with very best wishes

£120 signed

Brendan Kennelly

Brendan O'Connor

ECHINODERMATA

INTERNATIONAL ECHINODERM CONFERENCE

9 TH.

PROCEEDINGS OF THE FIFTH INTERNATIONAL ECHINODERM CONFERENCE
GALWAY / 24-29 SEPTEMBER 1984

Echinodermata

Edited by
BRENDAN F. KEEGAN & BRENDAN D.S. O'CONNOR
University College, Galway, Ireland

A.A. BALKEMA / ROTTERDAM / BOSTON / 1985

The texts of the various papers in this volume were set individually by typists under the supervision of each of the authors concerned.

ISBN 90 6191 596 1

Distributed in USA & Canada by: A.A.Balkema Publishers, P.O.Box 230, Accord, MA 02018

Printed in the Netherlands

Table of contents

Abstracts

2 *Crinoidea*

Abstract

3 *Echinoidea*

Abstracts

4 *Holothurioidea*

Abstracts

5 *Asteroidea*

Abstracts

6 *Ophiuroidea*

Abstracts

Foreword

The International Echinoderm Conference is now entrenched in the marine biological 'calendar of events'. This of itself testifies to the success of these meetings and the growing need to facilitate interchange between colleagues in a rapidly expanding research area. Echinoderms have "come into their own"!

Previous conferences were held in the U.S.A. (Washington, D.C. 1972 and Tampa Bay, 1981), Yugoslavia (Rovinj, 1975) and Australia (Sydney, 1978). Galway, in the quincentennial year of its city charter, welcomed 197 participants from 28 countries to the 1984 Conference.

These Proceedings contain abstracts of all papers/posters presented at the Conference, and complete papers from those submitted for publication and accepted on the recommendations of referees. The 'camera-ready' system has been employed throughout and most of the inclusions remain, substantially, in their final, submitted form. Whilst this system unfortunately has its typographical limitations, it is hoped that they in no way distract the reader. We thank the authors for their co-operation in this attempt to reduce costs and increase the speed of publication. Secretarial deadlines were met largely through the diligence and quiet efficiency of Sandy Lawson and Anne O'Higgins.

Where appropriate, referees were selected from amongst the actual participants. Anonymity was preserved and we endeavoured to make copy manuscripts available in advance of the actual presentations. This worked much better than expected; the referees responded with alarcrity and we very much appreciate their labours.

We thank Padraig Ó Céidigh, Professor of Zoology at University College, Galway, and Colm Ó hEocha, President of University College, Galway, for their opening and closing addresses respectively. Frank Rowe and Alain Guille spoke for the Conference at large in their memorable appreciations of Ailsa Clark and Gustave Cherbonnier.

We are grateful to all those who contributed to the working of the Conference: the chairpersons, whose exactitude gave synchrony to the parallel sessions; the projectionists whose task we did not envy; Tim Bowmer for his labour of love in fashioning our logo; the Organising Committee (researchers, technical staff and students) which unstintingly and enthusiastically framed the ethos of the meeting.

The National Board for Science and Technology was very gracious in making the r.v. Lough Beltra available for the collection of material at sea. As hosting institute, University College Galway accorded us every facility; we benefited particularly from the care and attention of personnel from the Buildings and Grounds Office.

In conclusion, we hope these Proceedings prove as useful as their predecessors. The next Conference will be held in Canada in 1987 and we wish the organisers every success.

Brendan F. Keegan
Brendan D.S. O'Connor

Organisation of the Conference

Organising Committee

Brendan F. Keegan, Chairman
Brendan D.S. O'Connor, Co-Chairman

Brendan Ball
Timothy Bowmer
Anne-Marie Brennan
David Burke
John Costelloe
Mark Costelloe
William Crowe
Patricia Dineen
James Dunne
John Galvin
Anthony Grehan
Peter Heffernan
John Kennedy
Albert Lawless
Yvonne Leahy
David McGrath
Brian Munday
Sinead Neiland
Anne O'Higgins
Ann Quinn
Mark White

Zoology Department
School of Marine Sciences
University College
Galway

Session Chairpersons

F.S. Chia (Canada)
E. Dukes (Ireland)
T. Ebert (U.S.A.)
R. Emson (U.K.)
L. Fenaux (France)
J. Ferguson (U.S.A.)
L. Fishelson (Israel)
C. Franzen-Bengtson (Sweden)
A. Guille (France)
M. Jangoux (Belgium)
M. Jensen (Denmark)
J. Lawrence (U.S.A.)
T. Motokawa (Japan)
D. Nichols (U.K.)
D. Pawson (U.S.A.)
D. Roberts (N. Ireland)
F. Rowe (Australia)
G. Sevastopulo (Ireland)
B. Thomassin (France)
T. Valentinčič (Yugoslavia)
P. Voogt (The Netherlands)
C. Walker (U.S.A.)

Plenary Lectures

M. Jangoux (Belgium)
J. Lawrence (U.S.A.)
B. O'Connor (Ireland)
N. Sloan (Canada)
K. Tenore (U.S.A.)

Dedication

These proceedings are dedicated to

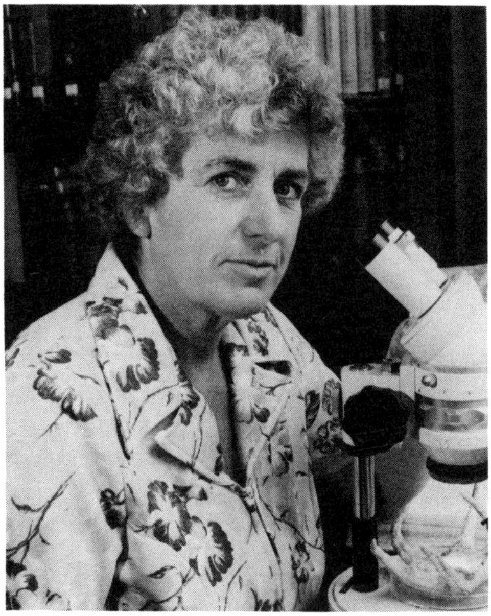

Ailsa McGowan Clark

and Gustave Cherbonnier

in recognition of their esteemed contributions to our knowledge and understanding of the Echinodermata.

1. General

Proceedings of the Fifth International Echinoderm Conference / Galway / 24-29 September 1984

Echinoderm role in the history of Phanerozoic tiering in suspension-feeding communities

WILLIAM I. AUSICH
The Ohio State University, Columbus, USA

DAVID J. BOTTJER
University of Southern California, Los Angeles, USA

ABSTRACT: Echinoderms have played a dominant role in the vertical ecological structure of benthic communities throughout the Phanerozoic. Despite their importance, echinoderms have only contributed to epifaunal tiering and not to infaunal tiering. The ecological importance of tiering for echinoderm communities is demonstrated by the iterative evolution within echinoderm classes for forms adapted to different tiering levels. In shallow subtidal settings, echinoderms dominated the higher tiers from the Cambrian through the Jurassic. Echinoderms have been important members of lower tiers throughout the Phanerozoic. The relative success of sessile suspension-feeding echinoderms has paralleled that of the epifaunal Paleozoic fauna.

1 INTRODUCTION

In 1982 Ausich and Bottjer presented the characteristic history of tiering among suspension-feeding organisms living on soft substrata in Phanerozoic shallow and epicontinental sea settings (Fig. 1). Tiering is the vertical ecological structure of organisms within a community, and this history represented the tiering capabilities of individual organisms. Accordingly, this history excluded reef settings, where tiering capability of individual organisms is enhanced by the vertical relief of the reef. Ausich and Bottjer (1982) included all body and trace fossils that have consistently been interpreted to result from the activity of suspension-feeding organisms.

The concept of vertical structuring of invertebrates in ancient benthic communities was first applied to the study of Mississippian crinoids (Lane 1963). Lane (1973), Sprinkle (1976) and Ausich (1980) expanded this concept with echinoderms, and Ausich and Bottjer (1982) elaborated further, naming this phenomenon tiering. The influence of echinoderms in the development of the tiering model reflects the significant role that echinoderms have played in the history of tiering among benthic invertebrates. Curiously, echinoderms have only occupied half of the potential tiering space. Echinoderms have been dominant as epifaunally tiered organisms, but no true infaunal suspension-feeding echinoderms are known to have evolved.

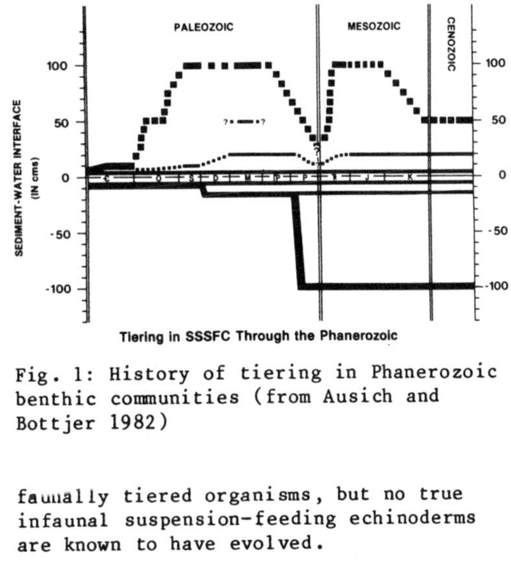

Fig. 1: History of tiering in Phanerozoic benthic communities (from Ausich and Bottjer 1982)

2 TIERING HISTORY

The history of tiering among benthic invertebrates corresponds in part to the three great "evolutionary faunas" of marine invertebrates defined by Sepkoski (1981), which include the Cambrian fauna, the post-Cambrian Paleozoic fauna and the modern fauna.

The Cambrian trilobite-dominated fauna lived very close to the sediment-water interface. In this fauna stalked echinoderms and sponges probably lived highest above the sea floor. Eocrinoids were the highest echinoderms, reaching approximately 10 cm above the substratum (Sprinkle 1973) (Fig. 1).

The evolution of the later Paleozoic fauna was an adaptive radiation of largely epifaunal suspension feeders, and a striking tiering structure developed relatively rapidly in geologic time. From the Early Ordovician to the Middle Silurian, crinoids reached 100 cm above the substratum (Watkins and Hurst 1977). This level persisted through the remainder of the Paleozoic, so that this epifaunal structuring typified the later Paleozoic fauna (Lane 1963, 1973; Ausich 1980). The depth to which infaunal suspension feeders burrowed was minor compared with epifaunal tiering through most of the Paleozoic. Typical limits of infaunal tiering were 12 cm below the bottom by at least the Devonian (Fig. 1) or as early as the Ordovician (Miller and Byers 1984) and 100 cm by at least the Permian (Hattin 1957).

The Mesozoic-Cenozoic modern fauna has largely been characterized by infaunal tiering, although the potential for both significant epifaunal and infaunal tiering was realized through the Jurassic (Fig. 1). Crinoids remained the highest suspension feeders until the end of the Jurassic when stalked forms migrated into deeper water in most areas. At that time sponges and alcyonarians inherited the high level suspension-feeding niches. In the fossil record bivalves and suspension-feeding arthropods have dominated the infaunal tiers of the modern fauna.

3 REPEATED CONVERGENT EVOLUTION

Epifaunal suspension feeding has been a very important habit among echinoderms (Fig. 2). All echinoderms that led a predominantly sessile existence are thought to have included only suspension

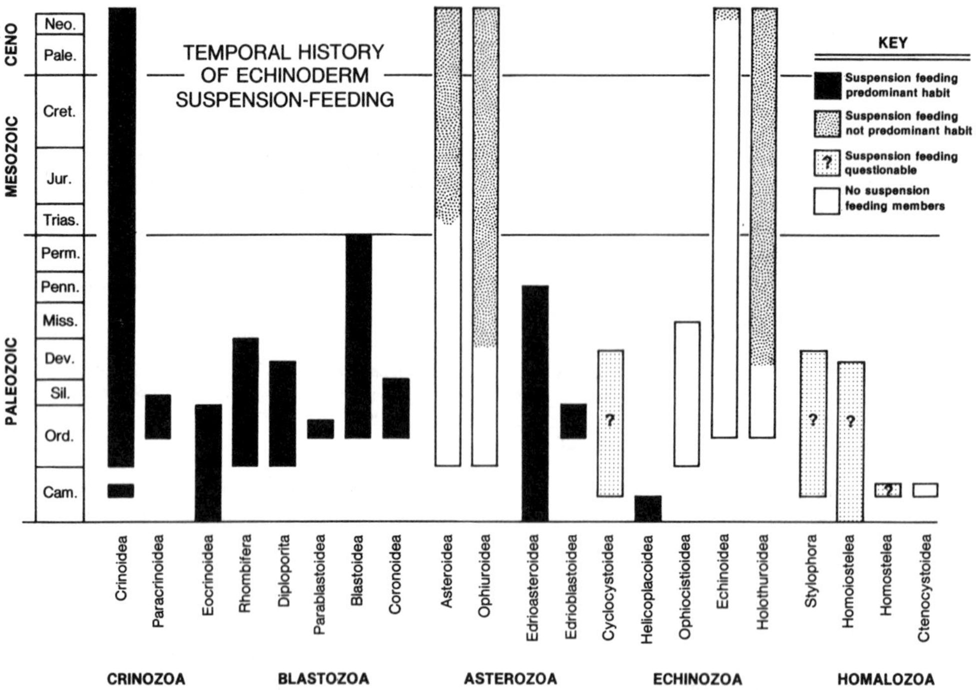

Fig. 2: Temporal history of suspension feeding among echinoderms, indicating relative importance of this trophic habit by class. Summarized from many sources including Jangoux (1982), Kolata and Jollie (1982), Massin (1982), Parsley (1980, de Ridder and Lawrence (1982), Smith and Paul (1982), Sprinkle (1980), Ubaghs (1981) and Warner (1982)

feeders (11 classes). Of the four extant non-crinoid classes, all have members that adopted a suspension-feeding habit. Only 2 of the 21 recognized classes, Ophiocistioidea and Ctenocystoidea, are thought not to have had the possibility of suspension-feeding members (Sprinkle 1980).

The importance of tiering for the ecological organization of echinoderms is underscored by the capability of members within a single class for iterative convergent evolution to varied suspension-feeding tiers. This is true for stalked echinoderms that primitively possessed a column and either progressively increased column height or secondarily reduced the column. Unstalked sessile suspension-feeding echinoderms also attained differing tiering levels by upward extension of their bodies.

Most classes that were predominantly suspension feeders adapted a variety of suspension-feeding levels. The strongest argument for this repeated evolution can be made by considering extremes of adaptation. Extremes are best displayed among stalked echinoderms, Crinozoa and Blastozoa. The dominant habit of stalked echinoderms has been an erect posture above the substratum. The extreme alternative to this erect habit is life on the substratum. Most classes of Blastozoa and both classes of Crinozoa have at least one or typically numerous members that have secondarily adapted a suspension-feeding habit in the lowest epifaunal tier (0 to +5 cm) immediately above the substratum (termed here "bottom dwellers"). Figures 3, 4 and 5 are non-comprehensive listings of bottom-dwelling "stalked" echinoderms. Included are a few eocrinoids, rhombiferans and paracrinoids; several blastoids; and many diploporitids and crinoids. Certain classes had numerous taxa specialized for bottom dwelling (e.g. diploporitids cemented to the substratum, Bockelie 1984) (Figure 3), whereas other classes had fewer representatives. All seven post-Cambrian orders of Paleozoic crinoids had forms interpreted to have been bottom dwellers, and five of seven post-Paleozoic crinoid orders had stemless or functionally stemless forms, with some of these adopting a pelagic habit.

From a constructional morphologic view, loss of the column would probably have been a rather minor developmental perturbation. However, the pervasive iteration for adaptation to the lowest epifaunal tier requires an ecological explanation of adaptability of echinoderms to various tiering levels.

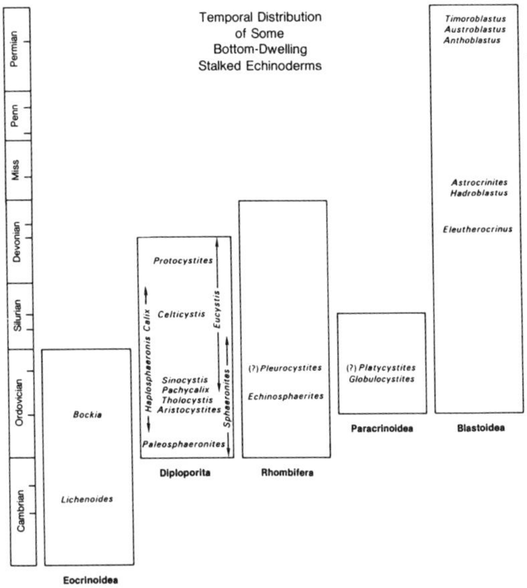

Fig. 3: Temporal history of non-crinoid bottom-dwelling blastozoans and crinozoans. This listing is not necessarily comprehensive. Summarized from numerous sources including Breimer and Macurda (1972), Bockelie (1979, 1984), Frest, Strimple and McGinnis (1976), Macurda (1977), Millendorf (1979), Moore (1968) and Paul (1967, 1973)

Adaptation to a bottom-dwelling habit was attained by numerous means. The following list of four adaptations for development of a bottom-dwelling habit is modified from Ubaghs (1978, p. T93-T94): 1, shorten column so as to be non-functional; 2, column discarded during ontogeny either completely or retaining a single proximal columnal (centrodorsal) or centrale; 3, cementation to the substratum; and 4, recumbent column. Again within many classes, two or more means were employed to adopt a bottom-dwelling habit.

Among echinozoans both edrioasteroids and holothurians developed different tier levels, although without a column tiering has apparently been limited to lower tier levels. The considerable mobility of ophiuroids has offered them the ability to climb onto other organisms (e.g., crinoids), thereby attaining much higher suspension-feeding levels than would otherwise be possible.

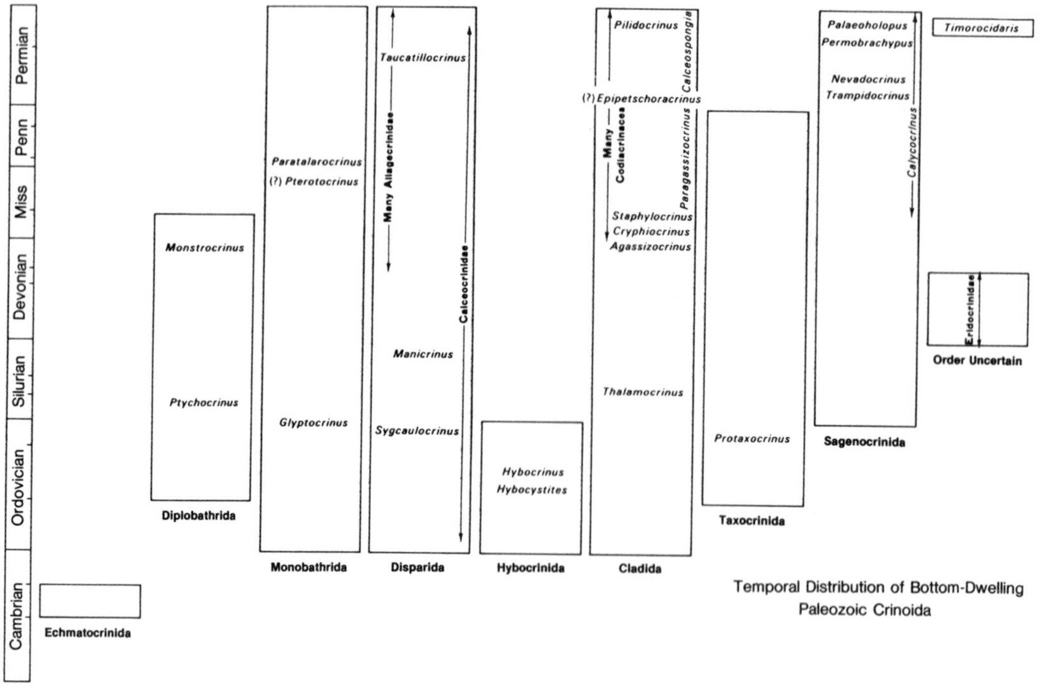

Fig. 4: Temporal history of bottom-dwelling Paleozoic crinoids. This listing is not necessarily comprehensive. Summarized from numerous sources including Brett (1980), Brower (1966), Chesnut and Ettensohn (1984), Ettensohn (1975, 1980), Frest and Strimple (1978), Lewis (1981), Moore and Teichert (1978), Strimple (1975), Strimple and Coney (1978) and Willink (1978)

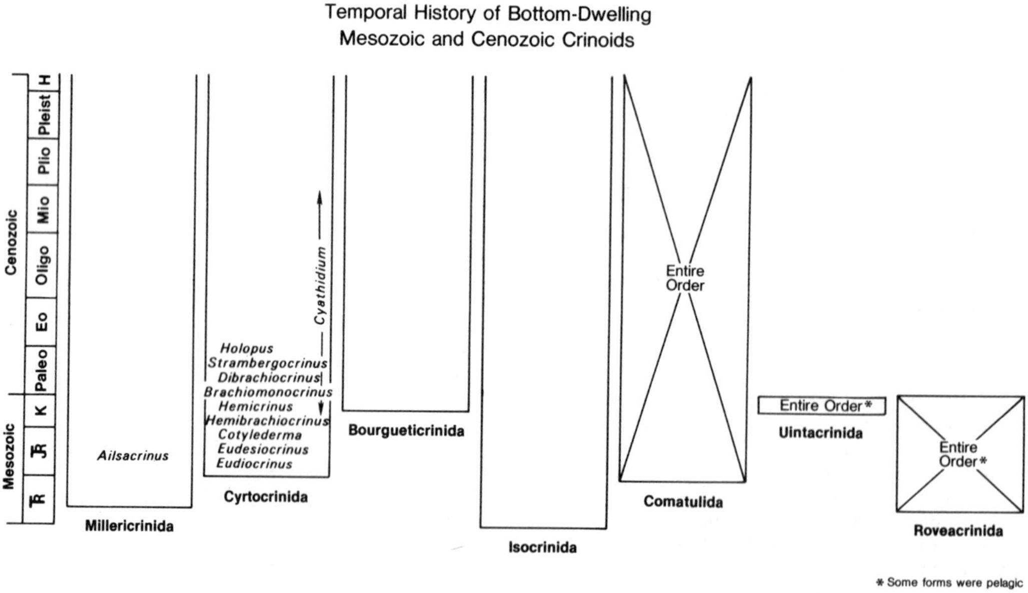

Fig. 5: Temporal history of post-Paleozoic bottom-dwelling crinoids. This listing is not necessarily comprehensive. Summarized from Moore and Teichert (1978) and Žitt (1979, 1983)

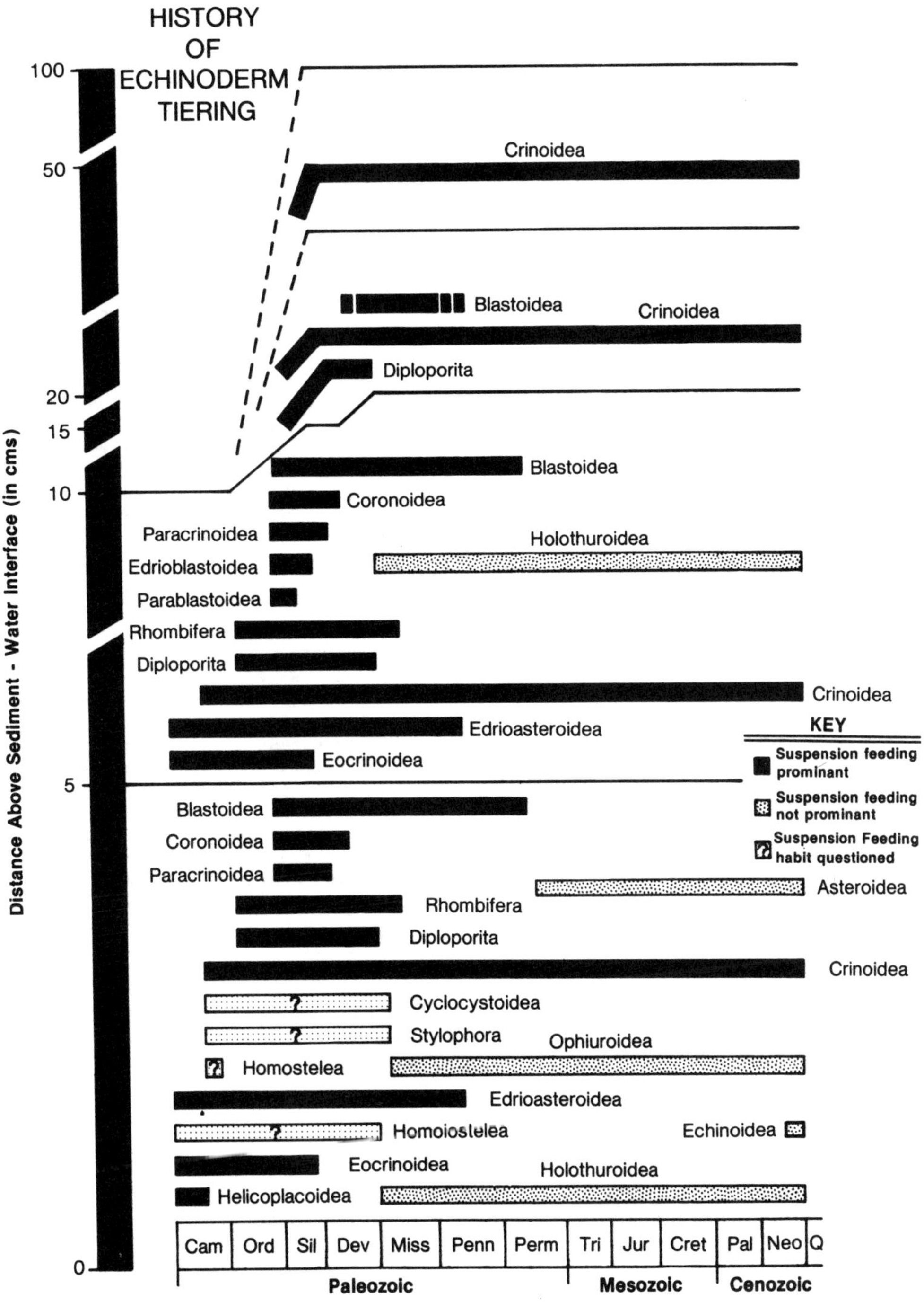

Fig. 6: Temporal history of Phanerozoic suspension-feeding echinoderms according to tier subdivisions

The repeated evolutionary development of echinoderms throughout the epifaunal tiering structure of benthic communities presents a compelling argument for the ecological importance of tiering and for the ecological influence on echinoderm evolution.

4 ECHINODERM TIERING HISTORY

The history of suspension-feeding echinoderms in tier subdivisions of Ausich and Bottjer (1982) is presented in Figure 6. Representatives of all suspension feeding echinoderm classes exploited the 0 to +5 cm tier, and many exploited the +5 cm to +10, 15 or 20 cm tier. Apparently only crinoids, blastoids and diploporitids attained positions within the +20 cm to +50 cm tier, and only crinoids became elevated to heights above 50 cm.

Studies to date indicate that all suspension-feeding echinoderms of the Cambrian were positioned within 10 cm of the substratum. In describing Cambrian faunas, Sprinkle (1976) divided echinoderm suspension feeders into two levels, "low" and "high". These levels correspond to the 0 to +5 cm and +5 to +10 cm epifaunal tiers. Within these Cambrian communities, edrioasteroids, helicoplacoids and perhaps homoiosteleans were reconstructed in the 0 to +5 cm tier; and crinoids, eocrinoids and perhaps stylophorans were within the +5 to +10 cm tier (Sprinkle 1976). In other settings some edrioasteroids may have reached the +5 to +10 cm tier.

From an ecological viewpoint, echinoderms played a significant role in the Ordovician adaptive radiation of the later Paleozoic fauna. From the Late Cambrian to the Middle Ordovician the number of echinoderm classes increased from 6 to 18. Considering only classes that are interpreted as predominantly suspension feeders, class diversity increased from 3 to 10; and generic diversity increased from 5 to 159 (generic estimates from Moore 1968, Moore and Teichert 1978). Absolute precision of these generic diversities is not important but these data do accurately reflect the order of magnitude shift in the importance of suspension-feeding echinoderms from the Cambrian fauna to the later Paleozoic fauna.

The vast increase in epifaunal echinoderms was ecologically accommodated by increased tiering complexity. By the Middle Ordovician more classes of predominantly suspension feeders had representatives in higher tiers: eight classes in the 0 to +5 cm tier, 10 classes in +5 to +10 cm tier and two classes in the +10 to +50 cm tier (Figure 6). Crinoids became established in higher level tiers during the Middle Ordovician.

By at least the Middle Silurian the maximum height of typical epifaunal suspension feeders was attained at +100 cm by crinoids. Crinoids dominated the upper tiers during the Middle Paleozoic and dominated all tiers by the Mississippian. Although subordinate to crinoids in upper tiers, diploporitids were in the +10 to +50 cm tier by the Middle Ordovician, and blastoids were probably in this tier by the Silurian. Diploporitids became extinct in the Devonian, and blastoids became extinct in the Late Permian.

Despite the progressive development of tiering heights from the Middle Ordovician to Middle Silurian, class level diversity of predominant suspension feeders decreased steadily from 10 in the Middle Ordovician to 2 by the Late Pennsylvanian (Fig. 7). This decline records a decreased number of classes with suspension-feeding capabilities, yet the generic diversity within these classes remained high. As echinoderm class diversity declined, the remaining classes expanded, both in diversity and intraclass tiering potential. New echinoderm suspension feeders included holothurians in the Early Devonian and ophiuroids in the Middle Devonian.

At the close of the Paleozoic, blastoids became extinct, leaving some suspension-feeding ophiuroids in the lowest tier, some suspension-feeding holothurians in both the 0 to +5 cm and +5 to +10 cm tiers, crinoids in all tiers and some

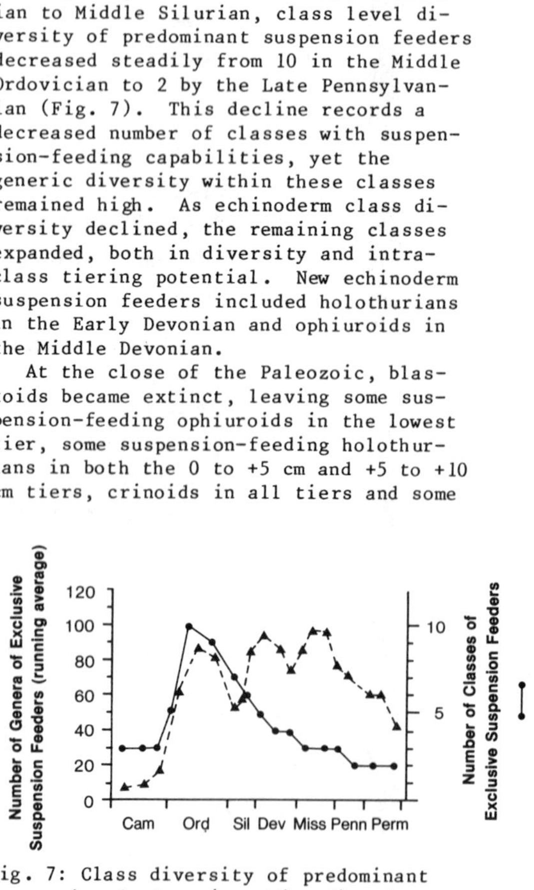

Fig. 7: Class diversity of predominant suspension feeders (see Fig. 2) and generic diversity within these classes (from Moore 1968, Moore and Teichert, 1978). Generic diversity is plotted as a three-value running average

asteroids that presumably developed the suspension-feeding habit during the Triassic.

Epifaunal tiering potential of echinoderms changed little during the development of the modern fauna. The epifaunal habit gave way to an infaunal life style (Vermeij 1977). Stalked crinoids were apparently not a part of shallow-water communities after the Jurassic (Meyer and Macurda 1977), so the maximum levels of tiering were not realized in shallow-water faunas. The only addition was the development of suspension feeding in echinoids during the later Cenozoic (Stanton, Dodd and Alexander 1979; De Ridder and Lawrence 1982).

5 DISCUSSION

Tiering, both above and below the sediment-water interface, has been an extremely important ecological factor among marine benthos. Echinoderms have played a fundamental role in the tiering of benthic communities throughout the Phanerozoic, and they have been key in the definition and development of this concept. The story of echinoderm tiering is exclusively one of epifaunal tiering; and to a large extent, complexity of tiering among echinoderms has paralleled the fortunes of the Paleozoic fauna of Sepkoski (1981). Complex epifaunally tiered communities, commonly dominated by echinoderms, persisted for approximately 260 million years during the Paleozoic and for perhaps approximately 100 million years during the early Mesozoic. The importance of stalked echinoderms in the history of the marine benthos is considerable.

A notable void in the history of tiering among echinoderms is the absence of infaunal suspension feeders. In today's oceans both the echinoid Dendraster and certain ophiuroids could be considered to be semi-infaunal suspension feeders. Although in both cases, the individual is partially buried in the sediment, and their suspension feeding apparatus is held erect above the substratum. Functionally these forms are epifaunal feeders. Infaunal suspension feeding may be a supplementary feeding habit in some heart urchins (Smith, 1984).

Echinoderms that participated fully in the infaunalization of the modern fauna are all mobile deposit feeders. Apparently the lack of infaunalization of functionally more sessile echinoderms or the lack of development of an effective means for continual access to water above the substratum (e.g., fused siphons of bivalves) precluded echinoderms from the infaunal suspension-feeding habit. The character of the modern fauna would have been radically altered had echinoderms developed inovations enabling them to exploit infaunal suspension feeding niches.

REFERENCES

Ausich, W. I. 1980. A model for niche differentiation in Lower Mississippian crinoid communities. J. Paleontol. 54:273-288.

Ausich, W. I. & D. J. Bottjer 1982. Tiering in suspension-feeding communities on soft substrata throughout the Phanerozoic. Science 216:173-174.

Breimer, A. & D. B. Macurda, Jr. 1972. The phylogeny of the fissiculate blastoids. K. Ned. Akad. Van Wet., Verh., Afd. Nat., Eerste Reeks, 26(3), 390p.

Brett, C. E. 1978. Attachment structures in the rhombiferan cystoid Caryocrinites and their paleobiologic implications. J. Paleont. 52:714-726.

Bockelie, J. F. 1979. Celticystis n. gen., a gomphocystitid cystoid from the Silurian of Sweden. Geol. Foren. Stockholmn Forhand. 101:157-166.

Bockelie, J. F. 1984. The Diploporita of the Oslo region, Norway. Palaeontol. 27:1-68.

Brower, J. C. 1966. Functional morphology of Calceocrinidae with description of some new species. J. Paleontol. 47:613-634.

Chesnut, D. R., Jr., & F. R. Ettensohn 1984. Life mode of Pterotocrinus in the Lower Pennington (Lower Carboniferous) of East-Central Kentucky. Geol. Soc. Amer. Abst. with Prog. 16: 128.

De Ridder, C. & J. M. Lawrence 1982. Food and feeding mechanisms: Echinoidea. In M. Jangoux & J. M. Lawrence (eds.), Echinoderm Nutrition, p. 57-115. Rotterdam, Balkema.

Ettensohn, F. R. 1975. The autecology of Agassizocrinus lobatus. J. Paleontol. 49:1044-1061.

Ettensohn, F. R. 1980. Paragassizocrinus: Systematics, phylogeny and ecology. J. Paleontol. 54:978-1007.

Frest, T. J. & H. L. Strimple 1978. Manicrinus (nov.), a cladid evolutionary homeomorph of the bottom-dwelling Hybocrinus, Brownsport (Silurian: Ludlow) of Tennessee. Southeastern Geol. 19:157-175, 150 p.

Hattin, D. E. 1957. Depositional environment of the Wreford Megacyclothem (Lower Permian) of Kansas. Kan. Geol. Survey Bull. 124.

Jangoux, M. 1982. Food and feeding mechanisms: Asteroidea. In M. Jangoux & J. M. Lawrence (eds.), Echinoderm Nutrition, p. 117-159. Rotterdam, Balkema.

Kolata, D. R. & M. Jollie 1982. Anomalocystitid mitrates (Stylophora-Echinodermata) from the Champlainian (Middle Ordovician) Guttenberg Formation of the Upper Mississippi Valley Region. J. Paleontol. 5:631-653.

Lane, N. G. 1963. The Berkeley crinoid collection from Crawfordsville, Indiana. J. Paleontol. 37: 1001-1008.

Lane, N. G. 1973. Paleontology and paleoecology of the Crawfordsville fossil site (Upper Osagian: Indiana). Univ. Calif. Pub. Geol. Sci. 99, 141 p.

Lewis, R.D. 1981. Archaetaxocrinus, new genus, the earliest known flexible crinoid (Whiterockian) and its phylogenetic implications. J. Paleontol. 55:227-238.

Macurda, D. B., Jr. 1977. Two Carboniferous blastoids from Scotland. Paleontol. 55:225-236.

Massin, C. 1982. Food and feeding mechanisms: Holothuroidea. In M. Jangoux & J. M. Lawrence (eds.), Echinoderm Nutrition, p. 43-55. Rotterdam, Balkema.

Meyer, D. L. & D. B. Macurda, Jr. 1977. Adaptive radiation of the comatulid crinoids. Paleobiol. 3:74-82.

Millendorf, S. A. 1979. The functional morphology and life habits of the Devonian blastoid Eleutherocrinus cassedayi Shumard & Yandell. J. Paleontol. 53:553-561.

Miller, M.F. & C. W. Byers 1984. Abundant and diverse early Paleozoic infauna indicated by the stratigraphic record. Geology 12:40-43.

Moore, R. C. (ed.) 1968. Treatise on Invertebrate Paleontology, Echinodermata 1, Part S. Lawrence, Kansas: Geol. Soc. Amer.

Moore, R. C. & C. Teichert (eds.) 1978. Treatise on Invertebrate Paleontology, Echinodermata 2, Part T. Lawrence, Kansas: Geol. Soc. Amer.

Parsley, R. L. 1980. Paracrinoidea. In T. W. Broadhead & J. A. Waters (eds.), Echinodermata Notes for a Short Course, p. 139-143. Dept. Geol. Sci., Univ. Tenn. Studies in Geol. 3.

Paul, C. R. C. 1967. The functional morphology and mode of life of the cystoid Pleurocystites, E. Billings, 1854. In N. Millott (ed.) Echinoderm Biology, p. 105-123. Symposia of the Zoological Society of London 20. London, Academic Press.

Paul, C. R. C. 1973. British Ordovician cystoids. Palaeontol. Soc. Monogr. 127, pt. 1, 64 p.

Sepkoski, J. J., Jr. 1981. A factor analytic description of the Phanerozoic marine fossil record. Paleobiol. 7:36-53.

Smith, A. 1984. Echinoid Palaeobiology. London, George Allen & Unwin.

Smith, A. B. & C. R. C. Paul 1982. Revision of the class Cyclocystoidea (Echinodermata), Phil. Trans. Roy. Soc. London 296:577-684.

Sprinkle, J. 1973. Morphology and evolution of blastozoan echinoderms. Mus. Comp. Zool. Spec. Pap. Cambridge, Mass.: Harvard Univ., 284 p.

Sprinkle, J. 1976. Biostratigraphy and paleoecology of Cambrian echinoderms from the Rocky Mountains. Brigham Young Univ. Geol. Stud. 23:61-73.

Sprinkle, J. 1980. An overview of the fossil record. In T. W. Braodhead & J. A. Waters (eds.), Echinodermata Notes for a Short Course, p. 15-26. Dept. Geol. Sci., Univ. Tenn. Studies in Geol. 3.

Sprinkle, J. (ed.) 1982. Echinoderm faunas from the Bromide Formation (Middle Ordovician) of Oklahoma. Univ. Kansas Paleontol. Contrib. Mono. 1, 369 p.

Stanton, R. J., Jr., J. R. Dodd & R. R. Alexander 1979. Eccentricity in the clypeasteroid echinoid Dendraster: environmental significance and application in Pliocene paleoecology. Lethaia 12:75-87.

Strimple, H. L. 1975. The rare inadunate crinoid genus Thalamocrinus. Southeastern Geol. 16:249-253.

Strimple, H. L. & C. Coney 1978. Paratalarocrinus, a new camerate genus from the Upper Mississippian of Virginia. J. Paleontol. 52:484-487.

Ubaghs, G. 1978. Skeletal morphology. In R. C. Moore and C. Teichert (eds.), Treatise on Invertebrate Paleontology, Part T, Echinodermata 2, p. T58-T216. Lawrence, Kansas, Geol. Soc. Amer.

Ubaghs, G. 1981. Réflexions sur la nature et la fonction de l'appendice articulé des carpoïdes Stylophôra (Echinodermata). Ann. de Paléontol. Invert. 67:33-48.

Vermeij, G. J. 1977. Mesozoic marine revolution: evidence from snails and predators and grazers. Paleobiol. 3:245-258.

Warner, G. 1982. Food and feeding mechanisms: Ophiuroidea. In M. Jangoux and J. M. Lawrence (eds.), Echinoderm Nutrition, p. 161-181. Rotterdam, Balkema.
Watkins, R. & J. M. Hurst 1977. Community relations of Silurian crinoids at Dudley, England. Paleobiol. 3:207-217.
Willink, R. J. 1978. Catillocrinids from the Permian of eastern Australia. Alcheringa 2:83-102.
Žítt, J. 1979. Štrambergocrinus gen. n. (Cyrtocrinida) from the Lower Cretaceous of Štramberk (Czechoslovakia). Časopis pro mineral. a geol. 24:237-247.
Žítt, J. 1983. Spoon-like crinoids from Štramberk (Lower Cretaceous, ČSSR). Sborn Narod. Muz. Praze, Acta mus. Nation. Pragae 39 B:69-114.

Proceedings of the Fifth International Echinoderm Conference / Galway / 24-29 September 1984

Ontogeny and functional morphology of two Ordovician calceocrinids

J.C.BROWER
Department of Geology, Syracuse University, N.Y., USA

ABSTRACT: Calceocrinids exploited a unique life style. The crown was movably hinged to a recumbent stem. Ontogenetic studies of two primitive Ordovician calceocrinids disclose solutions to the functional problems faced by these organisms. In order to open the crown into the feeding position, the hinge was either isometric or positively allometric relative to the effective weight of the crown. The food-gathering capacity of a crinoid can be estimated by the number of food-catching tube-feet multiplied by the width of the food grooves which reflects the number and size of the food particles that can be collected. The food-gathering capacity was subject to marked positive allometry compared to the volume of the crown.

INTRODUCTION

The mode of life of calceocrinids is unique among crinoids (e. g. Brower, 1977; Kesling & Sigler, 1969; Fig. 1). The stem functioned as a runner along the seafloor. The crown is joined to the stem and basal circlet with a movable hinge. Closing the hinge and depressing the stem elevated the crown into the feeding orientation where the crown was at roughly a right angle to the stem with the arms spread. The crown was raised by ligaments attached to the outer part of the hinge (Fig. 2). There are two possible orientations for the crown: facing upstream or downstream. I believe that calceocrinids are analogous to modern crinoids and they fed with the aboral parts of the arms facing the current. This position is universal in living crinoids (e. g. Meyer & Macurda, 1980) and was favored by Ausich (1980) for a Mississippian calceocrinid. Hydrodynamic considerations probably require that the aboral parts of the arms be located upcurrent (Rubenstein & Koehl, 1977; Warner, 1978). Once in the upright position, the hinge ligaments were probably locked to hold the crown in place with little expenditure of effort. This can be done by living echinoids and probably also by extant crinoids (D.L. Meyer, 1979, pers. comm.). The crown was closed by ligaments and/or muscles inserted on the inner margins of the basal and radial circlets (Fig. 2).

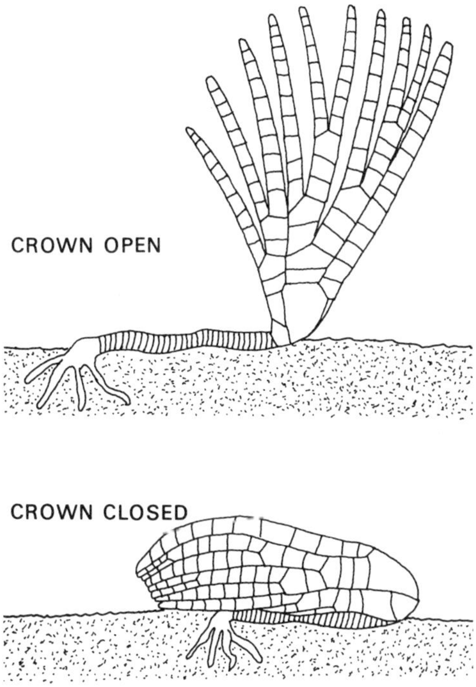

Figure 1. Living positions of calceocrinids.

This life style posed two functional problems for calceocrinids, namely the

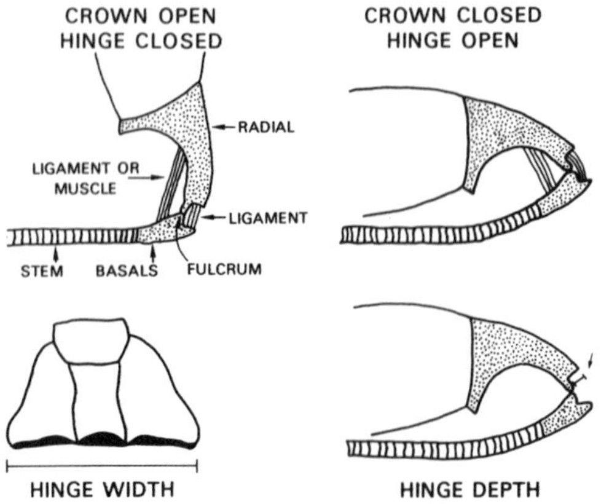

Figure 2. Functional morphology of the calceocrinid hinge.

operation of the hinge and food gathering. The study of allometry in two Middle Ordovician species, Cremacrinus ramifer (Brower, 1977), a small form, and Paracremacrinus laticardinalis Brower (1977), a large taxon, from the Bromide Limestone of Oklahoma, USA. illustrates the solutions evolved by the animals.

This research was supported by NSF Grant EAR-82-18374.

STATISTICAL METHODS

Allometry is defined as change of shape which is correlated with increase in size, (Gould, 1966). The relationships analyzed here are bivariate and most statistics are derived from the simple power function or allometric equation:

$$Y=aX^b$$

in which X and Y are the independent and dependent variables, respectively. The initial intercept a equals Y when X is unity. The exponent b is the ratio of the specific rates of change in Y and X and it reveals the changes in the shape ratio Y/X. A series of exponents for isometric change, in which the shape of the animal remains the same at all sizes, is useful for comparison with the exponents based on observed change. If the observed exponent exceeds the corresponding isometric value, the shape changes with positive allometry. If the observed exponent is less than the isometric figure, the animal exhibits negative allometry. Student's t tests were used to measure the statistical significance of observed versus isometric exponents and for comparison of various regression parameters. The regression lines were calculated with the reduced major axis algorithm.

PRIMARY MEASUREMENTS

The variables measured by paleontologists and biologists often only provide indirect representations of the parameters which they are intended to investigate (e. g. Gould, 1966). Results obtained from such studies can be misleading. This point is important, and this study is designed to work as closely as possible with the functional relationships of the animals. The following have been determined directly in mm and milligrams.

1. Volume of the crown. The crowns are preserved with little sediment in the arms (Fig. 3). The dorsal cup, anal tube and food grooves are either empty or filled with crystalline calcite. Calceocrinids are dense crinoids with comparatively small amounts of tissue. I assume that the volume of tissue is isometric relative to the total volume. This is supported by some data on modern crinoids (Brower, 1973).
2. Effective weight of the crown. The weight lifted when the crown was raised into the feeding position is the weight in water (original effective weight). The measured effective weight of the crown must exceed the original effective weight when the animal was alive, because the pores in the plates and other areas, which were originally filled with soft parts, are now replaced by calcite. Nevertheless, the original effective weight should be isometric with the measured effective weight.
3. Number of plates, including both brachials and ramulars, in the food-gathering system. 4. Length of the food gathering system. 5. Mean number of covering plates per brachial and ramular. 6. Average width of food groove. 7. Width and depth of hinge (Fig. 2).

Parameters derived from these variables are discussed in the text.

ONTOGENY OF THE HINGE

The first problem is unique to calceocrinids in that the ligaments attached to the hinge must have borne the weight of the crown. The forces that the hinge must overcome comprise the effective weight of the crown and the hydrodynamic drag between the water and the crown. The drag cannot be estimated owing to uncertainty about the drag coefficients and flow velocities. Nevertheless, it is possible to relate the allometry of the hinge to the effective weight of the crown. The area where the ligaments joined the hinge is approximated by the hinge width multiplied by its depth (Fig. 2). Assuming that both species had the same type of hinge ligaments, the strength of the ligaments is proportional to the area of the hinge where the ligaments were inserted. This is because the stresses generated by the hinge are roughly orthogonal to its area (Fig. 2). However, the hinge area provides only part of the story. The critical point is how much force can the hinge exert. The force will scale as the hinge moment which consists of the hinge area multiplied by the moment arm over which it applies leverage. When the hinge is opened and closed, the average moment arm is roughly half of the hinge depth (Fig. 2).

Analyzing hinge area (HA) versus effective weight (EW) resulted in the following equations.

$$(HA) = 0.01074(EW)^{0.6548} \quad (\textit{C. ramifer})$$

and

$$(HA) = 0.004913(EW)^{0.8471} \quad (\textit{P. laticardinalis})$$

The isometric exponent for this relationship equals 0.67 which would hold true for size increase with no change in shape. Comparison of the observed and predicted exponents indicates that the growth of C. ramifer is isometric whereas P. laticardinalis exhibits positive allometry. In C. ramifer, the hinge area is augmented at a rate that one would expect for an isometric crinoid but this growth rate is accelerated for P. laticardinalis. The differential growth of the two species is clearly reflected in the hinge shape (Fig. 3). The cup of all specimens of C. ramifer and young individuals of P. laticardinalis shows a rectangular or trapezoidal outline. In P. laticardinalis the width of the hinge increases more rapidly than the rest of the dorsal cup so the hinge of adult crinoids is extended into "ear" like projections. The shape of the hinge in C. ramifer is largely stabilized compared to the rest of the animal. The "ears" of P. laticardinalis probably have functional significance. When the crinoids were alive, the hinge lay along or close to the substrate. Adult crowns

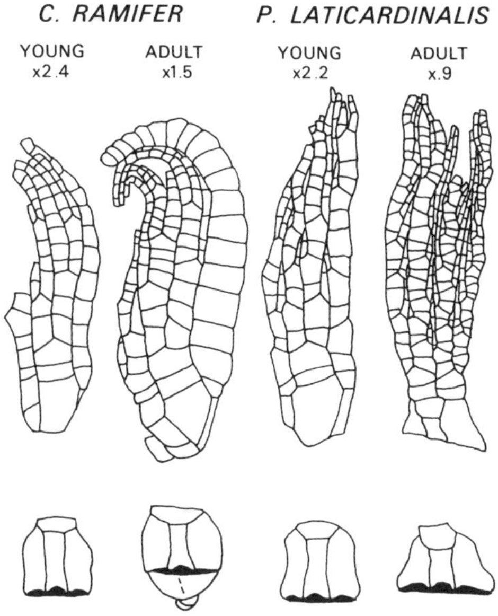

Figure 3. Growth sequences of Ordovician calceocrinids. Hinge is shown in black.

of P. laticardinalis are larger, higher and were probably more top heavy than mature individuals of C. ramifer. In effect the "ears" augmented the length of the hinge along the substrate which stabilized the crown, especially when the hinge was closed.

As discussed previously, the hinge moment should be proportional to the force exerted by the hinge to raise the crown. Figure 4 contains the graph of hinge moment (HM) versus effective weight (EW). The isometric exponent equals 1.0 and the allometric equations follow.

$$(HM) = 0.00009595(EW)^{1.051} \quad (\textit{C. ramifer})$$

and

$$(HM) = 0.00001877(EW)^{1.350} \quad (\textit{P. laticardinalis})$$

As before, C. ramifer is isometric whereas positive allometry is observed in P. laticardinalis. Student's t tests denote that neither the exponents or the initial intercepts of the two taxa are significantly different at a one- or two-sided risk level of 0.05. Statistically both the exponents and initial intercepts are homogeneous. Despite the lack of statistical significance, I consider the results to be meaningful.

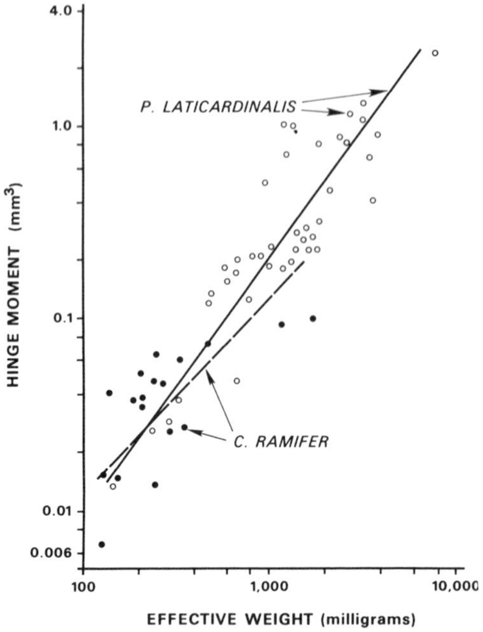

Figure 4. Graph of effective weight versus hinge moment. Number of specimens (N) and correlation coefficients (r): C. ramifer 20 and 0.696; P. laticardinalis 41 and 0.867.

Examination of Figure 4 discloses that young specimens of both taxa lie in the hinge moment:effective weight range of about 0.00011. However, the crinoids diverge with progressive age and larger specimens of P. laticardinalis are characterized by higher ratios of hinge moment:effective weight than equivalent sized C. ramifer. Thus the hinge moment of P. laticardinalis increases much more rapidly than that of the other species. Furthermore, this relationship is extrapolated to larger body sizes in P. laticardinalis than in C. ramifer where the adult crown effective weights comprise approximately 7800 and 1700 milligrams, respectively. The ratios of hinge moment:effective weight of the largest known specimens of P. laticardinalis and C. ramifer equal roughly 0.00043 and 0.00014, listed in the same order, so that adults of the former species possess comparatively larger hinge moments than those of the latter. This is interpreted as allometry related to size and habitat which allowed the mature P. laticardinalis to function in the same way as adult individuals of C. ramifer. Positive allometry is required for P. laticardinalis whereas isometry is adequate for C. ramifer owing to several reasons. Fully developed crowns of P. laticardinalis are larger, have higher effective weights, and almost certainly encountered larger drag forces than the adults of C. ramifer. This is because these animals exhibit larger crowns with more complexly branched arms (Fig. 3) and they lived in more agitated environments (Sprinkle & Longman, 1982). (The drag force on the crown is a function of the area of the arms and the square of the current velocity.)

This interpretation is supported by discriminant analysis on the two species using the effective weight (EW) and hinge moment (HM). The observed F ratio is significant at a two-sided risk level of 0.05 and it indicates that the two crinoids represent different populations. Furthermore both parameters are necessary to differentiate the two calceocrinids. The fact that the exponents and initial intercepts of the regression lines for the two species are not significantly different merely means that the separation involves an interaction between these parameters.

GROWTH OF THE FOOD-GATHERING SYSTEM

The second problem is common to many filter-feeders. The arms and their food-catching tube-feet must provide the entire volume of tissue with food. Four morphological parameters, length of stem, width of the food grooves, branch density (number of arm branches/area covered by the arms), and tube-foot spacing, are important in defining the feeding strategies of crinoids (Ausich, 1980; Meyer, 1979, 1982). These relate to level at which the animals feed, maximum size of food particles and the spacing of the food gathering net or filter.

Food gathering of the two calceocrinids was similar. Inasmuch as the stems ran along the sea floor (Brower, 1977), the arms filtered water at roughly the same elevations above the substrate (Fig. 3). The differences between the two species were tested by discriminant analysis based on the branch density, food-groove width, and tube-foot spacing. The F ratio is not significant at the 0.05 risk level, and the two crinoids overlap in the multivariate distribution of these variables. I conclude that P. laticardinalis and C. ramifer grade into each other relative to the size of the food particles collected (shown by tube-foot spacing and width of food-grooves) and packing of the food-gathering net or screen (branch density and tube-foot spacing). Sprinkle & Longman (1982) discussed the paleoecology of the Bromide Limestone. P. laticardinalis was found in an agitated area at a depth of 10 to 20 m. Conversely C. ramifer lived in quieter and deeper waters (10's of m). Regardless of the differences in body size and paleoecology of the two taxa, they were at least partly equivalent in food gathering. Surprisingly, the crowns of the two species contrast markedly in geometry. The larger form, P. laticardinalis, exhibits densely branched arms with numerous ramules whereas the smaller C. ramifer branches less frequently (Fig. 3).

The size of the food-gathering system can be approximated by its total length and number of plates, including both brachials and ramulars. The volume of tissue that must be supplied with food is assumed to be proportional to and isometric with the crown volume. The number of plates and the length of the arms should behave as linear dimensions so the appropriate exponents for isometry consist of 0.33. The observed exponents for the number of plates, 0.6058 for C. ramifer and 0.5607 for P. laticardinalis, significantly exceed the isometric figure. Consequently, the number of plates is augmented much more rapidly than if the crinoids were isometric. A similar situation is observed for the length of the food-gathering system. Both animals add new ramules throughout ontogeny (Fig. 3). This allows new plates to form at the growing tips of the arms and ramules, which in turn creates a plate supply rate that is multiplicative rather than additional. Two processes interact to dictate the growth of length of the arms, namely the plate supply rate and calcite accretion onto plates that formed earlier during ontogeny. Most of the increase of length can be attributed to the development of new plates. The exponents for the calceocrinids compare favorably with those given by Brower (1973) for Ordovician camerate crinoids with pinnulate arms. This demonstrates that having arms with ramules instead of pinnules does not necessarily impose a slower plate supply rate or a lower rate of increase of length of the food-gathering system. Although the differences between the exponents and initial intercepts of the two crinoids are not statistically significant, the regression lines of P. laticardinalis lie above those of C. ramifer. At any given crown volume, the food-gathering system of the former species is longer and possesses more plates than that of the latter.

Crinoids catch food particles with the tube feet and their number and spacing is obviously critical (e. g. Meyer, 1979, 1982). I estimated the number of food-catching tube-feet at one per covering plate for the calceocrinids, assuming that they were arranged as in the pinnules of Antedon bifida (Nichols, 1960). (Actually the tube-foot order is not important as long as it remains constant throughout growth.) The mean numbers of food-catching tube-feet per plate represent 6.9 and 11.4 for C. ramifer and P. laticardinalis, listed in the same order. The total number of food-catching tube-feet equals the average number per plate times the number of brachials and ramulars.

A reasonable estimate of the food-gathering capacity of a crinoid should be provided by the number of food-catching tube-feet multiplied by the width of the food grooves. The reasoning is simple: the number of tube-feet will be roughly proportional to the number of food particles that can be caught whereas the width of the food grooves restricts the size of the food stuffs that can be processed. As defined here, the food-gathering capacity ought to increase as a linear dimension squared, because the number of food-catching tube-feet and the width of the food-grooves represent linear dimensions. The amount of tissue that must be supplied with food should be isometric relative to the crown volume. Admittedly this ignores the soft parts in the stem. Luckily the Ordovician calceocrinids have short stems with small axial canals that contained little tissue. Omitting the stem will cost little in terms of information content.

Figure 5 shows the data for crown volume (CV) versus food-gathering capacity (FGC). The predicted exponent for isometry comprises 0.67. The equations follow.

$$(FGC) = 5.185(CV)^{1.016} \quad (\text{C. ramifer})$$

and

$$(FGC) = 12.16(CV)^{0.8769} \quad (\text{P. laticardinalis})$$

In both instances the food-gathering capacity is augmented more rapidly than if the animals were isometric, and

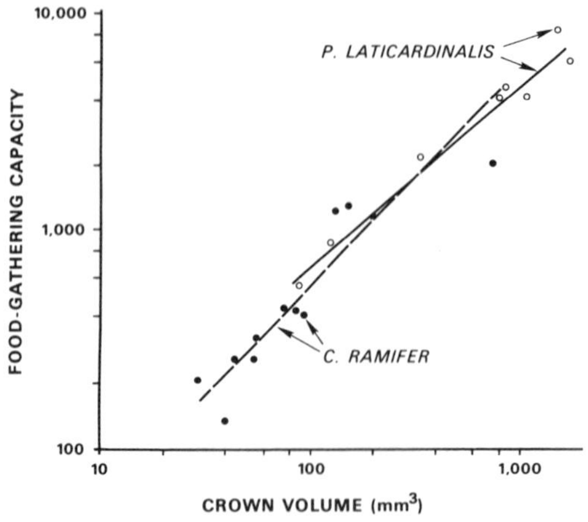

Figure 5. Graph of crown volume versus food-gathering capacity. Values of N and r: C. ramifer 12 and 0.919; P. laticardinalis 8 and 0.989.

Student's t tests denote that the patterns are statistically significant. If the food-gathering ratio (food-gathering capacity/crown volume) were constant throughout ontogeny, an exponent of 1.0 would be observed. The exponent of C. ramifer is not significantly larger than 1.0. For all practical purposes, the food-gathering ratio is independent of the size of the crinoid. Conversely the exponent of 0.8769 for P. laticardinalis is smaller than 1.0. Hence the food-gathering ratio declines in progressively larger crinoids and the decrease is statistically significant. Student's t tests were used to compare the exponents and the initial intercepts for the two calceocrinids. The differences are insignificant statistically. Thus the exponents and initial intercepts can be considered as homogeneous. I interpret this to mean that the two regression lines should be regarded as parallel and superimposed (Fig. 5). With respect to food-gathering capacity versus crown volume, the ontogeny of P. laticardinalis represents an extrapolation of that of C. ramifer into a realm of larger body sizes.

REFERENCES

Ausich, W.I. 1980. A model for differentiation in lower Mississippian crinoid communities. J. Paleontol. 54:273-288.

Brower, J.C. 1973. Crinoids from the Girardeau Limestone (Ordovician). Palaeontographica Americana. 7:261-499.

Brower, J.C. 1977. Calceocrinids from the Bromide Formation (Middle Ordovician) of Southern Oklahoma. Oklahoma Geological Survey Circ. 78:1-28.

Gould, S.J. 1966. Allometry and size in ontogeny and phylogeny. Biol. Rev. 41: 587-640.

Kesling, R.B. & J.P. Sigler. 1969. Cunctocrinus, a new Middle Devonian calceocrinid crinoid from the Silica Shale of Ohio. Univ. of Michigan Mus. of Paleont. Contr. 22:339-360.

Meyer, D.L. 1979. Length and spacing of the tube feet in crinoids (Echinoderamata) and their role in suspension feeding. Mar. Biol. 51:361-369.

Meyer, D.L. 1982. Food composition and feeding behavior of sympatric species of comatulids from the Palau Islands (western Pacific). In J.M. Lawrence (ed.), Int. Echinoderms Conf., Tampa Bay, p. 43-49. Rotterdam:Balkema.

Meyer, D.L. & D.B. Macurda Jr. 1980. Ecology and distribution of the shallow-water crinoids of Palau and Guam. Micronesica. 16:59-99.

Nichols, D. 1960. The histology and activities of the tube-feet of Antedon bifida. Quart. Jour. Microscopical Sci. 101:105-117.

Rubenstein, D.I. & M.A.R. Koehl. 1977. The mechanisms of filter feeding: some theoretical considerations. Amer. Nat. 111:981-994.

Sprinkle, J. & M.W. Longman. 1982. Echinoderm paleoecology. Univ. Kansas Paleont. Contrib., Monograph 1:68-75.

Warner, G.F. 1977. On the shapes of passive suspension feeders. In B.F. Keegan, P.O. Ceidigh, & P.J.S. Boaden (eds.), Biology of benthic organisms, p. 567-576. Pergamon.

Proceedings of the Fifth International Echinoderm Conference / Galway / 24-29 September 1984

Biostratigraphy and evolution of crinoid columnals from the Ordovician of Britain

S.K.DONOVAN
Department of Geology, Trinity College, Dublin, Ireland
Present address: NERC, Polaris House, North Star Avenue, Swindon, Wilts, UK

ABSTRACT: Only about 30 crinoid species based on calyces have been described from the Ordovician of Britain, yet crinoid columnals are present in many Ordovician beds. Columnals are relatively rare fossils before the Upper Ordovician, where they greatly increase in both numbers and diversity. Columnals can broadly be divided into two groups, meric (composed of more than one plate) and holomeric (columnals formed from a single ossicle). Only two British crinoids have tetrameric stems, Ramseyocrinus cambriensis (Lower Arenig) and Colpodecrinus forbesi (Caradoc). Bimeric and trimeric columnals are only known from the Ashgill. Pentameric columnals, and circular and pentagonal holomeric columnals, are common throughout the Ordovician. In the Upper Ordovician several new morphotypes appear, including columnals with a pentastellate lumen (Bystrowicrinus) and columnals with synarthrial articulating ridges (Ristnacrinus). The first cirriferous roots are Caradoc in age but Ordovician attachment structures are rare.

1 INTRODUCTION

Although the earliest pelmatozoans are Lower Cambrian (Sprinkle 1973), and Echmatocrinus brachiatus (the earliest crinoid) and the earliest true column are found in the Middle Cambrian (op. cit.: 177), stem ossicles (i.e. columnals) are very rare before the Lower Arenig (the only exception in the British Ordovician is the stem of a Tremadoc rhombiferan cystoid, Macrocystella mariae (Paul 1968)).

Crinoid columnals are common fossils in the British Ordovician, yet only about 30 crinoid species have been described from this interval, due mainly to the rarity of thecae. This poor record of complete specimens, which also generally have a limited geographic distribution, does not recommend the crinoids as potential zone fossils. However, columnal morphospecies often have short stratigraphic ranges and show "enough morphological characters to enable them to be identified and distinguished easily", a basic requirement for use in biostratigraphy (Clarkson 1979: 20). A preliminary stratigraphic scheme for columnal genera in the British Ordovician is outlined below. The only previous research of this kind was by Wright (1983), who considered columnals from the mid Caradoc strata of Snowdonia, North Wales.

2 MERIC COLUMNALS

It is convenient to divide columnals into two groups, meric (each columnal composed of more than one calcite plate) and holomeric (columnals formed from a single ossicle). Meric columns were an intermediate stage between primitive multiplated hohlwurzels and advanced holomeric stems (Sprinkle 1973: 37-40: Paul & Smith 1984: fig. 15). The known distribution of meric columnals in the British Ordovician is shown in Fig. 1 (1-3, 5, 6, 8). A single bimeric columnal has been found in the Rhiwlas Limestone of North Wales (Rawtheyan). Trimeric columnals are also rare and are limited to the mid Ashgill. Such columnals are known from the crinoid genus Ectenocrinus but the taxonomic affinities of the British specimens are still uncertain. Two crinoids with tetrameric columnals are known from the British Ordovician, Ramseyocrinus cambriensis from the Lower Arenig and Colpodecrinus forbesi from the Caradoc (Donovan 1983, 1984).

Pentameric stems are relatively common in the Arenig and Ashgill but are unknown from the Middle Ordovician and Caradoc, except in the Craighead Limestone of the North American faunal province (Paul 1976). However, Stukalina (1968) notes that the pentamerism of some columnals only becomes apparent on ground surfaces or in thin

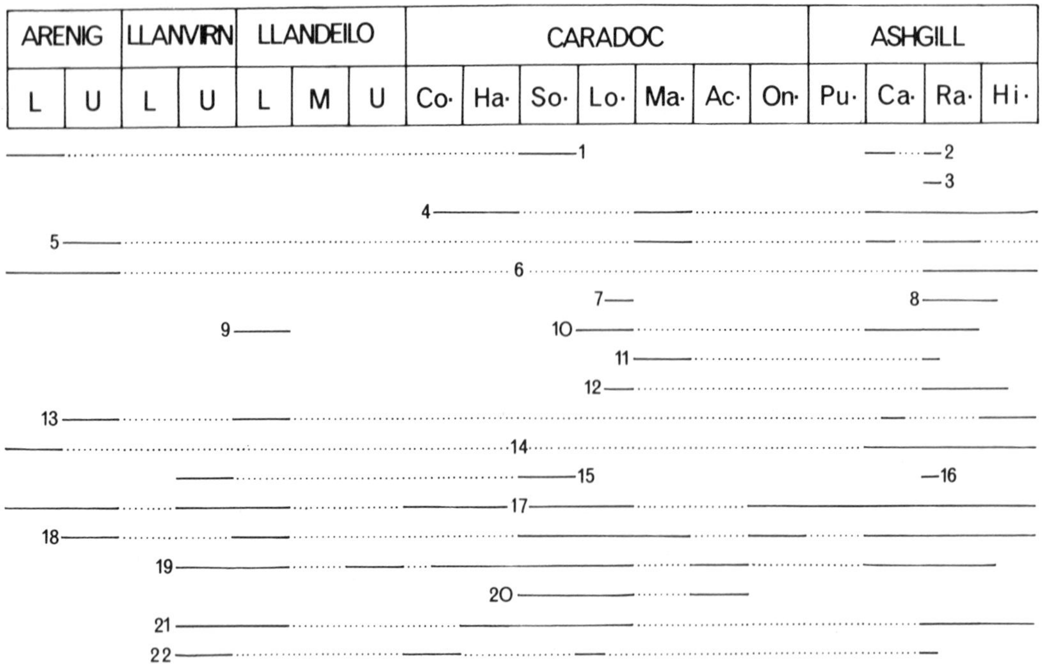

Fig. 1: Stratigraphic distribution of 22 columnal configurations in the Ordovician (excluding Tremadoc) of Britain. 1. Tetrameric. 2. Trimeric. 3. Bimeric. 4. Bystrowicrinus Yeltysheva. 5. Pentameric with a pentastellate lumen. 6. Pentameric with a large, pentagonal lumen. 7. Hemicyclocyclopa Wright. 8. Pentameric with a small lumen. 9. Bilaterally symmetrical columnals (cf. Fig. 2, 12). 10. Ristnacrinus Öpik. 11. Elliptical columnals with a pentagonal lumen. 12. Trigonocyclicus Yeltysheva & Stukalina. 13. Tetragonocyclicus Dubatolova. 14. Tetragonotetragonalis Yeltysheva. 15. Hexagonocyclopa Wright. 16. Circular columnals with a hexastellate lumen. 17. Cyclocyclicus Yeltysheva. 18. Pentagonocyclicus Yeltysheva. 19. Pentagonopentagonalis Yeltysheva. 20. Balacrinus Ramsbottom. 21. Cyclopentagonalis Yeltysheva. 22. Ramulicrinus Stukalina. (Solid line=definite presence, dotted line=inferred. L=Lower, M=Middle, U=Upper, Caradoc and Ashgill stages after Williams et al. 1972).

section. As the majority of columnals in the Ordovician of Britain are preserved as external moulds, these tests cannot usually be applied. Pentameric columnals with a pentastellate lumen are limited to the North American province until the Ashgill. Columns with large and small pentagonal lumina are differentiated in Fig. 1 as this is thought to indicate the amount of soft tissue in the axial canal. Columns with large axial canals are thought to be more primitive than those with narrow canals, because they probably contained organs which were concentrated in the cup of later pelmatozoans (Sprinkle 1973). Pentameric columnals with small lumina do not appear until the British Ashgill, whereas those with large lumina are known from the Arenig. In North America, pentameric columns with narrow lumina are known from at least the Upper Llandeilo, Bromide Formation (Warn 1982).

3 HOLOMERIC COLUMNALS

Most geologists intuitively think of a typical columnal as being circular with a circular lumen i.e. the morphogenus Cyclocyclicus. (The Cyclo- of this generic name describes the lumen shape and the -cyclicus the facet outline. The same rules apply to other Russian morphogeneric names). Ossicles of this configuration range throughout the Ordovician (Fig. 1; 17). However, gaps exist in the range of even this morphogenus (the numerous gaps shown in Fig. 1 emphasise the preliminary nature of this survey).

Of the bilaterally symmetrical columnals, Hemicyclopa (Wright 1983) is limited to the Longvillian and the columnal illustrated in Fig. 2 (12) is only found in the Llandeilo. Ristnacrinus is known from a number of European localities and has a total range from late Llandeilo to late Ashgill. Only

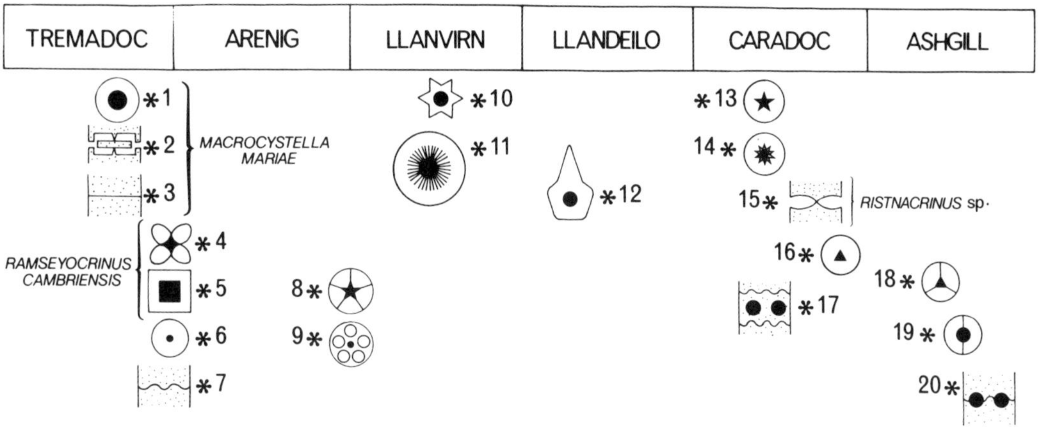

Fig. 2: First appearances of distinctive features of columnals in the British Ordovician. 1. Holomeric columnal (cystoid). 2. Proximal glyptocystitid articulation. 3. Synostosial articulation. 4. Tetramerism. 5. Holomeric, tetragonal columnal. 6. Holomeric, crinoid-like columnal. 7. Symplexial articulation. 8. Pentameric columnal with a pentastellate lumen. 9. Five-fold arrangement of ligament pits. 10. Hexagonal, holomeric columnal. 11. Broad, flange-like epifacet. 12. Holomeric columnal with bilateral symmetry. 13. Holomeric columnal with pentastellate lumen. 14. Ten-fold arrangement of ligament pits. 15. Synarthrial articulation. 16. Holomeric columnal with 3-fold symmetry. 17. Canaliculi. 18. Trimerism. 19. Bimerism. 20. Fossulae.

two elliptical columnals with pentagonal lumina are known. Trigonocyclicus columnals are moderately common in the Ashgill. Tetragonocyclicus and Tetragonotetragonalis both have long ranges with big gaps. Hexagonocyclopa and circular columnals with hexastellate lumina are both rare and do not overlap stratigraphically.

Columnals with pentagonal symmetry are common and have been divided into a number of morphogenera by previous authors. However, the existing morphogenera do not indicate the true diversity of these plates (e.g. see Wright 1983). The stratigraphic range of Balacrinus, the only 'true' genus included in this group, is quite short, yet those of the (artificial) morphogenera are noticeably longer.

First appearances of important features are summarised in Fig. 2. Obviously many morphological features first appear in the Tremadoc and Arenig, coinciding with the earliest appearances of certain crinoids and cystoids in Britain. A second group of new features occurs in the Upper Ordovician, coincident with the increase in columnal diversity and numbers.

4 ATTACHMENT STRUCTURES

Crinoid attachment structures are rare in the British Ordovician. Ramseyocrinus had a distal, planar coil (Donovan 1984) which was a much more open structure than those found in later crinoids (Brett 1981). Iocrinus brithdirensis (Lower Llanvirn) is the earliest example of a crinoid with a distal spiral coil holdfast (Bates 1965, pl. 45, fig. 5). A single specimen from the Caradoc has a pseudocirriferous coiled attachment, and an undescribed species from the Llanvirn has a pseudocirriferous dendritic radice. Straight cirriferous runners are first known from the Caradoc. True terminal discoidal and cirriferous holdfasts are not known until the Ashgill. However, as these attachment structures are much rarer fossils than columnals, it is not possible to propose a definite evolutionary or stratigraphic scheme yet. Known first appearances of different types of attachment differ between, say, Britain and North America (cp. Lewis 1982, for example).

5 INTERPRETATION

The obvious problem in basing a study in evolution solely upon fragmentary fossils is that it is not normally possible to determine whether similar units are derived from closely related or extremely dissimilar organisms e.g. the synarthrial ridge evolved in all four major crinoid subclasses and the glyptocystitid cystoids. Only gross conclusions can be made e.g. that the increase in columnal numbers and diversity

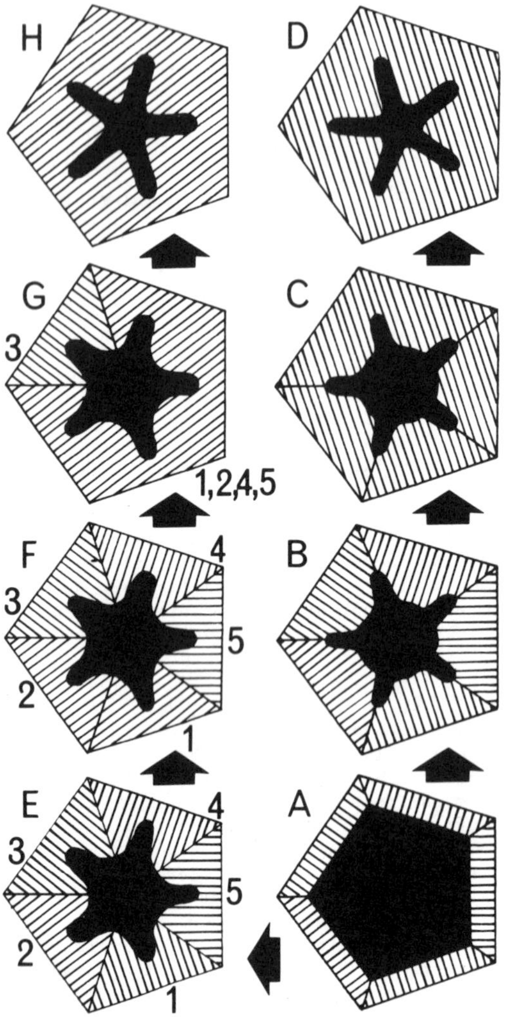

Fig. 3: Evolution of Bystrowicrinus-type columnals from a primitive, pentameric ancestor. Cleavage of meres shown diagrammatically by parallel lines.

in the Upper Ordovician probably reflects an associated increase in crinoid taxa.

Sprinkle (1973) and Stukalina (1966) have both discussed the early evolution of the pelmatozoan column on the basis of North American and Russian material, respectively. Sprinkle considered both blastozoan and crinozoan stems to have evolved from similar ancestral hohlwurzels (1973, fig. 14) and recognised the earliest holomeric columnals to be of blastozoan origin (the Middle Cambrian eocrinoid Eustypocrinus minor). The first crinozoans are Middle Cambrian, i.e. after the first blastozoans, and the evolution of the holomeric crinoid columnal was a relatively later event. Sprinkle's scheme of blastozoan columnal evolution may be incomplete because it does not recognise a pentameric intermediate between early hohlwurzels and Eustypocystis, whereas pentameric columns are found in later rhombiferans such as Caryocystites.

Stukalina (1966, fig. 1) recognised a number of functional changes in the evolution of holomeric from pentameric columnals; fusion of meres, reduction of axial canal diameter, development of stronger ligamentation and decrease in columnal height. Evidence for these changes is found in the British Ordovician.

The fusion of pentameres to form a holomeric columnal is accepted by Stukalina without detailed discussion of how this might have occurred (1966, 1967). In particular her evolutionary pathway III to IV (1966, fig. 1) is herein considered i.e. the evolution of a holomeric columnal with a pentastellate lumen from a primitive, pentameric ancestor. Stukalina's figure shows lumen rays coincident with columnal angles but these sometimes alternate. It is probable that both coincident and opposed lumen orientations evolved by similar pathways, rather than one from the other, although Stukalina has postulated that the latter may have occurred (1967, fig. 6).

Fig. 3 shows two hypothetical pathways whereby a pentameric columnal with a large, pentagonal lumen (A) could have evolved into columnals with pentastellate lumina which are either coincident with (B-D) or opposed to (E-H) the columnal angles. Either mechanism could have produced either lumen configuration. Both pentastellate lumen configurations existed and evidence for both mechanisms is known. It is therefore probable that pentastellate lumina evolved more than once during the Ordovician.

Consider the sequence A to D. A is a pentameric columnal with a large, coincident lumen. Meric sutures correspond to the columnal angles. The lumen decreases in size, becoming stellate as the diameter reduces mainly between the meric sutures (B). Separate meres still have different crystallographic orientations. If, after a further evolutionary step, the meres become crystallographically coincident, the overall appearance of the columnal will not have changed (C) but it would only require fusion of the meres, and a possible further decrease in lumen diameter, to produce a holomeric Bystrowicrinus-type columnal (D).

Alternately, in the sequence A, E to H, reduction in lumen diameter from A once again produces a pentastellate outline (E). In this example, meric sutures are situated between lumen rays. All meres have different crystallographic orientations and are numbered 1 to 5. An evolutionary step leads to, say, 1 and 2 having similar crystallographic orientations (F) and a further step fuses them. This is repeated twice more, so that 1, 2, 4, 5 are fused and separated from 3 by different crystal orientations and the meric sutures (G). Reorientation of the calcite in 3 precedes a final fusion event and a possible further lumen reduction (H).

These evolutionary sequences agree with the evidence of the fossil record (Fig. 2). Pentameric columns with large, pentagonal axial canals are known from the Lower Arenig of Britain and elsewhere e.g. Aethocrinus moorei Ubaghs (1969). The earliest pentameric columns with pentastellate lumina are Upper Arenig, from the Tourmakeady Limestone of Co. Mayo, Eire. Bystrowicrinid columnals first appear in the Caradoc and presumably evolved from pentameric columnals in the Middle Ordovician. Only a single Bystrowicrinus columnal has been found from south of the Iapetus Suture in pre-Ashgill rocks i.e. such columnals are almost exclusively limited to the North American faunal province until the Ashgill. Mixing of brachiopod and trilobite faunas of the South European and North American provinces increased through the Ordovician as the width of the Iapetus Ocean decreased (Cocks & Fortey 1982). The common occurrence of both pentameric and holomeric columnals with pentastellate lumina in the Ashgill of the South European province may be a further indication of the mixing of faunas as the Iapetus Ocean closed.

Tetragonal and trigonal columnals do not fit into simple evolutionary schemes such as that described above. Ramseyocrinus has both tetrameric and holomeric columnals in different parts of the stem (Donovan 1984). The distal, holomeric part of the stem is assumed to have a square lumen similar to that of the rest of the column. However, this indicates a long gap in the range of 'Tetragonotetragonalis' (Fig. 1; 14). Stems with five-fold symmetry which are xenomorphic sometimes have a proximal holomeric portion and are pentameric distally e.g. Euspirocrinus (Ubaghs 1978, fig. 45). This situation is reversed in the column of Ramseyocrinus i.e. the dististele is holomeric. For these reasons it is thought unlikely that Ramseyocrinus is directly related to any later tetragonal morphospecies.

Colpodecrinus is a camerate with a tetrameric stem and the Ashgill genus Xenocrinus has a holomeric, tetragonal stem (with a pentastellate lumen). Both are monobathrids. Colpodecrinus is known from both sides of the Iapetus Suture (Sprinkle & Kolata 1982: Donovan 1983) but Xenocrinus is limited to the North American faunal province. However, with the greater mixing of faunas during the Ashgill, it may be that some of the tetragonal plates from south of the Iapetus Suture, particularly in the Ashgill, are xenocrinids.

The appearance of trimeric columnals postdates that of holomeric, trigonal columnals, although Cyclotristellatopa B and Cyclocyclopa K are possibly trimeric (Wright 1983). Columnals with three-fold symmetry are rare and no evidence exists to suggest that trimeric and holomeric forms were ever associated together in the same stem or evolutionary lineage.

Other columnal morphologies have been considered by Stukalina and Sprinkle. The available information from the British Ordovician supports their schemes and they are accepted here as convenient models for use until further information is available.

6 ACKNOWLEDGEMENTS

I thank Drs C.R.C. Paul and E.P.F. Rose for critically reading an early draft of this paper, and Mr. D. J. McCabe for photographing the figures. This research was completed during the tenure of a NERC research studentship at the University of Liverpool, under the supervision of Dr. Paul. This paper was written during the tenure of a Royal Society ESEP fellowship at Trinity College.

7 REFERENCES

Bates, D.E.B. 1965. A new Ordovician crinoid from Dolgellau, North Wales. Palaeontology 8: 355-357.

Brett, C.E. 1981. Terminology and functional morphology of attachment structures in pelmatozoan echinoderms. Lethaia 14: 343-370.

Clarkson, E.N.K. 1979. Invertebrate palaeontology and evolution. London: George Allen & Unwin.

Cocks, L.R.M. & R. A. Fortey 1982. Faunal evidence for oceanic spearation in the Palaeozoic of Britain. J. geol. Soc. 139: 465-478.

Donovan, S.K. 1983. Tetrameric crinoid columnals from the Ordovician of Wales.

Palaeontology 26: 845-849.
Donovan, S.K. 1984. Ramseyocrinus and Ristnacrinus from the Ordovician of Britain. Palaeontology 27: 623-634.
Lewis, R.D. 1982. Holdfasts. In J. Sprinkle (ed.), Echinoderm faunas from the Bromide Formation (Middle Ordovician) of Oklahoma. Paleont. Contr. Univ. Kans. Monogr. 1: 57-64.
Paul, C.R.C. 1968. Macrocystella Callaway, the earliest glyptocystitid cystoid. Palaeontology 11: 580-600.
Paul, C.R.C. 1976. Palaeogeography of primitive echinoderms in the Ordovician. In M.G. Bassett (ed.), The Ordovician System: proceedings of a Palaeontological Association symposium, Birmingham, September 1974, p. 553-574. Cardiff: National Museum of Wales.
Paul, C.R.C. & A.B. Smith 1984. The early radiation and phylogeny of echinoderms. Biol. Rev. (in press).
Sprinkle, J. 1973. Morphology and evolution of blastozoan echinoderms. Cambridge, Mass.: Harvard Univ., Mus. Comp. Zool.
Sprinkle, J. & D.R. Kolata 1982. "Rhomb-bearing" camerate. In J. Sprinkle (ed.), Echinoderm faunas from the Bromide Formation (Middle Ordovician) of Oklahoma. Paleont. Contr. Univ. Kans. Monogr. 1: 206-211.
Stukalina, G.A. 1966. O printsipakh klassifikatsii stebley drevnikh morskikh liliy (sum.: Principles of classification of stems of ancient sea lilies). Paleont. Zh. 3: 94-102.
Stukalina, G.A. 1967. O taksonomicheskikh priznakakh segmentirovannykh stebley morskikh liliy (sum.: On taxonomic features of articulated stems of crinoids). Trudy vses. nauchno.-issled. geol. Inst. 129, Biostrat. Sbornik 3: 200-206.
Stukalina, G.A. 1968. K sistematike gruppy Pentamerata (Crinoidea) (sum.: On the systematics of the Pentamerata (Crinoidea) group). Paleont. Zh. 1: 81-91.
Ubaghs, G. 1969. Aethocrinus moorei Ubaghs, n. gen., n. sp., le plus ancien crinoide dicyclique connu. Paleont. Contr. Univ. Kans. Pap. 38: 1-25.
Ubaghs, G. 1978. Skeletal morphology of fossil crinoids. In R.C. Moore & C. Teichert (eds), Treatise on Invertebrate Paleontology, Part T, Echinodermata 2(1), p. T58-T216. New York & Lawrence: Geological Society of America.
Warn, J.M. 1982. Long-armed disparid inadunates. In J. Sprinkle (ed.), Echinoderm faunas from the Bromide Formation (Middle Ordovician) of Oklahoma. Paleont. Contr. Univ. Kans. Monogr. 1: 77-89.
Williams, A. et al. 1972. A correlation of Ordovician rocks in the British Isles. Spec. Rep. geol. Soc. Lond. 3: 1-74.
Wright, D.K. 1983. Crinoid ossicles in Upper Ordovician benthic marine assemblages from Snowdonia, North Wales. Palaeontology 26: 585-603.

NOTE

It should be pointed out that the Russian system of columnal morphogeneric terminology (Yeltysheva 1956) places emphasis on the lumen outline e.g. Pentagonocyclicus is a columnal with a pentagonal lumen and a circular facet outline. In contrast, the system proposed by Moore (1939: Wright 1983) places a description of the facet outline first e.g. Cyclopentagonopa is a circular columnal with a pentagonal lumen.

REFERENCES

Moore, R.C. 1939. The use of fragmentary crinoidal remains in stratigraphic paleontology. J. scient. Labs. Denison Univ. 33: 165-250.
Yeltysheva, R.S. 1956. Stebli morskikh lilii i ikh klassifikatsiya (sum.: Classification of sea lily stems). Vest. Leningr. gos. Univ. (Geol. Geog.) 12: 40-46.

Proceedings of the Fifth International Echinoderm Conference / Galway / 24-29 September 1984

Bone idle – A recipe for success?

R.H.EMSON
Zoology Department, King's College, Strand, London, UK

ABSTRACT: This paper takes a fresh look at the characteristic features of echinoderms, and speculates on the role that some may have had in the success of the group. The position, size and the method of formation of the skeleton and the advantages this gives, are emphasized as are the consequences of the substitution of collagenous tissues for muscles in certain circumstances. It will be shown that these features combined with the low level of metabolic activity exhibited by the group may have given the echinoderms competitive advantages over other animals at certain parts of the life history and possibly contributed to their success.

1 INTRODUCTION

Echinoderms are among the most familiar of marine organisms and a highly successful group in the marine environment. They form an important component of most marine communities and are often dominant either in number or in their effect upon the community. To many, however, they appear slow, clumsy, bizarre and apparently ill adapted compared with other groups. Even echinoderm biologists have referred to them in such terms as "elaborate and cumbersome". If they are indeed slow, clumsy, bizarre or ill adapted, then there must be compensating features which allow them to compete with and often dominate over representatives of other animal groups with similar life styles.

Few authors have speculated on the reasons for echinoderm success and even then most often in an oblique fashion. Among those authors who have commented, Nichols (1975) stresses their pentamerous symmetry, the unique skeleton and elaborate nervous system and the considerable regenerative powers.

The present paper is a speculative attempt to demonstrate the role that certain unique echinoderm characteristics may have played both in their early success and in ensuring their continued success to the present day. It is not an attempt to explain totally the success of this major group - this would clearly be an impossible task.

2. ADAPTATION, DESIGN AND CONSTRAINT

Several features of echinoderms are very characteristic of the group. Can these be shown to be significant for the initial and continuing success of the group ?

2.1 Nichols (1967, 1975) has claimed that pentamerous symmetry is a component of echinoderm success, the very fact of its persistence through evolutionary time suggests that it confers some advantage on the group. Attempts to discover the nature of the advantage remain, of necessity, highly speculative. Nichols' own well known suggestion is that a pentamerous arrangement of skeletal plates confers strength on a rapidly growing juvenile echinoderm, the pentamerous arrangement of plates being the minimum number which can produce a body with a circular cross section and yet have few lines of weakness. Nichols suggests that the advantage is transient and pentamerism retained in adult forms only because it occurs in structures which determine the adult body symmetry. It is difficult to see how, in this situation, pentamerous symmetry can be regarded as a reason for the wide success of this group. Stephenson (1967, 1974) seeks a different functional significance for pentamery, stating that there is a case for suggesting that an ancestral lophophorate echinoderm should have an odd number of arms, on functional grounds. He considers that three arms are not as satisfactory as five for the acquisition of food, the detection of enemies etc. and that the addition of further pairs of arms does not confer equivalent advantages. He further suggests that five is the most acceptable solution for ray number in ancestral echinoderms, on cost grounds, believing calcifying the arms to be very energy consuming and reduction of such use of metabolic energy to be a selective agent. Stephenson's ideas remain debatable but provide no support for a suggestion that pentamery is a strong reason for echinoderm success.

2.2 The form, position and magnitude of Echinoderm skeletons is one of the most characteristic features, and the group is unique among the major invertebrate phyla in having an endoskeleton. The internal nature of the skeleton is important as will be demonstrated later, but it is the magnitude of the skeleton and the implications of its size which I wish to consider first.

In all groups, save the holothurians, the skeleton is not only obvious but comprises a very significant component of the dry weight and volume of the body weight. This is after all why echinoderms are favourite ornaments and familiar fossils. Table 1 gives figures for the proportion of the dry weight of a number of common European and Caribbean echinoderms which is actually skeleton. It can be seen that in all cases the skeleton forms more than 50% of dry weight and approximately 25% of volume (since calcite has a density of 1.7-2.0 times that of water) and in some cases (e.g. Echinus, Astropecten) a much higher proportion of the whole.

The skeleton obviously has a supportive function and is also probably of defensive value for many echinoderms, but are there other ways in which possession of a skeleton gives the possessor an advantage over other invertebrates and thus contributes to the animals' short and long term success?

The adaptations of animals are basically to aid species in the struggle for survival either by increasing the chances of survival (reducing the chance of mortality), or optimization of growth as a result of feeding, or of enhancing reproductive effort. The presence of an inorganic skeleton may affect all three in a beneficial manner.

It is commonly believed that many marine invertebrates including echinoderms have survivorship curves of type 3 of Pearl (1928). For such animals, mortality is extremely high in the first weeks of life when individuals are small and those which grow and survive these first weeks have a good chance of achieving maturity. An animal which can rapidly increase in size will progress swiftly through this vulnerable stage and have a considerable selective advantage over slower growing forms.

For most animals size is totally dependent upon conversion of food absorbed into soft tissues, often rather inefficiently, and upon the creation of volume increasing spaces within the body by the separation of tissues. In echinoderms, increase in size involves both these processes, indeed the multiplication of coelomic systems is a characteristic echinoderm feature, but also results from the growth of an inorganic skeleton within the body.

This is an important difference because the laying down of a skeleton of this kind almost certainly requires the expenditure of relatively little energy compared with the fabrication of soft tissues. No actual work has apparently been carried out in this area and the suggestion is pure speculation. However, such physiologists as have been consulted were unanimous in agreeing that the energy costs of

Table 1. The proportion of the dry weight of a series of echinoderms which is inorganic skeleton.

a. European forms		%
Ophiuroidea	Amphiura filiformis	69
	Ophiothrix fragilis	79
	Ophiocomina nigra	60
Asteroidea	Crossaster papposus	75
	Asterias rubens	51
	Marthasterias glacialis	63
	Asterina gibbosa	64
Echinoidea	Psammechinus miliaris	85
	Echinus esculentus	87
b. Tropical forms		
Crinoidea	Cenocrinus asterius (pinnules)	85
Ophiuroidea	Ophiactis savignyi	71
	Ophioderma cinereum	71
	Ophiocoma pumila	78
	Ophiolepis paucispina	83
Asteroidea	Astrophyton muricatum	78
	Astropecten duplicatus	90
Echinoidea	Echinometra methaei	91
	Eucidaris tribuloides	91

precipitation and construction of an inorganic skeleton would be significantly less than those required for an equivalent amount of soft tissue. Thus an animal capable of forming such an inorganic skeleton within its body appears to have an energetic advantage over one which cannot. An animal whose body bulk is 33% (1/3) inorganic material could have used less energy reaching the same size than an animal constructed totally of organic materials, assuming that both have body cavities of approximately equal dimensions. To put it another way, two animals receiving the same rations and with similar capacity for converting food to tissue would be different in size; that animal with the inorganic skeleton being larger. The consequences for a series of hypothetical animals of a variety of shapes and with skeletal components comparable with modern echinoderms are shown in Fig.1.

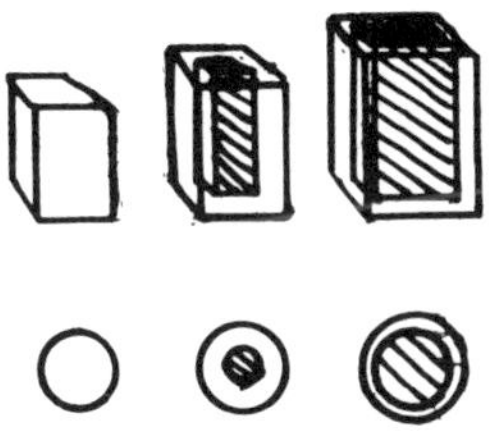

Figure 1.

Diagrams to show the increased size achieved through inclusion of a calcareous skeleton within the body.

As it has been noted that increased size gives the possessor a selective advantage, this would seem to be a clear example of the value of a low cost inorganic skeleton. Such a skeleton would be retained and the size increased up to a limit at which there was a selectively disadvantageous drawback to such a skeleton. The size of the skeleton in most echinoderms (e.g. Table 1) is both an indication of the importance of the tissue to the success of these animals and indirect support for the hypothesis that its production is not highly energy consumptive.

The advantage gained in size increase has other important implications. It has been suggested (e.g. Nichols, 1967) that the ancestral echinoderms were filter feeding lophophorates. The lophophore was depicted as having skeletal support. Successful filter feeders require a large surface area, and it will be obvious that an echinoderm will have, for a given organic tissue weight, a larger surface area.

Crinoids like many other filter feeders function by having a large filter held out in the water. The calcareous skeleton both strengthens such a fan and allows it to be larger (per unit assimilated energy) than those of animals without or can be manufactured at less cost than those of its competitors. Again, a self sustaining advantage is achieved, since as the animal grows it acquires more food and so grows proportionately more.

The previous two examples of possible considerable selective advantage imparted by possession of an internal skeleton relate to survival and growth. If growth to maturity is made more rapid by reason of possession of a calcareous skeleton, this would be of further potential significance. Calow (1983) points out that in several groups of invertebrates, somatic tissues are energetically more costly to fabricate than are gonad materials. Thus for an animal incorporating calcareous material, the size at which the switch to production of gametes occurs may be reached temporally earlier than for other species and if both continue to grow, more energy (relatively) would be available for gonad fabrication to the echinoderm.

If the hypothesis that the skeleton has been of fundamental importance in echinoderm success is accepted, it is easy to point to examples of filter feeding echinoderms in which the skeletal presence is put to maximum effectiveness. Examples can be found from each class. Among the ophiuroids, the gorgonocephalids are able to achieve the massive branched arm structure with a huge surface area because they consist of over 70% (Table 1)calcite. High surface area for deposit feeding is also the goal achieved by the highly successful sand dollars such as Mellita or Dendraster and is achieved by reason of the massive skeleton. Stalked crinoids are characteristic deep sea animals dependent on filter feeding. Their ability to succeed in a food deficient environment may depend on the ability, at little cost, both to have a large filtering apparatus and to raise it above the sea floor on the largely calcareous stalk.

The possession of an extensive internal skeleton may have provided the possessors with certain advantages, but had many consequences both advantageous and apparently disadvantageous to the animal. Positive features would include the fact that a skeleton would make echinoderms less vulnerable to predation, while, as it is internal, leaving the epidermis to develop other defensive attributes such as distasteful chemicals, and remain available as a sensory and absorptive surface. As the skeleton is formed of individual separate plates, the evolution and development of defensive spines would also be a relatively simple matter. The skeleton also provides an excellent attachment point for muscles, and allows the development of massive coelomic cavities, as in sea urchins, without the support of large amounts of soft tissues. The development of a variety of body forms is clearly made possible by the position and nature of the skeleton.

There are, however, serious problems associated with a skeleton of this kind which may have influenced the pattern of echinoderm evolution. These are overcome by some of the unique qualities of echinoderms. Obvious problems include the fact that echinoderms, because of skeletal bulk, are likely to be relatively inflexible and cumbersome and thus inevitably slow moving. Because the skeleton is internal and often complete, there may be problems of adequate transfer of nutrients from the gut to external surfaces. Even relatively slow movement should require considerable energy expenditure.

These would seem to be serious disadvantages limiting the kind of life styles the group can adopt successfully and also limiting the circumstances in which possession of a skeleton is an advantage. I believe that there can be no doubt that possession of an internal skeleton does limit echinoderms in certain ways, and I think it is also possible to show that echinoderms have developed certain other characteristics to the full in order to become competitive. Thus it seems that echinoderms have become in various ways highly energy efficient.

2.3 Fig. 2 represents in cartoon form one of the two features I would like to mention in support of both this argument and the title, which is that the general level of activity of echinoderms is low. Many, of course, are sedentary or semi-sedentary, often opportunistic feeders and capable of existing for long periods on minimal quantities of food. This is made possible by the fact that the metabolic requirements as measured by respiration rates are lower in echinoderms than in other groups. Lawrence and Lane (1982) and Schick (1983) conclude that this is

Figure 2.

"Echinoderms are mostly sedentary and hence they are more peaceful than other phylas" (from an undergraduate examination answer, quoted by Kenchington, 1983).

the case. Differences in oxygen consumption if total net weight is taken as the parameter of comparison, are quite pronounced because of the influence of the non-respiring skeleton, but even comparisons of ash free dry weight according to Schick show echinoderms to have a low level of oxygen uptake. This has been attributed by Lawrence and Lane (1982) to a low level of muscular activity in echinoderms, since muscle movement requires much energy. Recent studies have, apparently, shown that the nervous system of vertebrates consumes a large proportion of available energy. If this rule also applies to invertebrates, perhaps the low level of organization of the echinoderm nervous system and low level of utilization also means that it is less energy consuming than that of more cephalized competitors.

2.4 One way in which energy saving is achieved in an apparently unique way in echinoderms is that in various circumstances muscles can be substituted by collagenous tissues with the ability to vary their state. In the past few years the presence of collagenous tissues with this ability has been demonstrated in a number of echinoderms and is suspected in others (see Emson and Wilkie, 1980; Wilkie, 1984; Motokawa, 1984b for reviews).

Such tissues appear to substitute for muscles in the cirri of comatulid crinoids (Holland and Grimmer, 1981; Wilkie, 1983), the spine joints of sea urchins and starfish (Motokawa, 1982, 1983), the inter segmental connections of ophiuroid arms (Motokawa, 1984b, Wilkie, 1984), the body wall of holothurians (Motokawa, 1982, 1984a, 1984b; Wilkie, 1984) and possibly in sea-urchin tube feet and pedicellaria (Florey and Cahill, 1982; Hilgers and Splechtna, 1982).

The important feature of this change of state is that it seems likely that it is mediated or controlled by changes in the ionic composition of the ground substance surrounding the collagen fibres, causing this to become more or less fluid and thus allowing the fibres to move to a greater or lesser degree. Since the change is ionic, it is almost certain to require less energy to cause the change than would be required by a muscle. It seems possible that virtually no energy will be required to maintain this collagen in a locked condition, a possibility noted independently by Motokawa (1984a,b) and by Wilkie (1984) as well as the present author. Thus when collagen is acting in the same way as a tonic or catch muscle, less energy is used. It is striking that all the active effector organs of the outside of echinoderms have been shown to use such mechanisms. This of course could be making a virtue out of necessity since, as was pointed out above, most of the effectors are remote from the gut and nutrition consequently difficult. This economy must, however, contribute to the low energy requirements of most echinoderms, and thus to their efficiency.

Most present day holothurians appear to be a very obvious exception to the suggestion that a calcareous skeleton is an advantage. Holothurians, however, are advanced echinoderms which have adopted the veriform body shape and frequently the borrowing or crevice dwelling habit, for which elasticity and contractility are required.

It is striking that modern holothurians have developed the use of viscosity changes in collagen to a high degree, so that movements, changes of shape etc., are largely the result of collagen function.

It could be that evolution, beginning with a holothurian ancestor with an external skeleton with some flexibility provided by collagenous links, has worked to emphasize the latter at the expense of the former, resulting in the gradual increase of flexibility of holothurians and the reduction in the otherwise important skeleton.

2.5 Considerable regenerative ability and the nature of the nervous system have been mentioned above as factors contributing to echinoderm success and are clearly importand to modern echinoderms. There appears, however, to be no way in which the contribution they may have made to echinoderm success can be easily estimated and this will not be considered further here.

3 CONCLUSIONS

Echinoderms demonstrate many design features unique among invertebrates including pentamerism, a multiple coelomic system, extensive endoskeleton, external appendages and the apparently unique properties of the collagen. This paper has attempted to demonstrate how the presence of the endoskeleton may have been of selective advantage to early echinoderms in more ways than was previously realized and to indicate how its presence then imposed constraints on the subsequent design of echinoderms. The formation of the skeleton was suggested to be relatively cheap in terms of metabolic energy because the process was one of ionic precipitation. It was also apparent that in some echinoderm functions where collagen substitutes for muscle, ionic changes requiring little energy are also involved. It is plain that if the speculations made above are true, then an important component of the success of echinoderms is due to their ability to use inorganic ions to maximum effect. Evidently many other factors are also involved, but these properties seem to be particularly characteristic of echinoderms.

Hyman (1955) saluted the echinoderms as 'a noble group especially designed to puzzle the zoologist'. I salute them as a group which has brought economy of effort to a fine art. Bony, they certainly are, idle, they may be - being bone idle may have been their recipe for success!

4 REFERENCES

Calow, P. 1983. Energetics of reproduction and its evolutionary implications. Biol.J.Linn.Soc. 20:153-165.

Emson, R.H. & Wilkie, I.C. 1980. Fission and autotomy in echinoderms. Oceanogr. Mar. Biol Ann Revs 18:155-250.

Florey, E. & Cahill, M.A. 1982. Scanning electron microscopy of echinoid podia. Cell Tissue Res. 224:543-551.

Hilgers, H. & Splechtna, H. 1982. Zur Steuerung der Ablosung von Gift-pedizellarien bei Sphaerechinus granularis (Lam) und Paracentrotus lividus (Lam)(Echinodermata) Zool.Jb. Anat. 107:442-457.

Holland, N.D. & Grimmer, J.C. 1981. Fine structure of the cirri and a possible mechanism for their motility in stalkless crinoids (Echinodermata). Cell & Tissue Res. 214:217.

Hyman, L.H. 1955. The Invertebrates Vol. IV. Echinodermata. New York, McGraw Hill.

Kenchington, W. 1983. Out of focus. J.Biol.Education 17(3):179.
Lawrence, J.M. & Lane, J.M. 1982. The utilization of nutrients by post-metamorphic echinoderms. In Echinoderm Nutrition, M.Jangoux & J.M.Lawrence (eds), p.331-371. Rotterdam, Balkema.
Motokawa, T. 1982. Rapid change in mechanical properties of echinoderm connective tissues caused by coelomic fluid. Comp.Biochem.Physiol. 73C:223-229.
Motokawa, T. 1983. Mechanical properties and structure of the spine joint central ligament of the sea urchin Diadema setosum (Echinodermata, Echinoidea). J.Zool.Lond. 201:223-235.
Motokawa, T. 1984a. The viscosity change of the body-wall dermis of the sea cucumber Stichopus japonicus caused by mechanical and chemical stimulation. Comp.Biochem.Physiol. 77A, 3:419-423.
Motokawa, T. 1984b. Connective tissue catch in echinoderms. Biol.Revs. 59:546-549.
Nichols, D. 1967. The origin of echinoderms. Symp.Zool.Soc.Lond. 20:209-229.
Nichols, D. 1967. The rule of five in animals. New Scientist. 35:546-549.
Nichols, D. 1975. The uniqueness of echinoderms. Oxford Biology Readers Oxford University Press.
Pearl, R. 1928. The rate of living. New York, Knopf.
Schick, J.M. 1983. Respiratory gas exchange in echinoderms. Echinoderm Studies, 1:67-110. Rotterdam, Balkema.
Stephenson, D.G. 1967. Pentameral symmetry in echinoderms. Nature. 216:994.
Stephenson, D.G. 1974. Pentamerism and the ancestral echinoderm. Nature. 250:82-83.
Wilkie, I.C. 1983. Nervously mediated change in the mechanical properties of the cirral ligaments of a crinoid. Mar.Behav.Physiol. 9:229-248.
Wilkie, I.C. 1984. Variable tensility in Echinoderm Collagenous Tissues: A Review. Mar.Behav.Physiol. 11:1-34.

Proceedings of the Fifth International Echinoderm Conference / Galway / 24-29 September 1984

Echinoderm zonation in the Rockall Trough (NE Atlantic)

J.D.GAGE & MARGARET PEARSON
Dunstafnage Marine Research Laboratory, Oban, Argyll, Scotland, UK

D.S.M.BILLETT
Institute of Oceanographic Sciences, Wormley, Godalming, Surrey, UK

AILSA M.CLARK
British Museum (Natural History), London, UK

MARGIT JENSEN
Zoological Museum, Copenhagen, Denmark

G.L.J.PATERSON
British Museum (Natural History), London, UK

P.A.TYLER
Department of Oceanography, University College of Swansea, UK

ABSTRACT: In deep-sea sampling undertaken by the SMBA over the past 10 years much further information has been obtained on the distribution of echinoderms in the Rockall Trough. Study of the more abundant species has defined bathymetric distributions which reflect vertical zonation in the overall distribution of the echinoderm fauna. The most clearly expressed pattern was a transition zone around 1000 m depth. A further increase in the rate of first and last occurrences of species seems to occur around 2000 m for ophiuroids and holothurians. Another trend was a considerably broader bathymetric range shown by post-larval and juvenile stages compared to the adults of certain species that are thought to have indirect larval development.

1 INTRODUCTION

The programme of benthic sampling undertaken by the Scottish Marine Biological Association since 1973 has provided the first opportunity to expand our knowledge of the echinoderm fauna of the Rockall Trough since the exploratory sampling undertaken in the nineteenth century.

Although the SMBA programme has focussed mainly on quantitative and seasonal sampling at two fixed stations, sampling also has been undertaken elsewhere in the Rockall Trough in order to map distributional trends and to try to relate these to hydrographic features and bathymetry. Our initial emphasis in these studies has been on the Echinodermata because they are generally large enough to be conspicuous in deep-sea hauls and, perhaps because of this, are the best known of all the groups comprising the deep-sea fauna. The present paper considers the distributional trends shown by this important group with respect to bathymetry.

The positions of the sampling stations yielding echinoderm records are indicated on the bathymetric chart of the Rockall Trough area shown in Fig. 1. Further details of the bottom sampling stations worked are listed by Gage, Pearson, Clark, Paterson & Tyler (1983) and by Gage, Billett, Jensen & Tyler (in press). These two publications also give information on the kinds of gear employed and summarize previously known information on the distribution of the species identified, providing notes on their biology.

Samples have been taken at over a representative range of depths, and at various locations in the Rockall Trough, in order to describe the general distribution of the echinoderm fauna in relation to bathymetry. A concentration of depth-related sampling has taken place on a relatively gently-sloping portion of the continental slope and continental margin lying to the west of Barra. There was a further concentration of sampling effort around a permanent station in the deeper water of the southern Rockall Trough. Elsewhere, Fig. 1 shows a scattering of samples taken generally from intermediate depths at other areas of the Rockall Trough. The distribution of the samples with respect to depth is summarised in Fig. 2.

2 RESULTS

Fig. 3 shows a diagrammatic representation of the bathymetric distribution of the most abundant sea stars and brittle stars, while that of the sea urchins and sea cucumbers is shown in Fig. 4. Species occurring in less than four samples are excluded. The Crinoidea are also excluded because our records are still too sparse to present a clear picture of their depth-related

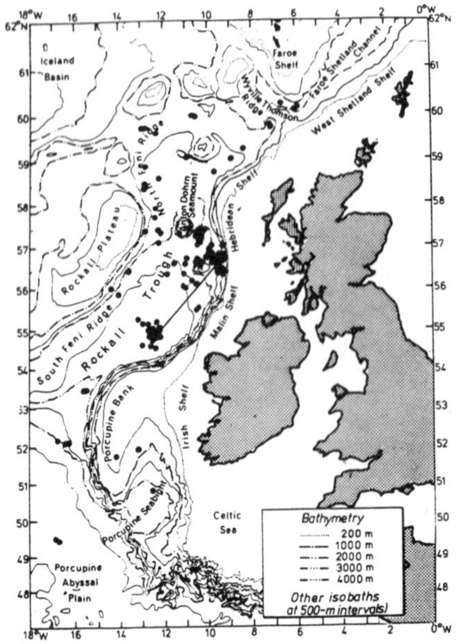

Fig. 1: Chart of the Rockall Trough showing sampling stations (filled circles). Only those lying within the Rockall Trough are used in the present analysis. Straight line shows track of bathymetric section depicted in Fig. 5.

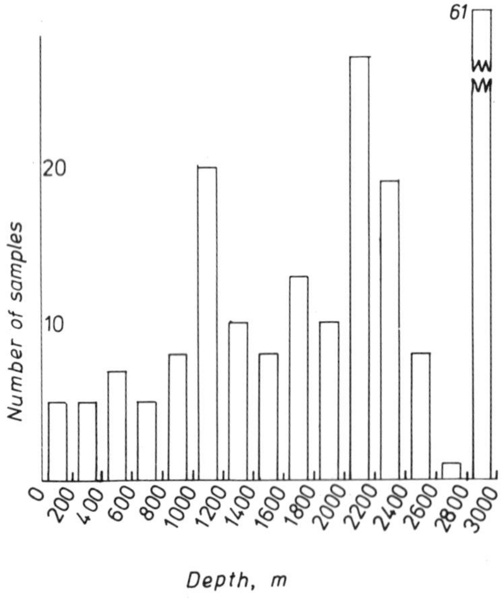

Fig. 2: Bathymetric distribution of samples.

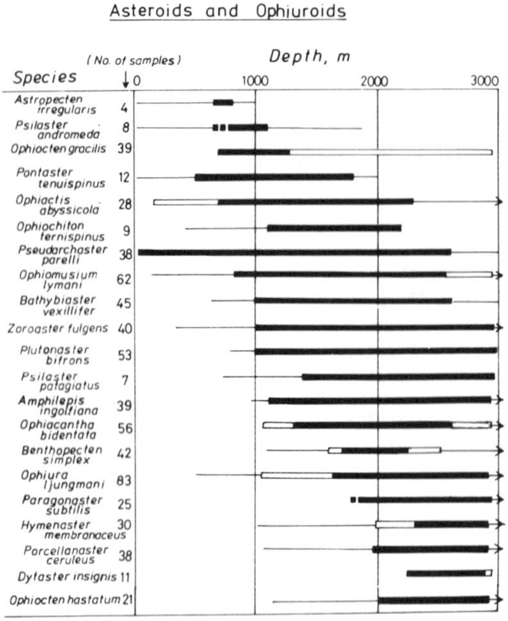

Fig. 3: Vertical range (filled bars) of the most abundant species of sea stars and brittle stars. The unfilled portions of the bars show the bathymetric distribution of post-larval and juvenile stages. Broken bars denote an upper or lower limit that, because of varying depth during the haul, was not clearly defined, the broken bar indicating the maximum possible limit. Thin bars show the maximum limits known for the species over its full range. The numeral after the species name gives the number of samples on which the vertical limits are based.

distribution. As is evident from Figs. 3 and 4, we have recovered juvenile stages of many species occurring over a wider range than that of adults.

The Coincidence-of-Range, or Transect Analysis, method described by Backus, Mead, Haedrich & Ebeling (1965) was applied to the bathymetric limits of adults shown in Figs. 3 and 4. This method employs presence/absence data in order to examine the distribution of the sum of first-plus-last captures of species at between-station intervals of a transect of stations arranged along some form of environmental gradient. In the present case this gradient is one of increasing depth. The observed value of the sum of

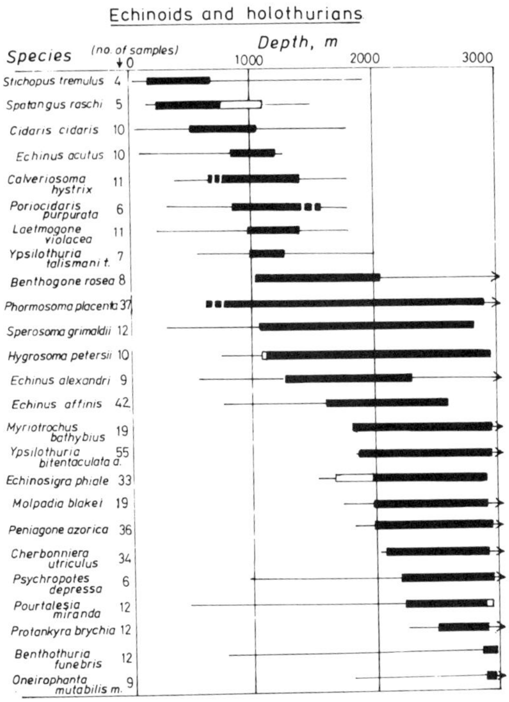

Fig. 4: Vertical range of sea urchins and sea cucumbers. Details as for Fig. 3.

captures is compared as a chi-squared statistic to that expected assuming a homogeneous distribution of the fauna. The chi-squared values are in this way used as an index of faunal change for each depth interval. Further details of the method, including a derivation of the distribution of expected values, is provided by Backus et al. (1965). The relative values of the index are plotted on a bathymetric profile (Fig. 5) whose track is shown on Fig. 1 and encompasses the concentration of sampling effort on the Hebridean slope and the Permanent Station.

The cumulative addition of all species ordered with respect to depth is shown in Fig. 6 for each of the four classes of Echinodermata considered.

3 DISCUSSION

The coincidence-of-range method represents an approach to zonation studies that is not based on the usual concept of bathymetric divisions based on the maximum rate of change between adjacent pairs of stations arrayed along the depth gradient (Ekman, 1953; Vinogradova, 1962; Menzies, George & Rowe, 1973). The latter approach permits objective methods of classification and ordination to be applied (e.g. Haedrich, Rowe & Polloni, 1980; Carney & Carey, 1982). However, these methods require the comparison of samples obtained with similar gear using standardized methods. Comparison of samples obtained with variety of gear, as employed in the present study, would present problems of serious bias in such an analysis. The coincidence-of-range method was first applied to the bathymetric gradient in the deep sea by Haedrich, Rowe & Polloni (1975). In this context the important statistics are the number of upper and lower depth limits that occur within each depth interval. In a situation with regions of faunal homogeneity separated by regions of transition, the occurrence of upper and lower limits should occur concurrently more often than expected on the basis of chance. Carney, Haedrich & Rowe (1983) point out that, unlike the simple presence or absence of a given species from a sample, the presence of an upper or lower limit within a certain part of the depth gradient is dependent on the intensity of sampling beyond the observed range. Hence, this approach will be sensitive to varying sampling intensity along the gradient. As such, the range coincidence method is best suited for heavily studied areas where the vertical range of species is well established.

The results from the analysis of coincidence-in-range (Fig. 5) indicate peaks in faunal change occur around the 1000-m and 2000-m depth levels. However, these depth levels also roughly correspond to peaks in sampling effort in relation to depth (Fig. 2). Although this correspondence must throw doubt on the validity of the results, it is felt that the level of sampling was probably sufficient at all depths to satisfactorily delimit the range of the species considered. Further evidence for the validity of zones of maximal faunal change occurring at around the 1000-m and 2000-m levels is provided by the species recruitment curves. These include all (non-crinoid) echinoderm records rather than the most abundant species. The curves for each of the echinoderm classes considered in general show similar trends in the rate of addition of new species with increasing depth. These show the maximum rate of faunal change occurs at around 1000 m. A more gradual gradient was evident

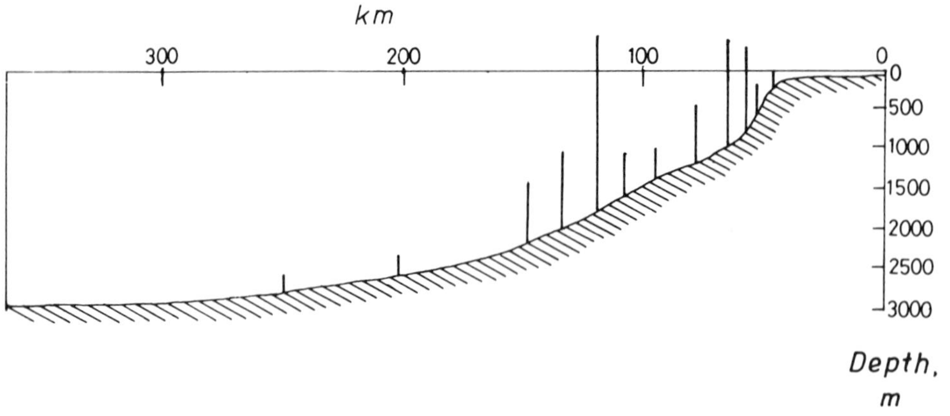

Fig. 5: Relative values of the chi-squared statistic of coincidence in range plotted on the bathymetric profile whose track is indicated in Fig. 1.

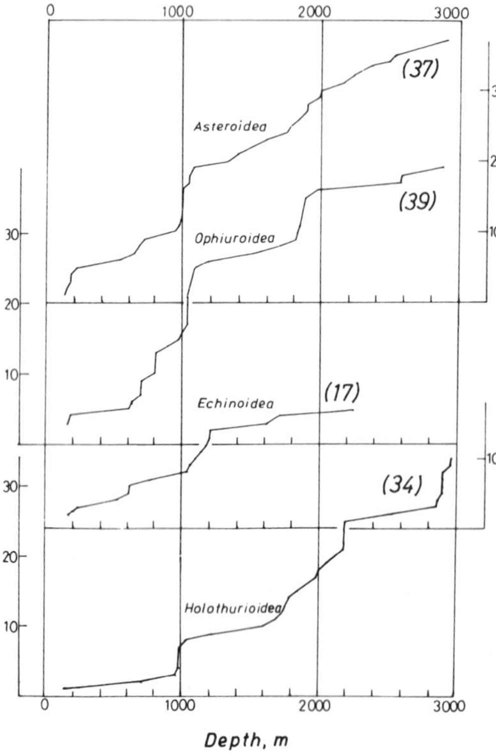

Fig 6: Species recruitment curves for all non-crinoid species encountered in the samples plotted against increasing depth. Total numbers of species are given in parenthesis.

around 2000 m, although more variation was shown between the four echinoderm groups at this depth. However, since these two depths also correspond to the peaks in sampling effort, the same doubts as to their validity must remain. A markedly asymptotic relationship of species to number of samples is evident from a plot (Fig. 7) from the large number of samples taken from a repeat station at about 2200 m depth lying to the E of the Anton Dohrn Seamount (see Fig. 1). We feel that since four samples was the smallest number providing records at a given depth interval (Fig. 2), it seems unlikely that the additional number of samples in the intervals at 1000-1200 m and 2000-2200 m would have yielded sufficient additional species to account for the observed large increments in new species.

3.1 Zonation patterns elsewhere

In general terms, the rich fauna of echinoderm species found in the Rockall Trough is similar to those found at equivalent depths elsewhere in the N E Atlantic. A zonation of the echinoderm fauna occurs that in general terms follows closely that described by Le Danois (1948) in his classic 'Les Profondeurs de la Mer'. On the basis of epibenthic faunal associations, he recognized three major zones. The first, the slope zone, extends from the shelf-slope break at about 200 to 1000 m depth and is subdivided into two subzones.

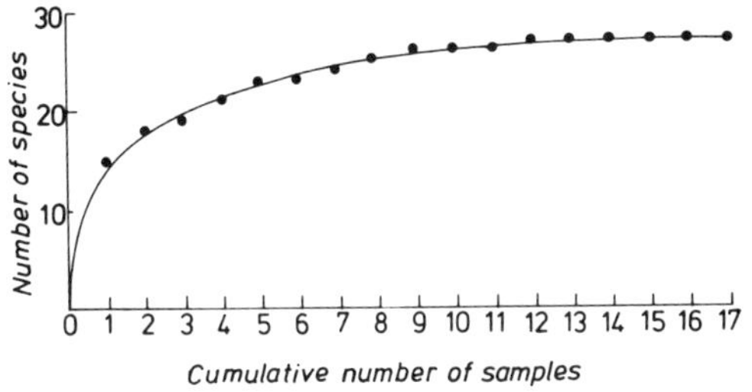

Fig. 7: Species recruitment curve for repeat station located E of Anton Dohrn Seamount.

This is followed by the semi-abyssal zone at 1000-2000 m, and by the abyssal zone that is subdivided into three levels, at 2000-3000 m, 3000-5000 m and that beyond 5000 m. The echinoderm components that Le Danois described for these faunal associations correspond closely to the areas between the areas of maximum change shown by the present study. Le Danois (1948) related his zones to the sediment characteristics found at different depths.

Comparison with the results obtained by the French BIOGAS investigations from sampling in the Bay of Biscay also show many similarities in terms of the bathymetric distribution of individual Asteroidea and Holothurioidea (Sibuet, 1977) and in Ophiuroidea (data of G.L.J.P.).

The present results for the mainly megafaunal Echinodermata are in general accord with the findings of previous investigations into the zonation of deep-sea fauna as summarised by Menzies et al. (1973) and Carney et al. (1983). These include the occurrence of a distinct fauna on the continental slope below the shelf/slope break at about 200 m, and an 'abyssal' fauna that starts at about 1000 m. The intervening area between these areas of relative faunal homogeneity has been termed the archibenthal zone of transition by Menzies et al. This transitional zone has been subdivided by some workers (Menzies et al., 1973; Mills, 1972; Haedrich et al., 1975, 1980) although the supporting data involve restricted subsets of the total benthic fauna that vary from study to study. The megabenthic echinoderm data of the study of Haedrich et al. (1980) in the N W Atlantic provides one the most promising sources for comparison with the present data. The species recruitment curve obtained by these workers indicates the rate of species increment to be greatest at 200-300 m with another, wider, zone of increment extending from 1200 to about 3000 m depth. For the epibenthic holothurians, Carney & Carey (1976) found a similar transitional zone corresponding to the shelf/slope break in the N E Pacific. This was followed by continuous replacement of species down to about 2000 m with no discernible boundaries. However, in terms of species of certain other groups encountered for the first time, Pearcy, Stein & Carney (1982) found evidence of both a transition at the shelf-edge and another in mid-slope at 1900-2200 m. In view of the disparity that is evident between the present results and those obtained in the N E Pacific and western N Atlantic it seems justified to look for more local rather than global factors such as the pressure gradient (Somero, Siebenaller & Hochachka, 1983) and the depth of the 10°C isotherm (Ekman, 1953) in order to account for deep-sea zonation patterns. The relationship of zonation to such factors as water-mass structure and the physico-chemical characteristics of the water column in the Rockall Trough must remain speculative. However, certain features that correspond to the zonal pattern observed may be mentioned. These include the occurrence of Mediterranean Water at 800-1200 m characterized by relatively high salinity and temperature, and low oxygen and the underlying layers of, firstly, cold Labrador Sea Water at 1200 to about 2000 m with high oxygen content, and high-silicate Northeast Atlantic Deep Water below about 2000 m (Ellett & Martin, 1973). The recent discovery of contour-following slope current of northward flowing water at the 200 to ca 1000 m depth level (Booth & Ellett, 1983) may also be an important factor in the establishment of the faunal zone observed between these depth levels.

3.2 Juvenile compared to adult range

The use of the fine-meshed epibenthic sled in our sampling has revealed a wider distribution of juveniles and post-larvae of many species, both with respect to their geographic and bathymetric range. Such early stages are not always easy to identify. Examples amongst the more abundant species are shown in Figs. 3 and 4. We suggest that probably planktotrophic or demersal lecithotrophic larval development occur in these species (see Tyler, Muirhead, Gage & Billett, this volume). Such larvae will be dispersed by currents to areas inimical to survival to adulthood. It is also interesting to note that for all those species that these authors indicate the large egg size to be present that is indicative of direct development, no such extended bathymetric limits were found.

3.3 Global versus local influences

Amongst many of the relatively well-known species a bathymetric range has been found in our samples that is somewhat narrower than that known over the full range of the animal (Figs. 3 and 4). However, this is not unexpected given the comparatively small area of the Rockall Trough compared to the known range of species. Such restricted ranges tend to support the view that local as well as global effects are important influences on benthic zonation patterns.

4 REFERENCES

Backus, R.H., G.L.Mead, R.L.Haedrich & A.W.Ebeling 1965. The mesopelagic fishes collected during Cruise 17 of the R/V Chain, with a method for analyzing faunal transects. Bull.Mus.comp.Harv. 134: 139-158.

Booth, D.A. & D.J.Ellett 1983. The Scottish continental slope current. Cont. Shelf Res. 2: 127-146.

Carney, R.S. & A.G.Carey,Jr. 1976. Distribution pattern of holothurians on the Northeastern Pacific (Oregon, U.S.A.) continental shelf, slope and abyssal plain. Thalassia. Jugosl. 12(1): 67-74.

Carney, R.S. & A.G.Carey,Jr. 1982. Distribution and diversity of holothuroids (Echinodermata) on Cascadia Basin and Tufts Abyssal Plain. Deep-Sea Res. 29: 597-607.

Carney, R.S., R.L.Haedrich & G.T.Rowe 1983. Zonation of fauna in the deep sea. In G.T.Rowe (ed.), The Sea, vol. 8, p.371-398. New York, Wiley-Interscience.

Gage, J.D., Margaret Pearson, Ailsa M.Clark, G.L.J.Paterson & P.A.Tyler 1983. Echinoderms of the Rockall Trough and adjacent areas I. Crinoidea, Asteroidea and Ophiuroidea. Bull.Br.Mus.nat.Hist. (Zool.) 45(5): 263-308.

Gage, J.D., D.S.M.Billett, M.Jensen & P.A.Tyler (in press). Echinoderms of the Rockall Trough and adjacent areas 2. Echinoidea and Holothuroidea. Bull.Br. Mus.nat.Hist. (Zool).

Ekman, S. 1953. Zoogeography of the sea. London, Sidgwick & Jackson.

Ellett, D.J. & J.H.A.Martin 1973. The physical and chemical oceanography of the Rockall Channel. Deep-Sea Res. 20: 585-625.

Haedrich, R., G.T.Rowe & P.Polloni 1975. Zonation and faunal composition of epibenthic populations on the continental slope south of New England. J.Mar. Res. 33: 191-212.

Haedrich, R., G.T.Rowe & P.Polloni 1980. The megabenthic fauna in the deep sea south of New England, USA. Mar.Biol. 57: 165-179.

Le Danois, E. 1948. Les profondeurs de la mer. Paris, Payot.

Menzies, R.R., R.George & G.T.Rowe 1973. Abyssal environment and ecology of the world ocean. New York, Wiley-Interscience.

Mills, E.L. 1972. R.R. Stebbing, The Challenger and knowledge of deep-sea amphipods. Proc.R.Soc.Edinb. 72B: 67-87.

Pearcy, W., D.Stein & R.Carney 1982. The deep-sea benthic fish fauna of the Northeastern Pacific Ocean on Cascadia and Tufts abyssal plains and adjoining continental slopes. Biol.Oceanog. 1: 375-428.

Sibuet, M. 1977. Repartition et diversite des Echinodermes (Holothurides-Asterides) en zone profonde dans le Golfe de Gascogne. Deep-Sea Res. 24: 549-563.

Somero, G.N., J.F.Siebenaller & P.W. Hochachka 1983. Biochemical and physiological adaptations of deep-sea animals. In G.T.Rowe (ed.) The Sea, vol. 8, p. 261-330. New York, Wiley-Interscience.

Vinogradova, N.G. 1962. The vertical distribution of the deep-sea bottom fauna in the abyssal zone of the ocean. Deep-Sea Res. 8: 245-250.

Proceedings of the Fifth International Echinoderm Conference / Galway / 24-29 September 1984

The echinoderm fauna of the Newfoundland continental slope

RICHARD L.HAEDRICH
Newfoundland Institute for Cold Ocean Science, Memorial University, St.John's, Canada

JOHN E.MAUNDER
Newfoundland Museum, St.John's, Canada

ABSTRACT: Deep ocean trawling between 200 and 2360 metres in the vicinity of Carson Submarine Canyon, Grand Banks of Newfoundland, reveals a moderately diverse assemblage of crinoids, asteroids, echinoids, and ophiuroids. No holothurians were taken. Vertical zonation is not pronounced. Faunal affinities are greatest with more northerly assemblages. When the fauna of this region is compared with that from the Hudson Canyon area, New York Bight, similarity, continuity, and coherence appear weak. The situation found in the echinoderms is not like that found in fishes. What deep ocean ecologists refer to as communities may be better considered random assemblages, predictable only in a very local sense.

1 INTRODUCTION

This paper is part of a series on the oceanic fauna of Newfoundland. Despite the fact that the region was one of the very first to be discovered and explored by Europeans, Newfoundland remains today one of the more poorly known areas of the earth. Only recently have modern syntheses concerning its biogeography begun to appear (e.g. South 1983).

Our oceanic work began in 1980 with a focus on the Carson Submarine Canyon (Fig. 1). Three cruises to the canyon have been conducted, two on the "Gadus Atlantica" in June 1980 and May 1981, and one on the "Oceanus" in September 1982. Both benthic and pelagic collections were made in the canyon and on and over the continental slope.

The present paper has several objectives. The first is to describe and characterize the echinoderm fauna of Newfoundland's continental slope. The affinities of this fauna, whether more strongly to the north or to the south, are of interest. The data will be compared with those from other areas to see if the general picture of distribution patterns developed elsewhere applies as well here. Of particular interest in this regard is vertical zonation of the fauna, and the continuity and coherence of faunal zones around the ocean margin. Finally, the results obtained from a consideration of the echinoderms will be compared with those already derived concerning the fishes of the region. The question is whether there is a consistent response or considerable variation from group to group. The premise is that one's view of conditions in the deep sea is very likely coloured by the taxa one considers.

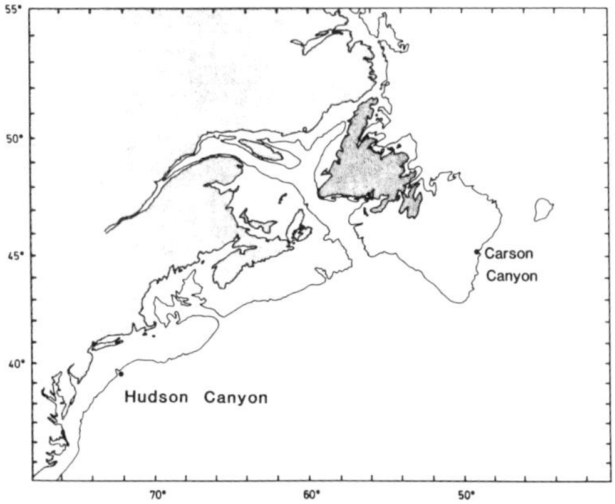

Figure 1. Trawling sites in the Northwest Atlantic: Carson Canyon off Newfoundland and Hudson Canyon off New England.

Table 1. Station data for bottom trawl collections on the continental slope off Newfoundland. Time = time on bottom; Species = number of echinoderm species.

Station	Cruise	Position	Date	Time	Depth, m	Species
814	GA 37	45-27N 48-33W	03/06/80	0550-0645	296-296	13
815	GA 37	45-26N 48-29W	03/06/80	0935-1000	730-730	11
816	GA 37	45-23N 48-29W	03/06/80	1231-1250	900-900	13
838	GA 37	45-18N 48-36W	04/06/80	0812-0832	1420-1425	15
839	GA 37	45-18N 48-36W	05/06/80	1140-1220	1050-1200	5
853	GA 37	45-18N 48-33W	05/06/80	1410-1430	1220-1280	5
865	GA 37	45-25N 48-36W	08/06/80	0110-0130	1050-1050	4
951	GA 51	45-28N 48-30W	15/05/81	1402-1422	596-606	17
952	GA 51	45-27N 48-28W	15/05/81	1700-1720	748-760	13
955	GA 51	45-31N 48-29W	15/05/81	2030-2050	200-209	9
957	GA 51	45-30N 48-29W	15/05/81	2245-2305	302-306	10
958	GA 51	45-29N 48-29W	16/05/81	0000-0020	407-418	10
959	GA 51	45-31N 48-22W	16/05/81	0820-0840	808-811	14
960	GA 51	45-31N 48-25W	16/05/81	1053-1117	498-500	15
961	GA 51	45-33N 48-23W	16/05/81	1225-1245	403-408	8
962	GA 51	45-33N 48-10W	16/05/81	1410-1430	1290-1320	16
963	GA 51	45-34N 48-07W	16/05/81	1605-1625	1490-1505	2
964	GA 51	45-36N 47-56W	16/05/81	1920-1940	1460-1490	10
1350	OC 126	45-29N 48-31W	11/09/82	1205-1235	280-285	10
1352	OC 126	45-18N 48-35W	11/09/82	1730-1800	1250-1450	12
1354	OC 126	45-21N 47-43W	12/09/82	0635-0710	2290-2360	8
1356	OC 126	45-18N 47-51W	12/09/82	1545-1615	2150-2280	8
1357	OC 126	45-24N 47-46W	13/09/82	0600-0635	1960-2010	9
1359	OC 126	45-33N 47-36W	13/09/82	1325-1345	1700-1720	7
1360	OC 126	45-33N 47-30W	13/09/82	1605-1630	1850-1875	13
1361	OC 126	45-41N 47-37W	13/09/82	1955-2015	1360-1360	9
1362	OC 126	45-15N 48-34W	14/09/82	0710-0730	1390-1395	2
1363	OC 126	45-17N 48-34W	14/09/82	1015-1035	1140-1270	9
1365	OC 126	45-08N 48-45W	14/09/82	1555-1615	910-1180	5
1366	OC 126	45-20N 48-34W	14/09/82	1953-2013	1150-1200	5
1367	OC 126	45-19N 48-37W	15/09/82	0730-0750	1150-1350	6
1368	OC 126	45-10N 48-33W	15/09/82	1135-1155	1325-1360	18
1372	OC 126	45-17N 48-34W	16/09/82	0645-0700	1110-1125	11
1373	OC 126	45-17N 48-26W	16/09/82	0955-1010	1700-1725	3
1376	OC 126	45-19N 48-34W	16/09/82	2105-2130	1080-1100	7

2 MATERIALS AND METHODS

The gear and sampling protocol are the same as those used in a similar program in the New York Bight (Haedrich and others 1980). A 12.5-m Gulf-of-Mexico shrimp trawl of 3.7-cm stretch mesh with a 2.5-cm heavy knotted liner in the cod end was used with steel V-doors and 20-m bridles. The speed of the ship during sampling was about 1½ knots. Material was fixed at sea in 10% buffered formalin, and was later transferred to alcohol in the laboratory.

Station data associated with the samples are given in Table 1. Positions of the stations are plotted on the chart of Figure 2. In this region, the coldest waters are associated with the Labrador Current which runs along the edge of the shelf break at depths of about 50-250 m (Table 2). Temperatures throughout the vertical range do not exceed 4° centigrade (Clarke and others 1980). Material from the study has been deposited in the Newfoundland Museum, St. John's. Individual species data by station are available from the authors on request.

To establish whether zonation was present, a variation of the chi-square test (Gardiner and Haedrich 1978) was employed. This method examines the distribution of the ends of individual species ranges, with the null hypothesis being that these occur at random. Faunal overlap between regions was measured using either percentage similarity, PS (the complement of the Bray-Curtis measure; Field and others 1982), or simple overlap. Diversity, H, was calculated using the information function (Haedrich and others 1980).

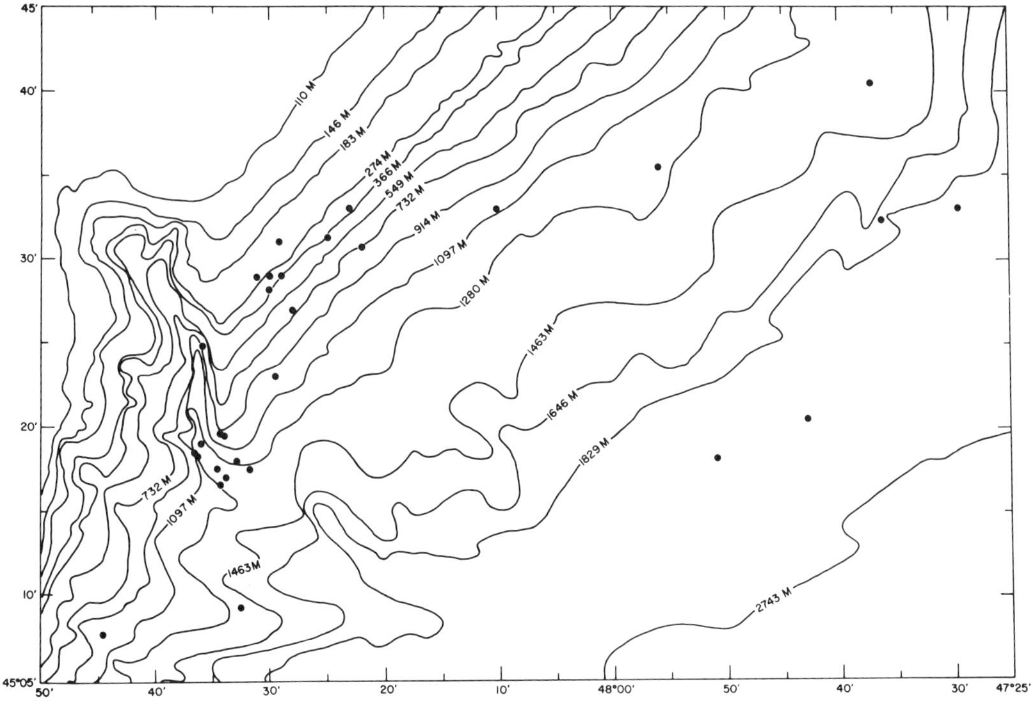

Figure 2. Positions of trawl stations and bathymetry in the Carson Canyon region.

Table 2. Hydrographic data for the Carson Canyon region. To 1000 m: Gadus Atlantica sta 966. Below 1000 m: Chain sta 1266 (Clarke and others 1980).

Depth, m	Temperature, degrees C	Salinity, ppt
0	3.6	32.71
50	0.5	33.26
100	0.8	33.86
200	2.5	34.52
300	3.1	34.69
400	3.7	34.81
500	3.8	34.85
600	3.8	34.86
700	3.7	34.87
800	3.6	34.88
900	4.0	34.95
1000	3.8	34.95
1500	3.6	-
2000	3.2	-
2500	2.9	-

3 RESULTS

3.1 Taxonomic composition of the fauna

In all, 49 species of echinoderms in 23 families were identified from the 35 successful trawls that fished between 200 and 2360 m depth (Table 3). The 10 top-ranking species (82% of total specimens) were Ophiura sarsii, Henricia sanguinolenta, Ophiopholis aculeata, Mediaster bairdii, Echinarachnius parma, Crossaster squamatus, Ophiacantha bidentata, Phormosoma placenta, Brisaster fragilis and Ophiacantha anomala.

3.2 Individual species accounts

In the list which follows, the numbers in brackets [ ] after the species name are number of stations, number of specimens, and vertical depth range in our samples. We report depth ranges conservatively as the deepest point reached by the shallow station to the shallowest point of the deep station. Where only a single station is involved, the depth reported is the range fished by the trawl at that station.

Crinoidea

Antedonidae - three species: Hathrometra sarsi (Duben & Koren 1846) [2, 9, 730-1325 m], ?Heliometra glacialis (Owen 1833) [1, 1, 296 m] and Trichometra cubensis

Table 3. Echinoderm species from the Newfoundland continental slope, and their occurrence in other lists. NF:Carson Canyon, Newfoundland; NE:New England slope (Haedrich and others 1980); GRN:Greenland, ICE:Iceland, SVL:Svalbard (compiled from Mortensen 1932, 1933, Einarsson 1948, Blacker 1957 and Paterson and others 1982); RKL:Rockall (Gage and others 1983). Entries are either numbers taken, +/- (presence/absence), or na = not applicable.

Species	NF	NE	GRN	ICE	SVL	RKL
Amphiophiura saurura	21	36	-	-	-	+
Amphiura fragilis	10	7	+	-	-	+
Bathybiaster vexillifer	1	51	+	+	+	+
Benthopecten simplex	23	-	-	-	-	+
Brisaster fragilis	151	1	+	+	-	na
Brisingella coronata	9	-	-	-	-	+
Ceramaster granularis	33	-	+	+	+	-
Crossaster papposus	11	-	+	+	+	-
Crossaster squamatus	180	-	+	+	+	+
Echinarachnius parma	260	-	-	-	-	na
Gephyrocrinus grimaldii	1	-	-	-	-	-
Gorgonocephalus lamarcki	31	-	+	+	+	-
Hathrometra sarsi	9	-	+	+	-	+
Heliometra glacialis	1	-	+	+	+	+
Henricia sanguinolenta	634	4	+	+	+	+
Hippasteria phrygiana	16	-	+	+	+	-
Homalophiura tesselata	2	1821	-	+	-	+
Leptychaster arcticus	1	-	+	+	-	-
Lophaster furcifer	5	-	+	+	+	-
Marginaster pectinatus	1	-	-	-	-	-
Mediaster bairdii	309	2	-	-	-	-
Ophiacantha anomala	151	-	+	+	-	-
Ophiacantha bidentata	180	4	+	+	+	+
Ophiacantha spectabilis	3	-	-	+	-	-
Ophiochiton ternispinis	1	-	+	-	-	+
Ophiochondrus sp.	1	na	na	na	na	na
Ophiocten gracilis	10	-	-	+	-	+
Ophiomitra hamula	1	-	-	+	-	-
Ophiomusium lymani	125	39679	+	+	-	+
Ophiopholis aculeata	362	35	+	+	+	-
Ophiopus arcticus	15	-	+	+	+	-
Ophioscolex glacialis	113	-	+	+	+	-
Ophiotjalfa vivipara	7	-	+	+	-	-
Ophiura mundata	1	-	-	+	-	-
Ophiura nitida	3	-	-	-	-	-
Ophiura robusta	57	-	+	+	+	-
Ophiura sarsi	759	22858	+	+	+	-
Phormosoma placenta	168	785	+	+	-	na
Plinthaster dentatus	3	-	-	-	-	+
Poraniomorpha hispida	9	-	+	+	+	-
Pseudarchaster parelii	16	11	+	+	+	+
Psilaster andromeda	16	334	+	+	-	+
Pteraster militaris	30	1	+	+	+	+
Pteraster obscurus	4	-	+	+	+	-
Radiaster tizardi	1	-	-	-	-	+
Solaster endeca	1	-	+	+	+	-
Strongylocentrotus drobachiensis	18	-	+	+	+	na
Trichometra cubensis	51	-	+	+	-	-
Zoroaster fulgens	9	1262	-	-	-	+

(Pourtales 1869) [6, 51, 811-1325 m]. Our H. glacialis is a small, damaged specimen.

Plicatocrinidae - one species: Gephyrocrinus grimaldii Koehler & Bather 1902 [1, 1, 1850-1875 m]. Clark (1923) gives the distribution of this species as equatorial mid-Atlantic to the Canary Islands and Madeira.

Asteroidea

Astropectinidae - three species: Bathybiaster vexillifer (Wyville Thomson 1873) [1, 1, 2290-2360 m], Leptychaster arcticus (Sars 1851) [1, 1, 596-606 m], and Psilaster andromeda (Muller & Troschel 1842) [9, 17, 408-1460 m].

Benthopectinidae - one species: Benthopecten simplex (Perrier 1881) [3, 23, 1875-2290 m].

Brisingidae - one species: Brisingella coronata (Sars 1871) [4, 9, 1380-2290 m].

Echinasteridae - one species: Henricia sanguinolenta (Muller 1776) [26, 634, 209-2290 m]. The confused taxonomy of this genus has been reviewed by Rasmussen (1965).

Goniasteridae - 5 species: Ceramaster granularis (Muller 1776) [8, 33, 296-1420 m], Hippasteria phrygiana (Parelius 1768) [5, 16, 403-808 m], Mediaster bairdii (Verrill 1882) [22, 309, 500-1960 m], Plinthaster dentatus (Perrier 1884) [3, 3, 1875-2150 m] and Pseudarchaster parelii (Duben & Koren 1846) [12, 16, 285-1960 m].

Poraniidae - two species: Marginaster pectinatus Perrier 1881 [1, 1, 596-606 m] and Poraniomorpha hispida (Sars 1872) [6, 9, 760-1325 m].

Pterasteridae - two species: Pteraster militaris (Muller 1776) [9, 30, 296-2150 m] and Pteraster obscurus (Perrier 1891) [2, 4, 285-307 m].

Radiasteridae - one species: Radiaster tizardi (Sladen 1882) [1, 1, 1325-1360 m]. Mortensen (1927) records this species from British seas only.

Solasteridae - 4 species: Lophaster furcifer (Duben & Koren 1846) [4, 5, 500-1290 m], Crossaster papposus (Linnaeus 1758) [5, 11, 209-407 m], Crossaster squamatus (Doderlein 1900) [8, 180, 209-596 m] and Solaster endeca (Linnaeus 1771) [1, 1, 900 m]. There is some question as to the validity of the species C. squamatus, although there is no doubt that two distinct Crossasters occur in our collections.

Zoroasteridae - one species: Zoroaster fulgens Wyville Thomson 1873 [6, 9, 1390-1960 m].

Echinoidea

Echinothuridae - one species: Phormosoma placenta Wyville Thomson 1874 [13, 168, 900-1960 m].

Hemiasteridae - one species: Brisaster fragilis (Duben & Koren 1844) [13, 151, 500-1460 m].

Scutellidae - one species: Echinarachnius parma (Lamarck 1816) [8, 260, 209-1000 m].

Strongylocentrotidae - one species: Strongylocentrotus drobachiensis (Muller 1776) [2, 18, 209-280 m].

Ophiuroidea

Amphiuridae - one species, Amphiura fragilis Verrill 1885 [5, 10, 296-1850 m]. This species is considered by Mortensen (1933) to be synonymous with A. denticulata Koehler 1898 described from the Grand Banks near the study area.

Gorgonocephalidae - one species: Gorgonocephalus lamarcki (Muller & Troschel 1842) [17, 31, 408-1490 m]. Most specimens were entangled with an alcyonarian (Nephthyidae) in an apparently parasitic relationship noted by others.

Ophiacanthidae - five species: Ophiacantha anomala Sars 1871 [7, 151, 606-1325 m], Ophiacantha bidentata (Retzius 1805) [18, 180, 209-2290 m], Ophiacantha spectabilis (Sars 1871) [1, 3, 1325-1360 m], Ophiochondrus sp. [1, 1, 1150-1350 m] and Ophiomitra hamula Mortensen 1933 [1, 1, 2150-2280 m]. Our Ophiochondrus specimen shows affinities with O. armatus (Koehler 1909), O. granulatus Koehler 1914 and O. convolutus Lyman 1869. It does not, however, agree precisely with descriptions of any one. Distinctive features include six shortish spines on each lateral arm plate (equal except for the slightly longer uppermost one), a disc completely covered with rough granules (including surface of radial shields), and supernumerary plates associated with the upper brachial plates extending a considerable way out onto the arms.

Ophiactidae - two species: Ophiopholis aculeata (Linnaeus 1767) [16, 362, 209-1420 m] and Ophiopus arcticus Ljungman 1867 [5, 15, 730-2290 m].

Ophiochitonidae - one species: Ophiochiton ternispinus Lyman 1833 [1, 1, 1850-1875 m].

Ophiomyxidae - one species: Ophioscolex glacialis Muller & Troschel 1842 [12, 113, 285-1420 m].

Ophiuridae - 9 species: Amphiophiura saurura (Verrill 1894) [10, 21, 900-2150

Table 4. Echinoderm species from the Newfoundland continental slope: number of individuals by depth zone. Shallow = 200-1200 m; middle = 1200-1800 m; deep = 1800-2400 m. Derived community parameters appear in Table 5.

Species	Shallow	Middle	Deep
Amphiophiura saurura	8	12	1
Amphiura fragilis	8	1	1
Bathybiaster vexillifer	-	-	1
Benthopecten simplex	-	-	23
Brisaster fragilis	54	97	-
Brisingella coronata	-	1	8
Ceramaster granularis	29	4	-
Crossaster papposus	11	-	-
Crossaster squamatus	180	-	-
Echinarachnius parma	260	-	-
Gephyrocrinus grimaldii	-	-	1
Gorgonocephalus lamarcki	15	16	-
Hathrometra sarsi	8	1	-
Heliometra glacialis	1	-	-
Henricia sanguinolenta	405	227	2
Hippasteria phrygiana	16	-	-
Homalophiura tesselata	1	1	-
Leptychaster arcticus	1	-	-
Lophaster furcifer	3	2	-
Marginaster pectinatus	1	-	-
Mediaster bairdii	166	140	3
Ophiacantha anomala	12	139	-
Ophiacantha bidentata	101	66	13
Ophiacantha spectabilis	-	3	-
Ophiochiton ternispinis	-	-	1
Ophiochondrus sp.	-	1	-
Ophiocten gracilis	2	1	7
Ophiomitra hamula	-	-	1
Ophiomusium lymani	5	68	52
Ophiopholis aculeata	350	12	-
Ophiopus arcticus	4	10	1
Ophioscolex glacialis	110	3	-
Ophiotjalfa vivipara	7	-	-
Ophiura mundata	-	1	-
Ophiura nitida	-	-	3
Ophiura robusta	45	12	-
Ophiura sarsi	752	7	-
Phormosoma placenta	11	136	21
Plinthaster dentatus	-	-	3
Poraniomorpha hispida	5	4	-
Pseudarchaster parelii	6	7	3
Psilaster andromeda	5	11	-
Pteraster militaris	16	12	2
Pteraster obscurus	4	-	-
Radiaster tizardi	-	1	-
Solaster endeca	1	-	-
Strongylocentrotus drobachiensis	18	-	-
Trichometra cubensis	3	48	-
Zoroaster fulgens	-	7	2
TOTAL	2624	1051	149

m], Homalophiura tesselata (Verrill 1984) [2, 2, 1125-1325 m], Ophiocten ?gracilis (Sars 1871) [5, 10, 308-2150 m], Ophiomusium lymani Wyville Thomson 1873 [15, 125, 1100-2290 m], Ophiotjalfa vivipara Mortensen 1913 [3, 7, 500-730 m], Ophiura ?mundata (Koehler 1906) [1, 1, 1325-1360 m], Ophiura nitida Mortensen 1933 [1, 3, 2290-2360 m], Ophiura robusta (Ayres 1851) [13, 57, 209-1420 m], and Ophiura sarsi Lutken 1855 [14, 759, 209-1420 m]. Our specimens of Ophiocten agree fairly well with the description and illustrations of O. gracilis as provided by Paterson and others (1982). However, with regard to dorsal scaling, primary rosette, and radial shields they resemble somewhat the recently described O. centobi Paterson, Tyler & Gage 1982. Our single Ophiura mundata is damaged and cannot be identified with full assurance. It has been previously recorded from Denmark Strait and Canaries/Azores.

3.3 Zonation of the fauna and community parameters

The analysis based on the chi-square test suggests that the fauna is zoned with depth. Figure 3 shows the occurence of newly encountered species (species recruitment curve) and the last occurences of previously encountered species with increasing depth. There is a step in both curves in the 1200-1400 m and 1800 m depth ranges, indicating that these are the regions of greatest faunal change. Faunal similarity (PS) between the shallow and middle regions is 35%, and between middle and deep regions is 33%. The number of specimens of each species by region is given in Table 4. Dominant species in the shallow region are Ophiura sarsii, Henricia sanguinolenta, Ophiopholis aculeata and Echinarachnius parma. Ranking species in middle depths are Henricia sanguinolenta, Mediaster bairdii, Ophiacantha anomala and Phormosoma placenta and in the deep region are Ophiomusium lymani, Benthopecten simplex, Phormosoma placenta and Ophiacantha bidentata.

Table 5 gives community parameters in the shallow, middle and deep regions. Species number was highest in the shallow area, but diversity as measured by the information function was about the same in all areas. Abundance declined with depth. We found no correlation between size of individual and depth of occurrence. Size frequency distributions were mostly unimodal.

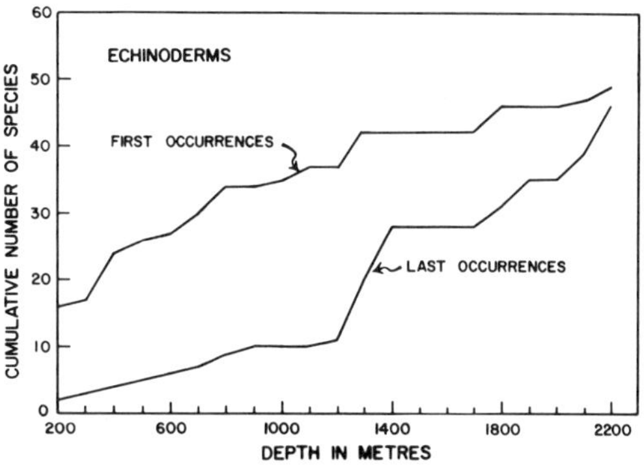

Figure 3. Rates of change in echinoderm species composition with depth increasing by 100-m increments, Carson Canyon region.

Table 5. Community parameters for echinoderm assemblages at shallow (200-1200 m), middle (1200-1800 m) and deep (1800-2400) depths, Newfoundland slope.

	Shallow	Middle	Deep
Species	36	31	20
Diversity, H	2.3	2.4	2.2
Specimens/hr	362	217	89
Stations	18	13	4

4 DISCUSSION

The echinoderm fauna off New England was sampled by Haedrich and others (1980) in the vicinity of Hudson Submarine Canyon (Fig. 1) using the same gear and approach as the Newfoundland study. At similar depths, 52 echinoderm species were taken. This is about the same as the number of species found off Newfoundland (49), but of these only 16 (33% of Newfoundland fauna) occurred in common. In contrast, 32 (65%) of the species recorded from Greenland, 34 (69%) of the species recorded from Iceland, and 21 (43%) of the species from the far northern island of Svalbard were found on the Newfoundland slope (Table 3). From the Rockall Trough, where the echinoids and holothuroids have not yet been reported, there are 20 (44%) species in common with Newfoundland. Thus the Newfoundland fauna has stronger affinities with faunas to the north and east, even across the North Atlantic, than it does with the fauna off southern New England. This is true despite the fact that considerable distance and a number of major ocean deeps separate Newfoundland from the other northern regions, whereas it is removed from New England by only about 1600 km of continuous and presumably fairly uniform continental slope.

Of the 16 species that co-occur off Newfoundland and off New England, 10 (63%) are more abundant to the north (Table 6), an observation that also underlines the northern affinities of this fauna. Three species (19%) that are clearly more abundant to the south are widespread forms that range into deep water: Homalophiura tesselata, Ophiomusium lymani and Zoroaster fulgens. The fourth, Ophiura sarsii, is a wide-ranging and very patchily distributed cool water species.

With respect to the vertical ranges of species taken off both Newfoundland and New England, there are no uniform trends (Table 6). Two species have shallower ranges in the Newfoundland samples: Pteraster militaris and Zoroaster fulgens. Three species occur deeper off Newfoundland than off New England: Brisaster fragilis, Henricia sanguinolenta and Psilaster andromeda. Not surprisingly, Arctic/boreal species tend to have greater vertical ranges off Newfoundland. Widespread species tend to have greater vertical ranges off New England, but Amphiophiura saurura is an exception. Individual species' responses within what is usually considered to be a rather homogeneous deep ocean fauna are clearly not the same from place to place.

Of the 15 echinoderm species found off New England only at depths in excess of 2400 m, none were found in Newfoundland. The essentially northern Newfoundland species simply do not extend farther south in the western Atlantic, and continued investigation will probably show the Grand Bank region to be an important faunal boundary at deep ocean depths. Such a conclusion is not without precedent. Cutler (1975) found evidence for abrupt changes in faunal composition along a short stretch of the slope at Cape Hatteras, and Markle and Musick (1974) showed that the fish fauna at 900 m changed markedly in a transect across Hudson Submarine Canyon.

The absence of holothurians in the Newfoundland collections is somewhat surprising. This cannot be attributed to sampling, for the same gear fished in the same way took numbers of these animals off New England (Haedrich and others 1980). Soft-mud substrates do seem to be more restricted off Newfoundland, but they were sampled by at least 7 trawls which captured 332 specimens of protobranch bivalves (Hutchings and Haedrich, 1984).

Off New England, at least five rather distinct zones were found on the continental slope. Off Newfoundland, there were only three, and these were not well-defined. Composition, in terms of both relative abundance and species assemblages, bore little similarity from zone to zone. This is best seen by comparing PS values between Newfoundland and New England (Table 7), where the mean PS = 11% and the maximum is 42%. The deeper regions of the slope and rise are more similar in composition than are the shallower slopes.

Table 6. Relative abundance and depth ranges of echinoderm species occurring off both Newfoundland and New England.

Species	% of fauna: NF	NE	Vertical range, m: Newfoundland	New England
Amphiophiura saurura	0.6	0.04	900-2150	1885-1967
Amphiura fragilis	0.3	0.007	296-1850	474-2995
Bathybiaster vexillifer	0.03	0.05	2290-2360	1352-2450
Brisaster fragilis	3.9	0.001	500-1460	256-273
Henricia sanguinolenta	16.6	0.004	209-2290	200-385
Homalophiura tesselata	0.05	1.9	1125-1325	260-2497
Mediaster bairdii	8.1	0.002	500-1960	750-950
Ophiacantha bidentata	4.7	0.004	209-2290	1554-1590
Ophiomusium lymani	3.3	41.8	1100-2290	260-2626
Ophiopholis aculeata	9.5	0.04	209-1420	474-529
Ophiura sarsi	19.9	24.1	209-1420	390-1575
Phormosoma placenta	4.4	0.8	900-1960	1250-2155
Pseudarchaster parelii	0.3	0.01	285-1960	393-2654
Psilaster andromeda	0.4	0.4	408-1460	360-520
Pteraster militaris	0.8	0.001	296-2150	2288-2497
Zoroaster fulgens	0.2	1.3	1390-1960	2080-2626

Table 7. Percentage overlap (PS) between echinoderm assemblages off Newfoundland and New England. Shallow: 200-1200 m; Middle: 1200-1800 m; Deep: 1800-2400 m.

Newfoundland:	Shallow	Middle	Deep
New England:			
Shelf (40-264 m)	1	1	0
Slope (283-650 m)	29	1	0
Slope (653-1290 m)	2	12	42
Slope (1380-1947 m)	1	7	25
Rise (2116-2481 m)	1	8	37

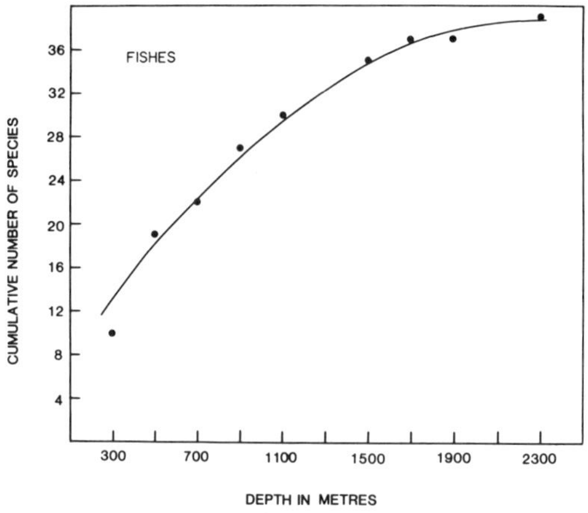

Figure 4. Cumulative first occurrences of demersal fishes with depth increasing by 200-m increments, Carson Canyon region.

In fishes, comparisons between Newfoundland and New England do not give the same picture as does consideration of the echinoderms. The New England fish fauna is zoned with depth, but the Newfoundland fauna shows only a gradual and continuous replacement of species across the slope (Fig. 4). There are about the same number of echinoderm species in the two places, but fish diversity is considerably higher off New England, with 72 species recorded vs 39 off Newfoundland. Affinities are to the south. The Newfoundland fauna has 24 species (61.5%) in common with New England, and 18 (46.1%) in common with the slope fauna between Iceland and Greenland (Haedrich and Krefft 1978). Species shared by Newfoundland and New England are almost all more abundant to the south, and their vertical ranges overlap broadly.

Between the data sets from Carson Canyon and Hudson Canyon one is struck more by differences than by similarities. The conclusion must be that the patterns which so many investigators have observed in the deep sea are predictive really only in rather local instances. But, on a more global scale, deep-sea species may be much more likely to occur in random assemblages and not in predictable and obligate community associations (Williams and others 1981, Grossman and others 1982).

REFERENCES

Blacker, R.W. 1957. Benthic animals as indicators of hydrographic conditions and climate change in Svalbard waters. Invest. Ser. II, volume XX, no. 10. Min. of Agric., Fish., and Food.

Clark, A.H. 1923. Crinoidea. Danish Ingolf Expedition 4(5) : 1-58.

Clarke, R.A., H.W. Hill, R.F. Reiniger & B.A. Warren 1980. Current system south and east of the Grand Banks of Newfoundland. J. Phys. Oceanog. 10(1): 25-65.

Cutler, E.B. 1975. Zoogeographical barrier on the continental slope off Cape Lookout, North Carolina. Deep-Sea Res. 22: 893-901.

Einarsson, H. 1948. Echinoderma. Volume IV, part 70, The Zoology of Iceland. Copenhagen: Munksgaard.

Field, J.G., K.R. Clarke & R.M. Warwick 1982. A practical strategy for analysing multispecies distribution patterns. Mar. Ecol. Prog. Ser. 8: 37-52.

Gage, J.D., M. Pearson, A.M. Clark, G.L.J. Paterson & P.A. Tyler. 1983. Echinoderms of the Rockall Trough and adjacent areas. I. Crinoidea, Asteriodea and Ophiuroidea. Bull. Br. Mus. nat. Hist. (Zool.) 45(5): 263-308.

Gardiner, F.P. & R.L. Haedrich 1978. Zonation in the deep benthic megafauna; application of a general test. Oecologica Berl. 31: 311-317.

Grossman, G., P.B. Moyle & J.O. Whitaker 1982. Stochasticity in structural and functional characteristics of an Indiana stream fish assemblage: a test of community theory. Amer. Nat. 120: 423-454.

Haedrich, R.L. & G. Krefft 1978. Distribution of bottom fishes in the Denmark Strait and Irminger Sea. Deep-Sea Res. 25: 705-720.

Haedrich, R.L., G.T. Rowe & P.T. Polloni 1980. The megabenthic fauna in the deep sea south of New England, U.S.A. Mar. Biol. 57: 165-179.

Hutchings, J.A. & R.L. Haedrich 1984. Growth and population structure in two species of bivalves (Nuculanidae) from the deep sea. Mar. Ecol. Prog. Ser. 17:

135-142.
Markle, D.F. & J.A. Musick 1974. Benthic fish associations on the continental slope of the Middle Atlantic Bight. Mar. Biol. 26: 225-233.
Mortensen, Th. 1927. Handbook of the Echinoderms of the British Isles. London: Oxford University Press.
Mortensen, Th. 1932. Echinoderms. The Godthaab Expedition 1928. Medd. om Gronland 79(2): 1-61.
Mortensen, Th. 1933. Ophiuroidea. Danish Ingolf Expedition 6(8): 1-121.
Paterson, G.L.J., P. Tyler & J.D. Gage. 1982. The systematics and zoogeography of the genus Ophiocten (Echinodermata: Ophiuroidea) in the North Atlantic Ocean. Bull. Br. Mus. nat. Hist. (Zool.) 43(3): 109-128.
Rasmussen, B.N. 1965. On taxonomy and biology of the North Atlantic species of the asteroid genus Henricia Gray. Medd. Danmarks Fisk. Havundersog. 4(7): 157-213.
South, G.R. (ed.) 1983. Biogeography and Ecology of the Island of Newfoundland. The Hague: W. Junk.
Williams, W.T., J.S. Bunt, R.D. John & D.J. Abel 1981. The community concept and the phytoplankton. Mar. Ecol. Prog. Ser. 6: 115-121.

Our work was supported by grant A-7230 from the Natural Sciences and Engineering Research Council, Canada. Contribution no. 67 from the Newfoundland Institute for Cold Ocean Science.

Proceedings of the Fifth International Echinoderm Conference / Galway / 24-29 September 1984

The energetic echinoderm

JOHN M.LAWRENCE
University of South Florida, Tampa, USA

ABSTRACT: The application of the concepts of energetics can be applied to the study of echinoderms at all levels of organization; molecular, tissue and organ, organism, and population. Such an approach is unifying and allows the transfer of information between levels. The approach also provides a qualitatively different type of information. Knowledge about the energetics of echinoderms is incomplete at any level and is completely absent for almost all groups, ways of life, and environments.

1 INTRODUCTION

"The line between ecology and physiology is ... difficult and perhaps ... impossible to draw with exactness... The physiologist seldom gets beyond considering an individual as his upper limit... In contrast, the ecologist usually regards an individual organism as his smallest unit..." (Allee et al. 1949)

In our attempt to understand the biology of echinoderms, we are involved in a quantification of various characteristics at all levels of organization and in both the static and active states. In recent years there has been an increase in quantification in terms of energy units. There are two major reasons for this.

First is the recognition that energetics is a basic characteristic of living systems. To perpetuate themselves through time, living systems must expend energy to maintain their organization and to carry out their processes. In addition, the deposition of matter in their body involves the deposition of energy. Quantification of these energetic phenomena can give insight into the system which quantification in other terms cannot. Representation in terms of energy gives another view of biological phenomena.

The second reason for the increased interest in the expression of the characteristics of the system in terms of energetic units is the potential to equate them over different levels of organization. It thus is possible to unify molecular, physiological, organismal, and population processes in a way not possible with other units. The significance of this is that knowledge of the suborganismal phenomena helps the understanding of the bases for organismal phenomena. Knowledge of the organismal phenomena gives meaning to the suborganismal phenomena and can help explain supraorganismal phenomena. It is the individual which is the fulcrum here. The significance of the organismal level of organization is expressed in the quotation above.

Despite the usefulness of the energetic approach to the analysis of biological systems such as echinoderms, one should always keep in mind that it is not the sole means of analysis which will give insight into the system. Energetics gives no information on size, shape, form, or numbers, and thus an analysis ignorant of these other ways of measurement is invariably lacking in completeness.

My intent here is to indicate the present state of information about the energetics of echinoderms at different levels of organization, to show how information can be integrated between the different levels of organization, and to indicate the directions in which I think such analyses can go in the future.

2 MOLECULAR LEVEL

At the molecular level, much further investigation is needed on the levels of activities of the various energy-producing metabolic pathways. Ellington (1982) concluded that the enzymes involved in glycolysis, the hexose monophosphate shunt, the Krebs cycle, and oxidative phosphorylation exist, but pointed out two major kinds of information that are needed: the actual rate of functioning of the systems and the way in which the pathways are regulated.

Regarding the activity of the entire pathways, almost all studies have been done with single or only a few enzymes of a particular tissue, so that comparisons of the relative importance of different metabolic pathways are difficult. Ellington & Hammen (1977) evaluated the activities of enzymes involved in the utilization of glucose by the longitudinal muscles of the holothuroid Sclerodactyla briareus according to physiological function or position at metabolic branchpoints, and showed the significance of malate dehydrogenase and lactate dehydrogenase which was related to the anaerobic production of energy and the ecology of the species. Smith (1983) showed that although the enzymes of the glycolytic pathway are present in the pyloric caeca of the asteroid Luidia clathrata, the pathway is probably not functional in channeling carbon units to the Krebs cycle for energy production, but probably functions in the transfer of carbon units with lipid anabolism and production of reduced nucleotides. The activities of these metabolic pathways can vary with the nutritional state of the individual, as found for various glycolytic enzymes in the pyloric caeca of the asteroid Echinaster (Watts & Lawrence 1982). The activities vary with other changes in physiological state such as reproduction in echinoids (Propp et al. 1983) and asteroids (de Mik-van der Plas 1981, Propp et al. 1983). Much remains to be done to clarify the pathways by which ATP is produced for energy requiring processes and reduced nucleotides for anabolism.

An important unresolved problem is the degree to which production of energy by different tissues is by aerobic or anaerobic processes. There has been no estimation of the relative contribution of aerobic and anaerobic metabolism to energy production. Anaerobic production of energy is implied by a number of observations (see Shick 1983). There is a general consensus that the body wall has primarily an aerobic metabolism, but this remains to be demonstrated. Giese et al. (1966) and Webster & Giese (1975) suggested that internal organs of echinoderms may rely considerably on anaerobic production of energy, and the calorimetric measurements (Shick 1983, Bookbinder & Shick, this volume) seem to substantiate this. The metabolic bases for this postulated anaerobiosis of internal organs remain to be demonstrated adequately. Ellington (1975) suggested that the Krebs cycle enzymes, usually thought to indicate aerobic metabolism, function instead in synthetic pathways in the pyloric caeca of the asteroid Asterias forbesi. The activity of the electron transport system for a variety of echinoid and asteroid tissues as indicated by the equivalent oxygen-acceptance by a substitute electron acceptor is greater than that measured directly (Monahan 1982). Either the potential for electron acceptance is much greater than actually used or there is considerable regulation of the rate of transfer of electrons into the system. There need to be more measurements of the relative degrees of aerobic and anerobic energy production by the various tissues of echinoderms. The possibility of anaerobic energy production by embryos and larvae has been little considered, but it is probably true that the embryos and larvae are aerobic systems. The molecular pathways for embryonic development are relatively well known (Ellington 1982), and are generally considered to involve combustion of lipids initially, (Lovtrup-Rein & Lovtrup 1980). Information about the metabolic pathways in larvae is non-existent. The amount of energy for metabolism should increase with an increase in larval size. In addition, the rate of energy metabolism, as indicated by the activity of the electron transport system, increases throughout the develoopment of the larva of Lytechinus variegatus (McEdward & Lawrence 1981).

It is known that the activity of various metabolic enzymes of echinoderm eggs and embryos are regulated by nucleotides and metabolites (Ellington 1982), but there have been no studies on regulation of the enzymes in tissues of post-metamorphic individuals. Another aspect of regulation of energy production at the molecular level is the role of hormones. Steroid hormones affect the deposition of nutrients in the pyloric caeca of asteroids (de Mik-van der Plas 1981). Estradiol stimulates oocyte growth and catecholamines inhibit it (Schoenmakers & Voogt 1980, Takahashi & Kanatani 1981; Khotimchenko 1982a,b, respectively). The significance of these studies is that it is now possible to do experimental studies that will give insight into the control of energetic processes associated with nutrition and reproduction. All studies done thus far have involved females. Studies on males need to be initiated as their energetics are not the same as that of females in some echinoderm species (see Lawrence, in press,a).

The effect of environmental variables on the production of energy at the molecular level has been little considered. The most obvious variables are oxygen, pressure, salinity and temperature. Even though the rate of oxygen consumption by echinoderms can remain constant over a considerable decrease in environmental oxygen level and although some species can survive several days under anaerobic conditions (see Ellington 1982, Lawrence & Lane 1982, Shick 1983), most echinoderm species are affected quickly by a lack of oxygen. This does not give any indication of the degree to which the total amount of energy production is based on aerobic or anaerobic metabolism, but simply that at least some tissues are highly aerobic. Anaerobic production of energy is implied by a number of reasons (see summary by Shick 1983). Ellington & Hammen (1977) demonstrated the enzymatic basis for anaerobic production of energy in the longitudinal muscles of Sclerodactyla dactyla and related it to the ability of this burrowing species to survive periods of hypoxia.

Regarding pressure, an individual echinoderm does not experience the range in pressure which could conceivably have an effect on their metabolism. Some species have extremely great vertical distributions in the ocean: the asteroid Pteraster pulvillus occurs from ca. 100 to 6000 m (Sladen 1888), and the holothuroid Elpidia glacialis, from ca. 200 to 19,500 m (Hansen 1975). There might be adaptations of the metabolic enzymes of individuals at the extremes of their vertical distribution.

Echinoderms are more euryhaline than usually thought. The energy budget of the asteroid Leptasterias hexactis is affected by environmental salinity levels (Shirley & Stickle 1982). The basis for this might be neuromuscular, with the effect being on the ability to move and feed. However, it might have a molecular basis in the cost of intracellular osmoregulation which occurs in echinoderms. Intracellular osmoregulation by Luidia clathrata does not involve a maintenance of constant intracellular ionic concentrations (Diehl 1983) and the metabolic machinery might be expected to be affected by changes. The activity of an important enzyme in the Krebs cycle (citrate synthase) is affected by physiological levels of ions which occur in the pyloric caeca of Luidia clathrata in different salinities (Marsh 1983). Unless the metabolic pathways were regulated to counter this effect, energy production would be different in the asteroid stabilized at different salinities.

Echinoderms exist over the entire biokinetic zone from several degrees below zero to ca. 35°C. Individuals of some species experience ranges which can affect their energy metabolism: Luidia clathrata can experience an annual temperature of 12 to 30°C (pers. obs.); the echinoid Strongylocentrotus droebachiensis, from 0 to 15°C (Percy 1973). Most studies show that the rate of aerobic metabolism of individuals varies with temperature in accordance with van't Hoff's law and that there is little or no seasonal acclimatization(see Lawrence & Lane 1982). Here it is important to keep in mind that most echinoderms show seasonal changes in their physiology associated with annual reproductive and nutritional cycles and thus are not physiologically the same at different seasons. Thus the concept of acclimatization must be applied with caution under such circumstances. Similarly, temperature affects the activity and consequently the energy intake by echinoderms. The nutritional condition affects the physiological state of the organs, and is known to affect the activity of metabolic enzymes (Watts & Lawrence 1982). Enzymes adapted to environmental temperatures are to be expected in species which live at widely disparate temperatures. There is no evidence that there are special metabolic systems or pathways associated with cold, polar species in contrast to temperate or tropical species that have a basis in temperature level per se. Seasonal temperature-related changes in enzymes associated with energy production do not appear to occur (Ellington & Lawrence 1974, Watts & Lawrence 1982). Temperature-related changes associated with energy production can occur however. The activity of the hexosemonophosphate shunt in the gut varies with the past temperature history in Strongylocentrotus droebachiensis (Percy 1971). It is possible that this is associated with quantitative changes in the metabolic pathways (either in the amount of enzyme or altered activity by regulation) rather than qualitative changes. It may also be related to temperature-induced changes in physiological state rather than a direct response to temperature. Certainly much anabolic metabolism occurs during low winter temperatures. Frequently, gametogenesis and even deposition of nutrient stores occurs at such times. Either little energy is required for such processes, or the enzymes must be functional at the low temperatures. Species of echinoderms which experience wide ranges of temperature usually are inactive at the seasonal low temperature. It is possible that there will be differences in the temperature sensitivity of energy-producing enzyme systems in the neuromuscular system and those of other tissues which remain active at low temperatures.

The major unanswered questions concerning energetics at the molecular level for echinoderms are the rates and sources of production of energy (ATP) and reduced nucleotides in the different tissues and in different physiological states, and the mechanisms of regulation of energy production.

3 TISSUE AND ORGAN LEVEL

Analysis of the organs of echinoderms is complicated by their lack of specialization. This makes it difficult to analyze the energetics of single cell-types or tissues. Despite this, the organs of the echinoderms do differ sufficiently to identify them in broad functions of protection, locomotion, digestion, nutrient deposition, and reproduction (Giese 1966) and to compare their productivity and production energetically.

The measurements of oxygen consumption by tissues, and thus of aerobic energy production, have been done at environmental levels of oxygen, and may not reflect actual rates under reduced oxygen levels in situ. There is limited knowledge of the relative contribution of aerobic and anaerobic metabolism to energy production by tissues. Shick (1983) showed that the rate of heat (energy) production by various internal tissues of echinoderms under anoxic conditions ranged from 69% to 81% of the normoxic rate. Bookbinder & Shick (this volume) found that anaerobiosis accounted for 76-92% of the total energetic heat production by the ovaries of Strongylocentrotus droebachiensis. These studies need to be extended to ascertain the rate of anaerobic energy production by the various tissues of echinoderms under essentially every physiological and ecological variable. Basically this means that at this time the energy requirements of the organism cannot be assumed to be equivalent to that indicated by oxygen consumption. Knowledge about the degree of anerobiosis will give direction to metabolic studies and are essential for the proper interpretation of the uses of energy as done with energy budgets for individuals.

The level of energy deposited in tissues and organs can be estimated by calorimetry or by use of conversion factors after organic analysis. Calorimetry is quicker, but I have preferred to do organic analysis because of the greater information content of the data. There is general agreement between energy levels measured by calorimetry and calculated by conversion (see Lawrence, in press,a). Knowledge about the amount of energy deposited in the various organic classes is helpful in various ways. Data on the relative levels of glycogen and lipid have implications for the type of anabolic and catabolic systems which exist, and in understanding the energy metabolism of carnivorous and herbivorous echinoderms. The information is useful in understanding trophic relations between echinoderms and their predators. E.g., how much and what is the organic basis of energy obtained by a predator on ophiuroid arm tips. Energy deposited in refractive protein would be resistant to digestion. Lane (in press) has postulated that the large theca of many Paleozoic crinoids contained the energy-rich gonads, and that the development of the large anal sac in inadunate crinoids was to separate them from other vital organs and thus make predation less lethal.

The level of energy in echinoderm tissues and organs varies with the amount and type of organic material present (Lawrence, in press,a). The level depends on both the level of inorganic material and the qualitative nature of the organic material. The level of energy in the body walls varies within and between classes. The body wall of echinoids and the rays of ophiuroids have low levels of energy; the body wall of asteroids and holothuroids vary in levels of energy according to the degree of calcification. Although the level of inorganic material is similar in the testes and ovaries, the level of energy varies as a result of the difference in organic composition. The deposition of intracellular triglycerides in eggs results in a higher energetic level in eggs than in sperm (see Lawrence & Lane 1982). As I have pointed out (Lawrence, in press,a), analysis of entire gonads does not necessarily give an indication of the level of energy deposited in the gametes due to the heterogeneous nature of the organ. Only at reproductive maturity when non-gametogenic tissue is minimal does analysis of the entire gonad give an indication of the level of energy deposited in the gametes. In general, the proximate composition of echinoderm eggs which show either direct or indirect development is simiilar, except for a lower level of soluble protein in large eggs (Turner & Lawrence 1979, Lawrence et al., in press). This means that the difference in egg size with development type does not involve a difference in energy deposition among the major organic components.

A low level of energy in a tissue or organ does not mean that there is little energy deposited there. The amount of energy (= production) depends on both the level of energy and on the degree of development of the organ (table 1). The development of the organ varies with the species, the stage of growth, and often the physiological state of the individual. E.g. the amount of energy in the spines of echinoids differs greatly according to the species; the amount of energy in the Aristotle's lantern varies with size and nutritional state of the individual (Ebert 1980, Black et al. 1982); the amount of energy in the pyloric caeca of asteroids varies with season. These estimates are indicative of the amount of production associated with the various organs of various species.

Within each class, the allocation of energy in the various body tissues or organs allows an evaluation of the energetic significance of the various components, i.e., how much energy does a species allocate to different parts of its body (fig. 1). There are major differences among the various species of ophiuroids in which the amount of energy in the arms and the disc. To what aspect of the biology of the species is this related? To the feeding type? To the habitat? Even with an inability to associate the amount of energy in the arm and disc with their functions, there are implications for understanding the biology of the species. The amount of energy deposited in the arm indicates that this amount of energy is not available for deposition in the disc. An amphiurid and a euryalid of similar disc sizes deposit much different amounts of energy in their arms.

Only asteroids among the echinoderms have a discrete nutrient-storage organ, the pyloric caeca, which is relatively energy-rich because of its high levels of lipid. How does the amount of energy deposited in an individual asteroid differ from other types of echinoderms because of the presence of this organ? What is the role of the body wall and the gonad as energy stores in non-asteroid echinoderms? What are the biological consequences for asteroids of the possession of energetically-dense storage organs? How does the amount of energy deposited in various species of asteroids differ between species with much different forms of arms which result in different relative sizes of pyloric caeca? How does energetic representation of echinoids differ according to the characteristics of the test, the spines, and the Aristotle's lantern?

Within a species, the energy deposited within an individual will vary with its size, and also with its nutritional and reproductive state. In asteroids, the nutritional state will affect the amount of energy deposited in the pyloric caeca as well as in the other somatic tissue as in other classes. Consequently, the energetic view of the asteroid in different nutritional states will depend considerably on the size of the pyloric caeca whereas it depends primarily on the body size alone in other classes. Even in good nutritional condition, the pyloric caeca of asteroids varies considerably in size with the seasonal reproductive state, and the amount of energy in an individual will differ as a result.

The development of the gonad to its reproductive maturity affects the amount of energy present in individuals of all classes. The maximal development of the gonad (= reproductive output) varies among and within classes (table 2), within an individual according to its body size, and sometimes even with sex (fig. 2). There may be space or area constraints of a particular body form which limits the maximal gonad size and thus energy deposition in the gonad. E.g., the greatly sunken marsupia of Abatus cordatus results in ovaries smaller than the testes of males (Magniez 1983). Contrast the coelomic volume of the flat rays of Anasterias to the round ones of Echinaster (Lawrence et al., in press). The difference in energetic reproductive output indicates the difference in the energetic requirements of the individual and its ability to use the energy for different purposes. Gonads are not homogeneous in regard to cell-type and even developing gametes vary with stage. Analysis of whole ovaries thus give no information about the energy content of eggs except for those situations in which only ova are present at reproductive maturity as with the spatangoid Abatus cordatus (Magniez 1983). The amount of energy in the small eggs with indirect development can range from 0.0104 $J.egg^{-1}$ to 0.0159 $J.egg^{-1}$, while large eggs with direct development can range from 2.5 $J.egg^{-1}$ to 159 $J.egg^{-1}$ (Lawrence et al., in press). These data have obvious implications in considering the analysis of the energetics of development and larval ecology. I.e., is the energy needed for development or to produce a certain sized newly-metamorphosed individual. The maximal energy content of the egg may represent the amount of matter necessary for a particular planktotrophic feeding type (Strathmann & Strathmann 1982). The maximal energy content may simply have to do with the limitations of the developmental processes, or associated with the physical characters of the egg such as the strength of the cell membrane. The minimal energy conent of a direct developing egg may be that of a functional, newly-metamorphosed juvenile. There is little decrease in

Table 1. Energy content of echinoderm body parts.

Species	Body part	kJ.ind^{-1}	Reference
Holothuroids			
Actinopyga mauritana	BW	1475.6	Lawrence & Guille unpub
Afrocucumis africanus	BW	2.1	"
Benthogone rosea	BW	55.4	Sibuet & Lawrence 1981
Chiridota rigida	BW	0.4	Lawrence 1980
Cladodactyla crocea	BW	12.1	Lawrence & Guille 1982
Eumolpadia violacea	BW	5.9	"
Holothuria atra	BW	106.3	Lawrence 1980
Holothuria difficilis	BW	2.0	"
Holothuria glaberrima	BW	73.7	Lawrence & Kafri 1979
Holothuria hilla	BW	12.6	Lawrence 1980
Holothuria leucospilota	BW	103.8	"
Paelopatides gigantea	BW	97.1	Sibuet & Lawrence 1981
Pentacucumis bovetensis	BW	0.6	Lawrence & McClintock unpub
Pseudocnus laevigatus	BW	0.9	"
Psychropotes longicauda	BW	272.0	Sibuet & Lawrence 1981
Stereoderma laevigata	BW	24.7	Lawrence & Guille 1982
Stichopus chloronotus	BW	242.7	Lawrence 1980
Trachythyone elongata	BW	5.9	Lawrence & Guille 1982
Trachythyone muricata	BW	2.5	"
Ophiuroids			
Amphiura antarctica	D	0.138	Lawrence & Guille unpub
	A	0.511	
	total	0.649	
Amphiura chiajei	D	0.197	"
	A	0.209	
	total	0.406	
Amphiura filiformis	D	0.138	"
	A	0.293	
	total	0.431	
Ophiacantha vivipara	D	1.640	"
	A	2.503	
	total	4.143	
Ophionotus hexactis	D	2.653	"
	A	1.167	
	total	3.820	
Ophiura ambigua	D	0.473	"
	A	0.485	
	total	0.958	
Ophiura brevispinus	D	1.256	"
	A	0.598	
	total	1.854	
Echinoids			
Abatus cordatus	T	4.0	Lawrence & Guille unpub
	S	0.9	
	total	4.9	
Brissopsis lyrifera	T & S	7.5	
Colobocentrotus atratus	T	10.9	Lawrence unpub
	S	11.3	
	AL	0.8	
	total	23.0	
Echinometra lucunter	T & S	79.5	"
	AL	14.2	
	total	93.7	

Table 1 (cont.).

Echinometra mathaei	T	134.3	"
	S	31.0	
	AL	6.2	
	total	171.5	
Echinostrephus aciculatus	T & S	4.6	"
	AL	1.6	
	total	6.2	
Echinothrix diadema	T & S	17.2	"
	AL	12.6	
	total	29.8	
Sterechinus diadema	T	34.7	Lawrence & Guille unpub
	S	15.5	
	AL	6.3	
	total	56.5	
Asteroids			
Anasterias perrieri	BW	87.9	Lawrence & Guille unpub
	PC	70.2	
	total	158.1	
Asterina frigida	BW	3.0	"
	PC	0.4	
	total	3.4	
Diplasterias meridionalis	BW	310.1	"
	PC	15.9	
	total	326.0	
Leptychaster k. kerguelensis	BW	21.3	"
	PC	13.8	
	total	35.1	
Luidia clathrata	BW	18.0	Lawrence 1973
	PC	21.3	
	total	39.3	
Pteraster affinis	BW	11.3	Lawrence & Guille unpub
	PC	21.3	
	total	32.6	

A: Arms; AL: Aristotle's lantern; BW: Body wall; D: Disc; PC: Pyloric caeca; S: Spines; T: Test.

energy content during the development of several direct-developing echinoderms (Lawrence et al. in press).

The rate of energy production (= productivity) of tissues and organs differ according to the same variables. Fuji (1967) did a thorough evaluation of the productivity of the somatic tissues. The annual productivity of 1-year-old individuals was ca. 50% of 4-year-old individuals due to the higher productivity of reproduction in the latter. Productivity associated with the test of the young individuals was 3-fold that of the large ones, but still amounted to 18% of the total productivity in the large individuals. These kinds of data give insight into the food requirements of individuals and the ways in which energy is used. Some studies have ignored the energy productivity of the test of adult echinoids, assuming it to be low simply because of the low level of energy present.

Comparative studies need to be done on the rates of energy production for the different body components with echinoderms of different body forms and sizes. E.g. there are no studies on rates of energetic production for any crinoid, ophiuroid, or holothuroid. One can ask: Does the rate of energy production differ for a suspension-feeding dendrochirote or a deposit-feeding aspidochirote? Does the rate of energy production for a suspension-feeding dendrochirote differ from that of a suspension-feeding crinoid? Even for asteroids and echinoids, there are major distinct sizes, forms, or functional types which can be expected

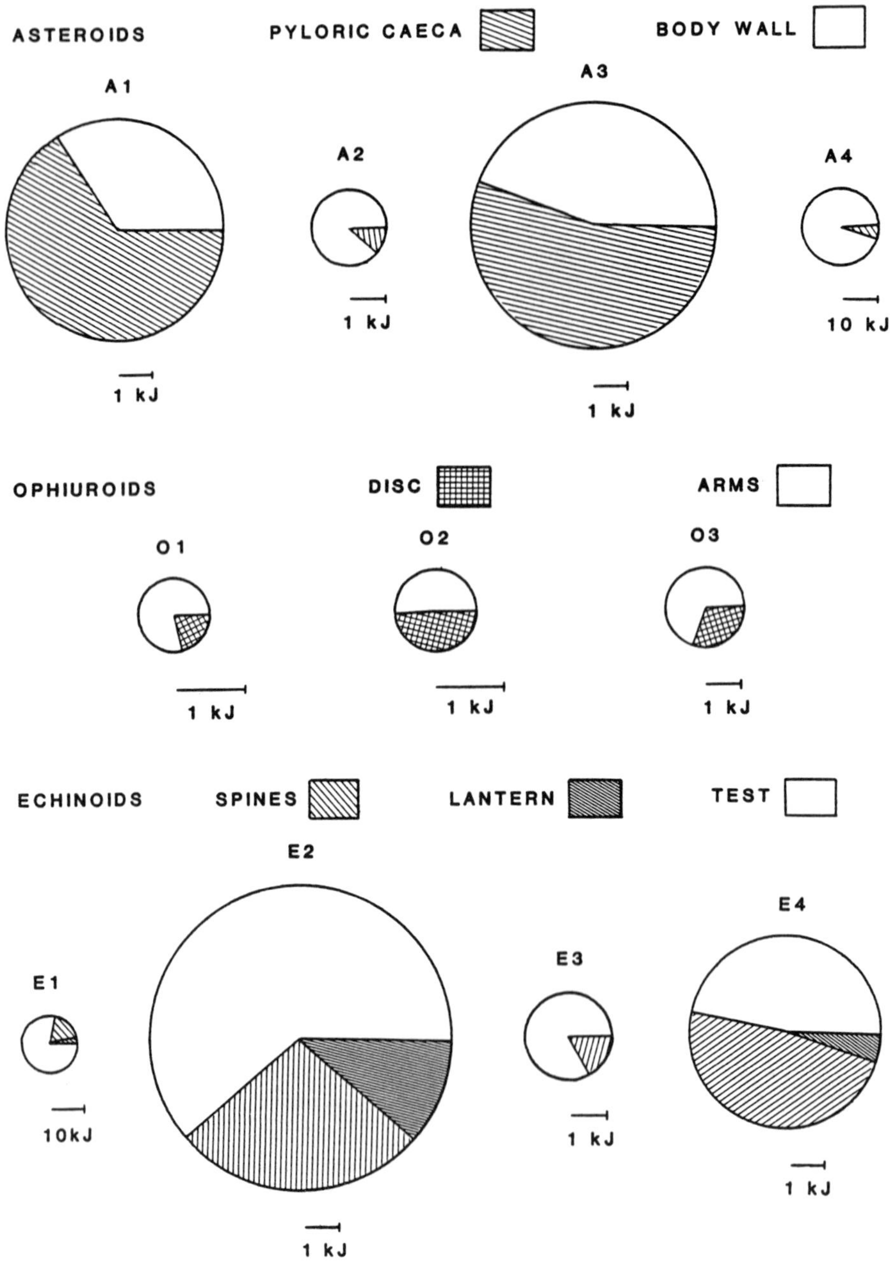

Figure 1. Allocation of energy to body components in asteroids (A1: Pteraster affinis, A2: Asterina frigida, A3: Luidia clathrata, A4: Diplasterias meridionalis); ophiuroids (O1: Amphiura antarctica, O2: Ophiura ambigua, O3: Ophionotus hexactis); and echinoids (E1: Echinometra mathaei, E2: Sterechinus diadema, E3: Abatus cordatus, E4: Colobocentrotus atratus). Prepared from unpublished data of Lawrence and of Lawrence & Guille. The area of the circle is propor tional to the number of kJ, with the bar indicating the calculated radius. The relative proportion of the total due to the different body components is indicated by the area of the sectors.

Table 2. Energetic reproductive output of echinoderms.

Species	kJ.ind^{-1}	Reference
Asteroids		
Echinaster I		Scheibling & Lawrence 1982
testes	10.3	
ovaries	21.8	
Echinaster II		
testes	8.7	Scheibling & Lawrence 1982
ovaries	11.6	
Leptasterias hexactis		Menge 1975
testes	2.6	
ovaries	1.9	
Luidia clathrata		Lawrence 1973
testes	4.2	
ovaries	4.7	
Pisaster ochraceus		calculated from Giese 1966
testes	134	
ovaries	209	

to have much different rates of energy production than those species which have been investigated. How does the rate of energy production of a spatangoid with a thin test compare with that of the clypeasteroid with a thick test studied by Lane (1977)? How does the rate of energy production by the carnivorous asteroids Pisaster and Asterias compare to the omnivorous Patiria or the deposit-feeding Ctenodiscus?

4 ORGANISMAL LEVEL

It is at the organismal level that information from the lower levels can be put together. At the basic level, one can compare the representation of a species in terms of amount of energy in contrast to weight as a result of eliminating the variables of water and inorganic material. These data can be useful in evaluating the importance of a species in the field in terms of energy. The few species of echinoderms for which such representation can be made do not begin to span the range of sizes which exist in each class. The energy contents of many more species of all classes throughout their entire range of individual sizes need to be ascertained. It is informative to relate the energy content of the body to the way of life and the functional morphology of a species. The largest echinoderms which ever existed were the extinct crinoids which reached over 30 m in length. This body size is beyond that possible for an errant echinoderm. An ophiuroid cannot be as large as an asteroid. The slender ophiuroid arm gives potential for many activities but cannot become as large as the asteroid if the activities are to be done. The ophiuroid and crinoid arms cannot develop a large coelomic cavity and lack the potential to evolve energy-rich pyloric caeca. All extant classes have representatives less than 10 mm in size.

The energy budget is a useful way of analyzing the energy requirements and uses by an individual. The budget ($C - F = A = R + U + P_r + P_s$, Ricker 1968) balances the energy intake and its use according to different physiological categories (C: consumption, F: feces, A: absorption, R: respiration, U: excretion and secretion, P_r: reproductive production, P_s: somatic production). Its use is possible because of the equivalency of energy in different forms. The value of its use lies in the possibility of ascertaining the effect of different levels of energy intake on the absolute and relative allocation of energy to various physiological processes. Essentially any variable (age, size, physiological state, environmental variables) can affect the physiological processes to varying degrees. Consequently, the preparation of an energy budget is an enormous task. There have been relatively few energy budgets prepared for echinoderms; most are incomplete in one or more factors or do not consider the effect of variables. The study of single factors of the energy budget is a valid approach, but one must be aware that internal checks are lost, and that an evaluation of the relative importance of the various factors cannot be done if only single factors are measured. The loss of internal checks is most important when more than one process can contribute to a factor. The two most important of these are C (which typically considers only ingestion and not uptake of

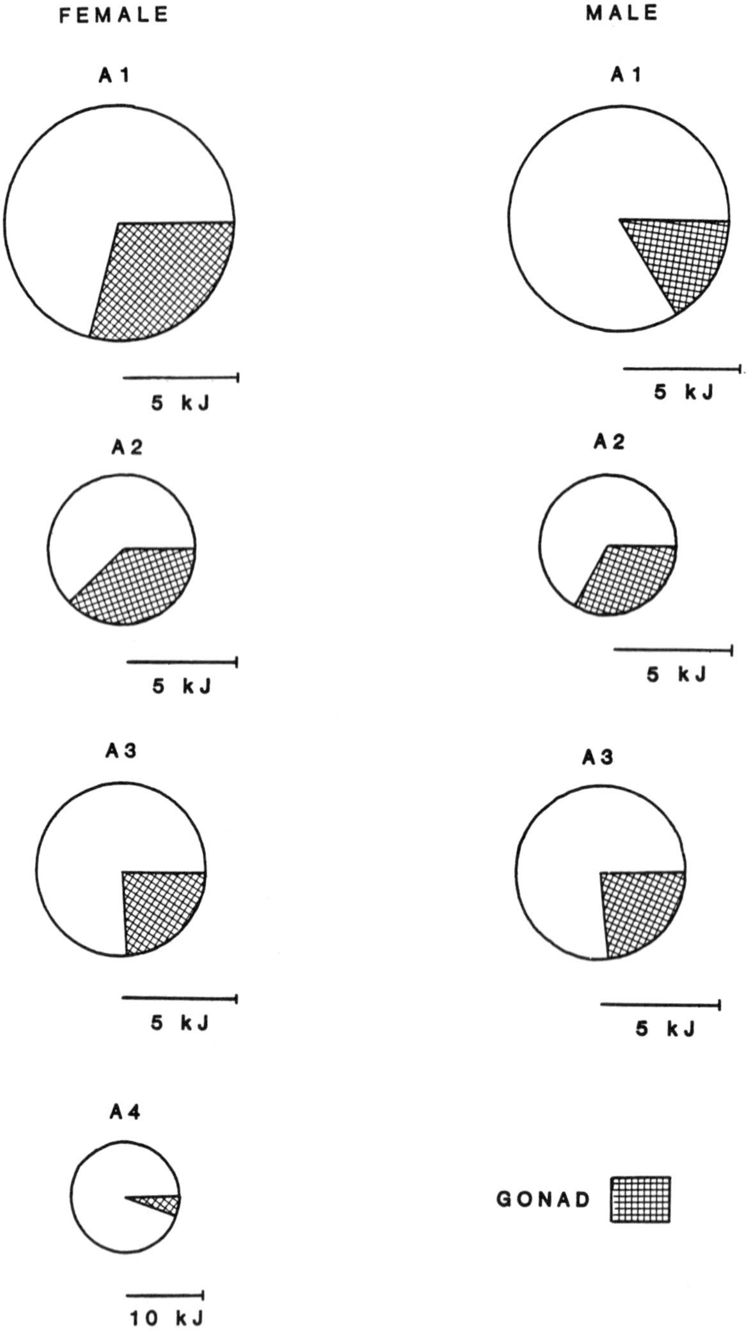

Figure 2. Allocation of energy to ovaries and testes of asteroids (A1: Pteraster affinis, A2: Asterina frigida, A3: Luidia clathrata, A4: Diplasterias meridionalis). Prepared from unpublished data of Lawrence and of Lawrence & Guille. The area of the circle is proportional to the number of kJ, with the bar indicating the calculated radius. The relative proportion of the total due to the gonads is indicated by the area of the sectors.

dissolved organic material, DOM), and R (which typically considers only aerobic production of energy. Estimates of the energetic equivalents of these physiological processes for echinoderms are few. Problems with balancing the energy budgets probably result to some degree from the failure to consider uptake of DOM and anaerobic respiration, but also from the fact that the individual components are physiological phenomena which respond to almost any variable. The best situation would be to measure all components of the energy budget with the same individuals to avoid the problem of group effects. Despite these problems, the studies that have been done probably do indicate in broad terms the energetics of the individual.

To measure C, one must know the rate of feeding and the absorption efficiency in terms of energy. These data are available for several species of echinoids (Fuji 1967, Miller & Mann 1973, Lilly 1975, Lane 1977, Greenwood 1980, Hawkins & Lewis 1982, Thompson 1982) and an asteroid (Shirley & Stickle 1982). C will vary with size of the individual which has ecological implications. A will vary with the amount of food consumed. The effect of the amount of food on the energy consumed is not only direct, but because retention time in the gut is affected. Absorption increases as the amount of food decreases. Thompson (1982) has shown the effect of food amount on gonadal development in Strongylocentrotus droebachiensis. No values are available for any species of the other classes. In evaluating the energy consumed, it is helpful to know the organic basis of the energy in the food. E.g., there is no correlation between the food preferences of S. droebachiensis and the energy level of algae (Larson et al. 1982). This is not surprising as the majority of the energy in algae is associated with structural carbohydrates which are not digested and absorbed (Lawrence 1976). Any adaptiveness in terms of energy acqusition must be sought in terms of the energy which is available. Knowledge of the relative amounts of energy consumed and absorbed in the different organic classes will also aid the investigation of the molecular processes of energy utilization. An aspect of asteroid feeding which affects the energy budget is the size of the prey. The size of prey (mussels) eaten by Marthasterias glacialis and Asterias forbesi increases with increasing asteroid body size (Penny and Griffiths 1984, Campbell 1983). Because the number of prey eaten.day^{-1} increases only slightly with size, the energy consumption is relatively constant. Individuals of different sizes select diffferently sized prey. Because of this, Penny & Griffiths concluded that the size of mussel eaten should differ when the size-frequency distribution of the prey differed. Campbell found that all but the largest Asterias forbesi selects that size of mussel which gives maximal energy intake. The energetics of feeding by carnivorous asteroids on units of food may be much different than that of herbivorous echinoids or of suspension-feeding echinoderms which have continuous ingestion.

The major lack in estimates of energy consumed is that of uptake of DOM which occurs in both pre- and post-metamorphic individuals (see Lawrence, in press,a). It has been reported that the uptake of free amino-acids alone could support 58% of the energy equivalent of aerobic respiration in the clypeasteroid Dendraster excentricus and 37% in the euechinoid Lytechinus variegatus (Stephens et al. 1978, Ferguson 1982a,b, respectively) Although these calculations indicate the significance of DOM in the overall energy budget, they do not consider the amount of energy produced by anaerobic metabolism or deposited in the tissue in growth. DOM seems to be important in the energy consumption of planktotrophic larvae (Dixit 1973, Manahan et al. 1983). The contribution of extra-intestinal uptake after external digestion (Pequignat 1970, 1972) has not been evaluated.

The aerobic rates of respiration R have been measured for numerous echinoderms (see Lawrence & Lane 1982) and the energetic equivalence of this can be calculated with some basic assumptions on the nature of the metabolites used. The validity of this assumption can be established by analysis of the body. Lawrence (1973) showed that the utilization of lipid in the pyloric caeca by the asteroid Luidia clathrata was close to that calculated by oxycalorific equivalents of the rate of aerobic respiration. Similarly, Turner & Rutherford (1976) calculated that the decrease in carbohydrate is sufficient to account for estimated embryonic energy requirements for the holothuroid Cucumaria curata. At the moment, however, the degree to which aerobic respiration completely reflects catbolism is a major question. The contribution of anaerobic respiration to the total respiration factor has not been included in any energy study involving echinoderms. The absolute energy of aerobic respiration increases with body size, but the relative amount decreases (see Lawrence & Lane 1982, Shick 1983). As the amount of energy involved in somatic production similarly decreases with size, it would seem that most of the energy of respiration of small individuals is associated with growth. As the amount of gonadal production of large individuals can equal the amount of somatic production of small individuals, the respiratory requirements of somatic production may exceed those of gonadal production. The molecular basis for this needs to be ascertained. Vahl (1983) has proposed that the increase in the rate of oxygen consumption which occurs after feeding by an asteroid is an indication of the specific dynamic action associated with growth. This assumes that most energy associated with growth is derived from the metabolic processes associated with aerobic respiration which occurs soon after feeding. There is some disagreement about the relation between the effect of reproductive state on the amount of energy associated with respiration (Lawrence, in press,a).

A direct relation between the rate of aerobic respiration and gonad size as suggested for Parechinus angulosis (Greenwood 1980) would imply considerable energy requirements for gametogenesis and an aerobic basis for it. This needs to be resolved. There seems to be little oxygen present in ripe echinoid ovaries (Webster & Giese 1975, Bookbinder & Shick, this volume). The energy of respiration is directly related to nutritional state, the amount falling quickly with decreased energy consumption. This may be related simply to the decreased energy allocation to growth, or a change in physiology to decrease energy utilization appropriately with decreased income. The energy of aerobic respiration of an individual echinoderm is directly related to temperature; the molecular basis for this has been indicated. One would assume that the energy of anaerobic respiration would be similarly affected. Echinoderms experiencing seasonal changes in temperature will have considerable changes in the amounts of their energy consumption allocated to respiration.

The primary excretory product of echinoderms is ammonium (see Jangoux 1982) and some studies on excretory energy loss in echinoderms have measured ammonium only. The energy loss from excretion $\underline{U}$ of ammonium is low due to its low energy content. Although less is excreted, primary amines can account for much of the energy loss. Energy loss from ammonium was no different from that of primary amines in Leptasterias hexactis (Shirley & Stickle 1982). The energy loss through nitrogen excretion could be expected to vary with the protein content of the food. No effect of protein level of food on nitrogen excretion was found in Luidia clathrata, but an increase associated with tissue catabolism was noted (Diehl & Lawrence 1979). There have been no estimations of the energetic loss by secretions such as mucus by echinoderms.

The amount of energy allocated to production $\underline{P}$ is of considerable interest because of its implications for physiological processes and for life-history strategies. Here is apparent the basic conflict of the benefits of growing ($\underline{P}_s$) as quickly as possible and those of reproducing ($\underline{P}_r$) as quickly as possible. The two processes are entangled in that size has a direct bearing on the reproductive output and effort.

The amount of energy ingested and absorbed by Strongylocentrotus intermedius increases as individuals increase in size, but the allocation of the energy absorbed shifts from somatic deposition to gonadal deposition was age and size (Fuji 1967). A similar shift occurs in the clypeasteroid Mellita quinquiesperforata (Lane 1977). It would be of considerable interest to see if the shift from somatic to gonadal production is similar in regular echinoids which Ebert (1975) categorized as rapid growers with a high mortality rate and those which are slow growers with a low mortality rate. He related the differential mortality to the difference in the amount of energy applied to reproduction ($\underline{P}_r$) and to maintenance ($\underline{P}$ + $\underline{R}$).

Differences in allocation of energy to reproduction associated with the change in sex of protandric species and those in which there is a sexual difference in reproductive output would be of considerable interest. Sequential hermaphrodites are invariably protandrous, a phenomemon probably associated with the energetics of reproduction. Ovaries are more energetically expensive to produce than testes (Lawrence 1976). Because of the difference in size, it is probable that a functional number of sperm can be produced by small individuals which could not produce a functional number of eggs. This differential allocation of energy to the ovaries and testes can also be seen in simultaneous hermaphrodites (Hendler 1979, Strathmann et al. 1984). Similarly, in gonochoric species, the size at which males produce testes is invariably less than that at which females produce eggs. A greater allocation of energy to the ovaries in gonochoric species occurs (fig. 2) and often seems associated with those which aggregate (e.g. Scheibling 1981). A very interesting situation is Leptasterias tenera in which about one-half of the testes of adult males are degenerate or absent (Hendler & Franz 1982). Are there different energy requirements of males and females of the same size in this species or in the ways

Table 3. Effect of size on yearly energy budgets and production ratios of echinoids (values in kJ).

Species	Size	C	R	P_s	P_r	P_s+P_r	P/R	P/C	Reference
Diadema	HD=15mm	58	8.8			3.7	41	6.2	Hawkins &
antillarum	HD=40mm	20'	114			21	18	10.1	Lewis 1982
	HD=60mm	317	198			47	24	15.0	
Mellita	1 g dry body wt	160	30	16	0	16	53	10.0	Lane 1977
quinquies-	10 g dry body wt	431	67	18	5	23	35	5.4	
perforata	30 g dry body wt	810	120	25	43	68	61	8.9	
Strongylo-	3.47 g dry body wt		8.8	7.3	0	7.3	83		Propp 1977
centrotus	8.01 g dry body wt		27	18	12	30	115		
droebach-	17.4 g dry body wt		64	17	32	49	75		
iensis	31.9 g dry body wt		107	8.7	58	67	63		
Strongylo-	HD=18mm	368		44	28	72		20	Fuji 1967
centrotus	6 g wet body wt								
inter-	HD=38 mm	709		25	48	73		10	
medius	19 g wet body wt								
	HD=54mm	925		15	63	78		8.5	
	48 g wet body wt								

Body sizes given are approximate. C: consumption; R: respiration; P_s: somatic growth; P_r: reproductive growth.

in which the same amount of energy is used for spawning?

The energetics of variations in reproductive types is of interest. Chia (1974) suggested that production of many small eggs in indirect development is energetically more expensive than the production of few large eggs with direct development, on the basis of comparison of reproductive effort in terms of gonad index. Such a comparison is crude except between species with similar body form and structure. Echinaster type I is larger, has a lower reproductive effort, and a greater reproductive output than Echinaster type I even though both produce large eggs with direct development (Scheibling & Lawrence 1982). Otteson & Lucas (1982) suggested that the energetic cost per reproductive unit in small asteroids which do not have planktotrophic development increases in the sequence 1) pelagic lecithotrophic development, 2) direct development, 3) direct development and brooding, 4) viviparity, and 5) asexual reproduction by fission. The energetic reproductive effort and output of the parent is of interest. Body size and energetics are not the only variables which affect the type of reproduction; the species population structure, and ecological physical and biological features are important.

The few energy budgets done thus far have all involved species which have annual reproductive cycles. In these species there are seasonal changes in the allocation of energy to the gonad. Whether this results in seasonal changes in the allocation of energy to somatic production is questionable. Numerous studies have shown a reciprocal relation between somatic and gonadal growth, but others have found no relation (see Lawrence & Lane 1982). This probably depends on food availability. There seem to be two thresholds involved: A minimal amount of food (energy) is necessary for maintenance. Below this first threshold, gonadal growth does not occur. Above this first threshold, the amount of food can support gonadal growth. Above a second threshold level of food, both gonadal and somatic growth can occur. Compare the use of energy by Strongylocentrotus intermedius (Fuji 1967) to that by Mellita quinquiesperforata (Lane 1977, table 3). S. intermedius consumes more energy than the clypeasteroid. Much more energy is associated with the test of S. intermedius. The body form of S. intermedius does not restrict the energy production in the gonad as in the clypeasteroid. The total energy of production in S. intermedius does not increase in size, as the decrease in P_s is countered by the increase in P_r as in M. quinquiesperforata. Thompson (1982) reported that the energy of production increased with production for Strongylocentrotus droebachiensis. His values for production are not comparable to those of Fuji, as Thompson considered gonadal production to be that of induced spawn while Fuji considered it to be the entire gonad. In addition, Thompson considered somatic production to be the residual gonad and gut (excluding the test), while Fuji considered it to be all non-gonadal tissue (including the test).

The ratio of energy of production to energy of consumption is an indication of the relative amount of the energy obtained which can be used for somatic or gonadal growth (table 3). Fuji reported that it decreased with increasing size for Strongylocentrotus intermedius, but Lane found a minimal rate for Mellita quinquiesperforata at an intermediate body size. In contrast, Hawkins & Lewis (1982) found a continued increase in P/C for Diadema antillarum. We do not know the basis for the differences found.

Ebert (1975, 1982) used the principles of the energy budget to evaluate evolutionary selection. He (1975) proposed that separation of regular echinoids into those which have high or low growth and natural mortality rates is the result of the way in which energy resources are allocated. Those species which grow rapidly and expend much of their energy resources on reproduction have less to spend on maintenance and have a higher probability of mortality. Although logical, there are some assumptions which are necessary for the concept to be true. The first is that energy consumption is necessarily limited and cannot support both growth and maintenance maximally at the same time. Ebert (1982) pointed out that there is a finite amount of energy which can be ingested, whether as a result of limitation of food availability or biological limitations of the individual. However, the consequences of the two kinds of limitations are distinct. With limited food availability, the individual must allocate the energy ingested to different processes. The degree to which there is a decrease in allocation of energy to a particular process indicates the degree the process is essential to the system. Ebert's proposals assume that echinoderms have no regulation over their use of resources which are limited. Numerous studies have shown that the rate of respiration, somatic and gonadal production, and size of individuals all decrease with low levels of energy consumption. There is no indication of increased mortality. Maintenance of the individual at the expense of somatic growth can be seen in Asterias rubens which can assume a "waiting stage" immediately after settling during which no growth occurs until food-energy supplies increase. Pisaster ochraceus use resources for regenerating arms at the expense of reproduction (Harrold & Pearse 1981). Similar maintenance of the individual at the expense of reproduction is seen in the failure to allocate energy to reproduction in certain populations of Echinocardium cordatum. When the amount of energy ingested is due to biological limitations of the individual, its allocation to different processes must indeed be in the overall best interests of the individual and population. It is less clear that there is a conflict between maintenance (sensu latu of Ebert), growth, and reproduction in this case as Ebert (1982) suggested. Ebert suggested that the energy deposited in the body walls of 17 species of echinoids was an allocation of energy directly related to natural survival. He did not demonstrate that allocation of energy to maintenance affected the allocation to reproduction. Certainly within a species, somatic and gonadal growth are not necessarily inversely related, indicating that energy consumption may be sufficient to support both simultaneously. The same may be true for the relation between the use of energy for maintenance and growth.

A basis for comparison of the energy budgets of other echinoderms with different body forms and life histories are glaringly absent. The only energy budget done on a non-echinoid is that for the carnivorous asteroid Leptasterias hexactis (Shirley & Stickle 1982). The gross production efficiency ($P.C^{-1}$) was 26%, not much higher than for the herbivorous Strongylocentrotus intermedius, but the net production efficiency ($P.A^{-1}$) was 40%, much higher than for S. intermedius. The findings indicate that for these two representatives of herbivory and carnivory, the amount of energy production per unit of energy consumed is similar, but that the carnivore makes much better use of the energy consumed through digestion and metabolism than the herbivore. This has significance in considering not only digestion, but the ability to use different types of absorbed metabolites.

5 POPULATION LEVEL

The application of the energetic concept to the population level involves the energetic characteristics of the individuals and the characteristics of the population (i.e., the size-frequency distribution of the individuals, their density, and their physiological state). It is possible to calculate the amount of energy which a population represents, the consumption of energy from the other components of the environment, the loss of energy as

Table 4. Caloric densities of echinoderms.

Species	$kJ.m^{-2}$	Habitat	Reference
Holothuroids			
Actinopyga mauritana	5670	reef flat	Lawrence 1981
Afrocucumis africana	103	"	"
Benthogone rosea	$4.1 \cdot 10^{-3}$	deep-sea	Sibuet & Lawrence 1981
Chiridota rigida	31	reef flat	Lawrence 1981
Cladodactyla crocea	172	subtidal	Lawrence & Guille 1982
Cucumaria sykion	55	rocky coast	Berry 1982
Eumolpadia violacea	38	subtidal	Lawrence & Guille 1982
Holothuria atra	117	reef flat	Lawrence 1981
Holothuria difficilis	62	"	"
Holothuria glaberrima	412	rocky coast	Lawrence & Kafri 1978
Holothuria hilla	108	reef flat	Lawrence 1981
Holothuria leucospilota	195	"	"
Paelopatides gigantea	$3.4 \cdot 10^{-3}$	deep-sea	Sibuet & Lawrence 1981
Pentacucomis bouvetensis	136	rocky coast	Lawrence & McClintock unpub.
Pseudocnus laevigatus	176	"	"
Psychropotes longicauda	$2.7 \cdot 10^{-3}$	deep-sea	Sibuet & Lawrence 1981
Stereoderma laevigata	115	subtidal	Lawrence & Guille 1982
Stichopus chloronotus	183	reef flat	Lawrence 1981
Ophiuroids			
Amphiura angularis	21.7	subtidal	Lawrence & Guille 1982
Amphiura antarctica	46.8	"	"
Amphiura chiajei	16.7	"	"
Amphiura filiformis	15.9	"	"
Ophiocantha vivipara	9.7	"	"
Ophionotus hexactis	81.9	"	"
Ophiura ambigua	14.5	"	"
Ophiura brevispina	18.8	"	"
Ophiura texturata	50.3	"	"
Echinoids			
Echinometra lucunter	2017	rocky coast	Lawrence & Kafri 1978
Mellita quinquiesperforata	235	sand	Lane 1977
Parechinus angulosus	2268	subtidal	Field et al. 1980
Asteroids			
Henricia ornata	18	subtidal	Field et al. 1980
Marthasterias glacialis	256	"	"
Patiria granifera	8	"	"

the heat energy produced during respiration and the deposition of the energy through growth, the transfer of energy out of the population through death and predation, excretions, and spawning. Most of the approaches thus far have been autecological rather than synecological.

The energetic densities have been estimated for a number of echinoderm species in a variety of habitats (table 4) and gives a much different view of the species than numerical density. These values are the maximal ones and reflect both the size of the individuals and their densities. It is assumed that the energetic density indicates the ability of the environment to support a particular species, but this is not always the case. E.g., all combinations of high and low densities and large and small individuals can be found in echinoid populations and seem to indicate alternate states of existence (Lawrence in press,b). The densities and sizes may have nothing to do with the available food energy, but result rather from biological (predation or competition) or physical factors. Small Echinometra mathaei were most dense (136 $individuals.m^{-2}$, 5838 $kJ.m^{-2}$) in burrows on an exposed flat covered with an algal mat on the windward side of a small cape on Reunion Island, while large individuals were much less dense in terms of numbers ($19.m^{-2}$) but similar in

Table 5. Population energy budgets of echinoids ($kJ.m^{-2}.y^{-2}$).

Species	C	$P_{somatic}$	$P_{gonadal}$	Total P	R	Reference
Diadema antillarum	3579			322	1680	Hawkins & Lewis 1982
Mellita quinquiesperforata						Lane 1977
1972	4620	172	21	193	598	
1973	2482	100	31	131	356	
Parechinus angulosus						Greenwood 1980
Robben Island	52717	1872	411	2283	7619	
Oatland Point	34050	1467	322	1789	8550	
Strongylocentrotus droebachiensis	4821	167	42	209	749	Miller & Mann 1973

terms of energy (4733 $kJ.m^{-2}$) on the protected side. It is probable that large individuals simply could not exist due to the physical effect of waves on the windward side (Lawrence 1980b). At this point, it seems that holothuroids and echinoids can exist in greater energetic densities than the other classes, but there are simply too few data to draw a strong conclusion. Certainly dense populations of large asteroids occur which would be expected to have a high energetic density as well. It is unlikely that the energetic density of ophiuroids would be as high despite the extremely high numerical density in which they can occur. Sibuet & Lawrence (1981) pointed out that although abyssal holothuroids were energetically less dense than shallow-water species, their individual body size and composition was similar. Presumably the lower levels of nutrients available in the deep sea (Khripounoff et al. 1980) reduces the density, but not the body size. Two distinct phenomena are involved.

There is a general inverse relation between numerical density and maximal caloric body size of echinoderms (Lawrence 1980b, Sibuet & Lawrence 1981, Lawrence & Guille 1982). The relationship does not seem dependent on trophic level or functional type. In contrast, there is a general direct relation between caloric density and maximal caloric body size (Lawrence 1980b, Lawrence & Guille 1982). This is probably based on a surface:volume relationship. An increase in surface allows a disproportionate increase in volume (biomass) so that a single large individual has more energy content than numerous smaller individuals occupying the same area.

Population energy budgets reflect the physiological state of the individuals, the demography of the population, and the environmental conditions (primarily temperature and food quality and quantity). They indicate the energy role of the population in the community. Estimations are available for several species of echinoids (table 5). The production efficiencies (table 6) show the differences in the way different species and different populations (in space and time) use energy.The effect of demographic changes on the energy budget of Mellita quinquiesperforata (Lane 1977) demonstrates clearly the importance of age in a population. With a massive settlement, a new cohort comprised 76% of the population initially, but only 35% in its second year. The different energy parameters, especially production, clearly demonstrate the effect of increasing age on the total population. The first year was more productive due to the greater density of individuals. Net growth efficiencies of growth, however, were higher in the second year due to the increased individual growth efficiencies which results with age. Individual production increased sufficiently to offset loss from mortality. The production:biomass decreased for the same reason.

The analysis of community structure requires ascertaining the abundance of different forms of the organisms which comprise the community. The contribution of various echinoderms to the energy content represented by invertebrates off the coast of southwest Africa is important and variable (Field et al. 1980). Rainer (1982) reported that the

Table 6. Annual population energetic production efficiencies of echinoids.

Species	P/R	P/C	P/B	Reference
Diadema antillarum	19	9.0		Hawkins & Lewis 1982
Mellita quinquiesperforata				Lane 1977
1972	32	4.2	3.9	
1973	37	5.3	1.6	
Parechinus angulosus				Greenwood 1980
Robben Island	30	4.3		
Oatland Point	21	5.3		
Strongylocentrotus droebachiensis	28	4.3	0.8	Miller & Mann 1973

asteroid Astropecten polyacanthus, the spatangoid Echinocardium cordatum, and an ophiuroid Amnphiura sp. accounted for 7.5, 13.8, and 8.8 % of the total community energy production.y^{-1}. Warwick (1982) produced production-rank curves for sublittoral macrobenthic communities. All members of the Amphiura community are deposit feeders and production is allocated relatively evenly among themselves in a lognormal fashion. The Modiolus community is almost completely dominated by a single suspension feeder, Ophiothrix fragilis. The basis of this was thought to be the unpredictable and fluctuating food supplies of suspension feeders such as the ophiuroids, resulting in unspecialized food requirements. Even though there is no trophic interrelationship within the suspension-feeding community, one species may outcompete the others for space. of an intertidal community at Kerguelen showed that holothuroids were relatively minor in importance (Lawrence & McClintock, unpub.). This study also showed that another way of describing a community, i.e., its diversity (H'), varies according to its calculation in terms of numbers of individuals or in terms of energy. The community was much less diverse when the amount of energy in its component species was considered than in terms of numbers of individuals. Once again, this probably indicates the role of area in determinging numbers of individuals, and of volume in determining energy.

6 SUMMARY AND PROJECTIONS

The energetric analysis of echinoderms at all levels of organization provide a different perspective which can give insight into their life histories. At no level of organization have sufficient studies been made to provide more than a suggestion for the role of energy in the echinoderm life histories. Specifically, more information is needed on metabolic pathways for energy production and deposition; the levels and amounts of energy associated with the different body compartments of echinoderms with different body forms and their variation with stage in life history (growth and reproduction); the ways in which the various components of the energy budget vary with stage in life history, with food supply, and environmental variables; and the way in which echinoderms contribute energetically to the structure and function of communities.

Acknowledgments

I thank James B. McClintock and Stephen A. Watts for their comments on the manuscript.

REFERENCES

Allee, W.C., O. Park, A.E. Emerson, T. Park, & K.P. Schmidt. Principles of animal ecology. W.B. Saunders Company, Philadelphia.

Berry, P.F. 1982. Biomass and density of detritovores on a littoral rocky reef on the Natal coast, with an estimation of population production for the ascidian Pyura stolonifera. Invest. Rep. oceanogr. Res. Inst. (53), 1-15.

Bookbinder, L.H. & J.M. Shick. (this volume). A respirometric and direct calorimetric study of ovary energy metabolism in Strongylocentrotus droebachiensis.

Campbell, D.B. 1983. Determination of the foraging strategy of Asterias forbesi (Echinodermata: Asteroidea). Ph.D. dissertation, Univ. of Rhode Island, Kingston.

Chia, F.S. 1974. Classification and adaptive significance of developmental patterns in marine invertebrates. Thalassia Jugosl., 10, 121-130.

de Mik-van der Plas, A.J. 1981. Pyloric caeca of the starfish Asterias rubens L., during the reproductive cycle. Doctoral dissertation, State Univ. of Utrecht, Utrecht.

Diehl, W.J. 1983. Patterns and mechanisms of isomotic intracellular regulation in Luidia clathrata (Say) (Echinodermata: Asteroidea) exposed to hypo- and hyperosmotic stress. Ph.D. dissertation, Univ. of South Florida, Tampa.

Diehl, W.J. & J.M. Lawrence. 1979. Effect of nutrition on the excretion rate of soluble nitrogen products of Luidia clathrata (Say) (Echinodermata: Asteroidea). Comp. Biochem. Physiol. 62A, 669-671.

Dixit, D.B. 1973. Uptake of amino acids and development in the sea urchin Strongylocentrotus purpuratus. Ph.D. diss ., Univ of California, Irvine.

Ebert, T.A. 1975. Growth and mortality of post-larval echinoids. Amer. Zool. 15, 755-775.

Ebert, T.A. 1980. Relative growth of sea urchin jaws: an example of plastic resource allocation. Bull. Mar. Sci. 30, 467-474.

Ebert, T.A. 1982. Longevity, life history, and relative body wall size in sea urchins. Ecol. Monogr. 52, 353-394.

Ellington, W.R. 1975. Glucose degradation and respiratory metabolism in starfish tissue. Fed. Proc. 34, 466.

Ellington, W.R. 1982. Intermediary metabolism. In: M. Jangoux & J.M. Lawrence (eds.). Echinoderm nutrition. pp. 395-415. A.A. Balkema, Rotterdam.

Ellington, W.R., & C.S. Hammen. 1977. Metabolic compensation to reduced oxygen tensions in the sea cucumber, Sclerodactyla briareus. J. comp. Physiol. 122, 347-358.

Ellington, W.R., & J.M. Lawrence. 1974. Seasonal temperature characteristics of supernatant malic dehydrogenase of adult and juvenile Mellita quinquiesperforata. Comp. Biochem. Physiol. 48B, 107-115.

Ferguson, J.C. 1982a. A comparative study of the net metabolic benefits derived from the uptake and release of free amino acids by marine invertebrates. Biol. Bull. 162, 1-17.

Ferguson, J.C. 1982b. Support of metabolism of superficial structures through direct uptake of dissolved primary amines in echinodersm. In: J.M. Lawrence (ed.). Echinoderms: proc. internat. conf., Tampa Bay. pp. 345- 351. A.A. Balkema, Rotterdam.

Field, J.G., C.L. Griffiths, R.J. Griffiths, N. Jarman, P. Zoutendÿk, B. Velimirov & A. Bowes. 1980. Variation in structure and biomass of kelp communities along the south-west Cape coast. Trans. Roy. Soc. S. Afr. 44, 145- 203.

Fuji, A. 1967. Ecological studies on the growth and food consumption of Japanese common littoral sea urchin, Strongylocentrotus intermedius (A. Agassiz). Mem. Fac. Fish. Hokkaido Univ. 15, 83-160.

Giese, A.C. 19667. On the biochgemical constitution of some echinoderms. In: R.A. Boolootian (eds.). Physiology of Echinodermata, pp. 757-796. Interscience Publ., New York.

Giese, A.C., A. Farmanfarmaian, S. Hilden & P. Doezema. 1966. Respiration during the reproductive cycle in Strongylocentrotus purpuratus. Biol. Bull. 130, 192-201.

Greenwood, P.J. 1980. Growth, respiration and tentative energy budgets for two populations of the sea urchin Parechinus angulosus (Leske). Est. Coast. Mar. Sci. 10, 347-367.

Hansen, B. 1975. Systematics and biology of the deep-sea holothuroids. Galathea Report. 13.

Harrold, C. & J.S. Pearse. 1980. Allocation of pyloric caecum reserves in fed and starved sea stars, Pisaster giganteus (Stimpson): somatic maintenance comes before reproduction. J. exp. mar. Biol. Ecol. 48, 169-183.

Hawkins, C.M. & J.S. Lewis. 1982. Ecological energetics of the tropical sea urchin Diadema antillarum Philippi in Barbados, West Indies. Est. Coast. Shelf Sci. 15, 645-669.

Hendler, G. 1979. Sex-reversal and viviparity in Ophiolepis kieri, n.sp., with notes on viviparous brittlestars from the Caribbean (Echinodermata: Ophiuroidea). Proc. Biol. Soc. Wash. 92, 783-795.

Jangoux, M. 1982. Excretion. In: M. Jangoux & J.M. Lawrence (eds.), Echinoderm nutrition. pp. 437-445. A.A. Balkema, Rotterdam.

Khotimchenko, Y.S. 1982b. Effect of noradrenaline, dopamine and adrenolytics on growth and maturation of the sea urchin, Strongylocentrotus nudus Agassiz. Internat. J. Invert. Reprod. 4, 369-373.

Khotimchenko, Y.S. 1982b. Effect of adrenotropic substances on the growth and maturation of oocytes of the sea urchin, Strongylocentrotus nudus. Experentia. 38, 696-697.

Khripounoff, A., D. Desbruyeres, P.

Chardy. 1980. Les peuplements benthiques de la faille Vema: donnees quantitatives et bilan d'energie en milieu abyssal. Oceanol. Acta. 3, 187-198.
Lane, G.N. (in press). Predation and survival among inadunate crinoids.
Lane, J.M. 1977. Bioenergetics of the sand dollar Mellita quinquiesperforata (Leske 1778). Ph.D. dissertation. Univ. of South Florida, Tampa.
Larson, B.R., R.L. Vadas & M. Keser. 1980. Feeding and nutritional ecology of the sea urchin Strongylocentrotus droebachiensis in Maine, USA. Mar. Biol. 59, 49-62.
Lawrence, J.M. 1973. Level, content and caloric equivalents of the lipid, carbohydrate and protein in the body components of Luidia clathrata (Echinodermata: Asteroidea: Platyasterida) in Tampa Bay. J. exp. mar. Biol. Ecol. 11, 263-274.
Lawrence, J.M. 1976. Patterns of lipid storage in post-metamorphic marine invertebrates. Amer. Zool. 16, 747-762.
Lawrence, J.M. 1980a. Maximal numerical and caloric densities of the echinoids, Colobocentrotus atratus and Echinometra mathaei, at Boucan Canot, Reunion Island, Indian Ocean. Amer. Zool. 20, 847.
Lawrence, J.M. 1980b. Numbers and biomass of the common holothuroids on the windward reef at Enewetak Atoll, Marshall Islands. In: M. Jangoux (ed.), Echinoderms: present and past. pp. 201-204. A.A. Balkema, Rotterdam.
Lawrence, J.M. (in press, a). Bioenergetics of echinoderms. In: T.J. Pandian & F.J. J. Vernberg (eds.). Animal Energetics, Academic Press, N.Y.
Lawrence, J.M. (in press, b). Alternate states of populations of Echinometra mathaei (de Blainville) (Echinodermata: Echinoidea) in the Gulf of Suez and the Gulf of Aqaba. Proc. internat. Red Sea Conf.
Lawrence, J.M. & A. Guille. 1982. Numerical and caloric densities of ophiuroids and holothuroids (Echinodermata) from Kerguelen (South Indian Ocean) and Banyuls (Mediterranean Sea). In: J.M. Lawrence (ed.). Echinoderms: proc. internat. conf., Tampa Bay. pp. 313-317. A.A. Balkema, Rotterdam.
Lawrence, J.M. & J. Kafri. 1979. Numbers, biomass and caloric content of the echinoderm fauna of the rocky shores of Barbados. Mar. Biol. 52, 87-91.
Lawrence, J.M. & J.M. Lane. 1982. The utilization of nutrients by post-metamorphic echinoderms. In: M. Jangoux & J.M. Lawrence (eds.). Echinoderm nutrition. pp. 331-371. A.A. Balkema, Rotterdam.
Lawrence, J.M., J.B. McClintock & A. Guille (in press). Organic level and caloric content of eggs of brooding asteroids and an echinoid (Echinodermata) from Kerguelen (South Indian Ocean). Internat. J. Invert. Reprod. Develop.
Lilly, G.R. 1977. The influence of diet on the growth and bioenergetics of the tropical sea urchin, Tripneustes ventricosus (Lamarck). Ph.D. dissertation. Univ. of British Columbia, Vancouver.
Lovtrup-Rein, H. & S. Lovtrup. 1980. Energy sources in the sea urchin embryo: a critique of current views. Acta Embryol. Emorphol. Exper. n.s., 1, 93- 101.
Lowe, E.F. 1978. Relationships between biochemical and caloric composition and reproductive cycle in Asterias vulgaris (Echinodermata: Asteroidea) from the Gulf of Maine. Ph.D. dissertation, Univ. of Maine, Orono.
Magniez, P. 1983. Reproductive cycle of the brooding echinoid Abatus cordatus (Echinodermata) in Kerguelen (Antarctic Ocean): changes in the organ indices, biochemical composition and caloric content of the gonads. Mar. Biol. 74, 55-64.
Manahan, D.T., J.P. Davis & G.C. Stephens. 1983. Bacteria-free sea-urchin Strongylocentrotus purpuratus larvae: selective uptake of neutral amino-acids from sea water. Science. 220, 204-206.
Marsh, A.G. 1983. The effects of sodium, potassium, calcium, and magnesium on the activity of citrate synthase (EC 4.1.3.7) in Luidia clathrata (Say) (Echinodermata: Asteroidea). M.A. thesis, Univ. of South Florida, Tampa.
Menge, B. 1975. Brood or broadcast? The adaptive significance of different reproductive strategies in the two intertidal sea stars Leptasterias hexactis and Pisaster ochraceus. Mar. Biol. 131, 87-100.
Miller, R.J. & K.H. Mann. 1973. Ecological energetics of the seaweed zone in a marine bay on the Atlantic coast of Canada. III. Energy transformation by sea urchins. Mar. Biol. 18, 99-114.
Monahan, M.J. 1980. Respiratory electron transport system activity in echinoderm tissues (Echinodermata: Asteroidea and Echinoidea). M.A. thesis, Univ. of South Florida, Tampa.
Ottesen, P.O. & J.S. Lucas. 1982. Divide or broadcast: interrelation of asexual and sexual reproduction in a population of the fissiparous hermaphroditic seastar Nepanthia belcheri (Asteroidea: Astserinidae). Mar. Biol. 69, 223-233.
Penny, A.J. & C.L. Griffiths. 1984. Prey selection and the impact of the starfish Marthasterias glacialis

(L.) and other predators on the mussel Chloromytilus meridionalis (Krauss). J. exp. mar. Biol. Ecol. 75, 19-36.

Pequignat, E. 1970. Biologie des Echinocardium cordatum (Pennant) de la Baie de Seine. Forma Functio. 2, 121-168.

Pequignat, E. 1972. Some new data on skin digestion and absorption in sea urchins and sea stars (Asterias and Henricia). Mar. Biol. 12, 28-41.

Percy, J.A. 1971. Thermal acclimatization and acclimation in the echinoid, Strongylocentrotus droebachiensis (O.F. Muller 1776). Ph.D. dissertation, Memorial University, Newfoundland.

Percy, J.A. 1973. Thermal adaptation in the boreo-arctic echinoid Strongylocentrotus droebachiensis (O.F. Muller, 1776). II. Seasonal acclimatization and urchin activity. Physiol. Zool. 46, 129-138.

Propp, M. 1977. Ecology of the sea urchin Strongylocentrotus droebachiensis of the Barents Sea: metabolism and regulation of abundance. Mar. Biol. Vlad. (1), 39-51.

Propp, M.V., V.I. Ryabushko, A.A. Zhuchikhina & L.N. Propp. 1984. Seasonal changes in respiration, ammonia and phosphate excretion, and activity of carbohydrate-metabolism enzymes in four echinoderm species from the Sea of Japan. Comp. Biochem. Physiol. 75B, 707- 711.

Rainer, S.F. 1982. Trophic structure and production in the macrobenthos of a temperature Australian estuary. Est. Coastal Mar. Sci. 15, 423-441.

Ricker, W.E. (ed.). 1968. Methods for assessment of fish production in fresh waters. I.B.P. Handbook No. 3. Blackwell, Oxford.

Scheibling, R.E. 1981. The annual reproductive cycle of Oreaster reticulatus (L.) (Echinodermata: Asteroidea) and interpopulation differences in reproductive capacity. J. exp. mar. Biol. Ecol. 54, 39-54.

Scheibling, R.E. & J.M. Lawrence. 1982. Differences in reproductive strategies of morphs of the genus Echinaster (Echinodermata: Asteroidea) from the eastern Gulf of Mexico. Mar. Biol. 51-62.

Schoenmakers, H.J.N. & P.A. Voogt. 1980. Stimulation of vitellogenesis in Asterias rubens. In: M. Jangoux. Echinoderms: past and present. p. 411. A.A. Balkema, Rotterdam.

Shick, J.M. 1983. Respiratory gas exchange in echinoderms. In: M. Jangoux and J.M. Lawrence. Echinoderm Studies I. pp.67- 110. A.A. Balkema, Rotterdam.

Shirley, T.C., & W.B. Stickle. 1982. Responses of Leptasterias hexactis (Echinodermata: Asteroidea) to low salinity. II. Nitrogen metabolism, respiration and energy budget. Mar. Biol. 69, 155-163.

Sibuet, M., & J.M. Lawrence. 1981. Organic content and biomass of abyssal holothuroids (Echinodermata) from the Bay of Biscay. Mar. Biol. 65, 143-147.

Sladen, W.P. 1889. Report on the Asteroidea collected by H.M.S. Challenger during the years 1873-1876. Rept. Scient. Results Voyage Challenger, Zool. 30.

Smith, L.A. 1983. Glycolytic activity in pyloric caeca of Luidia clathrata (Say) (Echinodermata: Asteroidea). M.S. thesis. Univ. of South Florida, Tampa.

Stephens, G..C., M.J. Volk, S.H. Wright & P.S. Backlulnd. 1978. Transepidermal accumulation of naturally occurring amino acids in the sand dollar Dendraster excentricus. Biol. Bull. 154, 335-347.

Strathmann, R.R. & M.F. Strathmann. 1982. The relationship between adult size and brooding in marine invertebrates. Am. Nat. 119, 19-101.

Strathmann, R.R., M.F. Strathmann & R.H. Emson. 1984. Does limited brood capacity link adult size, brooding, and simultaneous hermaphroditism? A test with the starfish Asterina phylactica. Am. Nat. 123, 796-818.

Takahashi, N. & H. Kanatani. 1981. Effects of 17-beta estradiol on growth of oocytes in cultured ovarian fragments of the starfish Asteri na pectinifera. Dev. Growth Differ. 23, 565-570.

Thompson, R.J. 1982. The relationship between food ration and reproductive effort in the green sea urchin Strongylocentrotus droebachiensis. Oecologia. 56, 50-57.

Turner, R.L. & J.M. Lawrence. 1979. Volume and composition of echinoderm eggs: implications for the use of egg size in life history models. In: S.E. Stancyk (ed.). Reproductive ecology of marine invertebrates. pp. 25-40. Univ. of South Carolina Press, Columbia.

Turner, R.L. & J.C. Rutherford. 1976. Organic, inorganic and caloric composition of eggs, pentaculae and adults of the brooding sea cucumber Cucumaria curata Cowles (Echinodermata: Holothuroidea). J. exp. mar. Biol. Ecol. 24, 49-60.

Vahl, O. 1983. The relationship between specific dynamic action (SDA) and growth in the common starfish, Asterias rubens L. Oecologia. 61, 122-125.

Warwick, R.M. 1982. The partitioning of secondary production among species in benthic communities. Netherlands J. Sea Res. 16, 1-16.

Watts, S.A. & J.M. Lawrence. 1982. Activity and kinetic characteristics of triosephosphate isomerase from the pyloric caeca of Echinaster sp. (Echinodermata: Asteroidea) exposed to different temperatures. Comp. Biochem. Physiol. 73B, 881-885.
Webster, S.K. & A.C. Giese. 1975. Oxygen consumption of the purple sea urchin with speci reference to the reproductive cycle. Biol. Bull. 148, 165-180.

"Echinoderms are mostly sedentary and hence they are more peaceful than other phylas" (from an undergraduate examination answer, quoted by Kenchington, 1983)

Proceedings of the Fifth International Echinoderm Conference / Galway / 24-29 September 1984

Catch connective tissue: The connective tissue with adjustable mechanical properties

T.MOTOKAWA
Department of Biology, The University of the Ryukyus, Nakagusuku, Okinawa, Japan

ABSTRACT: The catch connective tissue is defined as the collagenous connective tissue whose mechanical properties can be changed rapidly (in seconds to minutes) under nervous control. This tissue is found in all the supportive systems of echinoderms. When the tissue is in a catch state, the supportive systems become hard and immovable so that they maintain the animal posture and serve as protective "shells". When it relaxes, the supportive systems become flexible so as to allow deformation and locomotion of the animal. Further softening allows for autotumy (defence) and for fission (reproduction). The echinoderms' success has very likely depended on the development of catch connective tissues.

1 DEFINITION OF CATCH CONNECTIVE TISSUE

Echinoderms have developed a connective tissue whose mechanical properties can be changed rapidly. I have studied 6 kinds of such connective tissues from 12 species out of 4 classes of the phylum Echinodermata. Some common features of these connective tisses have been revealed:

1. They are collagenous connective tissues with or without muscles in them.
2. They change their mechanical properties spontaneously and in response to stimulation of various modalities.
3. The change is reversible except at autotomy and fission.
4. The change is rapid: it completes in the order of seconds or minutes.
5. The change in mechanical properties occurs mainly in viscosity; the change in elasticity is small.
6. The viscosity is controlled by nerves.
7. Neurosecretory-like cell processes with electron-dense granules are always found in the tissue.
8. The viscosity seems to be controlled through the ionic environment in the tissue.
9. The tissues are found in the supportive systems of echinoderms. They are used in control of the body tone, in defence including autotomy, and in fission.

I propose to call this connective tissue with adjustable mechanical properties "the catch connective tissue": I define it as the collagenous connective tissue whose mechanical properties can be changed rapidly (in seconds to minutes) under nervous control. I will briefly review how these tissues are used in the supportive systems of echinoderms, and discuss the possible contribution of catch connective tissues in echinoderms' success.

2 THE ROLE OF CATCH CONNECTIVE TISSUES IN SUPPORTIVE SYSTEMS

The main role of animal connective tissues is in support. Supportive systems must be rigid enough to bear the body weight and to withstand the forces imposed by the environment. The supportive system must be flexible; however, efficient movements are impossible if the body is so rigid as to be nondeformable. These requisites for a supportive system seem incompatible. In echinoderms, however, the supportive systems meet these requirements by being able to change the mechanical properties of the catch connective tissues. Echinoderms have three kinds of supportive systems: hydrostatic, articulated, and imbricate skeleton. Catch connective tissues are found in all these supportive systems; they function in the maintaining the tone of the body.

2.1 Animals with hydrostatic skeleton: sea cucumbers

Aspidochirotid holothurians have a large coelom which is surrounded by a thick body wall (Fig. 1a). The body wall is mainly composed of a connective tissue dermis, which is a catch connective tissue. The isolated dermis changes its mechanical properties in response to various kinds of stimuli: photic, chemical, electrical, and mechanical stimulations are effective (Serra-von Buddenbrock, 1963; Stott et al., 1974; Motokawa, 1981, 1982a,b, 1984a,b). When an animal is mechanically irritated, it stiffens the body wall. When it pushes itself through a narrow crevice, the body wall softens and becomes deformable. Although there is no in vivo measurements, the dermis is most possibly responsible for these changes.

Another instance of a hydrostatic skeleton in echinoderms is found in tube feet. The connective tissue layer in the wall of echinoid tube feet has been suggested to be a catch connective tissue (Florey & Cahill, 1977).

2.2 Animals with articulated skeleton: sea lilies, brittlestars, and sea urchins

The arms of crinoids and ophiuroids consist of series of ossicles which are articulated and bound with ligaments and muscles (Fig. 1b). These ligaments are catch connective tissues (Wilkie, 1978a,b, 1983; Motokawa, 1984c). When they are stiff or in catch , the joints are immobilized so that the posture of the animal is maintained. When the ligaments are soft or relaxed, the joints are free to move so that the animal can flex and move the body.

Sea urchins have spines which articulate on the tubercles of their tests (Fig. 1c). There are two kinds of ligaments at a spine joint; both are catch connective tissues (Takahashi, 1967; Motokawa, 1983). One ligament, the catch apparatus, controls the movability of the joint and thus maintains the posture of the spine. The other ligament, the central ligament, limits the excursion of spines and thus prevents disarticulation.

Some sea stars and brittlestars have articulated spines. Their joint ligaments change the mechanical properties (starfish ligament: Motokawa, 1982b; brittlestar ligament: unpublished observation).

In summary, the catch connective tissues control the movability of the joints and thus control the posture of animals.

2.3 Animals with imbricate skeleton: starfish

The body wall of some starfish consists of imbricate ossicles, which are bound together with muscles and dermis. The merit of this arrangement is that although the body is covered with rigid plates it remains flexible: the animal can deform the body by sliding the adjacent plates past one another. Starfish have remarkable abilities to change body shape and to make the body stiff. These abilities can be explained if we assume that the connective tissues between the plates are catch connective tissues (Wainwright, 1982): when the connective tissues are soft the plates can slide and thus the body is flexible; when the connective tissues become stiff the plates are frozen at that position and thus the posture of the animal is maintained. This mechanism is very probable because of the wide occurrence of catch connective tissues in echinoderms, although the stiffness change of the connective tissue between the starfish ossicles has not yet been shown by mechanical tests.

Imbricate skeletons are found not only in modern asteroids but also in some extinct echinoderms such as helicoplacoids and edrioasteroids (see U133 and U238 of Moore, 1966). The catch connective tissues have quite possibly controlled the flexibility of the bodies of these animals.

3 AUTOTOMY AND FISSION

Autotomy is found in every class of echinoderms (Emson & Wilkie, 1980). Crinoids, ophiuroids, and sometimes asteroids autotomize their arms when they are grasped by the arms. The connective tissue ligaments, which bind the ossicles, become soft enough to lose tensile strength at autotomy (Wilkie, 1978a; Holland & Grimmer 1981). These responses have defensive roles. The most active defensive reaction is the attack autotomy of echinoid pedicellariae. The globiferous pedicellaria detaches from the test after it has bitten the enemy; it keeps on biting after autotomy. The connective tissue ligament which connects the pedicellaria to the test softens to break at autotomy (Hilgers & Splechtna, 1982). In sea cucumbers evisceration is an autotomy phenomemon. When an animal is strongly poked or when aquarium conditions are poor, a part of the body wall softens to break, and through the rupture the viscera are

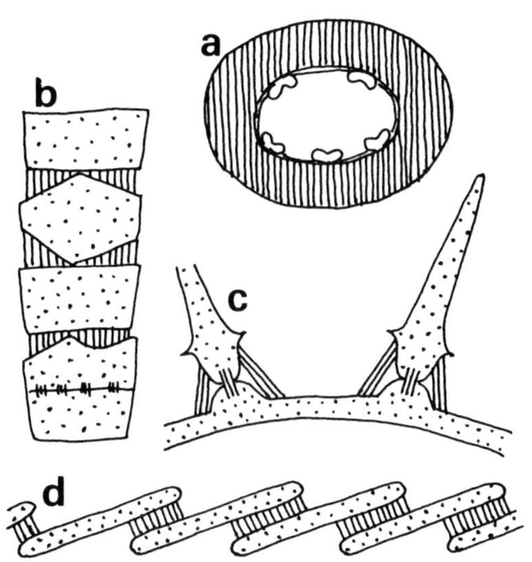

Figure 1. Disposition of catch connective tissues in the supportive systems of echinoderms. Stripes indicate catch connective tissue; stipples indicate the ossicle. a) hydrostatic skeleton: cross section of a sea cucumber. b) articulated skeleton: crinoid arm joints. c) articulated skeleton: longitudinal section of echinoid spine joints. The catch connective tissue which surrounds the joint is the catch apparatus, the tissue disposed at the centre is the central ligament.
d) imbricate skeleton: cross section of starfish integument.

extruded.

The stiffness change of holothurian dermis depends on the intensity of mechanical stimulation: when the stimulation is weak the response is stiffening; at a strong stimulation the response is softening and it is often followed by evisceration. The intervertebral ligament of a brittlestar has been shown to reversibly change its stiffness (Motokawa, 1984c). Therefore it is a simpler and thus a more probable interpretation to consider that the softening at autotomy as an extreme case of softening in the connective tissue catch pheomenon rather than to postulate two separate mechanisms, connective tissue catch and autotomy.

The catch connective tissues become extremely soft to allow a part of the body to break off in the event of an emergency. This ability to divide the body into parts may have led the echinoderms to employ fission: asexual reproduction by fission is observed in a number of echinoderms (Emson & Wilkie, 1980). In some coral reef asteroids, fission is suggested to be the main reproductive method (Rideout, 1978).

4 THE SECRET OF ECHINODERMS' SUCCESS

The Echinodermata is one of the most successful animal phyla: echinoderms are conspicuous large animals in the sea; they are dominant among the fauna of the deep sea floor. Although their distribution is restricted to the sea, there is no doubt that they constitute a significant part of the total standing crop of the earth, because the most of the surface of the earth is covered with sea water. Why are echinoderms so successful? A contributory reason may be that they have developed the catch connective tissues.

The success of an animal group depends on how well protected it is from predation. There are two trends in avoiding predation: one is to develop a protective shell around the body, and the other is to develop a high locomotory ability with which the animal can escape from predators. Let me call the animals with the former strategy the "defense-oriented animals", and those with the latter the "locomotion-oriented animals". Because the defense-oriented animals are covered with rigid, thick, and thus heavy shells, their locomotory activities are quite limited; they sometimes lead sedentary lives. The examples are shelled molluscs and corals. The locomotion-oriented animals must develop not only powerful muscles but also light and flexible supportive systems. Their integuments are therefore soft and pliant, and thus they have little defensive function. These two groups have different supportive systems: the defense-oriented animals have rigid and inflexible supportive systems whereas the locomotion-oriented animals have flexible ones. These animals fulfil only one of the two requisites for supportive systems (i.e. rigid and flexible, see Section 2).

In contrast to these animals there are certain animals which have supportive systems with both rigid and flexible characteristics. With these supportive systems, they can undertake extensive locomotory activities without sacrificing the defensive function of integuments. Examples are found among the arthropods and echinoderms, the two major successful animal groups.

The arthropods developed a chitinous

cuticle. The cuticular exoskeleton is strong and rigid, and thus it provides an effective defensive shell. In addition, it is light so that locomotion is not hindered. The cuticle is flexible at joints where appendages for locomotion are connected. In some instances the active change in mechanical properties of cuticle can be observed (Reynolds, 1975; Tychsen & Vincent, 1976). It is widely accepted that the arthropod cuticle is the key to their present success (Currey, 1970).

Echinoderms have developed a dermal system in which ossicles are bound together by catch connective tissues. Because the catch connective tissue rapidly changes its viscosity by 100-1000 fold (Motokawa, 1984d), the rigidity of the dermal system can be adjusted in order to meet the mechanical environment and the needs of the animal: the system can become stiff to be a defensive "shell" against predators and environmental mechanical stress; it can be stiffened to maintain the posture of the animal; it can become soft to allow the animal to change the body shape so as to hide into a narrow crevice; it can become soft to allow the animal to flex the body for locomotion; it can become very soft to enable the animal to divide the body at autotomy and fission.

The catch connective tissues soften to make the body flexible to improve locomotion efficiency: sea cucumbers make the dermis soft and generate peristaltic waves; sea urchins make the catch apparatus soft and move the spines for walking; brittlestars make the intervertebral ligaments soft and generate propagating waves on the arms when they move; featherstars make the brachial ligaments soft and swim by flapping movement of the arms. Many echinoderms walk with tube feet. The rigidity and flexiblity of tube feet may be controlled by the catch activity of the connective-tissue wall.

This remarkable connective tissue has enabled echinoderms to live exposed on the sea floor. Other locomotion-oriented animals with locomotory ability similar to echinoderms (such as annelids), however, are usually found hidden in the substrate, because they have no stiff integuments to protect them against more active predators. Echinoderms have integuments which can make defensive barriers when necessary: the imbricate skeleton of asteroid body wall can become an excellent armour; the arrays of echinoid spines which are held rigid and straight by catch connective tissues makes a barrier against attack; body walls with catch connective tissues and ossicles (which are large in ophiuroids and crinoids while very small in holothuroids, probably compensated for by the thickness of the body wall) become rigid protective integuments. Moreover when these stiff integuments are inadequate to resist the attacks of powerful enemies, the echinoderm makes the integument very soft to autotomize a part of the body so that they can sacrifice that part to the enemy as food. The animal can escape while the enemy is eating the autotomized part (Aldrich, 1976). This ability to live safely exposed enables the echinoderms to feed freely irrespective of their feeding habits: rheophilic suspension feeders can stand exposed in the currents; starfish can attack clams in an exposed position; sea urchins can graze algae on the open sea floor.

The catch connective tissues are used in maintenance of posture in echinoderms. This is unusual: in other animals the posture is maintained by muscles. The catch connective tissues probably consume less energy than muscles, because crinoids and ophiuroids engaged in suspension feeding (and thus with ligaments in catch state) have lower metabolic rate than those of other animals (LaBarbera, 1982). The economical echinoderm body plan may also have contributed to their success.

The success of echinoderms in the sea is very likely dependent on the catch connective tissues. This very dependence has been suggested to be the reason echinoderms have failed to invade fresh water (Eylers, 1982). The catch connective tissue is quite sensitive to the ionic environment in the tissue. This dependence has led to the proposal of an "ionic environment hypothesis" for the viscosity change mechanism of the catch connective tissues (see review by Motokawa, 1984c). In fresh water, echinoderms may have difficulties in maintaining the ionic environment of the catch connective tissues, and thus they may lose the ability to control the mechanical properties of their supportive systems.

Because this unique connective tissue is so deeply involved in the life of echinoderms, I believe that the study of this tissue provides answers to why echinoderms have such unique and different features from other animal phyla (Nichols, 1975).

5 ACKNOWLEDGEMENTS

I wish to thank Professor David Nichols for reading the manuscript.

6 REFERENCES

Aldrich, J.C. 1976. The spider crab Libinia emarginata Leach, 1815 (Decapoda Brachyura), and the starfish, an unsuitable predator but a cooperative prey. Crustaceana, 31:151-156.

Currey, J. 1970. Animal skeleton. London: Arnold.

Emson, R.H. & I.C.Wilkie 1980. Fission and autotomy in echinoderms. Oceanogr. mar.Biol.ann.Rev. 18:155-250.

Eylers, J.P. 1982. Ion-dependent viscosity of holothurian body wall and its implications for the functional morphology of echinoderms. J.exp.Biol. 99:1-8.

Florey, E. & M.A.Cahill 1977. Ultrastructure of sea urchin tube feet. Evidence for connective tissue involvement in motor control. Cell Tissue Res. 177:195-214.

Hilgers, H. von & H.Splechtna 1982. Zur Steuerung der Ablösung von Giftpedizellarien bei Sphaerechinus granularis (Lam.) und Paracentrotus lividus (Lam.) (Echinodermata, Echinoidea). Zool.Jb. Anat. 107:442-457.

Holland, N.D. & J.C.Grimmer 1981. Fine structure of syzygial articulations before and after arm autotomy in Florometra serratissima (Echinodermata: Crinoidea). Zoomorph. 98:169-183.

LaBarbera, M. 1982. Metabolic rates of suspension feeding crinoids and ophiuroids (Echinodermata) in a unidirectional flow. Comp.Biochem.Physiol. 71A:303-307.

Moore, R.C. (ed.) 1966. Treatise on invertebrate paleontology, part U, Echinodermata 3, vol.1. Lawrence: Geol. Soc. Amer. and Univ. Kansas.

Motokawa, T. 1981. The stiffness change of the holothurian dermis caused by chemical and electrical stimulation. Comp.Biochem.Physiol. 70C:41-48.

Motokawa, T. 1982a. Factors regulating the mechanical properties of holothurian dermis. J.exp.Biol. 99:29-41.

Motokawa, T. 1982b. Rapid change in mechanical properties of echinoderm connective tissues caused by coelomic fluid. Comp.Biochem.Physiol. 73C:223-229.

Motokawa, T. 1983. Mechanical properties and structure of the spine-joint central ligament of the sea urchin, Diadema setosum (Echinodermata, Echinoidea). J.Zool.Lond. 201:223-235.

Motokawa, T. 1984a. The viscosity change of the body-wall dermis of the sea cucumber Stichopus japonicus caused by mechanical and chemical stimulation. Comp.Biochem.Physiol. 77A:419-423.

Motokawa, T. 1984b. Viscosity increase of holothurian body wall in response to photic stimulation. Comp.Biochem. Physiol. in the press.

Motokawa, T. 1984c. Connective tissue catch in echinoderms. Biol.Rev. 59:255-270.

Motokawa. T. 1984d. Viscoelasticity of holothurian body wall. J.exp.Biol. 109:63-75.

Nichols, D. 1975. The uniqueness of the echinoderms. London: Oxford Univ. Press.

Reynolds, S.E. 1975. The mechanical properties of the abdominal cuticle of Rhodnius larvae. J.exp.Biol. 62:69-80.

Rideout, R.S. 1978. Asexual reproduction as a means of population maintenance in the coral reef asteroid Linckia multifora on Guam. Mar.Biol. 47:287-295.

Serra-von Buddenbrock, E. 1963. Études physiologiques et histologiques sur le tégument des holothuries (Holothuria tubulosa). Vie Milieu, 14:55-70.

Stott, R.S.H., H.R.Hepburn, I.Joffe, & J.J.A.Heffron 1974. The mechanical defensive mechanism of a sea-cucumber. S.Afr.J.Sci. 70:46-48.

Takahashi, K. The catch apparatus of sea-urchin spine II. Responses to stimuli. J.Fac.Sci.Univ.Tokyo,Sec.IV, 11: 121-130.

Tychsen, P.H. & J.F.V.Vincent 1976. Correlated changes in mechanical properties of the intersegmental membrane and bonding between proteins in the female adult locust. J.Insect Physiol. 22:115-125.

Wainwright, S.A. 1982. Structural systems: hydrostats and frameworks. In C.R.Taylor, K.Johansen & L.Bolis (eds.), A companion to animal physiology, p.325-338. Cambridge: Cambridge Univ. Press.

Wilkie, I.C. 1978a. Arm autotomy in brittlestars (Echinodermata: Ophiuroidea). J.Zool.Lond. 186:311-330.

Wilkie, I.C. 1978b. Nervously mediated change in the mechanical properties of a brittlestar ligament. Mar.Behav.Physiol. 5:289-306.

Wilkie, I.C. 1983. Nervously mediated change in the mechanical properties of the cirral ligaments of a crinoid. Mar.Behav.Physiol. 9:229-248.

Proceedings of the Fifth International Echinoderm Conference / Galway / 24-29 September 1984

Evolution and adaptation in some Caribbean Oligo-Miocene Clypeasters

R.H.PODDUBIUK
Department of Geology, Bedford College (University of London), UK

ABSTRACT: The echinoid genus Clypeaster appears to have reached the Caribbean area only in mid-Oligocene times. Its apparent explosive radiation there during mid-Tertiary times is, however, a human artefact resulting mainly from underestimation of the morphological variation within Clypeaster species.The Miocene species C.concavus and C.cubensis appear to be ancestral to the distinctive Recent Caribbean type species C.rosaceus. Trends within this lineage, particularly variation in test shape, test construction and tuberculation pattern, indicate progressive adaptation to an epifaunal mode of life, possibly prompted by the Miocene expansion of seagrass communities in the Caribbean.

1 INTRODUCTION

Clypeaster is one of the commonest and most diverse post-Eocene echinoid genera of shallow tropical seas. Over 400 nominal species have been described, grouped into a variety of subgenera defined mainly on the basis of gross test morphology and now recognised as of little taxonomic value (Durham 1966).The genus itself is, however, clearly distinguished from other members of the Order Clypeasteroida, aborally by the simple pseudocompounding of plates within its petals, adorally by the discontinuity of its interambulacra near the peristome and the presence of simple food grooves flanked by irregularly distributed accessory tube feet.

In addition to its purely numerical importance and the interesting biological diversity this reflects, Clypeaster is, as a result of its high preservation potential, a macrofossil of some biostratigraphical, biogeographical and occassionally even sedimentological importance. It appears to have evolved in the Tethyan region about the end of the Middle Eocene. The oldest recorded species (C.marbellensis,C.boussaci, C.sayni, C.alzadai and C.moianensis) occur in the Bartonian/Auversian of south-west Europe. Dispersal across the Atlantic did not take place until the Oligocene, the earliest New World species known, C.rogersi, ranging down into strata of probable early Chattian age (Cooke 1959).Transatlantic gene flow proved insufficient to prevent rapid allopatric speciation of the Caribbean populations. No species of Clypeaster seems to have maintained an amphiatlantic distribution but several distinct Caribbean mid-Tertiary species appear to have close affinities with contemporaneous or earlier Old World taxa (e.g. C.batheri with C.solanoi, C.oxybaphon with C.pulvinatus) suggesting a number of independent east to west crossings.

2 CARIBBEAN MID-TERTIARY CLYPEASTERS

Published records suggest that the arrival of Clypeaster in the New World was followed by a tremendous adaptive radiation. Over sixty species have been recorded from Upper Oligocene and Lower Miocene strata of the West Indies alone, just one of six subprovinces into which the Miocene Caribbean province can be divided (Woodring 1974). Re-examination of type material, in conjunction with statistical studies on newly collected population samples, indicates that known Upper Oligocene through Middle Miocene Clypeasters from Caribbean islands east of Cuba should be grouped into no more than seven species: C.batheri Lambert,1915; C.caudatus Jackson,1922; C.concavus Cotteau, 1875; C.cubensis Egozcue y Cia, 1897; C. julii Roman, 1952; C.oxybaphon Jackson,1922, pars.; C.oxybaphon Jackson, 1922, sensu stricto (Poddubiuk in prep; Poddubiuk & Rose in press). The names given are generally the oldest for which material figured or referred to in the original description is still available. Names of inadequately described species for which no type material is available are ignored since objective assessments of their discretness and relationships are no longer possible. Although a number of additional valid species have certainly been recorded from Cuba it is clear that well over two-thirds of all described West Indian fossil Clypeaster species are synonyms. The principal reason for this has been the tendency in the past to define Clypeaster species on the basis of quantitative differences in a small number of variables (such as relative petal length) without examining the extent of their intraspecific variation. This worldwide problem in the taxonomy of the genus was compounded for Caribbean mid-Tertiary species by two other factors:

1. Most early students worked in Europe on small collections sent to them by amateurs and often had to describe species from single isolated specimens or not at all
2. The long and diverse colonial history of the region has resulted in type specimen repositories being scattered across the Caribbean, the USA, and Europe. As a consequence, past systematists have often had to rely on incomplete descriptions with poor, even inaccurate, illustrations or no illustrations at all.

Reliance on quantitative diagnostic characters has also hindered evolutionary studies on Clypeaster more directly. Such characters are particularly subject to parallel evolution and, in the absence of a continuous gradational sequence, may not allow objective assessments of ancestry.

3 ANCESTRY OF CLYPEASTER ROSACEUS

Clypeaster rosaceus is perhaps the most distinctive representative of the genus in the Caribbean region. Though the type species, it is morphologically and ecologically rather atypical. Most members of the genus are shallow burrowers (Seilacher 1979). Clypeaster rosaceus, while still capable of burrowing under laboratory conditions (R. Mooi pers. comm.), is essentially a surface dweller in nature, apparently with a preference for seagrass beds in some areas (Kier & Grant 1965;Seilacher 1979). Surface dwelling Clypeasters have evolved several times independently (C.julii of the Caribbean Upper Oligocene being an earlier, and unrelated, epifaunal form), but C.rosaceus seems to be exceptional in that there is some evidence both for how and why it evolved. The pattern of its evolution can be discerned with some confidence because, in addition to quantitative changes in morphology two qualitative innovations distinguish it from its possible Lower Miocene ancestors. First, its test has a distinct double wall construction well illustrated by Mortensen(1948) and Seilacher (1979). Second, its genital pores are normally well separated from the madreporite. It is interesting that both these changes seem also to have taken place independently during the evolution of some other large and inflated Clypeasters, such as C.altus. In the Caribbean lineage the aquisition of the former feature distinguishes C.cubensis from C.concavus, while that of the latter distinguishes C.rosaceus from C.cubensis. Although the relationships between these species are best represented cladistically, their wide dispersal, their essentially sequential time ranges and the overlaps in gross morphology between each pair suggest that they may form part of a Darwinian lineage.

3.1 Clypeaster concavus Cotteau, 1875

Description: Medium to large; ambital outline oval to subpentagonal. Aboral surface low marginally with moderately elevated petaloid area centrally; anterior margin considerably thicker and more tumid than posterior. Adoral surface moderately concave. Petals lanceolate with moderate length, width and inflation; all petals slightly open; anterior petal longer and narrower than paired petals. All five genital pores in contact with stelliform madreporite. Test wall solid and of moderate thickness.

For quantitative information see figures and table below. For full redescription and synonymy see Poddubiuk(in prep.).

Distibution: The known stratigraphic range is from late Lower to early Middle Miocene. Geographic distribution was wide, including Anguilla, Tintamarre, St. Martin, Puerto Rico, Dominican Republic, Haiti, Cuba, Florida, Panama and perhaps Brazil.

Remarks: This widespread species displays considerable morphological variation and has an extensive synonymy (Poddubiuk in prep.). A specimen from the type locality originally described as a separate species, C.antillarum, illustrates that even early examples could attain a large sixe and in the process acquire some superficial similarities to C.cubensis.

3.2 Clypeaster cubensis Egozcue y Cia, 1897

Description: Large; ambital outline oval or occasionally subpentagonal. Aboral surface moderately and evenly inflated. Adoral surface moderately to highly concave. All petals similar in shape, lanceolate, slightly open, long with moderate to high width and inflation. All five genital pores in contact with madreporite. Double wall test construction.

For quantitative information see figures and table below.

Distribution: Known stratigraphic range is Middle and early Upper Miocene. Known geographic distribution includes Cuba, Puerto Rico and, as ´C.kugleri´, Trinidad.

Remarks: The holotype of this species could not be located. It was, in any case, poorly preserved adapically, according to the orignal description. A succinct illustrated description of Puerto Rican specimens ascribed to the species may be found in Gordon(1963) whose figured specimen would probably make the best neotype. They may be distinguished from Clypeaster rosaceus by their narrower and more closed petals, as well as by the contact between their genital pores and madreporite. Clypeaster caudatus Jackson, 1922, known only from two specimens collected in the Dominican Republic, appears to be distinguishable from C.cubensis only by its somewhat lower test and slightly greater petal closure(fig.2; table 1). The statistical significance of these differences cannot reliably be judged on the basis of such a small sample.

3.3 Clypeaster rosaceus Lamarck, 1801

Three subspecies are known: Clypeaster rosaceus rosaceus, the living form; C. rosaceus dalli from the Plio-Pleistocene of Florida; and C. rosaceus duchassingi from the ?Pliocene of Guadeloupe. All are characterized by a double-walled test construction and separation of genital pores from the madreporite. Although the species may extend back into the Miocene, with the possible exception of some poorly dated Venezualan material recorded by Cooke(1961), published Miocene records seem to be due to misidentification of C.cubensis. The best published description of C.rosaceus rosaceus is given by Mortensen(1948). Kier(1963) discussed similarities between the subspecies rosaceus and dalli, concluding that the latter could be distinguished by its greater width. Clypeaster rosaceus duchassingi, originally described as a separate species by Michelin(1861), is currently known only from a single specimen now in the Museum National d´Histoire Naturelle,Paris. It can be distinguished from the other subspecies by its wider, more open and almost perfectly flush petals (see table 1).

4 TRENDS DURING THE EVOLUTION OF C.ROSACEUS

As well as the two qualitative changes discussed above many quantitative morphological changes took place during the evolution of C.rosaceus. Perhaps the most obvious of these is an increase in test size, using as a measure either an overall size criterion based on length, width, and height or simply test length (Fig.2).The greatest differences occur between early and late populations of C.concavus, but there is no evidence to suggest that they are due to differ-ences in population structure or environmental stunting. Relative test height is very variable in all three species under consideration, but both its mean and maximum values consistently increased to reach their highest levels in the Plio-Pleistocene subspecies C.rosaceus dalli (Fig.2). Height of the peristome above the base of the test also increased, even relative to test height. It is normally less than 50% of test height in early C. concavus and over 60% in C.rosaceus. Relative test width does not appear to have changed significantly, though C.rosaceus dalli was slightly broader than its ancestors. Less easily quantifiable changes in overall test shape were the loss of a distinctly elevated petaloid area as overall inflation of the test increased, and decreasing anterior to posterior differences in aboral test camber, margin thickness and tumidity (Fig.1). As the external test form changed so did the pattern of internal supports. In C.concavus these are concentrated close to the ambitus, while in C.cubenis and C.rosaceus they are best developed about midway between the pole and ambitus of the test (compare Fig.1d to Fig.2 of Gordon[1963] and Fig.1 of Seilacher[1979] or Pl.1,figs.3&4 of Mortensen[1948]). The thickness of the test itself also changed drastically from a maximum of approximately 2mm in early C.concavus (<2.5% of test length) to over 7mm in Recent C.rosaceus (>6% of test length).All the petals increased in size relative to test length and became more open while the distinctions between the anterior petal and paired petals became less marked (Table 1). The maximum width of the petals shifted adambitally, occurring some 60% of the petal length away from the ocular pore in C.concavus but typically over 70% away in C.rosaceus. There was also an increase in the width of the poriferous zones relative to test length and in the similarity of adradial and perradial respiratory tube foot pores. Petal inflation was very variable, particularly from one population to another. There does, however, appear to have been a trend towards greater symmetry about the perradial

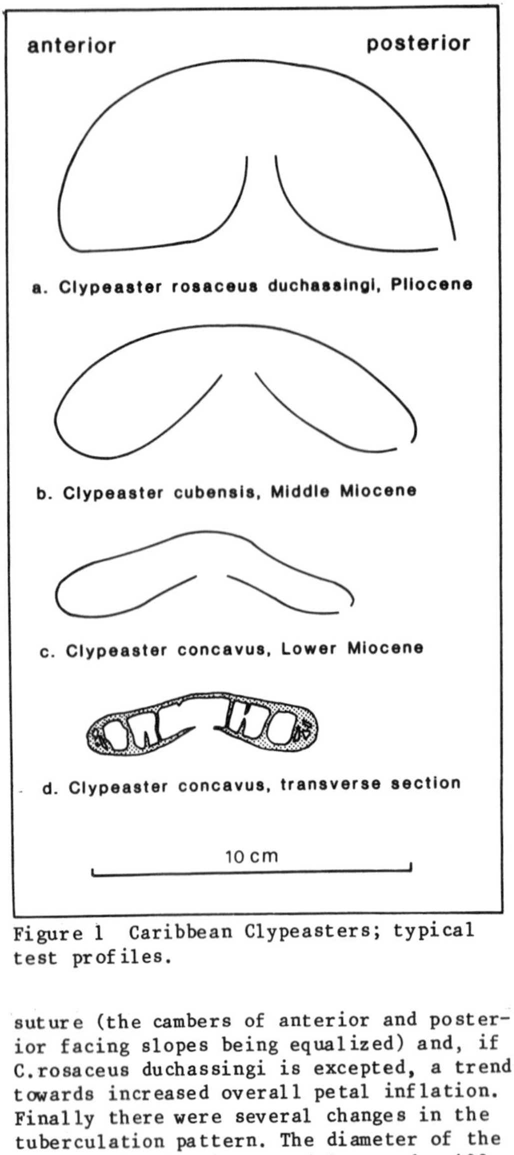

Figure 1 Caribbean Clypeasters; typical test profiles.

suture (the cambers of anterior and posterior facing slopes being equalized) and, if C.rosaceus duchassingi is excepted, a trend towards increased overall petal inflation. Finally there were several changes in the tuberculation pattern. The diameter of the aboral tubercles increased from under 400 microns in the figured cotype of C.concavus to over 700 microns in typical C.rosaceus. Simultaneously aboral tubercle packing density decreased from 250 to 300 tubercles per square centimetre in C.concavus to under 120 in C.rosaceus. The pattern of variation in tubercle size and packing density also changed. In particular the tubercles at the anterior margin of the test,much larger and more closely packed than those elsewhere in C.concavus, became almost indistinguishable from those of the lateral margins in C.rosaceus. Concurrently the tubercles of the lateral margins lost the clear posterior areole enlargement present in C.concavus, while the tubercles of the flattened adoral margin became larger and more densely packed.

5 FUNCTIONAL SIGNIFICANCE OF THE TRENDS

The morphological changes that took place during the evolution of C.rosaceus all indicate, or are consistent with, progressive adaptation from an infaunal to an epifaunal mode of life. The most obvious trends, increasing test height and reduction in anterior-posterior differentiation, involve a loss of streamlining. The zone of particularly large and densely packed tubercles with deep radially symmetrical areoles that characteristically covers the thick tumid margin of shallow sand burrowers such as C.concavus was lost, sediment loosening spines no longer being required. The spines of the aboral canopy, freed from the size restrictions imposed by their original sediment shielding function, became larger but more sparse. Loss of areole enlargement posteriorly in tubercles on the lateral margins of the test indicates the decreasing importance of their spines in test locomotion as the animal became epifaunal. In the same way the elevated petaloid area, generally interpreted as a ´snorkel´ in shallow burrowers (Seilacher 1979) was lost as the animal came to spend most of its time outside the sediment. A higher and more regularly domed test had positive structural advantages. Both these changes increased the uniformity of stress distribution over the test surface and the efficiency of its transference to the substratum. The thicker double-wall test construction and larger spines further increased the strength of the test, particularly its shock resistance, and perhaps also its effectiveness as a defense against predators. The evenness of the thickening both aborally and adorally indicates its primary function was probably not increased weighting of the animal for improved stability; for this purpose the added test weight would be best distributed around the ambitus, as in forms such as C.oxybaphon sensu lato. This circumferential distribution maximizes increases in the moment required to lift the edge of the test from the substrate. The divergence of the genital pores and their eventual separation from the madreporite reduced the weakness of the apical area, dispersing the ocular and genital perforations through the test over a wider area. The changes in internal support pattern appear, as Seilacher (1979) has stated, to reflect changes in the internal tethering pattern of a pneu associated with rounding of the test margin and deepening of the oral concavity.They do not seem to reflect design changes in a

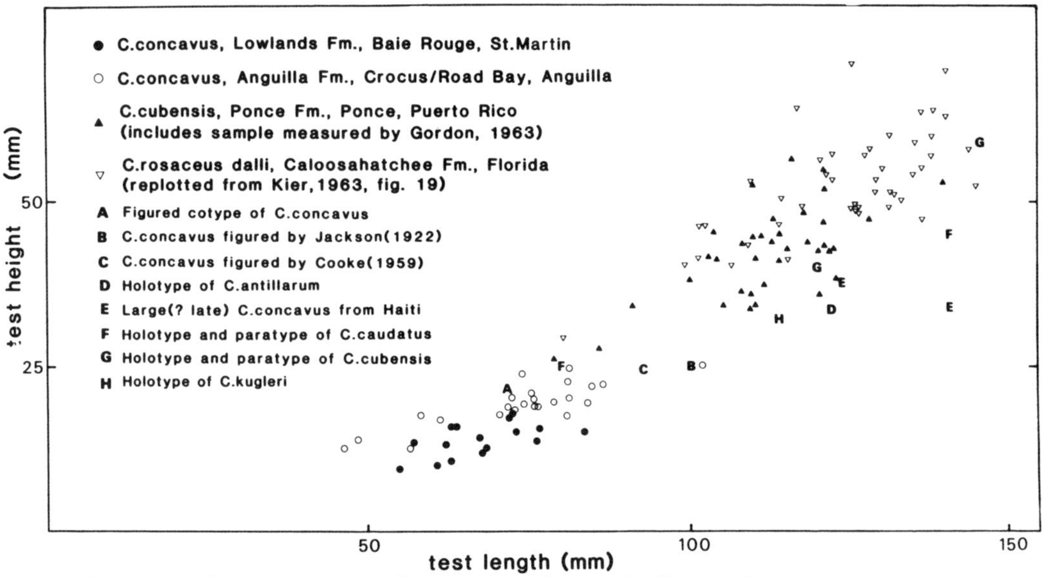

Figure 2 Test height and length in samples of some Caribbean fossil Clypeasters.

bracing system against external stresses. Changes in petal characteristics fall in two groups, those that increase their radial symmetry and those that increase their relative size. Return to approximately radial symmetry as far as the respiratory apparatus is concerned is a fairly common phenomenon among surface dwelling echinoids with an infaunal ancestry. In this case it involved increasing similarity between the different petals, more even petal inflation, spreading of the poriferous zones by broadening and opening of the petals, and increasing similarity of the adradial and perradial tube foot pores. These changes probably reflect the greater spatial homogeneity of an open water environment, in which resticted orientations of the gas exchange structures are no longer advantageous, and the need for forced flow of the external respiratory medium over them is reduced. The increasing relative size of the respiratory apparatus, due to increases in both the length and width of the poriferous zones relative to test length, seems at first to be inconsistent with adaptation to an epifaunal mode of life, in which one might expect gas exchange to be facilitated. It may perhaps be partly accounted for by increased test size and the resultant reduction in surface area to volume ratio. However this is not normally such a serious problem for echinoids as it is for other animals, much of the respiratory activity of a sea urchin being associated with its body wall (Giese 1976). Other factors may have been an increase in the relative volume of respiring tissue associated with thickening of the test wall, an increase in tissue activity, or adaptation to a surface environment abnormally low in oxygen such as a seagrass meadow at night. Too

Table 1 Petal characteristics of Clypeaster concavus, C.cubensis and C.rosaceus

Species		Specimen Number	Petal I			Petal II			Petal III		
			L*	W▽	C□	L*	W▽	C□	L*	W▽	C□
concavus	[Anguilla]	USNM 516746	34.8	20.7	8.5	31.5	19.8	7.6	37.3	19.1	6.4
concavus	[Anguilla]	USNM 214154	35.5	21.0	20.6	30.5	20.0	15.4	35.5	18.8	17.6
concavus	[Haiti]	ANSP 14825	31.6	19.6	19.0	28.8	19.2	13.6	35.2	18.5	23.1
cubensis		POD 17	43.5	24.0	21.0	39.2	23.9	16.4	45.5	21.7	18.0
rosaceus	duchassingi	MNHN no no.	42.5	24.3	32.1	40.0	20.3	19.3	40.3	21.3	39.6
rosaceus	dalli	BMNH E73324	42.3	27.6	32.5	36.8	24.3	32.7	44.0	25.1	35.1
rosaceus	rosaceus	POD 1455	42.6	23.7	37.2	37.8	23.5	27.1	41.4	23.4	38.4

* Petal length measured along perradial suture from the ocular pore to the point at which the line joining the distalmost respiratory pores of each poriferous zone crosses the suture, expressed as a percentage of test length.

▽ Maximum external width of petal expressed as a percentage of test length.

□ External width of petal at its distal end as a percentage of max. external petal width.

little is known about the respiratory economics and physiology of C.rosaceus to distinguish the relative importance of the various possibilities. Kier & Grant (1965) suggested that the deep oral concavity of C.rosaceus was an adaptation to the handling of seagrass debris as food. Progressive deepening of the aboral concavity, and a tendency for it to become restricted to the peristomial region, may thus reflect adaptation to surface feeding on such material. Brasier(1976) argued on foraminiferal evidence that seagrasses did not reach the Caribbean region until the early Miocene. Their absence from the Lesser Antilles was even accepted by Eva(1980) and is confirmed by the author's sedimentological studies in the eastern Caribbean (Poddubiuk in prep.). Their arrival may have triggered evolution of the distinctive C.rosaceus morphology by reducing the amount of ecospace available for shallow burrowers and providing a large new source of organic debris. Increasing body size may then reflect, as it seems to in other lineages (Hallam 1975), improved adaptation to a stable environment.

6 CONCLUSIONS

Clypeaster evolved in the Tethyan region during the Eocene but did not disperse to the New World before Oligocene times. The number of Caribbean mid-Tertiary species belonging to this genus has been vastly exaggerated, due mainly to a poor understanding of intraspecific variation. Clypeaster rosaceus appears to have evolved from an ancestor resembling C.concavus via a form resembling C.cubensis. The morphological trends within this lineage indicate progressive adaptation from a shallow burrowing to an epifaunal mode of life, perhaps in response to the expansion of seagrass communities in the Caribbean during the early Miocene.

7 ACKNOWLEDGEMENTS

I gratefully acknowledge the award of a research studentship and finance for Caribbean fieldwork from the Natural Environment Research Council. Numerous people kindly allowed access to collections and arranged loans of material, particularly Dr G.Davies (Academy of Natural Sciences, Philadelphia), Dr R.P.S.Jefferies(British Museum,Natural History, London), Dr P.Jung(Naturhistorisches Museum,Basle), Dr P.M.Kier(Smithsonian Institution,Washington), Dr E.Kirsteur(American Museum of Natural History, New York, Dr D.Pajaud(Université Pierre et Marie Curie, Paris), Dr A.Prieur(Université Claude Bernard, Lyon) and Dr J.Roman(Muséum National d'Histoire Naturelle, Paris). I would also like to thank all those who contributed to this paper by their comments and encouragement, particularly my supervisor Dr E.P.F.Rose,Dr M. Telford and Mr R.Mooi.

REFERENCES

Brasier, M.D. 1975. An outline history of seagrass communities. Palaeontology 18: 681-702.

Cooke, C.W. 1959. Cenozoic echinoids of Eastern United States. U.S.Geol.Surv. Prof. Pap. 321: 1-106, Pls. 1-43.

Cooke, C.W. 1961. Cenozoic and Cretaceous echinoids from Trinidad and Venezuela. Smithson. misc. Collect. 142(4): 1-35.

Durham, J.W. 1966. Clypeasteroids. In R.C. Moore(ed.), Treatise on Invertebrate Paleontology, p.U450-U491. Geol. Soc. America & Univ. of Kansas Press.

Eva, A.N. 1980. Pre-Miocene seagrass communities in the Caribbean. Palaeontology 23: 231-236.

Giese, A.C. 1976. Physiology of the echinoderm body wall. Thalassia Jugoslavica 12: 153-163.

Gordon, W.A. 1963. Middle Tertiary echinoids of Puerto Rico. J. Paleontology 37: 628-642.

Hallam, A. 1975. Evolutionary size increase and longevity in Jurassic bivalves and ammonites. Nature. London 258:493-496.

Jackson, R.T. 1922. Fossil Echini of the West Indies. Publ.Carnegie Inst. Washington 306.

Kier, P.M. 1963. Tertiary echinoids from the Caloosahatchee and Tamiami formations of Florida. Smithson. misc. Collect. 145(5) 63p.

Kier, P.M. & R.E. Grant 1965. Echinoid distribution and habits, Key Largo Coral Reef Preserve, Florida. Smithson. misc. Collect. 149(6):1-68.

Michelin, H. 1861. Monographie des Clypeastres fossiles. Mem.Soc.geol.Fr. ser.2 7: 101-148 pls. 9-36.

Mortensen, T. 1948. A Monograph of the Echinoidea IV.2 Clypeasteroida. Copenhagen; C.A.Reitzel.

Poddubiuk, R.H. in prep. PhD thesis. London.

Poddubiuk, R.H. & E.P.F.Rose in press. Relationships between mid-Tertiary echinoid faunas from the central Mediterranean and eastern Caribbean. Ann. Geol. Hell., tome hors serie.

Seilacher, A. 1979. Constructional morphology of sand dollars. Paleobiology 5: 191-221.

Woodring, W.P. 1974. The Miocene Caribbean faunal province and its subprovinces. Verhandl. Naturf. Ges. Basel 84: 209-213.

Proceedings of the Fifth International Echinoderm Conference / Galway / 24-29 September 1984

Slow evolution in the Holectypidae, a family of primitive irregular echinoids

EDWARD P.F.ROSE & JANE B.S.OLVER
Department of Geology, Bedford College (University of London), UK

ABSTRACT: The family Holectypidae includes some of the earliest and most primitive irregular echinoids. Two genera, Holectypus and Caenholectypus, are particularly common and widespread. They form a well-known Jurassic-Cretaceous lineage which, despite high recorded species diversity, persisted for over 100 million years with but little structural change. Other commonly supposed holectypids (Temnholectypus, Coptodiscus, Discholectypus, Metholectypus) have contrastingly low species diversity and restricted occurrence. They are relatively poorly known and of disputed holectypid status. The family Holectypidae thus essentially comprises a single morphologically conservative lineage, strikingly similar to the family Pygasteridae (Plesiechinus-Pygaster lineage) in both tempo and mode of evolution. "Conservatism" is a feature of these primitive Eognathostomata which contrasts with the more rapid and diverse morphological changes seen in other and later irregular echinoids. Holectypids and pygasterids were deposit-feeders but still relatively unspecialized as burrowers; they had therefore effectively achieved only the first of the two major adaptive advances which facilitated diversification of the irregulars as a whole.

1 INTRODUCTION

The order Holectypoida includes most, arguably all, of the currently-known earliest and most primitive "irregular" echinoids. Like many other echinoid orders, its earliest members are recorded from the Lower Jurassic, which seemingly marks the time of most rapid evolutionary change in echinoids (Kier, 1973; Rose & Olver, 1980). However, unlike the other widely recognised orders of early irregular echinoids (Pygasteroida, Galeropygoida, Disasteroida), it has living representatives (Echinoneus, Micropetalon). These preserve features of primitive irregular echinoid morphology and ecology (Rose, 1978) significant for functional interpretation of the initial irregular echinoid adaptive radiation (Smith, 1981; 1984).

The order and its constituent families are as yet poorly defined (Rose, 1982). However, two sub-orders are broadly recognisable: (1) Holectypina (with type family Holectypidae, genotype the Jurassic-Cretaceous genus Holectypus), an exclusively Jurassic-Cretaceous group of holectypoids, typically with primitive features such as jaws in the adult, perignathic girdle, a notched peristome, and apical system with five genital plates; (2) Echinoneina (with type family Echinoneidae, genotype the Oligocene-Recent genus Echinoneus), an essentially Cretaceous-Cenozoic group of holectypoids with advanced features (synapomorphies) such as lack of jaws and perignathic girdle in the adult, an unnotched peristome, a tetrabasal or monobasal apical system, and (occasionally) petaloid ambulacra.

The family Holectypidae is the type family for the order, containing the earliest holectypoids (sensu Wagner & Durham, 1966) and representing the ancestral stock from which all other holectypoids directly or indirectly arose. Yet most of the supposed members of this family are poorly known and in consequence its composition and characteristics lack clear definition.

2 FAMILY HOLECTYPIDAE LAMBERT, 1899

Created as a subfamily of the family Pygasteridae by Lambert (1899), this group was more fully defined by Lambert & Thiéry (1911) to accommodate the genera Pygaster (including Plesiechinus), Pygastrides,

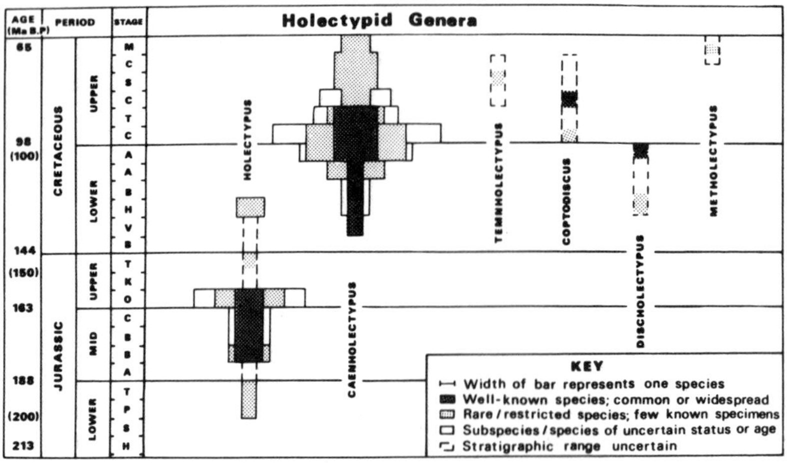

Fig. 1 Stratigraphic range and species diversity of holectypid genera; species number/range after Lambert & Thiéry (1911), Kier & Lawson (1978), with taxonomic/stratigraphic amendment, excluding three rare Caenholectypus species (aftabensis, decoratus, montalvensis) of imprecise U. Cretaceous range. Absolute ages (non-linear scale) and stratigraphic stages standardized after Harland et al. (1982).

Anorthopygus, Pileus, Holectypus (including Coenholectypus), Temnodiscus (= Temnholectypus), Coptodiscus, and Discholectypus. Characteristic features of the family Pygasteridae were "ambulacres simples; peristome avec scissures distinctes"; of the subfamily Holectypidae additionally "pas de cloisons internes". In discussion, Lambert & Thiéry (1911:281) grouped their genera in two series, one with periproct in contact with the apical system and one with apical system and periproct separated, thus in essence distinguishing pygasterids from holectypids respectively.

Mortensen (1948) reviewed the classification of these echinoids and followed Hawkins (1920) in recognising the Pygasteridae and Holectypidae as separate families, grouped together within the order Holectypoida. He assigned Pileus, together with Pygaster and Plesiechinus (as separate genera), to the Pygasteridae, distinguishing this group with supposed non-crenulate tubercles from the Holectypidae with crenulate tubercles. Five genera were assigned to the Holectypidae: Holectypus (including Temnholectypus), Caenholectypus (sic), Discholectypus, Anorthopygus, and Coptodiscus. Pygastrides, defined originally from a single Recent specimen, was excluded as merely the early post-larval stage of the cassiduloid Conolampas sigsbei.

In the Treatise on Invertebrate Paleontology, a separate order Pygasteroida, Durham & Melville, 1957, was adopted for the single family Pygasteridae (Pygaster, Plesiechinus, Pileus). Within the order Holectypoida, Wagner & Durham (1966) created a new family Anorthopygidae for Anorthopygus (alone), and redefined the Holectypidae to comprise Holectypus (including Temnholectypus), Coenholectypus (sic), Coptodiscus, Discholectypus, and Metholectypus.

Pileus was excluded from the Pygasteridae (as a probable Echinoneid) by Smith (1981), and the family therefore restricted to the two genera Plesiechinus and Pygaster. Primitive pygasteroids at least are now known to have crenulate tubercles, and there is little doubt that the Pygasteridae and Holectypidae are more closely related (Smith, 1981; Rose, 1982; Smith, 1984) than inferred by Mortensen (1948) or Durham & Melville (1957). However, possession of a supramarginal periproct in contact with the apical system clearly distinguishes pygasterids from all supposed holectypids, for these have submarginal or marginal periprocts separate from the apical system.

The separate family status for Anorthopygus has not been challenged, for although this genus has typical (primitive) holectypinid features such as a perignathic girdle (Kier, 1974, p.60) and notched peristome, it is quite distinct from other holectypoids in its large supramarginal/marginal (often oblique) periproct and ethmolytic apical system (Kier, 1974:48); its ambulacra are of simple primary plates, with uniserial pore-pairs; and unlike pygasterids and holectypids, its interamb tubercles are not arranged in regular series.

From the literature, it thus appears that

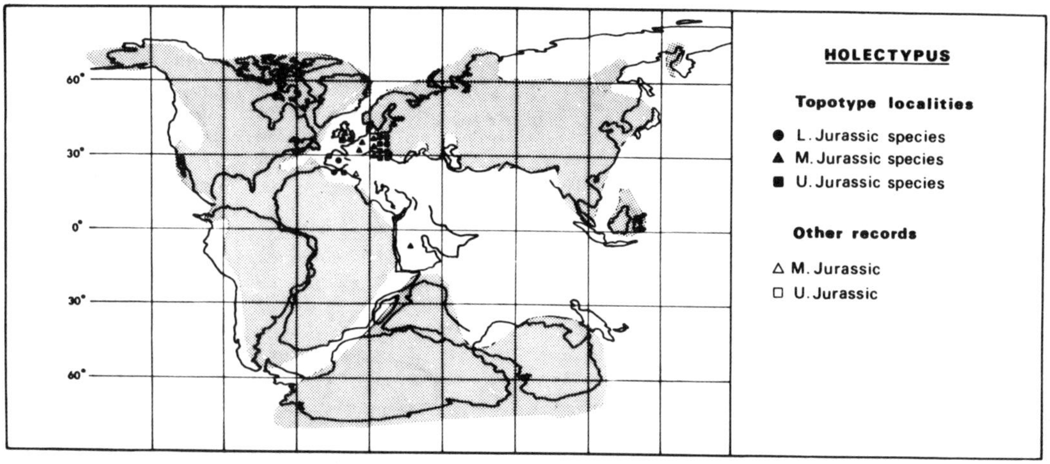

Fig. 2 Geographic range of Holectypus in Jurassic. Species as for fig. 1; mid-Jurassic continental reassembly and palaeogeography after Smith et al. (1977), Rose & Olver (1980). Marine areas: blank; land : stippled.

the family Holectypidae may comprise up to six genera, whose supposed close similarity and affinity with Holectypus are commonly indicated by their names: Holectypus, Caenholectypus, Temnholectypus, Coptodiscus, Discholectypus, Metholectypus (fig. 1).

3 HOLECTYPUS DESOR, 1842

Ascribed species: Type Discoidea depressa Agassiz, 1839 (= Echinites depressus Leske, 1778), common and widespread in NW Europe, Bajocian-Oxfordian; Lambert & Thiéry (1911) list fourteen other Jurassic species, and three more have been described since (Kier & Lawson, 1978; Kier, 1972). Additionally, an imperforate genital 5 occurs in the Hauterivian H. minutus Gauthier (Lambert, 1920) and H. almeidae Rey (1966), seemingly the only Cretaceous and thus the last true Holectypus. (Engel (1964) admits that the type of his supposed U. Cretaceous species H. boschmai may be a Jurassic H. corallinus).

Diagnostic features: Test small to medium (c. 20-45 mm adult diameter), rarely large (100 mm diameter), hemispherical, oral side flat or concave; apical system central, small, ethmophract, with five genital plates but only four gonopores (genital 5 imperforate), sometimes one or more complemental plates; ambs half to one third interamb width, plates low, simple primaries adapically, orally higher, every third plate enlarged; poriferous zones very narrow, with isopores similar throughout or slightly larger aborally with slightly elongate outer pore; interambs of low plates aborally, becoming very high adorally; tubercles small, crenulate, perforate, similar on ambs and interambs, appearing progressively to form orderly rows as plate length increases towards ambitus; (spines short, slender, with close-set longitudinal ribs, axis hollow); peristome central, large, almost circular, with well developed buccal notches; perignathic girdle of narrow amb processes and moderately developed interamb ridges; lantern well developed, teeth with lateral flanges; periproct inframarginal or marginal, large, radially elongate.

Discussion: Few of the twenty supposed species of Holectypus are well known; most are based on incomplete, 19th century definition; H. oblongus is erroneously based on a crushed specimen of H. depressus, and several other species may prove to be synonymous with one another.

In his description of the perignathic girdle of H. hemisphaericus, Hawkins (1934) observed that the girdle processes were "butressed by carinae which pass from a point about half-way up the process to taper very gradually into the test surface, disappearing about midway between the girdle and the ambitus", but these butresses are never greatly developed. Mortensen (1948: 29) therefore comments "Holectypus very closely resembles Discoidea... but the absence of internal radiating walls distinguishes Holectypus very clearly from Discoidea". Kier (1974:59) notes an evolutionary trend towards an increase in size of the interamb ridges within the holectypoids and pygasterids from Jurassic to Cretaceous species, but "well developed butresses" are generally accepted as characteristic of the family Discoididae in contrast to the "poorly developed internal butresses" of the Holectypidae (Wagner & Durham, 1966).

Occurrence: Fig. 2. Very rare and restricted in geographic distribution on first appearance in the L. Jurassic (H. hians, Pliensbachian, Algeria; H. conquensis, Toarcian, E. Spain). Individuals common, species more numerous and widespread throughout the M. Jurassic, and particularly subsequent Oxfordian stage, when Holectypus is widely recorded from shallow-water limestones across NW Europe, the circum Mediterranean countries, and the Middle East, extending S to Madagascar and E to India. At the end of its range, again seemingly rare and restricted to W Mediterranean, last recorded as H. afer (Tithonian, Algeria) and finally the Cretaceous H. minutus (S. France), and H. almeidae (Portugal).

4 CAENHOLECTYPUS POMEL, 1883

Ascribed species: Type Holectypus macropygus Desor, 1842 (= Discoidea macropygus L. Agassiz, 1840), widespread in Europe and N. Africa, Valanginian-L. Aptian; Lambert & Thiéry (1909-25) list 22 other Cretaceous "Holectypus" species (excluding subspecies and presumed synonyms); Kier & Lawson (1978) add 17 taxa. All are Caenholectypus except H. almeidae, H. minutus, H. boschmai, probably "C". macrostomus, possibly "C". cortesei (? = Discoidea pulvinata Desor, fide author).

Diagnostic features: Closely similar to Holectypus, but genital 5 perforate rather than imperforate.

Discussion: The perforate genital 5 marks a clear if minor morphological distinction between Caenholectypus and Holectypus; the two have very different stratigraphical ranges; each has a large number of ascribed species. It is therefore convenient to follow Pomel (1883), Mortensen (1948), and Wagner & Durham (1966) in maintaining them as separate genera, rather than follow Lambert & Thiéry (1911) in merging Caenholectypus in synonymy with Holectypus. Caenholectypus has sometimes been grouped as a subgenus of Holectypus (e.g. Cooke, 1955; Kier, 1974); this emphasises the close similarity and inferred affinity between the two but introduces a cumbersome nomenclature.

Since Caenholectypus occurs later in time than Holectypus but is so closely similar to it, direct descent is inferred.

Pomel spelt the name as Coenholectypus when defining the taxon (1883:75), but listed it as Caenholectypus in the index to his work (p.125). Each spelling is used only once; either would be appropriate, from Gk koinos = common, to indicate that species and/or individuals are common, or Gk kainos = new, to indicate that they are more recent than true Holectypus. Article 32(b) of the International Code of Zoological Nomenclature (Stoll, et al. 1961) rules that "If a name is spelled in more than one way in the original publication, the spelling adopted by the first reviser is to be accepted". Cooke (1955: 94) is the first reviser (sensu Article 24), since he explicitly places Coenholectypus in synonymy as a "Typographical error for Caenholectypus". Earlier authors commonly adopted the spelling Coenholectypus, but their usage does not constitute revision in the strict sense of the Code. Wagner & Durham (1966) maintained Coenholectypus, but on the criterion of page precedence rather than first revision, and therefore erroneously.

"Specific distinctions in Caenolectypus are based primarily on the size and location of the periproct and on the size and shape pf the peristome" (Cooke, 1955). Many of the ascribed species have been or probably should be placed in synonymy, and the stratigraphic ascriptions of Lambert & Thiéry (1909-25) are now often misleading.

Occurrence: Fig. 3. The earliest known species (C. macropygus) occurs in the Valanginian of Europe and N. Africa; wide geographic distribution (southern N. America, northern S. America, Europe, N. Africa, Japan) and increased species diversity was achieved by the Albian; seemingly even higher species diversity in the Cenomanian (particularly in N. Africa and the Middle East); fewer, but still widespread if imprecisely dated species range through the Senonian of N. America, Europe, N. Africa and the Middle East. The latest reported Coenholectypus are from the Maastrichtian (Shiranish Marls of Kurdistan; Mari Hills of Baluchistan; ?Tripolitania).

5 TEMNOHOLECTYPUS LAMBERT & THIERY, 1925

Ascribed species: Holectypus circularis Cotteau & Gauthier, 1895, from the Upper Cretaceous (Senonian) of Iran.

Diagnostic features: "Differs from Holectypus in the presence of numerous, regular pits on the test sutures" (trans. from Lambert & Thiéry, 1911:280).

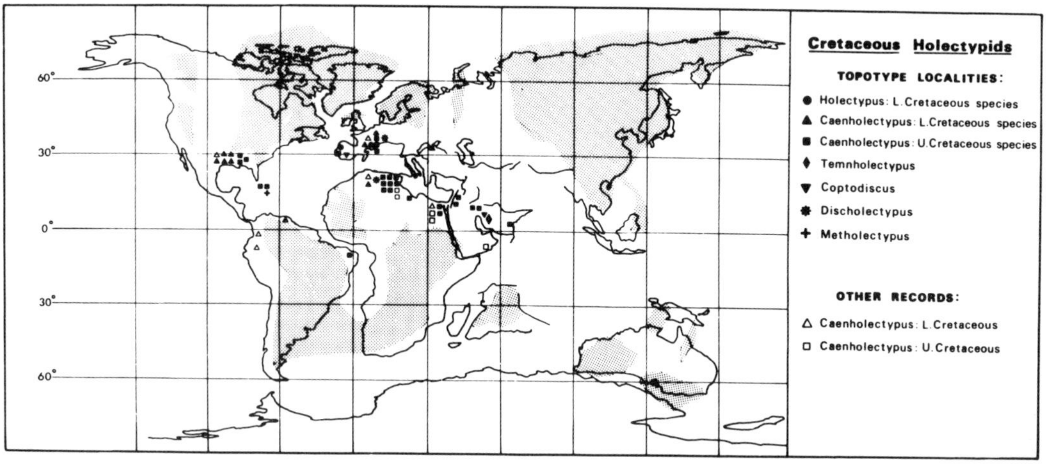

Fig. 3 Geographic range of "holectypids" in Cretaceous (cf. Fig. 2); mid-Cretaceous continental reassembly after Smith et al. (1977), Smith (1984).

Discussion: Temnholectypus, a replacement name for the preoccupied Temnòdiscus Lambert & Thiéry, 1911:280 (non Koken, 1896), is a monospecific taxon defined to accommodate T. circularis. Although supposedly characterised by fine, punctiform pits on the sutures of the oral side (Mortensen, 1948), Hawkins (1914) has shown similar pits to exist in several holectypoids, including Holectypus depressus and Caenholectypus serialis. Mortensen (1948) therefore concludes that the feature is not of generic significance. Temnholectypus is therefore a junior synonym, but since T. circularis has five gonopores, it is a synonym of Caenholectypus, not Holectypus as listed by Mortensen (1948) and Wagner & Durham (1966). Although thus included in Caenholectypus, C. circularis is distinguished from other species of that genus in its inflated test and very small periproct.

Occurrence: Described only from the Senonian of Iran. However, the Museum National d'Histoire Naturelle, Paris, contains specimens from N. Africa. Specimens in the British Museum (Natural History), presented by the Iraq Petroleum Company, from the Shiranish Marl of Kurdistan have been wrongly identified (R.G.S. Hudson, in MS) as H. cf. circularis; only one specimen (E. 44390) is a holectypid (Caenholectypus). This lacks the inflated shape and test pitting characteristic of C. circularis, and is too poorly preserved for further identification, but is important as one of the last known holectypids.

6 COPTODISCUS COTTEAU & GAUTHIER, 1895

Ascribed species: Type Coptodiscus noemiae Cotteau & Gauthier, 1895: U. Cretaceous (Senonian), Iran; also C. mengaudi Lambert 1919: U. Cretaceous (U. Cenomanian), Santander (Spain).

Diagnostic features: Similar to Caenholectypus, except for deep transverse depressions on test (amb and interamb) surface; on interambs pits in transverse sutures adapically, except near centre of each plate where suture is not depressed, thus producing longitudinal ridge extending from apical system to margin; marginally pits slight or absent but again strongly developed towards peristome.

Discussion: C. noemiae has been fully redescribed by Kier (1972) (from over 30 well-preserved specimens). Kier notes that distinctive pits, in both plates and plate sutures, which characterise the genus occur on both oral and aboral surfaces, not just the aboral surface as wrongly emphasised by Lambert & Thiéry (1911) and Mortensen (1948). Kier describes internal butresses (believed absent in Holectypus and Coenholectypus) and therefore transfers Coptodiscus from the Holectypidae to the Discoididae. However, these "butresses" are but weak ridges and ambulacral in position, quite different from the strongly developed interamb walls of Discoides (our observations.

Kier (1974:72) tables data showing that

and Caenholectypus (C. transpecosensis), is generally smaller in test size, but has similar percentage of interamb plates relative to amb plates. Mortensen (1948:24) notes that in Coptodiscus there is only a single series of periproctal plates, differing from the irregular polygonal periproctal plates of Holectypus.

The species name has been variously spelt C. noemiae (Cotteau & Gauthier, 1895; Lambert & Thiéry, 1911; Mortensen, 1948), C. nominae (Kier, 1972:69; 1974:72), and C. nomiae (Kier, 1972:190, 192).

Occurrence: C. noemiae is well described from the Senonian of Iran (Cotteau & Gauthier, 1895), Campanian of Oman (Lees, 1928), and Campanian of Saudi Arabia (Kier, 1972). It thus has a distinctive stratigraphic and geographic range within the U. Cretaceous of the Middle East. C. mengaudi is in contrast poorly known, and differs strikingly in earlier age and western rather than eastern Tethyan occurrence.

7 DISCHOLECTYPUS POMEL, 1883

Ascribed species: Type Holectypus meslei Gauthier, 1876; L. Cretaceous (Albian?), Algeria; also D. guebhardi Lambert, 1920; L. Cretaceous (Hauterivian), S. France.

Diagnostic features: Test large (some 43 mm diameter), hemispherical, oral surface flat; apical system with 5 gonopores (genital 5 perforate); ambs narrow, plates not simple primaries but fused in arbaciod/holectypoid pattern on both apical and oral surfaces (pairs of plates united by tubercle, each pair comprising a large plate joined to a smaller adapical plate, plate pairs separated from each other by a single plate of intermediate size); pore-pairs in single linear series adapically, offset in groups of two on oral surface; interambs with moderately large tubercles, about 5 in transverse series per interamb plate, forming distinct rows of tubercles radiating from apical system; peristome central, small (some 6 mm diameter); periproct inframarginal, distant from peristome, small for size of test.

Discussion: Lambert & Thiéry (1911) record an imperforate genital 5 and non-crenulate tubercles, errors corrected by Lambert (1920) and Mortensen (1948) respectively. Mortensen describes the amb plates as in groups of three, but the middle amb plate in each group to be the largest rather than the adoral plate as claimed by Lambert & Thiéry. British Museum (Natural History) specimen E.4096 (Albian, Algeria) reveals plate structure as figured by Gauthier (1876) and described above (following terminology of Smith (1984)).

The genus has been variously placed incertae sedis (Hawkins, 1912), with Lanieria (Hawkins, 1913), close to Holectypus (Lambert & Thiéry, 1911), or close to the Echinoneids (Hawkins, 1920). In its large test size and hemispherical test shape Discholectypus is clearly very different from all typical species of Holectypus and Caenholectypus, which are almost invariably smaller and flatter. It also differs fundamentally in its adapical plate structure and adoral pore arrangement. D. guebhardi is poorly known (from a single specimen); D. meslei lacks "cloisons internes" (Pomel, 1883:75), but gross test morphology and amb structure are far more similar to the discoidid Camerogalerus than they are to Holectypus.

Occurrence: L. Cretaceous of W. Mediterranean; uncommon.

8 METHOLECTYPUS HAWKINS, 1923

Ascribed species: Type M. trechmanni Hawkins, 1923: Cretaceous, Jamaica

Diagnostic features: Test small (22.8 mm diameter), subglobular; apical system very small, with 4 gonopores, genital 5 imperforate; ambs seemingly of simple primary plates throughout, with narrow poriferous zones of uniserial isopores, not grouped in arcs of 3 adorally; peristome small (5.5 mm diameter), central; periproct moderately large (length 6.5 mm, width 4 mm), radially elongate, inframarginal, close to peristome; tubercles on ambs and interambs of lower surface enlarged and more numerous relative to those of upper surface. (Internal test features unknown).

Discussion: Metholectypus is known only from the type species, and that from a single specimen (E.16642) too poorly preserved to reveal all significant test features. The genus is therefore known only from the uppermost Cretaceous, and from the extreme western end of the Caribbean island of Jamaica. Mortensen (1948) notes that it differs from Lanieria Duncan, 1889, only in possession of an imperforate rather than perforate genital 5; he therefore follows Lambert & Thiéry (1925) in placing Metholectypus as a junior synonym of Lanieria.

Lanieria is known from only two species:

the type L. lanieri (Cotteau, 1881) also from the uppermost Cretaceous and from the western Caribbean area (Cuba, Mexico), and L. uvaldana Cooke, 1953, from the Campanian of Texas, U.S.A. Hawkins (1913) redescribed L. lanieri, and his observations can be confirmed from one cotype (lectotype of Hawkins) (E.4570) and one topotype (E. 4571). Additional topotype specimens (E. 11691-2) not seen by Hawkins reveal that the peristome margin is notched by buccal slits and that the amb plates although simple primaries adapically become less regular near the peristome, occasionally in groups of 3 as figured (1913:201, fig.B) for Coenholectypus cubae Hawkins, 1913. The holotype (E.11516) and paratype (E. 11517) of C. cubae are similar in morphology to L. lanieri but for their hemispherical rather than subglobular shape; they have identical provenance. Jeannet (1936) has demonstrated a wide range of variation within a single large (188 specimen) sample of L. lanieri that embraces features of amb structure and test profile supposedly diagnostic of C. cubae. He therefore convincingly places C. cubae (and consequently the genus Hawkensia Lambert, 1928, erected for this species) as a junior synonym of L. lanieri.

The primary amb plates which supposedly distinguish M. trechmanni from L. lanieri are neither clearly proved throughout the ambs of M. trechmanni nor necessarily atypical of L. lanieri (Jeannet, 1936, fig. iii). The two are distinguished clearly on gonopore number, but in morphology (test size, shape, and gross structure), and geographic and stratigraphic distribution, Metholectypus resembles Lanieria rather than Holectypus. Lanieria has consistently been ascribed to the Discoididae (Hawkins, 1913; Mortensen, 1948; Wagner & Durham, 1966), but as Jeannet (1936) points out, the supposed radiating walls on which this ascription is primarily based are but feebly developed.

Occurrence: Barrettia Limestone, Houghton Hall, Green Island, Jamaica. Assigned an undivided Cretaceous age by Hawkins (1923), but the horizon dated as U. Cretaceous (Senonian-Maastrichtian) by Reed (1949: 255), and subsequently mapped (as the Green Island Formation) as of Campanian-Maastrichtian boundary age (McFarlane, 1977).

9 CONCLUSION

Discholectypus and Metholectypus are probably discoidids, and Temnholectypus is a synonym of Caenholectypus. Coptodiscus, even if a holectypid, is very narrow in its geographic, stratigraphic, and species range. The family Holectypidae thus contains only two genera of major importance: Holectypus (20 described species) and Coenholectypus (some 40 described species). These two genera represent a lineage that persisted for over 100 million years, through much of the Jurassic and Cretaceous periods. The lineage achieved great success, evidenced not only by its long stratigraphic range but also its high reported species diversity, wide geographic occurrence, and locally great abundance of individuals. However, it remained a conservative stock, with but minor variation in test morphology. The significant evolutionary change that did occur took place in the apical system, a change reflected in the generic distinction of the ancestral Holectypus (with 4 gonopores) from its descendant Caenholectypus (5 gonopores). Kier (1974:47) points out that a similar evolutionary trend occurred in the Cretaceous Discoididae, early (Aptian-Turonian) species of Discoidea possessing 4 gonopores, later (Cenomanian-Senonian) species 5 gonopores. The trend seemingly reflects inherent parallelism within the Holectypina.

The holectypid pattern of evolution is strikingly similar to that revealed by the family Pygasteridae, a group of similar, but slightly earlier origin. That family too comprises only two important genera: (1) Plesiechinus (eight described species, of total recorded range L. Jurassic: Sinemurian to M. Jurassic: Bathonian); (2) Pygaster (some twenty described species, range M. Jurassic: Bajocian to U. Cretaceous: Cenomanian). These two genera also represent a lineage that persisted for some 100 million years, although with less success than the succeeding holectypids if judged on species and individual diversity and geographic range. Significant evolutionary change was again associated with the apical system, if not gonopore number: the periproct elongated as its base migrated posteriorly, the periproct top and adjacent apical system narrowing behind it.

Both holectypids and pygasterids as here defined are similar in their retention of primitive features presumably inherited from a regular echinoid ancestry: a notched peristome and (inferred) functional jaws in the adult; ordered tuberculation; test which is radially symmetrical except for the periproct. Their similarities in morphology, stratigraphic range, and evolutionary history indicate close affinity and probable common origin (Smith, 1981; Rose, 1982; Smith 1984) rather than the separate origins previously advocated (Durham & Melville, 1957; Wagner & Durham, 1966).

Holectypids and pygasterids both represent conservative stocks of primitive irregular echinoids (of the superorder Eognathostomata Smith, 1981). They are examples of clades which persist for long intervals of geological time but which exhibit little net evolutionary change, an evolutionary pattern considered by Stanley (1979:122 et seq.) to support the "punctuated equilibrium" model of evolutionary change. Diversification did occur in the Eognathostomata, but not until the Cretaceous, when most of its new species, genera, and families first occur. Even so, the number of families and therefore the range of new morphologies remains low compared with those of the other superorder, the Microstoma Smith, 1984. The superorder Microstoma seemingly embraces more radical stocks, which more rapidly achieved advanced morphological features (such as early loss of jaws in the adult, and greater division of labour in both spines and tube feet), diversified more widely (into several orders by the early Cretaceous), and finally achieved the wide range of structural features which makes the Microstoma almost the only irregular echinoid superorder to survive and flourish at the present day.

The similar 100 million year success but ultimate extinction of both pygasterids and holectypids may reflect their similar palaeoecological adaptation and restriction. Smith (1980), from similarities of aboral tubercle arrangement, inferred that Plesiechinus and Holectypus both had a similar (semi-infaunal) mode of life, in coarse substrates. Previously, Smith (1978) had inferred that Plesiechinus fed using only its lantern (and perhaps peristomial tube feet) to ingest substrate particles of suitable size; and that it did not burrow, but only covered itself with coarse substrate material, probably excavated from beneath the animal and thus producing a shallow depression into which it could nestle. From a later study of functional morphology, Smith (1984) concluded that Holectypus "fed using only its lantern to gather food, scooping up bottom sediment fairly unselectively and producing a fairly copious discharge of faecal material"; and that it burrowed vertically down into (only) coarse sands or shell gravels, using its oral spines to excavate material from beneath itself, lacking both unidirectional movement and a high metabolic rate and so lacking the capability to plough through sediment collecting food. That both pygasteroids and holectypids fed by ingesting coarse detritus like the Recent holectypoid Echinoneus (Rose, 1978) is clear from the relatively large size of peristome and especially periproct, as well as spine and tube foot characteristics. That neither pygasterids nor holectypids were specialized as burrowers is evidenced by Smith's observation of their low aboral tubercle density (and consequent sparse aboral spine cover), radially orientated adoral tubercles (indicating lack of unidirectional movement), and their lack of specialized test structures and effector organs. They had therefore effectively achieved only the first of the two major adaptive advances which facilitated diversification of the irregulars as a whole. This restriction is reflected in their primitive ordered tuberculation, which contrasts with the disordered tuberculation seen in most later and all Recent irregular echinoids, and whose loss represents a significant event in irregular echinoid adaptive radiation.

ACKNOWLEDGEMENTS

Dr A.B. Smith, Dr R.P.S. Jefferies (British Museum, Natural History); Dr J. Roman (Muséum National d'Histoire Naturelle, Paris); Dr D. Pajaud (Université de Pierre et Marie Curie, Paris) all helpfully provided access to specimens in their care. Specimens cited here are in the BM(NH) unless otherwise indicated. We thank the Central Research Fund Committee of the University of London for funding fieldwork.

REFERENCES

Cooke, C.W. 1955. Some Cretaceous Echinoids from the Americas. U.S.G.S. Prof. Paper 264E: 87-112.

Cotteau, G.H. & V. Gauthier 1895. Mission scientifique en Perse par J. de Morgan. 3, Etudes géologiques échinides fossiles. Paris: A. Burdin. 142p., 16 pls.

Durham, J.W. & R.V. Melville 1957. A classification of echinoids. Jour. Paleontology 31: 242-272.

Engel, H. 1964. On two new species of Holectypus Desor from the Senonian of south Limburg near Maastricht, Netherlands. Zool. Meded., 39: 235-9.

Gauthier, V. 1876. In G.H. Cotteau, P.A. Peron & V. Gauthier Echinides fossiles de l'Algérie. Fasc. 3, Etages Urgo-Aptien et Albien. Paris: Masson édit. 90p.

Harland, W.B. et al. 1982. A geologic time scale. Cambridge: C.U.P. 131p.

Hawkins, H.L. 1912. Classification, morphology and evolution of the Echinoidea Holectypoida. Proc. Zool. Soc. Lond. (1912): 440-497.

Hawkins, H.L. 1913. On Lanieria, Duncan. Geol. Mag. 50: 199-205.

Hawkins, H.L. 1914. Some problematical structures in the Holectypoida. Geol. Mag. 51: 1-5.

Hawkins, H.L. 1920. The morphology and evolution of the ambulacrum in the Echinoidea Holectypoida. Phil. Trans. Roy. Soc. Lond., B, 209: 377-480, pl.61-69.

Hawkins, H.L. 1923. Some Cretaceous Echinoidea from Jamaica. Geol. Mag. 60: 199-216, pl.IX.

Hawkins, H.L. 1934. The lantern and girdle of some Recent and fossil Echinoidea. Phil. Trans. Roy. Soc. Lond., B, 233: 617-649, pl.68-70.

Jeannet, A. 1936. Encore Lanieria lanieri (d'Orb.) Cotteau. Observations nouvelles. Eclog. geol. Helvetiae, 29: 581-598.

Kier, P.M. 1972. Tertiary and Mesozoic Echinoids of Saudi Arabia. Smithson. Contrib. Paleobiol.: 10, 242p., 67pls.

Kier, P.M. 1973. The Echinoderms and Permian-Triassic Time. In A. Logan & L.V. Hills (eds.) The Permian and Triassic Systems and their Mutual Boundary. Mem. Canad. Soc. Petrol. Geol. 2: 622-629.

Kier, P.M. 1974. Evolutionary Trends and their Functional Significance in the Post-Palaeozoic Echinoids. Jour. Paleontology 48 (supp.): Paleont. Soc. Mem. 5, 95p.

Kier, P.M. & M.H. Lawson 1978. Index to living and fossil echinoids 1924-1970. Smithson. Contrib. Paleobiol.: 34, 182p.

Lambert, J. 1899. Etude sur quelques Echinides de l'Infra-Lias et du Lias. Bull. Soc. Sci. hist. & natur. Yonne, 53: 3-57.

Lambert, J. 1920. Notes sur quelques Echinides du Crétacé Inférieur de la Provenence. Notes provençales du Dr A. Guebhard. no.11. p.1-22. 2pls.

Lambert, J. & P. Thiéry 1909-25. Essai de nomenclature raisonée des Echinides. Chaumont: Lib. Ferriere. 607p., 15pls.

Lees, G.M. 1928. The Geology and Tectonics of Oman and parts of Southeastern Arabia. Quart. Jl. geol. Soc. Lond., 84: 585-670, pls.41-51.

McFarlane, N. 1977. Jamaica-Geology. 1: 250,000 map. Kingston, Jamaica.

Mortensen, T. 1948. A monograph of the Echinoidea. Vol.4, No.1 Holectypoida, Cassiduloida. Copenhagen: C.A. Reitzel.

Pomel, A. 1883. Classification méthodique et genera des Echinides vivants et fossiles. Algiers: A. Jourdan. 131p.

Reed, F.R.C. 1949. The geology of the British Empire. London; Arnold, 764p.

Rey, J. 1966. Sur un échinide nouveau de l'Hautérivian portugais. Bull. Soc. d'Hist. nat. Toulouse, 102: 295-300.

Rose, E.P.F. 1978. Some observations on the Recent holectypoid echinoid Echinoneus cyclostomus and their palaeoecological significance. Thalassia Jugoslavica, 12: 299-306.

Rose, E.P.F. 1982. Holectypoid echinoids and their classification. In J.M. Lawrence (ed.) Echinoderms: Proc. Internat. Conf., Tampa Bay, p.145-152, Rotterdam: Balkema.

Rose, E.P.F. & J.B.S. Olver 1980. Australian Jurassic Echinoids and their biogeographical significance. In M. Jangoux (ed.) Echinoderms: Present and Past. p.47-52, Rotterdam: Balkema.

Smith, A.B. 1978. A comparative study on the life-styles of two Jurassic irregular echinoids. Lethaia, 11: 57-66.

Smith, A.B. 1980. The structure and arrangement of echinoid tubercles. Phil. Trans. Roy. Soc. Lond., B, 289: 1-54.

Smith, A.B. 1981. Implications of lantern, morphology for the phylogeny of post-Palaeozoic echinoids. Palaeontology, 24: 779-801.

Smith, A.B. 1984. Echinoid Palaeobiology. London: G. Allen & Unwin. 190p.

Smith, A.G., A.M. Hurley & J.C. Briden 1977. Phanerozoic paleocontinental world maps. Cambridge: C.U.P.

Stanley, S.M. 1979. Macroevolution: Pattern and Process. San Francisco: W.H. Freeman.

Stoll, N.R. et al. (eds) 1961. International Code of Zoological Nomenclature. London: ITZN. 176p.

Wagner, C.D. & J.W. Durham 1966. Holectypoids. In R.C. Moore (ed.) Treatise on Invertebrate Paleontology. Pt.U Echinodermata 3. p.440-450. Lawrence: Kansas.

Proceedings of the Fifth International Echinoderm Conference / Galway / 24-29 September 1984

Preliminary analysis of distribution patterns of Australia's non-endemic, tropical echinoderms

F.W.E.ROWE
Australian Museum, Sydney, Australia

ABSTRACT: Over 1100 species of echinoderms are now known to occur in Australian waters. A tropical element accounts for at least two thirds of this number. Analysis of 460 non-endemic, tropical species shows at least six different distribution patterns. Some 35% of these species have an Indo-west Pacifid distribution; about 31% are distributed only through the East Indian region; approximately 16% are shared with the East Indian region and Indian Ocean while 11% are shared with the East Indian region and the west Pacific; 4% and 2% are shared with the Indian Ocean and west Pacific respectively. These patterns of distribution can best be explained as the result primarily of vicariance rather than as a direct result of dispersalism.

I INTRODUCTION

Following over 30 years of studying the Australian echinoderm fauna, H.L.Clark (1946) published a monograph detailing over 100 fossil and 788 extant species, together with his interpretation of the composition and origin of that fauna. Clark excluded the fossil records in his considerations of the origin of the Australian fauna because the species were imperfectly known at that time. The greatest majority of the extant species were recorded from littoral or shallow-water (30 m) habitats. Clark (1946) concluded that this fauna: comprised a majority of tropical or tropical-derived forms; that in support of Hedley (1904, 1926) four distinct zoogeographic provinces could be recognized around the Australian coast by definition of their respectively high levels of endemic species; that there was no question the Australian echinoderm fauna had originated "...from the north, and apparently the East Indian region...". Clark considered this fauna had extended south along the northwest coast (Dampierian Province) to the south coast (Flindersian Province), and south along the Queensland coast (Solanderian Province) to the south east coast (Peronian Province). He saw little "mingling" of the faunas across the northern or southern provinces, respectively, nor any justification for Whitley's (1932) division (on the basis of fish distributions) of the Solanderian Province into those species associated with the Great Barrier Reef (Solanderian) and those species associated with the mainland (Banksian Province). Regarding extra-Australian relationships Clark concluded: that there was practically no evidence of any common origin or close relationship between the Australian fauna and that of eastern Africa, Madagascar and Mauritius; that comparisons between the Australian fauna and the faunas from the Arabian and Sri Lankan (Ceylon) areas showed connection, rather, through the East Indian fauna and that comparison of the Australian fauna with the southwestern Pacific fauna showed a less close relationship than that with the East Indies, though he conceded that the southwest Pacific fauna, at that time, was imperfectly known. With respect to similarities between the faunas of South Africa, Australia and New Zealand, Clark (1946) concluded that this had resulted either from transportation on flotsam, rather than by any land bridge connection by which latter means he would have expected greater commonality of species, or that the commonality was due to invasion of South Africa and New Zealand from the north, which fauna would have originated from the East Indian region. Although he acknowledged Wegener's theory of continental drift, Clark concluded: "Only a desperate desire to validate Wegener's

Table 1. Numbers and percentages of Australian non-endemic tropical echinoderms shared with other Indo-west Pacific areas: IndoP = Indo-west Pacific; EI = East Indies, Philippine Islands and southern Japan; IO = Indian Ocean; P = Islands of the western Pacific Ocean; CRI = Crinoidea; AST = Asteroidea; OPH = Ophiuroids; ECH = Echinoidea; HOL = Holothurioidea.

	IndoP	EI	EI+IO	EI+P	IO	P	TOTAL
CRI	11	34	3	10	-	-	58
AST	23	44	16	15	2	2	102
OPH	47	29	22	11	11*	3	123
ECH	30	16	16	5	3	1	71
HOL	49	21	16	12	3	4	106
TOTAL	160	144	73	53	19	10	460
%	34.8	31.4	15.9	11.5	4.2	2.2	100

* A number of formerly endemic Australian species have been recorded from Madagascar by Cherbonnier and Guille, 1978.

Table 2. Numbers and percentages of Australian non-endemic tropical echinoderms; shared with other Indo-west Pacific areas; A - P = divisions of IndoP as delimited by Clark and Rowe (1971) (fig. 1); other abbreviations as Table 1.

a) IndoP.

	A	B	C	D	E	F	G	H	I	J	K	L	M	N	O	P
CRI	4	4	5	5	1	0	3	6	11	11	11	11	11	11	11	1
AST	14	15	15	12	9	6	3	12	14	10	19	23	19	20	23	10
OPH	28	19	36	23	15	7	6	19	26	25	39	47	37	35	47	13
ECH	24	23	26	20	17	5	6	21	21	22	30	30	26	28	29	16
HOL	24	27	34	26	19	8	4	27	27	32	46	49	47	29	49	20
TOTAL	94	88	116	86	59	26	22	85	99	100	145	160	140	123	159	60
%	59	55	73	54	37	16	14	53	62	63	91	100	88	77	99	38

b) EI.

	J(pt)	K	L	M	N
CRI	5	32	34	16	13
AST	7	19	44	19	5
OPH	4	28	29	17	9
ECH	1	14	16	3	5
HOL	1	17	21	8	5
TOTAL	18	10	144	63	37
%	13	76	100	44	26

c) EI + IO

	A	B	C	D	E	F	G	H	I	J	K	L	M	N
CRI	-	-	-	-	-	-	-	1	3	2	3	3	2	2
AST	4	2	6	5	7	5	1	5	12	13	15	16	10	9
OPH	3	3	13	6	3	5	1	7	8	7	20	22	15	7
ECH	4	4	5	7	6	5	4	6	13	10	16	16	13	14
HOL	1	-	3	4	2	3	3	1	12	9	13	16	7	7
TOTAL	12	9	27	22	18	18	9	20	48	42	64	73	47	39
%	16	12	37	30	25	25	12	27	66	58	88	100	64	53

d) EI + P

	J(pt)	K	L	M	N	O	P
CRI	4	10	10	9	7	10	-
AST	2	12	15	11	11	15	2
OPH	3	11	11	9	7	11	-
ECH	2	4	5	5	5	5	1
HOL	-	10	12	10	4	12	2
TOTAL	11	47	53	44	34	53	5
%	21	89	100	83	64	100	9

e) IO

	A	B	C	D	E	F	G	H	I	J	L
CRI	-	-	-	-	-	-	-	-	-	-	-
AST	1	2	1	-	1	-	-	1	1	-	2
OPH	-	1	8	3	1	1	1	-	1	-	11
ECH	-	-	1	2	-	-	-	-	2	-	3
HOL	1	-	-	-	1	1	2	-	1	-	3
TOTAL	2	3	10	5	3	2	3	1	5	-	19
%	10	15	52	26	15	10	15	5	26	-	100

f) P.

	L	O	P
CRI	-	-	-
AST	2	2	1
OPH	3	3	-
ECH	1	1	1
HOL	4	4	-
TOTAL	10	10	2
%	100	100	20

Table 3. Percentages of non-endemic tropical Australian species shared between localities on the Australian coast and each of the biogeographic areas: IndoP., EI+P., EI., EI+IO; for localities 1-17 see Fig. 2.

Localities	17	16	15	14	13	12	11	10	9	1	2	3	4	5	6	7	8
IndoP/160spp	1	4	15	37	46	72	72	71	71	77	83	74	63	22	8	4	1
EI+P/53spp	6	6	8	22	20	54	46	46	46	56	79	66	54	14	10	2	-
EI/144spp	1	1	9	15	23	51	44	38	35	42	50	40	34	16	7	2	2
EI+IO/73spp	5	6	15	32	51	80	70	64	62	62	62	56	40	30	6	1	1

hypothesis can find in the echinoderm fauna of Australia any evidence that is at all closely allied to that of South Africa or even of New Zealand".

H.L.Clark's (1946) monograph has served as a standard work on Australian echinoderms. Some 250 papers have since been published on extant Australian echinoderm species and over 80 on fossil species. Of these about 50% of papers dealing with extant forms describe collections and/or new taxa, 20% are popular accounts of the fauna and some 30% (published as government reports or in scientific journals) have dealt with the problem of Acanthaster planci and its effect on the Great Barrier Reef.

Few authors since H.L.Clark (1946),have discussed biogeographic relationships of Australia's tropical echinoderm fauna. Ekman (1953); 1967) following a re-analysis of Clark's (1946) work concluded that the fauna comprised essentially a northern tropical/subtropical element and a southern temperate element, with boundaries occurring approximately in the region of Abrolhos Islands (Geraldton on the mainland) in the west and Port Jackson (Sydney) in the east. He did not support the recognition of a Solanderian Province but concluded the Dampierian Province appeared to be a more or less independent region. Ekman believed the Indo-Malay region to be the faunistic center of the Indo-west Pacific. Endean (1957) showed that Whitley's mainland Banksian Province was not justified but formed part of a single tropical Australian Province. On the other hand, in recognizing both a "reef" and a "mainland" element in the Queensland fauna, Endean supported the identity of the Solanderian Province, as restricted by Whitley (1932), since he knew of few reef species which had been recorded west of Torres Strait. Interestingly, Endean shows that for mainland species the greatest percentage of species is shared with the East Indies whereas an analysis of the reef species shows the greatest percentage is shared with the East Indies and west Pacific. He concluded "mainland" species in Queensland arrived from the East Indies through the Torres Strait, whereas "reef" species arrived mainly from eastern New Guinea and from the Coral Sea. Endean also recorded a number of species occurring in both reef and mainland habitats. Briggs (1974) supported Ekman's views of the importance of the Indo-Malay region faunistically and recognized the Dampierian Province on the Australian tropical mainland. Gibbs, et al (1976) added some 22 new echinoderm records from the Queensland coast, including 15 from Australia for the first time. They referred to the Great Barrier Reef Province, defined by Maxwell (1972), to include the whole of Queensland's shelf region, including the reefs. Marsh (1976) analyzed the asteroid fauna of Western Australia and concluded not only that endemism was sufficiently high to justify recognition of the Dampierian Province but that a far larger group of Indo-Pacific "reef" species occurs in northwestern Western Australia than had previously been suspected. However, later, Marsh and Marshall (1983) were less committal about the validity of a northwestern Province, concluding that although 13% endemic species (excluding holothurians) occurred there, this figure may be "subject to correction in the light of future taxonomic and distributional studies".

1.1 Palaeogeographic considerations

The palaeogeographic history of the Indo-west Pacific and Australasian regions has recently been described by a number of authors including McKenzie and Sclater (1976), Knox (1980), Powell et al (1981), Adams (1981) and Zinmeister (1982), and need only be summarized here. The Australian plate has generally been considered to have separated from Antarctica about 55 mya (Knox, (1980), Powell et al (1981), Zinmeister (1982)). However, Ludbrook (1978) suggested initial rifting between the two continents began during the Cretaceous. More recently, Cande and Mutter (1982) concluded that separation of Australia from Antarctica began, at least west of 130° E, between 112-90 mya. The Tasman Basin, to the east of Australia, ceased opening about 55-56 mya (Zinmeister, 1982). The Australia plate commenced its collision with southeastern Asia between 15-10 mya and is continuing its northward progression (Powell, et al 1981). The Indian plate separated from Gondwanaland in the early Cretaceous and by 55-45 mya collided with the southern margin of Eurasia (McKenzie and Sclater (1976); Powell et al (1981)). The movements of Africa and Arabia finally severed the connection between the Indian and Atlantic Oceans by the middle Miocene (Adams, 1981). Volcanic islands and island chains have been present in the Pacific Basin at least since the mid Mesozoic and certainly since the Cretaceous (Kay, 1980; Springer,1982).

During Pleistocene glacial periods of lowered sea levels, large parts of the continental shelf regions surrounding the East Indies and northern Australian continent would have been exposed. This would have had the effect of almost isolating the Indian from the Pacific Ocean in the tropics except for a seaway through the Timor Trough (Sea). Extinctions, particularly of local shelf faunas would have affected faunal diversity because of changes in area of the continental shelves, leading to new faunal provinces (Wise and Schopf, 1981).

1.2 Hydrological considerations

Prior to and during the early part of the Tertiary, the Tethys Sea extended between Laurasia and Gondwanaland with water currents moving in a westward direction. This seaway gradually became restricted by the movement of the various continental plates with resultant modification of oceanic currents into a number of major gyres (van Andel, 1979). Present day water currents throughout the Indo-west Pacific region are detailed by various authors (see Fairbridge (ed.), 1966, Wyrtki 1961, 1973).

1.3 Analysis of non-endemic tropical species

Analysis of these species shows six distribution patterns within the Indo-west Pacific region (fig. 1 tables 1, 2e-f). Distribution records from, and division of the Indo-west Pacific (fig. 1) are taken principally from Clark and Rowe (1971). Records from the Bay of Bengal east of 90°E are included with the East Indies in tables 2 and 4. In each of the tables, percentage values (Marsh and Marshall, 1983) of Australian species shared with other areas A-P (fig. 1), (Clark and Rowe, 1971) are given.

Analysis of the tropical, non-endemic Australian species on a Class by Class basis shows the following distribution patterns:

Holothurioidea: IndoP and EI+P are dominantly represented by families of aspidochirotes and include the apodid chiridotids and the prominent synaptid genera Synapta, Euapta and Opheodesoma. EI and EI+IO are dominated by dendrochirotids molpadiids and the synaptid genera Synaptula and Protankyra.

Echinoidea: Most families are well represented throughout the IndoP (Table 1), however, temnopleurids and a number of irregular urchins tend to be restricted to the EI or EI+IO.

Crinoidea: It is difficult to recognize any family distributional trends, however, it is pertinent that three times as many

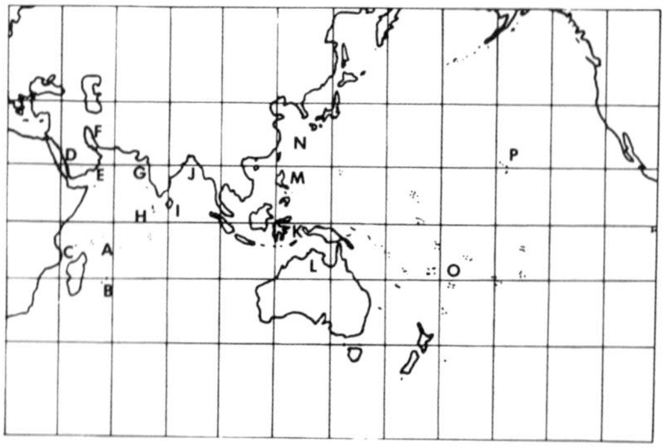

Figure 1. Division of Indo-west Pacific (after Clark and Rowe, 1971)

species are recorded from EI than from any of the other regions (Table 1).

Asteroidea: Like the Crinoidea the vast majority of asteroids (two to three times as many) are recorded from EI (Table 1). There are no clear family patterns but the genera Anthenea, Goniodiscaster, Halityle and Nepanthia are more or less restricted to EI, whilst those showing bias to the EI+IO include Luidia, Astropecten, Stellaster and Nardoa.

Ophiuroidea: The greatest majority of ophiuroids are wide-spread in the IndoP with the families Ophiocomidae and Ophiothricidae dominating. This latter is reflected also in the E I+P distributions. On the other hand distributions in the EI and EI+IO have a greater representation of euryalids, amphiurids and ophiurids.

Using Endean's (1957) and Marsh and Marshall's (1983) works as guides, a high percentage (80%) of the Australian species with IndoP and EI+P distributions can be considered "reef" species. Conversely those Australian species with EI and EI+IO distributions have similarly high percentages of "mainland" species.

The percentages of non-endemic, tropical Australian species records shared between localities on the Australian coast and each of the biogeographic regions, IndoP., EI+P., EI., and EI+IO., are tabulated fig. 2., table 3). These distributions appear directly related to the extra-Australian distributions (tables 2a-d) with the highest percentages of IndoP and EI+P occurring in northeastern Australia, while the highest percentages of EI and EI+IO occur in northwestern Australia.

1.4 Discussion

Analysis of 460 Australian non-endemic tropical echinoderm species supports the recognition of two faunal suites, a "reef"

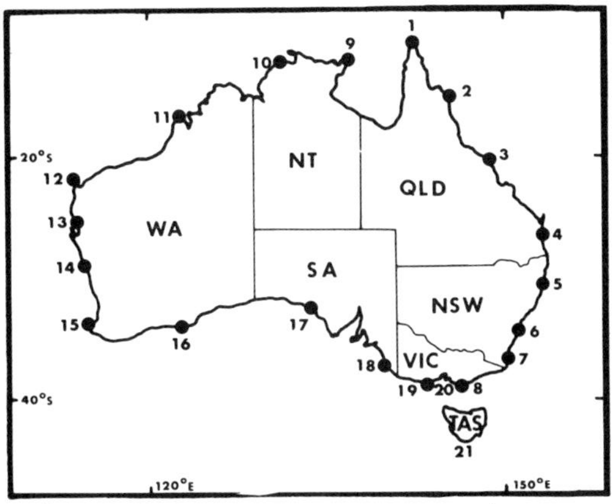

Figure 2. Localities on Australian coast: 1=Torres Strait (including Marray Ids); 2=Cape Flattery; 3=Bowen; 4=Double Island Point; 5=Coffs Harbour; 6=Port Jackson (Sydney); 7=Bermagui; 8=Wilson's Promontory (east(; 9=Cape Arnheim; 10=Darwin; 11=Cape Leveque; 12=North West Cape; 13=Shark Bay; 14=Geraldton; 15=Cape Naturaliste; 16=Esperence; 17=Ceduna; 18=Robe; 19=Cape Otway; 20=Wilson's Promontory (west); 21=Tasmania (incl. islands of Bass Strait).

suite and a "mainland" suite (Endean,1957). Four, if not six, faunal elements overlap along the northern Australian coast: 1. Indo-west Pacific; 2. Western Pacific/East Indies; 3. East Indies; 4. East Indies/Indian Ocean; 5. Indian Ocean; 6. Western Pacific. The Indo-west Pacific, Western Pacific/East Indies, and Western Pacific elements are represented by "reef" species whereas species in the remaining three areas are represented in the majority by "mainland" species. However, the Australian species shared with the East Indies/Indian Ocean may require further analysis since it may reflect both faunal suites, the "reef" species comprising the majority of those extending to the islands of the western Indian Ocean, whilst the "mainland" species may be more restricted in the eastern Indian Ocean. I believe too few species and too little is known of the distribution of the species recorded respectively from the Indian Ocean and Pacific Ocean (Table 2., e-f) to comment further on those relationships. The apparent absence of these species from the EI region, particularly when a number of species extend across the northern Australian coast, is difficult to account for.

Analysis of the distribution of the 460 non-endemic species along the northern Australian coast (table 3) is supportive of the North West Cape, Western Australia and Double Island Point, Queensland, as critical geographical end points beyond which many tropical species do not extend (Endean, 1957; Marsh and Marshall, 1983). Marsh and Marshall (1983) have pointed out that there is an abrupt change of habitat from the sheltered, turbid waters of Exmouth Gulf to high energy oceanic waters and coral reefs on the west side of the Cape. They considered NW Cape marks the end of range of many "mainland" species. It is evident from table 3 that this applies also to many "reef" species on both sides of the Australian continent despite the availability of tropical water currents to carry them further south (Leeuwin current-western Australia; East Australian current-eastern Australia).

There can be little support for considering the regions west of and southeast of Torres Strait as separate biogeographic provinces. Marsh and Marshall (1983) record a 13% endemism among echinoderm species (excluding holothurians) from northwestern Australia. Inclusive of holothurians I can confirm this percentage not only for the northwest but also the northeast coast of Australia. I agree with Marsh and Marshall (1983) that many of these endemic species are in need of taxonomic confirmation. Those parts of the coastline do, however, reflect relationships with the Pacific "reef" species and East Indian "mainland" species as might be expected, considering their topography and relationship to present-day oceanic currents and other hydrological conditions. Endean's conclusion that the reef species in Queensland have originated from the West Pacific, via oceanic currents from the Coral Sea, whilst mainland species have a strong affinity with those of the East Indies is substantially supported. Ladd (1960) concluded that the Indo-Pacific molluscan fauna had its origins among the Pacific islands and was spread westward by prevailing currents. Briggs supports Ekman's (1967) view of the East Indies as the "centre of origin" for Indo-Pacific marine faunas. The distribution patterns of Australian non-endemic tropical echinoderm species shows that the formation of the East Indies has had a profound effect on the distribution of echinoderms in the Indo-west Pacific. Further, the more recent effects of Pleistocene glacial periods and continental plate movements on eustatic sea levels have contributed to the disruption of formerly widespread tropical biotas, with both extinctions and speciation occurring. That the EI region forms a "centre of origin" from which species have been distributed, despite the presence of oceanic counter currents in the Pacific (counter currents in the Indian Ocean also flow from west to east, therefore originate from the western Indian Ocean and move towards the EI) is difficult to support. Rather the formation of the EI, including northern Australia, particularly with its extensive areas of continental shelves, has provided an ideal environment for major speciation of "mainland" species, as well as for the accumulation of other species brought to it by prevailing oceanic surface currents. The distribution of the Indo-west Pacific species reflects antiquity, I believe. Indeed, tropicopolitan species (e.g. Holothuria (Thymiosycia) impatiens, H. (T.) arenicola and Linckia guildingi signify, in my view, ancient species representing disjunct remains of pan-Tethyan distributions (Rowe, in press). Although fossil representatives of these species have not been recognized, at least some 25 extant IndoP echinoid species, which also occur in Australia, are recorded in fossil deposits. Fourteen of these extend to the Miocene while Prionocidaris bispinosus is recorded from the Eocene and Echinoneus cyclostomus is recorded from Oligocene. Fossil data indicate a minimum age for a species, though the rate of speciation may be as short as 10^3 years (Wise and Schopf 1981). Nonetheless, extensive distribution of at least 15 species would have been possible prior to the final restriction of the Tethyan Seaway. Indeed one species (Echinoneus cyclostomus) has a tropicopolitan distribution, whilst 7 species have an IndoP distribution, 6 have an EI+IO distribution and one (Clypeaster virescens) an EI distribution. The interpretation of the antiquity of Indo-west Pacific species supports the views of McCoy and Heck (1976) who contend that formerly widespread Tethyan taxa of some corals, seagrasses and mangroves have survived in the Caribbean, whilst others have survived in refuges in the tropical east Pacific. Similarly, Shuto (1983) has concluded that for species of the gastropod Murex, the more widely distributed species in the Indo-west Pacific are older than those with restricted distributions in the region. Three hypotheses might be proposed to explain the restriction of species to the East Indies/Australian region. Firstly, it may reflect a relatively young fauna formed <u>since</u> the development of the region beginning 15-10 mya. Distribution northward to Japan and southward along the coastlines of Australia is not difficult to account for, considering prevailing present-day ocean currents. Secondly, the fauna may have originated more than 15 mya, having its seat of origin in a combination of faunas from the continental shelf regions of the southeastern part of the Asian Plate (M of Fig. 1) and the northern (possibly principally the northwestern) coastline of Australia. Thirdly, if "mainland" species have originated, together with "reef" species in the western Pacific (Ladd, 1960), then the East Indies region might be reflecting an area, which in having extensive continental shelf, has permitted greater speciation and retention of mainland species to occur as well as supporting an element of reef species. The region is able to support such a large number and diversity of species because of its own structural diversity and its position in tropical latitudes.

Surprisingly little is known about premetamorphic life spans in echinoderms. According to Thorson (1961) the majority of echinoderm larvae have relatively short (4-6 weeks) premetamorphic lives. Yamaguchi (1977) found different larval swimming behavior between coral reef species (negatively geotaxic) and mainland (=continental) species (positively geotaxic). He suggested this might be reflected in geographic distribution patterns. Clearly knowledge of echinoderm larval behavior, which is important in any interpretation of biogeographic relationships of echinoderms, is limited. Similarly, also, fossil evidence is restricted. However, on the basis of the distribution patterns of Australian non-endemic tropical echinoderms and taking into account palaeogeographic events and their effect on the development of present-day ocean currents it is suggested that:

1. A suite of "reef" species and a suite of "mainland" species can be recognized in the North Australian fauna (Endean, 1957).
2. Division of the northern Australian coast into 2 or more biogeographic provinces cannot be justified on distribution patterns of echinoderms.
3. There are 4 if not 6 overlapping faunal, distribution related elements occurring on the tropical coast of Australia.
4. The distribution patterns do not support the hypothesis that the Indo-Malay region (Ekman 1967) is a "centre of origin" for an Indo-west Pacific fauna.
5. The East Indies/North Australian re-

gion has, by virtue of its latitudinal position and structural complexity permitted enhanced speciation, particularly of "mainland" species.

6. The formation of the East Indian Archipelago/Northern Australian region has had a profound disruptive effect on distributions of fauna in the Indo-west Pacific.

7. The origin of the Indo-west Pacific fauna is debatable but Ladd's (1960) view of a Pacific origin is considered feasible.

2 ACKNOWLEDGEMENTS

I extend my thanks to Miss Anne Hoggett, Australian Museum and Dr. D.L.Pawson, Smithsonian Institution for critically reading the manuscript. I also thank Harriet Greenblatt, Smithsonian, for typing the manuscript.

REFERENCES

Adams, C.G. 1981. An Outline of Tertiary Palaeogeography In Cocks, L.R.M. (ed.) The evolving earth, pp. 221-235. British Museum (Natural History). Cambridge Univ. Press.

Briggs, J.C. 1974. Marine zoogeography, 475 pp., New York:McGraw-Hill.

Cande, S.C. and J. C. Mutter. 1982. A revised identification of the oldest seafloor spreading anomalies between Australia and Antarctica. Earth and Planetary Science Letters 58:151-160.

Cherbonnier G. and Guille, A. 1978. Faune de Madagascar. 48. Echinodermes:Ophiurides. 253 pp. Paris.

Clark, A.M. and Rowe, F.W.E. 1971. Monograph of shallow-water Indo-west Pacific Echinoderms. 238 pp. London.

Clark, H.L. 1946. The echinoderm fauna of Australia Publs. Carnegie Inst. No. 566: I-566.

Ekman, S. 1967. Zoogeography of the sea (ed.), London:Sidgwick and Jackson. (reprinted from 1953).

Endean, R. 1957. The biogeography of Queensland's shallow-water echinoderm fauna (excluding Crinoidea) with a rearrangement of the faunistic provinces of tropical Australia. Aust. J. Freshwat. Res. 8:233-273.

Hedley, C. 1904. The effect of the Bassian isthmus upon the existing marine fauna:a study in ancient geography. Proc. Linn. Soc. NSW 28:876-883.

Hedley, C. 1926. Zoogeography. In The Australian Encyclopaedia. Vol. II, M-Z. pp. 743-744, Sydney, Angus and Robertson.

Kay, E.A. 1980. Little worlds of the Pacific. An essay on Pacific Basin biogeography, 40 pp. Hawaii, Lyon Arboretum.

Knox, G.A. 1980. Plate tectonics and the evolution of intertidal and shallow-water benthic biotic distribution patterns in the southwest Pacific. Palaeogeography, Palaeoclimatology, Palaeoecology Special Issue 31(2-4):267-297.

Ladd, H.S. 1960. Origin of the Pacific Island Molluscan fauna. Amer. J. Sci. 258-A:137-150.

Ludbrook, N.H. 1978. Australia. In Moullade, M. and A.E.M. Nairn (eds.) The Phanerozoic geology of the world II. The Mezozoic, pp. 209-249. Amsterdam, Elsevier.

Marsh, L.M. 1976. Western Australian Asteroidea since H.L. Clark. Thalassica Jugslavica 12(1):213-225.

Marsh, L.M. and J.I. Marshall 1983. Some aspects of the zoogeography of northwestern Australian echinoderms (other than holothurians). Bull. Mar. Sci.33(3): 671-687.

Maxwell, W.G.H. 1972. The Great Barrier Reef-Past, Present and Future. Qld. Nat. 20 (4-6):65-78.

McCoy, E.D. and K.L. Heck. 1976. Biogeography of corals, seagrasses and mangroves:an alternative to the center of origin concept. Syst. Zool. 25(3):201-210.

McKenzie, D.P. and J.G. Sclater. 1976. The evolution of the Indian Ocean. In Continents adrift and continents aground. Sci. Amer. (Readings) 1976:139-148.

Potts, D.C. 1983. Evolutionary disequilibrium among Indo-Pacific corals. Bull. Mar. Sci. 33(3):619-632.

Powell, C.McA., B.D. Johnson and J.J. Veevers. 1981. The early Cretaceous breakup of Eastern Gondwanaland, the separation of Australia and India, and their interaction with southeast Asia. In Keast, A. (ed.), Ecological biogeography of Australia, pp. 17-29. Hague, Junk.

Shuto, T. 1983. Larval development and geographical distribution of the Indo-west Pacific Murex. Bull. Mar. Sci. 33(3): 536-544.

Springer, V.G. 1982. Pacific Plate biogeography, with special reference to shorefishes. Smithson. Contr. Zool. No. 367: 1-182.

Thorsen, G. 1961. Length of pelagic larval life in marine bottom invertebrates as related to larval transport by ocean currents. Amer. Assoc. Adv. Science no. 67: 455-474.

van Andel, J.H. 1979. An eclectic overview of plate tectonics, paleogeography and palaeoceanography. In Gray, J. and A.J. Boucot (eds.). Historical biogeography, plate tectonics and changing environ-

ment. Proc. 37th Ann. Biol. Colloquium and Selected Papers. pp. 9-25, Corvallis:Oregon State Univ.
Whitley, G.P. 1932. Marine zoogeographical regions of Australasia. Aust. Nat.:166-167.
Wise, K.P. and T.J.M.Schopf, 1981. Was marine faunal diversity in the Pleistocene affected by changes in sea level. Paleobiology 7(3):394-399.
Woodland, D.J. 1983. Zoogeography of the Siganidae (Pisces): an interpretation of distribution and richness patterns. Bull. Mar. Sci. 33(3):713-717.
Wyrtki, K. 1961. Physical oceanography of the South eastern Asian waters Repts. Naga Exped., 2. La Jolla, Calif.,Scripps Inst. Oceanogr.
Wyrtki, K. 1973. Physical Oceanography of the Indian Ocean. In Zeitschel, B.(ed). The Biology of the Indian Ocean, pp.18-36. Berlin:Springer-Verlag.
Yamaguchi, M. 1977. Larval behaviour and geographic distribution of coral reef asteroids in the Indo-west Pacific. Micronesica 13(2):283-296.
Zinsmeister, W.J. 1982. Late Cretaceous-early Tertiary molluscan biogeography of the southern circum-Pacific. J. Palaeo. 56(1):84-102.

Proceedings of the Fifth International Echinoderm Conference / Galway / 24-29 September 1984

Quantitative distribution of echinoderms (Holothuroidea, Asteroidea, Ophiuroidea, Echinoidea) in relation to organic matter in the sediment, in deep sea basins of the Atlantic Ocean

MYRIAM SIBUET
Institut Français de Recherche pour l'Exploitation de la Mer, Centre de Brest, France

ABSTRACT : The quantitative distribution of echinoderm populations in the deep sea has been poorly investigated while as part of the benthos, the echinoderms play a significant role both in reworking the sediment and recycling the organic matter.
Intensive sampling conducted using a similar strategy in several abyssal Atlantic basins, between 1900 and 5250 m depth, allowed a comparison of the echinoderm populations. This study points out : - a great variability of the mean density of the echinoderms as a function of depth and of geographic area (from 1 ind./10^4 m^2 to 15000 ind./10^4 m^2), - a variable dominance of echinoderm classes, - an exceptionnally high abundance of holothuroids which dominates all other taxa in the Norwegian Sea, - a relatively small variation of the concentration of organic matter in the sediment between the different basins prospected which mostly shows the result of bioturbation and consumption by the fauna, - a significant correlation between holothuroids density with organic carbon content and terrigenous fraction of the sediment at tropical stations, where trophic input estimated by sediment traps, correlates better to the abundance of some echinoderm populations.

1 INTRODUCTION

The echinoderms can be considered as a well adapted group in the deep sea with respect to species diversity and the existence of strictly abyssal genera and families (e.g. Madsen, 1961 ; Hansen, 1975 ; Sibuet, 1976, 1977, 1980).
The quantitative distribution of echinoderm populations, as well as for other phyla of the megafauna in the deep sea, has been investigated in only a few cases (e.g. Grassle et al., 1975 ; Conan et al., 1981 ; Smith and Hamilton, 1983).
Due to the difficulties encountered in quantitative estimation of the megafauna (Rice et al., 1982), particularly the problem of sampling a sufficiently wide area with minimum sampling bias, little is known about the size of the populations and the variation in megafaunal density and biomass (Haedrich et al., 1980 ; Sibuet and Lawrence, 1981 ; Ohta, 1983).
During the past years the study of well defined areas in several Atlantic basins has been conducted by IFREMER (CNEXO) using the same strategy of intensive sampling (by cores, trawls, bottom photographic systems) usually controlled by acoustic navigation (Khripounoff et al., 1980 ; Rowe and Sibuet, 1983 ; Sibuet et al., 1984).
The fauna collected by trawls or observed by underwater photographic surveys, allows us to estimate the abundance of the echinoderm populations which live in several deep sea locations of the Atlantic Ocean. The results are presented in terms of density (number of individuals/10^4 m^2).
Their analysis gives informations on :
- the relative importance of each of the four classes commonly sampled : Holothuroidea, Asteroidea, Ophiuroidea and Echinoidea,
- the contribution of the echinoderms to the total density of the epibenthic megafauna (swimming animals excepted),
- the variation of the abundance of the four classes as a function of depth and in relation to the amount of organic matter which occurs on the upper part of the sediment. This analysis is extended to a discussion of the dependance of these benthic groups on food resources.

2 MATERIAL AND METHODS

The exploration of selected areas between 1900 and 5250 m depth in the Atlantic

Table 1 : Location of stations in the Atlantic Ocean and chemical analyses for dry weight sediment. a) Rowe and Gardner (1979) type of sediment trap (Sibuet et al., 1984) b) IFREMER sediment trap (Khripounoff et al., in press

Geographic area (Name of cruise)	Station code	Depth (m)	Ca Co_3 (%)	C. org. (mg/g) $\overline{M}$	C. org. (mg/g) σ(n-1)	C. org. flux (mg/m²/day)
Northern Bay of Biscay (BIOGAS Cruise)	G1	2 100	51-58	3,82	0,38	2,92 (b)
	G2	3 000	45-60	3,3	0,25	
	G3	4 200	42-50	4,02	0,34	4,28 (b)
	G4	4 750	35-40	3,9	0,1	2,31 (b)
Southern Bay of Biscay (BIOGAS Cruise)	G5	4 450		6	0,09	
	G6	1 900		8,5	0,13	
Porcupine (INCAL)	P1	2 650		4,8		
	P2	4 820		3,1	0,5	
Norwegian basin	NN	2 600-3 600		6,5	2,1	
Lofoten basin	NL	2 900-3 200		5,42	1,5	
Greenland basin (NORBI)	NG	2 500-3 700		6,62	2,5	
Cape basin	WB	4 650	50	3,42	0,04	
Angola basin (WALVIS)	WC	5 250	70	1,71	0,09	
Demerara plain (DEMERABY)	DA	4 420	10	6,05	0,58	15,9 (a)
	DB	4 850	30	4,54	0,13	9,8 (a)
Vema fracture zone (VEMA)	V	5 100	25-32	5,52	0,38	
Cape Verde basin (SEABED)	CV	5 190	45-50	2,46	0,14	3,29 (a)

Ocean (Fig. 1) has been principally conducted on board the R.V. "J. Charcot" during the following cruises : Biogas (1972-74, 1979-1981) in the Bay of Biscay or Golfe de Gascogne (Laubier and Sibuet, 1977, 1979) ; Norbi, 1975 in the Norwegian Sea (Dahl et al., 1977) ; Incal, 1976, partly in the Porcupine Abyssal Plain ; Vema, 1977 in the Vema Fracture Zone (Khripounoff et al., 1980) ; Demeraby, 1981 in the Demerara Abyssal Plain (Sibuet et al., 1984) ; Seabed, 1981 in the Cape Verde Basin (Sibuet, 1982).

Sampling of the scattered megafauna has been done with a large beam trawl of 5 meters width. 2 to 9 trawls were conducted at each station. The distance trawled on the bottom is obtained with the help of the acoustic navigation used since 1977. In a single region of about 30 km^2 between four navigation transponders, all the biological tools (trawls, photographs obtained with troïka and RAIE System), can be accurately followed in real time (Rowe and Sibuet, 1983). Although the trawl is not a truly quantitative gear, the mean density has been computed from the total area sampled by trawls. Long series of photographs (300 to 500) have been obtained at some stations. The photographs are obtained sequentially but there is no overlap between them. Knowing the altitude of the camera and the optical characteristics, the area photographed can be estimated. These data give a good evaluation of the density of large epibenthic echinoderms such as holothuroids and echinoids (Sibuet and Lawrence, 1981 ; Sibuet et al., 1984).

The environmental parameters determined in that study are the amount of organic-carbon and $CaCO_3$ measured in the superficial layer (2 cm) of non disturbed sediment collected by a Reineck or Usnel boxcore (e.g. Khripounoff et al., 1980 ; Sibuet et al., 1984 ; Khripounoff et al., in press).

Estimates of organic flux (in mg/m^2/day) of dry sediment have been obtained in certain areas by the deployment of two types of sediment trap. In the Demerara and Cape Verde basins the model of Rowe and Gardner (1979) was deployed (Sibuet et al., 1984) ; in the bay of Biscay the model constructed by IFREMER was deployed (Khripounoff et al., in press) at near 200 m from the bottom.

Table 2 : Density estimates (nb ind./10^4m²) of abyssal echinoderms. Numbers between brackets are example of densities of epibenthic holothuroids, asteroids and echinoids deduced from photograph analysis

Station code	Nb of trawls	Area sampled 10^4m²	Nb individuals/10^4m²							
			Holothuroidea		Asteroidea		Ophiuroidea		Echinoidea	
			$\overline{M}$	σ(n-1)	$\overline{M}$	σ(n-1)	$\overline{M}$	σ(n-1)	$\overline{M}$	σ(n-1)
G1	9	5,6	224 (422)	168	45 (50)	22	653,6	327	155 (124)	84,1
G2	3	1,8	53	23	46 (40)	20	97,1	82,4	20,2 (8)	8,9
G3	5	3	49,8	43	14,2	12,9	8,9	10,6	1,6	1,1
G4	7	6,7	22	12	14,8	11,3	20,6	20,3	0	
G5	4	4,9	38 (20)	46	2	2	52	82	0	
G6	3	2,4	146	114	4,6	1,5	33,7	8,1	20,2	2
P1	2	1,8	34,2	18	1119	58	285	238	34,2	1,6
P2	2	1,8	363	210	3,3	3	965	616	1,8	0,7
NN	6	7,4	5779	5281	17,6	20	0		7,1	10
NL	4	4,9	4971	1410	1,6	2	0,8		25,3	20
NG	4	1,3	7232	1023	244	119	0		432	280
WB	4	1,9	20,8	13	6,8	5	55,2	22	17,7	8,3
WC	4	2,3	5	0,9	0,5	1	1	2	2,6	1,3
DA	6	4,3	64 (88)	32	31,6 (4)	5,7	0,4	0,7	41,1	18,2
DB	8	6,2	6,6 (19)	3,4	3,5	1,8	6,7	3,5	0,4	1,1
V	3	3	31,4	9,8	7,2	3,48	3,6	1,6	5,6	2,6
CV1	3	2,9	1,2	1,6	0		0		0	

3 RESULTS AND DISCUSSIONS

The data obtained in each area of the Atlantic Ocean and considered here are shown in Table 1, which summarizes the environmental and sedimentological parameters and in Table 2 which shows the results of the mean density estimated for each class of echinoderm. Since the samples were obtained using the same methods a comparison between the stations is possible.

3.1. Geographical variations and relative importance of the echinoderm population

A comparison of the mean density of each class in the different areas prospected has been achieved. As the deep-sea biological productivity which depend of trophic resources is a parameter particularly difficult to obtain in the deep-sea environment, the study is here restricted to the analysis of the abundance in terms of number of individuals/10^4 m². This, of course, is not directly comparable to the biomass variation as the size of the individuals of distinct species of echinoderms varies as well as their life span. The standard deviations of the mean density show intra-station variations (Table 2) which can be explained by the heterogeneity of the faunal distribution. The high standard deviations can be due to the patchiness of the fauna or to the low number of samples taken in some areas. Results show a strong evidence of an aggregated dispersion pattern (__ > 1) at a large scale (i.e. the area sampled by each trawl is near 600 m²). This is particularly evident in

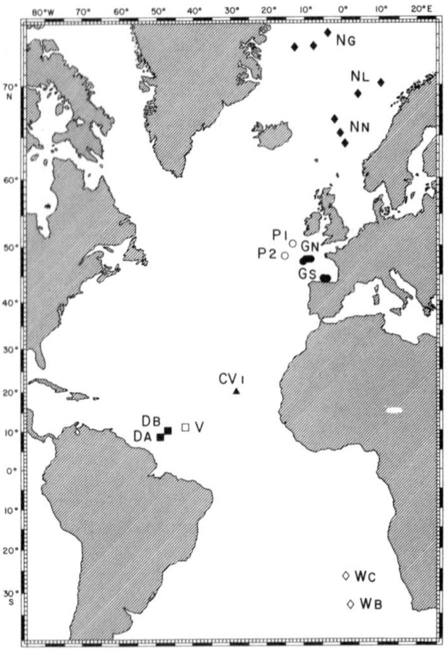

Fig. 1 : Map of the Atlantic Ocean showing location of study sites

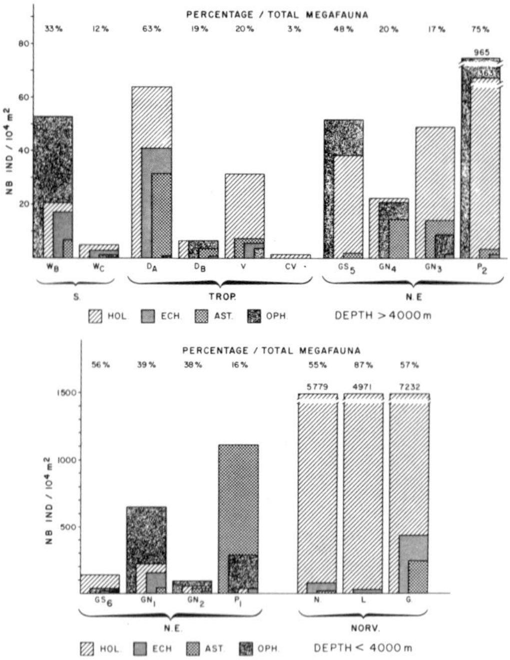

Fig. 2 : Quantitative distribution of the echinoderms (Holothuroidea, Echinoidea, Asteroidea, Ophiuroidea) at different stations located at more or less than 4 000 m depth

well sampled area where 9 trawls were taken (e.g. St G1). Moreover the level of magnitude of the mean density, deduced by trawls and by the more reliable photographic analyses, is the same in the examples given for the holothuroids, asteroids and echinoids (Table 2).

The stations located at more than 4000 m are distinguished in Fig. 2 from those distributed between 2 000 m and 3 700 m. Greater geographical variations of the mean density occur for stations situated at less than 4 000 m than for the deep stations : the density of each class does not exceed 64 ind./10^4 m^2 in the deep stations (excepted in the Porcupine abyssal plain with 363 holothuroids/10^4 m^2 and 965 ophiuroids/10^4 m^2 whereas in the shallower stations the range of the maximum density of one class lies between 97 and 7232 ind./10^4 m^2.

Whatever the depth range or the geographical area there is no general preponderance of one of the four class. The holothurians are always present and the dominant class in 10 of the 17 studied stations. The Norwegian sea is remarquable with the presence of very dense populations of the small species Elpidia glacialis, Theel, with a maximum of 15833 individuals/10^4 m^2 at 3 600 m in the Norwegian basin. Represented essentially by species of the order Elasipoda (like the dominant species, Benthogone rosea Koehler, in the Bay of Biscay, and Benthodytes typica Theel, in the Demerara basin (Bisol et al., 1984) and the family Synallactidae), the holothurians are mainly deposit feeders. The ophiuroids sometimes predominate in the deep stations such as Porcupine Plain (P2), South of the Bay of Biscay (G5) and Cape Basin (WB). In all cases the dominant species belong to the family Ophiuridae and show the predominance of deposit feeding species with increasing depth (e.g. Paterson et al., in press).The ratio between the density of the four classes of echinoderms and the total density of the epibenthic megafauna (swimming animals excepted, as their quantification is not available) is evaluated and expressed in terms of percentage. This shows (Fig. 2) a great variation in the importance of the echinoderms populations in the deep sea. In a few number of stations (6) they

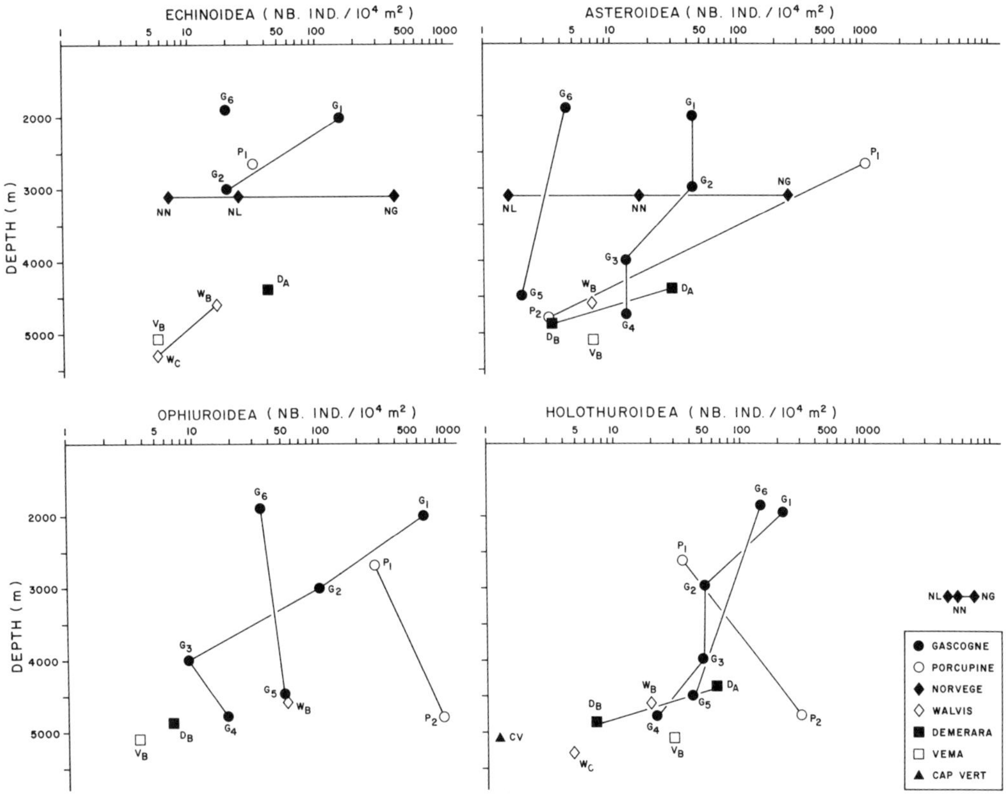

Fig. 3 : The mean density of echinoderm classes in the studied stations located at more than 2 000 m depth

represent more than 50 % of the total density, namely in the Norwegian sea, the southern Bay of Biscay, the Demerara Basin near the Amazone Cone (4420 m), and the Porcupine Abyssal Plain (4820 m). In the Cape Verde Basin, the holothuroids are the only echinoderm class present and represent only 3 % of the total megafauna density. This is an exception and in all other areas the four echinoderm classes considered contribute significantly to the total evaluated megafauna density.

3.2. Density variations as a function of depth

The change of abundance of the four classes with respect to depth range (1 900 m to 5 250 m) is shown in the graphs (Fig. 3). For the holothuroids a general decrease of the density with increasing depth is observed. An inverse relationship is only observed in the Porcupine Abyssal Plain, as the highest holothuroid abundance occurs at 4 820 m. The Porcupine Abyssal Plain has been probably under sampled. However, data obtained by Billett and Hansen (1982) and Rice et al. (1982) indicate that in the Porcupine Abyssal Plain a great density of the small species, Kolga hyalina, can occur at 4 820 m although differences of density appear through time in 1978 and 1979 at the same place. The tendance for some species to be highly aggregated imply an heterogeneity of distribution either in space or time. A significant negative correlation (r = - 0,89) is obtained between the density of the holothurians and the depth if we consider 12 stations (Porcupine and Norwegian Sea excepted). Greater variations of the density of the three other classes exist for the same depth range. For the asteroids, the decrease with depth is more accentuated in some geographical areas than others, with the highest decrease in the Porcupine Plain. For the ophiuroids no general tendancy is shown and in some cases (Porcupine Plain, southern Bay of Biscay) the density increases with depth. The scarcity or lack of echinoids is to be noted for the deepest stations (David and Sibuet, in press).

3.3. Density variations in relation to organic carbon and terrigenous matter in the sediment

The percentage of organic carbon and calcium carbonate of dry weight sediment has been estimated in the superficial layer collected with a minimum of disturbance by box core. The complementary fraction of the $CaCO_3$ in the sediment comprises mostly terrigenous material and a generally small amount of Opalin silica (< 10 %) (G. Auffret, personal communication). The organic carbon content in the sediment varies between 0.17 % and 0.85 %, whereas the $CaCO_3$ content varies between 10 and 70 % (Table 1). Only in some areas the rate of sinking particles has been achieved. In three tropical stations (DA, DB and CV) the results of the sediment trap (Sibuet et al., 1984) show a great variation of the organic carbon flux 3.29 mg/m²/day in the Cape Verde Basin to 15.9 mg/m²/day in the Demerara Abyssal Plain near the Amazone Cone. In the Bay of Biscay, the organic carbon flux, measured with another type of sediment trap and consequently not directly comparable, presents small variations between 2.92 mg/m²/day at 2 100 m, 4.28 mg/m²/day at 4 200 m and 2.31 mg/m²/day at 4 750 m (Table 1). The flux extimates (in mg/m²/day) are deduced from deployment over short periods, 10 days in the Demerara Abyssal Plain, 26 days in the Cape Verde Basin and one month in the Bay of Biscay. Seasonal variations cannot be excluded.

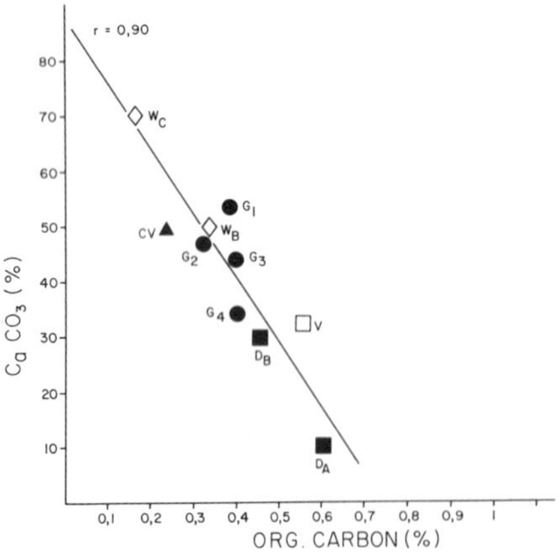

Fig. 4 : Relationship between organic-carbon and $CaCo_3$ (percentages of dry weight sediment) in the superficial sediment sampled in the study sites and regression line for all combined data

Figure 4 shows a significant (at 0.1 %) inverse linear relationship between the percentage change in organic carbon and Ca CO_3 (r = - 0.90). The present day negative correlation observed in the studied stations of the North and South Atlantic oceans has been independently established through time in the DSDP (Deep-Sea Drilling Project) cores. For example, Gardner et al. (1984) have clearly ascertained this correlation on the Walvis Ridge on the basis of 500 analyses of $CaCO_3$ and organic carbon content obtained simultaneously over a period of 2.5 M years. Consequently, this would mean that the negative correlation of $CaCO_3$ and organic carbon content could be valid both through time and space.
From the observed negative correlation, a positive correlation can be deduced between the complementary fraction of the $CaCO_3$ (mostly terrigenous material which can indicate the continental influence) and the organic carbon.
The relationship between the density of the fauna and the estimated environmental parameters is now examined figures 5 and 6, the density of each class is plotted in terms of Log.Nb.ind/10^4 m² against the amount of organic carbon in the sediment. Each geographic areas has been analyzed separately in order to respect regional characteristics or local conditions.

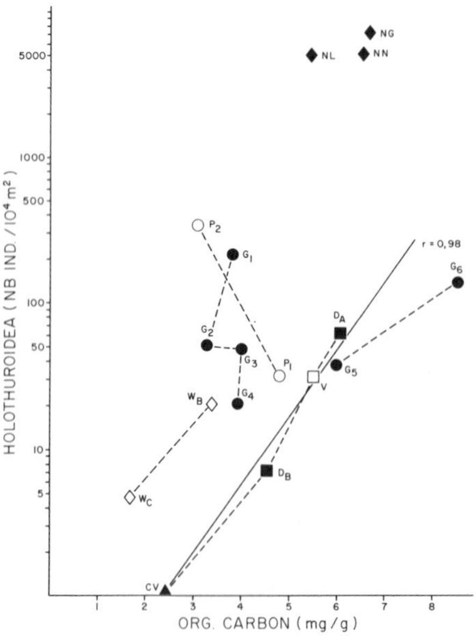

Fig. 5 : Relationship between organic-carbon in the sediment and the mean density of Holothuroidea with regression line for the tropical stations

A significant correlation exists in figure 5 for the tropical stations between the holothuroid density and the organic carbon which varies by a factor of 2.45 (r = 0.98). The holothuroid abundance is also highly correlated (r = 0.99) to the organic carbon flux which waries by a factor of 4.8 between the station CV in the Cape Verde Basin and the station DA in the Demerara Abyssal Plain near the Amazone cone. The influence of the organic carbon in the distribution of the holothuroids is here evidenced in stations located more or less in the same depth range (4 400 to 5 200 m). The organic carbon flux certainly gives a better information about the trophic input than the residual organic carbon in the sediment. Indeed, it was shown by Sibuet et al. (1984) that 85 % of the organic carbon which falls down is oxidized at the sediment-water interface through consumption by benthic fauna, microbial activity as well as chemical diagenesis. This means that only the great variation of trophic input are reflected by the analysis of organic compounds in the sediment. Moreover, the holothuroids abundance depends here of continental influences. In station DA near the Amazone Cone, the high levels of trophic input, of organic carbon and also of terrigenous material are related with the high density of holothuroid for that depth.

In the South Atlantic stations (WB and WC) and in the South of the Bay of Biscay (G5 and G6) the same relationship between the organic carbon content and the holothuroid density is shown. The Southern Bay of Biscay presents very high level of organic carbon which can be explained by the proximity of the continental land mass and by the concomitant levels of terrestrial input (e.g. plant debris were present in some sample). Curiously, there is no clear evidence of the trophic effect regarding our results on the density of the holothuroids or on the total megafauna (Sibuet and Segonzac, in press). Moreover, some extreme situations are shown in the Southern stations with the examples of azoïc samples (Laubier and Sibuet 1977, 1979). The variability in our biological data can results from the instability of the environment which can be largely influenced by turbidity currents.

In the northern stations of the Bay of Biscay (G1 to G4) the complexity of the environment (Auffret, in press) also explain the absence of great variations between the stations in the particulate flux or in the amount of the residual organic matter measured in the sediment (Table 1). Although there were no significant variations in the available organic matter measured between 2 000 m (G1) and 4 400 m (G4) the density of the holothuroids decreases by a factor of 10. The decrease of the density can be due to significant variation with time of the particulate flux not registered in the water-sediment interface. Moreover, recent bottom observations done with the submersible Cyana (cruise BIOCYAN in 1983)in station G1 has been made by the author who has observed a light agglutinate brown or dark green material similar to the one observed in the water column during the descent of Cyana. This material has been generally observed in holes and craters of the sea bottom. This observation either indicates an heterogeneity of the organic matter in the superficial film of the sediment water-interface not evidenced by box core samples or indicate seasonal arrival of organic matter. It can be compared to the results on seasonal sedimentation of phytoplankton observed in the deep-sea by Billett et al., 1983 in the Porcupine Sea-bight and already highlighted in subtropical and tropical regions by Deuser and Ross (1980) and Honjo (1982).

In the Porcupine Sea-bight (P1) and in the Porcupine Abyssal Plain (P2) the inverse relationship shown in figure 5 requires more studies to be understood, but could be due to temporal variations (Billet et al., 1983) which explain the heterogeneity of the distribution of the fauna.

The exceptionnal holothuroid richness in the Norwegian Sea is not related to a higher amount of organic carbon which of course reflects only in a small range of variation the variability of the trophic resources. The dominant species Elpidia glacialis is of small size (< 25 mm) and the biomass do not reach a high level, nevertheless the productivity is not negligible. This extreme situation can be perhaps explained by an easier colonization of some species in the deep-sea in cold water area.

The relationship of the Echinoidea and the Asteroidea are almost similar, with, in each region, a decrease of the density in relation to the decrease in organic matter (Fig. 6). The absence of these classes in some deep sea stations does not allow an evaluation of the correlation coefficient.

For the Ophiuroidea an inverse relation appears as in three areas (Porcupine, Southern Bay of Biscay and Demerara), their abundance decreases with an increase in the organic carbon in the sediment.

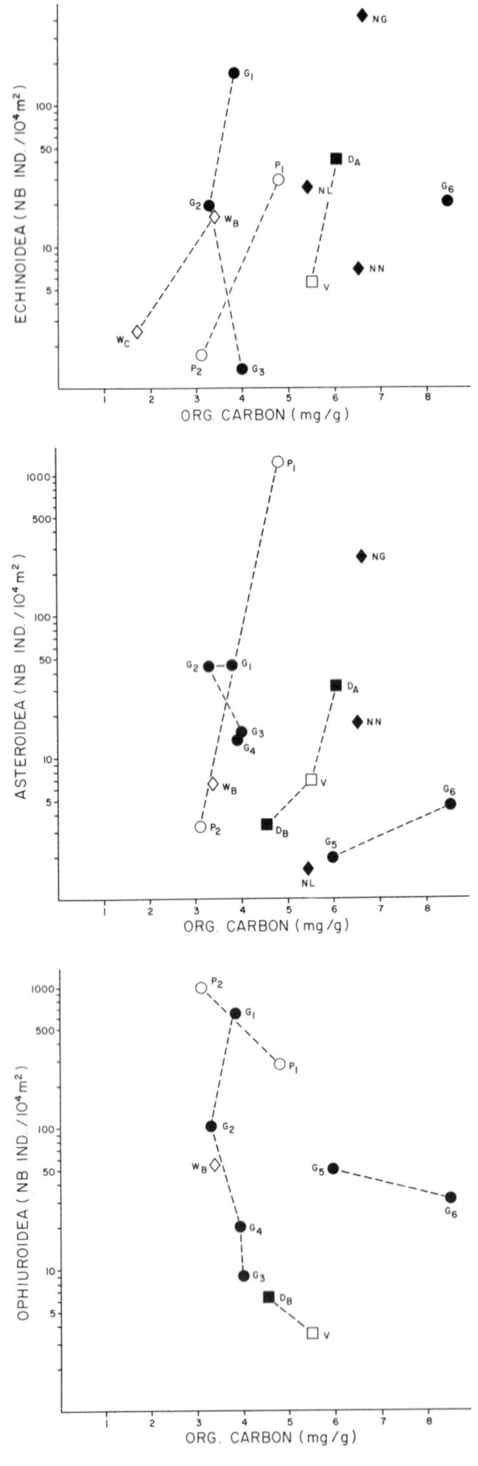

Fig. 6 : Relationship between organic-carbon in the sediment and the mean density of the Echinoidea, Asteroidea, Ophiuroidea

4 CONCLUSION

Knowledge of the distribution of the benthic megafauna is essential to an understanding of biological activity (bioturbation, recycling of organic matter) at the water-sediment-interface in the deep-sea ecosystem.

Echinoderms can contribute significantly in the deep sea community by their abundance, biomass, nutrition and respiration (e.g. specific studies by Khripounoff and Sibuet, 1980 ; Sibuet and Lawrence, 1981 ; Conan et al., 1981 ; Rice et al., 1982 ; Sibuet et al., 1982 ; Smith C.R. and Hamilton, 1983 ; Smith K.L., 1983 ; Sibuet, in press). Their importance cannot be discounted in the overall energy budget for deep sea benthic boundary layer communities (Smith K.L., 1983).

Density estimates of holothuroids, asteroids, echinoids, ophiuroids, collected or photographed over the past years in several deep sea Atlantic basins allowed comparison between 17 intensively prospected stations. These results shows :

- no general preponderance of one of the four classes but a regular presence only for the Holothuroidea ;
- a variable but not negligible contribution of the four classes to the total megafauna density (3 % to 87 % of the total evaluated megafauna and a mean of 38 % between the 17 studied stations) ;
- a general decrease in density with increasing depth except for the ophiuroids ;
- a significant negative correlation between the organic carbon and $CaCO_3$ content in the sediment and consequently a positive correlation between the organic carbon and the complementary fraction of $CaCO_3$ (mainly terrigenous matter in the studied stations) ;
- significant correlations of the holothuroid and asteroid abundance with the organic carbon flux, the organic carbon in the sediment and the terrigenous matter in the tropical Atlantic stations located within the same depth range (> 4 000 m). They explain that the trophic demands of these detritivores echinoderms are more satisfied even at great depth in an area influenced by continental trophic input ;
- that in complex environment like in the Bay of Biscay and the Porcupine Abyssal Plain the residual organic carbon presents no clear relationship with the change of abundance of echinoderms.

This highlights, for example, the need of informations about temporal variation in the type of sedimentation presently not reflected in the measured organic carbon from the sampled superficial sediment.

5 REFERENCES

Auffret, G.A., Environnement morphologique et sédimentologique des stations Biogas 1 à 4 (marge Nord du Golfe de Gascogne). In : Peuplements profonds du Golfe de Gascogne (IFREMER ed.) (in press).

Billett, D.S.M. and B. Hansen, 1982, Abyssal aggregations of Kolga hyalina Danielssen and Koren (Echinodermata : Holothurioidea) in the north east Atlantic Ocean : a preliminary report. Deep-Sea Res. 29 : 799-818.

Billett, D.S.M., R.S. Lampitt, A.L. Rice, R.T.C. Mantoura, 1983, Seasonal sedimentation of phytoplankton to the deep-sea benthos. Nature, Vol. 302 : 520-522.

Bisol, P.M., Costa, R. and Sibuet, M. 1984, Ecological and genetical survey of two deep-sea holothurians : Benthogone rosea and Benthodytes typica. Mar. Ecol. Prog. Ser., 15 : 275-281.

Conan, G., Roux, M. and Sibuet, M. 1981. A photographic survey of a population of the stalked crinoid Diplocrinus (Annacrinus) wyvillethomsoni (Echinodermata) from the bathyal slope of the Bay of Biscay. Deep sea Res., 28 : 441-453.

Dahl, E., Laubier, L., Sibuet, M. and Stromberg, Y.O. 1977, Some quantitative results on benthic communities of the deep Norwegian Sea. Astarte, 5 : 61-79.

David, B. and Sibuet, M. Les echinides dans le Golfe de Gascogne : distribution et diversité. In : Peuplements profonds du Golfe de Gascogne (IFREMER ed.) (in press).

Deuser W.G. and E.H. Ross, 1980, Seasonal change in the flux of organic carbon to the deep Sargasso-Sea. Nature, Vol. 283 : 364-365.

Gardner, J.V., Dean, W.E. and Wilson, C.R. 1984, Carbonate and organic-carbon cycles and the history of upwelling at deep sea drilling project site 532, walvis Ridge, South Atlantic Ocean. Hay, W.W., Sibuet, J.C. et al. Init. Repts. DSDP, 751 : Washington US.

Grassle, J.F., Sanders, H.L., Hessler, R.R., Rowe, G.T. and Mc Lellan, T. 1975, Pattern and zonation : a study of the bathyal megafauna using the research submersible Alvin. Deep sea Res., 22 : 457-481.

Haedrich, R.L. Rowe, G.T. and Polloni, P.T. 1980, The megabenthic fauna in the deep sea South of New England, USA. Mar. Biol., 57 : 165-179.

Hansen, B. 1975, Systematics and biology of the deep sea holothurians. Galathea Rep., 13 : 1-262, 14 pls.

Honjo, S. 1978, Sedimentation of materials in the Sargasso Sea at 5 367 m deep station. J. Mar. Res., 36 : 469-492.

Honjo, S. 1982, Seasonality and interaction of biogenic and lithogenic particulate flux at the Panama Basin. Science, Vol 218, n° 4575 : 883-884.

Khripounoff, A. and Sibuet, M. 1980, La nutrition d'echinodermes abyssaux, alimentation des holothuries. Mar. Biol., 60 : 17-26.

Khripounoff, A., Desbruyères, D. and Chardy, P. 1980, Les peuplements benthiques de la faille Vema : données quantitatives et bilan d'énergie en milieu abyssal. Oceanol. Acta, 3, 2 : 187-198.

Khripounoff, A., Crassous, P., Desbruyères, D. and Le Coz, J.R. Etude du flux organique particulaire et de ses transformations à l'interface eau-sédiment dans l'écosystème profond du Golfe de Gascogne. In : Peuplements profonds du Golfe de Gascogne (IFREMER ed.) (in press).

Laubier, L., M. Sibuet, 1977, Résultats des campagnes Biogas (3 août 1972 - 2 novembre 1974). Résultats des campagnes à la mer n° 11. Publ. CNEXO, 57 p.

Laubier, L. and Sibuet, M. 1979, Ecology of the benthic communities of the deep North-East Atlantic, in : The deep sea ecology and exploitation. Ambio Spec. Rep., 6 : 37-42.

Madsen, F.J. 1961, On the zoogeography and origin of the abyssal fauna in view of the knowledge of the Porcellanasteridae. Galathea Rep., 4 : 177-218.

Ohta, S. 1983, Photographic census of large-sized benthic organisms in the bathyal zone of Suruga bay, Central Japan. Bull. Ocean. Res. Inst., Univ. Tokyo, N° 15, 244 p.

Paterson, G.L.J., Lambshead, P.Y.D. and Sibuet, M. The Ophiuroidea fauna : its distribution and diversity. In : Peuplements profonds du Golfe de Gascogne (IFREMER ed.), (in press).

Rice, A.L., Aldred, R.G., Darlington, E. and Wild, R.A. 1982, The quantitative estimation of the deep sea megabenthos, a new approach to an old problem. Oceanol. Acta., 5, 1, 63-72.

Rowe, G.T., Gardner, N.D. 1979, Sedimentation rates in the slope water of the Northwest Atlantic Ocean measured directly with sediment traps. J. Mar. Res., 37 : 581-600.

Rowe, G.T. and Sibuet, M. 1983, Recent advances in instrumentation in deep sea biological research, in : The Sea, Vol. 8, edited by G. Rowe, J. Wiley and Sons, Inc., Publ., 81-95.

Sibuet, M. 1976, Le genre Hymenaster (asterides) dans l'océan Atlantique. Bull. Mus. Nat. Hist. nat. Paris, 256 : 304-324.

Sibuet, M. 1977, Répartition et diversité des echinodermes (holothurides - astérides) en zone profonde dans le Golfe de Gascogne. Deep sea Res., 24 : 549-563.

Sibuet, M. 1980, Adaptation des echinodermes à la vie abyssale. Echinoderms : present and past. M. Jangoux (ed.). AD. Balkema/Rotterdam, p. 233-240.

Sibuet, M. 1982, Résultats de la campagne de biologie benthique dans le bassin du Cap Vert - Seabed 2. Rapport CEA, TMC/14 575, 102 p.

Sibuet, M. Les invertébrés detritivores dans l'écosystème abyssal. Sélection de la nourriture et régime alimentaire chez les holothuries. Oceanis, in press.

Sibuet, M. and Lawrence, J. 1981, Organic content and biomass of abyssal holothuroids (Echinodermata) from the Bay of Biscay. Mar. Biol., 47 : 337-347.

Sibuet M. and M. Segonzac, in press, Abondance et répartition de l'épifaune mégabenthique. In : Peuplements profonds du Golfe de Gascogne (IFREMER ed.) (in press).

Sibuet, M., Khripounoff, A., Deming, A., Colwell, J., Dinet, A. 1982, Modification of the gut contents in the digestive tract of abyssal holothurians, International Echinoderms Conference, Tampa-Bay, edited by J.M. Lawrence, A.A. Balkema, Rotterdam, 421-428.

Sibuet, M., Monniot, C., Desbruyères, D., Dinet, A., Khripounoff, A., Rowe, G. and Segonzac, M. 1984, Peuplements benthiques et caractéristiques trophiques du milieu dans la plaine abyssale de Demerara dans l'Océan Atlantique. Oceanol. Acta., 7 n° 3 : 345-358.

Smith, C.R. and Hamilton, S.P. 1983, Epibenthic megafauna of a bathyal basin of southern California : patterns of abundance biomass and dispersion. Deep sea Res., 30 N° 9A : 907-928.

Smith, K.L. 1983, Metabolism of two dominant epibenthic echinoderms measured at bathyal depths in the Santa Catalina Basin. Mar. Biol., 72 : 249-256.

(Contribution IFREMER n° 29)

Acknowledgments : this research would not have been possible without the excellent cooperation of M. Segonzac, J. Galeron and P. Briand of the Centre National de tri biologique CENTOB - IFREMER - BREST.
I would like to thank Dr L. Laubier (IFREMER) and Dr D. Billett (I.O.S. Wormley) for their helpful comments on the manuscript, V. Martin (IFREMER) for the illustrations of this paper.

Echinoderm fisheries of the world: A review

N.A.SLOAN
Department of Fisheries and Oceans, Fisheries Research Branch, Pacific Biological Station, Nanaimo, British Columbia, Canada

ABSTRACT: Echinoderm fisheries represent ancient inshore fishery resources of some Asian and Mediterranean countries. Furthermore, import demand in Asia has stimulated harvesting abroad, in both developed and underdeveloped countries. Landings are now approaching 70,000 MT annually worldwide. Little is known of the biology of most harvested species. I provide what information is available on the landings, fisheries biology and management of sea stars, sea cucumbers, and sea urchins. Sea stars are occasionally fished for reduction to animal feed stocks and the curio trade but they may arouse more fisheries concern as pests feeding on shellfish. Sea cucumbers are harvested over a vast area of the tropics for export to Asia although by far the largest fishery occurs in Japan for the domestic market. Sea urchins dominate echinoderm landings, especially from Japan where there have long been many localized, cooperative-run fisheries. Chile and the U.S.A. (Pacific) report significant urchin landings as well. The fishery biology of urchins has not been well reported outside of Japan. Urchins have, however, attracted intense scientific interest because of their profound importance to benthic algal communities as grazers. In temperate areas the potential of urchins to influence other fisheries such as abalone, lobster, kelp and kelp-associated fin-fish has stimulated discussion of multispecies approaches to resource management.

INTRODUCTION

For thousands of years man has harvested echinoderms and annual world landings now approach 70,000 MT. Information on the fisheries for sea stars, sea cucumbers, and sea urchins is scattered and has never been reviewed on a world-wide basis. I discuss landings of these three groups, their biology and fisheries of the most prominent species. Where management occurs it is discussed.

SEA STAR FISHERIES AND PEST CONTROL

Reported world sea star (asteroid) landings are small, unreliable and strongly dominated by the fishery for Asterias rubens in Denmark (Table 1). The Danish Limfjord beamtrawl fishery occurs, ice permitting, from November to April. Sea stars are processed and exported to West Germany as an additive to fin-fish meal for poultry feed stocks (V. H. Jacobsen pers. comm.). Landings have decreased from a high of 1,100 MT in 1982 (unreported in FAO stastictics - Table 1) and only occur when other, more profitable fisheries are not operating. No management of the Danish fishery is exercised.

There have been attempts to harvest Asterias spp. along the N.E. coast of the United States for meal production (Burkenroad 1946) and for biological control as they damage oyster beds (Loosanoff 1958, 1961). Burkenroad (1946) concluded, however, that the use of sea stars in poultry feed stock supplement, or fertilizer, production was uneconomical due to high production costs and low product quality. Loosanoff (1961) confirmed the uneconomical nature of sea star fisheries in the United States. He cited irregularity of supply, low market demand and poor meat quality compared to finfish meal, despite improving processing technology. Rather than fishery harvesting, Loosanoff (1961) stressed the need for sea star eradication by harvesting and destruction or in situ chemical destruction by the use of "quicklime" (calcium oxide).

It is unlikely that sea stars will become the object of appreciable new meal fisheries although their predatory activities may be important in the negative sense to other

Table 1. World echinoderm landings (live weight) according to Class and major producing nations or areas from FAO Yearbook of Fisheries Statistics (Vols. 52 & 54).

	Year				
	1978	1979	1980	1981	1982
World sea star landings in metric tonnes (MT)	3977	1705	303	164(1)	--(2)
Percent of total landed by: Denmark	89.7(3)	98.8	100	98.0	--(2)
World sea cucumber landings (MT)	13475	13600	13544	12885	13371
Percent of total landed by: Japan	75.4	68.9	66.2	62.8	63.0
South Korea	18.6	24.4	23.2	27.9	28.1
W.C. Pacific(4)	4.2	5.4	9.0	7.8	7.9
W. Indian Ocean(4)	1.8	1.3	1.6	1.5(1)	1.0(1)
World sea urchin landings (MT)	41817	47796	48028(5)	51572(5)	47560(5)
Percent of total landed by: Japan	62.0	55.6	50.5	46.6	54.7
Chile	16.5	27.7	28.3	30.0	25.5
U.S.A. (Pacific)	19.0	15.0	20.2	22.0	18.6
World echinoderm landings total(6) (MT)	62220	65967	65288	70663	66649

(1)Signified incomplete estimate by FAO.

(2)Landings in Denmark missed by FAO for 1982, see text.

(3)This year 9.8% of world landings reported from West Germany and the Netherlands.

(4)Landings of tropical sea cucumbers are often recorded from dried product (Trepang) and are thus underestimated by an order of magnitude compared to live weight (C. Conand pers. comm.).

(5)Original FAO estimate missing U.S.A. landings which have been included here.

(6)World total includes some unspecified echinoderm landings by FAO.

fishable stocks, such as subtidal clams (Sloan & Robinson 1983) and sea urchins (Duggins 1983) in some areas.

More recently mariculturists have expressed concerns that Asterias spp. recruitment on mussel culture strings suspended from rafts could be a "serious problem" in certain areas of the N. E. United States coast (Incze & Lutz 1980). In Newfoundland, Sutterlin et al. (1981) and O'Neill et al. (1983) examined this problem experimentally in the field and the laboratory. They concluded that Asterias spp. predation could be beneficial to mussel production by their preferential cropping of smaller-sized mussels, allowing larger mussels more room to grow and thus improving the yield of marketable, older mussels. In the first year when both mussels and sea stars settled simultaneously there was a superabundance of small mussels available for the sea stars. In their second to third years sea stars showed a strong preference for small and newly recruited mussels. Only 4% of two-year mussels were eaten compared to 13-100% of the one-year mussels available at the same time.

Two other sea star fisheries warrant mention; firstly the harvesting of Asterias spp. as biological supply for biology teaching purposes throughout the U.S.A. This unreported fishery is certainly lucrative. Secondly, there are poorly documented sea star curio fisheries throughout the world whose retail value in sea-side shops is high. Curio fishing for Oreaster reticulatus has devastated accessible populations throughout the Caribbean and train-car loads of hapless stars go to processors (R. E. Scheibling pers. comm.).

SEA CUCUMBER FISHERIES

Introduction

Commercially exploited sea cucumbers (holothuroids) are sluggish, epibenthic, shallow-water and, with a few exceptions, deposit-feeders. Their cigar-shaped bodies have thick body walls which are eaten raw or dried, smoked or salted for the Asian specialty food market (Mottet 1976a). Common colloquial terms for the dried product are Trepang (Malaysian) or Beche-de-mer (French). Mottet (1976a) provides a recommended review of the Egnlish and Japanese literature on sea cucumber fisheries and, especially, processing. I update her review and discuss specific contributions to the fisheries biology and management of this group.

World landings have been dominated by Japan and South Korea with minor contributions from many developing countries in the west central Pacific and western Indian Oceans (Table 1). Almost all commercial species are aspidochirote sea cucumbers of the Families Holothuriidae and Stichopodidae. At least 9 species, not including Stichopus japonicus from Asia nor Parastichopus spp. from the western coast of North America, are harvested in appreciable numbers from the shallow tropical waters of developing countries (Table 2). Although numerous commercial species are listed in Panning (1944) and Daniel and Halder (1974), many are either rarely harvested or not mentioned in the recent fisheries literature. The N.E. Pacific fisheries for Parastichopus californicus and P. parvimensis from California to British Columbia will be mentioned only briefly here as they are covered in detail by Sloan (1985).

The suspension-feeding dendrochirotes Cucumaria japonica and C. frondosa (Family Cucumariidae) respectively form a very minor component of the human food fishery in Japan (Mottet 1976a) or were the subject of a discontinued fish meal fishery in Atlantic Canada (Sutterlin & Waddy 1975) and will not be discussed further.

Stichopus japonicus

Stichopus japonicus has been the subject of Japanese (Choe 1963) and Soviet (Levin 1982) monographs which, combined with the rest of the Japanese literature (see Mottet 1976a), comprises the most detailed information on any sea cucumber species. S. japonicus is an ubiquitous, shallow-water species along the Asian coast from 35° N to at least 44° N latitude, throughout Japan north of Kagoshima Bay to the southern coasts of Sakhalin and Kunashir Islands in the U.S.S.R. (Levin 1982). It has long been exploited as a fishery resource in U.S.S.R., China, Japan and North and South Korea. The landings of S. japonicus from Japan and South Korea alone account for 91.8% of the world total sea cucumber landings between 1978 and 1982 (Table 1).

Table 2. Major commercial sea cucumber species of the world from the recent literature and discussions with C. Conand.

Family Holothuriidae
Actinopyga echinites
A. lecanora
A. mauritiana
A. miliaris
Holothuria (Halodeima) atra
H. (Metriatyla) scabra
H. (Microthele) fuscogilva
H. (Microthele) nobilis
Family Stichopodidae
Parastichopus californicus
P. parvimensis
Stichopus japonicus
Thelenota ananas

In Japan the raw, undried body wall or processed gonads and alimentary canal are consumed. Improving the natural yield of Stichopus japonicus by management, habitat enhancement and mariculture through research has been discussed in the Japanese literature for some time (Mitsukuri 1897, 1903; Suguri 1965 in Mottet 1976a). Mitsukuri, who undertook biological studies at the request of the Imperial Japanese Government in the 1890's, mentioned the post spawning inactivity and crypsis or "summer sleep" of S. japonicus in late July. During the warm summer months sea cucumbers stopped feeding and their guts degenerated. He suggested that breeding occurred in the third summer of their 5-year life spans and that recruitment took place only in cryptohabitats among the bases of algae, invertebrate colonies or in crevices. Mitsukuri (1903) went on to make two recommendations based upon his research: the establishment of "breeding reserves" where no fishing would take place and the building of stone reefs in the shallows to supply cryptohabitat for adults in summer, and as a refuge for recent recruits. He then proposed that local catches be monitored and regulated, to supplement supply from the enhancement procedures, on a site-specific basis. Interestingly, he discovered that some of his recommendations had already been implemented in fishing villages a century before his time. Tanaka (1958a, b) confirmed the spent gonads and degenerated gut of S. japonicus during its late summer-early autumn "aestivation", whose onset he suggested was temperature-cued. Between November and June the gonads developed following commencement of feeding (Tanaka 1958b).

Suguri (1965 in Mottet 1976a) mentioned that one prefecture had alternated fishing years for at least a century and he listed another 25 prefectures which had seasonal closures within the maximum range of March to November. The first part of the closure (March-June) was for protection of brood stock and the last part because of low body wall quality due to summer "aestivation". Suguri recounted the use of stone reefs and the deployment of brushwood spat collectors for juveniles in the shallows where they recruit prior to migrating to deeper areas as adults. Success in enhancing larval recruitment appeared site-specific and thus varied greatly. The use of stone reefs and breeding reserves has not, however, been widely used in Japan. Other enhancement techniques such as relocating juveniles to ongrowing sites have been successful (Tauti & Matsumoto 1954; Suguri 1965). Suguri suggested, however, that enhancement techniques have been insufficiently evaluated to compare their relative merits.

Levin (1982) reported that the Bay of Peter the Great S. japonicus fishery off Vladivostok had mean annual landings of 180 MT from 1970-1978; considerably lower than earlier landings such as 550 MT in 1919 and 500 MT in 1929. Early Soviet sea cucumber fishing, by dredge or spear, was regulated by 1900. It was, however, not until 1923 that the "three field system" management, in which fishing grounds were partitioned and harvested areas allowed three years to recover, was implemented. Levin suggested that 6 to 10 years may be required for natural restocking of harvested areas. Even in the early days of the fishery a summer closure was maintained.

The Soviets now have a diving fishery managed by a reported quota of 242 kg per boat per day with a summer closure to promote reproduction and prevent the harvest of substandard (non-feeding and post-spawned) individuals. The body walls are processed for export to Asian countries although some local consumption occurs of muscle bands from the inside of the body wall. Levin (1982) suggested mariculture,

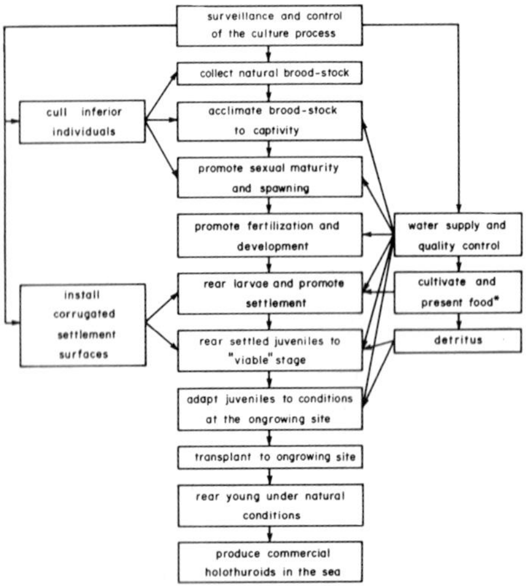

Figure 1. Organization of mariculture protocol for Stichopus japonicus translated from Levin (1982). *Levin listed food components as microflora, bacteria, dissolved organic matter and yeast.

first attempted in 1937, could be employed to enhance natural harvest. He claimed mariculture technology was sufficiently advanced to declare a goal of 1,000 MT production by 1990. Whether this projection was for the U.S.S.R. alone was not clear. Figure 1 illustrates his mariculture protocol for Stichopus japonicus. Levin (1982) also supported the use of Mitsukuri's stone reefs to mitigate the "limiting" factor of insufficient habitat for S. japonicus culture and/or enhancement in some areas. As part of this mariculture effort Rakov (1982) has studied age as well as growth in cages at different depths over 2-3 years. Estimated life span of S. japonicus was increased to 9-10 years. Most active growth apparently occurred in the warm summer months which, however, coincides with the above-mentioned cessation of activity and feeding as well as gut degeneration (Mitsukuri 1903; Tanaka 1958a, b). The success of reef development by Chinese fisheries workers was cited but Levin did not indicate whether the reefs were being widely used in the U.S.S.R.

The Chinese have been active in both use of hatchery-reared stock to increase sea cucumber productivity of sheltered bays (Gentle 1979) and transplanting stocks from northern China Sea areas to culture facilities in southern China (Xiamen) (Shuxu & Gongchao 1981).

The extensive Japanese literature

included in Mottet (1976a) and Levin's (1982) monograph suggest that enhancement of natural S. japonicus stocks by habitat improvement, transplantation and laboratory-based mariculture will become relatively more important in the future.

A trial export fishery for Stichopus mollis is currently underway in New Zealand (M. F. Barker pers. comm.).

N. E. Pacific Fisheries

Two Parastichopus spp. are fished along the Pacific coast of North America: P. californicus (by diving) in British Columbia; Washington State and California and P. parvimensis (by diving and trawling) in southern California (see review of Sloan 1985). There is little market for the relatively thin body walls and only the five muscle strips from inside the body wall are used; for both domestic consumption and export to Asia. Landings were first recorded in 1971 (3.5 MT), remained low until 1976 (6.7 MT), grew gradually to 153 MT in 1981 and increased strongly to 625 MT in 1983. Washington state dominated landings from 1977 to 1981, California in 1982 when total landings were low and British Columbia recorded 430 MT (69% of total landings) in 1983. The fishery is still characterized by incomplete market development which imparts instability and causes non stock-related fluctuations in landings. Large unfished stocks of P. californicus occur in the central and northern coasts of British Columbia and perhaps S. E. Alaska as well.

In British Columbia, sea cucumber landings were first recorded in 1980 and management has been pre-emptive to control fishing until the biology of local stocks is more thoroughly understood. Most of the coast is under closure with unrestricted harvesting allowed in a few south coast areas whose landings are monitored by fishermen's mandatory log books. Some progress in fisheries-related research has been made; after summer spawning and an approximate 70-day larval life, P. californicus recruit in nursery areas such as mats of filamentous algae (J.L. Cameron pers. comm.). P.V. Fankboner (pers. comm.) has found that individuals are harvestable at their approximate size of sexual maturity ($\approx$0.5 kg), which may be reached in 4 years. Also, he suggests that in early autumn feeding stops, body wall biomass declines and the gut atrophies, rather than being expelled as previously believed. It is not known yet whether the muscle strips are affected during these events. Landings have been reported from Washington State since 1971. That fishery has few management restrictions.

The California fishery has grown since its first landings record of 1981. Fishing occurs mostly in the Santa Barbara channel (southern California) where P. parvimensus dominates the catch. There are no restrictions on harvesting for licensed fishermen. P. parvimensus spawns from May to June and settlement occurs between September and October among kelp holdfasts for protection from fish predators (Muscat 1983). Muscat has also reported on morphometrics, distribution, activity patterns and ecological effects of feeding by P. parvimensus; all of which will aid future fisheries-related concerns for the species in southern California.

Tropical sea cucumber fisheries

Numerous sea cucumber species are harvested for export as trepang to Hong Kong, Singapore, Malaysia and Thailand from shallow tropical waters in many countries over a vast area from the east African coast to the central South Pacific (Hornel 1917; Panning 1944; Sachithananthan 1972; Anonymous 1979; Gentle 1979). The entire catch, however, represents less than 10% of recent reported world landings (Table 1). The scattered nature of this fishery has almost certainly led to underreporting of landings and confusion as to which species are harvested. Nonetheless, descriptive literature on fishing and processing techniques is widely available (Adithiya 1969; Beardsley 1971; Mottet 1976a; Crean 1977; Howell & Henry 1977; Parrish 1978; Conand 1979; Gentle 1979; Gentle & Conand 1979).

With the exception of Conand's (1981, 1982, 1983) work, research on the fisheries biology of harvestable tropical species is singularly lacking (Gawel 1981). Indeed, with few exceptions such as the histological monograph on Holothuria (Metriatyla) scabra (Bai 1980), the general biology of tropical species remains underworked although they are conspicuous and important components of many shallow-water communities (Bakus 1973; Massin 1982). There are about 9 major commercial tropical species (Table 2) among which Holothuria (Microthele) nobilis and H. (M.) fuscogilva are the most valuable per unit weight followed by Thelenota ananas and Actinopyga spp. although H. (M.) Scabra may dominate landed weight (Gentle & Conand 1979; C. Conand pers. comm.). All commercial species are deposit-feeders which assimulate bacteria and nitrogenous compounds of organic matter associated with sediments they ingest (Moriarty 1982). No stock assessments are available and none of the various species appears to be managed. A characteristic of the fishery is that accessible beds are rapidly depleted and local fishing interest wanes unless some control of the harvest is exercised (Carleton 1983).

Conand (1981, 1982) has reported on the morphometrics, reproductive cycle and size at sexual maturity of four of the most valuable commercial species in New Caledonia. Such data are essential if size limits and seasonal breeding closures are to be considered as management criteria in the future. Furthermore, there are indications that gonad development certainly requires nutrients from the viscera and perhaps the body wall (Krishnan 1968; Lawrence & Lane 1982) which could decrease product quality.

With the exception of asexually reproducing H. (Halodeima) atra, recruitment of tropical species is very poorly understood (Ebert 1983). Gentle (1979) has suggested that juveniles are highly cryptic, possessing camouflage coloration which they lose as adults who spend their lives exposed. Recent recruits apparently use the bases of turf-like seaweeds as nursery areas (Gentle 1979).

Sea cucumber growth is not well documented and Conand (1983) was unsuccessful using most traditional study methods, except tagging and recapture, on commercial species. There are no hard parts useful for age determination and body morphometrics are unreliable. Growth and size analysis of some species such as H. (Halodeima) atra is further complicated by the presence of sexual and asexual (transverse binary fission) modes of reproduction (Ebert 1978, 1983; Harriott 1982).

Recent, renewed interest has been shown in sea cucumber fisheries (Gawel 1981; Conand 1983) despite their general disorganization and incomplete market development (Carleton 1983). This chronic lack of organization at both production and marketing levels has robbed the fishery of its potential value and vitality. Asian buyers have encountered low product quality which has decreased prices (Carleton 1983) and thus the fishery has attracted insufficient attention to stimulate fisheries biology and management research. Potential for mariculture of tropical holothurians is not discussed in the recent literature (Brewer & Corbin 1984).

SEA URCHINS

Introduction

All commercially exploited sea urchins (echinoids) are epibenthic, shallow-water grazers. Sea urchins are harvested according to season for their gonads, which are highly prized in Asian and Mediterranean countries. Mottet (1976b) has reviewed the English and Japanese literature on edible sea urchin biology with some mention of management. I update her review on contributions relating to sea urchin fisheries biology and discuss some examples of their management. The fisheries and management of N.E. Pacific Strongylocentrotus franciscanus resources between California and British Columbia are mentioned here but discussed in detail elsewhere (Sloan 1985). Sea urchins are enormously important as grazers of algae in certain temperate rocky subtidal communities throughout the world and there is an extensive literature on the effects of their grazing (Lawrence 1975; Lawrence & Sammarco 1982). Some discussion of recent concerns about kelp-associated fisheries being influenced by urchin grazing activities (Pringle et al. 1980, 1982) is also provided here.

World sea urchin landings have been dominated by Japan although the proportion contributed by Chile and the United States (Pacific coast) has been increasing (Anonymous 1980; Table 1). Reported landings consist almost entirely of species in the Families Echinidae and Strongylocentrotidae although species, most of them tropical, in 10 other families are considered edible (Mottet 1976b). Japan also dominates the world as an importer of sea urchin roe from at least 13 countries in five continents. Chile is the most diversified exporter; selling roe in four continents (Wilson & Gorham 1982).

Atlantic Europe and Mediterranean fisheries

Sea urchins have been a prized delicacy in Italy since Roman times and have been harvested in Brittany since the Iron Age (1000 B.C.) (Allain 1972). The most prominent commercial species is Paracentrotus lividus which is harvested in S.W. Ireland, from Brittany through the Bay of Biscay to Portugal, North Africa to Egypt, Malta and Italy (Allain 1972). Other (large) species such as Echinus esculentus and E. melo are fished mostly for the curio trade and, although these small fisheries are locally destructive and of concern to conservationists (Allain 1972; Nichols 1978, 1979, 1981), they will not be mentioned further.

The P. lividus fishery in Brittany is the best documented European urchin resource (Allain 1972, 1975, 1978). In 1968, northern Brittany supplied half of France's production but by 1970-71 the fishery was in severe decline due to overfishing and stock assessments revealed depleted populations on all grounds. In Brittany the limited accessible subtidal grounds were all well known and the intertidal stocks fully exploited. Thus, there have been significant management problems with the fishery.

Fishing methods are controlled and commercial diving for urchins is prohibited throughout France (Allain 1975). In Brittany, for example, only certain gear types such as particular dredges, rakes and entanglement mops are permitted. The main Breton gear-type is a chain-weighted entanglement mop or "faubert" which is dragged over subtidal bedrock or boulder substrates less than 5 m deep. This is very destructive in boulder fields where the rocks are overturned, crushing urchins and disrupting their cryptohabitat. As intertidal harvesting is intense the only habitat refuges for P. lividus are bedrock crevices in subtidal areas shallower than 5 m, and all subtidal areas deeper than 5 m where, however, algal food availability is less. Size is set by market demand so fishermen keep all P. lividus $>$37.5 mm test diameter. Allain (1975) claims urchins above this size represent 66% of the population. Growth is considered slow and harvestable size is reached in 3-4 years of an estimated potential 10-year life span (Allain 1978). Allain (1975) reports that juveniles are scarce, implying low recruitment. The literature is unclear as to the size at sexual maturity of P. lividus. This may be confounded by differences in maximum size according to different areas in Brittany (Allain 1975), as in the Mediterranean (Régis 1980).

Allain mentioned other problems such as the nearshore pollution in the Brest area, rendering local P. lividus stocks unfishable, and lobster restocking in some urchin grounds which had greatly increased urchin mortalities from lobster predation.

Allain (1975) mentioned that the Breton fishing community rather than the government was active in proposing mitigation and enhancement initiatives for local P. lividus stocks. These measures included permanent area closures as reserves, ensuring production of juveniles, area restocking, control of predators such as the sea star Mathasterias glacialis and banning use of destructive gear such as the "faubert". Allain concluded that as recruitment of P. lividus appeared low, growth was slow and value increased continuously with size, mariculture would be a less attractive alternative to enhancement of stocks in nature through management changes and the cooperation of harvesters.

The Mediterranean P. lividus fishery is poorly documented but at least along the French coast stocks are overfished and there appears to be little management of this ancient fishery (J.-M. Pèrèz pers. comm.). Régis (1980) suggested that culture of Mediterranean P. lividus throughout its entire life cycle would be "unrealistic" but rather ongrowing natural urchin stocks in controlled, natural habitat conditions could be possible. She suggested examining the growth characteristics, in relation to roe production, on a site-specific basis to ascertain mariculture site characteristics affecting urchin gonads. Site criteria were the nature of food resources; accessibility for fishing and occurrence of appreciable natural recruitment.

There has been a diving export (to Paris) fishery for P. lividus off S. W. Ireland since 1948 and there are now some 20 active divers as well as rakers working the intertidal rock pools. (J. Arthur pers. comm.). Harvesting is unregulated and landings are not fully recorded. Market-demand sets a minimum size limit of 50 mm test diameter. This resource has been overexploited and there is a need to describe the biology of local stocks and implement some harvesting controls (J. Arthur pers. comm.).

Atlantic North America: Strongylocentrotus droebachiensis fisheries and multispecies interactions

The harvest of S. droebachiensis along the Atlantic coast of North America is very limited. No Canadian landings have been reported recently and the American fishery, perhaps 10 MT per year, is for supply of local ethnic markets such as the Italian communities of New York and Boston (Scattergood 1963; R. J. Miller pers. comm.). First American landings were recorded from Connecticut in 1929. These were supplanted by the Maine small boat, dip net fishery which averaged $\approx$3.2 MT annually between 1933 and 1963 (Scattergood 1963). I could not find recent data on this December to April New England area fishery.

In order to stimulate urchin fisheries interest in Canada's Maritime provinces, especially Newfoundland, Miller & Bishop (1973) reported on urchin densities, mid December to March roe season, harvesting with entanglement mops and product storage and shipping. Little urchin fishery development has, however, occurred in Newfoundland although Keats et al. (1983, 1984) have recently discussed holding S. droebachiensis for short periods under enhanced feeding conditions to improve roe yield. Kramer (1980) reported on the possibility of establishing a fishery in the St. Lawrence estuary. He recommended further stock assessments along the Quebec coast, fishing and processing technique refinements, market survey and roe quality assessments. Himmelman et al. (1983a) reported more detailed stock assessments in the St. Lawrence. Areas producing commercial sized urchins were limited, experienced low recruitment and were susceptible to overfishing. I know of no western Atlantic American sea urchin stock assessment surveys.

Although S. droebachiensis is not the subject of an appreciable fishery, its destructive grazing has created "barren grounds" along much of the eastern Nova Scotia coast, stimulating much fisheries-related research and debate. Fisheries ecologists are united in recommending a strong multispecies kelp ecosystem research approach to understanding the factors controlling nearshore productivity and management of kelp-associated lobster and finfisheries (Mann 1982; Pringle et al. 1980, 1982). Urchin grazing plays a pivotal role in these concerns. Ecologists differ, however, as to whether there is sufficient evidence to support theories concerning the origin, effects and longevity of urchin-dominated barrens.

Mann and his co-workers have proposed that reduction of lobster stocks due to environmental changes and overfishing released lobster predation pressure on urchins which led to the increased abundance of urchins. Thus, abundant urchins destructively overgrazed attached algae, creating barren grounds - a new, apparently stable, long term community structure. The decreased productivity and habitat shelter of the barrens aggravated poor lobster recruitment and lobster stock decline. They cited circumstantial evidence, field observations and experimental results in support of these hypotheses (Mann 1977, 1982; Wharton & Mann 1981; Bernstein et al. 1983; Mann et al. 1984). A related hypothesis from Harding et al. (1983) proposes that an environmentally-caused lobster recruitment failure initiated the decline of the lobster fishery and that lower lobster numbers caused an ecosystem "imbalance" in which "uncontrolled" urchins destructively grazed kelp.

Mann (1977) suggested that the eastern Nova Scotia barrens were new, "stable" changes involving a "long-term decrease" in productivity which urchins were capable of maintaining (Chapman 1981). Moreover, urchins appeared well adapted to occupying in abundance these habitats of relatively low productivity (Johnson and Mann 1982). Reportedly stable S. droebachiensis barrens have also been recorded from Norway (Hagen 1983). In the St. Lawrence estuary where lobsters are not present, Himmelman et al. (1983b) suggested the urchin-dominated habitats were stable as in some other habitats with few lobsters. Wharton & Mann (1981) stressed not implying permanency of barrens, but rather longer duration of barrens than previously cited. A change in the cycle, towards higher productivity, may be at hand due to recent mass urchin mortalities along 440 km of each Nova Scotia (Miller & Colodey 1983) attributed to disease outbreak associated with high sea surface temperatures (Scheibling & Stephenson 1984). Surveys of areas after the urchin mortalities revealed 2 to 14 times the algal cover compared to areas containing abundant urchins (Moore & Miller 1983). These recovering areas are, however, still much less vegetated than areas which have been without urchins for many years.

Pringle et al. (1982) do not feel there are sufficient data to support the Mann hypothesis. They conceded that significant decline of algal abundance has occurred in some Nova Scotia areas although they feel much of the remaining coast is too poorly documented historically. They feel that there is insufficient evidence to identify urchins as the sole agents of kelp forest declines and question the long term nature of barren grounds. Most importantly, they reject the role of lobsters as key predators of urchins and thus their influence on urchin abundance. Pringle et al. (1982) suggested that overexploitation since the 17th century of demersal fish (which at some stage in their lives eat urchins) such as haddock, cod, halibut, and flounder initiated a transformation from complex, stable kelp bed ecosystems to a simple, much less stable barrens of today off Nova Scotia. This elevated the ecological importance of lobsters and crabs in this unstable system in which a single-year recruitment failure of lobsters, along with strong fishing pressure, allowed "wild oscillations" of stocks. Thus, they proposed that localized population increase followed by overgrazing by urchins was due to removal of a suite of carnivores among which lobsters were only one.

Active S. droebachiensis control measures seem unlikely to succeed in view of the size of affected areas. Short term measures such as the use of quicklime to eradicate urchins, although locally effective (North 1976), appear too expensive to be widely used (Bernstein & Welsford 1982). Pringle et al. (1982) recommended a more indirect method of changing lobster fishery regulations to promote lobster recruitment thus eventually decreasing urchin densities through natural lobster predation. Until lobster stocks are enhanced, however, "economical restoration methods", i.e., urchin eradication of some sort, will be required in the barren grounds (Pringle et al. 1982). A good mitigation opportunity may be at hand in the form of naturally occurring disease-related mass urchin mortalities.

Pacific North America: Strongylocentrotus franciscanus fisheries

Since 1971 landings of the 'red' sea urchin Strongylocentrotus franciscanus have been reported by diving fishermen from British Columbia, Washington State, California and Baja California, Mexico; largely for export of fresh roe to Japan (see review of Sloan 1985). Alaskan resources are as yet undeveloped (Wilson & Gorham 1982). Due to their smaller size, neither the 'purple' urchin S. purpuratus nor the 'green' urchin S. droebachiensis are fished in appreciable quantities although abundant in some areas. Total landings of red urchins have increased dramatically from $<$ 1.0 MT in 1971 to 11,600 MT in 1981 and have declined to 8,590 MT in 1983. Since 1973 the southern California fishery has accounted for 85 to 96% of the total landings each year. With the exception of southern California, landings from other fisheries have fluctuated irregularly due to a combination of local stock depletions or underdeveloped marketing. Stock in central and northern areas of both British Columbia and California remain underutilized. This is due mainly to transport difficulties of fresh roe from these relatively remote (from international airports) areas to the Tokyo market and limited good weather conditions.

Management of the British Columbia urchin fishery has been pre-emptive to curtail widespread harvesting until more is known about the biology of local stocks. For example, recruitment is a concern as it appears to be considerably lower ($\approx$5.5% of the population per year) than among southern California stocks. Harvesting is monitored by mandatory log books and controlled by a suite of area and seasonal closures, landings quotas and a 100 mm test diameter size limit (which is reached in 4-5 years). Washington State also uses area and seasonal closures as well as size limits, which differ according to area. The southern California fishery is relatively unregulated with market demand controlling fishing effort. This lack of government intervention is due in part to lack of concern over recruitment, which is relatively high ($\approx$24% of the population per year) compared to northern stocks and in some areas urchins are considered pests to harvestable kelp, Macrocystis pyrifera, beds and their associated shell-and fin-fisheries. The Mexican fishery is not well documented but is apparently run by local co-operatives (M. J. Tegner pers. comm.). A multispecies management approach involving S. franciscanus and other kelp forest resources on the Pacific coast is discussed in detail elsewhere (Tegner 1980; Dayton & Tegner 1984; Sloan 1985).

Pacific South America: Loxechinus albus fishery

Chile has a large and growing urchin fishery whose landings are dominated by Loxechinus albus (Hancock 1969; Castilla & Becerra 1975; Table 1). Peru has no export fishery although L. albus are consumed locally (R. M. Perez pers. comm.). Similarly, Colombia has only some artisanal harvesting (F. H. Pineda Polo pers. comm.). Chilean stocks are known to be locally depleted by overharvesting (Deppe and Viviani 1977; Moreno & Sutherland 1982) and research is underway on the fisheries biology of L. albus (J. C. Castilla pers. comm.). In northern Chile the size limit is 80 mm test diameter versus 100 mm in central and southern Chile (Deppe & Viviani 1977). In 18 processing localities only 37.6% of the catch was of legal size, so there are management problems with the fishery.

Chilean kelp forests, although dominated by the same macroalga, Macrocystis pyrifera, as Pacific North American kelp forests, have few "organizational features" in common with their North American counterparts (Moreno & Sutherland 1982). Experimental observations in kelp forests between 39° and 55° latitude South have revealed that L. albus did not appreciably influence M. pyrifera settlement and growth nor algal community ecology (Castilla & Moreno 1982; Moreno & Sutherland 1982). These urchins fed on drift algae rather than grazing on attached plants. Furthermore, Castilla & Moreno (1982) proposed that urchin populations were not regulated by vertebrate or invertebrate predators but rather by a "complex of biological factors" as well as winter storms. For example, although L. albus is certainly eaten by the large Chilean sea star Meyenaster gelatinosus (Dayton et al. 1977) this predation pressure is much less significant than the predation of Strongylocentrotus franciscanus by another large sea star, Pycnopodia helianthoides, in Alaska (Duggins 1983).

Besides being a world leader in urchin landings, Chile has many other rocky shore fisheries (Hancock 1969; Castilla & Becerra 1975) such as the world's largest gastropod fishery for the muricid Concholepas concholepas (J. C. Castilla pers. comm.) and a major tunicate fishery for Pyura chilensis (Sloan 1985). Multispecies management as discussed previously for complexes containing kelp, urchin, abalone and lobster could be useful in managing Chile's extensive shellfisheries.

Japanese fisheries

Japan leads the world in reported landings of sea urchins (Table 1). The main species harvested are Strongylocentrotus intermedius north of central Honshu through Hokkaido, Heterocentrotus pulcherrimus and Pseudocentrotus depressus from central Honshu through Kyushu and Tripneustes gratilla south of Honshu through Okinawa (T. Motokawa pers. comm.). There are approximately 11 other minor commercial species as well (Matsui 1966; Fuji 1967). Fishing is widespread with at least 26 prefectures reporting landings which are dominated by strongylocentrotid species from northern Japan, especially Hokkaido (Kinoshita 1955, 1958). Japanese grounds have been thoroughly harvested for many years with few virgin beds remaining, and their urchin industry has a history of enhancement and resource management (Kinoshita 1958; Kawamura 1966; Matsui 1966; Fuji 1967; Mottet 1976b; Kito et al. 1980). Fishery regulation and enhancement by cooperatives (Sato 1980) maintains the urchin industry. Japanese inshore fisheries management is extremely decentralized with local fishermen's cooperatives controlling resources within their coastal areas (Mottet 1976b; Asada et al. 1983). Cooperative members have equal access to commonly-controlled fishing grounds whose regulations vary according to local concerns (Asada et al. 1983).

Despite Fuji's (1967) claim that the ecology of Japan's main commercial urchin species was insufficiently understood, more fisheries biology-related research has been done on echinoderms in Japan than any other country. Reproduction, growth, and feeding are well understood for urchins but there remain weaknesses in knowledge of recruitment, population dynamics (growth and mortality at every stage of life) and kelp forest ecology as a determining factor in urchin populations (Fuji 1967).

All authors agree that good feeding conditions must occur for gonad growth. Thus, many enhancement possibilities are considered in the light of optimal feeding criteria (Nagai & Kaneko 1975; Hayakawa & Kittaka 1984) during the seasonal growth phases of urchin gonads. These conditions vary locally throughout Japan and must be applied on a site-specific basis. The decentralized nature of Japanese coastal fisheries necessitates consensus and voluntary restraint for enhancement of urchin resources (Matsui 1966). The following interrelated goals, including their benefits and examples, have been identified from urchin harvest regulation and stock enhancement considerations by Kawamura (1966), Matsui (1966), Ishiwata et al. (1977), Mottet (1976b, 1980), Kito et al. (1980) and Sato (1980):

(1) Habitat improvement through use of artificial stone reefs for more substrate for algal food species and more cryptohabitat for urchins. For example, the introduction of goggles for free divers in Fukui Prefecture in 1902 led to stock declines and stone reefs were introduced in 1910. By 1935 further measures of urchin transplants were implement to increase production, see below.

(2) Urchin food enhancement at chosen sites through artificial "afforestation" of kelp which also supplies kelp spore recruits. Using (edible) kelp culture long line techniques, urchins can be ongrown to adequate size and be harvested. This, in turn, decreases urchin numbers sufficiently to permit subsequent kelp recruitment for the next urchin crop.

(3) Transplantation and ongrowing of adults for increased gonad growth in good feeding localities. Best areas are kelp beds $\approx$5 m deep and away from exposure, pollution, and osmotic stress from fresh water run off. Timing is essential and transplantation should occur at the end of the local fishing season and after spawning when feeding and gonad growth will be rapid. Some sites could support two transplantings a year, for example, in September/October for harvest in April/May, immediately followed by a second transplant for an August harvest. Sites must be regularly monitored for their algal cover and urchin gonad yield. High density culture under continuous surveillance in coastal enclosures such as shallow bays has also been proposed.

(4) Identification of optimal harvestable urchin size (age) for maintaining high gonad yield per individual from a stock of maximum size. Optimal size (and age) at

harvest varies between species and within species according to area. S. nudus and S. intermedius live to 7-8 years but best gonad growth and quality in relation to size occurs at approximately 5 years old. For S. nudus gonads increase 70% in size at 4-5 years old compared to only 30% increase at 6-7 years old. Older, less productive urchins also have a higher natural mortality than younger adults. Crowding on grounds must be prevented as it causes food shortages (lowering gonad yield per individual) and decreases kelp food recruitment.

(5) Closures by area and/or season permit grounds to revegitate, urchins to recruit and adults to immigrate from adjacent unharvested areas. Seasonal closures in shallows where harvesting is intense may allow juveniles to recruit in these preferred nursery habitats. Rotating grounds on a 3-year basis also aids post-harvest recovery but must not be delayed as to permit too many older, less productive urchins to occur in the harvestable population. Seasonal closures which, differ according to locality, prevent wasteful harvest of poor quality, post-or actively-spawning urchins, and aid successful breeding of remaining stocks.

(6) Fishing effort control by limiting harvest methods to prevent stock depletion and permit maximum numbers of cooperative members to participate in the fishery. Scuba diving appears prohibited in all but the least accessible harvesting areas. Free diving, wading, and small boat grappling or raking seem the dominant methods. One prefecture even prevents free divers from using wet suits!

Four additional goals of enhancing recruitment by deploying spat collectors, introducing artificially-reared urchin spat, urchin predator control and post-harvest stock assessments have been identified but are not well documented.

Japan is advanced in roe processing technology such as partial freezing and deoxygenization (Uchiyama et al. 1984) which increases its storage life. This could eventually permit exploitation of grounds previously too distant for transport of quality product to the fresh market.

The influence of urchin grazing on Japanese kelp forests has been studied since the early 1900's (see Hagen 1983 for references). The creation and persistence of crustose coralline algae barrens or "Isoyake" by urchins is a current research topic (Noro et al. 1983). Kito et al. (1980) proposed that overgrazing of adult and recruiting kelp by urchins created and maintained barrens. Environmental perturbations such as abnormally high sea surface temperatures were suggested by Noro et al. (1983) to initiate local kelp forest declines whose recovery would be aggravated by urchin grazing on kelp recruits. Urchins and abalone in such barrens may be numerous but they are small, presumably due to starvation (Kito et al. 1980).

Urchin grazing and multispecies management of harvestable kelp - associated fin - and shell-fish resources has received little unified discussion in the Japanese literature, perhaps because of the decentralization of coastal fisheries. Urchin removal has been successfully used for enhancement of kelp (Noro et al. 1983). Kelp enhancement has in turn been related to increases in urchin and abalone populations (Kito et al. 1980).

Australian, New Zealand and tropical urchin fisheries

The potential for urchin fisheries in temperate southern Australian waters has long been considered. Landings figures are not available and the potential fishery is still in the development stage. There was some early concern that urchins along the Victoria coast were competing with the valuable abalone stocks (Ward 1975) and Shepherd (1973) suggested the multispecies-ecosystem approach for coastal Australian resources. O'Connor and his colleagues (1975, 1976, 1977) examine the reproductive cycles and population dynamics of four New South Wales species with a view to fishery development. They discussed nearshore mariculture of urchins under enhanced food and perhaps thermal effluent conditions for maximizing gonad growth. No landings have, however, been reported from the southeast Australian coast (P. Sluczanowski pers. comm.).

The Heliocidaris erythrogramma resources of Tasmania have been assessed in detail (Dix 1977a, b, c). The resource was carefully evaluated from the harvesters' viewpoint and a small scale diving fishery appeared possible. Stock assessments, fishing effort and gonad yield surveys were completed and further information on growth and mortality will be forthcoming if the fishery develops sufficiently to warrant it. Dix found little concern for urchin/abalone competition as urchins occupied more sheltered rocky shores whereas abalone preferred exposed shores. A fishery has, however, not yet been established (A. J. Harrison pers. comm.).

New Zealand workers are exploring possibilities of an export fishery for the endemic Evechinus chloroticus (M. F. Barker per. comm.) and there is an artisinal fishery among native Maoris (J. H. Annala pers. comm.). The biology of E. chloroticus is quite well known (Dix, 1972; Choat & Schiel 1982) and sufficient stocks occur, thus an export potential for this species exists depending on market development.

There are many artisinal urchin fisheries along the coasts of tropical countries whose landings go largely unrecorded and for which little management is exercised such as the Tripneustes esculentus (=ventricosus) fishery in Barbados (Lewis 1958) and the Tripneustes gratilla and Heterocentrotus mammillatus fisheries in S. Pacific islands (B. A. Thomassin pers. comm.). Hong Kong has a substantial Anthocidaris crassispina fishery whose landings reached 2,000 MT in 1980 (S. T. Chiu pers. comm.). There is, as well, widespread urchin harvesting for the shell-craft curio trade which is poorly recorded. These local fisheries could represent potential for increased economic enhancement of coastal communities provided there is organization for export markets as for tropical sea cucumber fisheries mentioned above.

ACKNOWLEDGEMENTS

Special thanks to Drs. P. A. Breen, G. S. Jamieson, Z. Kabata, B. F. Keegan and D. L. Pawson for information and assistance with early drafts. The following colleagues very kindly supplied valuable information: J. H. Annala (Wellington), J. Arthur (Galway), M. F. Barker (Dunedin), J. L. Cameron (Vancouver), J. C. Castilla (Santiago), S. T. Chiu (Hong Kong), C. Conand (Brest), P. V. Fankboner (Vancouver), A. J. Harrison (Hobart), V. H. Jacobsen (Charlottenlund), R. J. Miller (Halifax), T. Motokawa (Okinawa), A. M. Muscat (Santa Catalina), D. O. Parker (Long Beach), J.-M. Pèrèz (Monaco), R. J. Perez (Lima), F. H. Pineda Polo (Cali), R. E. Scheibling (Halifax), J. R. Sibert (Noumea), P. Sluczanowski (Adelaide), M. J. Tegner (La Jolla), and B. A. Thomassin (Marseille).

I would greatly appreciate correspondence from colleagues concerning corrections and additions to this review for a future revision.

REFERENCES

Adithiya, L. A. 1969. Bech-de-Mer in Ceylon. Spolia Zeylanica 31: 405-411.

Allain, J. -Y. 1972. La pêche aux oursins dans le monde. Pêche Merit. No. 1133: 625-630. (Can. Fish. Mar. Serv. Transl. Ser. No. 4259-1978.)

Allain, J. -Y. 1975. Structure des populations de Paracentrotus lividus (Lamarck) (Echinodermata, Echinoidea) soumises a la peche sur les cotes nord de Bretagne. Rev. Trav. Inst. Pêches Marit. 39: 171-209. (Can. Fish. Mar. Serv. Transl. Ser. No. 3961-1977.)

Allain, J. -Y. 1978. Age et croissance de Paracentrotus lividus (Lamarck) et de Psammechinus miliaris (Gmelin) des côtes nord de Bretagne (Echinoidea). Cah. Biol. Mar. 19: 11-21.

Anonymous. 1979. Overseas prospects for Australian beche-de-mer. Aust. Fish. 39(9): 45-47.

Anonymous. 1980. Sea urchin catch soars in Chile. Fish. News Int. 19: 10.

Anonymous. 1983. 1981 Yearbook of Fishery Statistics, Vol. 52. FAO, Rome, 357 p.

Anonymous. 1984. 1982 Yearbook of Fishery Statistics, Vol. 54. FAO, Rome, 393 p.

Asada, Y., Y. Hirasawa & F. Nagasaki. 1983. Fishery management in Japan. FAO Fish. Tech. Pap. 238: 26 p.

Bai, M. M. 1980. Monograph on Holothuria (Metriatyla) scabra Jaeger. Mem. Zool. Surv. India 16: 1-75.

Bakus, G. J. 1973. The biology and ecology of tropical holothurians. In O. A. Jones & R. Endean (eds.). Biology and Geology of Coral Reefs, Vol. II, Biology I. p. 325-367. New York: Academic Press.

Beardsley, A. J. 1971. "Beche-de-Mer" fishery for Truk? Comm. Fish. Rev. 33: 64-66.

Bernstein, B. B., S. S. Schroeter & K. H. Mann. 1983. Sea urchin (Strongylocentrotus droebachiensis) aggregating behavior investigated by a subtidal multifactorial experiment. Can. J. Fish. Aquat. Sci. 40: 1975-1986.

Bernstein, B. B. & R. W. Welsford. 1982. An assessment of feasibility of using high-calcium quicklime as an experimental tool for research into kelp bed/sea urchin ecosystems in Nova Scotia. Can. Tech. Rep. Fish. Aquat. Sci. 968: 51 p.

Brewer, W. A. & J. S. Corbin. 1984. Aquaculture development for the Pacific Islands. In C. Gopälakrishnan (ed.). The emerging marine economy of the Pacific, p. 153-175. Boston: Butterworth.

Burkenroad, M. D. 1946. General discussion of problems involved in starfish utilization. Bull. Bingham Oceanogr. Coll. 9: 44-58.

Carleton, C. 1983. Miscellaneous marine products in the South Pacific: A review of available marketing information. 62 p. Commissioned by the Forum Fisheries Agency of the South Pacific, Honiara, Solomon Islands.

Castilla, J.C. & R. Becerra. 1975. The shellfisheries of Chile: An analysis of the statistics 1960-1973. In J.C. Valle (ed.), Proc. Int. Symp. Coastal Upwelling, p. 61-90, Santiago: Editorial Universitaria.

Castilla, J.C. & C.A. Moreno. 1982. Sea urchins and Macrocystis pyrifera: Experimental test of their ecological relations in southern Chile. In J.M. Lawrence (ed.), Proc. Int. Echinoderms Conf., Tampa Bay, p. 257-263. Rotterdam: Balkena.

Chapman, A.R.O. 1981. Stability of sea urchin dominated barren grounds following destructive grazing of kelp in St. Margaret's Bay, eastern Canada. Mar. Biol. 62: 307-311.

Choat, J.H. & D.R. Schiel. 1982. Patterns of distribution and abundance of large brown algae and invertebrate herbivores in subtidal regions of northern New Zealand. J. Exp. Mar. Biol. Ecol. 60: 129-162.

Choe, S. 1963. Biology of the Japanese common sea cucumber, Stichopus japonicus Selenka. Tokyo: Kaibundo, 226 p. (In Japanese).

Conand, C. 1979. Beche-de-Mer in New Caledonia: Weight loss and shrinkage during processing in three species of holothurains. Fish. Newsl. S. Pac. Comm. 19: 14-18.

Conand, C. 1981. Sexual cycle of three commercially important holothurian species (Echinodermata) from the Lagoon of New Caledonia. Bull. Mar. Sci. 31: 523-543.

Conand, C. 1982. Reproductive cycle and biometric relations in a population of Actinopyga echinites (Echinodermata-Holothuroidea) from the Lagoon of New Caledonia, western tropical Pacific. In J.M. Lawrence (ed.), Proc. Int. Echinoderms Conf., Tampa Bay, p. 437-442. Rotterdam: Balkema.

Conand, C. 1983. Methods for studying growth in holothurians (Beche-de-Mer), and preliminary results from a Beche-de-Mer tagging experiment in New Caledonia. Fish. Newsl. S. Pac. Comm. 26: 31-38.

Crean, K. 1977. Some aspects of the Beche-de-Mer industry in Ongtong Java, Solomon Islands. Fish. Newsl. S. Pac. Comm. 15: 36-48.

Daniel, A. & B.P. Halder. 1974. Holothuroidea of the Indian Ocean with remarks on their distribution. J. Mar. Biol. Ass., India 16: 412-436.

Dayton, P.K., R.J. Rosenthal, L.C. Mahen & T. Antezana. 1977. Population structure and foraging biology of the predaceous Chilean asteroid Meyenaster gelatinosus and escape biology of its prey. Mar. Biol. 39: 361-370.

Dayton, P.K. & M.J. Tegner. 1984. The importance of scale in community ecology: A kelp forest example with terrestrial analogs. In P.W. Price, C.N. Slobodchikoff & W.S. Gaud (eds.), A New Ecology: Novel Approaches to Interactive Systems, p. 457-481. New York: J. Wiley and Sons.

Deppe, R. & C.A. Viviani. 1977. La pesqueria artesanal del erizo comestible Loxechinus albus (Molina) (Echinodermata, Echinoidea, Echinidae) en la region de Iquique. Biol. Pesp. Chile 9: 23-41.

Dix, T.G. 1972. Biology of Evechinus chloroticus (Echinoidea: Echinometridae) from different localities, 4. Age, growth, and size. N.Z.J. Mar. Freshwat. Res. 6: 48-68.

Dix, T.G. 1977a. Survey of Tasmanian sea urchin resources. Tasmanian Fish. Res. 21: 1-14.

Dix, T.G. 1977b. Prospects for sea urchin fishery in Tasmania. Aust. Fish. 36(7): 18-22.

Dix, T.G. 1977c. Reproduction in Tasmanian populations of Heliocidaris erythrogramma (Echinodermata: Echinometridae). Aust. J. Mar. Freshwat. Res. 28: 509-520.

Duggins, D.O. 1983. Starfish predation and the creation of mosaic patterns in a kelp-dominated community. Ecology 64: 1610-1619.

Ebert, T.A. 1978. Growth and size of the tropical sea cucumber Holothuria (Halodeima) atra Jager at Enewetak Atoll, Marshall Islands. Pacific Sci. 32: 183-191.

Ebert, T.A. 1983. Recruitment in Echinoderms. In M. Jangoux & J.M. Lawrence (eds.), Echinoderm Studies, I, p. 169-203. Rotterdam: Balkema.

Fuji, A. 1967. Ecological studies on the growth and food consumption of Japanese common littoral sea urchin, Strongylocentrotus intermedius (A. Agassiz). Mem. Fac. Fish. Hokkaido Univ. 15: 83-160.

Gawel, M. 1981. Marine resources development planning for tropical Pacific islands. In E.D. Gomez et al. (eds.), Proc. 4th Int. Coral Reef Symp. Vol. 1, p. 247-251. Quezon City: University of the Philippines.

Gentle, M. 1979. The fisheries biology of Beche-de-Mer. South Pac. Bull. 29: 25-27.

Gentle, M. & C. Conand. 1979. Beche-de-Mer of the tropical Pacific. S. Pac. Comm. Handbook No. 18: 29 p.

Hagen, N.T. 1983. Destructive grazing of kelp beds by sea urchins in Vestfjorden, northern Norway. Sarsia 68: 177-190.

Hancock, D.A. 1969. The Shellfisheries of Chile. Instituda de Fomenta Pesquero, Santiago No. 45: 94 p.

Harding, G.C., K.F. Drinkwater & W.P. Vass. 1983. Factors influencing the size of American lobster (Homarus americanus)

stocks along the Atlantic coast of Nova Scotia, Gulf of St. Lawrence, and Gulf of Maine: A new synthesis. Can. J. Fish. Aquat. Sci. 40: 168-184.
Harriott, V. 1982. Sexual and asexual reproduction of Holothuria atra Jaeger at Heron Island reef, Great Barrier Reef. Aust. Mus. Mem. 16: 53-66.
Hayakawa. Y. & J. Kittaka. 1984. Simulation of feeding behavior of sea urchin Strongylocentrotus nudus. Bull. Jap. Soc. Sci. Fish. 50: 233-240.
Himmelman, J.H., Y. Levergne, F. Axelsen, A. Cardinal & E. Bourget. 1983a. Sea urchins in the St. Lawrence Estuary: Their abundance, size-structure, and suitability for commercial exploitation. Can. J. Fish. Aquat. Sci. 40: 474-486.
Himmelman, J.H., A. Cardinal & E. Bourget. 1983b. Community development following removal of urchins, Strongylocentrotus droebachiensis, from the rocky subtidal zone of the St. Lawrence Estuary, eastern Canada. Oecologia (Berl.) 59: 27-39.
Hornel, J. 1917. The Indian beche-de-mer industry. Madras Fish. Bull. 11: 119-150.
Howell, R.M. & M. Henry. 1977. Dried sea cucumber processing. Rep. Mar. Res. Develop., Truk District, Trust Territory of the Pacific Islands, 17 p.
Incze, L.S. & R.A. Lutz. 1980. Mussel culture: An east coast perspective. In L.S. Incze & R.A. Lutz (eds.), Mussel Culture and Harvest: A North American Perspective, p. 99-140. Amsterdam: Elsevier.
Ishiwata, N., H. Fushimi, K. Maekawa & T. Nanba. 1977. Studies concerning the fishery biology of the sea urchin, Hemicentrotus pulcherrimus (A. Agassiz), in Kaji, Fukui Prefecture I. Sea urchin fishery. La mer (Bull. Soc. franco-jap. d'océanogr.) 15: 146-150.
Johnson, C.R. & K.H. Mann. 1982. Adaptations of Strongylocentrotus droebachiensis for survival on barren grounds in Nova Scotia. In J.M. Lawrence (ed.), Proc. Int. Echinoderms Conf., Tampa Bay, p. 277-283. Rotterdam: Balkema.
Kawamura, K. 1966. Ecological studies and some discussion about the methods of conservation of the sea urchin, Strongylocentrotus nudus on the coast of Urakawa, Hokkaido. Hokkaidoritsu Suissan Shikenjo Hokoku 5: 7-30. (Can. Fish. Mar. Serv. Transl. Ser. No. 3881-1976).
Keats, D.W., D.H. Steele & G.R. South. 1983. Food relations and short term aquaculture potential of the green sea urchin (Strongylocentrotus droebachiensis) in Newfoundland. Mar. Sci. Res. Lab. Memorial Univ. Tech. Rep. 24: 24 p.
Keats, D.W., D.H. Steele & G.R. South. 1984. Depth-dependent reproductive output of the green sea urchin, Strongylocentrotus droebachiensis (O.F. Müller), in relation to the nature and availability of food. J. Exp. Mar. Biol. Ecol. 80: 77-91.
Kinoshita, T. 1955. Essays on cultivation of shallow sea, No. 4, Part 10: Facts about sea urchins. Hokusuishi Geppo 12: 24-26. (Can. Fish. Mar. Serv. Transl. Ser. No. 4289-1978.)
Kinoshita, T. 1958. The present day sea urchin fishery in Hokkaido. Hokusuishi Geppo 15: 505-507. (Can. Fish. Mar. Serv. Transl. Ser. No. 4305-1978.)
Kito, H. S. Kikuchi & N. Uki. 1980. Seaweed as nutrition for seabed marine life technology for artificial marine forests. In Int. Symp. Coastal Pacific Mar. Life, p. 55-66. Bellingham: Western Washington University.
Kramer, D.E. 1980. Outlook for establishing a fishery for the green sea urchin (Strongylocentrotus droebachiensis) in the St. Lawrence River. Can. Ind. Rep. Fish. Aquat. Sci. 114: 15 p.
Krishnan, S. 1968. Histochemical studies on reproductive and nutritional cycles of the holothurian, Holothuria scabra. Mar. Biol. 2: 54-65.
Lawrence, J.M. 1975. On the relationships between marine plants and sea urchins. Oceanogr. Mar. Biol. Ann. Rev. 13: 213-286.
Lawrence, J.M. & J.M. Lane. 1982. The utilization of nutrients by post-metamorphic echinoderms. In M. Jangoux & J.M. Lawrence (eds.), Echinoderm Nutrition, p. 312-371. Rotterdam: Balkema.
Lawrence, J.M. & P.W. Sammarco. 1982. Effects of feeding on the environment: Echinoidea. In M. Jangoux & J.M. Lawrence (eds.), Echinoderm Nutrition, p. 499-519. Rotterdam: Balkema.
Levin, V.S. 1982. Japanese sea cucumber. Vladivostok: U.S.S.R. Academy of Sciences, Vladivostok: 191 p.
Lewis, J.B. 1958. The biology of the tropical sea urchin Tripneustes esculentus Leske in Barbados, British West Indies. Can. J. Zool. 36: 607-621.
Loosanoff, V.L. 1958. Underwater studies of starfish behavior and evaluation of control methods. Bull. Bur. Comm. Fish., Biol. Lab., Milford, Conn. 22: 1-5.
Loosanoff, V.L. 1961. Biology and methods of controlling the starfish, Asterias forbesi (Desor). U.S. Fish Wildl. Serv., Fish. Leaflet No. 520, 11 p.
Mann, K.H. 1977. Destruction of kelp-beds by sea-urchins: A cyclical phenomenon or irreversible degradation? Helgoländer

Wiss. Meeresunters 30: 455-467.
Mann, K.H. 1982. Kelp, sea urchins and predators: A review of strong interactions in rocky subtidal systems of eastern Canada, 1970-1980, Neth. J. Sea Res. 18: 414-423.
Mann, K.H., J.L.C. Wright, B.E. Welsford & E. Hatfield. 1984. Responses of the sea urchin Strongylocentrotus droebachiensis (O.F. Müller) to water-borne stimuli from potential predators and potential food algae. J. Exp. Mar. Biol. Ecol. 79: 233-244.
Massin, C. 1982. Effects of feeding on the environment: Holothuroidea. In M. Jangoux & J.M. Lawrence (eds.), Echinoderm Nutrition, p. 493-497. Rotterdam: Balkema.
Matsui, I. 1966. The propagation of sea urchins. Nippon Suisan Shigen Hogo Kyokai Booklet 12: 105 p. (Fish. Res. Bd. Canada Transl. Ser. No. 1063-1968.)
Miller, R.J. & C.A. Bishop. 1973. A sea urchin fishery for Atlantic Canada. Fish. Mar. Serv. Biol. Station, St. John's, Nfld., Circular 19: 1-9.
Miller, R.J. & A.G. Colodey. 1983. Widespread mass mortalities of the green sea urchin in Nova Scotia, Canada. Mar. Biol. 73: 263-267.
Mitsukuri, K. 1897. On changes which are found with advancing age in the calcareous deposits of Stichopus japonicus, Selenka. Annot. Zool. Jap. 1: 31-42.
Mitsukuri, K. 1903. Notes on the habits and life-history of Stichopus japonicus Selenka. Annot. Zool. Jap. 5: 1-21.
Moore, D.S. & R.J. Miller. 1983. Recovery of macroalgae following widespread sea urchin mortality with a description of the nearshore hard-bottom habitat on the Atlantic coast of Nova Scotia. Can. Tech. Rep. Fish. Aquat. Sci. 1230: 94 p.
Moreno, C.A. & J.P. Sutherland. 1982. Physical and biological processes in a Macrocystis pyrifera community near Valdivia, Chile. Oecologia (Berl.) 55: 1-6.
Moriarty, D.J.W. 1982. Feeding of Holothuria atra and Stichopus chlorontus on bacteria, organic carbon and organic nitrogen in sediments of the Great Barrier Reef. Aust. J. Mar. Freshwat. Res. 33: 255-263.
Mottet, M.G. 1976a. The fishery biology and market preparation of sea cucumbers. Washington Dept. Fish. Tech. Rept. 22: 57 p.
Mottet, M.G. 1976b. The fishery biology of sea urchins in the Family Strongylocentrotidae. Washington Dept. Fish. Tech. Rep. 20: 66 p.
Mottet, M.G. 1980. Factors leading to the success of Japanese aquaculture with an emphasis on northern Japan. Washington Dept. Fish. Tech. Rep. 52: 83 p.
Muscat, A.M. 1983. Population dynamics and the effect on the infauna of the deposit-feeding holothurian Parastichopus parvimensus (Clark). Ph.D. Dissertation, University of Southern California, 328 p.
Nagai, Y. & K. Kaneko. 1975. Culture experiments on the sea urchin Strongylocentrotus pulcherrimus fed on artificial diet. Mar. Biol. 29: 105-108.
Nichols, D. 1978. Can we save the sea urchin? Diver (October) 26-27.
Nichols, D. 1979. A nationwide survey of the British sea urchin, Echinus esculentus. In J.C. Gamble & J.D. George (eds.), Progress in Underwater Science, Vol. 4 (N.S.), p. 161-187. Plymouth: Pentech Press.
Nichols, D. 1981. The Cornish sea-urchin fishery. Cornish Stud. 9: 5-18.
Noro, T., T. Masaki & H. Akioka. 1983. Sublittoral distribution and reproductive periodicity of crustose coralline algae (Rhodophyta, Cryptonemides) in southern Hokkaido, Japan. Bull. Fac. Fish., Hokkaido Univ. 34: 1-10.
North, W.J. 1976. Aquacultural techniques for creating and restoring beds of giant kelp, Macrocystis sp. J. Fish. Res. Board Can. 33: 1015-1023.
O'Connor, C. 1975. A study of the reproductive periodicities and population dynamics of some New South Wales sea urchins with a view to open-sea culture. In G. Persoone & E. Jaspers (eds.), Proc. 10th European Mar. Biol. Symp., Ostend, Vol. 1, p. 283-301. Wetteren: Universa Press.
O'Connor, C. & J.C. Mulley. 1977. Temperature effects on periodicity and embryology, with observations on the population genetics, of the aquacultural echinoid Heliocidaris tuberculata. Aquaculture 12: 99-114.
O'Connor, C., G. Riley & D. Bloom. 1976. Reproductive periodicities of the echinoids of the Solitary Islands in the light of some ecological variables II. Superficial and histological changes in the gonads of Centrostephanus rodgersii (Clark), Phyllacanthus parvispinus (Tenison-Woods), Heliocidaris tuberculata (Clark), and Tripneustes gratilla (Linneus), and their relevance to aquaculture Thalassia Jugo. 12: 245-267.
O'Neill, S.M., A.M. Sutterlin & D. Aggett. 1983. The effects of size-selective feeding by starfish (Asterias vulgaris) on the production of mussels (Mytilus edulis) cultured on nets. Aquaculture 35: 211-220.
Panning, A. 1944. Die Trepangfischerei. Mitt. Hamburg Zool. Mus. Inst. 49: 2-76.
Parrish, P. 1978. Processing guidelines for beche-de-mer. Aust. Fish. 10(10): 26-27.

Pringle, J.D., G.J. Sharp & J.F. Caddy (eds.). 1980. Proceedings of the workshop on the relationship between sea urchin grazing and commercial plant/animal harvesting. Can. Tech. Rep. Fish. Aquat. Sci. 954: 273 p.

Pringle, J.D., G.J. Sharp & J.F. Caddy. 1982. Interactions in kelp bed ecosystems in the northwest Atlantic: Review of a workshop. Can. Spec. Publ. Fish. Aquat. Sci. 59: 108-115.

Rakov, V.A. 1982. Growth rates and duration of life in the sea cucumber Stichopus japonicus in the Posset Bay. Biol. Morya No. 4: 52-54. (Can. Transl. Fish. Aquat. Sci. No. 5089-1984).

Régis, M.-B. 1980. Etude des possibilités d'élevage des oursins reguliers en fonction de la valeur de certains indices physiologiques. Oceanologica Acta 3: 7-15.

Sachithananthan, K. 1972. South Pacific islands beche-de-mer fishery. FAO, Rome DP. RAS. 69. 102.11: 32 p.

Sato, S. 1980. A general concept of systems engineering in coastal areas. In Int. Symp. Coastal Pacific Mar. Life, p. 31-43. Bellingham: Western Washington University.

Scattergood, L.W. 1963. The sea urchin fishery. U.S. Fish Wildl. Serv. Fish. Leaflet No. 511: 1-5.

Scheibling, R.E. & R.L. Stephenson. 1984. Mass mortality of Strongylocentrotus droebachiensis (Echinodermata: Echinoidea) off Nova Scotia, Canada. Mar. Biol. 78: 153-164.

Shepherd, S.A. 1973. Competition between sea urchin and abalone. Aust. Fish. 32 (6): 4-7.

Shuxu, X. & G. Gongchao. 1981. Experiments on southward transplantation and artificial breeding of the sea cucumber (Stichopus japonicus). J. Fish. China 5 (2): 147-153.

Sloan, N.A. 1985. World jellyfish and tunicate fisheries and the N.E. Pacific echinoderm fishery. In G.S. Jamieson & N. Bourne (eds.), North Pacific workshop on recent advances in stock assessment and management of invertebrates. Can. Spec. Publ. Fish. Aquat. Sci. (in press).

Sloan, N.A. & S.M.C. Robinson. 1983. Winter feeding by asteroids on a subtidal sandbed in British Columbia. Ophelia 22: 125-140.

Sutterlin, A., D. Aggett, C. Couturier, R. Scaplen & D. Idler. 1981. Mussel culture in Newfoundland waters. Mar. Sci. Res. Lab. Memorial Univ. Tech. Rep. 23: 82 p.

Sutterlin, A.M. & S. Waddy. 1975. Harvesting, processing, and composition of sea cucumber meal as a possible substitution for herring meal in trout diets. Can. Fish. Mar. Serv. Tech. Rep. 531: 8 p.

Tanaka, Y. 1958a. Seasonal changes occurring in the gonad of Stichopus japonicus. Bull. Fac. Fish. Hokkaido Univ. 9: 29-36.

Tanaka, Y. 1958b. Feeding and digestive processes of Stichopus japonicus. Bull. Fac. Fish. Hokkaido Univ. 9: 14-28.

Tauti, M. & I. Matsumoto. 1954. On the propagation of Stichopus japonicus Selenka around Iejima Island in Hyogo Prefecture. Bull. Jap. Soc. Sci. Fish. 20: 386-387.

Tegner, M.J. 1980. Multispecies considerations of resource management in southern California kelp beds. Can. Tech. Rep. Fish. Aquat. Sci. 954: 125-143.

Uchiyama, H., K. Kakuda & Y. Uchida. 1984. Partial freezing as a new method for long period preservation of sea urchin gonad. Bull. Jap. Soc. Sci. Fish. 50: 839-843.

Ward, J. 1975. Sea urchins - a new Victoria fishery? Aust. Fish. 34 (1): 16.

Wharton, W.G. & K.H. Mann. 1981. Relationship between destructive grazing by the sea urchin, Strongylocentrotus droebachiensis, and the abundance of American lobster, Homarus americanus, on the Atlantic coast of Nova Scotia. Can. J. Fish. Aquat. Sci. 38: 1339-1349.

Wilson, J.R. & A.H. Gorham. 1982. Alaska underutilized species, Vol. III: Sea Urchin. Alaska Sea Grant Rep. 82-7: 92 p.

Proceedings of the Fifth International Echinoderm Conference / Galway / 24-29 September 1984

Evolutionary history of Jurassic Collyritidae (Echinoidea, Disasteroidea) in the Paris Basin (France)

J.THIERRY
Institut des Sciences de la Terre, Dijon, France

ABSTRACT: The evolutionary history of the Collyritidae of the Paris Basin, from the Upper Bajocian to the Middle Oxfordian (Middle and Upper Jurassic), is considered with respect to several data: changes in body shape and anatomy in space and time; sediment characteristics; stratigraphical and geographical distribution of populations; paleogeography and its variations in time; relative importance of the taxon in relation to other fossil echinoid groups. The morphological variations appear closely related to the grain size of the sediment and its carbonate content. Furthermore, this echinoid family is restricted to shallow but open and quiet environments; it is totally missing from very shallow low energy and sheltered environments and from high energy ones as well as from quiet and deeper environments. The settling occurs in several stages correlated with transgressive periods. Except in a few cases, collyritids occur to the exclusion of other regular or irregular echinoids.

A INTRODUCTION

Since the fundamental monographs of Cotteau (1849-56, 1867-74), Cotteau and Triger (1855-69), Valette (1906-07) and Mercier (1932) some regions of the Paris Basin have been known to contain a rich and diversified echinoid fauna. Among the various groups which are present, the Holectypoidea, Cassiduloidea and Disasteroidea are particularly frequent. Amongst the Disasteroidea, the family Collyritidea shows several interesting features, which are the subject of this paper. During the period under consideration, from the Upper Bajocian where they appeared in small numbers to the Middle-Upper Oxfordian boundary where they were driven out of the basin, a time space of a little more than 25 million years, the story of the Collyritidae may be considered under the following five headings : stratigraphical and geographic distribution of populations; characteristics of the sediments inhabited; palaeogeography and its variation in time; body shape and anatomy changes in space and time and their relationship to environmental parameters; relative importance of the Collyritidae within the total echinoid fauna.

B MATERIAL AND METHODS

Slighty more than 2000 specimens, distributed between about 65 samples, have been collected in classical or new sections throughout the Paris Basin. All the associated macrofauna (ammonites, brachiopods, bivalves, gastropods, etc... and of course other echinoids) has also been collected to give an idea of the fossil assemblages.

Among these taxa, ammonites play a special part; through them, each sample can be accurately located in time in terms of the standard Jurassic ammonite zonal scheme. Thus, changes of body shape of Collyritids in time can be assessed with precision.

Sedimentological parameters of the rocks from each of the localities sampled have been measured. The grain size and the ratio between silt-clay and carbonates appear to be important data for explaining variability of the test shape and the collyritid settlement. Sedimentological data are, of course, closely related to the palaeogeography of this west European region. Thus, geographical changes in test shape can be interpreted according to general environments.

In addition, the variability of each sample (including ontogenetic variation), and comparison between different samples, have been made by means of classical mor-

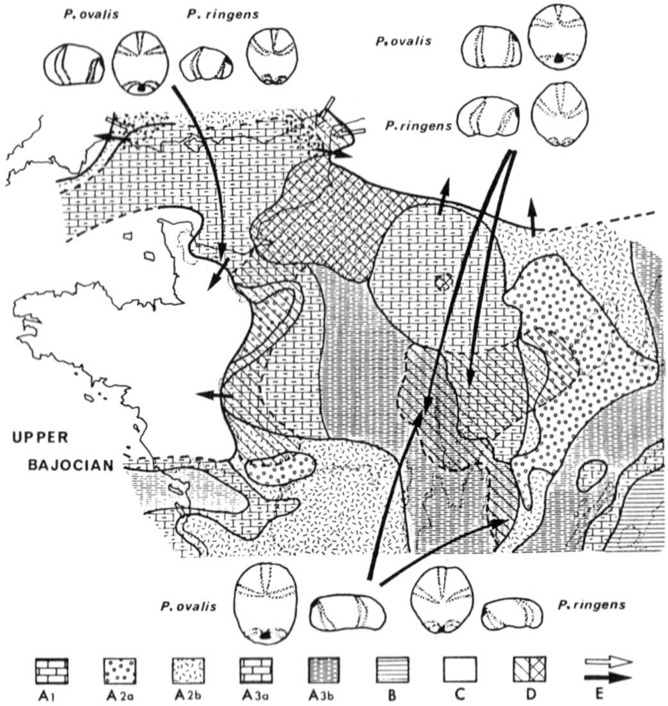

Fig. 1. Paleogeography of the Paris Basin and distribution of Pygorhytis ringens and Pygomalus ovalis in Upper Bajocian.
A- Carbonatic platforms, 1- Low energy and very shallow sheltered internal platforms; 2- Very high energy and shallow barriers or platform; 2b- Medium energy and shallow platform margins ; 3- Low energy and deep platforms: a- with carbonates predominant, b- with fine and clayish sediments.
B- Deep oceanic basin.
C- Supposed emerged areas.
D- Areas with reduced (oblique lines) or without (oblique cross lines), sediments.
E- White arrows with stipples: sands or silts; black arrows: transgression.

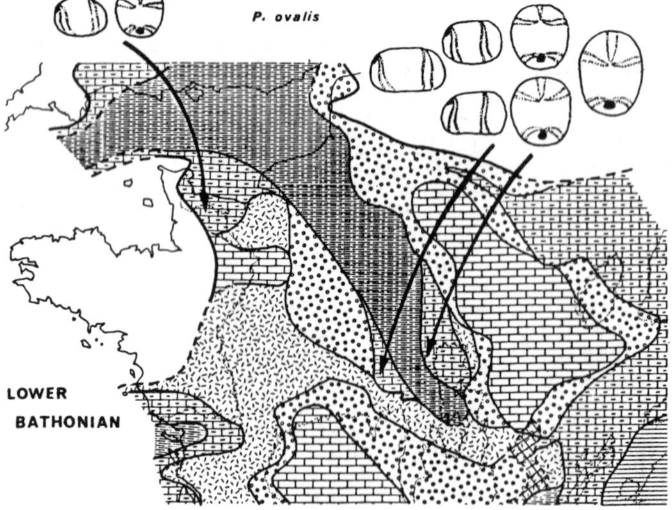

Fig. 2. Paleogeography of the Paris Basin and distribution of Pygomalus ovalis in Lower Bathonian (same legend as in fig. 1). Note the importance of superficial platforms.

phological studies, supported by statistical methods (multivariate analysis). Preliminary results (Thierry 1974a, 1974b) are here strongly corroborated because of the large numbers of specimens collected.

C SEDIMENTARY ENVIRONMENTS AND PALAEOGEOGRAPHY OF THE PARIS BASIN DURING THE MIDDLE AND UPPER JURASSIC

1. Relevent sedimentary and environmental conditions

In the Paris Basin, the Collyritidae almost always occur in medium or fine-grained carbonate rocks. The carbonates are more often wackestones or grainstones, particularly in the Bajocian-Bathonian-Callovian and Oxfordian; they are biomicrites or biosparites resulting from the lithification of an original calcareous medium or fine sand (grains are between 250µm dans 700µm). Where the matrix is fairly abundant, it is usually either micritic or slightly clayish, and occasionally is ferruginous. In rare cases, oolites (oobiomicrite) or quartz sand (sandy biomicrite) are present; their size reaches 1mm or more. Except when oolites or quartz are present, the grains are fairly well sorted. The carbonate content is always high, between 80 to 90% from the Bajocian to the Callovian; in the Oxfordian it is sometimes below 70% of the total sediment.

All these sedimentary data show that the Collyritidae were confined to shallow, but open, quiet environments, where carbonate sediment (bioclasts, lithoclasts, pellets, rare oolites and micritic matrix) outweighed fine or coarse clastic material (fine quartz, silts and clays). These enronments occured in moderately shallow external platforms; they extend either offshore of an emergent area or around the margins of bioclastic, oolitic or coral reef barrier.

It is notable that the Collyritidae are totally excluded from very shallow low energy and sheltered environments (internal platforms confined regions and lagoons with abundant fine or muddy carbonates) and from high energy ones (bioclastic, oolitic or coral-reef barriers with medium- to coarse-grained sediments). They are also very rare or absent in very quiet and deeper environments (external platforms with abundant detrital mud and fine sand). All these facies are extensively developed in the Paris Basin during the period under consideration.

2. Major Palaeogeographic features

The Paris Basin is considered to have been a marine epicontinental basin; it opened to the south-east into the Tethyan realm through the Jura Mountains and the Alps. In the east it led into the Swabian basin and to the south-west and north-west into the Atlantic realm through the Channel and the Poitou trough. It is bounded to the north by the London-Brabant-Ardennes high, to the east by the Armorican land and to the south by the Massif Central high.

From Upper Bajocian to Upper Oxfordian times the major palaeogeographic evolution of the basin involved two transgressive periods separated by, and also followed by, regressions.

D THE ADVENT AND EVOLUTIONARY HISTORY OF COLLYRITIDS IN THE PARIS BASIN

1. From Upper Bajocian to Middle Bathonian: advent

Pyrorhytis ringens and Pygomalus ovalis appear in the Upper Bajocian (Parkinsoni zone) and are known only in Normandy and Burgundy (Nièvre and Saône-et-Loire). They are found together in medium grained marly limestones (fig. 1). They are rare and the samples studied are small. On the whole they are small (20 or 25mm), round or slightly oval and gibbous. The distance between the bivium and the trivium is important relative to the body lenght; the peristome is central or subcentral; the periproct is supra-marginal and joined with the bivium.

No very important morphological or environmental changes occur in the Lower and Middle Bathonian (fig. 2 and 3) except for a slight increase of the body length in P. ovalis. P. ringens vanishes in the Lower Bathonian.

In the upper part of the Middle Bathonian, because of the extensive spreading of unfavourable environments (high energy barriers or internal platforms), the Collyritidae disappeared completely. Certainly, they existed elsewhere in Europe, but not within the Paris Basin. We must point out that rare specimens of Holectypus depressus may sometimes be found with Bajocian collyritids.

2. From Upper Bathonian to Middle Callovian: large-scale colonisation and important morphological changes

Pygomalus analis is widely distributed during the Upper Bathonian (fig. 4) and Collyrites elliptica during the Callovian (fig. 5-6) had an even greater geographical range. The distinction between these two

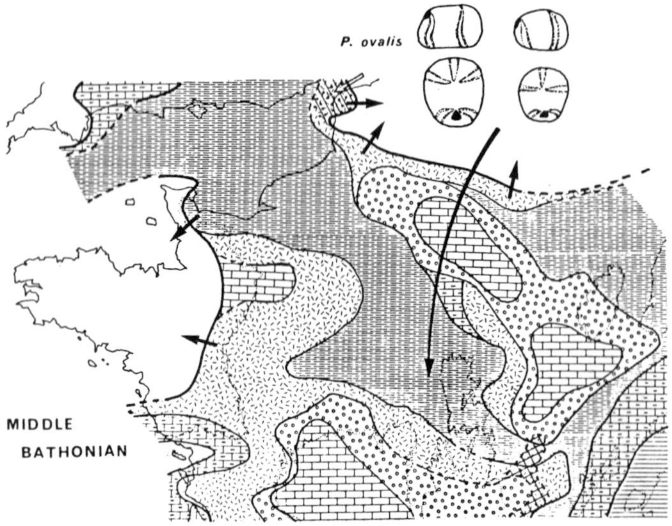

Fig. 3. Paleogeography of the Paris Basin and distribution of Pygomalus ovalis in Middle Bathonian (same legend as in fig. 1). Note the maximum extension of superficial platforms.

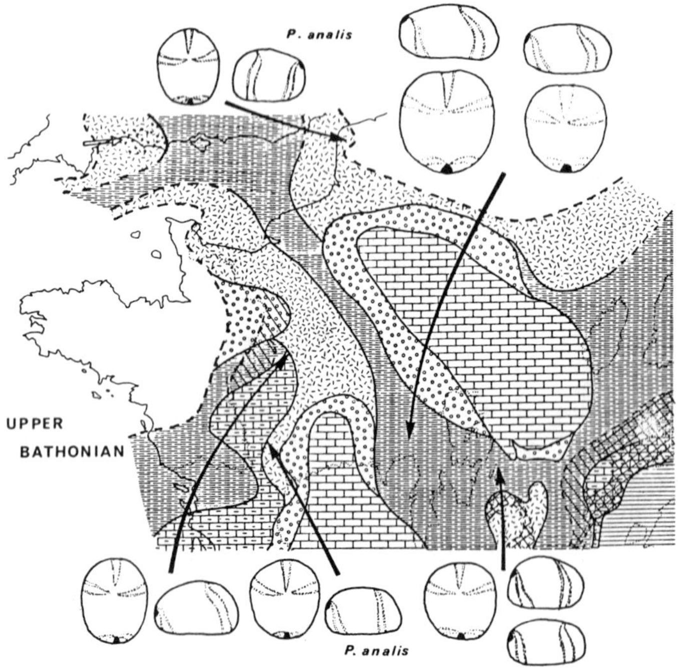

Fig. 4. Paleogeography of the Paris basin and distribution of Pygomalus analis in Upper Bathonian (same legend as in fig. 1). Note the beginning of the restriction of the superficial platforms and the spreading of Collyritidae.

taxa is very easy: the bivium, joined to the periproct in the former, separates from it in the latter at the bathonian-Callovian boundary. This important morphological break is accompanied by significant changes of the body shape discussed below:

- The antero-posterior length of the test increases: from about 30 or 35mm in Upper Bathonian (Retrocostatum and Discus zones) P. analis to 40 or 50mm in Lower-Middle Callovian C. elliptica (from the Macrocephalus to Coronatum zone); gigantic specimens in the Middle Callovian can have a length between 70 and 90mm.
- The bivium and trivium become closer relative to the total length of the test. During this shifting, the bivium separates

from the periproct. This has a major consequence upon the posterior position of the apical plates (Jesionek-Szymanska, 1956, 1959, 1962, 1963).

- The periproct moves backward into the posterior ambulacrum; from a supra-marginal position in P. analis, it becomes marginal in C. elliptica.

- The peristome shifts gradually anteriorly; from a subcentral position in the samples of P. analis, it moves towards the anterior margin in C. elliptica.

- The test, fairly round in the oldest collyritids such as P. ovalis, becomes more elongate in P. analis and C. elliptica, and the widest part of the test is found progressively nearer the anterior end.

- The height of the test is variable in all samples studied, but this variation is not distributed at random. The prevalence of gibbous tests is strongly related to coarsening of the sediment at some places in the Paris Basin. For instance, this occurs in oobiomicritic Lower and Middle Callovian sediments in North and South Burgundy and in sandy biomicritic sediments of Middle Callovian age in the western part of the basin (fig. 6). Some authors have created new species for this particular morphology, e.g. C. dorsalis and C. faba. I think that it is simply an ecomorph or variant.

All these morphological changes of the test must be considered with respect to the important palaeogeographic and sedimentary changes that occured in the Paris Basin from the Upper Bathonian to the Middle Callovian. This time corresponds to a transgressive period when large and superficial low or high energy carbonate platforms were gradually replaced by deeper and low to middle energy platforms. At the same time, coarse to medium grain bioclastic limestone gave way to fine-grained and marly or muddy limestone except in some places cited above. All these environmental changes seem to have been favourable to the Collyritidae because they attained their maximum distribution during this time.

A general increase in size is common in lineages; one interprets this as an adaptation of the organisms to a favourable environment. The progressive migration of the periproct towards the posterior margin and the gradual forward movement of the peristome are pronounded trends during the evolution of irregular echinoids; they must be correlated with the development of a stongly unidirectional movement and a greater adaptation to sediment swallowing (Smith, 1984). In fact, these changes coincide in the Paris Basin with the maximum spread of fine sediment (silty carbonate of wackestone to mudstone type) on quiet and moderately deep external platforms. The relationship between organism and sediment is seen also in the tendency for the collyritids to develop either flattened or gibbous tests depending on the grain size of the sediment. This confirms the observations made on the living species Echinocardium cordatum (Higgins 1972, 1974, 1976; Laurin, Marchand and Thierry, 1979).

As a last comment, we can point out that in all sections, Collyritidae are accompanied by an abundant and diversified benthic (bivalves, brachiopods, gastropods) and necto-benthic or nectonic (ammonites, belemnites) fauna, which corroborates the largely open water nature of the environments. For instance, soft bottom bivalves are common (mud stickers, edgewise and cup shaped recliners). Except in very few cases where collyritids occur there are no other regular or irregular echinoids belonging to the Cidaridae, Holectypidae, Nucleolitidae, Clypeidae or Disasteridae. One exception to this pattern is the occurence of a few specimens of Echinobrissus clunicularis and Hyboclypeus gibberulus together with Collyrites elliptica in medium to coarse grain size muddy limestones in Nievre and Sarthe. Conversely, in Boulonnais, collyritids are known only from less than five specimens in the whole of the Bathonian and Callovian, while other irregular echinoids can be collected by the hundreds.

3 FROM UPPER CALLOVIAN TO MIDDLE-UPPER OXFORDIAN: THE EXCLUSION OF THE COLLYRITIDS

The major morphological trends and environmental changes described above carry on during this third period. Collyrites (or Cardiopelta?) bicordata and C. acuta show the maximum shifting forward of the peristome and downward movement of the periproct; the antero-posterior lengthening stops and in some smaller adult specimens is, in fact, reduced. On the whole, flattened specimens are more common that gibbous ones, except in some rare outcrops in the south of the Paris Basin (C. conica).

But two, new and important morphological features appear as early as the Upper Callovian (Athleta and Lamberti zone). First, a notch and a groove become visible on the anterior margin and ambulacrum. This is more pronounced in the Middle-Upper Oxfordian (Transversarium and Bifurcatus zones). At the same time, the peristome is

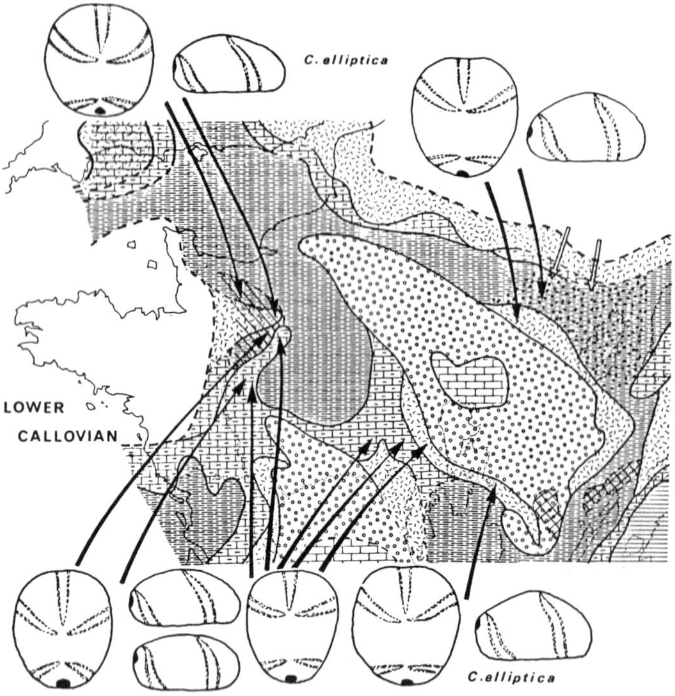

Fig. 5. Paleogeography of the Paris Basin and distribution of Collyrites elliptica in Lower Callovian (same legend as in fig. 1). Note the restriction of superficial platforms and the maximum spreading of Collyritidae.

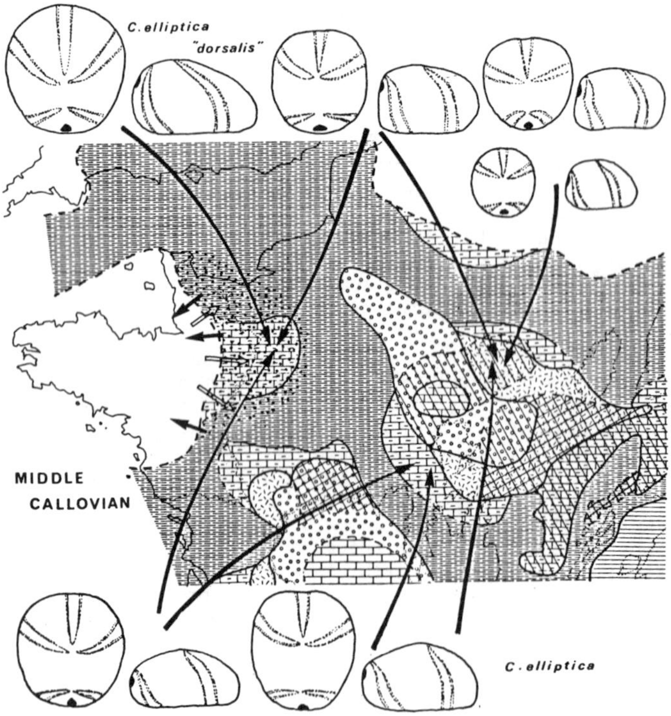

Fig. 6. Paleogeography of the Paris Basin and distribution of Collyrites elliptica in Middle Callovian (same legend as in fig. 1; same remark as in fig. 5).

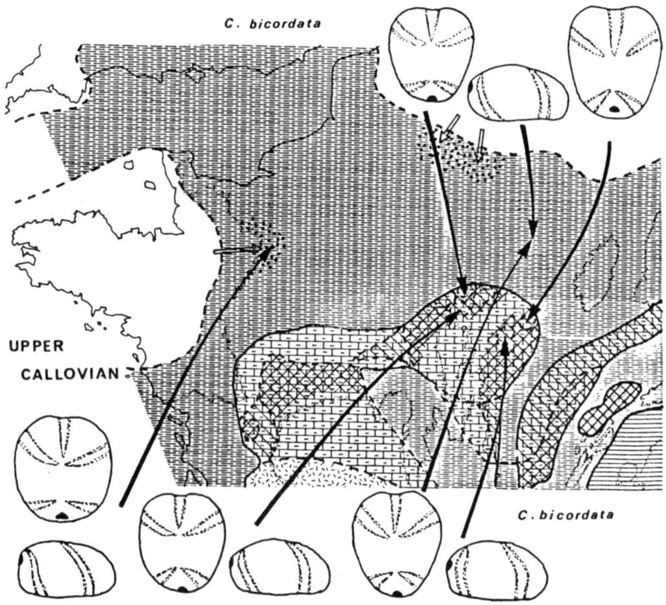

Fig. 7. Paleogeography of the Paris basin and distribution of Collyrites (or Cardiopelta?) bicordata in Upper Callovian (same legend as in fig. 1). Note the quite total disappearance of superficial platforms and the begining of the restriction of the Collyritidae distribution.

greatly shifted forward and is connected with the anterior notch through a very slight inferior groove. Second, the test becomes heart-shaped; some late Upper Callovian and Middle Oxfordian specimens may have a narrow (or acuminated) back part. They are the forerunners of the Upper Oxfordian species C. capistrata, which is not studied here because it is unknown in the Paris Basin.

All these species are found in very fine grained carbonate sediment or in siltstones and claystones (mudstones). Once more, morphological changes and environmental conditions are highly correlated. Here we can say that the changes certainly correspond with the progressive adoption of a more fully infaunal mode of life that the first Collyritidae. One can point out also that the anterior groove is deeper on flattened specimens, in fine or clay-rich sediments; the contrary was observed in living Echinocardium cordatum (Higgins, op. cit.).

If we consider now the geographical distribution of collyritids in the Upper Callovian (fig. 7, 8, 9) we can observe that they are more and more restricted to the south of the Paris basin. Only this region retained favourable conditions. In the north, a new development of high energy, superficial platforms drove out the collyritids. In the Upper Oxfordian they completely disappeared from the Paris Basin except for its south-east margin (Jura mountains) where rare specimens can be collected with the first Disaster.

CONCLUSIONS

The present work has been carried out on only a restricted set of morphological features; furthermore only the Collyritidae of the Paris basin have been considered. Nonetheless, one must try to emphasize some results.

The systematics of Middle and Upper Jurassic collyritids must be revised (Mintz 1966, 1968). Even if there is no doubt about the distinction between Pygorhytis and Pygomalus or Pygomalus and Collyrites and between species in these genera, there is no obvious morphological break between Collyrites and Cardiopelta. The validity of the latter genus is doubtful; the anagenetic evolutionary patterns observed do not permit easy distinction of each taxon (Thierry 1974a, b; 1982). Conversely, the Collyritidae can be used in biostratigraphy (Thierry, 1984), if an accurate study of their variability is made.

The second important point is about morphological adaptations versus environmental changes; here,only those affecting overall body shape are considered. We can prove that Collyritidae of Upper Callovian and

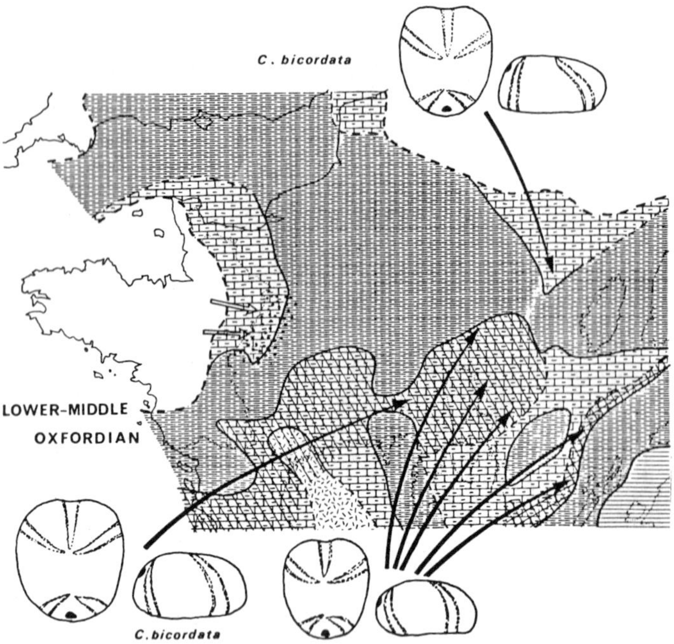

Fig. 8. Paleogeography of the Paris Basin and distribution of Collyrites (or Cardiopelta ?) bicordata in Lower-Middle Oxfordian (same legend as in fig. 1). Note the only existence of Collyritidae in the East part of the Basin.

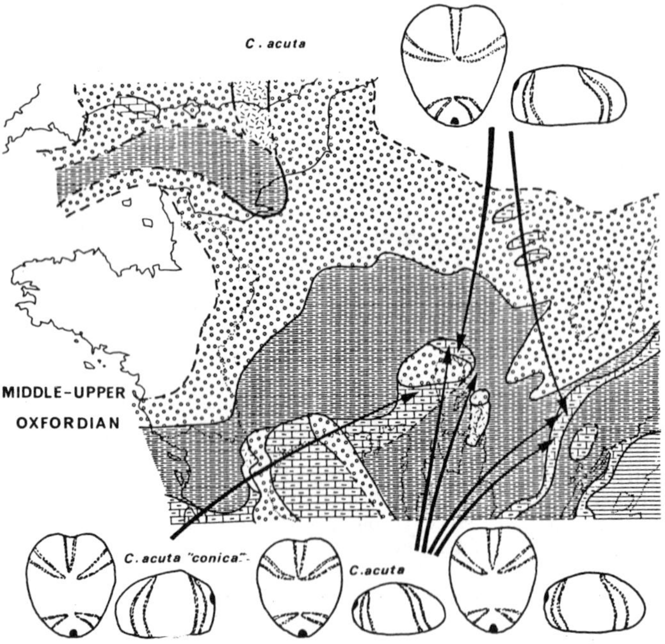

Fig. 9. Paleogeography of the Paris Basin and distribution of (Cardiopelta acuta) in Middle Upper Oxfordian (same legend as infig. 1). Note the restricted area with Collyritidae and the new development of superficial platforms on quite the whole Paris basin.

Oxfordian are better adapted to living infaunally in fine sediments than their predecessors. The development of a sunken frontal ambulacrum, correlated with the shifting forward of the peristome and the backward migration of the periproct, coincides with the extention of the areas of fine-grained sedimentation. These adaptations still exposed by authors (Kier, 1974; Smith, 1984) are also marked in other morphological features as plastron, tube feets and spines (Jesionek-Szywanska, 1963; Smith, 1984). They must be the subject of another paper.

REFERENCES

Cotteau, G. 1949-56. Etudes sur les Echinides fossiles du département de l'Yonne. 1. Terrain jurassique.

Cotteau, G. 1967-74. Description des animaux invertébrés, terrain jurassique. Paléontologie Française. 9. Echinides irréguliers.

Cotteau, G. & Triger, B. 1855-69. Echinides du département de la Sarthe considérés au point de vue zoologique et stratigraphique. Paris.

Higgins, R.C. 1972. Localised variation in the population structure of Echinocardium cordatum (Pennant)(Echinoidea, Spatangoidea) in Tasman Bay. New Zealand Ocean Inst. Rec. 1 : 127-134.

Higgins, R.C. 1974. Specific status of Echinocardium cordatum, E. australe and E. zealandicum (Echinodea: Spatangoida) around New Zealand, with comments on the relation of morphological variation to environment. Journal Zool. 173, 4: 451-475.

Higgins, R.C. 1976. Observations on the morphologiy of Echinocardium cordatum (Echinoidea, Spatangoidea) from diverse geographical areas. Journ. Zool. 177, 4: 507-516.

Jesionek-Szymanska, Obserwacje had morfologia Pygomalus analis (Agassiz)(Echinida Disasteridae). Acta Paleont. Polonica. 1, 1: 4-68.

Jesionek-Szymanska, W. 1959. Remarks on the structure of the apical system of irregular echinoids. Acta Paleont. Polon. 3: 339-351. 3.

Jesionek-Szymanska, W. 1962. Contribution à l'étude du genre "Orbignyana" Ebray, 1860 (Echinida, Disasteridae Gras, 1848): Bull. Mus. hist. Nat. 34, 2: 184-187.

Jesionek-Szymanska, 1963. Echinides irreguliers du Dogger en Pologne. Acta Paleont. Polonica. 3, 3: 293-414.

Kier, P.M. 1974. Evolutionary trends and their functional significance in the post-Paleozoic echinoids. Jour. Pal., 48, 3: part. I of II.

Laurin, B., Marchand, D., Thierry, J. 1979. Variations morphologiques du test chez Echinocardium cordatum (Pennant): étude qualitative et quantitative de cinq échantillons de Bretagne et de Normandie. Bull. Soc. Géol. Norm. et Amis du Museum du Havre. 65, 4: 89-104.

Mercier, J. 1932. Etude sur les Echinides du Bathonien de la bordure occidentale du Bassin de paris. Mém. Soc. Linn. Normandie. 2.

Mintz, L.W. 1966. The origins, phylogeny, descendants of the Echinoid family Disasteridae, A. Gras, 1848. Dissertation Doctor Philosophy, University of California, Berkeley.

Mintz, L.W. 1968. Echinoids of the Mesozoic families Collyritidae, d'Orbigny 1853 and Disasteridae, A. Gras, 1848, Journ. Pal. 42, 5: p. 1272-1288.

Smith, A. 1984. Echinoid Palaeobiology. Special Topics in Palaeontology. G. Allen and Unwin Ltd. London.

Thierry, J. 1974a. Etude quantitative et dynamique des Collyritidae (Echinoidea) du Jurassique de Bourgogne. Bull. Soc. Geol. Fr. 7, 16: 385-395.

Thierry, J. 1974b. Les Collyritidae (Echinoidea, Atelostomata, Disasteroidea) du Jurassique de Bourgogne (Evolution, Taxonomie, Biostratigraphie). 2e R.A.S.T., Pont à Mousson, Nancy.

Thierry, J. 1982. Coupure, continuité et discontinuité en paléontologie: un exemple chez les Collyritidae, oursins irréguliers du Jurassique. 9e R.A.S.T., Paris.

Thierry, J. 1984. Middle and Upper Jurassic biozonation Collyritidae (Irregular Echinoids) in the Paris Basin (France). Internat. Symp. on Jurassic Stratigraphy, Erlangen, September 1984 (in press).

Valette, A. 1906-07. Echinides fossiles de l'Yonne. Bull. Soc. Sc. Hist. Nat. de l'Yonne. 60: 9-191.

Proceedings of the Fifth International Echinoderm Conference / Galway / 24-29 September 1984

Gametogenic strategies in deep-sea echinoids and holothurians from the N.E. Atlantic

P.A.TYLER & A.MUIRHEAD
Department of Oceanography, University College, Singleton Park, Swansea, UK

J.D.GAGE
Scottish Marine Biological Association, Oban, Argyll, Scotland, UK

D.S.M.BILLETT
Institute of Oceanographic Sciences, Wormley, Godalming, Surrey, UK

ABSTRACT: The gametogenic biology of ten echinoids and eight holothurians from deeper than 1000m in the N.E. Atlantic have been examined. In the echinoid species Echinus affinis, E. acutus norvegicus, E. alexandri and Cidaris cidaris a maximum egg size of 110µm diameter is produced, suggestive of planktotrophic development. Samples from a time-series suggest a distinct seasonality of reproduction in Echinus affinis and from the limited evidence available we may infer a similar strategy for the other species. In the five echinothurid species examined, and in Poriocidaris purpurata an egg from 1100µm to 1500µm diameter is produced. In these species nutritive tissue is well developed. Phagocytes breakdown unspawned oocytes and may translocate the products to developing oocytes. Of the holothurians an intermediate egg size of c. 350µm is found in Laetmogone violacea, Peniagone spp., Ypsilothuria bitentaculata attenuata and c. 200µm in Cherbonniera utriculus. Alternatively large eggs, in excess of 650µm, are found in Benthogone rosea, Deima validum and Oneirophanta mutabilis. In the last two species we have found no evidence of spermatogenesis although half the population are undergoing oogenesis, and in the remaining half no evidence of gametogenesis has been found. It is inferred that these specimens are males. The selection of reproductive strategies within each of these species is discussed within the context of the ecology of the deep-sea.

1 INTRODUCTION

Echinoderms form an important element of the deep sea megabenthos in the N.E. Atlantic. An unique opportunity to study reproduction in this deep-sea group has arisen from the long-term sampling programme undertaken by the Scottish Marine Biological Association, the Institute of Oceanographic Sciences and University College, Swansea.

A study of reproduction of deep sea species of asteroids and ophiuroids has shown that there are a variety of gametogenic strategies within these groups varying from the production of larger eggs and continuous development to the production of many small eggs and distinct interspecific synchrony in the seasonality of gametogenic development, albeit in a limited number of species (Tyler, Pain & Gage, 1982; Tyler et al. 1982). Since the completion of our studies of the asteroids and ophiuroids we have concentrated on reproduction in the echinoids and holothurians from the same geographic area. Once again we observe a variety of reproductive strategies from continuous development (Tyler & Gage, 1984a) to seasonality of development (Tyler & Gage, 1984b). This paper compares the reproductive strategies observed in 10 echinoids and 8 holothurians collected in the Rockall Trough, Porcupine Seabight and Porcupine Abyssal Plain at depths between 1000 and 4500m. All species are benthic with the exception of Peniagone diaphana which is benthopelagic (Barnes et al., 1976; Billett, Hansen & Huggett (this Vol.)). Little was known previously about the reproductive ecology of these deep-sea echinoderms apart from Hansen's (1975) observations of the elasipodid holothurians.

2 MATERIALS AND METHODS

All the specimens examined were collected within the Rockall Trough as part of the Scottish Marine Biological Association/ University College Swansea Deep Sea Programme (Gage et al., 1980; Gage & Tyler, 1982) or in the Porcupine Seabight/Abyssal Plain as part of the Institute of Oceanographic Sciences programme. The depth

Table 1. Gametogenic parameters in deep-sea echinoids from the N.E. Atlantic.

	May Oocyte Size (μm)	Dominant Oocyte Size (μm)	Yolk Type	Seasonality	Nutritive Tissue	Phagocytosis	Depth (m)	Total No. Examined
Cidaris cidaris	110	Varies with Season	PAS +++	+++	Well Dev.	-	508-1500	8
Poriocidaris purpurata	1500	**	PAS +++	-	Well Dev.	+	990-1145	6
Echinus affinis	110	Varies with Season	Fine granular PAS +++	+++	Well Dev.	-	1785-2300	303
Echinus alexandri	110	**	-	?+	Well Dev.	-	1632	5
Echinus acutus var. norvegicus	110	**	-	?+	Well Dev.	-	1075	10
Phormosoma placenta	1100	<300	Very granular PAS +++	-	Well Dev.	+	616-2910	179
Calveriosoma hystrix	1250	*	Vacuolated PAS+ve	-	Well Dev.	+	640-1075	42
Araeosoma fenestratum	1250	<300	Vacuolated PAS+ve	-	Well Dev.	+	631	20
Hygrosoma petersii	1150	<300	PAS +++	-	Well Dev.	+	990-2965	10
Sperosoma grimaldii	1100	<300	PAS +++	-	Well Dev.	+	1210	20

*Very variable.

**Too few specimens

PAS = Periodic acid Schiff

+ve = Positive

+++ = Strongly positive

?+ = Seasonality expected but too few specimens.

ranges at which individual species were collected are given in Tables 1 and 2 and further zoogeographic details are given in Gage et al. (in press).

All specimens examined were treated in a similar way. The test height and diameter of sea urchins, and the length of holothurians were noted. The gonads were removed, processed to paraffin wax, sectioned at 7μm and stained with haemalum and eosin, PAS, Toluidine blue or Methyl green/pyronin. The details of the gametogenic cycle were determined microscopically and the presence of any reproductive seasonality was determined by the use of oocyte/size frequency figures (Grant & Tyler, 1983).

3 OBSERVATIONS AND RESULTS

3.1 Echinoidea

The gametogenic biology observed in deep-sea echinoids from the N.E. Atlantic (Table 1) showed the variation in strategy observed previously in deep-sea asteroids (Tyler et al., 1982). The dominant reproductive strategy seen in all the echinothurid species examined and in Poriocidaris purpurata was the production of numerous large eggs, from 1100μm maximum diameter in Phormosoma placenta to 1500μm maximum diameter in Poriocidaris purpurata. The development of the oocyte followed a classical pattern. Oogonia within the nutritive tissue developed to young primary oocytes, recognised by their basophilic cytoplasm and large nucleus. Vitellogenesis commenced at about 250μm egg diameter in Poriocidaris and between 350 and 500μm in the echinothurids. The yolk structure appeared to be variable in the different species. In Poriocidaris purpurata, Phormosoma placenta, Hygrosoma petersii and Sperosoma grimaldii the yolk was very granular and PAS-positive whilst in the closely related Calveriosoma hystrix and Araeosoma fenestratum the cytoplasmic structure was more vacuolar and was less PAS-positive than in the other echinothurids. This suggests a biochemical variation in the carbohydrate yolk structure between these species. It is of interest that one specimen of Hygrosoma petersii showed intra-gonadal hermaphroditism, a condition rarely encountered in echinoids.

The nutritive tissue in both Poriocidaris purpurata and the echinothurids was dense and well structured. This tissue became more vacuolated as oocytes in the immediate vicinity developed, suggesting a transfer of stored energy during the process of vitellogenesis. There was no obvious seasonality of reproduction in any of these species.

An alternative reproductive strategy was seen in a population of Echinus affinis sampled at 2200m in the Rockall Trough. In this echinoid the oocyte development began in November/December of each year while the ovary contained oocytes from the preceding generation. The oocytes undergo vitellogenesis during the summer months and spawnout in the following January/February suggesting an oocyte development time of 14 months and a strict annual seasonality of reproduction. The small egg size of c. 110μm indicates a planktotrophic larva and a discrete recruitment of Echinus postlarvae was observed in April/May of each year (Tyler & Gage, 1984b). Although the number of specimens examined was very limited a similar reproductive strategy was seen in E. acutus var. norvegicus, E. alexandri and Cidaris cidaris. In the last species the larva is well known (Prouho, 1888) and is recorded in the North Atlantic plankton from February to June (Mortensen, 1901) suggesting inter-specific synchrony of reproduction with Echinus affinis.

In all the males examined conical headed spermatozoa, typical of echinoids, were observed. These developed seasonally in the Echinus species but were present the year round in the echinothurids and Poriocidaris.

3.2 Holothurioidea

The reproductive strategy of deep-sea holothurians showed a more subtle degree of variation than that observed in echinoids. In contrast to the pentamerous arrangement of the other echinoderm classes, the holothurian gonad is a paired structure lying in interradius CD (sensu Hyman, 1955).

The dominant holothurian order in the deep-sea is the Elasipodida. To date we have examined six species from three different families (Table 2). In the Laetmogonidae Laetmogone violacea had a compact nodose ovary containing numerous eggs up to 400μm diameter. In vitellogenic oocytes the yolk appeared reticulate and moderately PAS-positive. Most oocytes would appear to be spawned and phagocytosis is rare. The confamilial Benthogone rosea had a maximum egg size of 750μm diameter and the yolk was finely granular and PAS-positive. Phagocytosis of unspawned oocytes was common. Male L. violacea produced a small spermatozoon with a 2μm head whilst the spermatozoon of B. rosea

Table 2. Gametogenic parameters in deep-sea holothurians from the N.E. Atlantic.

	May Oocyte Size (µm)	Dominant Oocyte Size (µm) *	Yolk Type	Fecundity	Phagocytosis	Seasonality	Depth (m)	Total No. Examined
Ypsilothuria bidentaculata attenuata	350	*	PAS +++	Low	Rare	-	>1800	225
Cherbonniera utriculus	200	<100		Low	Rare	?		10
Laetmogone violacea	400	<150	Reticulate PAS+ve	Mod	Rare	-	225-1804	274
Benthogone rosea	750	*	Fine granular PAS+ve	Mod	Common	-	1103-4100	376
Deima validum	700	<400	Fine granular PAS+ve	Mod	Limited	-	4050-4505	22
Oneirophanta mutabilis	950	<400	Coarse PAS+ Vacuolated	Mod	Limited	-	2900-6000	156
Peniagone azorica	c.300	*	Vacuolated PAS (±)	Mod	Common	-	1991-3463	113
Peniagone diaphana	c.300	*	Vacuolated PAS+ve	Mod	Common	-	1900-5600	30

*Variable.

PAS = Periodic acid Schiff

+ve = Positive

± = Weakly positive

Low = <100 eggs/individual

Mod = >100-10000 eggs/individual

had a head diameter of 5-7µm.

We have had the opportunity to study only a limited number of specimens of the Deimatidae. In both Oneirophanta mutabilis and Deima validum the production of oocytes was a simple process with very little connective or nutritive tissue present and only a limited amount of phagocytosis of spent oocytes. A maximum egg size of between 700 and 950µm is produced although most of the eggs are smaller than 400µm diameter. In any population sampled, oogenesis occurred in about half the specimens but in the other specimens gametogenesis was absent. We suggest that these holothurians are males in a reproductively inactive phase. This is in contrast to other deep-sea echinoderms that breed continually in which the males are usually at constant maximum development.

In the Elpidiidae we have examined the benthic Peniagone azorica and the benthopelagic Peniagone diaphana (Barnes et al., 1976; Billett, Hansen & Huggett (this Vol.)). Oocyte development to the maximum size of c. 350µm is quite simple. However both the structure of the supporting tissue in the ovary and breakdown process of unspawned oocytes are complicated. If the oocyte is not released it undergoes phagocytosis within a strongly β-metachromatic cell membrane. As this internal material was phagocytosed and transported away the β-metachromatic membrane collapses in on itself and the end product was a series of "sacs" within the connective tissue of the gonad wall.

We have also taken the opportunity of examining the dactylochirotid Ypsilothuria bitentaculata attenuata (not Y. talismani (Tyler & Gage, 1983; Gage et al., in press)), and the molpadid Cherbonniera utriculus. Ypsilothuria bitentaculata attenuata produced up to 50 oocytes at any one time with a maximum diameter of 350µm. There is no synchrony of gamete development although oocyte development may be synchronous within individual tubules of the ovary. Cherbonniera utriculus, which grows to a maximum size of only 6mm (Sibuet, 1979) produced very few eggs up to 200µm diameter.

4 DISCUSSION

The gametogenic strategies observed in deep-sea echinoids and holothurians from the N.E. Atlantic demonstrate the same high degree of variability that was observed in the asteroids and ophiuroids (Tyler, Pain & Gage, 1982; Tyler et al., 1982).

The dominant strategy observed is that of the production of a large egg, indicating direct development with no obvious reproductive seasonality. This is seen in the echinothurid sea urchins, in Poriocidaris purpurata and in the elasipodid holothurians Oneirophanta, Deima and Benthogone. Of these species the most remarkable feature is the apparent lack of male development in Oneirophanta and Deima. This may be a result of inadequate temporal sampling or spawning on capture. However even in the latter case we would expect to see some relict spermatozoa. Hansen (1975) did not find males within his samples although he recognized a post-spawning O. mutabilis affinis male. We also exclude the possibility of hermaphroditism, parthenogenesis or parasitic castration in both Oneirophanta mutabilis and Deima validum. We do not observe any seasonality or brooding in Oneirophanta mutabilis mutabilis as suggested by Hansen (1968, 1975) for O. mutabilis affinis.

The remaining holothurians examined, with the exception of Cherbonniera utriculus, produced an egg of 350-400µm suggesting some form of abbreviated development although this remains speculative at present. Eggs of this size have been found in several species of Elasipodida and Dactylochirotida. It obviously has adaptational advantages in the deep-sea as those species exhibiting it are found in large numbers. Non-feeding pelagic larvae might be expected to be formed in the benthopelagic P. diaphana, while the distribution of L. violacea is related to ocean currents suggesting that a pelagic larva is formed (Hansen, 1975).

In Cherbonniera utriculus the maximum egg size observed was only 200µm. Billett and Hansen (1982) observed an egg of similar size in Kolga hyalina and interpreted this as indicating abbreviated development. However, extrapolation from data for reproduction in shallow water holothurians (Tanaka, 1958; Green, 1978) would suggest that an egg size of 200µm is indicative of planktotrophic development.

Strong evidence for planktotrophic development is seen in Cidaris cidaris and Echinus affinis. For Cidaris cidaris the larva is known; however, this species lives on the upper slope and extends into relatively shallow water. This could not be said of Echinus affinis taken at 2200m. Tyler and Gage (1984b) have speculated at length on the reproductive strategy of this species and concluded that E. affinis and other synchronously breeding deep-sea echinoderms undergo their gametogenic and spawning processes as a function of the sinking of the products of surface primary production. This provides a food source for developing larvae and a labile food source for the adult undergoing vitellogenesis to produce the subsequent generation of eggs (Tyler et al., 1982; Billett et al., 1983).

ACKNOWLEDGEMENTS

We wish to thank Dr. M.B. Collins in whose department the work was carried out; the Master and Crew of R.R.S. Challenger for their assistance at sea. The Scottish Marine Biological Association is Grant-aided by the Natural Environment Research Council. This research was carried out during the tenure of N.E.R.C. Grant GR3/4131 to PAT and under contract to the Department of the Environment as part of its radioactive waste management research programme. The results may be used in the formation of Government policy, but at this stage, they do not necessarily represent Government policy.

REFERENCES

Barnes, A.T., L.B. Quetin, J.J. Childress & D.L. Pawson 1976. Deep-sea macroplanktonic sea cucumbers; suspended sediment feeders captured from deep submergence vehicle. Science, 194: 1083-1085.

Billett, D.S.M. & B. Hansen 1982. Abyssal aggregations of Kolga hyalina D & K (Echinodermata: Holothurioidea) in the N.E. Atlantic Ocean: a preliminary report. Deep-Sea Res. 29:799-818.

Billett, D.S.M., B. Hansen & Q. Huggett (This volume). Pelagic Holothurioidea of the N.E. Atlantic.

Gage, J.D., R.H. Lightfoot, M. Pearson & P.A. Tyler 1980. An introduction to a time-series of abyssal macrobenthos: methods and principle sources of variability. Oceanol. Acta. 3:169-176.

Gage, J.D. & P. Tyler 1982. Growth and reproduction of the deep-sea brittlestar Ophiomusium lymani Wyville Thomson. Oceanol. Acta. 5:73-83.

Gage, J.D., D.S.M. Billett, M. Jensen & P.A. Tyler 1984. Echinoderms of the Rockall Trough. II. Echinoidea and Holothurioidea. Bull. Br. Mus. nat. Hist. (In Press).

Grant, A. & P.A. Tyler 1983. The analysis of data in studies of invertebrate reproduction. II. The analysis of oocyte size/frequency data, and comparison of different types of data. Int. J. Invert. Rep. 6:271-283.

Green, J.D. 1978. The annual reproductive cycle of an apodous holothurian Leptosynapta tenuis: a bimodal breeding season. Biol. Bull. Mar. Biol. Lab., Woods Hole, Mass. 154:68-78.

Hansen, B. 1968. Brood protection in a deep sea holothurian Oneirophanta mutabilis Theel. Nature, 217:1062-1063.

Hansen, B. 1975. Systematics and biology of the deep-sea holothurians. I. Elasipoda. Galathea Report, 13:1-262.

Hyman, L. 1955. The Invertebrates. IV. Echinodermata. New York: McGraw Hill Book Co. 763pp.

Mortensen, T. 1901. Echinodermlarven. Nordisches Plankton III(a), 1-30.

Prouho, H. 1888. Recherches sur le Dorocidaris papillata et quelques autres Echinides de la Mediterranee. Arch. Zool. exp. et gen. Ser. 2, 5:213-380.

Sibuet, M. 1979. Cherbonniera utriculus gen. nov. sp. nov. petite Holothurie (Molpadonia, Molpadiidae) des vases abyssales du Nort Est Atlantique. C.r. hebd. Acad Sci. Paris, 279:1443-1445.

Tyler, P.A., S.L. Pain & J.D. Gage 1982. Gametogenic cycles in deep-sea phanerozoan asteroids from the N.E. Atlantic. In J.M. Lawrence (ed.), Proceedings of the International Echinoderm Conference, Tampa Bay, p.431-434.

Tyler, P.A., A. Grant, S.L. Pain & J.D. Gage 1982. Is annual reproduction in deep-sea echinoderms a response to variability in their environment? Nature, 300:747-750.

Tyler, P.A. & J.D. Gage 1983. The reproductive biology of Ypsilothuria talismani (Holothuroidea: Dendrochirota) from the N.E. Atlantic. J. Mar. Biol. Ass. U.K., 63:609-616.

Tyler, P.A. & J.D. Gage 1984a. The reproductive biology of echinothurid and cidarid sea urchins from the deep sea (Rockall Trough, N.E. Atlantic Ocean). Mar. Biol., 80:63-74.

Tyler, P.A. & J.D. Gage 1984b. Seasonal reproduction in Echinus affinis (Echinodermata: Echinoidea) in the Rockall Trough, N.E. Atlantic Ocean. Deep-Sea Res. 31:387-402.

Proceedings of the Fifth International Echinoderm Conference / Galway / 24-29 September 1984

The paleobiogeography of Irish and British Lower Carboniferous blastoids

J.A.WATERS
West Georgia College, Carrollton, USA

G.D.SEVASTOPULO
Trinity College, Dublin, Ireland

ABSTRACT: Twenty species of blastoids belonging to twelve genera have been identified from the Lower Carboniferous (Dinantian) strata of Ireland and Great Britain. One genus and three species are new. Fragmentary material may represent an additional three genera. Most of the genera are monospecific and many are either rare or geographically restricted. Most have short geological ranges and are contained within one or two of the Dinantian stages.

1 INTRODUCTION

Blastoids are a class of stemmed echinoderms which are known from the Middle Ordovician to the Upper Permian. They are, in general, uncommon fossils in Europe and seldom rivalled the crinoids either in diversity or abundance.

Our purpose in this paper is to comment on stratigraphical and paleobiogeographic distribution of Irish and British Carboniferous blastoid faunas. The faunas are grouped according to their ages, using as a framework the stages proposed for the Dinantian (Lower Carboniferous) by George et al. (1976) and finer subdivisions based on conodont zones established by Groessens (1977). Waters and Sevastopulo (in press) provide a more comprehensive review of the stratigraphic distribution of Irish blastoids and give details of many of the localities mentioned in this account.

2 DISTRIBUTION OF BLASTOIDS

2.1 Courceyan Stage (Siphonodella and Polygnathus communis carina Zones)

Rocks of early Courceyan (Siphonodella Zone) age are distributed in much the same pattern as the slightly younger rocks shown in Figure 1. A single, minute, indeterminate specimen from Blind Strand, Seven Heads, County Cork is the only blastoid reported from them. This lack of records probably results mainly from collection failure, but it is possible that the very shallow near shore environments represented in the Siphonodella Zone were not favourable habitats for blastoids.

The distribution of lithofacies in the early part of the succeeding carina Zone is shown in Figure 1. The most prolific blastoid faunas of this age are known from limestones and shales of the Hook Head Formation at Hook Head, County Wexford. Orophocrinus pentangularis has been found at several horizons through the lower 120 metres of the formation and a new genus of orbitremitid, which is currently under study, occurs within the lower 50 metres. Two other genera, represented by material inadequate for formal description, have been found above and below a 25m thick, shallow water dolomite, 124m above the base of the formation. All the blastoids occur with a diverse fauna of brachiopods, bryozoa, tabulate and solitary rugose corals and crinoids, mostly camerates of the genera Dialutocrinus and Platycrinites. Blastoids formed a minor component of the benthos of an offshore shelf, in moderately shallow water but beneath fair weather wave base. Blastoids are apparently absent from the upper part of the Hook Head Formation which is interpreted as being of deeper water origin.

Elements of the Hook Head blastoid faunas have been found at Dungarvan to the west, and at Granny to the north of Hook Head. The Dungarvan fauna consists of Orophocrinus cf. pentangularis, in rocks of similar facies to the Hook Head occurrences but possibly of slightly younger age.

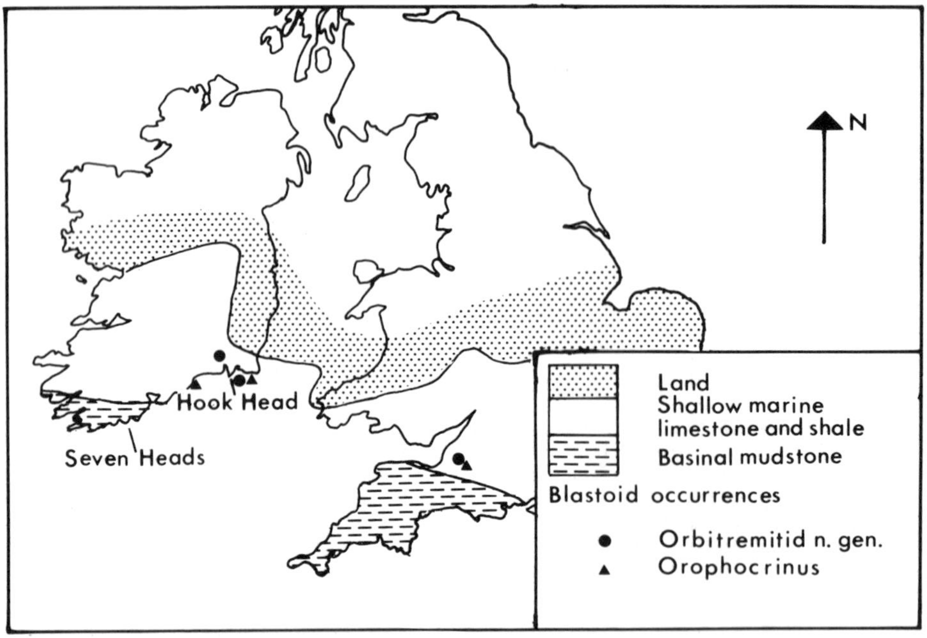

Figure 1: Sketch palaeogeographic map showing the distribution of blastoids in Courceyan (carina Zone) times.

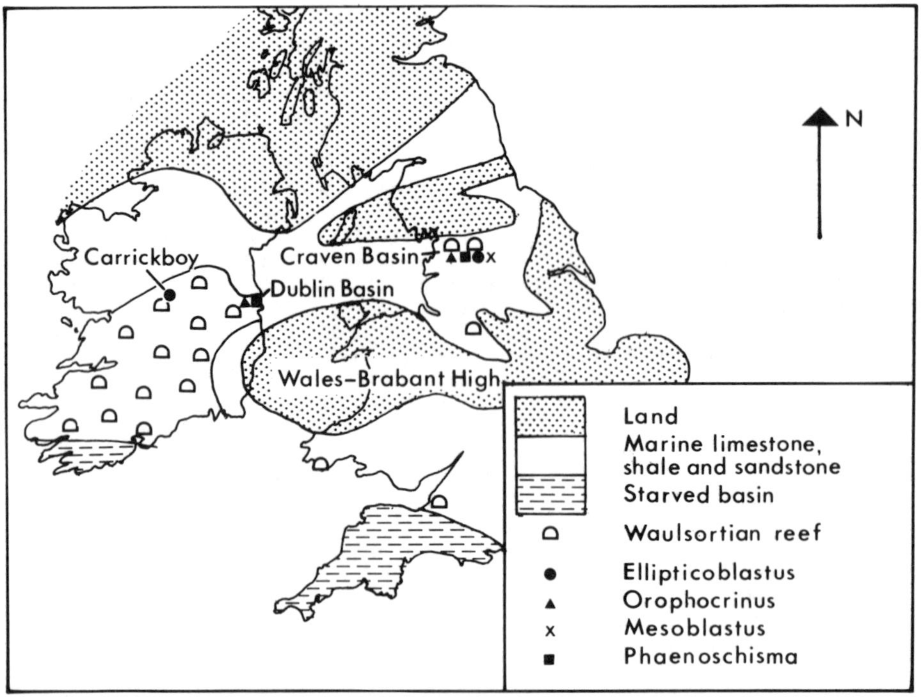

Figure 2: Sketch palaeogeographic map showing the distribution of blastoids in early Chadian times. Distribution of Waulsortian reefs is generalized.

The undescribed orbitremitid occurs at Granny at the same horizon and in rocks of the same facies as at Hook Head.

Only one blastoid has been formally described from rocks of Courceyan age in Britain. Orophocrinus orbignyanus is known from Waterlip Quarry, Shepton Mallet, Somerset (Bather 1912) which is cut in the Black Rock Limestone, probably at a level equivalent to some of the section at Hook Head. Based on our observations of Bather's specimens at the British Museum we believe they may be juveniles of O. pentangularis rather than O. orbignyanus. Macurda (1965) has noted the similarities between these two species. In addition, four specimens of the undescribed orbitremitid genus have been collected from just below the Black Rock Limestone at Windsor Hill Quarry, one mile northwest of Shepton Mallet. Reference to studies of the section at Windsor Hill by Butler (1973) suggests that the Irish and Somerset occurrences of the orbitremitid genus are at almost identical horizons and in similar lithofacies.

As shown in Figure 1 all the blastoids of Courceyan (carina Zone age) occur in a similar palaeogeographic setting, an offshore shelf bordering the large continental area which had come into being during the Devonian and which was progressively submerged by the Carboniferous sea. The lack of records from further west in Ireland is ascribed to collection failure; in southern and eastern England Carboniferous rocks are buried under a thick Mesozoic cover.

2.2 Courceyan (Scaliognathus anchoralis Zone) and Chadian Stages

During the late Courceyan (anchoralis Zone) and the early part of the Chadian, much of Ireland was covered by limestones of the Waulsortian complex (Figure 2). The complex consists of bank facies, exceptionally pure micritic limestones, some with stromatactis cavities, which formed topographic highs on the sea floor that are often referred to as reefs; and interfingering and laterally equivalent off-bank facies. They differ from the bank limestones in being clearly bedded, lacking stromatactis and in containing chert and argillite. Echinoderms appear to be more abundant in the flanks of individual reefs or in the off-reef facies. The depth at which the Waulsortian reefs grew is difficult to guage but there is evidence in Belgium that they developed in relatively deep water (below the photic zone).

In Britain the Waulsortian is not nearly as widely developed as in Ireland; the principal occurrences of Waulsortian are in the Craven Basin, where the reefs are Chadian in age; in gulfs on the northern side of the Wales-Brabant High; in south Wales; and in southwest England (Figure 2).

Most of the blastoid faunas known are of Chadian age; all of them are associated with reefs; and almost all of them are from the Craven and Dublin Basins. In the Craven Basin, the fauna consists of the fissiculates Orophocrinus pentangularis, O. verus and Phaenoschisma acutum and the more common spiraculates, Ellipticoblastus ellipticus, Mesoblastus angulatus and M. elongatus. Each taxon has a limited stratigraphic and geographic range among the various knolls. For instance, blastoids are very rare at Coplow Knoll, Clitheroe, but the nearby and slightly younger Salthill Knoll (Miller and Grayson, 1972) has yielded relatively large populations of Mesoblastus and Ellipticoblastus. Phaenoschisma and Orophocrinus are also known from Salthill but they are not as abundant. Large populations of Ellipticoblastus and some additional specimens of Orophocrinus are also known from the Calamine Mines of the Duke of Buccleugh near Whitewell.

Faunas associated with reefs in the Dublin Basin are of latest Courceyan and early Chadian age. They comprise Orphocrinus celticus, O. praelongus, O. verus, Phaenoschisma acutum and Hadroblastus hibernicus. This fauna is completely represented at Feltrim Hill, north of Dublin (Hudson et al. 1966); various components of it have been found at other localities in the Dublin Basin. In comparison with faunas from the Craven Basin, the absence of spiraculates is striking. However, Ellipticoblastus ellipticus occurs near the top of the reef exposed at Carrickboy Quarry, County Longford, to the west of Dublin. The horizon at which it has been found is judged to be of Chadian age. There are no confirmed records of Mesoblastus in Ireland.

The factors affecting the distribution of blastoids between the various reefs of the Dublin and Craven Basins are not known. Biostratigraphical correlations between the reefs are not sufficiently accurate to disprove the possibility that some of the differences are a consequence of the restricted ranges of particular taxa in time, but it seems likely that ecological controls were more important. The blastoids occur normally within shales in the upper part of a reef, as at Carrickboy, or in beds immediately overlying a reef,

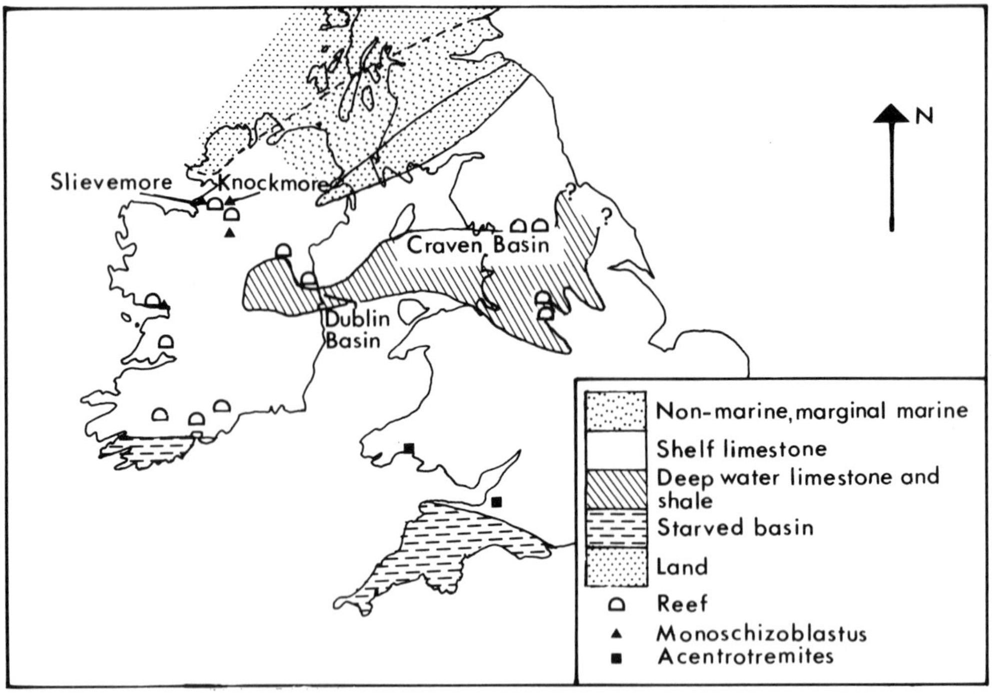

Figure 3: Sketch map showing distribution of blastoids during Asbian times.

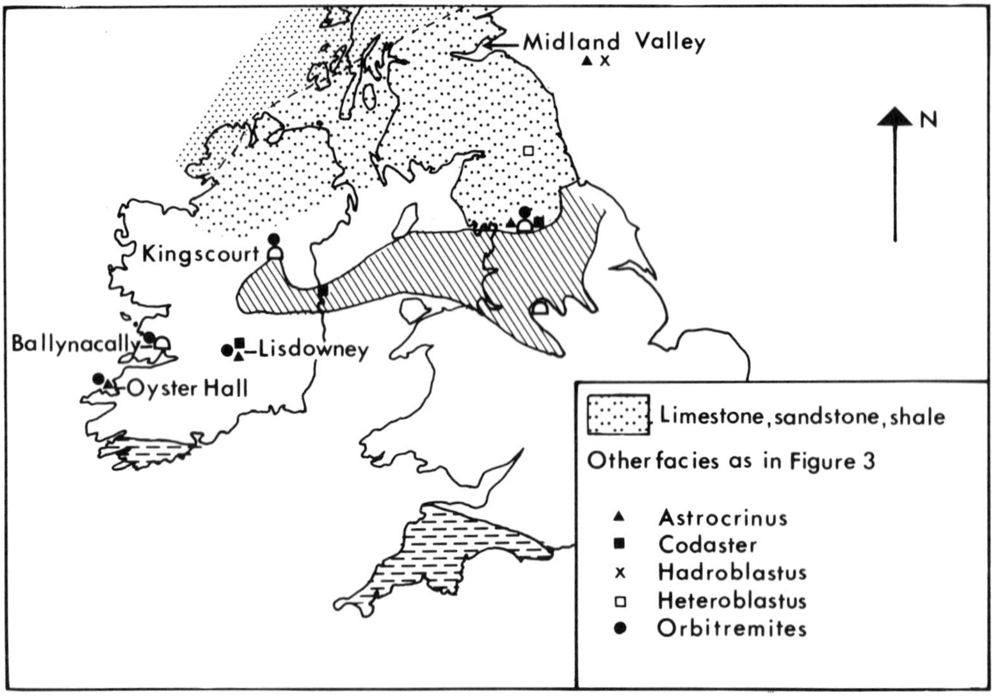

Figure 4: Sketch map showing distribution of blastoids during Brigantian times.

as at Salthill, Clitheroe and at Feltrim. Hudson et al. (1966) have documented the rich fauna from the mudstone overlying the reef at Feltrim. Blastoids form only a very small percentage of the benthos which was dominated by corals, brachipods and crinoids. The crinoid fauna, judged by the number of thecae identified, was made up mainly of camerates, particularly Sampsonocrinus and Platycrinites spp.; however, several inadunates have been identified from washed samples of the mudstone, and the original fauna may have contained camerates and inadunates in approximately equal numbers. Rarer flexible crinoids are also present. Collections of blastoids made recently from Salthill, Clitheroe are from beds overlying the reef which are extremely rich in crinoid material. As at Feltrim, the numbers of preserved thecae are heavily biased in favour of camerates, particularly species of Amphoracrinus and Platycrinites, but inspection of washed samples suggests that the original fauna contained as many inadunates as camerates, both in terms of the number of taxa and the number of individuals. A few flexibles are also present. Although blastoids are more abundant than at Feltrim, they still formed only a small proportion of the population of echinoderms. Detailed palaeoecological studies of other localities which have yielded blastoids have not been made, but it is clear that at all of them, blastoids are rare relative to other echinoderms.

A few records of blastoids are included in faunal lists of various localities in the Waulsortian of Ireland made in the last century, and some of the species known from the Craven Basin are also recorded from Derbyshire and Westmoreland in England. The concentration of localities with blastoids in the Craven and Dublin Basins (Figure 2) is, therefore partly due to collection bias, but the close association of blastoid faunas with Waulsortian reefs is unlikely to be an artefact of the data. When new Waulsortian sites with blastoid faunas are discovered it may be possible to determine the factors controlling the distribution of such genera as Ellipticoblastus and Mesoblastus.

2.3 Arundian and Holkerian Stages

Arundian and Holkerian blastoids are extremely rare. Only one indeterminate spiriculate from Arundian shales of northwest Ireland has been recorded (Waters and Sevastopulo, in press). Arundian and Holkerian crinoids are also very rare in the British Isles; this is most probably a reflection of the paucity of facies in which echinoderms are likely to have been preserved, since well-washed crinoidal calcarenites were extensively developed during both stages.

2.4 Asbian Stage

Blastoids are very unevenly distributed in rocks of Asbian age (Figure 3). The fissiculate Acentrotremites ellipticus, represented by three specimens from Wrington, Somerset (Bather 1912) and Kidwelly, Carmarthenshire, is the only taxon recorded from Britain.

In contrast, many localities in Ireland have yielded faunas of Asbian age. They are almost all in the northwest, in the Glencar and Dartry Limestones and equivalent formations. The faunas are almost all monotypic assemblages of the spiraculate Monoschizoblastus rofei. The most prolific fauna known to us is from thin crinoidal limestones with shale partings on the flank of a large Asbian reef near Knockmore Cliff, County Fermanagh. We have collected several hundreds of specimens from this locality. They are associated with crinoids, including diverse poteriocrinine inadunates and the camerates Amphoracrinus and Platycrinites. Counts of echinoderms from washed shale suggest that the blastoids are the most abundant elements of the fauna. However, M. rofei is not restricted to reefal or peri-reefal environments and in most localities where it occurs it is abundant. Considering how abundant it is in northwest Ireland, its apparent absence from other areas of Ireland and Britain, particularly those where reefs are developed, is surprising.

The youngest Asbian blastoid fauna known from northwest Ireland is contained in a thin fossiliferous nodular clay within the Meenymore Formation exposed at Slievemore, County Sligo. The clay has yielded several tens of specimens of a new species of Hadroblastus, several specimens of Orbitremites sp., three specimens of Astrocrinus sp. and a single high basal cone belonging to yet another unidentified taxon. The associated fauna is very rich and includes numerous inadunate crinoids and microcrinoids, disarticulated asteroid, cyclocystoid and ophiuroid ossicles, abundand cephalopods and gastropods, several trilobite taxa and several braciopods. Corals and cryptstomes bryozoans are notably rare. Almost all of the fossils including the blastoids are very small. The Meenymore Formation throughout northwest Ireland shows evidence of a very shallow water origin (Brandon, 1977). On Slievemore, the fossiliferous clay contains bioclasts with probable algal coats and overlies limestones which include algal laminates. We interpret the small size of the fossils and the scarcity of corals and cryptstome bryozoa as an indication of environmental stress in a very shallow water habitat.

2.5 Brigantian Stage

Blastoid faunas of Brigantian or probable Brigantian age are known from widely separated localities (Figure 4). Among the best known are large populations of Orbitremites derbiensis, Codaster acutus and rare specimens of Astrocrinus tetragonus which occur on the shelf margins of the Craven Basin, associated with reefal buildups. They have been well documented by Hoysey (1954). A population of Orbitremites derbiensis has been found at Kingscourt, north of Dublin, in what we believe is a similar reefal buildup. The original occurrence of Codaster acutus on the shelf to the north of Dublin basin is signalled by the discovery of a single specimen at the base of a coarse limestone, full of allochthonous shelf components in the basinal Brigantian succession at Loughshinny, north of Dublin. Orbitremites also occurs in association with reef limestone near Ballynacally, County Clare.

In many instances, however, Brigantian blastoid faunas are not associated with reefs. For example, a prolific fauna from near Lisdowney, County Kilkenny has been found in shallow water limestones with shale partings. The most abundant blastoid is a new series of Orbitremites represented by several hundred specimens. It is morphologically similar to O. derbiensis, but much smaller. Codaster acutus is moderately abundant, while Astrocrinus tetragonus is rare. Blastoids and microcrinoids are the most abundant elements of the fauna, and typical late Dinantian poteriocrinine inadunate crinoids are less diverse than is normally the case in the Brigantian. Elements of this fauna have been found near Oyster Hall, County Kerry.

Blastoids from the Midland Valley of Scotland occur in shales and limestones of the Lower Limestone Group. The fauna consists of Astrocrinus, which is known from several localities, and Phaenoschisma (?) benniei, which is less common (Macurda 1977). They are associated with crinoid faunas dominated by advanced poteriocrinine inadunates.

Heteroblastus cumberlandi is a spiraculate blastoid known only from a single locality, near Hexham, Northumberland, where it is abundant, but usually fragmented. It occurs in a dark grey shale containing abundant siderite nodules with a fauna dominated by molluscs. Although this assemblage is unusual for most Dinantian blastoids, it is the same as reported by Macurda and Mapes (1982) for Pennsylvanian blastoids from the midcontinent of the United States.

3 SUMMARY

Most of the blastoid genera we have studied have short geological ranges and many were geographically restricted. We believe this endemism was caused by the patchy distribution of favourable environments for the blastoids (which seem to change with time) rather than geographical barriers inhibiting migration into the various depositional basins.

Courceyan (Tournaisian) blastoids generally are found along the southern margin of a large land mass called the Wales-Brabant High. Although locally abundant, the blastoids were a minor element of a shallow water echinoderm community dominated by camerate crinoids.

Chadian (Lower Visean) blastoids are best known from the Craven basin in Great Britain and its extension, the Dublin basin, in Ireland. The blastoids were associated with the flank facies of Waulsortian mound buildups and lived in rather deep water. Although taxonomically diverse in some faunas, blastoids were relatively rare and were minor members of a crinoid community which included numerous camerates.

Arundian and Holkerian (Middle Visean) blastoids are very rare, possibly because of the lack of suitable facies for their preservation.

Asbian and Brigantian blastoids occurred in a wide range of environments, and some taxa, for instance, Orbitremites and Astrocrinus have an extensive geographical distribution. Others, such as Monoschizoblastus, are geographically restricted. The blastoids were locally very abundant and in many cases were the dominant element of the fauna. They are usually associated with advanced poteriocrinine inadunate crinoids. This change in blastoid abundance and dominance is striking and is also seen at an approximately equivalent stratigraphical level in the United States. In the U.S. the change can be correlated with patterns of extinction among crinoid groups (Waters et al. 1982) but the pattern in Great Britain and Ireland is more tenuous. Even though most of the Lower Carboniferous crinoid and blastoid genera are endemic to either Europe or North America, there is a suggestion of a common evolutionary - paleoecological transition that significantly changed the composition of Dinantian echinoderm communities.

4 ACKNOWLEDGEMENTS

J. A. Waters' contribution to this project was funded by grants from the Eppley Foundation for Scientific Research, the American Philosophical Society and a West Georgia College Faculty Research Grant. The support of all these sponsors is gratefully acknowledged. Grainne Moloney and Bonita Stewart typed versions of the manuscript.

REFERENCES

Bather, F. A. 1912. Two blastoids (Crophocrinus and Acentrotremites) from Somerset. Proc. Bristol Nat. Soc. 4th series. 3:47-50.

Brandon, A. 1977. The Meenymore Formation - an extensive intertidal evaporitic formation in the Upper Visean (B_2) of north-west Ireland. Rep. Inst. Geol. Sci. No. 77/23. 14 p.

Butler, M. 1973. Lower Carboniferous conodont faunas from the eastern Mendips, England. Palaeontology 16:477-517.

George, T. N. et al. 1976. A correlation of Dinantian rocks in the British Isles. Geol. Soc. London Spec. Rept. No. 7:1-87.

Groessens, E. 1977. Distribution de conodonts dans le Dinantien de la Belgique. International Symposium on Belgian Micropalaeontological limits. Namur 1974, No. 17:1-193.

Hudson, R. G. S. et al. 1966. A detailed account of the fauna and age of the Waulsortian reef knoll limestone and associated shales, Feltrim, Co. Dublin. Scient. Proc. R. Dublin Soc. Series A. 2:251-277.

Joysey, K. A. 1954. On the geologic distribution of Carboniferous blastoids in the Craven area, based on a study of their occurrence in the Yoredale series of Grassington, Yorkshire. Q. J. geol. Soc. London, III:209-224.

Macurda, D. B., Jr. 1965. The functional morphology and stratigraphic distribution of the Mississippian blastoid genus Orophocrinus. J. Paleontol. 39:1045-1095.

Macurda, D. B., Jr. 1977. Two Carboniferous blastoids from Scotland. Paleontology 20:225-236.

Macurda, D. B., Jr. & R. Mapes 1982. The enigma of Pennsylvanian blastoids. Proc. Third N. Amer. Paleon. Conv. 2:343-345.

Miller, J. & R. F. Grayson 1972. Origin and structure of the Lower Visean "reef" limestones near Clitheroe, Lancashire. Proc. Yorks. geol. Soc. 38:607-638.

Waters, J. A., T. W. Broadhead & A. S. Horowitz 1982. The evolution of Pentremites (Blastoidea) and Carboniferous crinoid community succession. In J. M. Lawrence (ed.), Proc. Int. Echinoderm Conf. 133-138. Rotterdam, Balkema.

Waters, J. A. & G. D. Sevastopulo (in press). The stratigraphical distribution and palaeoecology of Irish Lower Carboniferous blastoids. Ir. J. Earth Sci.

Echinodermata (Canary Islands) – Provisional check-list

J.J.BACALLADO
Departamento de Zoologia, Universidad de La Laguna, Tenerife, Canary Islands
E.MORENO
Centro de Tecnología Pesquera, Taliarte, Telde, Gran Canaria, Canary Islands
A.PÉREZ RUZAFA
Departamento de Ecología, Universidad de Murcia, Spain

This work forms part of a more generalized research project on the marine benthos of the Canary Archipelago, a project financed by the Government of the Autonomic Community of the Canary Islands. It has been in progress since 1978. The checklist presented here is based on material collected by the authors along the whole length of the Canary coast, together with that obtained from dragging and dredging on the adjacent submarine banks. The Canary Islands lie opposite the north-west coast of Africa, at a minimum distance, from the southeast of Fuerteventura to Cape Juby, of approximately 100 km. They are situated between longitude 27°37' and 29°23' N and latitude 13°20' and 18°10' W, on the southern edge of the temperate zone of the Northern Hemisphere. The islands rise steeply from the continental edge of Africa, with depths of more than 2000 m separating individual islands. Their volcanic nature is indicated by the absence of notable insular platforms; the marked inclination of the sea bed results in the predominance of hard substrates. Areas of sediments, principally detrital sands, clays and organic shingles, are relatively scarce. Tides are typically oceanic with a semi-diurnal regime. The Islands lie on the route of the descending arm of the Gulf Stream, known as the Canary Current, which flows in a south-southeasterly direction, transporting relatively cold water from higher latitudes. In winter, this current has its maximum velocity of 65 cm/sec. The surface water temperature normally ranges between 16-17°C in winter and 22-23°C in summer. These values are higher than those of the adjacent African coast due to the upwelling that occurs there. The annual values of salinity (surface) are between 36 and 37%.

CHECK-LIST

Crinoidea

Antedon bifida (Pennant)
Generally infralittoral, under stones, in fissures, on sponges, etc.: circalittoral in the community of D. ramea.

Holothurioidea

Mesothuria intestinalis (Ascanius & Rathke)
Bathyal, between 300 and 1445m., on muddy bottoms.

Mesothuria connectens (Perrier)
Bathyal, on bottoms between 1000 and 2000m.

Mesothuria verrilli (Théel)
Bathyal and abyssal, on bottoms between 500 and 4200m.

Paelopatides gigantea (Verril)
Bathyal, on bottoms between 1100 and 4000m.

Bathyplotes pourtalesi (Théel)
Bathyal, on bottoms between 134 and 1477m.

Stichophus tremulus (Gunnerus)
Circalittoral and bathyal, on bottoms between 20 and 1960m.

Eostichopus regalis (Cuvier)
Infralittoral and circalittoral, on detrital bottoms (5 to 400m).

Holothuria (Platyperona) cf. parvula (Selenka)
Mesolittoral, on rocky bottoms.

Holothuria (Platyperona) sanctori Delle Chiaje

Infralittoral, on rocky bottoms.
Holothuria (Lessonothuria) poli Delle Chiaje
Infralittoral, on sandy bottoms with beds of phanerogams.
Holothuria (Lessonothuria) arguinensis Koehler & Vaney
Infralittoral, on rocky bottoms
Holothuria (Vaneyothuria) lentiginosa Marenzeller
On soft bottoms between 100 and 250m.
Holothuria (Holothuria) tubulosa Gmelin
Infralittoral, on mud-sand bottoms and in submarine algal beds.
Holothuria (Holothuria) mammata Grube
Infralittoral, on rocky bottoms.
Holothuria (Holothuria) dakarensis Panning
Mesolittoral and infralittoral on rocky bottoms.
Holothuria (Holothuria) helleri Marenzeller
Circalittoral, on mud-sand bottoms and on coral (in the community of D. ramea).
Peniagone ferruginea Grieg
Bathyal, on bottoms between 2800 and 3000m.
Benthodytes lingua Perrier
Bathyal, on bottoms between 860 and 2200m.
Benthodytes typica Théel
Bathyal and abyssal. Benthopelagic.
Psychropotes depressa (Théel)
Bathyal, on bottoms between 1915 and 3510m.
Psychropotes semperiana Théel
Bathyal and abyssal.
Psychropotes longicauda Théel
Bathyal and abyssal.
Aslia lefevrei (Barrois)
Rocky infralittoral zone.
Pawsonia saxicola (Brady & Robertson)
Mesolittoral and rocky infralittoral zone on algae.
Neocnus incubans Cherbonnier
Mesolittoral and infralittoral on Cystoseira.
Thyone fusus (O.F. Müller)
Infralittoral
Molpadia musculus Risso
Muddy bottoms between 35 and 5025m.
Euapta lappa (J. Mueller)
Rocky infralittoral zone.
Leptosynapta inhaerens (O.F. Müller)
Infralittoral, in mud-sand bottoms.

Asteroidea
Astropecten aranciacus (Linnaeus)
Between 1 and 100m., normally on sandy bottoms and in beds of marine phanerogams.
Astropecten irregularis serratus Müller & Troschel
From the intertidal zone down to depths of 1500m., on mud-sand bottoms.
Astropecten hermatophilus Sladen
At 100m., on mud-sand bottoms.
Tethyaster subinermis (Philippi)
Between 80 and 90m., on sandy bottoms.
Luidia ciliaris (Philippi)
In the community of D. ramea
Chaetaster longipes (Retzius)
Between 50 and 1000m., on rocky substrates.
Ophidiaster ophidianus (Lamarck)
Between 1 and 1000m., on rocky bottoms next to sand or mud bottoms.
Hacelia attenuata (Gray)
In shallow caves and in the community of D. ramea.
Narcissia canariensis (d'Orbigny)
Generally found on infralittoral rocky or sandy bottoms, rarely at greater depths.
Asterina gibbosa (Pennant)
Between 1 and 500m., on diverse substrates.
Allopatiria ocellifera (Gray)
On mud-sand bottoms.
Henricia sanguinolenta (Müller & Troschel)
Between 1 and 1000m., on rocky bottoms.
Echinaster sepositus (Retzius)
At 10m., on sandy and rocky bottoms.
Marthasterias glacialis (Linnaeus)
Inhabits the tidal zone and infralittoral rocky bottoms.
Coscinasterias tenuispina (Lamarck)
Inhabits the tidal zone and infralittoral rocky bottoms.

Ophiuroidea
Astropartus mediterraneus (Risso)
Circalittoral, in the community of D. ramea.
Ophiacantha setosa (Retzius)
Circalittoral, in the community of D. ramea.
Amphiura chiajei Forbes
Circalittoral, in calcareous concretions.

Amphipholis squamata (Delle Chiaje)
Infra- and circalittoral in interstices, among sponges, rhizoides of algae and marine phanerograms and in the community of D. ramea.
Ophiactis savignyi Müller & Troschel
Infralittoral, on mud-sand bottoms; among sponges on rocky bottoms.
Ophiactis virens (M. Sars)
Mesolittoral, under stones.
Ophiothrix fragilis (Abildgaard)
Infralittoral, under stones.
Ophiocomina nigra (Abildgaard)
Infralittoral, on rocky bottoms.
Ophioderma longicaudum (Retzius)
Meso-, infra- and circalittoral, on rocky and sandy bottoms, under stones.
Ophiolepis paucispina (Say)
Mesolittoral and infralittoral, on rhizoides and between stones.
Ophiopsila aranea Forbes
In beds of D. ramea, between 50 and 100m.

Echinoidea
Cidaris cidaris (Linnaeus)
Circalittoral, in the community of D. ramea and on mud-sand bottoms.
Stylocidaris affinis (Philippi)
Circalittoral, in the community of D. ramea.
Diadema antillarum (Philippi)
Very abundant in the infralittoral zone; Circalittoral, rare in the community of D. ramea.
Centrostephanus longispinus (Philippi)
Circalittoral, in the community of D. ramea; rare.
Arbacia lixula (Linnaeus)
Infralittoral on rocky bottoms.
Genocidaris maculata A. Agassiz
On the sea-bed between 200 and 300m.
Sphaerechinus granularis (Lamarck)
Mesolittoral and infralittoral, on rocky bottoms and in beds of marine phanerogams.
Paracentrotus lividus (Lamarck)
Mesolittoral and infralittoral, on shallow rocky bottoms.
Echinocyamus pusillus (O.F. Müller)
Infralittoral, on mud-sand bottoms; circalittoral in the community of mäerl.
Echinocardium cordatum (Pennant)
Infralittoral, on sandy bottoms.
Schizaster canaliferus (Lammarck)
Infralittoral, in semi-dark caves; very rare.
Brissopsis atlantica Mortensen
Infralittoral on mud-sand bottoms.
Brissus unicolor (Leske)
Infralittoral on mud-sand bottoms.
Plagiobrissus costai (Gasco)
Infralittoral on mud-sand bottoms.

ACKNOWLEDGEMENTS

We thank the Canarian Government for making this study possible. Also, we would like to express our thanks to T. Cruz, G. Pérez Dionis and J.M. Hernández Otero, for supplying us with additional material.

Proceedings of the Fifth International Echinoderm Conference / Galway / 24-29 September 1984

Stability and change in the history of sea stars

DANIEL B.BLAKE
University of Illinois, Urbana-Champaign, USA

All known post-Paleozoic sea stars differ from all known Paleozoic representatives in the structure of the ambulacral column; differences include placement of the water vascular system and nature of articular arrangements. These ambulacral differences are complex enough to imply a strictly monophyletic origin for the younger fauna from some as yet unrecognized Paleozoic lineage. Post-Paleozoic families are long-ranging because many taxonomically diverse surviving families were established by the Jurassic, and few Mesozoic families are known to have suffered extinction. Endurance of Paleozoic lineages is less clear. Many major sea star constructional types ("major" types are those with numerous representatives) are known from both faunas, originated in the Ordovician and Jurassic, and survived through long intervals of Phanerozoic time. The fossil record, with its variety of constructional types in the Ordovician, suggests an early Paleozoic establishment of diverse life habits, then continuity until widespread extinction around the end of the Paleozoic. Extinction was followed in the Mesozoic by renewed diversification into all of the old, but no new general life habits. The younger sea stars appear to have replaced rather than displaced the older, regardless of any possible superiority of the post-Paleozoic construction. The post-Paleozoic again has been a time of evolutionary stability, perhaps engendered by the broad range of environmental tolerances and feeding types present in the group. Basic sea star organization thus appears capable of enduring all but major environmental stresses (such as those around the end of the Paleozoic) but the lack of new organizational patterns following widespread extinction suggests the presence of strong constraints on adaptive potential. According to some authors, significant changes took place during the Paleozoic/Mesozoic transition in the nature of predatory, burrowing and boring faunas. Sea star constructional patterns (and therefore inferred life modes) appear unaffected by changes around them, although differences in relative importance of different life habits would remain undetected from the sketchy fossil record.

Proceedings of the Fifth International Echinoderm Conference / Galway / 24-29 September 1984

Dynamic connective tissues: The echinoderm paradigm

J.P.EYLERS
Biology Department, Mercer University, Macon, Ga., USA

The "catch apparatus" is by now a well established phenomenon in all classes of echinoderms and a close parallel with the process of autotomy has been noted. The results of stress relaxation experiments conducted on body wall tissues from several species of holothurians can be summarized as follows. First, during short term relaxation the tissue behavior can be modelled as two parallel Maxwell elements, each consisting of an elastic component and a viscous component in series. Second, the principle mechanical changes associated with the state of "catch" occur in the viscous component of the second Maxwell element. Third, interstitial calcium ion concentration appears to have a role in regulating the "catch" state. A component based model of the tissue must account for the mechanical behavior of three phases: collagen fibers, matrix, and interstitial water. Collagen fibers store elastic energy by straining when the tissue is deformed, but stored elastic energy will quickly dissipate through viscous shearing of the interveining matrix. These two effects combined result in behavior similar to a Maxwell element, the elastic component derived from the fibers and the viscous component from matrix shear. The mucopolysaccharide matrix is a network of randomly coiled polymers cross-linked to each other at discrete bond sites. When the tissue is deformed the matrix undergoes a reduction in conformational entropy which manifests itself as an elastic restoring force, a phenomenon known as entropy elasticity. Stretched polymers within the matrix cause increased forces at bond sites. Random thermal motions regularly cause bonds to break, resulting in an exponential loss of stored energy as stretched polymers recoil. This process, called chemorheology, occurs in series with entropy elasticity, and together they are modelled as a Maxwell element. Viscous resistance to internal water flow is the third mechanical element of the system. Since the tissue is of constant volume, interstitial fluid must move about when the tissue is deformed. The degree of permeability of the matrix and the rate of deformation determine the stress induced in the fluid. In sum, the tissue behaves approximately like two Maxwell elements and one viscous element all in parallel. Since strain rate is zero during stress relaxation the fluid stress component is zero and the tissue will behave like two parallel Maxwell elements. In addition to conforming with the known facts, this model also makes specific, testable predictions about the relationships among fiber density, fiber-matrix bonding, polymer chain dynamics, matrix cross-link strength, and the mechanical behavior of the tissue. If we interpret the viscous behavior of the second Maxwell element as reflecting matrix bond strength, the model implies that the "catch" mechanism involves ion dependent cross-links in the matrix phase of the tissue which can be modulated by regulating interstitial calcium concentration.

Preliminary report on the reproduction of mediterranean echinoderms collected between 0 and 1000 meters, during the 'ECOMARGE' programme

J.G.FERRAND
Laboratoire de Biologie Cellulaire et Animale, UER Sciences, Université d'Orléans, France

C.VADON, J.P.FERAL & A.GUILLE
Muséum National d'Histoire naturelle, Laboratoire de Biologie des Invertébrés Marins et Malacologie, Paris, France

The "ECOMARGE" programme consists of a coordinated study of the benthic slope ecosystem. This interdisciplinary programme is carried out by ten French oceanographic laboratories. The subject of our participation is the comparison of the evolution of biological processes (reproduction and nutrition) from 0 to 1000 meters depth in the Mediterranean (Banyuls/mer) and in the Atlantic ocean (Arcachon) at the same latitude (43-44°N) and in relation to the environmental factors. Since June 1983, our research concerns the study of the breeding cycles of three Echinoderm species chosen because of their broad bathymetric range and of their high population density. The figures 1, 2 and 3 show the stages found in four groups of males and females sampled at 3 seasons of the year at different depths. These data are not yet sufficient for establishing breeding cycles. However, the histological study of their gonads allowed us to find five gametogenetic stages (tab. 1).

Figure 1. Gametogenetic stages of Ophiura texturata. See table 1 for legend.

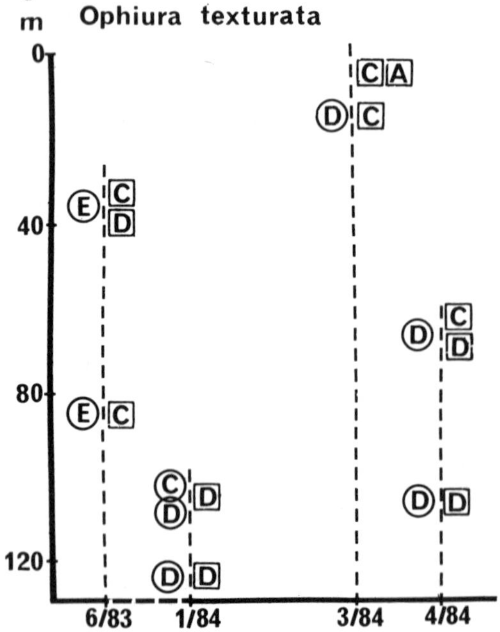

Table 1. The different gametogenetic stages observed.

"O"	unsexuable	
	♂	♀
Recovering spent "A"	arrest of spermatogenesis	previtellogenic oocytes
Early growth "B"	beginning of spermatogenesis	beginning of vitellogenesis
Late growth "C"	complete spermatic columns with first spermatozoa	large oocytes and smaller oocytes
Maturity "D"	large mass of spermatozoa	only large oocytes
Shedding "E"	empty gonads or mass of non-ejected spermatozoa	relict oocytes

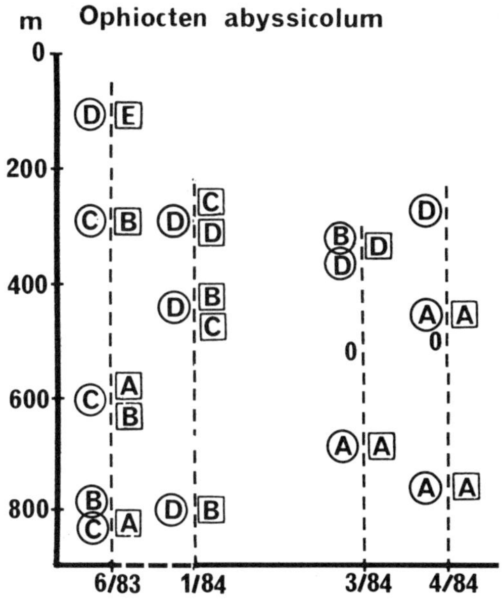

Figure 2. Ophiocten abyssicolum, same legend as fig. 1.

Concerning Ophiura texturata and Brissopsis lyrifera, the specimens from the upper levels seem to be in quasi-synchronism with those found at deeper levels. Thus, it is very probable that no relation exists between breeding cycles and depth. Concerning Ophiocten a abyssicolum, a difference of gametogenetic development between deeper and shallower water specimens is noticeable at spring time: at this season, the genital products of the deeper water specimens are already released, and a new gametogenesis appears to be initiated; the shallower water specimens are still finishing their maturation and have not yet released their gametes. The seasonality of the reproduction of deep-sea ophiuroids may occur even in the deeper individuals of species presenting a broad bathymetric range.

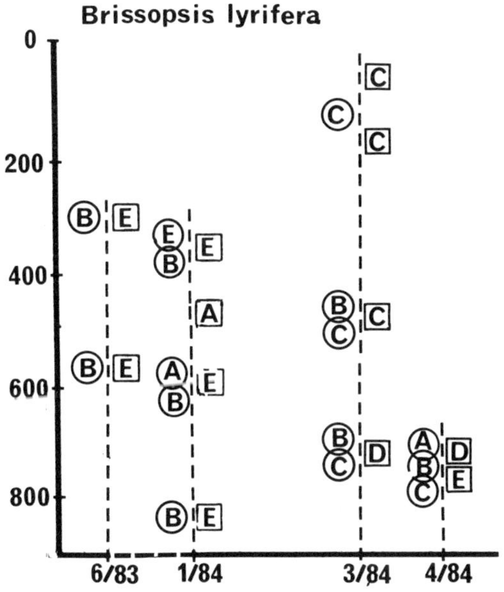

Figure 3. Brissopsis lyrifera, same legend as fig. 1.

Proceedings of the Fifth International Echinoderm Conference / Galway / 24-29 September 1984

Middle Cambrian echinoderms from Sweden

C.A.M.FRANZEN-BENGTSON
Paleontologiska institutionen, Uppsala Universitet, Sweden

The lower Middle Cambrian Oelandicus shales at Norrtorp, south central Sweden have yielded a widely diversified fauna. The echinoderms are represented by Cigara Barrande, edrioasteroids and stylophoran carpoids. Five specimens of Cigara show a long, slender, plated stele. The proximal part of the stele forms a hollow cone of thin, fairly large convex plates gradually passing into the tapering base of the calyx. The distal part is composed of two series of alternating ossicles. The edrioasteroids represent a new species of Stromatocystites. The carpoids may be a new species of Ceratocystis.

Proceedings of the Fifth International Echinoderm Conference / Galway / 24-29 September 1984

Echinoderms of the coral reef of New Caledonia

A.GUILLE, G.CHERBONNIER, J.P.FERAL & C.VADON
Muséum National d'Histoire naturelle, Laboratoire de Biologie des Invertébrés Marins et Malacologie, Paris, France

P.LABOUTE & J.L.MENOU
Centre ORSTOM, Nouméa Cedex, New-Caledonia

M.JANGOUX & C.DE RIDDER
Université Libre, Faculté des Sciences, Laboratoire de Zoologie, Brussels, Belgium

D.MEYER
University of Cincinnati, Dept. of Geology, Ohio, USA

Regular investigations, over a six year period, by a team of divers from the ORSTOM center at Noumea (New Caledonia), produced a collection of 236 species of Echinoderms for the lagoon, the reef and its external slope to a depth of 60m. This comprises: 31 species of comatulid, 60 holothuroids (with 14 new species), 35 echinoids (1 new species), 50 asteroids (3 new species) and 60 ophiuroids (1 new species. Each species was photographed in colour in its natural habitat. Together with new data on taxonomy, particularly that of the holothuroids, the records include information recorded during day and night dives. The composition of the echinoderm fauna of the coral reef of New Caledonia shows its closed relationship with the fauna of the Great Barrier of Australia. Some species described from Micronesia are here recorded for the first time. Most of the species are common in the tropical indo-west-pacific zone. Some of the new species have since been taken again in the Polynesia area.

Proceedings of the Fifth International Echinoderm Conference / Galway / 24-29 September 1984

Biotic diseases of echinoderms

MICHEL JANGOUX
Laboratoire de Biologie marine, Université Libre de Bruxelles, Belgium

Although a great deal of echinoderm parasites have been described for more than one century, there is little information on the host-parasite relationships or on the effect of parasitism on the echinoderm life-cycle. (It is noteworthy that, while often intensively parasitized, echinoderms never behave as parasites). Thus far, most biotic diseases of echinoderms are caused by animal agents, and records of infections by microorganisms and algae are very few. More than 400 animal agents infesting echinoderms have been recorded. About one-third of these live either on or in holothuroids which appear to be very suitable hosts for parasites. As a rule, turbellarians, gastropod molluscs and copepod crustaceans are the more numerous. However, sporozoans, myzostomids and ascothoracid cirripedes are also well represented. About one half of the recorded agents live either attached to the echinoderm body wall or sheltered within the body cavity. Although conspicuous body-wall deformations or castrating effects may occur, it must be emphasized that most diseases caused by animal agents do not produce major consequences for the echinoderms and are rather well tolerated. Actually, one may consider each echinoderm as an animal substrate on or in which various organisms live, the echinoderm and its associates together forming a well-balanced biological complex. Calculation of a pathogenic index (i.e. number of species of animal agents/estimated number of recent host species x 100) for each echinoderm group shows that the two main recent classes have a very different sensitivity to animal pathogens, the index values being 12.7 for Echinozoa and 2.8 for Asterozoa.

Proceedings of the Fifth International Echinoderm Conference / Galway / 24-29 September 1984

Life modes of cornute and mitrate carpoids

DENNIS R.KOLATA
Illinois State Geological Survey, Champaign, USA

Cornute and mitrate carpoids (class Stylophora) are enigmatic, extinct echinoderms whose phylogenetic relationships, functional morphology, and life modes have been subjects of much debate. Because neither mouth nor anus can be identified with certainty, there is no general agreement on even the basic elements of orientation (dorsal/ventral; posterior/anterior). Interpretations of these animals must take into account: (1) body form, (2) specialized skeletal structures, (3) the presence of only one body opening common to all species, and (4) cornutes and mitrates are closely related, and, therefore, the basic organ systems were arranged in very similar ways in the two groups. On the basis of these constraints, it is hypothesized that in all stylophorans the main thecal orifice was the site of both mouth and anus, and that a protractible feeding structure was present. Life modes, however, were distinct in the cornutes and mitrates. Evidence for a sluggish epifaunal life habit in the cornutes is seen in the flattened asymmetrical body, the presence of downward projecting spines and knobs on the marginals in many species, and the specialized thecal pores and slits (interpreted here as respiratory structures) that most investigators believe faced away from the substrate. The aulacophore is a long, slender, relatively inflexible appendage. In contrast, the mitrates were adapted to an infaunal life mode. They have a streamlined, generally symmetrical body composed of relatively large thecal plates that lack knobs and protuberances. The aulacophore was more flexible and muscular than that of the cornutes. Spines and blades on the lower surface of the stylocone and proximal part of the distal aulacophore are inferred to have pulled the animal backwards (aulacophore first) into the substrate. In many species, transverse cuesta-like ribs on the theca, marginal denticles, and forward imbrication of thecal plates would serve to grip the substrate and inhibit removal by either predators or currents. The mitrates probably lived near the surface (e.g., upper three or four centimeters of sediment) protracting the feeding structure upward into the water column. The anterior spines present in many species are well placed for a tactile and/or protective function.

Proceedings of the Fifth International Echinoderm Conference / Galway / 24-29 September 1984

Deep-sea echinoderms: A time and motion study

R.S.LAMPITT & D.S.M.BILLETT
Institute of Oceanographic Sciences, (NERC), Wormley, Godalming, Surrey, UK

Aspects of the ecology of deep-sea echinoderms can be deduced from photographs taken sequentially from a camera mounted on an epibenthic sledge or other towed gear. This transect photography provides information on population density, spatial distribution, method of locomotion, feeding behaviour, faecal density (in the case of holothurians), burrowing behaviour, and the degree to which echinoderms bioturbate the sediment surface. Photographs taken between 400 and 4100m in the Porcupine Seabight off the southwest coast of Ireland have provided information on these features of echinoderm life. The population density of the common echinoderms visible in photographs rarely exceeds 1000 individuals per hectare (e.g. Phormosoma placenta, Benthogone rosea) apart from some exceptional cases, such as Elpidia echinata, 17,400 individuals per hectare at 1640m depth, and Kolga hyalina, 500,000 individuals per hectare at a depth of 3680m. K. hyalina has a patchy distribution and may occur in dense aggregations. The degree of aggregation is not constant and variations may be due to the seasonal deposition of phytodetritus on the seabed. Although most epifaunal echinoderms crawl over the seabed with remarkably little disturbance of the sediment surface, some species, such as Bathybiaster vexillifer, plough through the upper centimetre of the sediment, and other species cause disturbance while burrowing. Other information on echinoderm behaviour can be obtained using time-lapse photography. Bathysnap, a time-lapse photographic and current meter system, takes photographs of the same area of seabed at intervals from 4 minutes to 8 hours, for periods lasting from a few hours to several months. Data are now available on the rate of movement of several species. For instance, Bathybiaster vexillifer at a depth of 2000m may move at 100cm h^{-1} but also has been seen to remain stationary for up to 5½ days. The average speed for B. vexillifer is therefore much slower. Maximum rates of 310cm h^{-1} for Hymenaster membranaceus and 89cm h^{-1} for Benthogone rosea, have also been calculated. Aggregations of K. hyalina have been seen to move as a "herd" passing through the field of view (about 1m width) within 16 hours, and leaving the sediment surface strewn with their faeces.

Proceedings of the Fifth International Echinoderm Conference / Galway / 24-29 September 1984

Echinoderms of the 'Cantabrico 83' expedition off Asturias (N. Spain)

ALICIA LÓPEZ-IBOR
Catedra de Invertebrados, Facultad de Biología, Universidad Complutense, Madrid, Spain

During the 'Cantabrico 83' expedition off the coast of Asturias, 21 echinoderm species (1 crinoid, 3 holothuroids, 3 ophiuroids, 10 asteroids, and 4 echinoids) were recorded from depths of 42 to 728m. The only such previous expedition was the 1968 cruise of the 'Thalassa' which collected off Galicia and, to a lesser extent, also off Asturias. Within the area investigated, the continental slope is the most interesting where the echinoderm fauna is concerned.

In the Mediterranean, which is more shallow, species diversity is not so great and the numbers of individuals for some 'endemics' are much less than for mediterranean-atlantic species. It should be pointed out that littoral (0-100m) atlantic species can range more deeply within the Mediterranean, a phenomenon which may be related to factors such as higher temperatures and salinities.

In the Bay of Biscay, and often in the "Mar Cantabrico", boreal or semi-boreal species occur which have extended their range southwards.

SPECIES LIST

The following is an alphabetical listing of the species recorded:

Anseropoda placenta
Astropecten irregularis
Astropecten platyacanthus
Brisingella coronata
Chaetaster longipes
Cidaris cidaris
Echinus acutus
Laetmogone violacea
Leptometra celtica
Nymphaster arenatus
Ophiactis balli
Ophiothrix lutkeni
Ophiura ophiura
Phormosoma placenta
Plutonaster bifrons
Pontaster tenuispinus
Porania pulvillus
Psilaster andromeda
Spatangus purpureus
Stichopus regalis
Stichopus tremulus

Proceedings of the Fifth International Echinoderm Conference / Galway / 24-29 September 1984

Palaeoecology of a late Jurassic echinoderm community from the Swiss Jura mountains

C.A.MEYER
Geological Institute, University of Bern, Switzerland

INTRODUCTION

Isolated starfish- and ophiuroid-ossicles are quite common in clayey sediments of middle Oxfordian age (HESS,1975),although complete specimens are very rare.An exception is the excellently preserved asteroid-brittle star-decapod community in fine grained,silty sandstone which is described in this paper. The Effingen beds containing this echinoderm community are situated in the Northern part of the Weissenstein-anticline near Solothurn (southern part of the Swiss Jura mountains;MEYER,1984).

MATERIAL AND METHODS

The material consists mainly of a large slab (1x1m) excavated during a two week field trip in summer 1979 sponsored by the Museum of Nature (Solothurn).The slab was subsequently prepared using KOH-flakes to digest clay minerals,air-tool to mechanically disaggregate the sample and airbrasive for the final cleaning.

RESULTS

The sequence is dominated by calcite cemented, fine grained,silty sandstone (up to 60% quartz,mean grain size 0.04mm) and consists of four different sedimentary units detected by radiography and two units interpreted to have been formerly present.The first unit contains the ophiuroids Ophiomusium gagnebini (THURMANN) and Sinosura wolburgi HESS and the crustacean Eryma sp.. The second unit bears the same animals and the astropectinid sea star Pentasteria (P.) longispina HESS.
The mean density of Eryma sp. ranges from 0.08 Ind.m^{-2} (unit I) to 0.017 Ind.m^{-2} in unit II.Such values agree well with those mentioned in literature for recent relatives (CHAPMAN & RICE,1971). P.longispina has a mean density of 3.3Ind.m^{-2}; some areas are very poorly populated (0 - 2 Ind.m^{-2}),while others are densly crowded (up to 12 Ind.m^{-2}). Again,similar dispersal patterns are also observed in modern asteroids (WARNER,1979).Population density of O.gagnebini ranges between 330 - 1800 Ind.m^{-2} in the first unit and decreases to 0 - 200 Ind.m^{-2} in the second unit,whereas the density of S.wolburgi remains very low (0.3-0.16 Ind.m^{-2})throughout the units. Recent ophiuroid populations show a very low density compared with fossil populations.An exception to this are the suspension-feeding species Amphiura filiformis and Ophiotrix fragilis in European waters where they form dense aggregations (2000 Ind.m^{-2};WARNER,1979).The occurence of comparable dense patches in the studied fossil population and a current speed of 20-60cm sec^{-1},estimated from the sedimentological record, suggest that ophiuroid aggregational behaviour can be traced back at least to Mesozoic times.
Size-frequency diagramms of the brittle stars and starfishes display a distribution suggesting different year groups.Juvenile individuals are almost completly absent indicating a seasonality of the annual recruitment. Regeneration rate in the starfish is very high (30%),while ophiuroids show a very low rate with arms

regenerated (0.05%).These arms show signs of predation either by fish or crustaceans.

DEPOSITIONAL HISTORY

After a long period of clay sedimentation,currents deposit large quantities of fine sand,which allows the brittle star-decapod community to settle(unit I).According to sedimentary structures (cross lamination, ripple marks) current speed did not exceed 20-60cm sec^{-1}.The overlying unit II is characterized by a uniform,low sedimentation rate and the appearance of sea stars.Studies on population dynamics and longetivity indicate that the life span and the sedimentation time for the first two units lies between 10 and 20 years. The succeeding column of sediment (7-20cm) is not sufficiently thick enough to prevent the endobenthic animals from digging themselves out. Death by intoxication of the pore water seems improbable because of the low pyrite content (5%).Several storm layers are developed higher up in the sequence.This suggests that a storm layer was once interbedded between unit II and the following unit V which rapidly covered the animals and prevented them from escaping.This storm bed was subsequently evenly eroded without leaving any traces,except for a thin argillaceous lamina which can be explained as suspended material settled out after the erosional event (unit IV).The following arguments supply this interpretation: the sea stars died while moving (fulcral ridge of the ambulacrum;position of the supramarginal spines : REESE, 1966);the internal cavities of the sea stars are filled with clay,as are the Erymid burrows.The next unit (V) shows layers with ripple marks and different types of trace fossils (e.g. escape traces,attributed to a burrowing sea anemone).At the top of this unit, a new settlement of juvenile starfish and crustaceans is developed.The following "hummocky cross stratification" caused by a heavy storm (WALKER,1979),finally buried the new settlers.Within this storm layer,partly disaggregated and wrinkled echinoderms can be found. Several regional outcrops contain the same lithological unit with an identical trace fossil assemblage, but no other echinoderm community could be found.The higher quartz content of these beds (up to 80%) the reduced original pore space and therefore, the lower organic content probably prevented the establishment of another echinoderm community due to unfavorable nutritional conditions.

Considering palaeoecological and sedimentological evidence,the studied echinoderms lived in the middle sublitoral between 50 and 100m waterdepth.

REFERENCES

CHAPMAN.C.J. & RICE,A.L.(1971): Some direct observations on the ecology and behaviour of the Norway lobster Nephrops norvegicus(LEACH).- Marine biol.10,321-329.

HESS,H.(1975): Die fossilen Echinodermen des Schweizer Jura.- Veröff. nathist.Mus.Basel,Nr.8.

MEYER,Ch.A.(1984): Palökologie und Sedimentologie der Echinodermenlagerstätte Schofgraben(mittleres Oxfordien,Weissenstein,Kt.Solothurn).- Ecl.geol.helv.77/3 (in press).

REESE,E.S.(1966): The complex behaviour of echinoderms.In: BOOLOTIAN,R.A.(Ed.): Physiology of Echinodermata(p.157-218).-Interscience,New York.

WALKER,R.G.(1979): Shallow marine sands.In: WALKER,R.G.(Ed.): Facies models(p.75-89).-Geosci. Canada,Repr.ser.1.

WARNER,G.F.(1979): Aggregations in Echinoderms.In:LAARWOOD,G. & ROSEN,B.R.(Ed.): Biology and Systematics of colonial organisms (p.375-396).- Syst.Assoc.Spec.Vol. 11,Academic Press,London.

Proceedings of the Fifth International Echinoderm Conference / Galway / 24-29 September 1984

The Irish echinoderm fauna

B.O'CONNOR, B.F.KEEGAN, T.BOWMER & J.COSTELLOE
Department of Zoology, School of Marine Sciences, University College, Galway, Ireland

The Echinodermata rank amongst the better documented phyla of the Irish Marine Fauna, due to their large size and the long history of dredging expeditions and surveys off our coasts. To date, the following 73 species have been recorded from the inshore area, i.e. up to and including the 200 metre isobath:

Antedon bifida (Pennant)
Leptometra celtica (Barrett and McAndrew)
Pontaster tenuispinis (Düben and Koren)
Astropecten irregularis (Pennant)
Luidia ciliaris (Philippi)
Porania pulvillus (O.F. Müller)
Asterina gibbosa (Pennant)
Asterina phylactica Emson and Crump
Anseropoda placenta (Pennant)
Pteraster pulvillus Sars
Crossaster papposus (L)
Solaster endeca (L)
Henricia oculata (Pennant)
Henricia sanguinolenta (O.F. Müller)
Stichastrella rosea (O.F. Müller)
Asterias rubens L
Leptasterias mulleri (Sars)
Marthasterias glacialis (L)
Ophiothris fragilis (Abildgaard)
Ophiothrix lutkeni Wyville Thompson
Ophiocomina nigra (Abildgaard)
Ophiopsila aranea Forbes
Ophiopsila annulosa (Sars)
Ophiactis balli (Thompson)
Ophiopholis aculeata (L)
Amphiura chiajei Forbes
Amphiura filiformis (O.F. Müller)
Amphiura securigera (Düben and Koren)
Amphiura incana Lyman
Acrocnida brachiata (Montagu)
Amphiphois squamata (delle Chiaje)
Ophiura ophiura (Lamarck)
Ophiura albida Forbes
Ophiura sarsi Lutken
Ophiura affinis Lutken
Mesothuria intestinalis Ascanius
Parastichopus tremulus (Gunnerus)
Holothuria forskali delle Chiaje
Leptopentacta elongata (Düben and Koren)
Paracucumaria hyndmani (Thompson)
Pawsonia saxicola (Brady and Robertson)
Aslia lefevrei (Barrois)
Ocnus lacteus (Marenzeller)
Ocnus brunneus Forbes
Ocnus planci (Barrois)
Thyone fusus (O.F. Müller)
Thyone roscovita Herouard
Pseudothyone raphanus (Düben and Koren)
Neopentadactyla mixta (Östergren)
Thyonidium pellucidum (Fleming)
Thyonidium commune (Forbes)
Psolus phantapus (Strüssenfelt)
Psolus squamatus Koren
Leptosynapta inhaerens (O.F. Müller)
Leptosynapta bergensis (Östergren)
Leptosynapta minuta (Becher)
Labidoplax digitata Montagu
Labidoplax media Ostergren
Labidoplax thompsoni (Herapath)
Cidaris cidaris (L)
Psammechinus miliaris (P. Müller)
Echinus esculentus L
Echinus acutus L
Echinus elegans Düben and Koren
Paracentrotus lividus (Lamarck)
Echinocyamus pusillus (O.F. Müller)
Spatangus purpureus (O.F. Müller)
Spatangus raschi Loven
Echinocardium cordatus (Pennant)
Echinocardium flavescens (O.F.Müller)
Echinocardium pennatifidum Norman
Brissopsis lyrifera Forbes

Recent studies have had the new dimension of in situ observation and experimentation. These have provided insight into the importance of many echinoderms, even to the extent of their having dominant roles in regulating community organisation. A detailed inventory is being prepared which will include notes on the taxonomy, nomenclature, biology, distribution and zoogeography of these species.

Contribution No. 230 from the School of Marine Sciences, University College, Galway

Proceedings of the Fifth International Echinoderm Conference / Galway / 24-29 September 1984

Echinosphaerites (Rhombifera) and its community relationships from the Middle Ordovician of Bohemia, Czechoslovakia

RONALD L.PARSLEY
Department of Geology, Tulane University, New Orleans, La., USA

RUDOLF PROKOP
Department of Paleontology, Narodni Muzeum, Prague, CSSR

Echinosphaerites (Wahlenberg 1818) is one of the most widely distributed of all "cystoids". This rhombiferan is found in the Middle Ordovician of North America, northwestern Europe, Bohemia (? African plate-Gondwanna), Yunan, China and extends well into the Upper Ordovician strata of Bohemia. Echinosphaerites typically has a large globular polyplated theca, four to six basals, humatirhombs across all facets of thecal plates, two to five brachioles (some branch to produce up to seven or eight), and a very thin, rarely preserved (? five part) column. Echinosphaerites infaustus (= Arachnocystites Neumayr 1889) from Bohemia has a robust six part column adjoining six basals. There are three unbranched brachioles, with small oval platform; humatirhombs average nine to ten fistulipores (maximum fourteen). Juveniles (= Deutocystites) often with large rounded plates that retain their shape in ontogeny and subsequently were surrounded by several irregular series of polygonal plates. Anal pyramid and gonopore well developed; hydrophore, hitherto unknown in Caryocystitide, on posterior side of oral platform. The primitive nature of column and brachioles is offset by the more advanced nature of humatirhombs (see Bockelie 1981) giving a good example of mosaic evolution. Thecal plates have large cores and well developed (? seasonal) growth rings which probably indicate Bohemia's cooler waters and higher Ordovician paleolatitude than that of North America and/or Western Europe. The best documented Middle Ordovician locality in Bohemia is in the grewacke-like Zahorany (D_4) Formation, Zahorany locality, southwest of Prague. Here the community was dominated by "vertically stratified" echinoderms. Dendrocystites sedgwicki (Homoiostelea), "Hemicystites" bellulus (Edrioasteroidea), Cardiocystis (rare, Eocrinoidea) were recumbent on the bottom; Codiacystis (Diploporida) without a column, sat on the bottom with brachioles directed upwards; Aristocystites (Diploporida), also without a column, sat on the bottom but bent over so that brachioles fed at or near water-sediment interface; Cheirocrinus, Echinosphaerites, and (rare) Rhombifera (all Rhombifera) are stalked and fed up to several hundred millimeters off the bottom; and (rare) Polycrinus romulatus (Crinoidea), on columns up to one quarter to one half meter long and with very long arms relative to thecal length, were the emergents in this community. Very large trilobites and conularids are numerous; pelecypods, gastropods and brachiopods are normal to reduced in size relative to their more tropical counterparts.

The early radiation of echinoderms

C.R.C.PAUL & A.S.SMITH
Department of Geology, Liverpool University, UK

Fossil evidence suggests that the water vascular system of echinoderms evolved before stereom which, in turn, evolved before pentamery. Cladistic analysis of Lower Cambrian genera reveals that the carpoid/echinoderm dichotomy was the first evolutionary divergence. Helicoplacoids are stem-group echinoderms with multi-layered spiral plating and three biserial ambulacra with tube feet and multi-plated cover plate sheets, which radiated from a lateral mouth. Camptostroma is the most primitive pentameral echinoderm and probably ancestral to all living groups. It had multi-layered plating, five ambulacra in a 2-1-2 arrangement and the anus, gonopore and hydropore in the CD inter-ray. The position of the latter is fundamental in orientating echinoderms. Camptostroma gave rise to the Eleutherozoa via Stromatocystites and the Pelmatozoa via Kinzercystis. Stromatocystites had a flat theca with single-layered, tesselate plating. Kinzercystis had an elongate stalk and biserial brachioles derived from aligned cover plates in Camptostroma. Lepidocystis had both brachioles and free terminal portions of its ambulacra as true arms. Lepidocystis probably gave rise to Echmatocrinus with multiple, free, uniserial arms and thence to the crinoids. Kinzercystis probably gave rise to Gogia and the systoids s.l. (= blastozoans).

Proceedings of the Fifth International Echinoderm Conference / Galway / 24-29 September 1984

An annotated atlas of abyssal echinoderms from the Clipperton-Clarion Fracture Zone, equatorial eastern north Pacific

DAVID L.PAWSON
National Museum of Natural History, Smithsonian Institution, Wash., USA

ERIC J.FOELL
Deepsea Ventures Inc., Gloucester Point, Va., USA

A photographic and free-fall grab-sampling survey of an area rich in manganese nodules in the Clipperton-Clarion Fracture Zone of the equatorial eastern north Pacific Ocean, in depths of 4,500-5,200 meters, has revealed the presence of a conspicuous epibenthic invertebrate megafauna of more than 70 species. Approximately 38 are echinoderms: 6 asteroids, 2 ophiuroids, 3 crinoids, 2 echinoids and 25 holothurians. Most of these echinoderms are illustrated by means of seafloor photographs. For each species, where appropriate, comments are given on systematic status, geographic and bathymetric distribution, and living habits. Approximately 33% of the echinoderms of this poorly explored area are apparently restricted to the eastern Pacific in the broad sense; 30% are Atlantic and Indo-Pacific in distribution; 13% are cosmopolitan; 13% are Indo-Pacific; approximately 11% are known so far only from the Pacific Ocean. At least six species are new. The fauna includes at least one and perhaps two species of large, hitherto undescribed, swimming holothurians, which appear to spend a considerable time on the seafloor, presumably feeding.

Proceedings of the Fifth International Echinoderm Conference / Galway / 24-29 September 1984

The origin and early evolution of *Balanocrinus* (Crinoidea: Articulata)

MICHAEL J.SIMMS
Department of Geological Sciences, University of Birmingham, UK

The genus Balanocrinus is a common Jurassic crinoid recognised primarily on the pattern of the columnal articulations. Superficially similar columnals are found as early as the Middle Triassic and have been taken as ancestral. However, strong similarities in morphology and columnal microstructure exist between Lower Jurassic Chladocrinus and true Balanocrinus but are absent in the supposed Triassic ancestor. Chladocrinus and Balanocrinus share an identical shape and arrangement of stereom lumina on columnal latera, identical arrangement of brachial articulations, and similar very long cirri. Four species, representing a monophyletic lineage, can be unequivocally assigned to Balanocrinus up to the close of the Domerian (fig.1); B. quiaiosensis Loriol, 1891 (Oxynotum-Taylori Subzone), B.subteroides (Quenstedt, 1858)(Taylori-Figulinum Subzone), B.gracilis (Charlesworth, 1847) (Capricornus-Stokesi Subzone) and B. sp.nov. (Subnodosus-Apyrenum Subzone). 'Balanocrinus' antiquus Loriol, 1887, from the Hettangian, is considered a remnant of a Triassic lineage. Balanocrinus quiaiosensis is a small species showing strong morphological similarities to juvenile Chladocrinus. Derivation of Balanocrinus from Chladocrinus, by paedomorphosis, during the mid-Sinemurian seems probable. Subsequent evolution fits the punctuated equilibria model, with prolonged periods of evolutionary stasis punctuated by brief speciation events which correlate with rapid and widespread facies changes. No significant changes have been detected within species but several inter-specific trends are apparent. These include a progressive increase in maximum size, nodal spacing and robustness of the cirri.

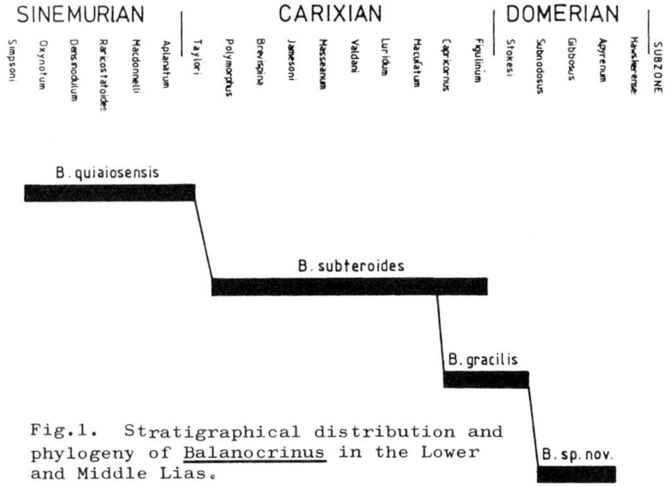

Fig.1. Stratigraphical distribution and phylogeny of <u>Balanocrinus</u> in the Lower and Middle Lias.

Detritus as food for echinoderms

K.R.TENORE
Chesapeake Biological Laboratory, University of Maryland, USA

Sediment detritus, as either a major or supplemental food resource, can be important to echinoderm growth. The degree of importance cannot necessarily be determined by bulk measures of carbon, nitrogen, or calories but by measuring nutritional components such as essential amino acids and fatty acids that vary with source and state of decomposition of the detritus pool. Historically, the sense of Liebig's law of limiting factor has led to identifying the nutrient that limits the growth of a consumer. In detritus-based food chains, the food resource is a heterogeneous pool composed of various sources of detritus (fecal pellets, seaweed and vascular plant-derived particulates) and associated microbes. The contribution of these sources change with ecosystem habitat and seasonally. The various components differ quite dramatically in essential nutrient and caloric content and provide a composite nutritional source to the deposit feeder. Thus, one must evaluate the overall food potential in the context of different components of the detritus pool providing different parts of the overall nutritional requirements of the deposit feeder. A "balanced diet" can be the result of a mixed food supply.

Predation influences the microhabitat distribution of ophiuroids and echinoids in the rocky subtidal zone of the Gulf of Maine, USA

J.D.WITMAN
University of New Hampshire, Durham, USA

The distribution, abundance, and size structure of Ophiopholis aculeata and Strongylocentrotus droebachiensis was quantified along a depth gradient from 8 to 33 m at exposed subtidal sites. With the exception of urchin feeding aggregations at 10-12 m depth, Ophiopholis and Strongylocentrotus were significantly larger and more abundant inside subtidal mussel beds (Modiolus modiolus) than on the substratum outside the beds at all depths sampled. The hypothesis that the mussel beds function as spatial refuges from predation was tested by controlled field experimentation. None of the ophiuroids inside the mussel beds were eaten in two trials conducted with Ophiopholis tethered to positions inside and outside experimental mussel beds. In striking contrast, ophiuroids outside the mussel beds were devastated by predatory lobsters (Homarus americanus) and flounder (Pseudopleuronectes americanus). There was a marked temporal pattern to the predator attacks as lobsters preyed on exposed ophiuroids at night and flounder preyed on exposed ophiuroids during the day. A trial conducted with tethered Strongylocentrotus (20-30mm test diameter) yielded a similar result: all urchins outside the beds were consumed by predators while none were consumed inside the mussel beds. Crabs (Cancer borealis) and lobsters accounted for most of the night predation. During the day, exposed urchins were preyed on by the cunner, Tautogolabrus adspersus. The observed distribution and abundance patterns of ophiuroids and echinoids reflect differential survival inside and outside the mussel beds. Open, rocky substrata with little structural complexity represent high risk habitats where the probability of death from predation is high. Thus, O.aculeata and S. drobachiensis avoid predation by occupying structurally complex mussel beds where they attain high population densities.

2. Crinoidea

Proceedings of the Fifth International Echinoderm Conference / Galway / 24-29 September 1984

Bathymétrie et variabilité morphologique chez les Pentacrinidae (Echinodermes – Crinoïdes pédonculés) du Pacifique occidental

JEAN PAUL BOURSEAU & MICHEL ROUX
Département des Sciences de la Terre, Université Claude Bernard, Lyon I et Centre de Paléontologie stratigraphique et Paléoécologie, associé au CNRS, Villeurbanne, France

The study of recent stalked crinoid populations (family: Pentacrinidae), and the accurate data provided by some authors, allow a detailed analysis of bathymetric distribution for the different species of this family in the western Pacific area. Principal morphological variations of the skeleton linked to this factor are examined. Intraspecific polymorphism in regard to external stem morphology and arm organization is considered for numerous species in that geographic province. The number of taxa could be reduced because some of them are probably ecophenotypal and geographical varieties. Species regroupings are proposed for the Pacific area, especially for members of the genus Metacrinus, Stem and arm morphological variations linked to bathymetry, are interpreted in terms of adaptative strategy (r and k selection). Biogeographical repartition of these species is analysed from that point of view. Such an approach has been conducted to examine stalked crinoid species of genera other than Metacrinus, living in other geographical areas (Atlantic Ocean). Taxonomy, paleoecology and biostratigraphy of fossil stalked crinoids must be reconsidered and discussed in the application of these results.

1 INTRODUCTION

L'analyse de la distribution bathymétrique des espèces actuelles de crinoïdes pédonculés (figure 1) montre une distribution globale bimodale (Roux, 1982). Le premier mode (strate supérieure) se situe entre 200 et 500 mètres de profondeur. Il est dominé par les Pentacrines (Isocrinida, Pentacrinidae). Le second (strate inférieure) se place entre 1 500 et 3 000 mètres de profondeur. Il comprend principalement des représentants du sous-ordre des Bourgueticrinina (Bourgueticrinidae, Bathycrinidae). Entre 600 et 1 500 mètres de profondeur, le nombre d'espèces demeure toujours faible ; en revanche, l'abondance absolue des individus de chaque espèce peut être relativement élevée.

Si la province biogéographique du Pacifique occidental montre clairement cette distribution bimodale, la province de l'Atlantique occidental tropical et celle de l'Atlantique Nord-Est présentent un mode nettement dominant situé soit dans la strate supérieure (Atlantique Ouest) soit dans la strate inférieure (Atlantique Est). Les cri-

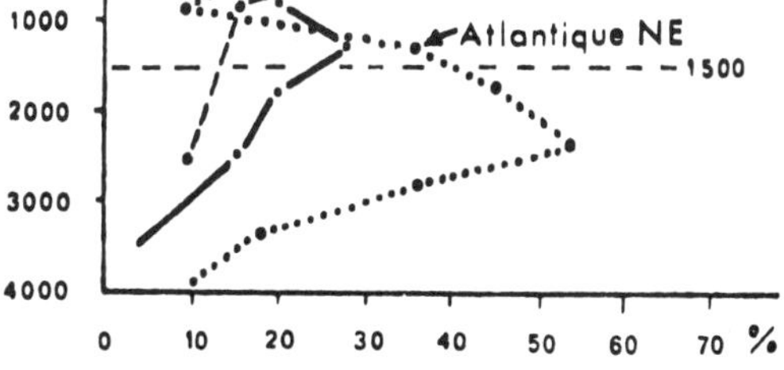

Fig. 1 : Répartition bathymétrique des Crinoïdes pédonculés dans les trois principales provinces actuelles. En ordonnées : bathymétrie en mètres ; en abscisses : diversité spécifique.

noïdes pédonculés étant des filtreurs efficaces dont le mode de sustentation est très sensible aux variations d'énergie du milieu, cette différence de part et d'autre de l'Atlantique a été interprétée comme la conséquence soit d'un flux trophique trop faible se raréfiant dès les profondeurs moyennes (à l'Ouest), soit de la présence de courants d'upwellings saisonniers (à l'Est).

Dans le Pacifique occidental, les populations de Pentacrines de la strate supérieure ont été analysées récemment à la suite des campagnes Musorstom I-II à l'Ouest des Philippines et de la campagne Corindon dans le détroit de Macassar (Roux, 1981 ; Bourseau & Roux, à paraître). Ces résultats s'ajoutent aux données très précises sur des espèces du Sud du Japon fournies par T. Gislen (1922) et à celles concernant une population recueillie près de Tahiti d'une espèce par ailleurs connue aux Philippines (Roux, 1980). Cet ensemble nous permet une analyse détaillée de la distribution bathymétrique des Pentacrines et des principales variations de la morphologie du squelette qui y sont liées.

2 REPARTITION BATHYMETRIQUE DES PENTACRINIDAE DANS LE PACIFIQUE OCCIDENTAL

La figure 2 illustre la répartition bathymétrique des espèces de Pentacrines recueillies au cours des campagnes Musorstom I-II et Corindon. La distribution de chaque espèce observée dans le secteur des Philippines est plus restreinte que celle qui apparaît lorsqu'on additionne les données issues des diverses régions du Pacifique occidental. Ceci pourrait résulter d'un échantillonnage insuffisant. Il est plus probable d'envisager des fluctuations de la position bathymétrique des espèces, selon les lieux en fonction des conditions écologiques qui varient considérablement dans un domaine aussi vaste.

Les plus grandes Pentacrines recueillies par les campagnes Musorstom et Corindon vivent à une profondeur relativement faible (126-230 mètres). Saracrinus superbus est souvent associé dans une même station soit à Metacrinus musorstomae soit à S. angulatus. M. interruptus n'est associé à aucune autre espèce de Pentacrine et semble occuper une niche écologique différente (énergie du milieu plus faible). M. serratus et Diplocrinus sibogae, espèces de taille moyenne, vivent entre 300 et 600 mètres, dans des stations différentes. Dans le détroit de Macassar, le créneau bathymétrique compris entre 600 et 800 mètres est occupé par une seule espèce M. wyvillei représentée à une station par un nombre d'individus très élevé. Les espèces les plus profondes sont M.

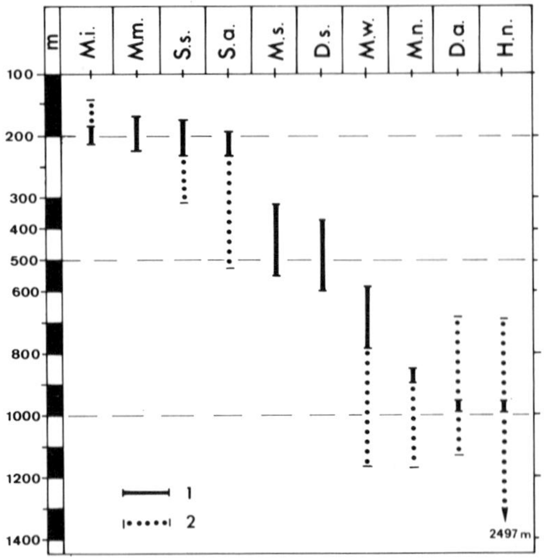

Fig. 2 : Distribution bathymétrique des espèces de Pentacrines recueillies lors des campagnes Musorstom et Corindon. 1 - profondeur des stations Musorstom et Corindon ; 2 - extension bathymétrique d'après les campagnes antérieures ; M.i. : Metacrinus interruptus ; M.m. : Metacrinus musorstomae ; S.s. : Saracrinus superbus ; S.a. : Saracrinus angulatus ; M.s. : Metacrinus serratus ; D.s. : Diplocrinus sibogae ; M.w. : Metacrinus wyvillei ; M.n. : Metacrinus nodosus ; D.a. : Diplocrinus alternicirrus ; H.n. : Hypalocrinus naresianus.

nodosus, D. alternicirrus, Hypalocrinus naresianus. Cette dernière est souvent représentée dans les stations par un grand nombre d'individus alors que les exemplaires appartenant aux deux premières sont rares. La station n° 56 de la campagne Musorstom II a fourni, en un seul trait de chalut, cinquante spécimens d'H. naresianus pour un seul exemplaire de Diplocrinus alternicirrus.

La manière de présenter une zonation bathymétrique dépend beaucoup de la conception large ou étroite de la notion d'espèce que l'on adopte. L'observation des très nombreux individus récoltés sur les stations Musorstom et Corindon conduit à penser qu'il existe une très large variabilité des espèces, voire un polymorphisme intraspécifique portant sur la morphologie du pédoncule et sur l'organisation de la couronne brachiale. A l'inverse de la figure 2, une conception large de la notion d'espèce a été adoptée pour la figure 3 qui porte sur l'ensemble des Pentacrines du Pacifique. Les regroupements d'espèces proposés s'appuient sur une étude détaillée des articulations des bras et du pédoncule. Ils tien-

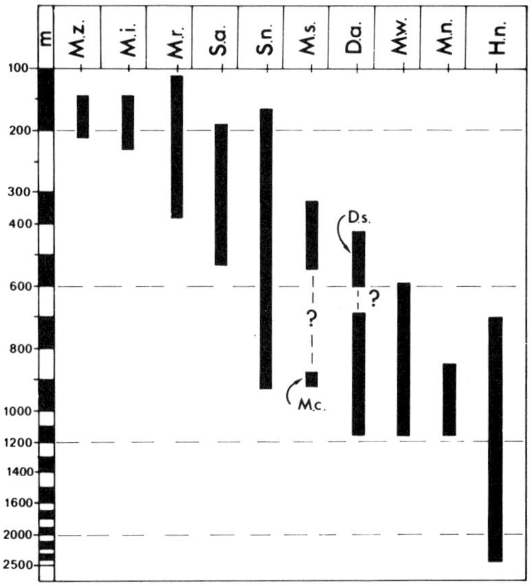

Fig. 3 : Distribution bathymétrique des Pentacrines du Pacifique dans l'hypothèse d'un nombre d'espèces réduit à dix. M.z. : Metacrinus zonatus ; M.i. : Metacrinus interruptus (incluant M. musorstomae) ; M.r. : Metacrinus rotundus (incluant M.c. : M. costatus) ; M.w. : Metacrinus wyvillei ; M.n. : Metacrinus nodosus ; S.a. : Saracrinus angulatus (incluant toutes les formes affines, voir texte) ; S.n. : Saracrinus nobilis (incluant S.s. : S. superbus et S. varians) ; D.a. : Diplocrinus alternicirrus (incluant D.s. : D. sibogae) ; H.n. : Hypalocrinus naresianus.

nent compte aussi de la présence très fréquente, dans une même station, d'espèces en fait morphologiquement très proches (Bourseau & Roux, à paraître). Ce point de vue aboutit à une réduction considérable du nombre d'espèces. A.H. Clark (1923) en distinguait vingt-trois. Une position maximaliste ramène ce nombre à dix. M. musorstomae peut être interprété comme un écophénotype de M. interruptus. L'organisation de sa couronne ne présente pas de différence supérieure à celle qui existe entre les spécimens du locus typicus (Philippines) et ceux du Japon décrits par T. Gislen (variété géographique ?). De même, M. rotundus regrouperait les formes affines M. multisegmentatus et M. cyanus. S. angulatus inclurait S. acutus, S. cingulatus, S. batheri, S. suluensis qui entrent dans le même champ de variabilité. S. nobilis, espèce eurybathe, montre une variabilité morphologique très importante qui se manifeste aussi au cours de l'ontogenèse. Elle pourrait inclure des formes allant de S. superbus à S. varians. M. serratus est à rapprocher de M. costatus en raison de leurs très fortes affinités morphologiques. Le hiatus bathymétrique qui sépare les deux taxons pourrait résulter d'une lacune d'information. On retrouve ce problème chez Diplocrinus avec D. alternicirrus et D. sibogae. Les autres espèces (M. wyvillei, N. nodosus, H. naresianus) sont clairement distinctes.

Ces dix espèces se relaient assez bien avec la profondeur pour préserver une diversité minimale de trois espèces différentes par niveau bathymétrique. On atteint cinq espèces vers 200 mètres et six entre 800 et 1 200 mètres. Au-delà, il ne subsiste qu'une seule espèce : H. naresianus.

Cette présentation conduit à penser qu'au sein de chaque espèce, il pourrait y avoir des variations morphologiques liées à la bathymétrie. Parmi les formes que nous avons regroupées au sein de S. nobilis, S. superbus, pourrait représenter le morphotype dominant autour de 200-250 mètres alors que S. varians serait celui inféodé plus particulièrement aux sites plus profonds. De même chez Diplocrinus, D. sibogae, taxon mal connu, toujours trouvé vers 500 mètres, pourrait être un morphotype de faible profondeur de D. alternicirrus espèce montrant par ailleurs une vaste répartition géographique accompagnée d'une variation morphologique importante. Le nombre de cirres est le plus fréquemment de cinq sur un spécimen typique des Philippines (locus typicus) ; les verticilles incomplets semblaient pourtant le caractère le plus symptomatique de l'espèce. H. naresianus paraît présenter un nombre de bras variable en fonction de la profondeur : les spécimens en ayant plus de dix apparaissent uniquement dans la partie supérieure de la zone de répartition bathymétrique.

L'espèce la plus eurybathe (groupe de S. nobilis) est aussi celle qui a la plus large répartition géographique (du Japon à la Nouvelle Zélande). Chez Metacrinus, ce sont les espèces situées aux pôles de la répartition bathymétrique du genre qui ont la plus large dispersion géographique : M. rotundus, espèce la moins profonde, est connue du Japon au Sud-Est de l'Australie ; M. wyvillei, espèce la plus profonde, s'étend également du Japon aux îles Kermadec. La seule espèce de Pentacrines commune aux Philippines et au Pacifique Central (Tahiti), D. alternicirrus (inclus D. sibogae) est aussi relativement eurybathe.

A l'inverse, certaines espèces plus sténobathes ont une extension géographique plus restreinte. M. zonatus est limité à la mer des Célèbes. M. interruptus et S. angulatus sont connus de l'Indonésie au Sud du Japon. M. serratus n'a été récolté qu'aux environs des Philippines.

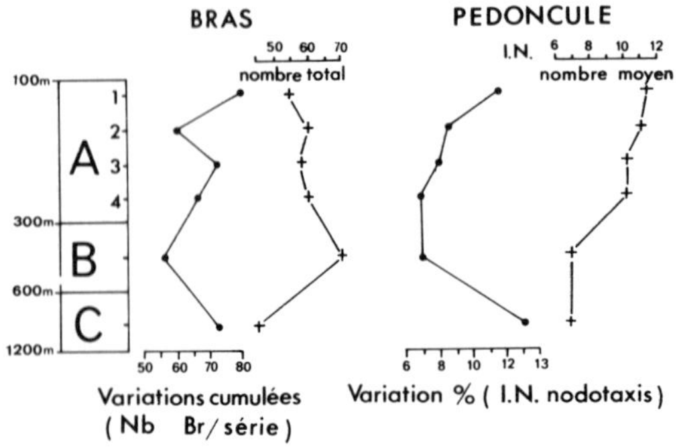

Fig. 4 : Variabilité morphologique chez Metacrinus en fonction de la profondeur. Les espèces ont été ordonnées en fonction de la profondeur minimale connue. Pour A_3 et A_4, cette profondeur est identique ; c'est alors le degré d'énergie du milieu (plus faible pour A_4) qui a été utilisé. A - tranche de 100-250 mètres ; A_1 : M. rotundus ; A_2 : M. interruptus (Japon) ; A_3 : M. musorstomae ; A_4 : M. interruptus (Philippines). B - tranche de 300-550 mètres illustrée par M. serratus. C - tranche de 500-1 200 mètres illustrée par M. wyvillei. I.N. : internodales ; Nb Br/série : nombre de brachiales par série.

3 VARIABILITE MORPHOLOGIQUE ET BATHYMETRIE

3.1 Analyse chez le genre Metacrinus

Le genre Metacrinus présente la plus forte diversité spécifique et la plus large répartition bathymétrique. L'analyse n'a porté que sur les espèces recueillies en nombre suffisant pour que les données soient statistiquement significatives ainsi que sur les caractères du squelette les moins dépendants de la croissance et les plus aisément quantifiables (figure 4).

Le nombre de bras croît avec la profondeur jusqu'à la tranche bathymétrique comprise entre 300 et 600 mètres. Au-delà de ce seuil, il chute brutalement pour l'espèce la plus profonde M. wyvillei. La variabilité de l'organisation de la couronne brachiale est la plus importante pour les taxons situés aux bornes de la répartition bathymétrique : M. rotundus et M. wyvillei. M. serratus qui illustre la tranche comprise entre 320 et 550 mètres de profondeur montre une variabilité très faible dans l'organisation de sa couronne brachiale (tableau 1)

Le pédoncule présente lui aussi avec la profondeur une diminution régulière du nombre moyen d'internodales par nodotaxis. Le changement le plus important (passage de plus de dix internodales à sept internodales) est réalisé à partie de 300 mètres de profondeur (M. serratus, M. wyvillei). La variabilité de ce caractère baisse avec l'accroissement de la bathymétrie jusqu'au seuil de 600 mètres puis elle redevient plus élevée avec l'espèce la plus profonde (M. wyvillei). On remarquera que c'est encore pour les espèces situées aux bornes de la tranche de répartition bathymétrique (tableau 1) que cette variabilité est la plus importante (M. rotundus, M. wyvillei).

	I B_r	II B_r	III B_r	IV B_r	Nb IN
M. rotundus (1)	16,3	25,1	19,9	23,9	11,6
M. interruptus (1)	7,1	11,9	22,0	19,6	8,7
M. musorstomae (2)	7,3	19,5	20,0	25,3	8,1
M. interruptus (2)	8,1	19,3	17,5	21,9	6,9
S. superbus (2)	8,1	16,3	17,6	20,4	7,2
S. angulatus (2)	4,4	16,9	15,4	20,5	12,5
M. serratus (2)	7,9	13,7	17,9	16,9	6,8
M. wyvillei (3)	11,8	13,5	22,3	23,7	13

Tabl. 1 : Comparaison des coefficients de variation (exprimés en %) des principaux caractères des espèces de Metacrininae étudiées. Les espèces sont ordonnées en fonction de leur répartition bathymétrique (en haut, les moins profondes, en bas les plus profondes). Nb IN : nombre d'internodales par nodotaxis mature - X Br : nombre de brachiales par série. (1) Japon, d'après les données de T. Gislen (1922) et de M. Roux (1981) ; (2) Philippines, d'après J.P. Bourseau & M. Roux (à paraître) ; (3) détroit de Macassar, d'après J.P. Bourseau & M. Roux (à paraître).

3.2 Analyse sur l'ensemble des Pentacrines

Elle porte sur les caractères communs à toutes les Pentacrines et sur les espèces pour lesquelles nous disposons d'une information satisfaisante (figure 5).

La longueur maximale des bras décroît avec la profondeur jusque vers 800 mètres. Au-delà de ce seuil, elle réaugmente légèrement. Le nombre de bras décroît régulièrement avec la profondeur à partir de 300 mètres. Globalement, les espèces vivant à des profondeurs inférieures au seuil de 600 m ont plus de cinquante bras tandis qu'au-delà ce nombre devient nettement plus faible. H. naresianus qui occupe la zone la plus profonde dans la strate des Pentacrines voit son nombre de bras se réduire au minumum de dix.

Au niveau du pédoncule, le diamètre maximal ne diminue de façon significative que pour l'espèce la plus profonde. La valeur maximale est atteinte pour la forme S. superbus

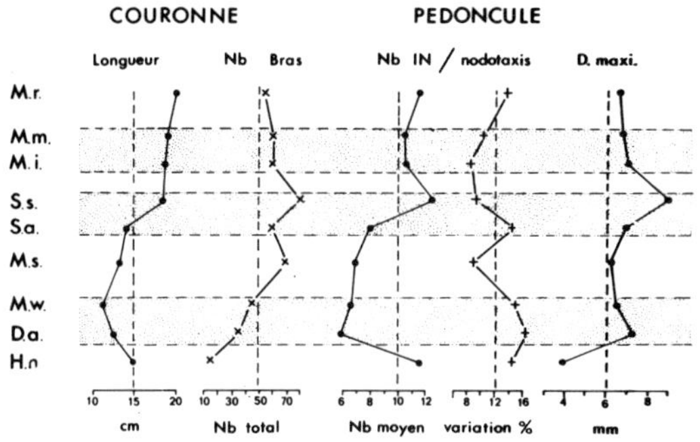

Fig. 5 : Variation des principaux caractères de la couronne et du pédoncule des Pentacrines du Pacifique en fonction de la profondeur. De haut en bas, les espèces sont ordonnées par créneau bathymétrique de profondeur croissante. Les bandes en pointillés correspondent aux espèces ayant le même créneau bathymétrique (Initiales des espèces citées dans le texte et dans la légende des figures 2 et 3). Pour la longueur et le nombre de bras de la couronne, ce sont la longueur maximale et le nombre le plus élevé de bras qui ont été utilisés. Pour D. alternicirrus, les valeurs sont tirées de M. Roux, 1980.

du groupe de S. nobilis qui vit entre 200 et 300 mètres de profondeur. Le nombre moyen d'internodales par nodotaxis diminue progressivement avec l'accroissement de la bathymétrie, évoluant d'un pôle à onze internodales pour les espèces de faible profondeur vers un pôle à six internodales pour les espèces vivant entre 350 et 900 mètres. Pour l'espèce qui se maintient au-delà de 1 200 mètres (H. naresianus), on passe brusquement à un nombre élevé de près de douze internodales. La variabilité de ce caractère pour les taxons recensés est encore une fois plus importante pour les espèces situées aux bornes de la strate à Pentacrines (M. rotundus la moins profonde ; M. wyvillei, D. alternicirrus, H. naresianus, les plus profondes). Elle est la plus faible pour les espèces vivant dans la tranche bathymétrique comprise entre 300 et 600 mètres.

Sur la plupart des courbes, les deux espèces de Saracrinus analysées s'écartent sensiblement de la tendance générale de modification de la morphologie avec la profondeur. En revanche si on prend les valeurs moyennes entre les deux espèces une telle irrégularité dans les courbes disparaît. Cette approximation peut être envisagée car les populations analysées ont été recueillies dans les mêmes stations. Chacune des espèces du genre Saracrinus paraît s'être différenciée vers un des pôles du gradient de variation morphologique possible dans un même biotope.

4 INTERPRETATION

Avec la profondeur, la nourriture se raréfie. La capacité de filtrage de la couronne dépend du nombre de bras et de leur longueur. Par ailleurs, les principaux besoins énergétiques se situent dans les bras (muscles, glandes génitales notamment). La réduction de la longueur des bras avec la profondeur chez Metacrinus aboutit à limiter les besoins énergétiques, l'essentiel de la capacité de filtration étant préservée grâce au maintien d'un grand nombre de bras. De ce point de vue, l'optimum semble réalisé chez M. serratus qui vit entre 300 et 600 mètres. En revanche, pour les deux espèces les plus profondes (D. alternicirrus, H. naresianus) la réduction des besoins énergétiques s'effectue par une diminution considérable des divisions brachiales, tempérée par une légère réaugmentation de la longueur des bras qui permet le maintien d'un bon niveau de filtration.

L'interprétation des variations morphologiques du pédoncule avec la profondeur est moins évidente. Elle pourrait être recherchée au niveau des facteurs hydrodynamiques et de leurs implications morphofonctionnelles.

Le modèle antérieurement proposé (Roux, 1982) s'appuie sur la notion de stratégie adaptative. A l'approche de conditions instables, la stratégie r (opportuniste) tend à dominer. Morphologiquement, elle se traduit généralement par une forte variabilité et une spécialisation faible. Pour les Pentacrines, deux bornes d'instabilité limitent leur répartition bathymétrique : une borne supérieure (vers 100 mètres) correspond au seuil de vulnérabilité hydrodynamique ; une borne inférieure (vers 250 mètres) représente le seuil de vulnérabilité trophique. Nos résultats confirment ce modèle : les espèces vivant vers les bornes de la répartition bathymétrique présentent la variabilité la plus élevée. La zone la plus stable pour ces animaux paraît se situer entre 300 et 600 mètres avec développement de la stratégie de type k : optimum de capacité de filtration et faible variabilité de M. serratus. L'analyse morphologique fournit des résultats parfaitement complémentaires des aspects biogéographiques de la stratégie adaptative (eurybathie ou sténobathie, dispersion géographique). Les stratèges de type r les plus typiques sont M. rotundus, M. wyvillei et S. nobilis. Ces espèces ont la plus large extension géographique ; S. nobilis paraît l'espèce la plus eurybathe et la plus polymorphe.

Il sera intéressant de comparer les résultats obtenus sur les Pentacrines du Pacifique occidental avec les données issues d'autres provinces. D'ores et déjà, nous pouvons constater que, dans l'Atlantique oriental, l'unique espèce vivant le plus souvent entre 1 200 et 2 000 mètres (Diplocrinus wyvillethomsoni) a les caractéristiques d'un stratège de type r avec une forte variabilité morphologique (de 11 à 30 %) (Roux, 1976). Les caractères de la couronne et du pédoncule s'intègrent fort bien dans les courbes de la figure 5, entre D. alternicirrus et Hypalocrinus naresianus. Malgré un matériel disponible abondant (voir Meyer et al., 1978), des analyses biométriques assez précises n'ont pas encore été menées sur les Pentacrines de l'Atlantique tropical occidental.

La plupart des caractères morphologiques analysés sont étudiables chez les fossiles. Outre les applications paléoécologiques que l'on peut en espérer, nos résultats doivent inciter à la prudence dans la définition des taxons fossiles et leur utilisation biostratigraphique.

5 REFERENCES BIBLIOGRAPHIQUES

Bourseau, J.P. & M. Roux. Résultats des campagnes Musorstom II et Corindon - Echinodermes : Crinoîdes Pentacrinidae (à paraître).

Clark, A.H., 1923. A revision of the recent representatives of the crinoid family Pentacrinidae with the diagnoses of two new genera. Journ. Washington Acad. Sc. 13 (1) : 8-12.

Gislen, T. 1922. The crinoids from Dr. S. Bock's expedition to Japan - 1914. Regiae Soc. Sc. Uppsaliensis. 5 (6) : 1-183.

Meyer, D.L., Messling, C.G. & Macurda, D.B., 1978. Zoogeography of tropical Western atlantic Crinoidea (Echinodermata). Bull. mar. Sc. 28 (3) : 412-441.

Roux, M., 1976. Aspects de la variabilité et de la croissance au sein d'une population de la Pentacrine actuelle : Annacrinus wyvillethomsoni Jeffreys (Crinoidea). Thalassia Jugoslavica. 12 (1) : 307-320.

Roux, M., 1980. Découverte de site à Crinoîdes pédonculés (genres Diplocrinus et Proisocrinus) au large de Tahiti. C.R. Acad. Sc. Paris. 290 (D) : 119-122.

Roux, M., 1981. Echinodermes : Crinoîdes Isocrinidae in Résultats des campagnes Musorstom - Philippines (18-28 mars 1976). Mém. ORSTOM. 91 : 477-543.

Roux, M., 1982. De la biogéographie historique des océans aux reconstitutions paléobiogéographiques : tendances et problèmes illustrés par des exemples pris chez les Echinodermes bathyaux et abyssaux. Bull. Soc. géol. France. 24 (5-6) : 907-916.

Proceedings of the Fifth International Echinoderm Conference / Galway / 24-29 September 1984

Post-spawning behavior and early development of the comatulid crinoid, *Antedon bifida*

MARIE-CHRISTINE LAHAYE & MICHEL JANGOUX
Laboratoire de Biologie marine, Université Libre de Bruxelles, Belgium

SUMMARY. Spawned females of **A. bifida** show a marked brooding behavior in protecting pinnule-borne eggs or embryos. Fully differenciated doliolariae hatch from the fertilization membrane ca 24 hours after fertilization. Swimming and pre-settling larval behaviors are described. It is suggested that the doliolaria apical tuft functions as a sensory organ that is able to test the quality of the substratum. Definitive larval settlement occur at the latest 5 days after fertilization.

INTRODUCTION.

Our knowledge of the development of the common European comatulid, **Antedon bifida** is based exclusively on nineteenth century studies (see, e.g., Carpenter 1876, Perrier 1886). No information occurs on the behavior of the precystidean larvae (i.e. doliolaria larvae). The aim of the present paper was to study the behavior and structure of **A. bifida** early larvae from both in vivo observations and SEM investigations. Special attention was drawn also to the peculiar post-spawning behavior shown by adult females of **A. bifida.**

MATERIAL AND METHODS.

Adult individuals of **Antedon bifida** (Pennant) were collected by scuba-diving at Morgat (Brittany , France). They were maintained in an open-circuit marine aquarium at the Biological Station of Roscoff.

SEM-observations were done on specimens (whole pinnules or larvae) fixed in Bouin's fluid for a 12 hours period. They were dehydrated by graded ethanol and dried by the critical point method using CO_2 as transition fluid. The pieces were mounted on aluminium stubs, coated with gold in a sputter coater and observed with a ISI DS-130 Scanning Electron Microscope.

OBSERVATIONS AND DISCUSSION.

1. Post-spawning behavior of female **A. bifida**

When mature the genital pinnules of **A. bifida** are conspicuously distended. Gametes are emitted through star-like gaps located on the side wall of each pinnule (one gap per pinnule). These gaps may be considered as temporary gonopores as they occur only at the time of spawning (Fig. 1). Spawning individuals induce the nearby ones to spawn in turn provoking a kind of chain reaction. Such epidemic spawning could be caused by the release by spawning individuals of a pheromone into the sea-water (Holland and Grimmer 1975).

Spawned eggs cluster to the pinnule outer surface owing to a particular binding substance secreted by the epidermis (Holland and Grimmer 1975) (Fig 2). Pinnule-bound eggs are fertilized in situ. Fertilization occur at the latest 48 hours after spawning. Most eggs however are fertilized within a 24 hours period.

Spawned females always show a protective behavior of their eggs whether or not they have been fertilized. Actually each spawned female draws its arms closer together, the pinnules folding back along the longitudinal axis of the arms. This behavior obviously should be considered

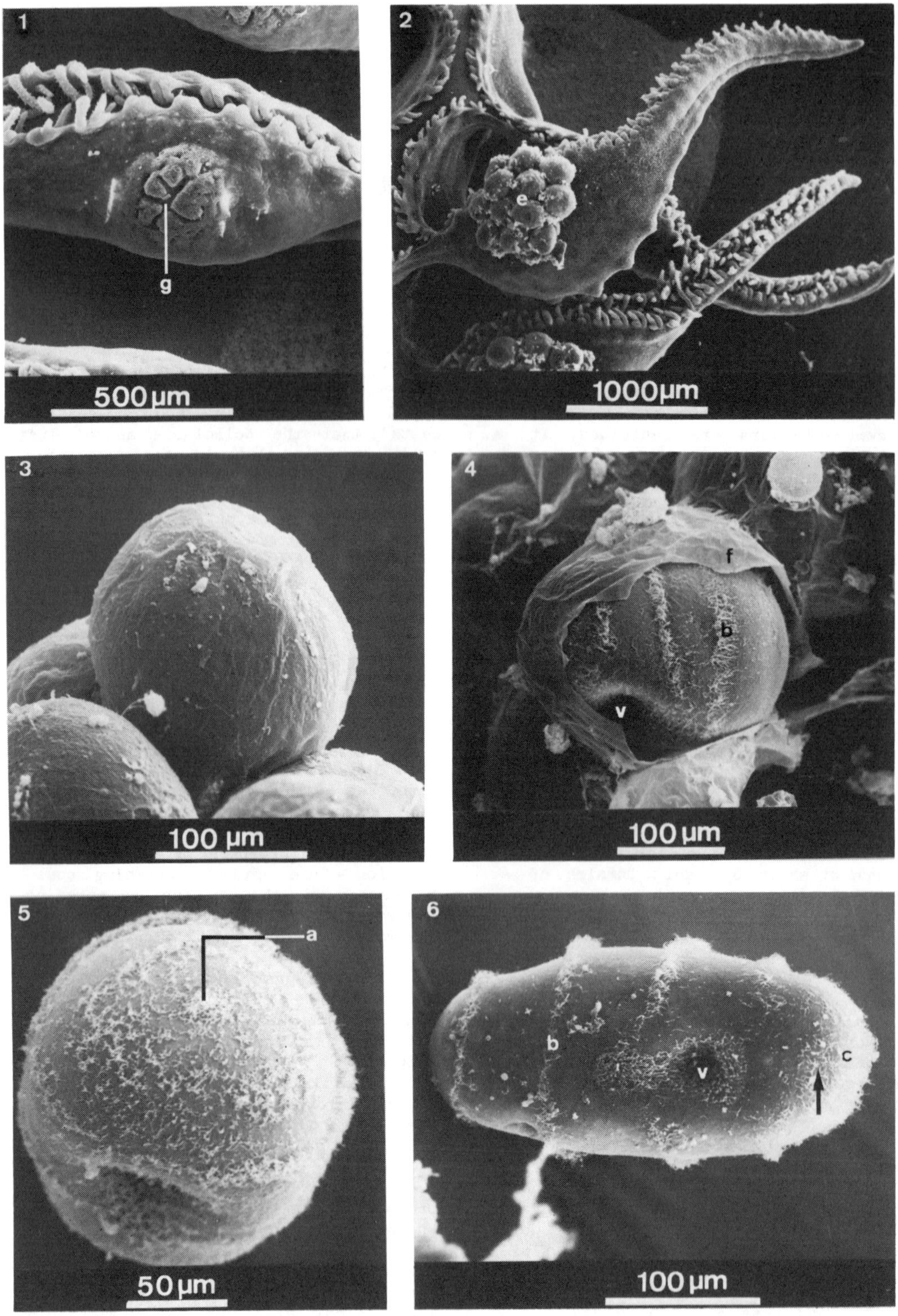
1
g
500 µm
2
e
1000µm
3
100 µm
4
f
b
v
100 µm
5
a
50 µm
6
b
v
c
100 µm

as a brooding behavior. It allows easy recognition of spawned females in the field. It should be noted also that, when brooding, female comatulids are almost insensitive to mechanical stimuli.

Aquarium observations of fertilized females showed that the brooding behavior occur until the larva hatches. The fate of unfertilized females is most peculiar as their egg-bearing genital pinnules tend to autotomize. This does not necessarily imply the death of the comatulid which can regenerate the autotomized genital pinnules.

2. The early larva of **A. bifida.**

The outer surface of the fertilization membrane of **A. bifida** is perfectly smooth (Figs 3, 4), being markedly different from the ornamented membrane previously described for several other comatulid species (see, e.g., Dan and Dan 1941, Holland and Jespersen 1973, Mladenov and Chia 1983). Hatching occur ca 24 hours after fertilization. Hatching larvae are rounded to slightly barrel-shaped. They are well-differenciated doliolaria having a ciliated and almost pyriform vestibular invagination, an anterior ciliary cap with a central apical tuft and four transverse ciliary bands (Figs 4, 5). Here also marked differences occur when comparing hatching larva of **A. bifida** to that of other species such as **Comanthus japonicus** and **Florometra serratissima** (see Holland and Kubota 1975, and Mladenov and Chia 1983, respectively). Hatching in these latter species seemingly occur with less differentiated larva, namely an uniformely ciliated larva having still a small blastopore at is posterior end.

Developing doliolaria of **A. bifida** progressively increase in size (Fig 6). About 48 hours after hatching an additionnal small depression -viz. the so-called adhesive pit- appears ventrally between the anterior ciliary cap and the first transverse ciliary band. That pit correspond to the future area of attachment of the doliolaria to the substratum.

The swimming behavior of doliolaria is summarized in Fig 7. The larva swim along two superposed helicoidal path while rotating clockwise along its longitudinal axis. Swimming movements of **A. bifida** doliolaria appear more complex than those of **F. serratissima** doliolaria, that latter swimming along a single sinusoidal path (Mladenov and Chia 1983).

Exploratory movements of the substratum (i.e. the pre-settling behavior of Mladenov and Chia 1983) occurred through the end of the third day of development. Pre-settling behavior may be observed for a period of ca 48 hours. During that time doliolariae usually are able to settle reversibly on different substrata. Before settling the larva always brushes the substratum using its apical tuft which seemingly functions as a sensory organ testing the quality of the substratum. Definitive settlement implies the secretion of an strongly adhesive cement by the larval adhesive pit.

Preliminary experiments showed that **A. bifida** doliolaria are able to discriminate between different substrata and that they never settle at random. Moreover they are markedly gregarious either in aquarium conditions or in the

Figure 1. SEM view of a mature genital pinnule of a female **A. bifida** showing the star-like gonopore (g).

Figure 2. SEM view of a genital pinnule just after spawning. e, eggs.

Figure 3. SEM view of **A. bifida** embryos. Note the smooth aspect of the fertilization membrane.

Figure 4 . Hatching of a doliolaria larva. b, ciliary band; f, fertilization membrane; v, vestibular invagination.

Figure 5. Upper view of a doliolaria larva. a, apical tuft.

Figure 6. Lateral view of a late doliolaria larva. b, ciliary band; c, ciliary cap; v, vestibular invagination.(The arrow indicates the place of the adesive pit).

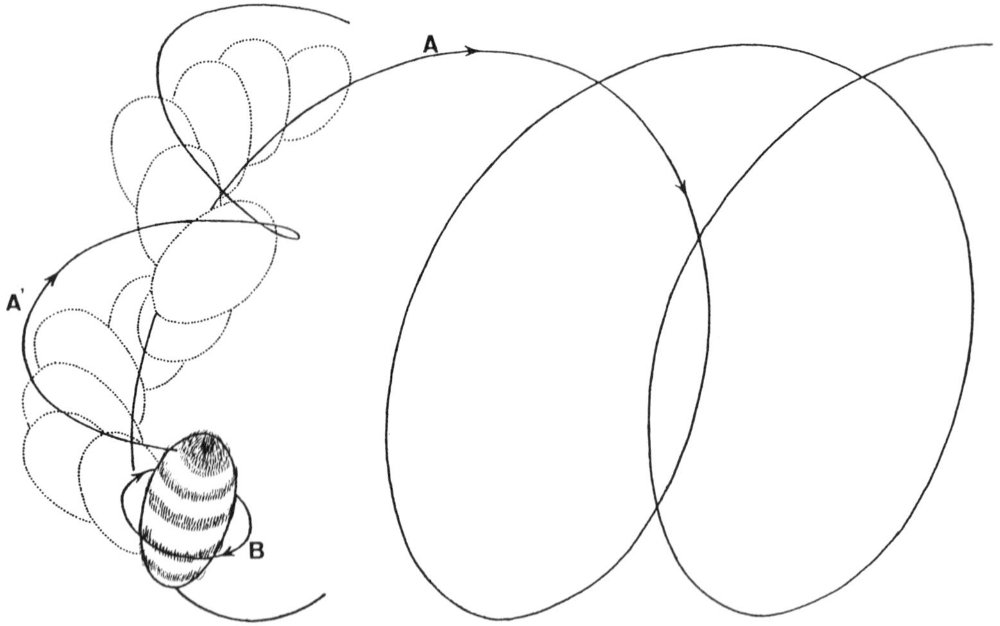

Figure 7. Swimming behavior of **A. bifida** doliolaria. A-A', helicoidal swimming; B, clockwise rotation along the larval longitudinal axis.

field (early cystidean larvae observed in situ were found most often close to or within an adult colony).

Acknowledgments. Work supported by I.R.S.I.A. grant 82085 to Marie-Christine Lahaye, and by F.R.F.C. grant 2.4525.80.

BIBLIOGRAPHY.

Carpenter, W.B., 1876, On the structure, physiology and development of **Antedon** (**Comatula** Lamk) **rosaceus**, Proc. roy. Soc. Lond., 24 : 211-231.

Dan, J. & K. Dan, 1941, Early development of **Comanthus japonicus**, Jap. J. Zool. 9 : 565-574..

Holland, N.D. & J.C. Grimmer, 1975, Epidermal mucus and the reproduction of a crinoid echinoderm, Nature, Lond., 225 : 223-224.

Holland, N.D. & A. Jespersen, 1973, The fine structure of the fertilization membrane of the feather star **Comanthus japonica** (Echinodermata: Crinoidea),Tissue Cell ,5: 209-214.

Holland, N.D. & H. Kubota, 1975, Correlated scanning and transmission electron microscopy of larvae of the feather star **Comanthus japonica** (Echinodermata : Crinoidea), Trans. Amer. microsc. Soc., 94 : 58-70.

Mladenov, P. & F.S. Chia, 1983, Development, settling behaviour, metamorphosis and pentacrinoid feeding and growth of the feather star **Florometra serratissima**, Mar. Biol., 73 : 309-326.

Perrier, E., 1886, L'organisation et le développement de la comatule de la Méditerranée **Antedon rosacea**, Nouv. Archs Mus. Hist. nat. Paris, 9 : 53-348.

Proceedings of the Fifth International Echinoderm Conference / Galway / 24-29 September 1984

Submersible observations of deep-water crinoid assemblages in the tropical western Atlantic Ocean

C.G.MESSING
Undergraduate Marine Science Program, University of Miami, Fla., USA

ABSTRACT: Observations from DSRV ALVIN and JOHNSON SEA-LINK in Bahamian waters, and the PC-8B off Jamaica reveal several distinct crinoid assemblages at different depths and on different substrates. South of Grand Bahama I., and off Jamaica in about 200m on sediment-veneered pavement and prominent outcrops, Cenocrinus asterius (Isocrinida) Leptonemaster venustus and Comactinia meridionalis hartlaubi (Comatulida) dominate. West of Little Bahama Bank in 280-400m on sediment-veneered pavement, abundant Endoxocrinus parrae and Neocrinus decorus (Isocrinida) appear with obviously different crown elevations; Democrinus brevis (Bourgueticrinida), normally associated with unconsolidated substrates, reaches densities of $5/m^2$ in 290m; an abundant comatulid fauna includes at least five species in four families. In the same area in 600-624m, short-stalked Diplocrinus maclearanus? reach densities of $15\text{-}20/m^2$ on lithoherms, whereas Isocrinus blakei (both Isocrinida) is restricted to intervening flats; the comatulid fauna is less diverse at this depth. Measured crinoidal contribution to the sediment reaches 11.6% by grain count but probably exceeds this in the densest aggregations. A stalked crinoid has been observed unattached and crawling.

1 INTRODUCTION

Ever since the great dredging expeditions of the last century, we have known that extant crinoids occur in a wide range of marine habitats, sometimes in enormous numbers (Agassiz, 1888). Structural diversity has suggested that different forms, e.g., bathycrinids, isocrinids, hyocrinids, exist on different substrates and under different conditions. It has only been in the last decade or so, however, that direct observations, in shallow water via SCUBA and at greater depths via manned submersibles, have begun to reveal details of crinoid posture and orientation, habitat and ecological relationships (Macurda and Meyer, 1974; Messing, 1984).

Although manned submersibles now allow us to examine many heretofore undredgable hard bottoms, most research into benthic communities below the limits of compressed air diving remains focussed on unconsolidated substrates.

During the last several years, a series of dives from the limits of SCUBA to over 1000m have allowed me to examine community structure and faunal relationships of a variety of hard bottom assemblages in general, and of crinoids in particular. Herein, I describe local and bathymetric distribution, substrate, water movement and faunal relationships of stalked and comatulid crinoids as observed from submersibles in the Straits of Florida and adjacent waters, and off the northern coast of Jamaica.

2 METHODS

Submersible dive data are listed in Table 1. All vehicles are equipped with hydraulically operated collecting arms of varying dexterity and function. DSRV ALVIN and PC-8B have simple claws. JOHNSON SEA-LINK has, in addition, a grab sampler and two suction hoses. Observations were recorded on audio and video tape, and via hand-held and externally mounted cameras. Most specimens have been deposited with the Division of Invertebrate Zoology, National Museum of Natural History, Washington,D.C.

3 OBSERVATIONS

Reef-dwelling comatulid crinoids such as Nemaster spp. and Analcidometra armata (Pourtalès) may occasionally be glimpsed during initial descent or final ascent, but because they are studied far more

Table 1. Summary of submersible dives contributing to observations.

Submersible	Dive Number	Date	Location	Maximum Depth
			Northwest Providence Channel	
ALVIN	843	11/1/78	26°23.6'N, 78°42.3'W	892m
"	844*	11/2/78	26°34.0'N, 78°11.9'W	961
"	846	11/3/78	26°25.7'N, 77°52.1'W	785
"	847*	11/4/78	26°12.6'N, 78°25.0'W	1277
"	848*	11/5/78	25°54.6'N, 77°59.0'W	998
			Straits of Florida	
"	1335	2/21/84	27°05.0'N, 79°20.0'W	635
JS-L	1355	6/13/83	26°38.5'N, 79°08'W	301
"	1356	6/14/83	" "	224
"	1358	6/15/83	" "	242
"	1359	6/15/83	" "	624
"	1361	6/16/83	" "	455
"	1362	6/17/83	" "	409
			Discovery Bay, Jamaica	
PC-8B	84-60	7/7/84	18°29.0N, 77°24.7'W	244
"	84-61	7/8/84	" "	195

*Dives unaccompanied by author but for which external camera rolls were examined in detail.

satisfactorily via SCUBA, they will not be discussed here.

The transition from the deep forereef escarpment to the deeper environments of the bank margin or island slope varies from place to place. Both the composition and upper limits of crinoid communities associated with these deeper environments vary as well. Off the northern coast of Jamaica, the vertical forereef escarpment, festooned with demosponges, sclerosponges, octocorals and antipatharians, ends abruptly at a mean depth of 122 m (Land and Moore, 1977). Below this the island slope, veneered with reef-derived and pelagic sediment, descends at a steep and locally variable angle. The relatively smooth slope is locally interrupted by crescentic embayments, shelf-like overhangs, promontories, vertical scarps, and flatter areas in which the sediment carpet is deeper and bioturbated. Along the upper reaches of the island slope, vertical and overhanging surfaces protected from sediment accumulation support attached organisms typical of the deep forereef above (e.g., sponges, antipatharians). Forms restricted to deeper water begin to appear here as well (e.g., branching and solitary ahermatypic corals) (see also Lang, 1974).

In contrast, the transition from deep forereef to deeper bottoms is more gradual along the western margin of the Little Bahama Bank. Although the base of the vertical escarpment may be recognizable, there is no obvious change in the substrate at this point. Rather, the irregular, rubble-like, cemented bottom gradually slopes away from the vertical; sediment accumulation increases with depth and flattening slope so that, by 170-180m, the bottom is sand with protruding bits of rubble very scattered. Above 150m, the hard substrate is carpeted with abundant encrusting sponges, scattered octocorals, antipatharians and a possible algal turf. No more than 10m deeper, however, exposed surfaces of cemented rubble appear strikingly barren. Arborescent anthozoans are very rare and the dominant organisms, where visible, appear to be encrusting sponges.

Along the southern margin of the bank in the Northwest Providence Channel (NWPC), the forereef slopes gradually away from the vertical to a cemented, rubble bottom with a veneer of coarse, reef-derived sediment. This bottom gradually becomes smooth with a coarse sediment veneer and little attached fauna. At least one large promontory-like outcrop was seen in each dive area.

Off Jamaica, comatulid crinoids, unobserved on the deeper reaches of the vertical forereef and scarce along the upper margin of the island slope, increase in numbers with depth and become among the most abundant epibenthic organisms by about 150m. A $6m^2$ transect along a relatively smooth, sediment-dusted portion of the slope using a $0.25m^2$ quadrat held against the bottom by the collecting arm gave a comatulid density of about $0.7/m^2$.

Figure 1. Stalked isocrinid, Cenocrinus asterius (L.), and monoplanar, fan-forming comatulid, Comactinia meridionalis hartlaubi (Messing), clinging to the ahermatypic coral Madracis myriaster (Milne Edwards & Haime) on a hard outcrop in about 200 m off Jamaica.

Distribution is patchy, however, with clusters in excess of $10/m^2$ occurring under overhangs and on vertical faces protected from sediment accumulation. Leptonemaster venustus A.H.Clark, a slender, ten-armed comasterid, appears to be the commonest feather star. The calyx is usually hidden in a crevice and the arms extended in a shallow, dish-like or deeper, conical, radial filtration fan. Occasional specimens were observed clinging close above the substrate to branches of arborescent anthozoans. A single specimen of Comactinia meridionalis (Agassiz), a more robust, ten-armed comasterid, was also collected.

In NWPC, comatulids do not appear until about 190m. Most abundant is Comactinia meridionalis hartlaubi (Messing), a large, yellow, robust, ten-armed comasterid that typically forms vertical, monoplanar filtration fans with 5-8 arms (fig.1). The calyx usually remains hidden in a crevice or among the basal branches of cnidarians such as the ahermatypic coral, Madracis myriaster (Milne Edwards & Haime). This feather star was also observed in this posture by D.B.Macurda,Jr. and D.L.Meyer during the 1973 NEKTON dive series off Jamaica (personal communication).

A large isocrinid, Cenocrinus asterius (L.), first appears in about 170m off Jamaica and 190m in NWPC, in both places on hard, sediment-veneered slopes, outcrops, and blocky promontories. This species is never abundant but is prominent wherever it occurs, singly, or in small groups. More than three are infrequently visible from the submersible at one time. The upright posture and parabolic filtration fan have been described by Macurda and Meyer(1974). One specimen was observed to rotate its crown from the vertical (in less than 0.1 knot current) to a 90° tilt in about 30 sec in response to a strong current generated by the submersible's screw. Off Jamaica, many specimens were also observed to be regenerating portions of one or two rays. C. asterius is rare in trawl collections partly because is often occurs on irregular bottoms but also because it appears to have a very restricted depth distribution. Although its depth limit was not observed off Jamaica (deepest observation = 244 m), it was not seen below 252 m in NWPC; most sightings fell between 190 and 220 m.

In contrast with the well-defined comatulid/Cenocrinus assemblage off Jamaica and in NWPC, few comatulids and no stalked crinoids were collected or observed in less than 198 m off the western margin of the Little Bahama Bank. In one area here, broad sand waves, occasionally veneered with fine ripple marks, appear in 170-190m. Although there was no active sediment transport or strong current activity at this time, attached organisms were absent and there was little evidence of bioturbation. Scattered rubble appears again in 170m and scattered attached fauna in 165m (Stichopathes sp.). On a subsequent dive in this general area, J.Miller and G. Hendler collected Leptonemaster venustus and Comactinia meridionalis on a steep, cemented rubble slope in 100-145m, and Nemaster discoidea(Carpenter) on a similar bottom in 120m(likely the deepest definite record for this reef-dwelling comasterid).

Below this barren area, the western margin of the Little Bahama Bank supports denser and more diverse crinoid assemblages than the other areas examined. From 200 to about 600 m, the bank margin drops away in an irregular series of near vertical scarps of a few centimeters to a few meters in height, separated by extensive, almost flat, lithified pavements dusted with sediment. Knoblike or buttresslike promontories up to several meters high project downslope to the west at intervals from the scarps. The frequency and regularity of their distribution along the bank margin are unknown. Especially in deeper water, there are extensive areas of deeper sediment, rippled or lineated by current movement or variously bioturbated. The most abundant attached organisms accompanying crinoids over this depth interval are stylasterine hydrocorals (+20/m^2 in places), solitary ahermatypic corals, antipatharians, paramuriceid and primnoid octocorals and, in the deeper reaches of this interval, lithistid, hexactinellid, geodiid sponges, and branching ahermatypic corals. Ophiuroids and echinoids are locally abundant.

From 200 to between 250 and 270m, the commonest crinoid is the slender bathycrinid, Democrinus brevis (A.H.Clark)(one was seen in 244m off Jamaica), accompanied by comatulids such as Comactinia meridionalis forming monoplanar, vertical fans with most of their ten arms. I saw no Cenocrinus asterius here, although it may have been observed on another dive. Here, the bottom is a mixture of pelagic and bank-derived sediment, apparently shallow enough for numerous (up to 20/m^2) antipatharian corkscrews (Stichopathes sp.) to attach to buried rubble or hardground.

Crinoid diversity and abundance subsequently increase dramatically. The isocrinids, Endoxocrinus parrae (Gervais) and Neocrinus decorus (Wyville Thomson), and the comatulids, Crinometra brevipinna (Pourtalès) (Charitometridae), Stylometra spinifera (Carpenter) (Thalassometridae), and Neocomatella pulchella (Pourtalès) (Comasteridae), all appear in numbers between about 260 and 290m. Unfortunately, some of my commentary did not successfully transfer to tape. I do not, therefore, have a record of the precise observed upper limit for all of these species in this area. The two isocrinids probably disappear above about 270m. A single S. spinifera was seen in 236m. Although not identified by me in shallower water on these dives, N. pulchella has been collected in much shallower water in this general area (143-210m; Messing, 1975; Meyer, et al., 1978).

In about 300-400m, these crinoids appear in a fairly typical aseemblage on almost flat, sediment-dusted, lithified pavements. The two isocrinids display typical parabolic filtration fans. Unlike C. asterius, in these, and especially in E. parrae, the cirri tend to fold down along the upright portion of the stalk (fig. 2). The two differ from each other in that, when tilted, the lower edge of the crown of E. parrae is raised off the bottom by about the equivalent of its crown diameter. In N. decorus, the lower edge of the slightly smaller, darker crown is raised off the bottom 2-3x the crown diameter. N. decorus thus usually overtops E. parrae. (The maximum stalk length of specimens of the former available to me at this time is 60cm; most, but not all, of this length is raised vertically.) N. decorus also differs from E. parrae in having shorter, finer cirri that just reach the length of an internode. Though common, individuals of these two species are rarely separated by less than 1-2 m. Despite weak current

Figure 2. Short-stalked Diplocrinus maclearanus (Wyville Thomson) (5 at right), long-stalked Endoxocrinus parrae (center), and comatulid Crinometra brevipinna on a sponge (lower left) on an irregular, lithified outcrop with coral and coral rubble [Lophelia prolifera (Pallas), rear] and primnoid octocorals (Candidella sp., left, rear). Figures 1 and 2 are composites taken from photographs provided by A.C.Neumann and D.L.Meyer.

speeds (0.1-0.3 knots from the SSE), most of the isocrinids exhibit the typical tilted crown response to unidirectional water movement. Similarly, D. brevis, which occurs in patches of up to 5/m^2 in 290m, also demonstrates a flow response; most specimens bend downcurrent. The weak observed current and upright posture of some specimens suggests, however, that the downcurrent bend is not necessarily a direct response to flow, but may be a longer term growth or acclimation response to a periodically stronger, unidirectional flow.

In 410m, on an open sediment bottom characterized by extremely sparse epifauna and scattered, broad, furrow trails probably made by paleopneustid echinoids, a single isocrinid (E. parrae?) was observed detached and crawling along the substrate. Although it did not move during the 5 min of observation, there was an obvious drag mark behind its stalk and it was observed to flex several arms adorally; subsequent aboral flexion would have driven it forward.

Most of the comatulids in this assemblage cling above the substrate to sponges, antipatharians, stylasterines, branching corals or octocorals. Crinometra brevipinna clings to the widest variety of perches and is also found directly on hard substrates. Its posture is a characteristic radial filtration cone or funnel, with the arms almost always curled in at the tips (fig. 2). Unlike the isocrinids, which exhibit a unidirectional flow response despite these weak flow conditions, C. brevipinna forms its funnel upright or tilted on both the upcurrent and downcurrent sides of its frequently planar, flabellate perches. Under stronger flow conditions (up to and exceeding 1.0 knots in about 600 m), C. brevipinna clings primarily to the downcurrent side of its perch. C. brevipinna includes 20 infrasubspecific, chiefly sculptural varieties that apparently intergrade significantly (Messing, 1975). However, between 300 and 400 m off the western edge of the Little Bahama Bank, two distinct color/sculptural forms occur; a strongly sculpted, yellow form and a smooth, grey and purple form. The two do not differ significantly in their habits, however, and their status remains to be worked out.

Stylometra spinifera is a bright orange comatulid with 30-40 arms and long slender, prehensile cirri that clings well off the bottom chiefly to the narrow branches of antipatharians and primnoid octocorals. Its posture is virtually identical to that of the isocrinids; it forms a parabolic filtration fan with the arms strongly recurved aborally into the current and its oral/aboral axis more or less parallel with the bottom.

Neocomatella pulchella, a comasterid with about 15-20 arms, forms dish- or shallow bowl-like radial filtration fans, and clings usually within a few cm of the substrate to the downcurrent side of sponges, octocorals, stylasterines, and branching ahermatypic corals.

A small antedonid with striking, black and white bands, tentatively identified as Coccometra nigrolineata(A.H.Clark), clings to hard, apparently sediment-free surfaces with its ten arms spread out in a horizontal, radial filtration fan. No more than 5-6 were observed during any 3-4 hr dive.

In 600-624 m, along the western margin of the Little Bahama Bank, another series of distinctive benthic assemblages appears. Unfortunately, I had no opportunity to observe the transition between these and the assemblages in 300-400m. Off West End, Grand Bahama I. (26°38.5'N, 78°08'W), substrates include relatively smooth, sediment-veneered hardgrounds, vertical and sloping scarps and isolated rock outcrops up to about 1m high, and rippled and current-lineated, pelagic sediment with and without scattered rubble and paleopneustid echinoid tests (Plethotaenia angularis). To the north, off Memory Rock (27°05'N, 79°20'W), similar substrates are accompanied by large, carbonate, whaleback ridges called lithoherms, up to 300m long and 50m high (Neumann, et al., 1977). The characteristic zonation of attached and semisessile benthic invertebrates on these lithoherms, apparently a response to variations in current velocity, will be described in a forthcoming paper (Messing, Neumann, and Lang, in prep.). In addition to crinoids, hard substrates in this depth range are dominated by primnoid and paramuriceid octocorals [e.g., Plumarella pourtalesii (Verrill), Candidella imbricata (Johnson), Swiftia sp.], sponges (e.g., Phakellia sp. fans, Isops sp., Rhizochalina sp., lithistids, and hexactinellids such as Farrea occa Bowerbank and Hyalonema sp.), ahermatypic corals [e.g., Lophelia prolifera (Pallas), Madrepora oculata L.], and stylasterine hydrocorals [e.g., Errina glabra (Pourtalès), Stylaster erubescens Pourtalès], with more scattered antipatharians (e.g., Stichopathes sp., Leiopathes sp.), octocorals (e.g., Corallium medea Bayer, Paragorgia johnsoni Gray, Isididae), and chitinous corkscrew tubes of the polychaete Eunice tibiana (Pourtalès) covered with zoanthids.

The most abundant crinoid in this depth range is Diplocrinus maclearanus (Wyville Thomson), a short-stalked isocrinid in which the crown is held adjacent to the substrate and the arms, forming a parabolic fan, are often longer than the stalk. The precise status of this species is unresolved. Rasmussen (1978) subsumed the genus under Endoxocrinus, but this species appears to be distinct from E. parrae. Sculptural varieties of unknown significance have been described for each as well. Pending detailed examination of western Atlantic isocrinids, short-stalked forms with about 20 arms are herein called D. maclearanus, and long-stalked forms with 30-40 arms, E. parrae. Both have long, robust cirri.

Diplocrinus maclearanus clings to lithified substrates, sponges, echinoid tests, and, occasionally, low stylasterine and scleractinian corals. On the downcurrent crests of the lithoherms mentioned above, D. maclearanus occurs (with abundant primnoids) chiefly along the margins of raised lithified crusts in aggregations of up to

15-20/m^2. Because much of the lithoherm surface is broken into horizontally linear or crescentic, lithified crusts separated vertically by poorly cemented (and often eroded) carbonate sediment, the crinoids and octocorals frequently appear to be "planted" in artificially regular rows although they also may line up on apparently smooth bottoms. Though currents in excess of 2 knots were recorded about 2 m above some lithoherm crests (about 20m above the surrounding flats), the tops of the tilted crowns of D. maclearanus were rarely more than about 10 cm above the substrate and well within the reduced velocity boundary layer. Under weak flow conditions (< about 0.3 knots), the parabolic fans of adjacent specimens orient either vertically or parallel to the bottom.

Endoxocrinus parrae is less common at this depth, but where it occurs, and in contrast with its situation in shallower water, it is the tallest isocrinid. Democrinus brevis is also less common here.

Isocrinus blakei (Carpenter) is a slender, usually ten-armed isocrinid that, unlike its local confamilials, does not form a parabolic fan. Instead, its stalk bends and its arms fan downcurrent in a conical or funnellike, radial fan (Macurda and Meyer, 1976). I. blakei appears to be restricted to low relief hardgrounds, bits of rubble and echinoid tests under weak flow conditions and is absent from the higher energy lithoherm crests. Its outcrops are often surrounded by rippled sediment though no active particle transport was observed during these dives. Individuals also appear directly on open sediment, but accompanying primnoid and paramuriceid octocorals suggest that they arise from buried rubble or echinoid tests.

Large, yellow Crinometra brevipinna is the most common comatulid here, forming the same conical or funnellike fans with incurled tips as in shallower water. Individual specimens appear every several meters whereas Neocomatella pulchella appears only at long intervals. The small antedonid, Trichometra cubensis(Pourtalès), clings to a number of slender perches, including antipatharians (9 were collected from a single bushy colony of Leiopathes sp.), primnoids, and Democrinus brevis stalks. Its arms are closely apposed at the bases and spread apart only slightly distally. The tips curl inward and the feather star appears as a bulbous funnel. Comatilia iridometriformis A.H.Clark, a small, yellow, ten-armed comasterid, occurs in dense but patchy aggregations of up to 60/m^2 on branching coral rubble and lithified crusts on the upcurrent ends of lithoherms (Messing, 1984).

Crinoid assemblages in about 250-1000m in NWPC exhibit reduced density and diversity relative to those in the Straits of Florida due, most likely, to the combined effects of steeper overall slope and reduced water movement. Current velocity records in NWPC are few, scattered, and, apart from on- and off-bank tidal exchange, suggest a slow, general flow from the east. No strong flow similar to the Florida Current exists, although, in agreement with most other flow observations made during these dives at all depths below the forereef, water movement is more or less parallel to the slope rather than up or down it. As a result of this reduced flow, horizontal and shallowly sloping surfaces tend to be more deeply blanketed with sediment than in the Straits of Florida (although sediment-veneered pavements and dusted ridges do occur here). Sediment-free vertical escarpments exceed 50m in places.

Slope steepness is also reflected in sediment composition. Along the northern margin of NWPC, a distinct, tongued boundary exists between sediment visibly dominated by coarse, reef-derived material and a much finer, pelagic sediment in about 650m. West of Grand Bahama I., pelagic foraminiferan and thecosome tests dominate the sediment in 400 m.

As in the Straits, isocrinids other than C. asterius first appear in about 280m and are accompanied by comatulids, but both remain sparse on most bottoms observed. Between 500 and 600m, a series of lithified, ridge-like shelves, 10-15m apart with a vertical relief of 0.5-1m, and running diagonally across the slope, bear numerous isocrinids and comatulids (including Crinometra brevipinna) accompanied by primnoids, other octocorals and antipatharians along their narrow crests. The vertical faces of these shelves are dominated by lithistid sponges. In about 680m, the cyrtocrinid, Holopus rangii d'Orbigny, was observed with lithistid sponges and sparse antipatharians on a sheer, jointed escarpment about 20m high.

In 780-850m, a series of elongated mounds more than 10m high are covered with a veneer of dead, branching, ahermatypic coral (including some intact, monoplanar colonies that have been knocked over) that becomes denser at and near the crests. No living colonies were observed and everything is covered with a dusting of sediment. Diplocrinus maclearanus is most common at and near the crests of these mounds; individuals occur every 2-3m^2, accompanied by comatulids, sponges, stylasterines, solitary corals, antipatharians and brisingid asteroids.

The deepest stalked crinoid observed during dives in NWPC was an unidentified isocrinid in 805m. Below this, from 836m to almost 1100m, appear occasional individuals of a slender, pinkish, ten-armed antedonid with long, fine cirri, probably Caryometra sp. This comatulid forms a deep, radial filtration bowl and, when disturbed by the collecting arm, actively swims. A single disturbed individual alighted on ALVIN's collecting basket and remained there for most of the ascent. Sediment-free or veneered bottoms in this depth range are relatively barren with respect to those in shallower water, but support sparse populations of sponges, antipatharians, brisingid asteroids and occasional actinians.

Between about 700 and 1000 m, a few pale, ten-armed comatulids with long, fine cirri were observed spread-eagled on open sediment. These may be Atelecrinus balanoides Carpenter, a particularly delicate form found in this area only on unconsolidated substrates between about 600 and 1500 m.

4 DISCUSSION

Observations from manned submersibles confirm conclusions drawn from dredged collections that crinoids occur on a wide variety of bottoms, that they are numerically important members of several benthic, hardground assemblages, that certain species are restricted to specific, localized habitats (e.g., Comatilia iridometriformis, Messing, 1984), and that some have bathymetric ranges of several hundred meters at least (e.g., Crinometra brevipinna, Democrinus brevis) whereas others have depth ranges of little more than 50m at most (e.g., Cenocrinus asterius). Macurda and Meyer (1974,1976) pointed out that stalked crinoids can be more accurate indicators of current direction in deep-water photographs than sedimentary features such as sediment shadows. Whereas isocrinids respond to weak flows (less than 0.3 knots), sedimentary features may represent responses of sediment to stronger flow conditions not currently active. The current series of observations indicates further that deep-water crinoids (like reef-dwelling comatulids; Meyer, 1973) may be indicators of current velocity as well. Though precise tolerances are not yet known, Isocrinus blakei, for example, appears to be restricted to a weaker flow regime than Diplocrinus maclearanus (although the latter occurs closer to the substrate). By contrast, all crinoids, as well as most other epifauna, are absent from bottoms that appear to be subject to particularly strong currents.

Repeatedly in different habitats, different species of crinoids are consistently raised to differing heights above the substrate (e.g., Neocrinus decorus vs. Endoxocrinus parrae in 300-400m; E. parrae vs. Diplocrinus maclearanus in 600m). Future examination of structural and dietary differences between members of these pairs represents an excellent opportunity to test the possibility that these crinoids are partitioning available resources--a proposal already put forward for multitiered assemblages of fossil crinoids (Ausich, 1980). The multi-tiered nature of contemporary, deep-water, hardground assemblages is not restricted to the crinoid component. For example, in 300-400m, N. decorus and E.parrae represent two tiers with the much smaller crowns of Democrinus brevis at about the same height as E. parrae. Below this, another tier includes comatulids, ophiacanthid ophiuroids, and flabellate paramuriceid octocorals. A still lower storey, no more than a few cm high, includes small stylasterines and solitary corals. On the lithoherms in about 600m, a low, recumbent tier of filterers below the "lawn" of Diplocrinus maclearanus and primnoids includes serpulid and other tubicolous polychaetes, encrusting hexactinellid sponges, and a pterobranch hemichordate, Cephalodiscus atlanticus Bayer. This fine encrustation is only recognizable from substrate samples examined in the laboratory. It is worth mentioning here because there is a considerable diversity of small, attached fauna on all of these hardgrounds that cannot be resolved by visual observation from the submersible (beyond, perhaps, recognition of a fine or filamentous turf, unless a close-up external camera is used) but that probably contributes significantly to community dynamics.

Several observations were unanticipated. It was not expected that Democrinus brevis would occur on lithified substrates. As a bathycrinid, most extant specimens have the terminal rootlets for anchoring in unconsolidated substrates that are typical of the family. The few specimens collected during these dives snapped off above their bases, suggesting that the attachment structure may be suitable for anchoring to hard bottoms (like the expanded base illustrated by A.M.Clark, 1977). Such an attachment may improve resistance to water movement relative to rootlets. These hard bottom forms reach more than twice the stalk length of D. brevis rooted in fine mud at a similar depth and possibly under a similar current regime on the Florida side of the Straits.

The observation of an isocrinid apparently crawling along the substrate was equally unexpected but illustrates the need for long-term, detailed analyses of these assemblages. How active are stalked (and unstalked) crinoids at these depths? Do any of them raise or lower their crowns in response to variations in current flow? What are their rates of turnover, growth and recruitment? By extension, and recognizing that these communities are dominated by dense assemblages of filterers at dysphotic and aphotic depths, what are the sources of food and how do they vary? Does the quantity or composition of particulate, suspended food vary with height above the bottom enough to contribute to multi-tiered resource partitioning, or does the height stratification of filterers reflect differing abilities to deal with shear stress? Finally, how may examination of these contemporary crinoid assemblages assist in interpreting fossil assemblages? Many of the latter are, of course, known only from dissociated ossicles. By comparing contemporary death assemblages (i.e., the contribution of extant crinoids to sediment) with living assemblages, parallels may be drawn that may be applied to the fossil record. For example, crinoid ossicles contribute 11.6% by grain count (7.8% by weight) to a gravelly sand sample from the rubbly nose of a lithoherm where living crinoids were few and scattered. Similarly, crinoid ossicles constitute fully 8.8% by weight of the chiefly foraminiferan/thecosome sand adjacent to a pavement with scattered isocrinids in between 300 and 400 m. The slow deposition rates usually attributed to pelagic sediments suggest that crinoids may be equally slow contributors to the sediment. Comparison of relative proportions of different ossicles in entire specimens and in the sediment may suggest, for instance, whether the stalks of extant crinoids continue to grow and shed terminal ossicles, or the arms and pinnules continually regenerate.

5 ACKNOWLEDGEMENTS

My participation in these submersible dives was made possible by the following: ALVIN: A.C.Neumann, University of North Carolina, Chapel Hill; JOHNSON-SEA LINK: J.C.Miller, Harbor Branch Foundation, Fort Pierce, Florida; PC-8B: J.Woodley & P. Janca, Discovery Bay Marine Laboratory, Jamaica, and K.Norton, Miami, Florida. My gratitude for their efforts runs rather deeper than the submersibles. I am also indebted to the following Smithsonian personnel for identification of non-crinoid invertebrates collected during these operations: F.M.Bayer (octocorals), S.D.Cairns (corals and stylasterines), K.Ruetzler (sponges), K.Fauchald (polychaetes), P. Kier, D.Pawson, G.Hendler and M.Downey (echinoderms).

6 LITERATURE CITED

Agassiz,A. 1888. Three cruises of the U.S. Coast and Geodetic Survey Steamer"Blake" in the Gulf of Mexico, in the Caribbean Sea, and along the Atlantic coast of the United States from 1877 to 1880. Bull. Mus.Comp.Zool.,Harvard 14,15.

Ausich,W.I. 1980. A model for niche differentiation in Lower Mississippian crinoid communities. J.Paleontol.54:273-288.

Clark,A.M. 1977. Notes on deep-water Atlantic Crinoidea. Bull.Brit.Mus. (Nat. Hist.)Zool. 31(4):159-186.

Land,L.S. & C.H.Moore,Jr. 1976. Deep forereef and upper island slope, North Jamaica. Amer.Assoc.Petr.Geol., Studies in Geol. (4):53-65.

Lang,J.C. 1974. Biological zonation at the base of a reef. Am.Scient. 62:272-281.

Macurda,D.B.,Jr. & D.L.Meyer. 1974. Feeding posture of modern stalked crinoids. Nature(London) 247:394-396.

Macurda,D.B.,Jr. & D.L.Meyer. 1976. The identification and interpretation of stalked crinoids (Echinodermata) from deep-water photographs. Bull.Mar.Sci. 26(2):205-215.

Messing,C.G. 1975. The systematics and distribution of the Crinoidea Comatulida (exclusive of the Macrophreatina) collected by the R/V Gerda in the Straits of Florida and adjacent waters. 296pp. Master's thesis, Univ. of Miami, Florida.

Messing,C.G. 1984. Brooding and paedomorphosis in the deep-water feather star Comatilia iridometriformis A. H. Clark (Echinodermata: Crinoidea). Mar. Biol. (Berlin) 80:83-91.

Meyer,D.L. 1973. Feeding behavior and ecology of shallow-water unstalked crinoids (Echinodermata) in the Caribbean Sea. Mar.Biol.(Berlin) 22(2):105-129.

Neumann,A.C.,J.W.Kofoed & G.H.Keller. 1977. Lithoherms in the Straits of Florida. Geology(Boulder,Colorado) 5:4-10

Rasmussen,H.W. 1978. Articulata, in Moore, R.C. & C.Teichert (eds.), Treatise on Invertebrate Paleontology, Part T, Echinodermata 2(3):T813-T928. Boulder, Colorado: Geol.Soc.Amer.

Proceedings of the Fifth International Echinoderm Conference / Galway / 24-29 September 1984

Comparison of reproduction in the Red Sea feather-stars *Lamprometra klunzingeri* (Hartlaub), *Heterometra savignii* (J.Müller) and *Capillaster multiradiatus* (L.)

J.L.RUTMAN & L.FISHELSON
Department of Zoology, George S.Wise Faculty of Life Sciences, Tel Aviv University, Israel

ABSTRACT: This paper summarises observations on reproductive activity of the feather-stars Lamprometra klunzingeri (Hartlaub), Heterometra savignii (J. Müller), and partly Capillaster multiradiatus (Linnaeus), from the Gulf of Aqaba (Red Sea). During this study gonads of nearly 1000 specimens of the three species were investigated. Along each arm, gonads of the same animal develop asynchronously, those in the lower quarter of the arm maturing first. During the reproductive cycle oogonia and spermatogonia differentiate from the innermost layer of the gonad wall. Six stages were defined in the reproductive cycle of each gonad. During a single reproductive cycle several spawns from each gonad are possible . Maturation of spermatozoa is gradual and liberated ripe sperm masses accumulate in the lumen of the testis. Egg maturation prior to spawning involves only 30% at the most of the oocytes present. Gonads under 50 um in diameter are usually inactive, and those with a diameter over 250 um are well advanced in the reproductive cycle. During reproduction, only 1.2% of female L. klunzingeri and 1.1% of female H. savignii held ripe eggs in the ovarian lumen. Males with ripe sperm masses comprised 25% in L. klunzingeri and 32% in H. savignii.

Reproductive activity of L. klunzingeri is continuous, without a definite seasonal pattern, whereas in H. savignii cessation of maturation occurs between April and June. Gonad structure and gametogenesis of C. multiradiatus resemble those of the other two species. The highest relative frequency of mature individuals was for L. klunzingeri in August (33%), for H. savignii in September (45%) and for C. multiradiatus in December (38%).

INTRODUCTION

The crinoids are the last investigated among the echinoderm classes, especially concerning their ecology and reproduction in natural habitats. The most important reason for this is that the majority of species dwells in deep waters. For this reason, papers on their biology or reproduction are mostly based on aquarium observations or collections of their planktonic larvae (see Blootian, 1966). The first sound descriptions of reproduction in a crinoid were given by Dana (1941) and by Fell (1966). Their reproduction in nature has been observed on only a few occasions (Fishelson, 1968). The anatomy of their reproductive system has been investigated for many years and most of the older papers are summarised by A.H. Clark (1921). Recently the description of the gonadal system of Comanthus japonicus was given by Dana and Kubota (1960); of Antedon bifida by Dimelow (1958) and of Nemaster rubiginosa by Holland (1971). In these species that the genital system consists of numerous small gonads situated on the secondary pinulli of the arms (Fig. 1 and 2), interconnected by a genital filament extending along the arm. The genital filaments from all the arms terminate in the central part of the calyx. From the observations in nature and in captivity, it is obvious that in the crinoids the sexes are separate, and no dimorphism is observed. In most cases fertilisation occurs in the surrounding waters. The first stage larvae-doliolaria, being free-swimming or brooded, pass through several morphological changes until settlement and development of the attached, final pentacrinoid stage (Fig. 3 and 4). In some species these stalked larvae were found attached to the adults (Mortensen, 1937; A.H. Clark, 1921; and Fishelson for Lamprometra and Antedon, unpublished). After maturation the calyx

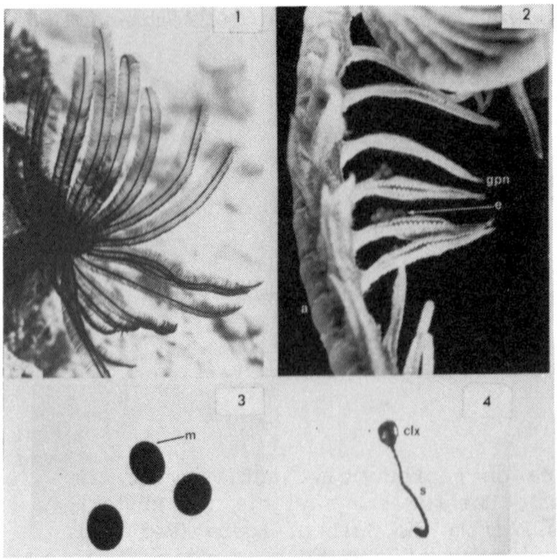

Figures 1-4: 1. Heterometra savignii in a feeding posture (1/4); 2. The swollen gonadial pinnuli and eggs (e) of Lamprometra klunzingeri; 3. Fertilized eggs of L. klunzingeri m - membrane; 4. Pentacrinid larva of L. klunzingeri two weeks old (length - 1.8 mm); clx - calyx; s - stalk.

of the pentacrinoid detaches from its stalk and a free-living juvenile crinoid is formed, that in some instances occupies habitats not occupied by their adults (Fishelson, 1974). In Lamprometra klunzingeri the development until maturation takes approximately 18 months (Fishelson, personal observation). This study follows the reproductive cycle of two, Red Sea crinoids, Lamprometra klunzingeri and Heterometra savignii, with additional information on Capillaster multiradiatus, all three inhabitants of the shallow subtidal in the Gulf of Aqaba, Red Sea (Rutman and Fishelson, 1971; Fishelson, 1974; Rutman, 1975). Most material included formed a part of the late J. Rutman's Ph.D. thesis.

METHODS

Fortnight and monthly samples of 30 to 35 individuals of the investigated crinoids were collected at random during 4 years. Parts of the crinoid arms were checked for gonad-development on the pinulli by means of a dissecting microscope and parts were fixed in Bouin's for histological investigation and in glutaraldehyde for electron-microscopy. The macroscopic observations allowed the segregation of the gonads into three categories: Young gonads of 70 um to 150 um diameter; medium gonads of 150 um to 250 um diameter and ripe gonads of over 250 um diameter. Serial histological sections, 5um and 8 um thick, were prepared from all these gonads, stained in Delafield's Hematoxylin-Eosin. Photographs were taken with a Zeiss camera on a Zeiss-Ultrascope.

HISTOLOGY OF THE REPRODUCTIVE SYSTEM

In general, the genital system in all three investigated species is similar, resembling that of Nemaster (Holland, 1971).

The histology of the gonads at various stages of development and the tissue-layer structure resembles the gonad of holothurians described by Atwood (1973). The most internal cell-layer of the gonad is the germinative one, formed by cuboidal cells of 6um-7um (Fig. 6). These cells line the gonad lumen (Fig. 5 and 7; Fig. 21 and 22). In males, with the onset of ripening (Fig. 24, 25, 26), almost all of these cells synchronously form spermatocytes, whereas in females the ripening is asynchronous, by cells situated on various sites of the germinative layer (Fig. 14 to 18). The maturation of oogonia and their liberation into the gonad lumen occurs shortly before reproduction (Fig. 10). Also the spermatids develop within the germinative layer, until the spermatocytes are ripe and liberated (Fig. 24-28). According to Ludwig and Hamann (1907) and Holland (1971) the source of the germinal cell layers is in the filament of germinal tissue situated within the arm, from where they migrate into the gonad. In our study we found oogonia and spermatogonia in the gonads only and no sign of translocation of these cells was observed. Thus it seems that repopulation of the gonad by sexual cells occurs from primordial germ-cells situated within the epithelium that lines the gonad lumen and not from outside. This layer becomes very poor in cells after a period of reproduction and processes of redevelopment are visible with onset of a new reproduction period.

The dimension of a single gonad changes during ripening, from 20um-50um in diameter at the non-reproductive stage to a maximum of 600 um during the peak of reproduction. Ripe gonads are spindle-like, with swollen central parts (Fig. 1 and 8), and in them the ripe, mature eggs are visible in the gonad lumen (Fig. 10). In females the color of such ripe gonads is yellow-orange and in males pale-

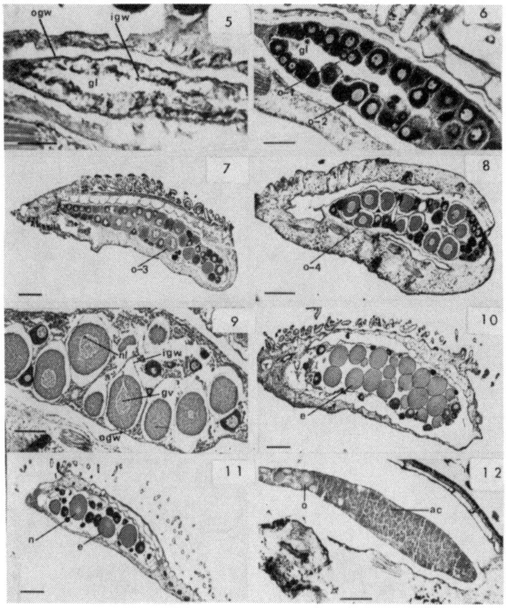

Figures 5-12: 5. The gonad at Stage I (sex cannot be identified); 6. Female gonad Stage II, with ovocytes of type 1 and 2 (O-1;' O-2); 7. Ovary Stage III with oocytes of stage 1, 2 and 3, the last one rich in yolk; 8. Ovary Stage IV, with oocytes stage 4; 9. End of Stage IV, beginning of eggs - ripening and liberation into the gonad lumen; 10. Egg in the gonad lumen before spawning; 11. Stage V - after partial spawning; 12. Stage VI - Degenerative gonad with accessary cells (ac). Abbreviations: e - egg; gl - gonad lumen; gv - germinal vesicle; igw - inner layer of gonad wall; n - nucleus; nl - nuceolus; o - oocyte; ogw - outer layer of gonad wall. (Marks: 5-50 um; 6, 9, 12 - 100um; 7, 8, 11 - 200um).

yellow. Along the same arms of an individual crinoid the gonads are in various stages of development: Generally the most proximate gonads are the most advanced in ripening, whereas those situated more distally along the arms, are less developed. During reproduction the liberation of gametes into the surrounding waters occurs sequentially, from the proximal toward the distal sections of the arms. This is shown in the the scatter diagram (Fig. 29) that summarises observations on correlations between the diameter of the gonads and their position on the arms of L. klunzingeri. As evident, the correlation coefficient (r) between the widest diameter of the gonad in the middle (m) part of the arms and those of the basal (b) parts, was 0.821 for females and 0.895 for males ($P < 0.01$); the r between b and distal (d) parts of the arms was 0.515 for females and 0.688 for males ($P > 0.01$). With the first liberation of gametes, gonopors are formed on the upper side of the pinuli wall (Fig. 30).

According to gametogenic activity, the gonads were divided as follows:

Stage 1: No gametogenesis observed. The diameter of the gonad is 30um-70um; the gonad wall is 25um to 50um thick. The narrow gonadal lumen is lined with a smooth layer of epithelium, with numerous cells in mitosis (Fig. 5 and 6).

Stage II: Onset of gametogenic activity; in females marked by occurrence of oogonia (7-10um in diameter) and oocytes (10-20um) (Fig. 6). The average maximal diameter of the gonads is 135um (± 56um). In more advanced gonads of this stage the oocytes attain 35u to 85um in diameter; in them the germinal vesicle is marked by its irregular membranes, along which microvesicules of yolk are situated (Fig. 15 and 16). Such lipid inclusions were also mentioned for Nemaster (Holland, 1971). In males of Stage II, the maximal diameter of the testis is 135 um (± 15 um); most germinal cells are in mitotic divisions; spermatogonia and a few spermatocytes are visible (Fig. 24).

Stage III: The average diameter of the female gonads is 257 um (± 67 um); of the male 207 um (± 78 um). The gonad lumen is well formed and the germinal layer outlining it forms lamellations protruding into the lumen (Fig. 7). The lamellation is especially strong in males, with numerous spermatids and spermatocytes hanging out into the gonad lumen (Fig. 26). The oocytes are over 100 um in diameter (Fig. 9 and 16) and around each of them a jelly coating layer develops.

Stage IV: The diameter of the female gonad is 343 um (± 75 um) and of the male 267 um (± 86 um). The maximal diameter of the oocyte is 200 um (Fig. 10 and 17). Ripe egg-masses are observed in the gonad lumen (Fig. 10). The large oocytes and the ripe eggs have irregular chorionic membranes, the protrusions of which penetrate into the jelly coats. In testes the central lumen is occupied by a mass of spermia (Fig. 27).

Stage V: The diameter of the female gonads are 428 um (± 81 um) and of the

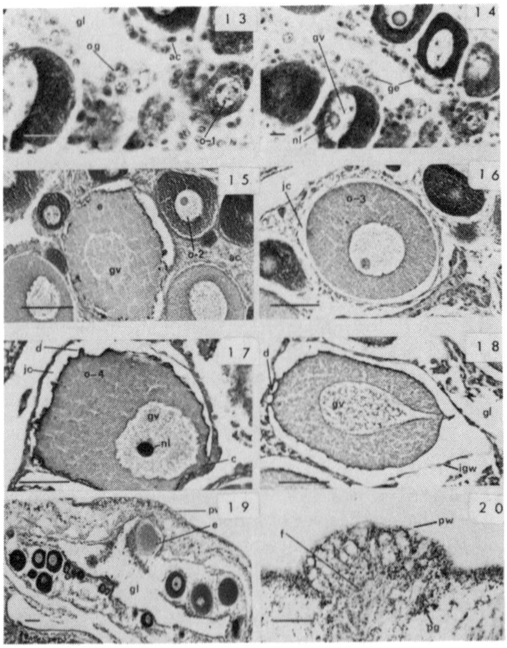

Figures 13-20: 13. Part of ovary with young oocytes (O-1) and oogonia (og); 14. Oocytes of stage 2; 15. Part of ovary with oocytes (O-2) and ripening eggs (e); 16. Oocyte stage 3 (O-3) surrounded by a jelly coat (jc); 17. Oocyte stage 4, only weakly attached to the germinal layer (c); 18. The ripening egg before liberation into the lumen; 19. An egg (e) leaving the gonad; 20. The tissue in the gonad opening. Abbreviations: f - fibroblasts; ge - germinal epithelium; og - oogonia; pg - pigment granules; pw = pinnule wall; d - dentetion of oocyte periphery. (Marks: 13, 14 - 10um; 5-20 - 50um).

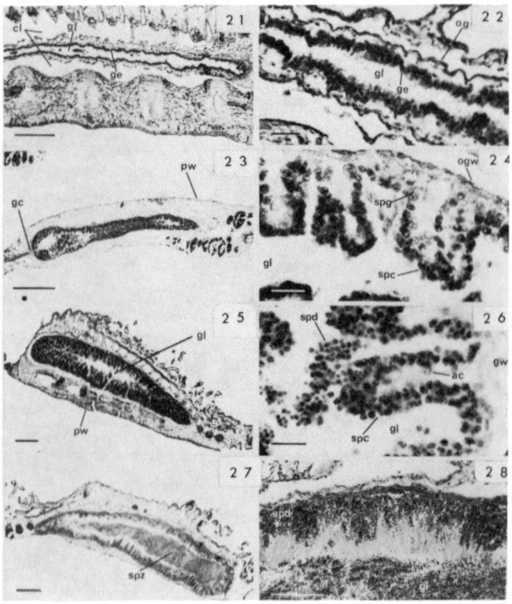

Figures 21-28: Testis: 21. Beginning of Stage II; 22. The same, enlarged; 23. End of Stage II; 24. The same enlarged, onset of folding of germinal tissue; 25. Stage III; 26. The same enlarged, showing ripe spermatocytes (spe) and spermatids (spd); 27. Stage IV, with a mass of spermia in the testis lumen; 28. The same enlarged, demonstrating ripe spermia. Abbreviations: cl - coelom; gc - genital cord; spg - spermatogonia; spz - spermatozoa. (Marks: 20 - 100um; 21 - 25um; 22-28 -50um; 24-26 - 10um; 25-27 - 200um).

males 285 um (± 76 um). Clear signs of post-spawnings, the gonad with remnants of egg and spermia (Fig. 12); open genital pores are visible (Fig. 30).

Stage VI: Post-reproductive gonads, with germinal tissue in degradation(Fig. 11 and 19). Numerous phagocytes are visible among the cells of the gametogenic layer. Parallel to this, gametogenetic activity within the gonads is renewed, marked by the occurrence of juvenile sexual cells.

In all the three investigated species of Red Sea crinoids, these six stages of gonadal development were found to be very similar.

COMPARISON OF THE ANNUAL CYCLES OF GONADS

To simplify the comparison of gonad development of L. klunzingeri and H. savignii the data from various years were compiled on a monthly basis. These are shown in Fig. 31, in which each graph represents the percent occurrence of a given gonadal stage in the total number of animals observed for a given month. Thus, it was demonstrated that for K. klunzingeri, the percent of individuals with Stage I gonads (first graph, upper left) is almost the same all during the year, exposing continuous gametogenesis of juvenile cells. In H. savignii, contrary to this, Stage I forms a peak in mid-summer (May-June), with a strong decrease in August. The opposite is evident from

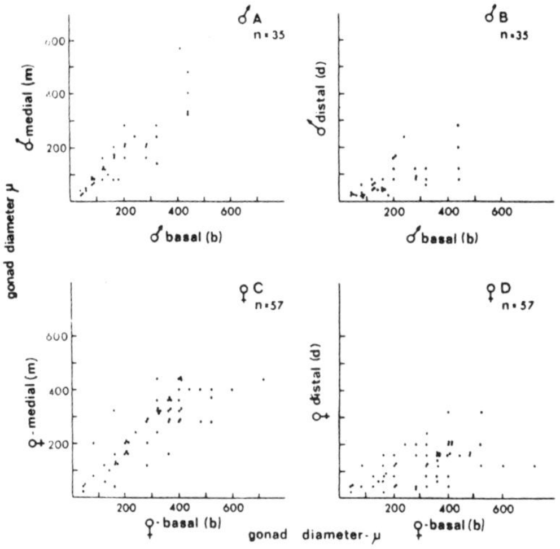

Figure 29: The diameter of gonads on different parts of the arms of L. klunzingeri: A and B - males; C and D - females. b - basal parts of arms; m - median parts; d - distal parts.

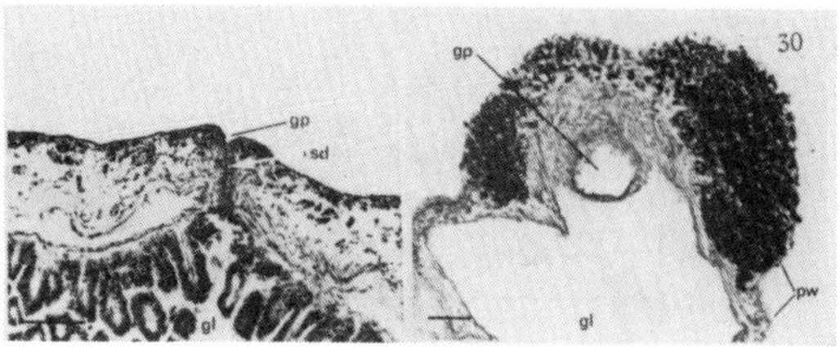

Figure 30: The gonopor (gp) of male (left) with sperm duct (sd) and the gonopor of the female (right) of L. klunzinger. (Marks - 40um).

the graphs of Stage III: The percent of L. klunzingeri in this category is much higher in summer than that of H. savignii. The dominance of L. klunzingeri in summer is also shown for Stage IV, where Stage IV in H. savignii dominates during the autumn and winter.

Although (except some months for H. savignii) in both species all stages of gonadial development are found in the populations during all months of the year, two different trends are recognised for the two species investigated: In L. klunzingeri ripe or post-spawned individuals are encountered during all months, whereas in H. savignii heterogeneity of the gonads is much lower and synchronization of activity much higher. Thus, for example, in H. savignii no ripe individuals were found from April until June (Fig. 31), although gametogenesis occured during the whole year. This means that no ripening occurs in the first summer months.

The seasonal and monthly separation between the two species becomes evident if we compare the time of highest dominance of ripe individuals of Stage IV: In L. klunzingeri, this stage dominates from May to July, whereas in H. savignii, they form the majority from September until December. In addition to this interspecific separation between the reproductive peaks of the two most common species, intraspecific differences in sexual ripening are found between male and females of the same species. Thus, for example, in males of L. klunzingeri Stage I occurs in May, June and October, whereas in females it occurs in June, October and November. Stage III dominates in females of the August sample and in males of November; Stage VI is found in 50% of the females from December and only in 16% of the males of the same month. Thus, it seems that although in general all the developmental stages are found in the gonad of the crinoids during all the months of the year, the percent of each of the stages in the total observed varies from month to month and between the sexes. The situation is slightly different in females and males of H. savignii. In males of this species Stage I of gonads are found in spring (February-March) as well as during winter (December) whereas in females these are observed only during summer months (May to July). Also Stage IV of the gonads that are observed in males during January-February, are not encountered in the females of this species at the same time. Thus, the sexual separation of gonadial stages within the population of H. savignii goes along a different pathway than L. klunzingeri. Additional differences were also found. In multiradiatus Stage I and II oocytes dominate during April and May, where ripe and postspawned Stage VI are found from September until December.

DISCUSSION

The gonads of the investigated crinoids are easily divided according to six stages of development: 1. Resting stage; 2. Beginning of gametogenesis; 3. Intensive gametogenesis; 4. Maturation; 5. Spawning; 6. Decrease of reproductive activity and phagocytosis of remnants. The same type of development has been described for many animal species, among them numerous echinoderms (Tanaka, 1958;

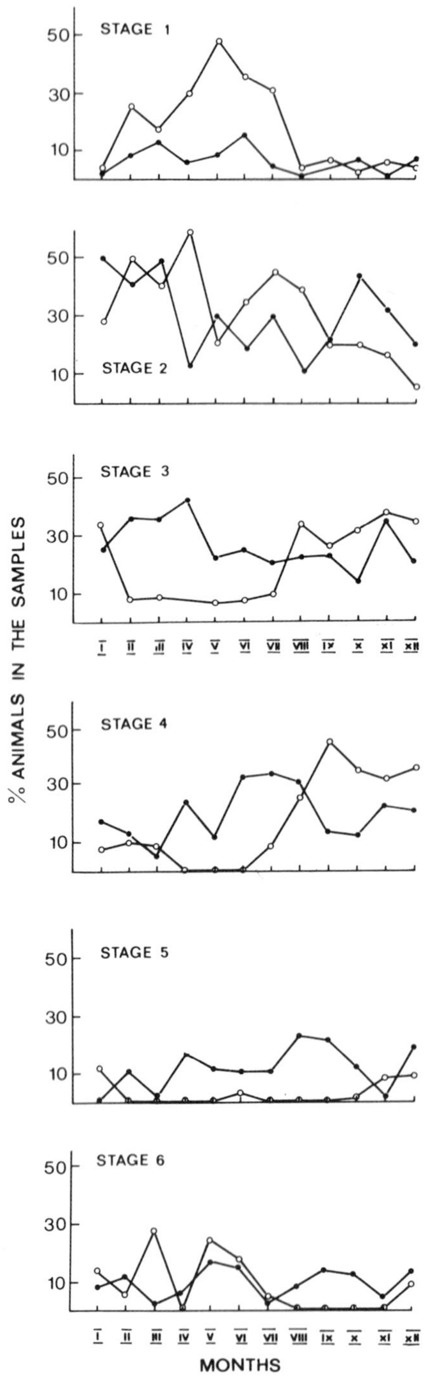

Figure 31: Stages of oogenesis in monthly samples of L. klunzingeri (N=583, black circles) and H. savignii (N=580, open circles).

Chia, 1968; Patent, 1969). In most of the described sequences, the gametogenesis is at its highest at Stages II and III, and as these stages are found also during Stages IV and V, the result is that in most species juvenile gonocytes are found year round, in the reproductive as well as non-reproductive months. Contrary to this, the ripening of oocytes and their liberation into the gonad lumen oocurs only shortly before spawning, as also demonstrated for other echinoderms (Dana, 1952; Holland, 1967; Holland and Rommel, 1969; Pearse, 1969; Fenaux, 1970). In the Japanese Comanthus, this lack of storage of ripe eggs within the gonad seems to be connected with the synchronised ripening (Holland, 1967). In our comatulids, the ripening is asynchronous, and in spite of this no prolonged storage of eggs was observed. Females of L. klunzingeri and H. savignii, as well as those of C. multiradiata, are able to spawn from the same gonad several times during the year. The same was also described for sea-urchins (Fuj, 1960; Holland and Giese, 1965; Holland and Rommel, 1969) as well as Gorgonocephalus caryi (Patent, 1969). Dan and Kubota (1960) noted that in Comanthus japonicus synchronisation of development exists between gonads situated on the same arm. This was not the case in the crinoids investigated by us. In our study such synchronisation was found only between gonads of various arms situated at the same distance from the calyx. The population of L. klunzingeri reproduces all the year round, whereas spawning individuals of H. savignii were encountered only from April to June. As the first species is a typical dweller of shallow-subtidal waters (Fishelson, 1974) compared to H. savignii and C. multiradiata which occupy slightly deeper sites, it is possible that the extensive reproduction of this is an adaptation to the more extreme disturbances frequent in this habitat.

REFERENCES

Atwood, D.G. 1973, Ultrastructure of the gonadal wall of the sea cucumber Leptosynapta clarki (Echinodermata: Holothuroidea). Z. Zellforsch., 141(3): 319-330.

Boolootian, R.A. 1966, Reproductive physiology, in: Physiology of Echinodermata ed. R.A. Boolootian, Interscience Publishers, pp. 561-613.

Chia, F.S. 1968, Some observations on the development and cyclic changes of the

oocytes of a brooding starfish Leptasterias hexactis. J. of Zool. (London) 154: 453-461.
Clark, A.H. 1921, A monograph of the existing crinoids. Smithsonian Inst., U.S. Nat. Mus., Bull. 82, 1(2), XXV+795 pp.
Clark, A.M. & Rowe, F.W.E. 1971, Monograph of shallow-water Indo-West-Pacific echinoderms. Trustees of the British Museum (Nat. Hist.) London. 238 pp., 31 pl.
Dan, K. 1952, Meiosis of the eggs of the crinoid Comanthus japonica. Annot. Zool. Japonenses, 25(1): 258-264.
Dan, K. & Kubota, H. 1960, Data on the spawning of Comanthus japonica between 1937 and 1955. Embryologia, 5(1): 21-37.
Dimelow, E.J. 1958, Some aspects of the biology of Antedon bifida (Pennant),with some reference to Neocomatella europa. Ph.D. Dissertation, Univ. Reading.
Fell, H.B. 1966, Ecology of crinoids, in: Physiology of Echinodermata, ed. R.A. Boolootian, Interscience Publishers,pp. 49-71.
Fenaux, L. 1970, Maturation of the gonads and seasonal cycle of the planktonic larvae of the ophiuroid Amphiura chiajei (Forbes). Biol.Bull., 138(3): 262-271.
Fishelson, L. 1968, Gamete shedding behavior of the feather-star Lamprometra klunzingeri in its natural habitat. Nature, 219 (5158): 1063.
Fishelson, L. 1974, Ecology of the Northern Red Sea crinoids, and their epi- and endozoic fauna. Mar. Biol., 26: 183-192.
Fuji, A. 1960, Studies on the biology of the sea urchin. I. Superficial and histological gonadal changes in gametogenic process of two sea urchins, Strongylocentrotus nudus and S. intermedius. Bull. Fac. Fish. Hokkaido Univ.,11(2): 1-14.
Holland, N.D. 1967, Gametogenesis during the annual reproductive cycles in a cidaroid sea urchin (Stylocidaris affinis). Biol. Bull. 133(3): 578-590.
Holland, N.D. 1971, The fine structure of the ovary of the feather star Nemaster rubiginosa (Echinodermata: Crinoidea). Tissue & Cell, 3(1): 161-175.
Holland, N.D. & Giese, A.C. 1965, An autoradiographic investigation of the gonads of the purple sea urchin (Strongylocentrotus purpuratus). Biol. Bull., 128(2): 241-258.
Holland, N.D. & Rommel, J.A. 1969, A frequency polygon study of oogenesis in the sea urchin Psammechinus microtuberculatus. Pubbl. Staz. Zool. Napoli, 37: 657-659.
Ludwig, H. & Hamann, O, 1907, Die Seelilien, in: Bronn's Klassen und Ordnungen des Thierreichs, 2 (Abt. 3, Buch 5.) Winter Verlag, Leipzig.
Mortensen, Th. 1937, Contributions to the study of the development and larval forms of echinoderms III. D. Kgl. Danske Vidensk. Selsk. Skrifter, naturv. og math. Afd. Ser. 9, 7(1): 1-65, 15 pl.
Patent, D.H. 1969, The reproductive cycle of Gorgonocephalus caryi (Echinodermata: Ophiuroidea). Biol. Bull., 136(2): 241-252.
Pearse, J.S. 1969, Reproductive periodicities of Indo-Pacific invertebrates in the Gulf of Suez. I. The echinoids Prionocidaris baculosa (Lamarck) and Lovenia elongata (Gray). Bull. Mar. Sci. 19(2): 323-350.
Rutman, J. 1975, The reproductive activity of the feather-star Lamprometra klunzingeri and Heterometra savignii from the Gulf of Eilat (Red Sea). Ph.D. Thesis, 99 pp. (in Hebrew).
Rutman, J. & Fishelson, L. 1969, Food composition and feeding behavior of the shallow-water crinoids at Eilat (Red sea). Mar. Biol. 3(1): 46-57.
Tanaka, Y. 1958, The seasonal changes occurring in the gonad of Stichopus japonicus Bull. Fac. Fish. Hokkaido Univ., 9(1): 29-36.

SEM and behavioural study of the feeding system of the comatulid, *Antedon bifida*

MARIE-CHRISTINE LAHAYE & MICHEL JANGOUX
Laboratoire de Biologie marine, Université Libre de Bruxelles, Belgium

Investigations on the feeding behaviour of Antedon bifida were done using a coloured suspension of food particles. SEM-observations were performed on pieces of comatulid arm or calyx, fixed in Bouin's fluid for a 12 hour period. The feeding system of adult A. bifida consists of ciliated grooves, lined by well-developed feeding appendages or podia bearing conspicuous papillae. Specialized feeding units, i.e. triplets of unequal-sized podia with an associated protective lappet occur all along the pinnular grooves. Both the primary and secondary podia of the feeding units are food-collecting appendages. Food particles are trapped on the mucus secretion of the podal papillae. Particles are transferred to the ciliary tracts of the grooves by the wiping action of collecting podia against the ciliary current. When active, tertiary podia show paddling movements against the ciliary current. Such movements increase mucus entrapment of food particles. Arm grooves are lined by triplets of equal-sized podia. These play an important role in regulating particle flow at the arm-pinnule junctions. Calycinal grooves are lined by single podia that act in guiding particle flow towards the mouth. Rate of particle transport down the arm groove is c. 4.3 cm/min. Typical pentacrine larvae have 25 podia topping the larval calyx, i.e. five triplets of radial podia (collecting podia) and five doublets of interradial podia (paddling podia). The feeding method of early pentacrinoid larvae is identical to that of the adult. Adult and larval feeding units are rather similar, being both made of collecting podia, paddling podia, protective structures (lappets or larval oral plates) and ciliary tracts.

Work supported by IRSIA grant 81382 to A. Lambert

3. Echinoidea

Proceedings of the Fifth International Echinoderm Conference / Galway / 24-29 September 1984

Reproduction and development in *Goniocidaris umbraculum*, a brooding echinoid

M.F.BARKER
Portobello Marine Laboratory, Dunedin, New Zealand

ABSTRACT: The reproductive biology of a population of the cidarid echinoid Goniocidaris umbraculum (Hutton) occurring on the outer continental shelf off Otago has been examined. Regular determination of the gonad index and histological examination of sectioned gonads show a clearly defined reproductive cycle. Gonads reach the peak of development in June-July each year and spawning is completed by August-September. Oocytes take two years to complete their development within the ovaries. Gonad indices of males are consistently higher than females. After spawning developing eggs are retained on the peristome, which sinks to accommodate up to 60 embryos brooded for approximately 8 weeks. The evolution and ecological significance of brooding in Goniocidaris umbraculum is discussed.

1. INTRODUCTION

The reproductive cycles and larval development of a large number of echinoids have now been described. Most species have been shown to have an annual reproductive cycle with spawning of a small (approximately 100 µm) egg which develops into a free swimming planktotrophic pluteus larva. Settlement on the bottom follows the period in the plankton. Other more direct forms of development occur less commonly in the echinoids. Lecithotrophic planktonic development from a large yolky egg is described for Peronella lesueuri (Mortensen 1921) and Heliocidaris erythrogramma (Williams and Anderson 1975). Brooding is uncommon in the class as a whole although it does occur commonly in many subantarctic and antarctic species (Hyman 1955, Magniez 1983). A detailed description of brooding is only available for two species, the cassiduloid Cassidulus caribbearum (Gladfelter 1978) and the spatangoid Arbatus cordatus (Mangiez 1980). The only data on the reproductive cycle of a brooding species is for Arbatus cordatus (Magneiz 1980, 1983).

Within the class Echinoidea the Cidaroida are the only living order within the primitive Subclass Perischoechinoidea. Although detailed information on reproduction is available for only a few cidaroids, all the reproductive strategies already mentioned are known to occur. The tropical species Eucidaris tribuloides has an annual cycle and larval development follows the normal indirect pattern with a pluteus larva (McPherson 1968). The deep water slope species Cidaris cidaris also has indirect development, (Mortensen 1927a) but does not appear to show reproductive seasonality (Tyler and Gage 1984). The gametogenic cycle described for Stylocidaris affinis (Holland 1967) and Prionocidaris baculosa (Pearse 1969) indicates an annual reproductive cycle but the diameters of ripe oocytes (80-100 µm) would imply that both species have planktotrophic development. Heliocidaris erythrogramma has an annual reproductive cycle (Dix 1977) with a large egg (400 µm) and lecithotrophic development (Williams and Anderson 1975). Tyler and Gage (1984) predict indirect development for the deep water slope species Priocidaris purpurata. Hyman (1955) reviewing information on brooding in echinoids notes that peristomial brooding is known for at least 9 species of cidaroids, most of these antarctic species.

Mortensen (1925) found that juvenile urchins were carried on the peristome of adult Goniocidaris umbraculum and in a later paper (1927b) included a brief description of a juvenile. Since these early papers by Mortensen there have been no further studies on reproduction in Goniocidaris umbraculum. The present investigation was undertaken to provide a full description of the reproductive cycle and to provide the first detailed description of brooding in a member in this interesting and primitive group.

2. MATERIALS AND METHODS

2.1 Reproductive cycle

Adult urchins were collected at approximately monthly intervals by Agassiz dredge from a station on the outer continental shelf off Otago (170°55'E, 45°46.8'S) at 95 m depth from May 1981 until December 1982. An initial survey had indicated that adult G. umbraculum were extremely plentiful at this station on a coarse shell/bryozoan rubble bottom. The test diameter was measured on all urchins collected and the presence of brooded eggs or juveniles noted. Fifteen or 16 larger urchins (test diameter 20-25 mm) were selected from each haul for subsequent dissection. Urchins were held in running seawater until they could be processed, usually within 24 hours. The heavy spines were cleaned of encrusting species (tubeworms, sponges, bryozoa etc.) and then each urchin was opened by an incision around the oral circumference of the test with a pair of heavy scissors. Next the 5 gonads were removed from each interradius, blotted dry and weighed. The remaining portion of the body was also blotted dry and then dried to constant weight at 60°C Gonad indices were calculated as the ratio

$$\frac{\text{wet gonad wt.} \times 100}{\text{dried body wt.}}.$$

Using the wet gonad overemphasises the weight of gonad relative to the body. This was necessary however in order to examine all 5 gonads histologically. Gonads were fixed in seawater formalin, dehydrated in alcohol and embedded in paraffin wax. The gonads were cut at 7-10 µm as a single mass and stained with haemotoxylin and eosin. For females the size frequency of oocytes was determined by measuring oocytes sectioned through the nucleolus in at least 5 or 6 animals. A total of at least 200 oocytes were measured for each sample but the small number and large size of individual oocytes often required measurement of all oocytes sectioned in the correct plane in each ovary mass. In male urchins only the stage of spermatogenesis was noted.

2.2 Spawning and juvenile development

Analysis of the very first sample of adult G. umbraculum collected in May 1981, revealed that urchins were nearly ripe. In early June a sample of 30-40 urchins was collected from the offshore station along with a quantity of the bottom material. These urchins and substratum were returned to the laboratory and maintained in a large (0.6 x 1.2 m) tank with a constant supply

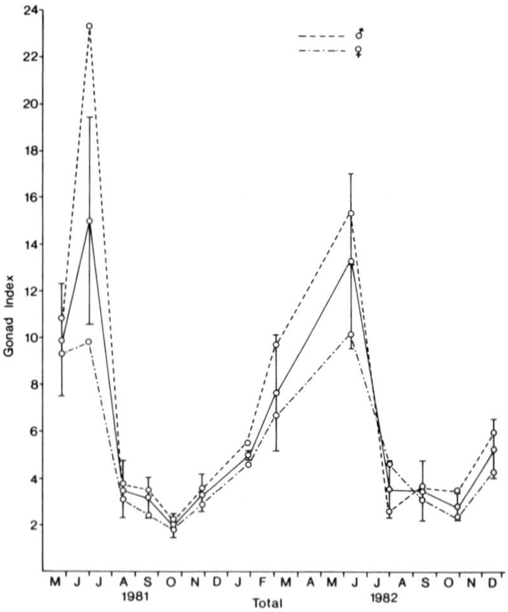

Figure 1. Gonad index cycle of Goniocidaris umbraculum: vertical lines represent 95% confidence limits of combined male and female indices (=0—0)

of running seawater. Every 1 or 2 days each urchin was checked for the presence of freshly spawned eggs. After spawning occurred numbered labels were attached to aboral spines of individual urchins. Eggs or developing embryos were periodically removed from the peristome area of these individuals and the stage of development noted before eggs were returned to the peristome. Spawned eggs were also collected from three females and kept in the main seawater circulation system in small perspex rings covered with bolting silk. Holes drilled in each side allowed the circulation of fresh seawater.

3. RESULTS

3.1 Gonad indices (Fig. 1)

In May 1981 when the study commenced the mean gonad index for both males and females was already high and peaked in early July. A sudden drop in the index between July and August 1981 indicated that spawning had taken place. The combined index dropped to a minimum value in October and then commenced to rise in late 1981-early 1982. Unfortunately bad weather and vessel breakdowns prevented samples being collected until early June when the index had risen to the highest value recorded for 1982. As in 1981 the index dropped sharply between

June and late July, reached a minimum value in October and had begun to rise when sampling was terminated in December. At the peak of the breeding season in June of both years the wide confidence limits indicate the presence of both ripe and recently spawned individuals in the population.

When male and female gonad indices are plotted separately (Fig. 1) males have a consistently higher index than females although on only two sampling dates (1.7.81 and 28.10.82) were the differences statistically significant ($P<0.001$ and $P<0.05$ respectively Mann-Whitney U test).

3.2 Oogenesis

Size frequency histograms of oocytes measured in sectioned ovaries are shown in Figure 2 and photomicrographs of developmental stages in Figure 3a-d. When sampling commenced in May 1981 two populations of oocytes were present, immature previtellogenic oocytes (0-200 µm) and larger vitellogenic oocytes (300-800 µm) (Fig. 3a). Nutritive phagocytes were also present. By early July numbers of mature oocytes (Fig. 3b) had increased. The disappearance of larger oocytes (500-800 µm) by mid August confirms the spawning date suggested by gonad indices. Numbers of quite large (300-600 µm) oocytes remain in ovaries (Fig. 3c). Over the period September to January these larger oocytes gradually increased in size and there was also a marked increase in the number of smaller (0-150 µm) previtellogenic oocytes (Fig. 3d). By June large oocytes were abundant while vitellogenesis had commenced in many of the smaller size classes. As in 1981 spawning occurred in the June-July period with considerable numbers of large (350-600 µm) oocytes remaining unspawned in October 1982. There was no indication of breakdown or resorption of these large vitellogenic oocytes in the months following spawning. The very large numbers of previtellogenic oocytes present in most 1981 ovaries sectioned were not seen in the latter part of 1982.

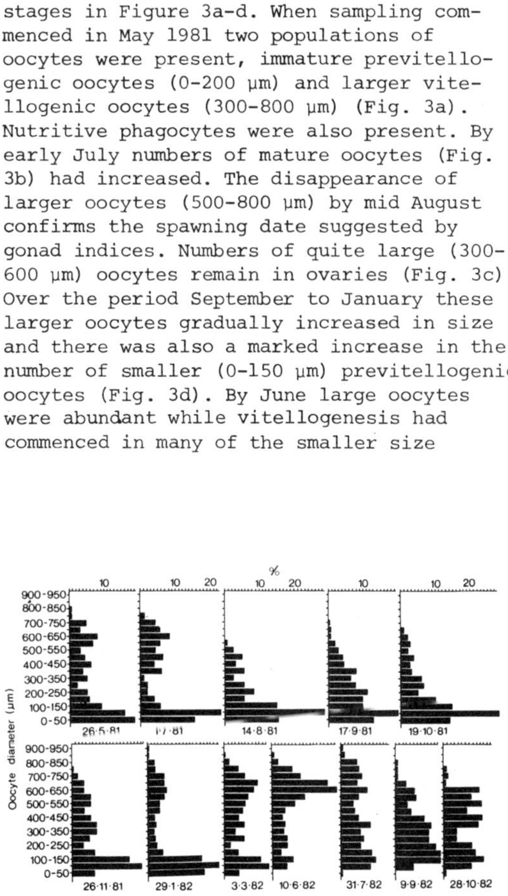

Figure 2. Mean oocyte size frequency distribution from sectioned ovaries of Goniocidaris umbraculum collected from May 1981 to October 1982.

3.3 Spermatogenesis (Fig. 3e-f)

When sampling commenced in late May 1981 testes were at an advanced stage of development. Spermatocytes and spermatids formed long columns from the germinal epithelium towards the testis lumen while mature spermatazoa were abundant within this luminal area (Fig. 3e). Nutritive phagocytes containing eosinophilic and basophilic globules were also abundant. By early July most testis lobes were packed with mature spermatozoa (Fig. 3f), although in some testes columns of spermatocytes and spermatids still formed a discrete layer. There was also an increase in numbers of nutritive phagocytes forming within cavities in the lumen and immediately inside the germinal epithelium. By mid August spawning was almost complete in all testes sectioned (Fig. 3g). Some residual spermatazoa were being phagocytized by abundant nutritive phagocytes. By mid September spermatogonia were present and spermatocytes formed conspicuous clusters immediately within the germinal epithelium (Fig. 3h). Nutritive phagocytes made up a large component of testis tissue. In the following months, October through to January, there was a gradual build up in the thickness of the germinal layer of spermatocytes and spermatids within testis lobes and by early March mature spermatazoa were again accumulating within the central lumens. Nutritive phagocytes were abundant during this period. By June testes were ripe. In the succeeding months the pattern was identical to that described for 1981 with complete spawning by late July and active spermatogenesis occurring with formation of clumps of spermatogonia against germinal epithelia in early September.

3.4. Spawning and juvenile development

3.41 Laboratory observations

In both years spawning occurred in laboratory-held urchins in mid-late August some 4-6 weeks after gonad indices indicate spawning had occurred in field populations.

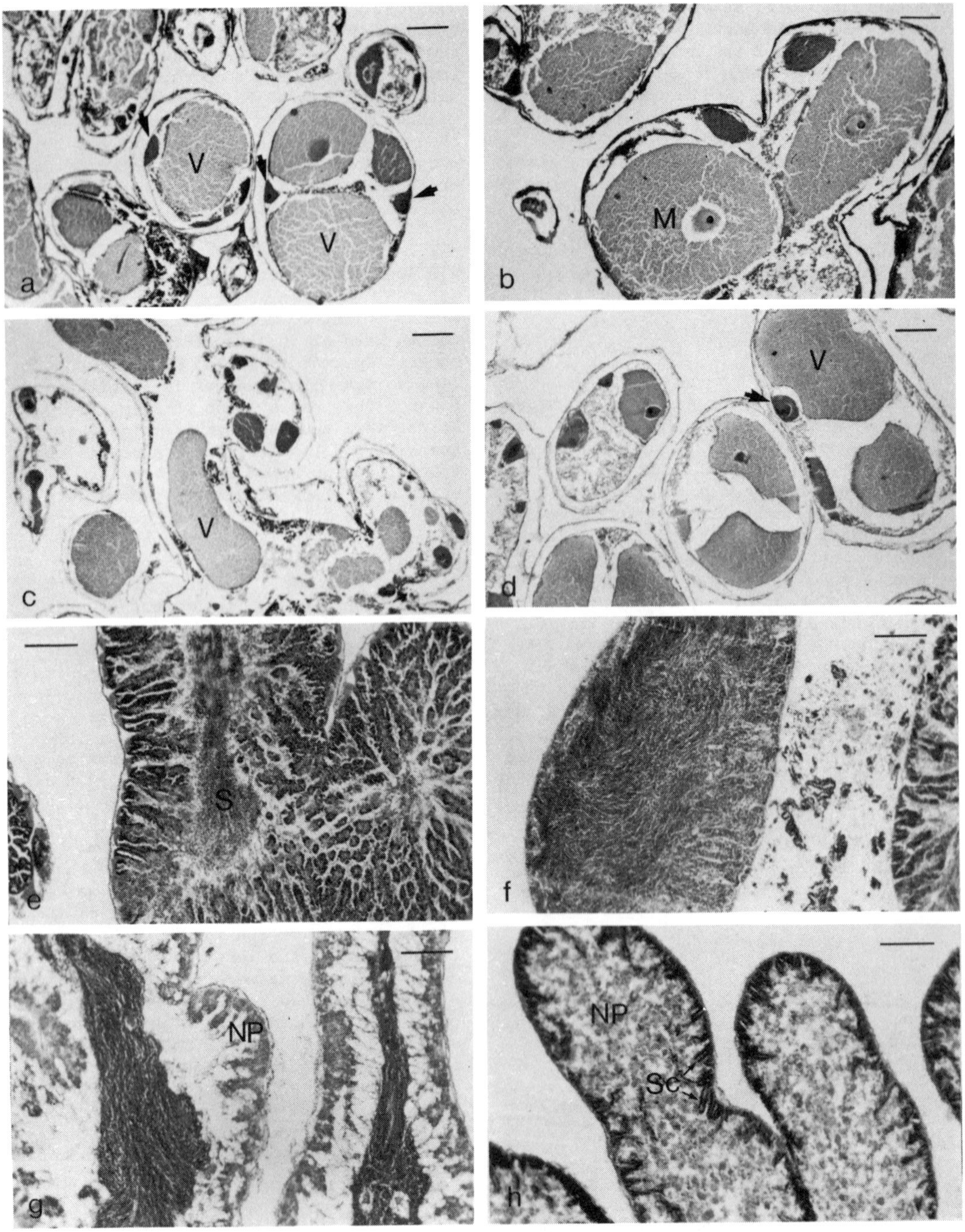

Figure 3. Histological structure of ovary and testis of Goniocidaris umbraculum. (a) Ovary 26.5.81 showing previtellogenic (arrow) and vitellogenic (V) oocytes. (b) Ovary 1.7.81 showing mature oocytes (M). (c) Ovary 14.8.81 showing unspawned vitellogenic oocytes (V). (d) Ovary 26.11.81 previtellogenic (arrow) and vitellogenic (V) oocytes present. Scale bar a-d = 200 µm. (e) maturing testis 26.5.81, spermatazoa (S) within lumen (f) Mature testis 1.7.81 (g) Spawned testis 14.8.81, note nutritive phagocytes (NP) (h) early testis development, 19.10.81 note spermatocytes (Sc) and abundant nutritive phagocytes (NP). Scale bar e-h = 100 µm.

The first indication of spawning was when several urchins partially buried themselves in the bryozoan/shell substrate (Fig. 4a). These animals were all found to be carrying eggs around the peristome. The remaining urchins in the tank were then closely monitored over the next few days and spawning was observed directly in two further females. Oocytes were released from the genital pores and were carried around from the aboral to oral region in the ambulacral areas. Movement from gonopore to peristome takes several minutes for each egg and is apparently accomplished by the action of cilia and tube feet and is aided by the formation of a channel created between the spines. The large (900-1000 μm) yolky oocytes are extremely buoyant, and float if separated from the female. The peristome appears to sink inwards and the eggs accumulate in this area their buoyant nature helping to hold them against the peristomial membrane (Fig. 4b). The oral spines are also used to cover the mouth area and this further assists in retaining eggs. Spawning was not synchronous from all 5 ovaries. Eggs were released intermittently from different ovaries and it took several hours for 50-60 eggs to accumulate within the peristome area. Spawning urchins were isolated from the main seawater system and kept for two days in filtered seawater at ambient temperature. Observations on eggs showed no further embryonic development. Females were then placed in a dilute sperm suspension for 5-10 minutes and observations continued for several hours. Within 2-3 hours the first cleavage had taken place.

Development is unusual and only a very brief description can be included here. The formation of a blastula took 5-6 days and invagination of an archenteron was first seen 6 days after fertilization. The gastrula stage lasted for 17-18 days at which time the blastopore closed. After 24-25 days 5 protuberances, primordial tube feet, appeared on one side of the embryo. At this time the embryo was very irregular in shape and approximately 1 mm in diameter. The tube feet gradually lengthened and after 30 days incubation further small protruberances, incipient spines, appeared close but peripheral to the developing tube feet. The spines also lengthened and as remaining yolk reserves were reduced embryos became more regular in shape. After 40-45 days the tube feet had developed suckers and were being used to move the embryos around. For the first time embryos were negatively buoyant. At approximately 47-55 days juveniles began to move from the peristome area to the bases of the spines on the oral side of the adult female urchin (Fig. 4c). At this stage embryos were lost from adult urchins in laboratory animals, but from field observations it is clear that in nature urchins may remain on the test of the mother while further tube feet, spines, the mouth and Aristotle's lantern develop (Fig. 4d). There was some variation in the stage of development within eggs brooded on particular female urchins. Occasionally some embryos were noted to be at the blastula or gastrula stage while others had well developed tube feet and were almost ready to leave the peristome area. Eggs collected from females as they were spawned and reared in vitro showed much greater synchrony in development. Asynchronous development to the degree outlined above results from several different spawnings by the parent female.

3.42 Field observations

The occurrence of brooded eggs or embryos around the peristome and juvenile urchins on the test for the different samples taken over the breeding season is shown in Table 1.

Only two urchins still carrying eggs or embryos around the peristome were retrieved from dredge trawls. For all others the highly buoyant eggs were probably lost from adults during dredge trawling or hauling. As juvenile urchins were able to cling to the adult much more effectively, many more adults were recovered with juveniles than with embryos attached. Juveniles first appeared on adults sampled in early-mid September in both 1981 and 1982. Laboratory studies indicate that these would have been at least 55-60 days old and spawned in early to mid July which agrees well with the drop

Table 1. The occurrence of brooded embryos and juveniles on adult *G. umbraculum* sampled during 1981 and 1982.

Date	sample size	Number adults brooding Eggs/ embryos	Juveniles
1.7.81	37	-	-
14.8.81	43	2	-
17.9.81	54	-	3
19.10.81	41	-	20
26.11.81	39	-	-
9.12.81	62	-	3
29.1.81	54	-	-
10.6.82	37	-	-
31.7.82	43	-	-
9.9.81	33	-	3
28.10.82	55	-	10
16.12.82	29	-	-

Figure 4. Brooding in Goniocidaris umbraculum. (a) Partially buried female, Aboral spines just visible (arrows) among bryozoan and shell rubble. (b) Oral view showing developing embryos (arrows) packed around the peristome. (c) juvenile 55 days old. (d) Juvenile urchins, approximately 3½ months old on adult spines. a & b scale bar = 5 mm., c & d scale bar = 0.5 mm.

in gonad indices. Adults carrying juveniles were most abundant in samples collected in mid-late October. By December 1981 only 3 urchins in a sample of 62 were still carrying juveniles and none were found in December 1982. This suggests juveniles stay with the parent for approximately 4 months.

The smallest urchins found to be carrying juveniles were 17.3 and 17.9 mm test diameter.

4. DISCUSSION

Goniocidaris umbraculum has a well defined annual breeding cycle as shown by seasonal variation of gonad indices. Histological evidence, however, shows that maturation from oogonium to mature oocyte takes about two years. This prolonged period for oogenesis is uncommon in echinoids. Presumably a prolonged growth phase is required for the accumulation of the considerable yolk reserves present in the large (800-1000 µm eggs. Both the brooding asteroids Leptasterias hexactis (oocyte diameter 800 µm, Chia 1968) and L. tenera (oocyte diameter 900-1000 µm, Worley et al. 1977) also take two years to complete oogenesis. Interestingly, Heliocidaris erythrogramma, with an oocyte diameter of 400 µm, completes oocyte development in a single year (Williams and Anderson 1975).

The environmental factors which synchronise the reproductive cycle of G. umbraculum are unknown. The exogenous controls on gametogenesis and spawning in marine invertebrates are still the subject of considerable debate. Seasonal fluctuations in sea temperature are often regarded as important (see Giese and Pearse 1974). At the sampling station for G. umbraculum off Otago the bottom temperature varied between 9.5°C in July and 13.5°C in February-March, i.e. an annual variation of only 4°C. While temperature may still have some influence in controlling the reproductive cycle, other factors such as photoperiod or the accumulation of nutrients may be more important. Photoperiod has recently been shown to be the primary exogenous control in the asteroid Pisaster ochraceous (Pearse and Eernisse 1982). Pearse (1969) suggested that fluctuations in accumulated food reserves were more important than temperature for synchronising reproductive periodicities in Prionocidaris baculosa.

Male gonad indices were consistently higher than female indices. That male urchins continue to maintain a higher index at the completion of spawning is a surprising result especially when the gametogenic cycle is considered, i.e. the retention in ovaries of quite large vitellogenic oocytes following the spawning season. A higher index in females than males, especially following spawning, would seem a more likely result. Higher male indices are unlikely to be due to males having a lighter test as there were no significant differences in the weight of dried evicerated male and female tests (P>0.5, Students t test) although there could be slight intersexual differences in water content of gonads. A more likely explanation for higher male indices could be the observed formation of large quantities of nutritive phagocytes in the testes shortly before and immediately after spawning. Although nutritive phagocytes were present in ovaries over this period they did not become abundant until January. A higher male gonad index does not necessarily imply a higher energy output. Magniez (1983) also found males to have a higher gonad index than females throughout the year in the brooding echinoid Arbatus cordatus. However the ovary had higher lipid and carbohydrate levels and overall a 40% higher calorific value than the testis.

In a description of gametogenesis in Stylocidaris affinis, Holland (1967) noted that the ovaries never contained ripe oocytes for an extended portion of the reproductive cycle. During spermatogenesis, spermatocytes accumulated and their maturation through to spermatazoa occurred only shortly before spawning. Holland suggested this pattern might represent a primitive feature of cidarids. This did not occur, however in Prionocidaris baculosa (Pearse 1969) and in the present study both mature spermatazoa and oocytes accumulated in gonads over several months of the reproductive cycle. Clearly more information on gametogenesis in cidarids is needed before such phylogenetic implications can be made.

The burying behaviour that takes place does not appear to have been described before in any epibenthic echinoderm. While brooding eggs or embryos, urchins do not feed. Movement of Aristotle's lantern may be sufficient to dislodge eggs from the peristome area. As urchins are not able to feed, burying may be a means of reducing the risks of attack or physical disturbance which would increase the risk of egg loss. Partial burying must provide a physically stable environment around the oral surface of the adult while early embryonic development is occurring. Presumably there is sufficient water passage through the very coarse bryozoan/shell substratum to allow the female to ventilate the eggs.

The origin of the brooding habit in G. umbraculum is of considerable interest. A number of subantarctic and antarctic species of cidarids brood their eggs on the peristome as described in the present study. If brooding in G. umbraculum is an archaic feature then the origins of this genus might be expected to lie in southern waters. Both Fell (1953) and McKnight (1969) believed that the genus Goniocidaris originated in or near Australasia during early or mid Oligocene times. Mortensen (1928) regarded the centre of distribution of modern species as the Indonesian and Australasian area, which suggests an Indo-Pacific origin. As indirect development is the likely form of development in tropical species, brooding in G. umbraculum would therefore seem to be a reproductive strategy which has evolved comparatively recently.

Brooding occurs in many species of starfish and is particularly characteristic of the genus Leptasterias (Hendler and Franz 1982, Himmelman et al. 1982). Several hypotheses have been advanced to account for brooding in this genus. Menge (1975) suggested that in L. hexactis brooding is an evolutionary response to competition induced small body size, the competitor in this case being the large broadcast spawner Pisaster ochraceus. Hendler and Franz (1982) however, believed that as most species of Leptasterias are cold water and slow growing, brooding and a two year period for oogenesis are cold adaptations rather than an evolutionary consequence of competition. Himmelman et al. (1982) also suggested that because the genus evolved in arctic or subarctic waters, brooding is probably a fixed trait.

At the station sampled on the Otago shelf G. umbraculum overlaps in distribution with the larger, broadcast spawning euechinoid Pseudechinus huttoni. Both of these species graze on the bryozoan Filicea elegans. Brooding in G. umbraculum could have evolved in response to competition with P. huttoni. However the overlap between these two species is only narrow, as G. umbraculum has an outer shelf-upper slope distribution and P. huttoni is mainly found on the mid-outer shelf. Competition, if it occurs, must only take place over a restricted area. It seems more likely that, as suggested for Leptasterias, brooding in G. umbraculum has evolved as a response to migration into southern waters which were considerably cooler than those in which the genus evolved. G. parasol, which is found in deeper water and has a more southerly distribution than G. umbraculum also broods around the peristome. (Pers.obs.).

In a variety of invertebrate taxa brooding seems to be a characteristic of small body size (Strathmann and Strathmann 1982). G. umbraculum is a small echinoid (test diameter 20-25 mm) and this may mean that only limited energy is available for reproduction despite the restricted dispersal that occurs with such a strategy. Clearly more information on such population parameters as growth and longevity is needed to come to valid conclusions as to the most energetically viable reproductive strategy for G. umbraculum.

REFERENCES

Chia, F.S. 1968. Some observations on the development and cyclic changes of the oocytes in a brooding starfish, Leptasterias hexactis. J.Zool. 154:453-461.

Dix, T.G. 1977. Reproduction in Tasmanian populations of Heliocidaris erythogramma (Echinodermata:Echinometridae). Aust.J. Mar.Freshwat. Res. 28:509-520.

Fell, H.B. 1953. The origins and migrations of Australasian echinoderm faunas since the Mesozoic. Trans.Roy.Soc.N.Z. 81:245-255.

Giese, A.C. & J.S. Pearse. 1974. Introduction: general principles, In A.C. Giese & J.S. Pearse (eds.). Reproduction of marine invertebrates. p.1-49. Academic Press, New York.

Gladfelter, W.B. 1978. General ecology of the cassiduloid urchin Cassidulus caribbearum. Mar.Biol. 47:149-160.

Hendler, G. & D.R. Franz. 1982. The biology of a brooding seastar, Leptasterias tenera, in Block Island Sound. Biol.Bull. 162:273-289.

Himmelman, J.H., Y. Lavergne, A. Cardinal, G. Martel & P. Jalbert. 1982. Brooding behaviour of the Northern sea star Leptasterias polaris. Mar.Biol. 68:235-240.

Holland, N.D. 1967. Gametogenesis during the annual reproductive cycle in a cidaroid sea urchin (Stylocidaris affinis). Biol.Bull. 133:578-590.

Hyman, L.H. 1955. The invertebrata. 4. Echinodermata. McGraw Hill:New York.

McKnight, D.G. 1969. An outline distribution of the New Zealand shelf fauna. Benthos survey, station list and distribution of the Echinoidea. Mem.N.Z. Oceanogr.Inst. 47:91 p.

McPherson, B.F. 1968. Contributions to the biology of the sea urchin Eucidaris tribuloides (Lamarck). Bull.Mar.Sci. 18: 400-443.

Magniez, P. 1980. Modalités de l'incubation chez Abatus cordatus (Verrill), oursin endémique des îles Kerguelen. In M.Jangoux (ed.), Echinoderms present and past, p.399-403 Rotterdam:Balkema.

Magniez, P. 1983. Reproductive cycle of the brooding echinoid Abatus cordatus (Echinodermata) in Kerguelen (Antarctic Ocean) changes in the organ indices biochemical composition and caloric content of the gonads. Mar.Biol. 74:55-64.

Menge, B.A. 1975. Brood or broadcast? The adaptive significance of different reproductive strategies in the two intertidal sea stars Leptasterias hexactis and Pisaster ochraceus. Mar.Biol. 31: 87-100.

Mortensen, T. 1921. Studies of the development and larval forms of echinoderms. G.E.C.Gad:Copenhagen.

Mortensen, T. 1925. Goniocidaris umbraculum A brood protecting species. N.Z.J.Sci. Technol. 8:192.

Mortensen, T. 1927a. Handbook of the echinoderms of the British Isles. Oxford University Press.

Mortensen, T. 1927b. On the postlarval development of some cidarids. K.danske Vidensk.Selsk.Skr.(Nat.Math.Afd. 8 raekke). 11:367-387.

Mortensen, T. 1928. A monograph of the Echinoidea. I. Cidaroidea. Reitzel, Copenhagen, 647 p.

Pearse, J.S. 1969. Reproductive periodicities of Indo-Pacific invertebrates in the Gulf of Suez. I. The echinoids Prionocidaris baculosa (Lamarck) and Lovenia elongata (Gray). Bull.Mar.Sci. 19:323-350.

Pearse, J.S. & D.J. Eernisse. 1982. Photoperiodic regulation of gametogenesis and gonadal growth in the sea star Pisaster ochraceus. Mar.Biol. 67:121-125.

Strathmann, R.R. & M.F. Strathmann. 1982. The relationship between adult size and brooding in marine invertebrates. Am.Nat. 119:91-101.

Tyler, P.A. & J.D. Gage. 1984. The reproductive biology of echinothurid and cidarid sea urchins from the deep sea (Rockall Trough, North-East Atlantic Ocean). Mar.Biol. 80:63-74.

Worley, E.K., D.R. Franz & G. Hendler. 1977. Seasonal patterns of gametogenesis in a North Atlantic brooding asteroid, Leptasterias tenera. Biol.Bull. 153:237-253.

Williams, D.G.C. & D.T. Anderson. 1975. The reproductive system, embryonic development and metamorphosis of the sea urchin Heliocidaris erythrogramma (Val). (Echinoidea:Echinodermata). Aust.J.Zool. 23:371-403.

Proceedings of the Fifth International Echinoderm Conference / Galway / 24-29 September 1984

Selection, storage and elimination of heavy sand particles by the juvenile sand dollar, *Dendraster excentricus* (Eschscholtz)

FU-SHIANG CHIA
University of Alberta, Graduate Studies, Edmonton, Canada

ABSTRACT: Juvenile sand dollar Dendraster excentricus selectively ingests iron oxide particles from the substrate. The selection is probably made by the buccal podia and at the floor of the buccal cavity and the selection is based on the density of sand particles or on the micro-flora on the sand surface, or both. Once ingested, the sand is stored in the intestinal diverticulum until the animal is two to three years old. There is some evidence that the sand may be dissolved in the diverticulum by organic acid or by iron-reducing bacteria. Structural and behavioural adaptations of the juvenile sand dollar are discussed.

1. INTRODUCTION

In a previous paper (Chia, 1973), I reported that the juveniles of the sand dollar, Dendraster excentricus, selectively ingest iron oxide particles from the substrate and store them in the intestinal diverticulum. The diverticulum, when filled with heavy sand, may function as a weight belt to assist the young animal to remain in a shifting sand environment. The sand particles are eliminated from the diverticulum when the animal grows to about 30 mm in diameter, at the age of approximately 2 - 3 years (Birkland and Chia, 1971).

The objective of this paper is to report the findings on the mechanisms and processes involved in the selection and elimination of iron oxide particles by the young animal.

2. MATERIALS AND METHODS

Juvenile sand dollars at the size of 1.5 to 30 mm in diameter were collected at different times during 1973 through 1984 from sandy beaches at 3 locations in the United States: Alki Point, Seattle; East Sound, Orcas Island; False Bay, San Juan Island. Some animals were brought back to The University of Alberta and kept alive for two months in an aerated aquarium at a temperature of 10°C. For histological studies of the diverticulum and mouth regions, specimens were fixed in Bouin's fluid. Wax sections were stained with Mason's trichrome or silver impregnation technique using the methods outlined in Drury and Wallington's text (1967). For analysis of the epipsamasmic diatoms, the methods of Hickman (1969) were used, and in this case the sand samples were collected in May and June, 1975, from East Sound, Orcas Island. The study of bacterial flora was conducted on the same sand samples; the culture medium and procedures adopted were those of Jobson (1971). X-ray radiograms were made from either live animals or dried specimens, on a medical X-ray machine.

3. RESULTS

3.1 Structure of the Diverticulum and Sand Storage

The morphology of the diverticulum in Dendraster excentricus and several other species of sand dollars was described by Mitchell (1972). The diverticulum originates from the small intestine at the point where the intestine loops 360 degrees, doubling back, leading to the rectum (Fig. 1). When viewed aborally, the diverticulum travels counterclockwise along the exterior border of the digestive tract and terminates blindly, almost at the point of its origin. In some cases, the diverticulum may extend further, much beyond its point of origin. A number of radiated pouches spring out from the diverticulum. Generally, there are four large pouches, extending to four of the five interambulacral regions; the posterior interambulacra is with-

Fig. 1 Diagrammatic representation showing the digestive tract and the diverticulum after the removal of the aboral half of the test of the juvenile sand dollar. A = Aristotle lattern, E = esophagus, S = stomach, I = intestine, R = rectum, D = diverticulum with pouches (P).

out a pouch as this space is occupied by the rectum. There are five pairs of small pouches located at the five ambulacral regions. In most cases, the four large pouches are filled with sand first, starting with the pouches at the anterior end of the animal.

Histologically, the diverticulum is an extension of the small intestine. The wall consists of, from inside to outside, ciliated columnar epithelium (sometimes it appears to be a syncytium), connective tissue fibers, longitudinal muscle, circular muscle, and the peritonium, which is a layer of ciliated cuboidal epithelium (Fig. 2). Compared to the wall of the intestine (Fig.3), the wall of the diverticulum is much thinner, and there are always large vacuoles found in or between the epithelial cells.

3.2 Site of Selection

There are two areas in which selection of sand particles can occur: an area around the mouth, outside the gut; and an area at the intestine-diverticulum joint, within the gut. If selection takes place outside the gut, only iron oxide particles are

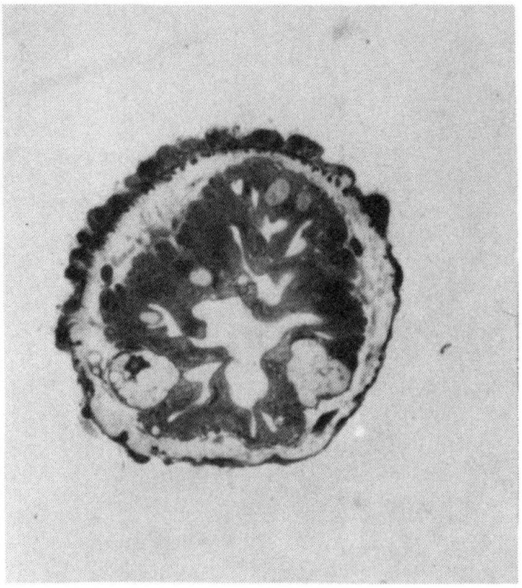

Fig.2 Cross section of the diverticulum. Note the vacuoles in the epithelium.

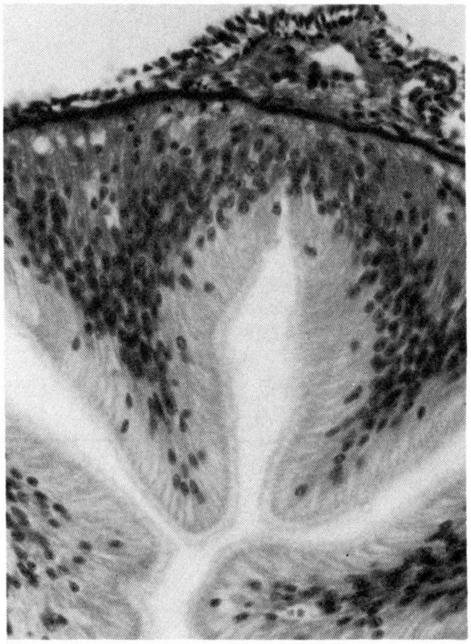

Fig.3 Portion of a cross-section of the intestine with a piece of the mesentery on the top of the Figure. The magnification of this figure is identical to that of Fig.2.

allowed to enter the gut; they will then be shunted into the diverticulum when it passes the joint. On the other hand, if selection occurs at the joint, one should expect to find mixed sand particles in the gut prior to the joint of the diverticulum.

We examined the gut contents from a number of juvenile sand dollars from all three locations and found only a small amount of sand particles, predominantly iron oxide. Furthermore, at the intestine-diverticulum joint we found no special structure such as a valve, nor was any sensory field obvious in that area. It is, therefore, concluded that selection of sand particles must take place outside the gut, and that the movement of iron oxide from the intestine to the diverticulum is of a passive nature.

If this conclusion is correct, one must ask where, outside the gut, the selection site is located. In a previous study on the feeding biology of Dendraster excentricus (Chia, 1969), I suggested that the selection or rejection of food was probably made by the buccal podia. Although the study was limited to adult animals, a similar mechanism may be used by juveniles in sand selection. After a detailed histological examination of the mouth region of the juveniles, with emphasis on locating nervous tissue, evidence suggests that selection may take place by both of the buccal podia as well as the buccal cavity site (Fig. 4).

The buccal cavity is bordered by both the nervous ring and radial nerve cords and the floor of the outer buccal cavity is a layer of folded columnar epithelium. Silver-stained nervous tissues are common beneath the epithelium. Moreover, the animal when freshly collected, retains a certain amount of food together with sand particles in this cavity. There is a narrow passage between the buccal cavity and the esophagus and this region may serve as the final selection site for iron oxide.

3.3 Nature of Selection

Since iron oxide is heavier than other sand particles in the substrate, it is possible that selection is dictated by the density of sand. On the other hand, selection may be based on the surface texture of the particles. However, I examined the surface of iron oxide and quartz (the major components of the substrates) with a scanning electron microscope and found no outstanding or consistent differences between these two particles.

It is also possible that sand selection by the juvenile sand dollar may be based on the differences of the microflora supported by various sand particles. If this is the case, selection is due to the detection of organic matter that may be related to feeding. In other words, the animal selected iron oxide over other particles because of

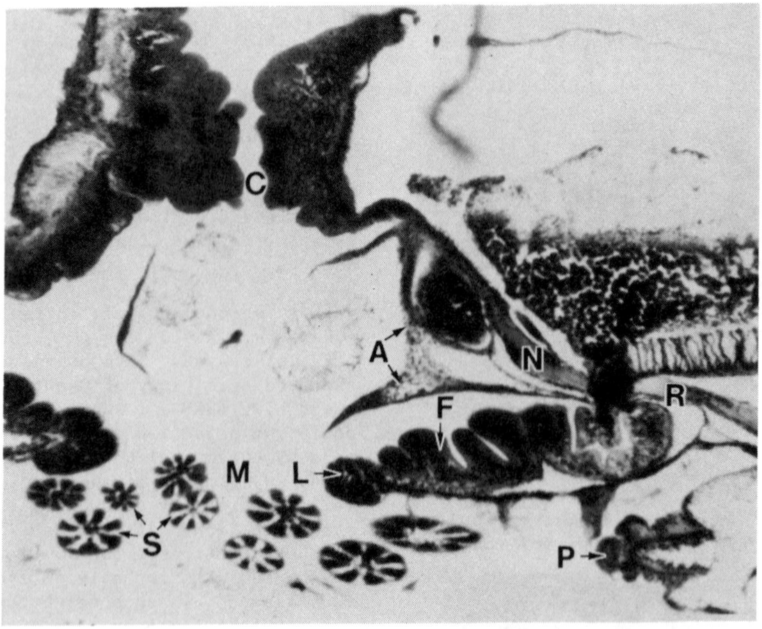

Fig.4 Oral-aboral section through the mouth region of a juvenile sand dollar, showing the buccal podium (P), spines (S), Lip (L), floor of the buccal cavity (F), radial nerve cord (R), nerve ring (N), the position of the mouth (M), the position of Aristotle lattern (A), the narrow canal (C) between the esophagus and the buccal cavity.

Table 1. Summary of mean number of potentially 8 species of diatom cells found on each sand particle as well as the mean number of diatom cells on a given surface area ($0.4mm^2$) of each type of sand particle. The total number of sand particles studied is 150 for each.

	Iron Oxide		Quartz & Feldspar	
Diatom Species	$\bar{X}$#/sand particle	$\bar{X}$#/$0.4mm^2$ surface area	$\bar{X}$#/sand particle	$\bar{X}$#/$0.4\ mm^2$ surface area
Fragilaria sp.A	1.97	11.25	11.00	25.00
Fragilaria sp.B	0.86	4.95	0.59	1.33
Fragilaria sp.C	0.30	1.72	0.58	1.31
Nictzchia sp.A	1.10	6.34	2.60	5.90
Nictzchia sp.B	0.50	2.88	2.30	5.20
Cocconeis sp.	1.96	11.41	1.90	4.30
Navicula sp.	0.70	4.03	1.30	2.75
Cyclotella sp.	0.72	4.2	0.8	1.81

specificity for the attached diatoms or bacteria on them. In examining this possibility, I studied the diatoms on two sand types, iron oxide and quartz. Eight species of five genera of diatoms are found on both sand types (Table 1). It is evident from Table 1 that there are more diatoms per particle on the quartz than on the iron oxide; however, this difference is less pronounced when the count is taken on a given unit area. What remains significant is the relative abundance of certain species on each of the substrates. For example there is more Fragilaria sp. on the quartz than on the iron oxide; the reverse is true for Cocconeis sp. The possible effect of these differences in sand selection should not be over-emphasized as the population of the attached diatoms must vary both in time and locality, and that the present result is limited to two samples collected in May and June, from one locality (Orcas Island).

A preliminary examination was conducted in an attempt to identify the bacteria from the two sand particles, quartz and iron oxide. After four days of incubation at room temperature, the B_{10} media with iron oxide show distinct whitish to bluish precipitates accompanied by a greening of the culture medium, indicating the presence of iron-reducing bacteria. On the other hand, the medium with quartz samples incubated under identical conditions turned black, indicating the presence of sulphite-reducing bacteria. There was a lack of sulphate-reducing bacteria in both sand samples and little or no black precipitate was found in the Butlin's medium. But again, the effect of bacteria on the selection of sand particles by the juvenile sand dollar cannot be assessed accurately based on this finding as the presence of bacteria on such sand samples may only be transitory.

3.4 Elimination of Sand

There are three possible ways in which sand particles can be eliminated from the diverticulum: 1) through the digestive system; 2) through the body wall; and 3) dissolution in situ by organic acid or by iron-reducing bacteria.

If the first instance is true, sand particles must be removed from the diverticulum into the small intestine, rectum, and then out via the anus. This would necessitate involvement of muscular contraction and/or ciliary activities. Although the epithelial lining of the diverticulum is ciliated and both circular and longitudinal muscle fibres are present in the diverticular wall, it does not appear to be strong enough to carry out this task.

The second possibility involves the rupture of the diverticulum, allowing the sand to enter the body cavity where it is expelled through the body wall. When in the coelom, certain coelomycytes may be involved in the transfer of sand particles to the skin surface, a mechanism known to occur in other echinoderms (Endean, 1966). If sand is eliminated in this way, it should be possible to find iron oxide particles in the coelom, especially when the animal is approaching a size of 30 mm in diameter. A dozen animals were dissected in an attempt to localize the sand particles in the coelom, without success.

Another method of detecting sand inside the body without dissection is to use the X-ray machine. This method met with considerable success (Fig. 5). At first, only the fixed and dried specimens were used, but later it was found that live specimens could be tested this way as well. Five animals were radiographed three times each at 10-day intervals. During the test period no shifting of sand particles in the diverticulum was detected. Moreover, among the 70 animals radiographed, no sand particles in the coelom were detected.

The third possibility is that sand is dissolved in situ by acid or iron-reducing bacteria. This hypothesis is based on the fact that iron-reducing bacteria have been found. Moreover, Howie (1974) was successful in using Thioglycollic acid (aqueous solution, ph = 2.2) to dissolve iron materials from the hematotic matrix on a fossil fish skull. In addition, the iron oxide particles in the diverticulum are small enough (300 - 400 um), and given a long period of time (a year or so), it is entirely possible that sand particles can be dissolved in situ by either acid or iron-reducing bacteria.

4. CONCLUSION AND DISCUSSION

This study suggests that the heavy sand particles are selected by buccal podia as well as in the buccal cavity based either on the density of the sand, or on the attached microflora, or both. Elimination of sand does not involve forcing the sand back to the intestine and expelling it through the anus. It is more likely that the sand is dissolved by organic acid or by iron-reducing bacteria within the diverticulum. We still do not know whether the animal simply stops ingesting sand particles when it is approximately 30 mm in diameter, or whether it continues to ingest but stops shunting the sand particles into the diverticulum.

Another interesting point is the behavioral and structural adaptations unique to this stage of ontogenetic development. The assumption is often made that once metamorphosed, the juvenile sand dollar will assume the lifestyle of the adult. This is hardly the case. As this study has demonstrated, storage of heavy sand particles in the diverticulum is a transitional

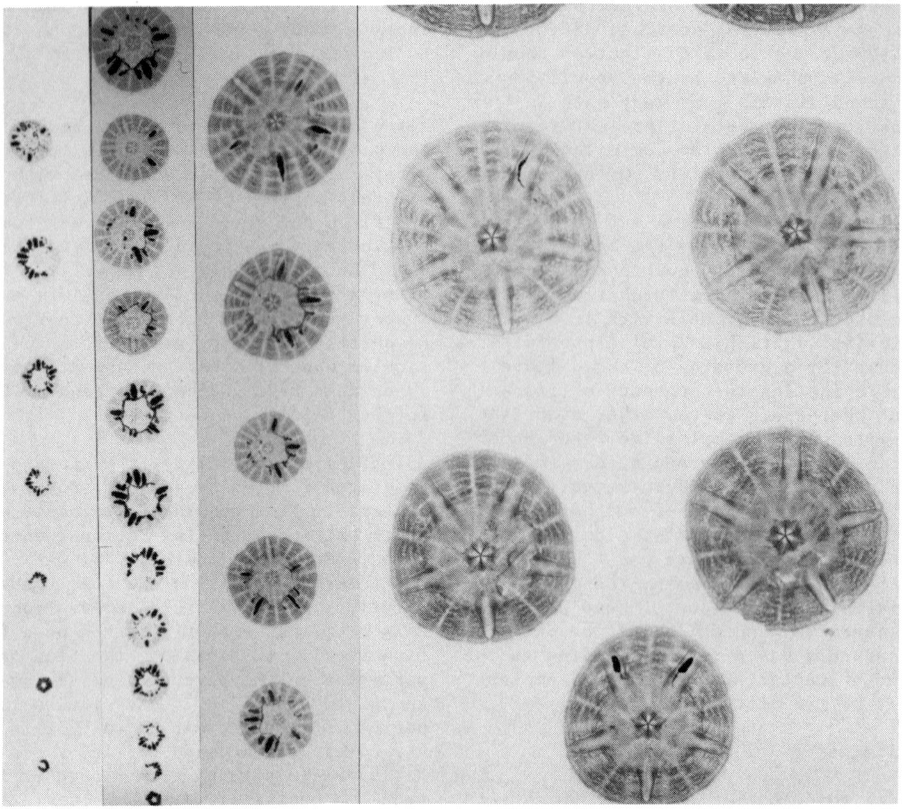

Figure 5. X-radiograms of juvenile sand dollars showing the iron oxide sand particles outlining the diverticulum in small specimens, but not in the large ones which are 30 mm or more in diameter.

phenomenon, limited to juveniles up to three years of age. Although I have reasoned that sand storage is adapted to increase the density of the animal in order to help it remain in a shifting environment (Chia, 1973), this may not be the complete story; in truth, we know very little about the life strategies of the juvenile sand dollar.

It should be noted that at Sydney Spit, Vancouver Island, the juveniles were found concentrated only in high intertidal pools, quite some distance from the adult populations. However, this was not the case at Alki Point, nor at East Sound. I have now examined two Australian sand dollars: Arachinoides placenta (L) and Laganum depressum Lesson. The juveniles of neither one of these species possess the intestinal diverticulum. We may reason that these two tropical sand dollars have not had to face the same problem or they may indeed share the same problem, but solve it differently.

Acknowledgement: This research was supported by a grant from the National Science and Engineering Research Council of Canada. I thank Mr. R. Seward, Mr. R. Koss, Mr. W. Andrias and Ms. B. Willen for technical assistance.

REFERENCES

Birkeland, C. and Chia, F.S. 1971. Recruitment risk, growth, age and predation in two populations of sand dollars, Dendraster excentricus (Eschscholtz). J. Exp. Mar. Biol. Ecol. 6 : 265-278.

Chia, F.S. 1969. Some observations on the locomotion and feeding of the sand dollar, Dendraster excentricus (Eschscholtz). J. Exp.Mar.Biol.Ecol. 3 : 162 - 170.

Chia, F.S. 1973. Sand dollar : a weight belt for the juvenile. Science 181 : 73 - 74.

Drury, R.A. and Wallington, E.A. 1967. Carleton's Histological Technique. Toronto: Oxford University Press. Fourth edition.

Endean, R. 1966. The coelomocyte and coelomic fluids in Boolootian (Ed.) physiology of echinodermata. John Wiley and Sons, New York. pp 822.

Hickman, M. 1969. Methods for determining the primary productivity of epipelic and epipsammic algal associations. Limnology and Oceanography. 14 : 936 - 941.

Howie, F.M.P. 1974. Introduction of Thioglycollic acid in preparation of vertebrate fossiles. Curator, 17 : 159 - 166.

Jobson, A.M. 1971. Bacterial metabolism of crude petroleum. M.Sc. thesis, Dept. of Microbiology. University of Alberta. pp. 132.

Mitchell, B.P. 1972. Rediscovery of Gregory's diverticulum in the scutellia sand dollar. M.Ed. thesis. Western Washington State College. pp.32.

Proceedings of the Fifth International Echinoderm Conference / Galway / 24-29 September 1984

Feeding biology of the short-spined sea urchin *Anthocidaris crassispina* (A.Agassiz) in Hong Kong

S.T.CHIU
University of Hong Kong

ABSTRACT: In Hong Kong, Anthocidaris crassispina commonly feeds on 9 algal species. In winter (December to April), urchins at all localities heavily graze upon Colpomenia sinuosa, Sargassum hemiphyllum and Ulva conglobata. During the warmer months, the urchins graze upon Sargassum spp. and Corallina spp. in localities where macrophytes are perennial. In areas experiencing a summer macrophytic decline, the urchins depend upon coralline (Corallina spp.), filamentous (Cladophora delicatula and Gelidium divaricatum) and encrusting (Lithophyllum) algae. Laboratory food preference experiments performed on 6 common algal species have shown that in terms of percentage total intake, Colpomenia sinuosa was most preferred, followed in order by Petalonia fascia, Sargassum hemiphyllum, Padina australis, Corallina pilulifera and Ulva conglobata. Consumption experiments have shown that urchins of 40-50 mm test diameter feed upon algae (daily percentage biomass) in the following order: Colpomenia sinuosa (5.7%), Ulva fasciata (3.5%), Padina australis (3.0%), Sargassum hemiphyllum (2.6%), Ulva conglobata (2.5%) and Corallina pilulifera (1.4%).

1 INTRODUCTION

The short-spined sea urchin Anthocidaris crassispina (A. Agassiz) is economically important regionally and is the only species that supports a local fishery. It is a benthic grazer of the lower intertidal and shallow sublittoral zone of rocky shores in Hong Kong. Previous studies have reported that regular echinoids preferentially feed on macrophytic algae (Fuji 1962, Himmelman and Steele 1971, Kawamura 1974, Vadas 1977, Mattison et al 1977, Larson et al 1980). The macroalgae of the Hong Kong seashore are characterised by seasonal fluctuations in species composition and abundance depending upon locality (Hodgkiss 1984).

A study was undertaken to determine the relationship between food availability, and the feeding biology of A. crassispina in Hong Kong. Laboratory experiments were also undertaken to determine food preferences and feeding rates upon common macroalgae. This programme is part of a general study of the ecology of A. crassispina in Hong Kong.

2 MATERIALS AND METHODS

2.1 Field studies

The general study area consists of 45 coastal sites spread throughout Hong Kong, and extending from areas in the east experiencing oceanic conditions, to sites in the west where estuarine conditions prevail (Watts 1973) (Fig. 1). Three stations (Lung Ha Wan in the east, Cape D'Aguilar to the south and Gemini Beach in the west) were chosen for a comparative study of the feeding biology of Anthocidaris crassispina.

The study was initiated in December 1982 and the results presented in this paper cover the period from July 1983 to July 1984. Sampling was carried out at monthly intervals. A 50 metre transect line was laid between permanent marker points, and the occurrence of algal species and their relative abundance determined for each metre interval. Distribution of urchins along the transect line was also recorded. At each station, 50 animals were carefully picked up and the presence of food items adhering to the oral

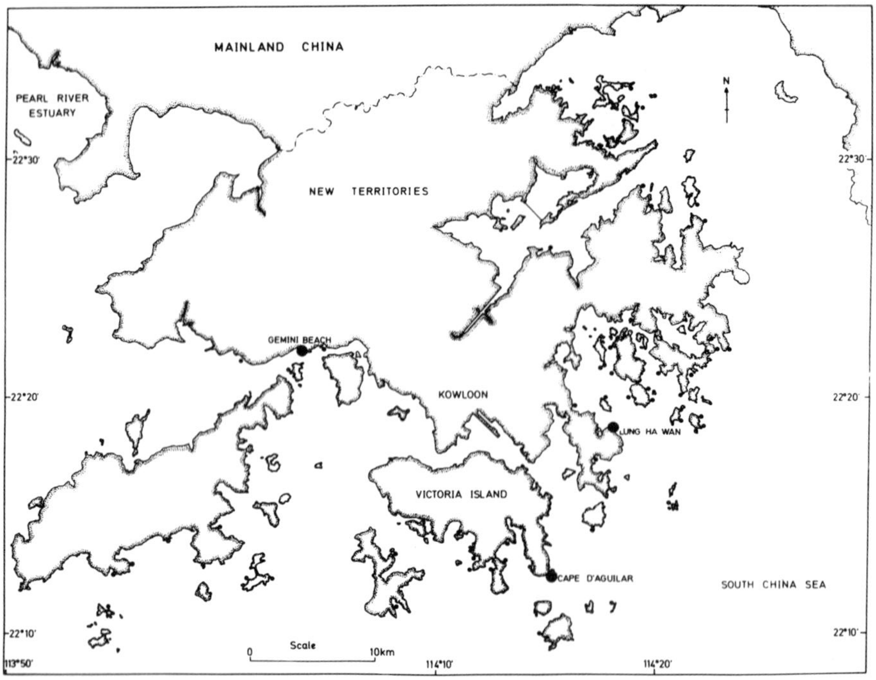

Fig. 1: Map of Hong Kong showing the study areas. Circles mark localities where Anthocidaris crassispina has been recorded

surface was noted. The process of ingestion of such items could normally be observed. The test diameter and height , and wet weight of these animals were recorded. Subsamples of these urchins were collected for gut contents analysis. Sample sizes varied from 9 to 21, taking into consideration the overall abundance of urchins at the particular site and the possible disturbance of the population by over-sampling. These animals were immediately injected with commercial grade 40% formaldehyde (1-2 ml).

Animals sampled were dissected, and the gut contents of each urchin assessed under a dissecting microscope. Food items were alloted points (1,2,4,8 and 16) according to volume. Hynes (1950) stated that plant material is not ingested in discrete units but is normally broken into pieces by the biting action of teeth. Accordingly, visual estimates of the volume of each food item is the most suitable scoring method.

2.2 Laboratory feeding experiments

Animals used in the laboratory experiments were collected from Lung Ha Wan where macrophytic algae are best represented (Table 1). The algae utilised in the experiments were those that were most frequently recorded in the gut and normally constituted at least 10 % of the diet. These include Colpomenia sinuosa(Roth) Derbes & Solier, Sargassum hemiphyllum(Turn) C.Agardh, Petalonia fascia (Muller)Kuntze, Padina australis Hauck, Sargassum augustifolium(Turn) C.Agardh, Ulva conglobata Kjellman, Ulva fasciata and Corallina pilulifera Postels & Ruprecht. Experiments were performed in an aquarium operating on a semi-open recirculating water system.

Three replicates of five urchins of similar weight (46.4± 3.1 g) and test diameter (46.5 ±0.8 mm) were each placed in one unit of the aquarium. The animals had previously been starved for 14 days. Of the 8 species of common algae, only 6 were used in this experiment as there were insufficient quantities of Ulva fasciata and Sargassum augustifolium for both sets of experiments. Test algae were blotted dry for 5 minutes before weighing and ≃ 50 g of each species was then placed in the test tanks. A control tank with algae but no urchins was established. In the following days, more algae were weighed and offered as required depending upon feeding rates so that the ur-

Table 1. The seasonal occurrence and relative abundance of algae at the three study sites. L.H.W. = Lung Ha Wan, C.D.= Cape D'Aguilar, G.B.= Gemini Beach
•••• Very abundant > 70 % cover per m^2
••• Common 30 - 70 % cover per m^2
•• Scarce < 30 % cover per m^2

	Winter (Dec. to April)			Summer (June to Oct.)		
	L.H.W.	C.D.	G.B.	L.H.W.	C.D.	G.B.
Phaeophyta						
Sargassum hemiphyllum	••••	••••		••••		
Sargassum augustifolium	•••					
Colpomenia sinuosa	••••	••••				
Pocockiella variegata	••••			••••		
Petalonia fascia	•••	••				
Scytosiphon lomentaria	••	••	•••			
Gymnogongrus flabelliformis	••	•••				
Padina australis	•••	••		••		
Chlorophyta						
Ulva fasciata	••	••	••			
Ulva conglobata	••	••				
Cladophora delicatula	•••	••••		•••	••••	
Rhodophyta						
Porphyra suborbiculata	••	•••	•••			
Gelidium divaricatum	••••	••••	•••	•••		
Corallina pilulifera	••••	••	•••	••••	•••	•••
Corallina sessilis	••••	••	•••	••••	•••	•••
Cyrtymenia sparsa		•••				
Lithophyllum	••••	••••	••	••••	••••	

chins could feed ad libitum.

After 14 days, the weight of remaining algae in all three replicates and the control was recorded. The difference in the final weight and the total weight of algae given was taken as the weight of algae ingested.

A second set of urchins were divided into 3 size groups based upon test diameter (i.e. 20-30 mm, 40-50 mm, 50-60 mm). Seven sets of experiments were established with seven species of algae. In each set, 6 animals of each size group were used. The urchins were starved for 14 days, and then offered 60 g blotted wet weight of one of the algal species. The urchins were allowed to feed on the alga for 14 days. Additional algae was offered depending upon the feeding rate. Actual intake (biomass) was then determined at the end of the experiment. No control was set up since weight changes of algae had previously been found to be negligible (<5%) under similar aquarium conditions. In addition, calorific values, organic content and the ash fraction of each algal species was determined.

3 RESULTS

3.1 Relative abundance and composition of algae

Table 1 shows the occurrence and relative abundance of macrophytic algae at the sampling stations.

A strong seasonal variation was observed at both Lung Ha Wan and Cape D'Aguilar with luxuriant growth of the phaeophytes Sargassum hemiphyllum, Colpomenia sinuosa, Petalonia fascia and Pocockiella variegata (Lamouroux) Papenfuss in the winter months (December to April). In the hotter summer months (June to October), only 3 species of phaeophytes persist at Lung Ha Wan, i.e. Sargassum hemiphyllum, Pocockiella variegata and Padina australis. No phaeophytes were recorded at Cape D'Aguilar during this time. The littoral zone of both areas is dominated by a turf of the calcareous alga, Corallina pilulifera and a complex of crustose red algae, "Lithophyllum", interspersed with the filamentous Gelidium divaricatum

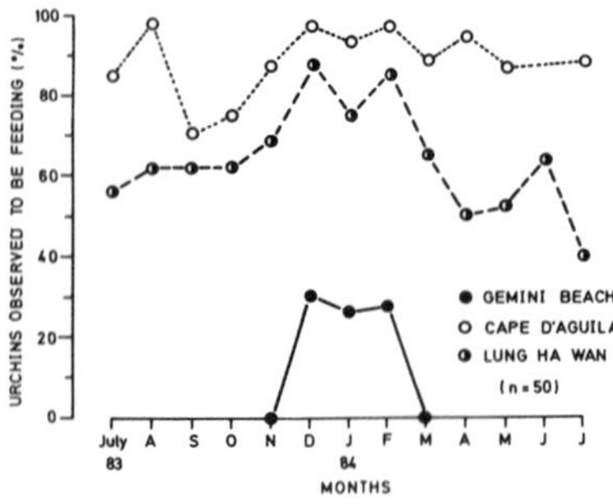

Fig. 2: Feeding activity of Anthocidaris crassispina in the study areas

(Martens) and Cladophora delicatula Mont..
Macrophytic algae are, however, poorly represented at Gemini Beach all year round. Sparse growths of Scytosiphon lomentaria (Lyngbye) C.Agardh, Porphyra suborbiculata Kjellman, Ulva fasciata and Gelidium divaricatum are recorded in winter only. Heavy sedimentation on the rock surfaces and poor visibility obscured accurate determination of the percentage coverage of crustose and calcareous red algae which were observed to be present in the winter months.

3.2 Feeding activity of Anthocidaris crassispina in the field

The population densities of A. crassispina were different at the three sites. The mean density of urchins at Cape D'Aguilar (6.2 per m^2) was higher than that of Lung Ha Wan (3.2 per m^2). Urchin density at Gemini Beach did not exceed 1.9 per m^2.

Figure 2 shows the actual observed incidences of food items being consumed by A. cra-ssispina in the field. At Lung Ha Wan, the frequency of animals observed to be ingesting food was very high in winter (65-95%) but decreased in the summer months (45-60%). This seasonal difference was even more marked at Gemini Beach where no animals were observed to be feeding except during the winter months of December to February. Even then, the actual numbers feeding was still comparatively low (30%). At, Cape D'Aguilar, the majority of urchins were observed to be feeding on all sampling dates throughout the year (75-100%).

3.3 Gut content analyses

Figures 3-5 show the monthly change in the composition of the gut contents in each study area. Distinct differences exist between the summer and winter diets.

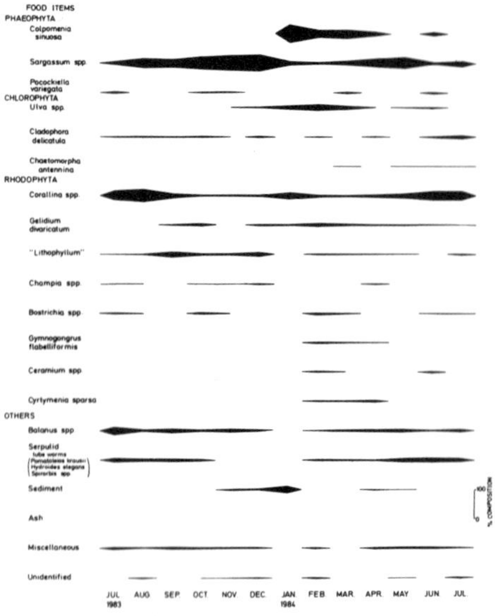

Fig. 3: Diagrammatic representation of the percentage composition of the gut contents of A. crassispina at Lung Ha Wan

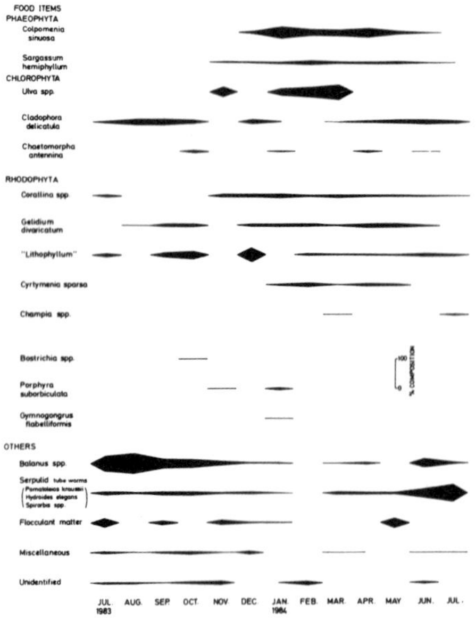

Fig. 4: Diagrammatic representation of the percentage composition of the gut contents of A. crassispina at Cape D'Aguilar

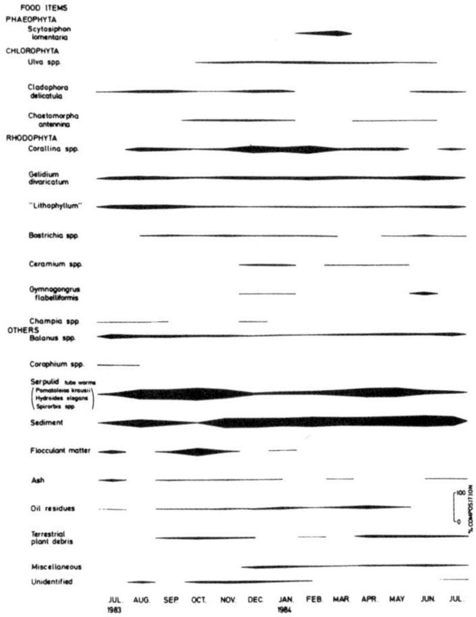

Fig. 5: Diagrammatic representation of the percentage composition of the gut contents of A. crassispina at Gemini Beach

At Lung Ha Wan, Corallina spp., barnacles and serpulid tube worms (Pomatoleios kraussii(Baird), Hydroides elegans(Haswell) and Spirobos spp.) comprise the main fraction of the diet (>60%) during the summer. Sargassum hemiphyllum is also recorded at this time, the percentage frequency ranging from 6 to 40%. In autumn, the percentage of Lithophyllum increased to between 9-18%. Miscellaneous food items, including the urchin's commensal pontoniine shrimp Athanas dorsalis Stimpson, the gastropods, Bittium sp., Diala sp. and Linopyrga sp., and charcoal ash (from beach camp fires) comprise less than 5% of the diet.

Progressing into the winter months, macrophytic algae become more important in the diet. The most abundant species recorded were Colpomenia sinuosa(20-60%), Sargassum spp.(8-50%), Pocockiella variegata(5%), Ulva spp.(1-20%), Corallina spp.(6-23%) and Gelidium divaricatum(4-10%). Of these, three are phaeophytes and Gelidium divaricatum is the only filamentous species.

At Cape D'Aguilar, in summer, barnacles (Balanus spp.), serpulid tube worms, Lithophyllum and Cladophora delicatula comprise the main fraction of the diet (65-90%). The encrusting Lithophyllum was again noted to be more frequently ingested from September to December (12-42%). With the exception of charcoal ash, miscellaneous food items at Cape D'Aguilar were identical to items recorded at Lung Ha Wan. During winter, a total of seven macroalgae were recorded in the diet, these include Colpomenia sinuosa(7-40%), Sargassum hemiphyllum(2-9%) and Ulva spp.(20-50%). These algae comprise 60-70% of the winter diet.

At Gemini Beach, serpulid tube worms and sediment constitute a steady 40-50% of the diet throughout the year. In winter, macrophytic algae were still poorly represented. The red calcareous algae Corallina spp.(8-20%) and some filamentous species such as Gelidium divaricatum(7-12%) and Cladophora delicatula(6%) were recorded in the diet. The only phaeophyte occurring in the winter diet was Scytosiphon lomentaria which is known to be associated with low salinity and is also regarded as a pollution marker (Morton and Morton 1983). Miscellaneous items in the diet were distinctly different from the other two sites and included such debris as poultry feathers, polyethylene twine, tissue paper and charcoal ash.

3.4 Food preference

Observations made during the food preference tests showed that when algae and urchins were placed in the same tank, feeding does not immediately commence. The urchins move around and frequently congregate at the air-water interface. After a few hours, however, the majority begin feeding. The most common first choice was Colpomenia sinuosa, although individual variation was apparent. Moreover, urchins were often observed to move away and feed on other algae without exhausting the algae that had previously been fed upon.

Table 2 summarises the results of the food preference tests. Of the six algae species offered, A. crassispina fed upon Colpomenia sinuosa and Petalonia fascia most readily. In order of preference, the other species are Sargassum hemiphyllum, Corallina pilulifera and Padina australis. Ulva conglobata was the least preferred.

Table 3 presents the consumption rates of different size groups of urchins. Small animals(20-30 mm) consumed a daily average of 4.84% of their biomass whereas larger specimens(50-60 mm) eat much less (1.40%).

The highest consumption rate, expressed as percentage biomass, was recorded for Colpomenia sinuosa for all three size groups (2.34-7.34%), and the lowest was for Corallina pilulifera (1.24-3.22%).

Urchins of the 20-30 mm size class also recorded high consumption values for Sargassum hemiphyllum(6.03%) and Ulva conglobata(5.54%). Urchins of the 40-50 mm size

Table 2. Results of the food preference tests on Anthocidaris crassispina as represented by average consumption rates per animal per day. Three replicates of 5 animals each. Duration of experiment = 14 days. Water temperature = 22 degrees Celsius. Salinity =33‰

Group No.	I		II		III	
	Wet weight(g) $urchin^{-1}$ day^{-1}	Total intake (%)	Wet weight(g) $urchin^{-1}$ day^{-1}	Total intake (%)	Wet weight(g) $urchin^{-1}$ day^{-1}	Total intake (%)
Colpomenia sinuosa	0.91	31.6	0.65	22.3	0.77	21.4
Sargassum hemiphyllum	0.95	19.5	0.61	21.1	0.52	14.4
Petalonia fascia	0.56	22.1	0.63	21.8	1.01	28.1
Corallina pilulifera	0.30	10.3	0.52	17.9	0.48	13.2
Padina australis	0.41	14.3	0.39	13.5	0.45	12.5
Ulva conglobata	0.06	2.2	0.10	3.4	0.37	10.3
Total	2.89	100	2.91	100	3.60	100

Table 3. Average daily consumption rates on individual algal species offered to different size groups of Anthocidaris crassispina.
No. of experimental animals in each size group =6. Duration of experiment =14 days.
Water temperature = 20-22 degrees Celsius. Salinity = 33‰

Urchin size group (Test diameter in mm)	Gp.1 (20 - 30)		Gp.2 (40 - 50)		Gp.3 (50 - 60)	
	Wet wt(g)	% biomass	Wet wt(g)	% biomass	Wet wt(g)	% biomass
Phaeophyta						
Colpomenia sinuosa	1.38	7.34±2.18	2.78	5.68±0.46	3.14	2.24±0.62
Sargassum hemiphyllum	0.88	6.03±2.60	1.04	2.60±1.11	1.31	1.41±0.10
Padina australis	0.80	4.85±1.69	1.37	2.99±0.25	1.15	1.31±0.15
Sargassum augustifolium	0.27	3.26±1.44	0.65	2.20±1.21	0.92	1.20±0.71
Chlorophyta						
Ulva fasciata	0.70	3.66±0.25	1.70	3.45±0.16	1.64	1.50±0.47
Ulva conglobata	0.59	5.54±1.86	0.87	2.50±0.28	0.69	0.89±0.01
Rhodophyta						
Corallina pilulifera	0.79	3.22±0.47	0.63	1.37±0.20	0.69	1.24±0.18

class, however, recorded higher intakes of Ulva fasciata(3.45%), followed by Padina australis(2.99%) and Sargassum hemiphyllum (2.60%).

The second highest intake of animals in the 50-60 mm size class was for Ulva fasciata(1.50%) followed by Sargassum hemiphyllum(1.41%) and Padina australis(1.31%).

Table 4 summarises the analysis of the nutrient composition of the algal species utilised in the food preference experiment. The phaeophytes tended to have higher calorific values (3.29-5.54 Kcal g^{-1}dry wt.) compared with the chlorophytes (1.38-1.43 Kcal g^{-1}dry wt.). Corallina pilulifera has a very low calorific value (0.05 Kcal g^{-1} dry wt.). Similarly, phaeophytes have a higher organic content (24.6-88.8% dry wt.) when compared with the chlorophytes (35.4-38.4% dry wt.). Corallina pilulifera has an organic content value of only 16.9% dry wt.. With the exception of Colpomenia sinuosa (69.4% dry wt.), the phaeophytes have a lower ash content (11.2-43.5% dry wt.) compared with the chlorophytes(59.3-61.7% dry wt.) and Corallina pilulifera (83.1% dry wt.).

Table 4. Results of the analysis of the nutritional composition of various algal species

	Calorific value (Kcal g^{-1} dry wt)	Organic content (% dry weight)	Ash content (% dry weight)
Phaeophyta			
Colpomenia sinuosa	5.54	24.6	69.4
Sargassum hemiphyllum	4.21	56.5	43.5
Padina australis	3.29	57.3	42.8
Sargassum augustifolium	4.05	88.8	11.2
Chlorophyta			
Ulva fasciata	1.43	35.4	59.3
Ulva conglobata	1.38	38.4	61.7
Rhodophyta			
Corallina pilulifera	0.05	16.9	83.1

4. DISCUSSION

Sea urchins are the dominant benthic herbivore of rocky shores in temperate zones. Urchin grazing is reported to be a major factor affecting the occurrence and community structure of macroalgae (Paine and Vadas 1969, Carpenter 1981, Cowen et al 1982, Hagen 1983). From the present study, the interaction between urchin grazing and algal abundance appears to present a different situation. The composition of the gut contents, being a reliable indicator of the natural food ingested by urchins, shows variation in the diet highly dependent on season, locality and nature of the substratum.

Anthocidaris crassispina at Lung Ha Wan, a typically food unlimited site with macrophytic algae all year round, exhibits selective feeding on Colpomenia sinuosa, Sargassum hemiphyllum and Corallina spp. in winter. In summer, with the absence of most macrophytic algae (Hodgkiss 1984), the urchin diet changes and comprises Corallina spp., Sargassum hemiphyllum and serpulid tube worms. The feeding upon serpulid tube worms, the main components of an 'under boulder community' (Morton and Morton 1983) reflects the 'summer habitat' of urchins where they hide under boulders to avoid exposure to strong illumination and very high surface temperatures especially during extreme low water spring tides.

Urchins at Cape D'Aguilar in the winter also selectively feed upon algae, and the presence of individual algal species in the gut is related to the duration of their occurrence on the shore. The record of Ulva fasciata and Ulva conglobata in the gut corresponds to the period that luxuriant growth of these species are recorded in situ from November to March. Barnacles, serpulid tube worms and the filamentous algae Cladophora delicatula and Gelidium divaricatum constitute the main items of the summer diet. Cladophora delicatula and Gelidium divaricatum are commonly found on the shells of barnacles and tube worms at Cape D' Aguilar.

Perhaps the most distinctly different site is Gemini Beach, situated close to the estuary of the Pearl River flowing down from Canton province, China. This station experiences large fluctuations in salinity (Fig. 6), and heavy sedimentation of the rock surfaces (5-10 mm). This results in sparse algal growth and non-occurrence of most algal species. The urchins in this area adopt a generalist mode of grazing (similarly expressed to a lesser degree at Cape D' Aguilar in the summer) and the main food items comprise sediment and serpulid tube worms. Polyethylene twine, oil residues, tissue paper and terrestrial plant debris are common constituents of the diet and originates from the nearby shipbuilding industries and densely populated residential estates. Nevertheless, in winter, when there is a sparse growth of Porphyra suborbiculata and Scytosiphon lomentaria, the urchins readily consume these algae.

Although a generalist mode of feeding in food limited areas has been reported for other urchin species (Don cited in Walker 1981, Carpenter 1981), the availability of suitable or preferred food has been shown to affect growth, reproduction and recruitment of urchin populations (Ebert 1968, Vadas 1977, Larson et al 1980). In Hong Kong, the spawning period of Anthocidaris crassispina (March to May) coincides with

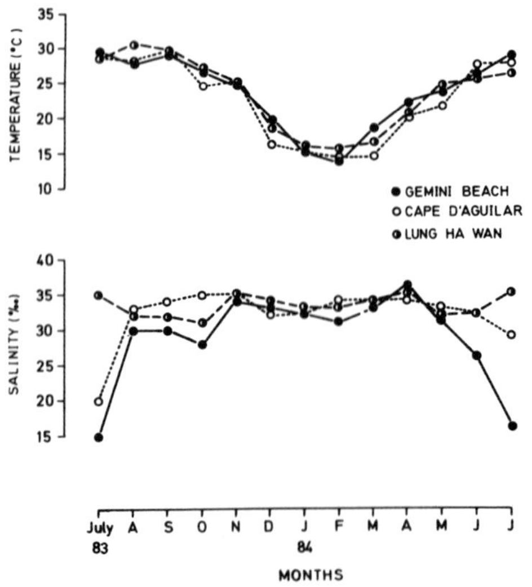

Fig. 6: Monthly variations in temperature and salinity recorded at the three study areas

the latter period of peak algal abundance, and juvenile urchin settlement is observed from April to June. The urchin population structures at the study areas (Fig. 7) show that the quality of a suitable food supply can be critical to successful recruitment. At Gemini Beach, urchin density is low (1.9 urchins per m^2) and no recruitment has been recorded during the two years study period. There is no previous data on the A. crassispina population around this area, and occu -rrences of juvenile recruitment prior to this study is unknown. The underlying causes for unsuccessful recruitment are not yet fully understood, but the limited food supply could be an important contributing factor considering the high juvenile feeding rates. The long term consequence of non-recruitment could also result in disappearance of A. crassispina from this area.

In terms of growth, the maximum sizes attained by a species might be expected to be greater in a food-unlimited environment (Ebert 1968). In Hong Kong, urchins in food abundant areas such as Lung Ha Wan, however, are seldom recorded to achieve a maximum potential size (Fig. 7) due to heavy fishing pressure. Due to poor underwater visibility and sparse population density, no fishing takes place at Gemini Beach. Here, therefore the urchins appear to attain a larger size.

Despite an ability to subsist on other food items, to achieve optimal growth and reproduction, A. crassispina demonstrates selectivity for certain food items. Although fifteen species of algae have been recorded in the gut contents, only five consistently comprise more than 40% of the diet. This high degree of selectivity for certain algae has also been demonstrated for other urchins (Fuji 1962, Kawamura 1974, Vadas 1977, Larson et al 1980).

Comparison of the food preference tests and the composition of algae in the natural diet provides interesting data. Colpomenia sinuosa is both the major component of the winter diet and the most preferred species in laboratory tests. It also has the highest consumption rate recorded for urchins of size ranging from 20 to 60 mm test diameter. Nevertheless, although Petalonia fascia is ranked as the second choice in experiments, it has never been recorded in the natural diet. Similarly, Padina australis was also never recorded from the gut contents. Their low relative abundance on rocky shores may account for the poor representation in the diet. On the contrary, Corallina pilulifera, ranked fourth in terms of preference, often accounts for up to 40% of an urchin's natural diet. It grows profusely as a coralline turf in the subtidal zone and its importance in the natural diet reflects the opportunistic mode of grazing adopted by A. crassispina when preferred food is unavailable.

In general, A. crassispina appears to prefer phaeophytes and rhodophytes to chloro -phytes. This may be an artefact of algal zonation. Phaeophytes and rhodophytes are the two most important groups of algae represented in the sub-littoral zone of exposed rocky shores which is the principal habitat of A. crassispina. On such shores, chlorophytes are mainly confined to the higher reaches of the eulittoral zone.

From the results obtained, gut contents analyses do accurately act as an indicator of food preference but also reflect the seasonal availability of food at each parti -cular locality. In fact, the actual distri -bution of A. crassispina in Hong Kong appears to be related to a food supply which is in turn, highly dependant upon season, locality and substratum.

5. ACKNOWLEDGMENTS

This study was carried out as part of the thesis requirement for the degree of Ph.D., at the University of Hong Kong. I express my sincere gratitude to Prof. B.S. Morton for his guidance and encouragement. I am also grateful to Prof. B. Lofts, Head of the Department of Zoology, the University of Hong Kong, for his support and permission to attend this conference. Participation

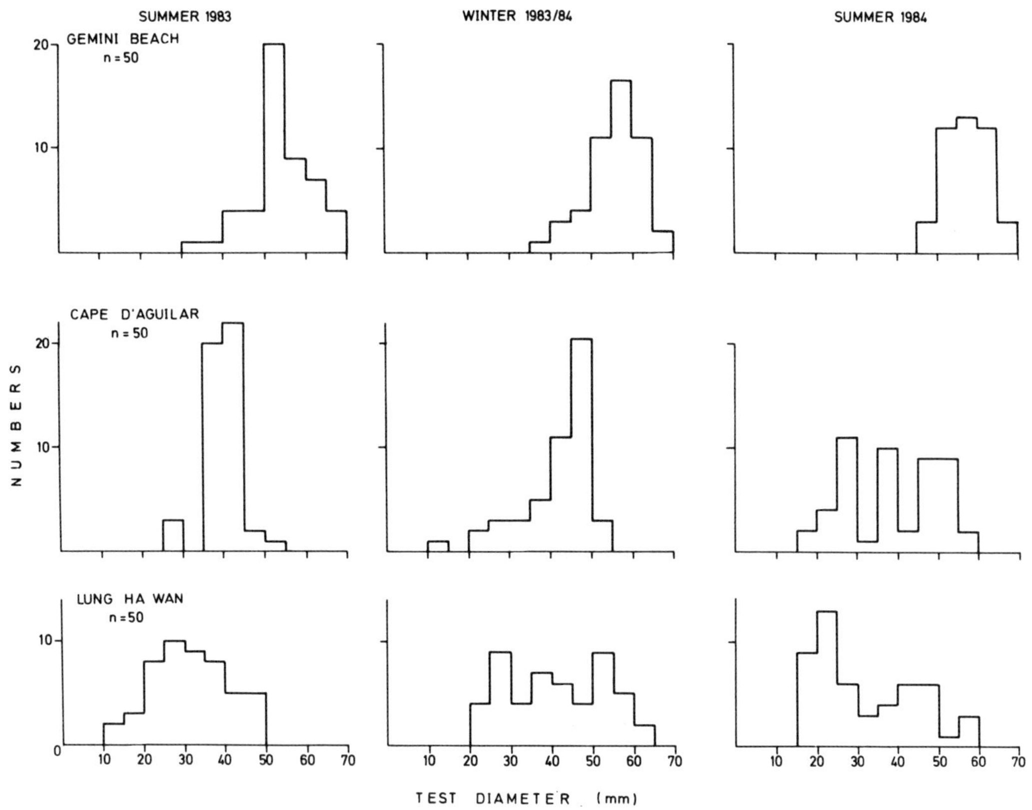

Fig. 7: Distribution by size class of Anthocidaris crassispina at the three study areas

at the conference was facilitated by a grant from Unesco, and I am indebted to Dr. J.R. E. Harger and Dr. D. Troost for their kind assistance.

6. REFERENCES

Carpenter, R.C. 1981. Grazing by Diadema antillarum (Phillipi) and its effects on benthic algal community. Journal of Marine Research, 39:749-765.

Cowen, R.K., Agegian, C.R. and Foster, M.S. 1982. The maintenance of community structure in a central California giant kelp forest. Journal of Experimental Marine Biology and Ecology, 64:189-201.

Ebert, T.A. 1968. Growth rates of the sea urchin Strongylocentrotus purpuratus related to food availability and spine abrasions. Ecology, 9:1075-1090.

Fuji, A. 1962. Ecological studies on the biology of the sea urchin. V. Food consumption of Strongylocentrotus intermedius. Japanese Journal of Ecology, 12:181-186.

Hagen, N.T. 1983. Destructive grazing of kelp beds by sea urchins in Vestfjorden, Northern Norway. Sarsia, 58:177-190.

Himmelman, J.H. and Steele, D.H. 1971. Foods and predators of the green sea urchin Strongylocentrotus droebachiensis in Newfoundland waters. Marine Biology, 9:315-322.

Hodgkiss, I.J. 1984. Seasonal patterns of intertidal algal distribution in Hong Kong. Asian Marine Biology, 1:78-84.

Hynes, H.B.N. 1950. The food of freshwater sticklebacks (Gasteroteus aculeatus and Pygosteus pungitius), with a review of methods used in studies of the food of fishes. Journal of Animal Ecology, 19: 36-58.

Kawamura, K. 1974. Fishery biological studies on a sea-urchin, Strongylocentrotus intermedius (A. Agassiz). Scientific Reports of Hokkaido Fisheries Experimental Station, 16:1-54.

Larson, B.R., Vadas, R.L. and Keser, M.

1980. Feeding and nutritional ecology of the sea urchin Strongylocentrotus droebachiensis in Maine, U.S.A.. Marine Biology, 59:49-62.
Mattison, J.E., Trent, J.D., Shanks, A.L., Akin, T.B. and Pearse, J.S. 1977. Movement and feeding activity of red sea urchins (Strongylocentrotus franciscanus) adjacent to a kelp forest. Marine Biology, 39:25-30.
Morton, B. and Morton, J. 1983. The sea shore ecology of Hong Kong. Hong Kong : Hong Kong University Press.
Paine, R.T. and Vadas, R.L. 1969. The effects of grazing by sea urchins, Strongylocentrotus spp., on benthic algal populations. Limnology and Oceanography, 14:710-719.
Vadas, R.L. 1977. Preferential feeding : an optimisation strategy in sea urchins. Ecological Monographs, 47:337-371.
Walker, M.M. 1981. Influence of season on growth of the sea urchin Evechinus chloroticus. New Zealand Journal of Marine and Freshwater Research, 15:201-205.
Watts, J.C.D. 1973. Further observations on the hydrology of the Hong Kong territorial waters. Hong Kong Fisheries Bulletin, Agriculture and Fisheries Department, Hong Kong. 3:9-35.

Proceedings of the Fifth International Echinoderm Conference / Galway / 24-29 September 1984

Effect of mechanical stress on the calcification pattern in regular echinoid skeletal plates

J.DAFNI
Hebrew University of Jerusalem, H.Steinitz Marine Biology Laboratory, Eilat, Israel

Abstract: The effect of tensile mechanical stress on the calcification pattern of isolated test plates of the echinoid Tripneustes gratilla elatensis was examined, using a radioisotope. Interambulacral plates, stretched in the horizontal direction (adradial-interradial) calcified at higher rates than the vertically stretched, or non-stretched plates. The ratio between the calcification rates at the upper and lower sutural margins and those of the horizontally directed (v/h ratio) was also affected in the treated plates: Plates stretched at the horizontal direction showed lower v/h ratios than the vertically stretched or the non-stretched. These results suggest that in situ the test plates are subject to mechanical stresses - tension along the longitudinal sutures and compression at the horizontal sutures.

1 INTRODUCTION

The hypothesis that peripheral growth and calcification of regular echinoid plates are controlled by a biomechanical effect - enhanced by sutural tensile stress and inhibited by compressional stress (Moss & Meehan 1968, Raup 1968, Dafni 1980,1983a), has never been tested. Heatfield (1970) and Meikle et al.(1979) proposed methods by which isolated skeletal parts (echinoid spines, mammalian cranial bones) could be used in radioactive in vitro experiments, to test the effects of chemical agents and mechanical forces. These methods were applied to test the above hypothesis.

2 MATERIAL AND METHODS

Five homologuous square pieces, each ca 2.5cm^2, were cut with a rotating disc, from the different superambital interambulacra of the coronal test of a living T. gratilla elatensis Dafni(1983b)(Fig. 1A), and cultured in 0.5 liter well-aerated seawater, at room temperature (25±1^{0}C) and continuous light. The explants were attached to a stretching device (Meikle et al. 1979), comprising of a Teflon cylinder (ϕ 20mm), upon which two movable pin-carrying half-cylinders moved, pulled apart by a 50-100g stain-

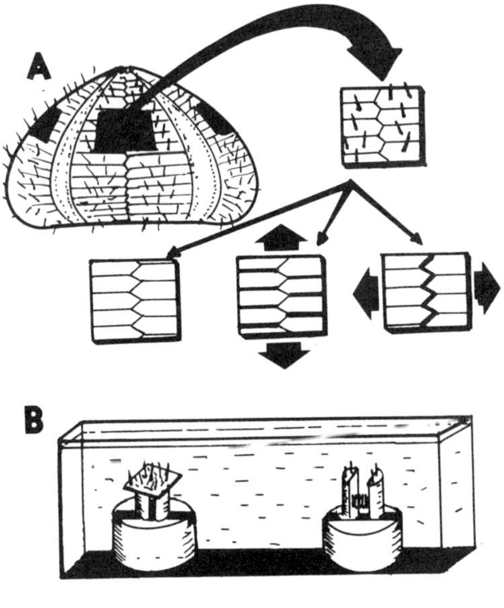

Fig. 1: Scheme of experimental procedure. A, skeletal squares removed from the superambital interambulacra and stretched in the horizontal direction (right), vertically (middle) or non-stretched. B, the stretching device, submerged in seawater; left - an explant is shown in stretched position.

Table 1. The effect of mechanical stretching of isolated T.g.elatensis test plates on (A) their calcification rates, and (B) on the ratio between vertical and horizontal sutural calcification rates (v/h), under different stretching conditions (vertical, horizontal stretching and non-stretched control). In parentheses - number of plates analysed for each treatment. The results were tested in a two-tailed Student's T-test.

A. Calcification rates (%.day^{-1}±SD×10^{3})

Individual no.	Parts analysed	Direction of stretching: Horizontal	Vertical	Control	Comparison between treatments†: Hor-vert	Vert-cont	Hor-cont
y-23A	Whole plates	---	30±10 (17)	22±5 (18)	---	1.39*	---
y-9	Sutural edges	104±55 (10)	---	64±24 (13)	---	---	1.63*
y-12	" "	108±26 (20)	82±21 (22)	74±11 (9)	1.32***	1.1(-)	1.46***
y-17	" "	64±14 (18)	68±13 (19)	---	0.94(-)	---	---
y-20	" "	87±47 (15)	49±26 (20)	42±20 (9)	1.77***	1.17(-)	2.07*

B. Effect on v/h ratio

Individual no.	Parts analysed	Horizontal	Vertical	Control	Hor-vert	Vert-cont	Hor-cont
y-9	Sutural edges	0.47±.19 (10)	---	1.11±.64 (13)	---	---	0.42***
y-12	" "	0.71±.22 (20)	1.09±.26 (21)	1.20±.34 (9)	0.65***	0.91(-)	0.59***
y-17	" "	0.73±.25 (18)	1.17±.48 (19)	---	0.63***	---	---
y-20	" "	0.12±.11 (15)	0.26±.28 (20)	---	0.46(-)	---	---
y-23	" "	0.47±.16 (19)	0.73±.42 (30)	---	0.65*	---	---

† Level of significance: (-), N.S.; *,$p<0.05$;**,$p<0.01$; ***,$p<0.005$

less steel spring (Fig. 1B). The explants were stretched either in the vertical (aboral-adoral direction, or horizontally. An impaled specimen without stretching served as a control. After incubation in $^{45}CaCl_2$ seawater for 12h the explants were transferred to non-radioactive seawater for 12h and then cleaned in NaOCl (5%). Determination of ^{45}Ca activity in the entire plates or at their periphery, along the sutural margins, followed the procedure described by Dafni & Erez (1982), and divided by the specific radioactivity of the incubation medium to obtain the absolute calcification rates (in % accretion per day).

3 RESULTS AND DISCUSSION

The skeletal explants maintained their vitality for at least two days, during which spines and pedicellaria were active and continued to move slowly, and the latters' jaws opened and closed frequently. Later the tissue vitality usually deteriorated. The low calcification rates, obtained during the first 24h (5-10% of the in vivo rates), are the unavoidable results of the physiological stress caused by both, the physical injury - which is known to reduce temporarily the calcification rates (Heatfield 1970) - and the use of seawater as culture medium. Nevertheless, a consistent pattern was shown, different than the patterns that characterize intact urchins (Dafni & Erez 1982)(Table 1): Explants of the urchin y-23A, tested to find whether vertical stretching enhances overall calcification, when compared to the non-stretched, showed that this was indeed the case. The stretched plates calcification rates were 1.39 higher than the control (Table 1A).

Analysis of the sutural edges provided additional information, mainly because both, absolute sutural calcification and the ratio between vertical (at the upper and lower margins) and horizontal calcification (vertical margins), the v/h ratio, were compared (Table 1A & B).

The plates of individual no. y-9, stretched only in the horizontal direction, showed increased sutural calcification, by a factor of 1.6. In other explants, horizontal and vertical stretching were compared with the former yielding calcification rates 1.3-1.7 greater than the latter (Table 1A).

As to v/h ratio, there was a nonsignificant difference between the vertically stretched plates and the controls (Table 1B), whereas horizontal stretching reduced this ratio significantly. This was due to the increase of the interradial (zigzag) suture calcification rates. It is noteworthy, however, that the v/h ratio of non-stretched sutures in this experiment were appreciably higher than the in vivo ratios (ca 0.3-0.4) (Dafni & Erez, in prep). This may suggest that the mere extraction of plates from the tight-fitting in situ mosaic affected both calcification rates and differential patterns.

These results support the hypothesis that in vivo the plates are under mechanical confinement (Raup 1968), which is more pronounced in the vertically directed sutural edges, due to the mechanical activity of the ambulacral tubefeet, that pull the entire test towards the substratum, pressing the vertical plate columns in this direction (Dafni 1980). Extracting the plates probably relieves this compression, causing increased v/h ratios even without the application of vertical stretching. Horizontal stretching possibly simulates the in situ tensile stresses in this direction, due to the expansion of the entire test (Moss & Meehan 1968) and to the spatial arrangement of the intestinal mesenterial threads (Dafni & Erez 1982), increasing sutural calcification and lowering v/h ratio.

It has been pointed out (Dafni, MS) that sutural gaps show often in rapidly growing sutures, and that skeletal trabeculae, that grow in these gaps, are abnormally long and slender. It seems logical that tensile stress, which pulls the plates apart, cause these gaps' formation, thus stimulating growth and calcification at the sutures, whereas compression goes together with the plates confinement.

Physical confinement is known to affect the growth and morphology of various cell aggregations and those of colonial organisms (Thompson 1942, Raup 1968, Stevens 1974). In the echinoid test stresses act in a more differential way, and a certain plate may be subjected to compression in one direction and tension in the other. This pattern may change appreciably as the plate shifts along the vertical column, during its ontogeny. It is also possible that mechanical "stress balance" is the main controlling factor that determines echinoid test morphology (Dafni MS).

5 ACKNOWLEDGEMENTS

The author wishes to thank Prof. F.D. Por and Dr. J. Erez for their guidance throughout this study. The H. Steinitz Laboratory staff provided help and encouragement.

6 REFERENCES

Dafni, J. 1980, Abnormal growth patterns in the sea urchin Tripneustes cf. gratilla (L.) under pollution (Echinodermata: Echinoidea). J. Exp. Mar. Biol. Ecol. 47: 259-279.

Dafni, J. 1983a, Aboral depressions in the tests of the sea urchin Tripneustes cf. gratilla (L.) in the Gulf of Eilat, Red Sea. J. Exp. Mar. Biol. Ecol. 67: 1-15.

Dafni, J. 1983b, A new subspecies of Tripneustes gratilla from the northern Red Sea (Echinodermata: Echinoidea: Toxopneustidae). Isr. J. Zool. 32:1-12.

Dafni, J. & J. Erez 1982, Differential growth in Tripneustes gratilla. In J.M. Lawrence (Ed.) Echinoderms: Proceedings of the International Conference, Tampa Bay, 71-75. Rotterdam, A.A.Balkema.

Heatfield, B.M. 1970, Calcification: effect of temperature and Diamox on incorporation of ^{45}Ca in vitro by regenerating spines of Strongylocentrotus purpuratus. Biol. Bull. 139:151-163.

Meikle, M.C., J.J. Reinolds, A. Sellers & J.T. Dingle 1979, Rabbit cranial sutures in vitro: A new experimental model for studying the response of fibrous joints to mechanical stress. Calcif. Tissue Int. 28:137-144.

Moss, M.L. & M.M. Meehan 1968, Growth of

the echinoid test. Acta Anat. 69:409-444.
Raup, D.M. 1968, Theoretical morphology of echinoid growth. J. Paleont. 42 (suppl. to no. 5): 50-63.
Stevens, P.S. 1974, Patterns in nature. New York, Penguin Books.
Thompson, D.W. 1942, On growth and form. 2nd edition. London, Cambridge University Press.

Proceedings of the Fifth International Echinoderm Conference / Galway / 24-29 September 1984

Significance of architectural patterns in the deep-sea echinoids Pourtalesiidae

B.DAVID
Institut des Sciences de la Terre, Dijon, France

ABSTRACT: The significance of architectural patterns among the deep-sea Pourtalesiidae is questioned. In the first part, a series of Pourtalesia and Echinosigra from 1mm to adult length are described and the main stages of architectural and morphological growth processes are compared. In the second part, the pattern of interrelationships within the family is determined from the study of various morphological features. From these two approaches, supported by the ecology of Pourtalesia, the direct adaptive significance of plate patterns is reconsidered.

1 INTRODUCTION

Echinoid growth allows great diversity in the construction of the adult test. The abyssal irregular sea urchins are a good illustration of that. An extraordinary shape, sometimes accompanied by an even more extraordinary plate array, is often observed. Among them, the familly Pourtalesiidae shows the greatest transformations of the test: "here Echinosigra paradoxa may well deserve the prize" (Mortensen, 1950). Each genus of this family combines at various degrees a more or less transformed architecture with a more or less unique shape. This double originality is not only an oddness of nature, but also represents the morphological and architectural designs of the most successful deep-sea group among the irregular echinoids. This leads us to ask a double question: how does this originality appear and how does it work within the Pourtalesiidea ?

How does it appear ? A dynamic analysis of the development of Pourtalesia and Echinosigra, chosen for example, would provide a large part of the answer. How does it work ? Ontogeny is once again a good key to interpret adult features to establish the main relationships within the family. Moreover, through these two questions, the knowledge of ontogeny could be a way to understand the ecological or adaptative significance of morphological or architectural structures.

Fig. 1. Location of the sampling stations. Black circles show INCAL stations where Echinosigra and Pourtalesia were found together; open circles INCAL stations where Echinosigra were found; black square BIOGAS station where Pourtalesia were found; black diamonds BIOGAS stations where Echinosigra were found.

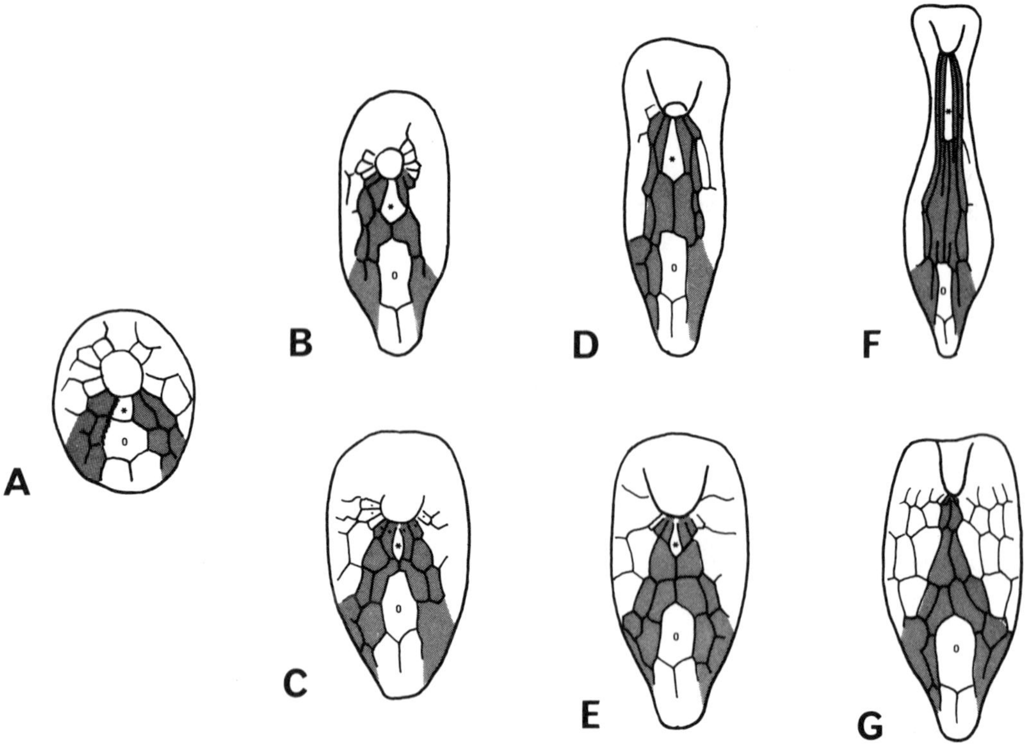

Fig. 2. Main ontogenetic stages of Echinosigra and Pourtalesia; oral sides. Posterior ambulacra I and V are stippled; labrum is shown by an asterisk and sternum by a circle.
A - Undetermined young sea urchin (length 1.2mm)
B - Young Echinosigra (length 4.3mm)
C - Young Pourtalesia (length 4.5mm)
D - Young Echinosigra (length 7.7mm)
E - Young Pourtalesia (length 7.0mm)
F - Adult Echinosigra (length 25mm)
G - Adult Pourtalesia (length 22mm)

2 MATERIAL AND METHODS

The family Pourtalesiidae includes 7 genera: Pourtalesia, Echinosigra, Helgocystis, Ceratophysa, Spatagocystis, Echinocrepis and Cystocrepis (after Mortensen, 1950). Except for the first two, they are more or less known, principally by works of Agassiz (1881, 1904), Loven (1883) Mortensen (1907).

The bulk of the material is made up of Pourtalesia and Echinosigra. Several hundred of these echinoids were dredged from the north-east Atlantic during the BIOGAS and INCAL cruises (CNEXO organizer). The two forms occur sympatrically in the Rockall Trough and separately on the northern Echinosigra) and southern (Pourtalesia) margins of the Bay of Biscay (fig. 1). Many of the specimens collected are juveniles and provide an excellent stock for an ontogenetic study. I do not have samples of the other genera, but a good bibliography exists about the austral forms Spatagocystis, Echinocrepis and Cystocrepis ; Helgocystis and Ceratophysa are hardly known.

The plate pattern was mainly investigated by camera lucida drawings under the binocular microscope. The observation of small tests was facilitated by putting a drop of acetone or alcohol on the specimen ; differential evaporation underlines the plate boundaries.

The introductory question was two-fold, therefore the research and the results are presented in two parts. Firstly, I will investigate and compare the ontogenies principally of Pourtalesia and Echinosigra. Secondly, I will try to find the relationships between all the genera of the family.

3 DEVELOPMENT OF THE TEST

3.1 Observation of the growth

It is possible to observe the plate array from sea urchins of 1 or 2mm length to the adults. Two examples are chosen to follow the architectural and morphological development: Pourtalesia from the Bay of Biscay and Echinosigra from Rockall and Porcupine. A detailed analysis of the dynamics of plate growth in Pourtalesia was made in a previous paper (David, in press) and Gage (1984) had given the ontogenetic series of Echinosigra. Thus the main growth stages will only be presented here with emphasis on comparison between the two genera.

The architectural or morphological growth is a continuous process. However, it is convenient to distinguish some main stages along this continuum.

Stage 1. Neither morphology nor architecture distinguishes Pourtalesia from Echinosigra less than 3 or 4mm length. The general body shape is oval and the plate array is almost regular (fig. 2A). The interambulacrum (IA) 5 is continuous. Its ventral part is made of 4 joined plates: labrum, sternum and 2 episternal plates. Laterally, at this IA, ambulacra I and V are a regular mosaic of alternating plates. All of the paired IA (1, 2, 3 and 4) begin with a small unpaired peristomial plate, a large unpaired plate follows in the second row and paired plates appear with the third row.

Stage 2. At 4 to 5mm length, the young Echinosigra become more elongated (fig. 2B) and are easily recognizable from Pourtalesia which are more shortened (fig. 2C). However, architectural developments are almost identical. The labrum and sternum are disconnected by the shifting growth of plates I.a.2 and V.b.2 which join together on the axis of symmetry and develop an important boundary. A single break exists in IA 5. In some specimens slight differences between the two genera could appear in the geometry of ambulacral plates: I.a.2 and V.b.2, and respectively I.b.2 and V.a.2 are more parallel and more elongated in Echinosigra than in Pourtalesia.

Stage 3. At 7 to 8mm, the differences are very important. The body shape of Echinosigra becomes more and more elongated, whereas the architecture retains the same aspect as in the former stage (fig. 2D). On the other hand, the body shape of Pourtalesia changes little, whereas the architecture adopts a new pattern. The external plates I.b.2 and V.a.2 rotate around their internal homologues and join together on the axis (fig. 2E). The labrum and sternum are separated by two rows of plates. A double break exists in IA 5.

Stage 4. Further development shows the culmination of these trends. Adult Echinosigra have a very elongated test made of very elongated plates and IA 5 has a single break (fig. 2F). Pourtalesia show a double break of IA 5 and have a lesser elongated test (fig. 2G). Thus, the plates are relatively short since the more advanced architectural development is itself a mechanism of longitudinal growth. At this stage, the labrum, often staying a small size, can move in the peristomial groove and is inconspicuous on the oral side.

3.2 Discussion

These observations of the growth lead to a new architectural diagram of Pourtalesia different from that described by Loven one hundred years ago (1883). The new hypothesis can be rigorously proven by studying the shape and growth relationships of certain plates as well as observing the water - vascular system (David, in press). Pourtalesia and Echinosigra may have a very high affinity.

The two genera have closely similar functional organizations: positions of peristome and periproct, and width of the IA 5 break (i.e. size of the sternum). They follow similar growth pathways. The body shape becomes more and more elongated. The peristome moves forwards step by step, and IA 5 is broken into two parts by the post-ambulacral plates rotating and shifting towards the longitudinal axis of symmetry. However, each genus make "a personal ontogenetic choice" to reach the same functional result and the architectural development of Pourtalesia balance the allometric morphological growth of Echinosigra. In other words, Pourtalesia is more advanced in architecture and Echinosigra in allometric morphology, while being closely alike in their structures.

As a whole, in comparison with other Holasteroids, all the Pourtalesiidae have similar architectural configurations with a break - single or double - of IA 5 (Ceratophysa is unknown)(fig. 2F-G and 3). The different adult plate patterns represent only variations along an unique architectural ontogenic process that related genera follow to varying degrees (as shown above between Echinosigra and Pourtalesia).

4 THE 7 GENERA : RELATIONSHIPS

This second part takes into account other features and constructs a pattern of rela-

	HELGOCYSTIS	CERATOPHYSA	POURTALESIA	ECHINOSIGRA	SPATAGOCYSTIS	CYSTOCREPIS	ECHINOCREPIS
Position of apical system	anterior	anterior	anterior	anterior	anterior	anterior	anterior
Genital plates array	close	close	close	close	close	close	close
Deep oral groove	yes	Y	Y	Y	Y	Y	Y
Vertical peristome	yes	Y	Y	Y	Y	Y	Y
Disjoined I.A. 5	yes	Y	Y	Y	Y	Y	Y
Bivium and trivium	disconnected	disconnected	disconnected	disconnected	disconnected	disconnected	juxtaposed
Aboral I.A. 5 plates	alternating	alternating	**facing**	facing	alternating	alternating	alternating
Peristomial ambulacral plates	biporous	uni- or biporous	uniporous	uniporous	uniporous	uniporous	uniporous
Hood	reduced	absent	small	small	**large**	very large	very large
Rostrum	large	large	large	large	small	absent	absent
Periproct	supramarginal	supramarginal	supramarginal	supramarginal	marginal	inframarginal	inframarginal
Sternum	?	?	large	large	small	small	small
Fasciole	yes	Y	Y	Y	Y/N	N	N
Size (mm)	100	98	30-50	30-40	48	92	100(200)
Genital pores	4	4	4	2	4	4	3
I.A. 5 break (architecture)	single ?	?	double	single	double	double	single

Table 1. Variation of the main features among the seven genera of Pourtalesiidae. Y = yes and N = no

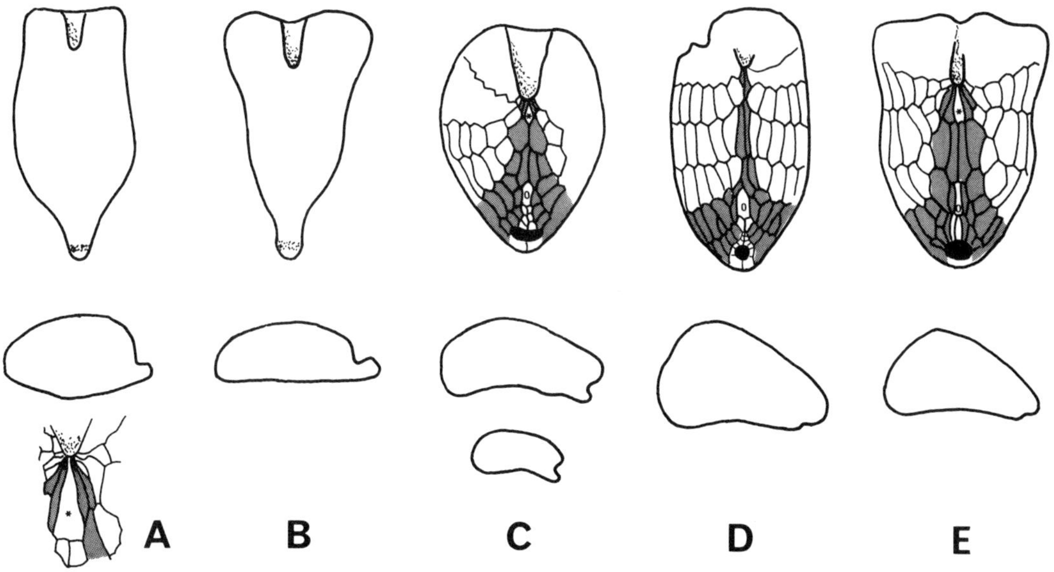

A B C D E

Fig. 3. Genera of the Pourtalessiidae; oral and lateral sides. Posterior ambulacra I and V are stippled; labrum is schown by an asterisk and sternum by a circle.
A - Helgocystis (length 90-100mm). Outlines of the test (modified from Agassiz, 1881) and detail of peristomial zone (taken from Loven, 1883).
B - Ceratophysa (length 98mm)(modified from Agassiz, 1881).
C - Spatagocystis (length 48mm). Oral side (modified from Agassiz, 1904); young and adult lateral views (taken from Agassiz, 1881).
D - Echinocrepis (length 80mm)(modified from Agassiz, 1881).
E - Cystocrepis (length 92mm)(modified from Agassiz, 1904).

tionships within the Pourtalesiidae. This system will show the distribution of plate patterns in the network of relationships.

The features of the test of the 7 genera are shown in figure 3 and the most relevant ones are indicated in table 1. These features fall into three categories :

1. Features which never change in the 7 genera. They are: an anterior apical system, with genital plates situated close together and even coalesced, a deep oral groove, a vertically placed peristome and a disjoined IA 5. These features attest to the homogeneity of the family. They represent advanced characteristics.

2. Features which change and are clearly apo- or plesiomorphic among the genera of Pourtalesiidae. These are significant in establishing the relationship among the genera. Their meaning is discussed below principally in comparison with other Holasteroids.

- The fact that the bivium and trivium are juxtaposed could be envisaged as an autapomorphy of Echinocrepis ; conversely,

Fig. 4. Morphological changes of hood and rostrum within Pourtalesiidae. A - Young Echinosigra - Pourtalesia; B - Helgocytis or middle sized Echinosigra - Pourtalesia; C - Adult Echinosigra - Pourtalesia; D - Middle sized Spatagocystis; E - Adult Spatagocystis ; F - Echinocrepis or Cystocrepis.

the fact that ocular plates I and V are disconnected from the anterior part of the apex is a symplesiomorphy shared by the other genera.

- A facing pattern of aboral plates of IA 5 is an apomorphy shared by Pourtalesia and Echinosigra; an alternating pattern is a plesiomorphy shared by other genera and also by typical Holasteroids.
- Uniporous peristomial ambulacral plates is an apomorphy shared by all the genera, except Helgocystis and occasionally Ceratophysa.
- In Echinosigra and Pourtalesia, the periproct rotates and becomes more and more vertical during growth (Fig. 4 A-B-C). This rotation induces, almost mechanically, the formation of the hood and rostum. In Spatagocystis, the hood becomes more important and the rostrum less important at the end of development(Fig. 4 D-E). This can be seen as a consequence of the continuation of the rotation process while the periproct attains an inverse position. Echinocrepis and Cystocrepis have an inframarginal periproct, a very prominent hood and a rostrum reduced to a slight projection of the ventral face (Fig. 4 F). This structure is the logical culmination of the previous growth trend, which can be interpreted as a progressive apomorphy. Thus,the extended hood, reduced rostrum and inframarginal periproct are advanced characteristics. The size of the sternum is linked with the oral shift of periproct, therefore a small sternum is apomorphic. In the same way, the subanal fasciole is linked with the existence of a conspicuous rostrum. All these features are morphologically correlated.

3. Features which seem to vary randomly: size, width of the apical separation of the bivium - trivium and number of genital pores (that they are limited to 2 in Echinosigra is due to the restricted space of the "head").

A cladogram can be drawn which summarizes the relationships within the family (Fig. 5). This scheme shows three distinct groups: 1) a primitive and paraphyletic group made of Helgocystis and Ceratophysa; 2) a derived and monophyletic group made of Pourtalesia and Echinosigra, two very closely related genera as shown above by their ontogenetic comparison ; 3) another derived group made of Spatagocystis, Cystocrepis and Echinocrepis which perhaps share more advanced characteristics. Within this last group Cystocrepis and Echinocrepis are closely related.

The cladogram describes theorical corroborated interrelationships. In this network, the distribution of architectural patterns is not congruent with that of

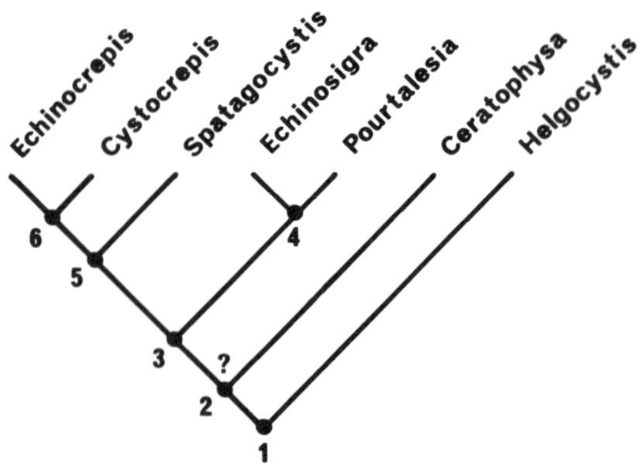

Fig. 5. Cladogram showing the most probable relationship among Pourtalesiidae.
(1) Anterior apical system ; genital plates close together ; deep oral groove ; peristome vertical ; IA 5 disjointed.
(2) Peristomial pores often single. This character is hardly defined and the node could be met with the last or with the next. It is difficult to make a choice and I would like to compromise, even though it gives an artificial picture.
(3) Peristomial pores always single ; susanal hood conspicuous.
(4) Facing dorsal plates of IA 5 (i.e. pairred pattern).
(5) Periproct open adorally ; well developed hood ; reduced rostrum; small sternum.
(6) Fasciole absent, subanal rostrum absent; inframarginal periproct.

other characters. Two solutions are possible.

1. Architecture is an apomorphy. Thus, the hypothesis must be changed and a new cladogram must be drawn.
2. Architectural similarities are the result of convergences and the cladogram does not need to be changed. The second solution agrees with usual cladistic methods since the hypothesis, suggested above, is the simplest explanation of the interrelationships. Although I do not especially agree with the parsimony criterion as a rule for choice of hypothesis, in this case, ontogenetic arguments (as uniqueness of architectural process) support it and architectural similarities can be considered as convergences.

5 CONCLUSIONS

More or less transformed plastronal architecture is a variable feature within the Pourtalesiidae. However, it merely represents deferred achievements of the same growth process. Between closely related genera, it leads to very similar functional results. Moreover, the pattern of relationships among Pourtalesiidae show that architectural similarities, such as between Pourtalesia and Cystocrepis or between Echinocrepis and Echinosigra, may be considered as convergences linked to the same regulation of the growth. From, this viewpoint, the more or less transformed architecture of a genus do not seem intrinsically adaptative, it is only a way to obtain the same functional result which is mainly the forward migration of peristome.

Now, I want to ask a last question : why such type of architecture ? This is a more "tricky point" with an underlying philosophical thought. The zoogeographical distribution of the genera agrees with the three subdivisions of Pourtalesiidae, but it does not provide any answer. On the other hand, the ecology of Pourtalesia can be discussed. Pourtalesia is a mobile epifaunal dweller. It mainly feeds on superficial deposits. Comparison with an "architecturally classic" sea urchin from the same environment shows that this one (Brissopsis in this case) is able to catch larger and more mobile prey such as worms or crustaceans (Billett. pers. comm.). Therefore Pourtalesia seems to have a lower ability to move and to catch mobile prey. From this point of view the architectural pattern of Pourtalesia, with a reduced sternum and few locomotory spines, seems to be linked with its lower mobility. One could suppose that the characteristics of the adaptative pressure of deep-sea environments allow the survival of slow-moving sea urchins. Thus these characteristics partly release the plastron from its necessary locomotory function. Other architectural patterns, without the classical large sternal plates, could be tried. The Pourtalesiidae pattern is perhaps the simpler : the nearest plates replace the sternum. This fascinating feature is no more than another solution drived by ontogeny: it is an epiphenomenon in regard to the migration of the peristome. But, the ontogenetic trend, once engaged, goes further and leads to a more or less transformed plate array whose direct adaptative significance can be challenged.

REFERENCES

AGASSIZ, A. 1881. Report on the Echinoidea dredged by H.M.S. Challenger, during the years 1873-1876. Rep. Scient. Resul. Voy. HMS Chall., Zool. 3: 1-321.

AGASSIZ, A. 1904. The panamic deep sea Echini. Mem. Mus. Compar. Zool. Harvard Coll., 31: 1-243 + atlas 110 pl.

DAVID, B.. Dynamics of plate growth in the deep-sea echinoid Pourtalesia miranda Agassiz: a new architectural interpretation (in press Bull. mar. Sc.).

GAGE, J.D. 1984. On the status of the deep-sea echinoids Echinosigra phiale and E. paradoxa. J. mar. biol. Ass. U.K., 64: 157-170.

LOVEN, S. 1883. On Pourtalesia a genus of Echinoidea. Kongl. Sven. Vetensk. Akad. Handling., 19: 1-95.

MORTENSEN, T. 1907. Echinoidea (part. 2). Danish Ingolf Exped., 4: 1-200.

MORTENSEN, T. 1950-1951. A monograph of the Echinoidea, t. 5 Spatangoida, vol. 1 (1950), p. 1-432 ; vol. 2 (1951), p. 1-193. Copenhagen: Reitzel.

Acknowledgements

I wish to thank Dr. J.M. Lawrence for critically reading this manuscript and Mrs K. Sebedio for translating it in english. Also many thanks to CENTOB and COB (Brest) for giving the deep-sea material and Mrs M.F. Jamelin-Crouzet for typing.

Proceedings of the Fifth International Echinoderm Conference / Galway / 24-29 September 1984

Food selection and absorption efficiency in the spatangoid echinoid, *Echinocardium cordatum* (Echinodermata)

CHANTAL DE RIDDER, MICHEL JANGOUX & EDWIN VAN IMPE
Laboratoire de Biologie marine, Université Libre de Bruxelles, Belgium

SUMMARY. The burrowing spatangoid, **Echinocardium cordatum**, is a selective detritus feeder. It passively traps detritus in the funnel that connects its burrow to the surface of the sediment. Estimation of the nutrient level (lipid, protein and carbohydrate) occuring in the gut content indicates that the main site for digestion and absorption of nutrient is the first digestive loop. The nutrients absorbed originate from detritus occuring in the sediment while organic matter adsorbed on sand grains is not used. Absorption efficiency of all investigated nutrient is given, with the highest value for the lipid fraction.

INTRODUCTION.

The burrowing spatangoid, **Echinocardium cordatum**, ingests both deep-seated and surface sediment. Surface sediment falls down the burrow's chimney reaching the upper part of the frontal ambulacrum. Sediment particles are then carried by the spines of the frontal ambulacrum close to the mouth where they are picked up by the peribuccal tube-feet (Nichols 1959, Buchanan 1966, Péquignat 1970). **E. cordatum** does not select the size of ingested particles, and is considered as a non-selective deposit-feeder (Lafon 1953, Nichols 1959). Occurrence of a chemical selection of food particles was never considered. However, surface sediment may be ingested exclusively (De Ridder 1982). Moreover, investigations have shown that starved individuals ingest distinctly more surface sediment than normally fed individuals (De Ridder and Jangoux 1985).

The purpose of this work was to determine the origin, nature and organic composition of the food of **E. cordatum**, to investigate whether or not food selection occurs, and to measure the absorption efficiencies for the protein, carbohydrate and lipid fractions present in the gut contents.

MATERIAL AND METHODS.

Individuals of **E. cordatum** were collected intertidally on a sandy beach at Le Home-Varaville (Normandy, France) in October-November 1980 and 1981 (mean size of specimens : 50.9 ± 1.7 cm length, 50.5 ± 1.4 cm width). Surface sediment (viz. top 0.5 cm sediment around the chimney opening) and deep-seated sediment (viz. sediment of the burrow's bottom) as well as intradigestive sediment collected from dissected individuals (viz. sediment of the first and second digesive loop, and from the rectum) were sampled. Granulometric and biochemical analyses were performed on each sample. Samples used for biochemical analysis were immediately frozen while others were preserved in 70 % ethanol. Sediment-associated detritus was characterized using a binocular microscope.

For the determination of sediment grain size, samples were washed in distilled water, treated with a weak solution of sodium hypochlorite for several hours in order to remove associated organic matter, washed again several times in distilled water, dried for 12 h at 100° C and subsequently weighed. Sediment samples were screened (mesh sieves in µm : 320, 265, 200, 160 and 100), the obtained fractions weighed and their respective percentage in the total sample weight calculated (mean values reported on cumulative curves).

For biochemical analysis lipid was firstly extracted using a 2/1 dichloromethane-methanol mixture. The protein fraction was then extracted using NaOH 1 N for 10 min.. Carbohydrate was analysed in both the proteic extract and remaining rinsed sediment. Lipid, protein and carbohydrate fractions were analysed using ,respectively, the methods of Marsh and Weinstein (1966), Lowry et al. (1961) and Dubois et al. (1956), as performed by Jangoux and Van Impe (1977).

RESULTS.

1.Sediment grain size distribution.

The sediment in which **E. cordatum** lives at Le Home-Varaville is composed mainly of fine sand grains (about 80 % of sediment dry weigth consists of grains whose size ranges from 100 to 160 µm). The size frequency distributions of each sediment sample - viz. surface sediment, deep-seated sediment, and intradigestive sediment - are similar (fig. 1). These results clearly indicate that **E. cordatum** is not selective with respect to sediment-particle size.

2.Levels of organic matter in the sediment.

The levels of organic matter in the surrounding and intradigestive sediment are given in fig. 2. Whatever the sediment origin, the level of associated organic matter always is very low (no more than 4 mg of organic matter per g of sediment). There are no significant difference in the level of organic matter in surface and deep-seated sediment. To the contrary, intradigestive sediment is conspicuously richer than that surrounding the burrow, with the highest value for the sediment in the first digestive loop. These changes concern both lipid and protein fractions while carbohydrate fraction seemingly does not fluctuate. These results clearly indicate that a chemical selection of food particles occurs.

3.Sediment-associated detritus.

Detritus occuring in coarsely particulate form in sediment samples

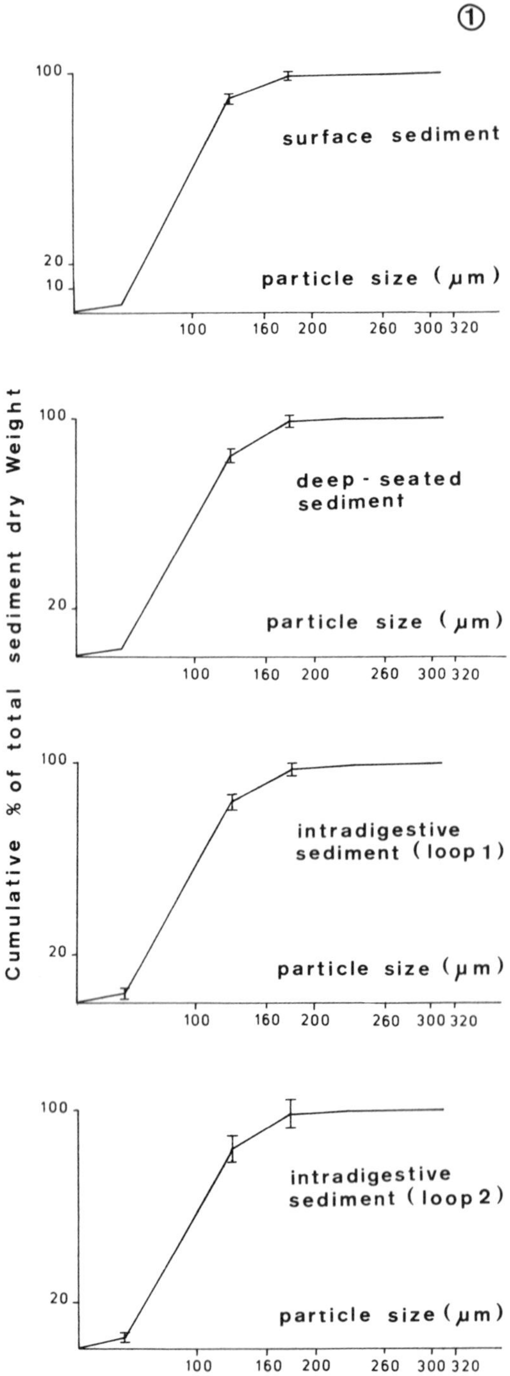

Figure 1. Particle size frequencies of surface, deep-seated and intradigestive sediment. (Mean values expressed in % of total sediment dry weight, for n=5, p=0.05).

include mostly organic-mineral aggregates, fecal pellets and fragments, vegetal fragments, animal fragments (chitinous molts, shells, dead colonies of Hydrozoa and Bryozoa, polychaetes tubes). In the studied area fecal pellets of the bivalve **Cardium edule** were easily recognisable and particularly abundant. To assess the availability of detritus to **E. cordatum**, the number of entire (unaltered) fecal pellets was determined in each sediment sample (observations realised under dissecting microscope). Fecal pellets of **C. edule** occured in all samples of intradigestive sediment as well as in surface sediment. The levels of fecal pellets in each sample are reported in fig. 3. Pellets are most numerous in the first digestive loop while their concentrations decrease progressively in the second loop and in the rectum. As is evident in figs 2 and 3, the change in number of entire fecal pellets along the gut length corresponds the gradient in decrease in organic matter. It should be noted, moreover, that an inverse increase of the number of altered pellets (viz. pellets in various stages of fragmentation) was also observed along the gut length.

4.The mobilisable sediment-associated organic matter.

The level of sediment-associated organic matter really available to the echinoids (i.e. the mobilisable organic matter) was tentatively estimated. The sediment column filling each digestive area - viz. first gut loop, second gut loop, rectum - was carefully removed and longitudinally separated in two almost equal parts. One part (untreated sediment) was immediately frozen as it is. The other part (treated sediment) was frozen after having been cleaned from unadsorbed organic matter, namely particles and suspended materials of detrital origin. (Particles were removed under a dissecting microscope before the sediment was rinsed repeatedly in order to eliminate suspended material).

The levels of organic matter in treated and untreated sediment are given in fig. 4. The level of mobilisable organic matter (Mo) may be easily calculated using the relation Mo = Uo - To, where Uo is the level of organic matter in untreated sediment, and To is the level of organic matter in treated sediment (i. e. the organic matter adsorbed on sediment grains). As

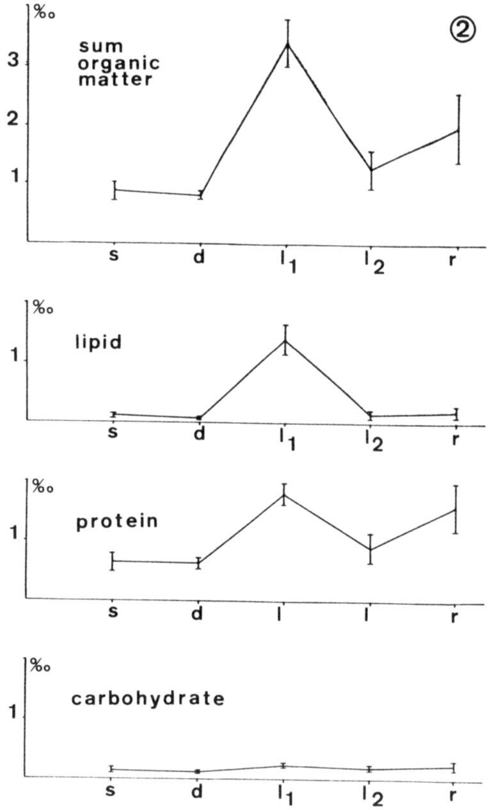

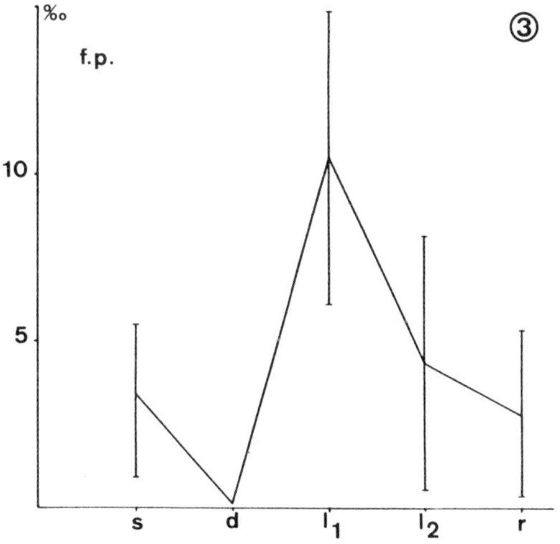

Figure 2 & 3. Levels of nutrient in mg/g sediment dry weight (fig. 2) and number of fecal pellets per g sediment dry weight (fig. 3) in surface (s), deep-seated (d), first (l_1) and second gut loop (l_2), and rectum (r) sediment. (Mean values for n=10, p=0.05).

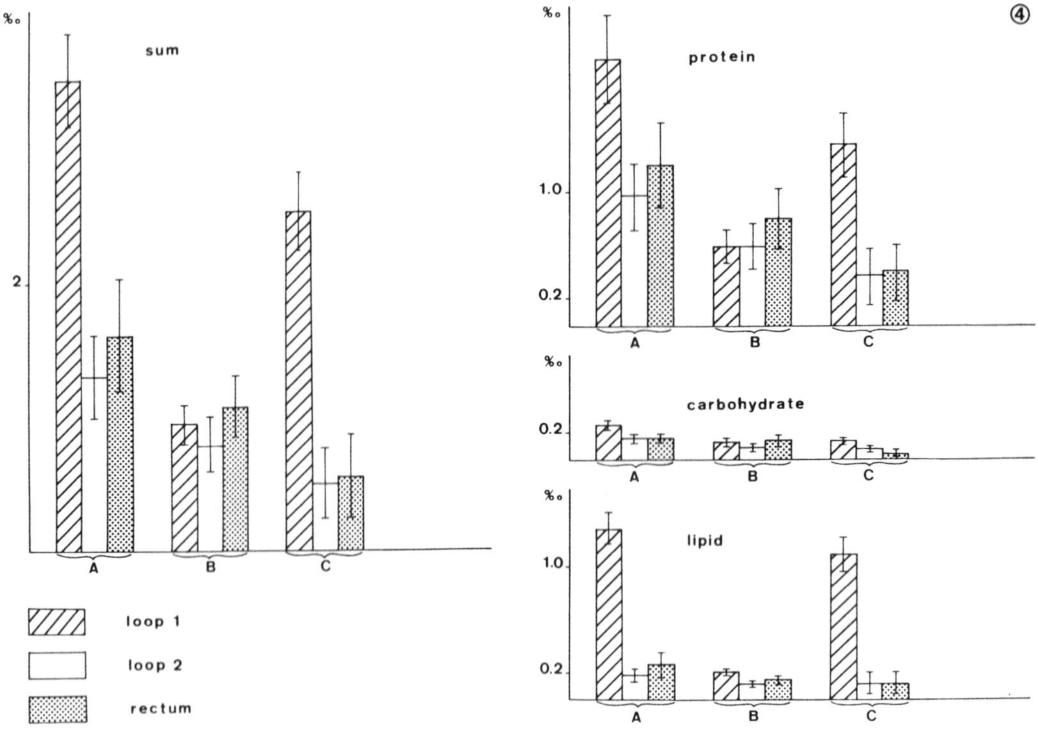

Figure 4. Evolution along the gut (loop 1, loop 2, and rectum) of the level of nutrient in untreated (A) and treated (B) sediment, and in the associated mobilisable organic matter (C). (Mean values expressed in mg/g sediment dry weight, for n=10, p=0.05).

seen in fig. 4 mobilisable organic matter represents about 75 % of the total sediment-associated organic matter. Moreover, it appears that the adsorbed fraction of organic matter (To) does not significantly change along the echinoid gut while the level of mobilisable organic matter (Mo) decreases markedly from the first to the second gut loop. As evident in fig. 4, there is, however, a slight increase in the protein level of the contents of the rectum when compared to the second gut loop. This increase seemingly only reflects sediment-adsorbed organic matter. It presumably results from the occurence of inorganic fragments from intestinal bacterial pellets that are expelled from the echinoid intestinal caecum (De Ridder 1982). These fragments presumably are rather rich in adsorbed organic matter.

5.Estimation of absorption efficiencies.

Absorption efficiencies were estimated adapting the relation of Lowe and Lawrence (1976),

$$A_N = 100 \times \left(\frac{N_{Ua}}{N_{Ta}} - \frac{N_{Ub}}{N_{Tb}} \middle/ \frac{N_{Ua}}{N_{Ta}} \right)$$

where A_N is the absorption efficiency for nutrient N, N_{Ua} and N_{Ub} are the level of nutrient in untreated sediment from the first loop (a) and either the second loop or the rectum (b), N_{Ta} and N_{Tb} are the level of nutrient in treated sediment from the first loop (a) and either the second loop or the rectum (b). This method is based on the assumption that the sediment-adsorbed nutrient fraction is not significantly absorbed during digestion (see fig. 4).

Table 1 : Absorption efficiencies (in % of nutrient in food ingested), absorption index (mg of nutrient absorbed per g dry weight of sediment ingested), and mg and kcal of nutrient absorbed per individual per day. (Calculation done by comparing the organic content in the first and the second gut loops).

Nutrient	Absorption efficiency	Absorption index	mg absorbed per individual per day	kcal absorbed per individual per day ($x 10^{-3}$)
Protein	51.3	1.02	9.30	52.5
Carbohydrate	10.0	0.02	0.18	1.0
Lipid	75.0	0.96	8.75	82.7
Sum		2.00	18.23	136.2

Absorption index I_N was calculated following Lowe and Lawrence (1976) by multipliing the absorption efficiency (A_N) for nutrient N by the level of nutrient N in the ingested sediment (N_{Ua}),

$$I_N = (A_N)\ (N_{Ua})$$

Absorption index were multiplied by the feeding rate of **E. cordatum** (i.e. 9.12 g dry weight sediment per individual per day; De Ridder and Jangoux 1985) to calculate the amount of each nutrient absorbed per individual per day. The caloric equivalent for each nutrient was calculated using conversion factors given in Winberg (1971). The obtained values are listed in Table 1.

As seen on Table 1, the absorption efficiency in the first gut loop is the highest for lipid (75 %), a lower value is observed for protein (51.5 %) while that of carbohydrate (10 %) is much less. As a consequence of the high level of protein in the ingested food, about equal quantities of protein (1.02) and lipid (0.96) are absorbed per g of sediment ingested, while the carbohydrates (0.02) are almost unused. Lipid, however, contributes more to the total amount of kilocalories absorbed than the protein (61% and 38.2%, respectively).

DISCUSSION.

As already stated by Lafon (1953) and Nichols (1959), **E. cordatum** is clearly not a selective deposit-feeder with respect to sediment grain size. Food selection occurs, however, since the level of organic matter associated to the gut content (i.e. the intradigestive sediment of the first gut loop) is about four times higher than that of surface sediment. Such an increase in the level of organic matter is due to a marked enrichment of gut sediment by detrital material, for instance bivalve fecal pellets.

De Ridder and Jangoux (1985) demonstrated that E. cordatum may ingest both surface and deep-seated sediment. That latter supposedly is picked up for some mechanical reasons as it contains almost no food (detrital particles). De Ridder and Jangoux hypothesized that a critical volume of intradigestive sediment is needed to allow the efficient shifting of the food bolus along the gut length. They suggested that the ingestion of deep-seated sediment actually contributes to gain that volume. Ingestion of that sediment had been infered, e.g., by Buchanan et al. (1980) when noted the accumulation of ferric phosphate within the connective tissue layer of the second gut loop. The most likely origin of that material is from pore water from the deep-seated sediment where reducing conditions prevail (here the iron would be in its dissolved ferrous form).

The question is how a burrowed **Echinocardium** concentrates detrital material in its gut content? It has been reported that the supply of food may occur as a result of the action of the elongated apical tube-feet (i.e. the funnel building tube-feet). These tube-feet either directly collect food particles from the sediment surface (Robertson 1871, Buchanan 1966, Péquignat 1970) or enhance the fall of particles in the funnel through their movements (Nichols 1959, De Ridder 1982). We have never seen the apical tube-feet collecting any kind of particle, although they seem to possess chemosensitive abilities, since they are more active in areas where surface sediment are rich in organic matter (Buchanan 1966, Péquignat 1970, De Ridder 1982). Whatever the chemosensory ability of apical tube-feet, the particular shape of the chimney opening has to be considered also. Actually this opening is typically funneled, and forms a perfect trap for detrital particles that are swept by local water currents (viz. wave actions and tidal currents). We suggest therefore that particle selection in **E. cordatum** could result from two different mechanisms that may act together, viz. a passive trapping of detrital particles through the funneled opening of the burrow, and a semi-active trapping of particles located around the funnel through the circular and repetitive movements of apical tube-feet (De Ridder 1982).

The first gut loop corresponds to the main site of digestion and absorption. This is conspicuously illustrated by the progressive increase of altered bivalve fecal pellets along the gut length. The change of level of nutrient from the first to the second loop clearly confirm that the first gut loop is the site for absorption. Comparison of untreated sediment ("natural sediment") and treated sediment (sediment where most detrital particles were removed), lead us to the conclusion that **E. cordatum** digests and absorbs only detrital material associated with the sediment while it does not use the organic matter adsorbed on sediment grains since there was no variation of the organic matter levels in the washed sediment.

That some deposit-feeding invertebrates are not sediment-size selective does not necessarily imply that they are non-selective for the food particles they ingest. This has been demonstrated in the present paper in the case of **E. cordatum**, but is also true for many others of deposit-feeding benthic animals (see, e.g. Powell 1977, Rijken 1979, Yingst 1982). As already noted by Lane and Lawrence (1982) classical measurements of nutrient absorption efficiencies cannot be performed with such organisms because the energetic value of their food (i.e. the surrounding surface sediment) will be as a rule lower than that of their feces. Such calculations would lead to absurd results, viz. negative absorbtion efficiencies. The absorption efficiency in selective deposit-feeders can be estimated only by analysing the associated nutrients of the sediment filling the proximal part of their gut, the nutrient levels being then compared with those obtained either from the analysis of the sediment filling the distal part of the gut or from the analysis of the feces (if feces are available). This was investigated on the gut content of **E. cordatum.** From the obtained results (see Table 1) it is obvious that the ability of **E. cordatum** to obtain nutrients from sediment-associated detritus is high.

ACKNOWLEDGMENTS.

We thank Dr. P. Le Gall, director of the marine Laboratory of Luc-sur-Mer (Normandy, France), for providing facilities.

BIBLIOGRAPHY.

Buchanan, J.B. 1966. The biology of **Echinocardium cordatum** (Echinodermata : Spatangoidea) from different habitats, J. mar. biol. Ass. U.K. 46 : 97-114.

Buchanan, J.B., B.E. Brown, T.L. Coombs, B.J.S. Pirie, & J.A. Allen 1980. The accumulation of ferric iron in the guts of some spatangoid echinoderms. J. mar. biol. Ass. U.K. 60 : 631-640.

De Ridder, C. 1982. Feeding and some aspects of the gut structure in the spatangoid echinoid, **Echinocardium cordatum** (Pennant), Proc. intern. Echinoderm Conf., Tampa Bay (J.M. Lawrence, ed.), pp 5-9. Balkema : Rotterdam.

De Ridder, C. & M. Jangoux 1985. Note sur l'alimentation et sur la durée du transit digestif chez l'oursin spatangide, **Echinocardium cordatum** (Pennant) (Echinodermata), Annls Inst. océanogr. 61 (in press).

Dubois, M., K.A. Gilles, J.K. Hamilton, P.A. Rebers and F. Smith 1956.

Colorimetric method for determination of sugars and related substances. Analyt. Chem. 28 : 350-357.

Jangoux, M. & E. Van Impe 1977. The annual pyloric cycle of **Asterias rubens** L. (Echinodermata : Asteroidea), J. exp. mar. Biol. Ecol. 30 : 165-184.

Lafon, M. 1953. Recherches sur les sables et quelques conditions de leur peuplement zoologique, Annls Inst. océanogr. 28 (3) : 113-161.

Lane, J.M. & J.M. Lawrence 1982. Food, feeding and absorption efficiencies of the sand dollar, **Mellita quinqui-esperforata** (Leske), Est., coast. and shelf Sci. 14 : 421-431.

Lowe, E.F. & J.M. Lawrence 1976. Absorption efficiencies of **Lytechinus variegatus** (Lamarck) (Echinodermata : Echinoidea) for selected marine plants, J. exp. mar. Biol. Ecol. 21 : 223-234.

Lowry, O.H., N.J. Rosebrough, A.L. Farr & R.J. Randall 1951. Protein measurement with the Folin phenol reagent, J. biol. Chem. 193 : 265-275.

Marsh, J.B. & D.B. Weinstein 1966. Simple charring method for determination of lipids, J. Lipid. Res. 7 : 574-576.

Péquignat, E. 1970. Biologie des **Echinocardium cordatum** (Pennant) de la Baie de Seine, Forma Functio 2 : 121-168.

Robertson, D. 1871. Notes on **Amphidotus cordatus** (Penn.), Quart. J microsc. Sci.11 : 25-27.

Rijken, M. 1979. Food and food uptake in **Arenicola marina**, Neth. J. Sea Res. 13 : 406-421.

Winberg, G.G. 1971. Methods for the Estimation of Production of aquatic Animals. 175 pp. Academic Press : N.Y.

Yingst, J.Y. 1982. Factors influencing rates of sediment by **Parastichopus parvimensis** (Clark), an epibenthic deposit-feeding holothurian Est., coast. and shelf Sci. 14 : 119-134.

Proceedings of the Fifth International Echinoderm Conference / Galway / 24-29 September 1984

Morphology of spines of *Heterocentrotus mammillatus* (Echinodermata, Echinoidae) and its ecological significance

A.DOTAN & L.FISHELSON
Department of Zoology, Tel Aviv University, Israel

ABSTRACT: The adaptative morphology and ecology of spines of the slate-pencil sea-urchin Heterocentrotus mammillatus (L.) was studied in the field in the Gulf of Aqaba (Red Sea). The digitiform spines of this sea-urchin show very clear concentric growth lines produced mainly by differetial deposition of organic material. Observations on marked small and medium-size individuals (D < 60 mm) show that the growth rate per increment of one ring is similar to the annual growth rate (p > 0.5). The study of growth of small individuals (D < 40 mm, age $\geq$ 4 years), and experiments on regeneration, enable the growth lines to be used as an index for age determination of H. mammillatus. Abrasion, breakage and shedding of whole spines may be easily detected by examining longitudinal sections of spines. This may enable a reasonable assessment of environmental stress on a population of H. mammillatus in a specific habitat. Special emphasis was placed on the understanding of form and function of spines situated on different parts of the test. The oral spines are short, flat and blunt and are used for anchorage and movement. On the other hand, the aboral spines are long, rounded and pointed. Their main function is defence. In situ regeneration experiments showed that truncated spines which grow below the ambitus, become shorter and wider than those of the upper part of the test. Sea urchins living along the reef flat, are more susceptible to mechanical stress by wave action and have thicker spines than those living in calm water, at a depth of 5 m and more.

1 INTRODUCTION

Calcification growth lines occur in various Echinoids. They are apparent in the lantern ossicles, the coronal and apical plates, the tubercles and the spines (Deutler 1926, Pearse and Pearse 1975). The coronal and apical plates of some species are deposited in regular annual cycles and the number of those was used as an indicator of age (Review by Pearse and Pearse 1975). Concentric rings of large, wedge-like calcite crystals, are deposited in the spines of many sea-urchins. These rings differ clearly from most of the calcareous mass of the spine, which consists of a labyrinthic network of trabecules (Ebert 1967, Weber 1969). Growth of most spines occurs at the periphery, with deposition of delicate trabeculae, followed by an external layer of wedges. In an intact spine, such growth starts at the base, near the milled ring and proceeds towards the distal tip of the spine (Heatfield 1971). In the large pencil-like spines of the genus Heterocentrotus growth lines are outlined by pigments, mainly naphthoquinones, incorporated in the crystals of the wedges (Vevers 1966).

There are two contradictory theories about the character of growth lines deposition in spines of sea-urchins. One theory argues that the growth lines are deposited periodically in a cycle controlled by environmental factors or by an endogenic rhythm. According to this approach, growth lines in spines can be used as "chronometers" (Moore 1966, Weber 1969). Contrary to this, Ebert (1967, 1968) asserts that growth lines are not regularly formed and additions are caused by damage to the spine and precipitation of new material during regeneration. Heatfield (1971) conducted a laboratory experiment in which he fractured spines of Strongylocentrotus purpuratus and examined their regeneration. He reported that

Table 1. Morphological characteristics of spines. (10 specimens from each habitat)
Flatness (Difference between width and height of cross-section) (mm)

		Reef slope	Reef flat	Fore reef
Above ambitus	Long radius	-0.1±0.52	0.7±0.79	0.3±1.03
	Short radius	0.2±0.56	0.2±0.81	0.2±1.03
Ambitus	long radius	1.6±1.10	3.1±2.40	2.0±1.52
	short radius	1.8±0.87	2.4±0.85	1.7±1.13
Below ambitus	long radius	3.2±1.20	4.1±1.27	3.3±1.42
	short radius	2.6±0.85	2.8±0.60	2.8±0.92
	ANOVA:	F	P	
	Habitats	2.02	0.153	
	Position	168.3	0.000	
	Radius	5.06	0.033	

Table 2. Morphological characteristics of spines (10 specimens from each habitat)
Pointedness (Differences between width at the base and width at the tip) (mm)

		Reef slope	Reef flat	Fore reef
Above ambitus	long radius	2.6±0.50	2.3±0.87	1.9±1.05
	short radius	1.9±0.88	1.4±0.82	2.3±0.90
Ambitus	long radius	1.5±1.05	0.3±1.88	0.3±0.69
	short radius	0.6±0.82	-0.2±0.85	1.0±1.12
Below ambitus	long radius	-0.3±1.09	-1.1±1.06	-0.9±0.80
	short radius	-0.4±0.70	-0.8±0.82	-0.6±0.57
	ANOVA:	F	P	
	Habitats	2.95	0.070	
	Position	164.1	0.000	
	Radius	1.48	0.235	

Table 3. Morphological characteristics of spines. (10 specimens from each habitat)
Average diameter (mm)

		Reef slope	Reef flat	Fore reef
Above ambitus	long radius	4.9±0.80	6.0±1.09	5.5±0.64
	short radius	3.9±0.26	4.8±0.77	4.3±0.48
Ambitus	long radius	5.6±0.72	6.4±0.69	5.7±0.69
	short radius	4.3±0.46	4.8±0.39	4.8±0.65
Below ambitus	long radius	5.0±0.49	5.2±0.27	4.4±0.78
	short radius	3.6±0.40	3.7±0.43	3.9±0.46
	ANOVA:	F	P	
	Habitat	6.38	0.005	
	Position	41.43	0.000	
	Radius	156.4	0.000	

repair of spines brings an increment of new growth lines in about 30-60 days. He also observed that new rings may be added regularly, even in the absence of a fracture and regeneration. A longitudinal section in the spine revealed traces of past injuries, either repair of a fracture or a replacement after complete loss of a spine. Ebert (1968) used the number of injuries in spines of individual sea-urchins as an indicator of the environmental stress caused by exposure to wave-action. Few studies mention the effect of environmental factors on the morphology of spines. Mortensen (1928) wrote that individuals of Eucidaris tribuloides which live in deep water, have longer spines than those living in shallow water. Cutress (1965) reported that when E. tribuloides was transferred from a small tank to a larger one, the spines became longer and sometimes thinner. No reason for this change was suggested. Regis (1981) found different growth rates in spines of Paracentrotus lividus living in different conditions of depth and wave action.

As indicated by Hyman (1955), the longest spines in most sea-urchins are situated near the ambitus and those of the oral and aboral sides are usually much shorter and sometimes also of different shape. Cutress (1965) showed that in E. tribuloides broken spines on plates above the ambitus do not change their morphological character after regeneration, while those near the peristome regenerated the "oral type" spines which are flat and short.

This paper summarizes a study on the morphology, regeneration and growth lines of the slate pencil sea-urchin Heterocentrotus mammillatus (L.) and the possible factors affecting them.

2. METHODS

The study was performed on individuals from natural populations. Most samples were collected near Na'ama Bay (North of Sharem al Sheikh) at the south-eastern coast of the Sinai Peninsula.

For morphological studies, the spines were filed longitudinally with a rotary whetstone to their median plane and their growth lines were examined with a dissecting microscope. Small spines were polished with carborundum powder. Assessment of growth lines and traces of fractures was based on examination of a series of interambulacral spines, from the ambitus downwards. From each specimen, 3 series of 4-5 spines each were examined.

We studied sea-urchins living in three zones of the coral reef: Zone 1: The fore-reef (n=112), Zone 2: The backreef 6 m from the reef edge (n=65) and Zone 3: The reef-slope, depth of 5 m and more (n=64). From these samples only a small number of specimens (8, 5 and 6 from the fore reef, back-reef, and reef slope respectively) were not taken into account because it was impossible to determine the precise number of lines on their spines or because contradictory counts were obtained. Living individuals of H. mammillatus were marked for growth studies by fastening textile tags to holes drilled in the spines. Animals whose body tests were 2 cm in diameter or greater, were marked during various months in 1981 and recollected in March 1982. The recapture rate was 29-56%.

An isolated patch reef of about 1.5 m^2 was cleared of all H. mammillatus and 15 unmarked individuals of 1 cm, 2 cm and 4 cm test diameter, were introduced. A year later these sea-urchins were collected and their growth was analyzed.

Regeneration experiments were conducted in May 1981 with marked individuals. Spines were truncated 1-1.5 cm above the milled ring. The broken parts of those spines were preserved and compared later with the regenerated spines after 10 months of regeneration.

The results of all our observations were analyzed with BMDP computer programs (Dixon and Brown 1982).

3 RESULTS

3.1 Morphology of spines in H. mammillatus

The morphology of the primary spines of H. mammillatus is variable: The longest spines at the ambitus or above it are usually longer than the oral-most ones. The aboral spines are pointed and round in cross section, while those on the oral side are blunt and flat. The "turning point" in this trend is at the ambitus (Tables 1-4). Comparison showed that spines of sea-urchins living on the reef-slope are thinner than those living in shallow habitats, that are more exposed to wave action. There are differences in flatness of cross-section and pointedness of spines in the different areas, but these are not statistically significant.

Table 4. Morphological characteristics of spines (10 specimens from each habitat) Relative length (length of spine/diameter of test)

		Reef slope	Reef flat	Fore reef
Above ambitus	long radius	1.24±0.128	1.21±0.096	1.23±0.121
	short radius	0.97±0.135	0.93±0.057	0.95±0.164
Ambitus	long radius	1.24±0.160	1.17±0.160	1.16±0.106
	short radius	0.93±0.144	0.87±0.090	0.96±0.097
Below ambitus	long radius	0.95±0.220	0.82±0.156	0.80±0.148
	short radius	0.59±1.129	0.56±0.154	0.59±0.136
	ANOVA:	F	P	
	Habitat	1.23	>0.25	
	Position	97.7	0.000	
	Radius	181.2	0.000	

Table 5. Percentage of damaged spines in samples of H. mammillatus (5-12 spines of each specimen)

	Reef slope (n=120)	Reef flat (n=70)	Fore reef (n=70)	Krushkal-Wallis test statistic
Major fracture	16 ±15.2	30 ±24.9	26 ±21.7	8.19*
Minor fracture	18 ±17.8	35 ±21.8	38 ±30.6	23.2**
Lost spine	5.1±13.9	5.3± 9.2	8.5±12.1	7.54*

Spearman Rank Correlation Coefficient

	Major spine	Minor spine	Lost spine
Major fracture	1		
Minor fracture	0.334*	1	
Lost spine	0.074	0.069	1

* Level of significance of difference < 0.05
** Level of significance of difference < 0.01

Table 6. Changes in dimensions of regenerated spines (10 months after fracture)

	Above ambitus (n=8)		Below ambitus (n=8)		T test	Krushkal-Wallis statistic test
	mm	%	mm	%		
Length of spine	2.1±3.40	7±9.9	-3.3±4.57	-8±13	2.7*	4.41*
Width of spine at the tip	-0.3±0.63	7±20	1.7±1.19	33±25	4.2**	8.96**

* Level of significance < 0.05
** Level of significance < 0.01

There are no differences in the length of spines in the three groups (Tables 1-4).

Comparison of disruptions in growth lines (see below) showed that spines of sea urchins on the reef-slope were less damaged than those of animals living on the reef table (Table 5). A distinction was made between major fractures, in which large parts of the spine was lost and several growth lines were found discontinued and minor fractures, where only the tip was damaged and only the external growth line was found disrupted. However, the general trend was found to be similar and there was a significant correlation between them. Complete loss of spines is a limited phenomenon and occurred mostly on the forereef. There is no correlation between a break of spines and loss of spines in the same animal.

3.2 Regeneration of spines

Regeneration studies show that the rate of repair of a broken spine is slow and it takes about 10 months for a truncated spine to reach its original length. After that period of time a new growth line was found in 10 out of 16 spines. In most cases, regenerating spines deposited new material from the plane of fracture distally and calcification proceeds proximally only for a few millimeters. Only in one case (out of 25), a broken spine added 11 mm of new band in the proximal direction while growing 13 mm distally.

10 months after truncation regenerated spines above the ambitus become longer and more pointed than the original, in contrast to those below the ambitus, that regenerated shorter and wider at the tip (Table 6).

3.3 Growth lines in spines

In most sections of spines of H. mammillatus, growth lines are prominent and easy to count. The number of growth cycles differs in spines along the interambulacral series. The 4-5 spines that are closest to the peristome and attached to the oldest plates, have the same number of cycles. The spine above them usually has one cycle less and spines that are situated further, have 2-3 lines less than the preceding one. Thus, there is a gradual decrease in cycles from the peristome upwards.

This regular arrangement of growth lines in the spines is complicated by some disruptions:

a. Near the base of numerous spines there are many dense lines following the external outlines of the spine in its early growth. In most of the upper spines, the lines look like round, bulb-like buttons that resemble secondary spines. This delicate structure is typical for original spines and do not appear in replacement spines. These basal dense lines do not mark growth cycles, they can already be seen in very small individuals that are less than one year old. Sometimes it is difficult to distinguish between these lines and the regular growth cycles, and in such cases it is impossible to determine accurately the number of cycles. This is especially problematic in small spines on the oral side.

b. When a spine grows rapidly, usually during the first cycles or during regeneration (Weber, 1969), the deposition of a growth band does not terminate "on time" and a new cycle begins before the preceding one is closed at the distal end of the spine. We counted these unfinished cycles as complete cycles.

c. The addition of a new cycle, that starts near the milled rings and proceeds to the tip of the spines, is a gradual one and because of this, many spines contain "unfinished" bands. In our study, such bands were counted as normal ones. Observations revealed that there is no synchronization in formation of new cycles, either within spines of a single sea-urchin or among various individuals of the same population. Moreover, no significant seasonal differences were found in the deposition of new cycles. Nevertheless, in most cases all the spines situated on the same level on the test contained the same number of growth lines. Rarely the deviation from the common number exceeded $\pm$ one line.

A high correlation was found between the diameter of the test and the number of growth cycles in the oldest spines of each individual (Fig. 1). Growth studies showed that the annual growth rate of medium sized sea-urchins (D=2-6 cm) is quite linear. Decline of growth rates was observed only in animals larger than 6 cm in diameter (Fig. 2). Comparison of the slope of test growth per increment of one growth cycle and the annual growth rate, reveals no difference (t test, $P > 0.5$). Although the growth rate of unmarked sea-urchins was studied only in a limited number of specimens, it seems that their growth rate is similar to that of marked animals.

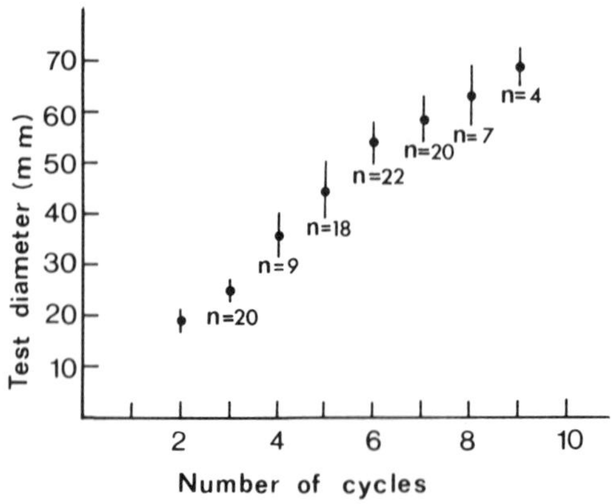

Fig. 1. Test diameter as a function of number of cycles in the spines of H. mammillatus (± standard deviation). For individuals with 3-6 cycles. Slope = 0.954±0.338

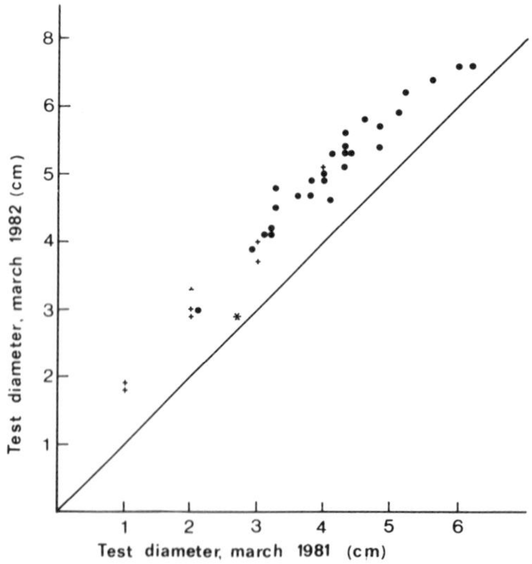

Fig. 2. Walford plot for growth of tests of H. mammillatus in a year. The diagonal line is zero growth. - marked individual, - marked individual infested with the parasitic gastropod Echineulima paluccia, -unmarked individuals (see text). For medium sized individuals (diameter=2-6 cm), annual growth = 0.981±0.205

4 DISCUSSION

Ebert (1967) used two facts to refute the hypothesis that regular and periodic deposition of growth bands occurrs in spines of Echinoids: 1. Spines always have growth lines on the outside. 2. In longitudinal sections the cycles are always distally terminated at a sharp discontinuity which suggested a break. This is not the case in the studied population of H. mammillatus. In this species many spines were found in the process of depositing a new cycle, usually without signs of a break. Heatfield's (1971) observation that addition of new bands in spines may take place at intervals in the absence of fracture, is supported in our study. The slow rate of regeneration, about 10 months compared with 2 months in Strongylocentrotus purpuratus (Heatfield 1971), explains why in our species this process cannot affect the periodic deposition of growth bands. The data on damaged spines show that the proportion of fractured spines in H. mammilatus is much smaller than in natural populations of S. purpuratus (Ebert 1968). This fact again emphasizes the minor influence of regeneration on the internal pattern of spine growth in Heterocentrotus.

Pearse and Pearse (1975), mentioned that two conditions are required to enable use of growth zones as "chronometers": 1. a correlation between differential deposition and seasonal changes. 2. the number of growth lines is independent of size. The second condition of Pearse and Pearse is fulfilled in this study. There is a considerable overlap among groups of sea-urchins that contain different numbers of lines. For example, the largest specimen found (D=8.2 cm) had 8 cycles, which is less than many much smaller individuals that have 9 or 10 cycles (Fig. 1). The first condition was not met in our study: We were not able to indicate the season and the environmental factors that induced new cycles. A similar problem was faced in many studies on growth bands in coronal plates (Markel 1975, Crapp and Willis 1975, Smith 1980). The evenness of the counts in each specimen suggest that, in spite of the vagueness concerning the formation of the growth lines, their deposition is periodic. Caution is certainly needed, but the results suggest that in the studied population, growth lines may be considered annual.

The morphological differences among primary spines of H. mammillatus are remarkable and may have important adaptive significance. The aboral spines that are long and pointed, function mainly for defence, while those from the ambitus orally, are used mainly for anchorage and movement. Spines on the oral side of the test are very short, thus aiding close attachment and preventing formation of space between the test and the substrate. The 4-5 spines above the peristome are usually of the same age (same number of lines) and it is evident that the difference in shape of the spines is caused by differential precipitation of material in accordance to the location of the spine. A different mechanism was found in cidarid spines that usually break near the base and regenerate in a different shape (Cutress 1965). The question of what controls the differential growth of spines, (factors of genetic or environmental nature) is still open.

Differences were found in the form of spines from different degrees of exposure to wave action, and thus the growth modulation of spines seems to be adapted to mechanical stress. Although some of the differences were not statistically significant, it seems that they show a real trend (e.g. Table 2). The different character of spines in different habitats indicates the plasticity of growth in Echinoid skeletal elements. Similar results were shown for the test (Dafni 1984, Regis 1981), the lantern (Ebert 1980, Black et al. 1982) and the spines (Cutress 1965, Regis 1981). Interestingly, the length of spines, which was reported to be sensitive to different environmental effects, was similar in all the studied habitats.

Assessment of number of repairs and regenerations of spines by examining longitudinal sections can be used as a convenient method for quantitative evaluation of environmental stresses on sea-urchins (Ebert 1968). The data on spine breaks is well correlated with intuitive scale of exposure to wave action. On the other hand, total loss of spines was not connected with spine breaks. Ebert (1968) stated that shedding of spines of S. purpuratus is influenced by strong wave action but in a earlier paper (1967) the same author reported that even when a spine is fractured near the milled ring, it regenerated a new shaft and did not fall off. In Cidaridae, shedding of spines was observed even without considerable damage (Markel and Roser 1983). Loss of spines may be induced by unfavorable physical conditions (revue by Cutress 1965) or by disease (Scheibling and Stephenson 1984). In our study these factors cannot explain the difference observed between the urchins in the forereef and the backreef at the same depth. Moreover, individuals of H. mammillatus living on the forereef receive better water circulation and are supposed to be less susceptible to problems of dessication, high water temperature etc. On the other hand, attempts of predation, mainly by trigger fish that usually attack the test, may be responsible for loss of spines, especially in the forereef where these fish are common.

5 ACKNOWLEDGEMENTS

The authors wish to acknowledge the Society for the Protection of Nature in Israel for the generous financial support and for providing space and facilities at the No'ama Field Studies Center.

REFERENCES

Black, R., Johnson, M.S. & Trendall, J.T. 1982. Relative size of Aristotle's lantern in Echinometra mathaei occurring in different densities. Mar. Biol. 71(1):101-106.

Crapp, C.B. & Willis, M.E. 1975. Age determination in the sea urchin Paracentrotus lividus with notes on the reproductive cycle. J. exp. mar. biol. ecol. 20:157-178.

Cutress, P.M. 1965. Observations on growth in Eucidaris tribuloides (Lamark) with special reference to the origin of the oral primary spines. Bull. mar. sci. 15(4):797-834.

Dafni, J. 1984. Growth and calcification in the sea urchin Tripneustes gratilla at Eilat. Ph.D. Thesis, Hebrew University of Jerusalem. In Hebrew with English summary.

Deutler, F. 1926. Uber das Wachstum des Seeigelskeletts. Zool. Jb. (Anat.) 48:119-200.

Dixon, W.J. & Brown, M.B. 1982. Biomedical computer programs, Univ. of California Press, Berkeley, California. U.S.A.

Ebert, T.A. 1967. Growth and repair of spines in the sea urchin Strongylocentrotus purpuratus (Stimpson). Biol. Bull. 11:141-149.

Ebert, T.A. 1968. Growth rates of the sea urchin Strongylocentrotus purpuratus related to food availability and spine adrasion. Ecology 49:1075-1091.

Ebert, T.A. 1980. Relative growth of sea

urchin jaw: an example of plastic resource allocation. Bull. mar. sci. 30:467-474.
Heatfield, B.M. 1971. Growth of the calcareous skeleton during regeneration of spines of the sea urchin Strongylocentrotus purpuratus (Stimpson): A light and scanning elecron microscope study. J. Morph. 134:57-89.
Hyman, L.H. 1955. The invertebrates. Vol. IV. Echinodermata. N.Y. Mcgraw-Hill.
Markel, K. 1975. Wachstum des Coronarskeletes von Paracentrotus lividus. Zoomorphologie 82:259-280.
Markel, K. & Roser, U. 1983. Calcite resurption in the spine of the Echinoid Eucidaris tribuloides. Zoomorphology 103:43-58.
Moore, G.P. 1966. The use of trabecular bands as growth indicators in spines of the sea urchin Heliocidaris erythrogramma. Aust. J. Sci. 29:52-54.
Mortensen, T. 1928. A monograph of the echinoidea. Vol. I. Cidaroida. Copenhagen: C.A. Reitzel.
Pearse, J.S. & Pearse, V.B. 1975. Growth zones in Echinoid skeleton. Am. Zool. 15:731-753.
Regis, M.B. 1981. Aspects morphometriques de la croissance de deux Echinoids du Golfe de Marseilles, Paracentrotus lividus (LMK) et Arabacia lixula (L). Cah. Biol. Mar. 22: 349-370.
Scheibling, R.E. & Stephenson, R.L. 1984. Mass mortality of Strongylocentrotus Droebachiensis off Nova Scotia, Canada. Mar. Biol. 78(2):153-164.
Smith, A.B. 1980. Stereom microstructure of the Echinoid test. Special papers in Paleonthology No. 25.
Vevers, H.G. 1966. Pigments in: R.A. Boolootian (ed.), Physiology of Echinodermata p. 73-85. N.Y. Interscience Publishers.
Weber, S.N. 1969. Origin of concentric banding in the spines of the tropical echinoid Heterocentrotus. Pac. Sci. 23:452-466.

Proceedings of the Fifth International Echinoderm Conference / Galway / 24-29 September 1984

The non-periodic nature of growth rings in echinoid spines

THOMAS A.EBERT
Department of Biology, San Diego State University, Calif., USA

ABSTRACT: Evidence is reviewed on growth rings in echinoid spines as indicators of age and the hypothesis is rejected that rings are periodic. Studies on Heterocentrotus mammillatus in Hawaii are consistent with the hypothesis that some if not all growth lines are formed in response to trauma.

1 INTRODUCTION

Internal spine morphology is a record of past events in a sea urchin's life, which may have included traumatic assaults (Ebert 1967) or, as has been proposed by others, records of periodic events (Moore 1966, Weber 1969). If growth-lines record periodic events they can be used to determine an animal's age and parameters of growth and survivorship functions.

The purpose of this paper is to review evidence concerning the nature of growth lines in echinoid spines to show that they can not be periodic and, secondly, to present new information on growth lines in spines of the slate-pencil sea urchin Heterocentrotus mammillatus to show that some, if not all, growth rings are a consequence of trauma.

2 REVIEW OF GROWTH LINES IN SPINES

Individual growth rates in sea urchins decrease with age and are very slow in old individuals. As a first approximation, sea urchin growth can be modeled using the Brody-Bertalanffy equation (Ebert 1975) although an additional shape parameter is needed to model more accurately the fact that some, and possibly all, sea urchins have an initial, rapid growth phase that is followed by very slow growth for a long period of time (Ebert 1982). If growth rings are indicators of age then the relationship between size and number of lines must have a shape like the Brody-Bertalanffy curve or the Richards function with a fractional negative shape-parameter (Ebert 1982).

Three published studies have included test size vs. growth-line data from spines: Heliocidaris erythrogramma in New South Wales, Australia (Moore 1966); Strongylocentrotus purpuratus in Oregon, USA (Ebert 1968); and Heterocentrotus trigonarius in Tahiti, French Polynesia (Weber 1969). The data in these papers together with unpublished data on H. mammillatus in Hawaii can be used to test whether the relationship between size and numbers of growth rings follows a reasonable growth curve.

2.1 Heliocidaris erythrogramma

Moore (1966) presents a double-logarithmic plot of volume (mm^3) vs. the average number of rings in nine primary spines for seventy animals that were collected on the east coast of New South Wales. Volume was estimated from the relationship:

$$V = 4/3\pi ab^2 \qquad 1)$$

where a is one half the diameter and b is one half of the height. This is a formula for estimating the volume of a prolate spheroid. H. erythrogramma is better described as an oblate spheroid so the published formula in Moore's paper may be a typographical error; however, I can not be sure and so I have used equation 1 to calculate volume for a sample (N = 358) of H. erythrogramma from Port Jackson, New South Wales. Parameters were estimated for a functional regression (Ricker 1975) between the logarithms of diameter (D) and volume (V):

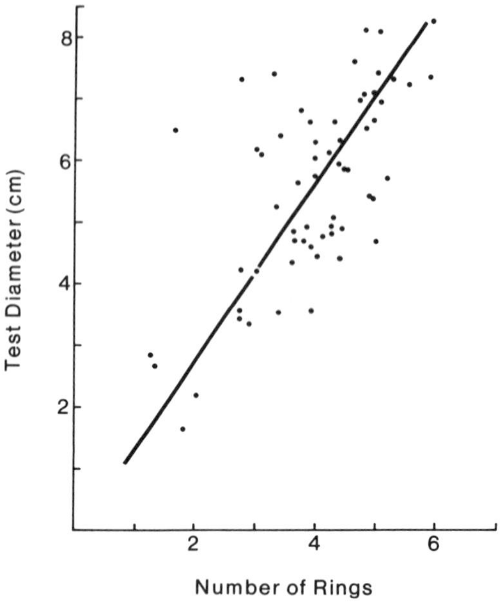

Fig. 1: Test diameter (D) as a function of average number of rings (R) in Heliocidaris erythrogramma (NSW, Australia); N = 61; replotted from Moore (1966); D = 1.40R.

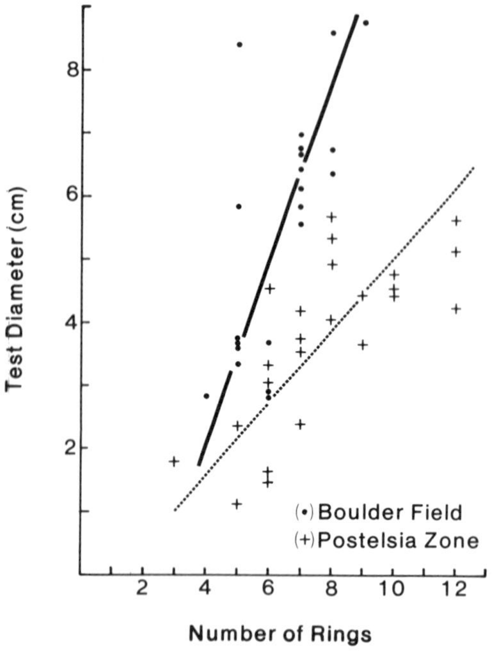

Fig. 2: Strongylocentrotus purpuratus, Oregon, USA (Ebert 1968); Postelsia Zone (N = 24): D = 0.57R - 0.70; Boulder field (N = 21): D = 1.48R - 3.94.

$$D = 0.250V^{0.310} \qquad 2)$$

Data values were extracted from Moore's Fig. 3 (N = 61 rather than 70: nine points must be coincident in his graph). Using equation 2, volumes were changed to test diameters and then plotted as a function of average number of rings in spines (Fig. 1). There is a fair degree of scatter but it is clear that large individuals have more growth lines in their spines than do small individuals. The relationship, however, is linear rather than a decreasing exponential growth curve.

2.2 Strongylocentrotus purpuratus

I have gone back to my original data file to obtain maximum number of growth rings in spines of a sample gathered in August 1964 on the south side of Sunset Bay, Oregon. Average number of rings in 2-5 primary spines was used previously (Ebert 1968). Individuals from two areas show different slopes for the linear regressions of test diameter vs. number of rings (Fig. 2). Animals from the Postelsia Zone have more rings at a particular size than do animals collected in the Boulder Field. Animals in the Boulder Field grow faster and achieve a greater maximum size than do animals in the Postelsia Zone so that number of rings vs. age is similar for both areas (Ebert 1968). As with H. erythrogramma, the relationship between diameter and number of rings is approximately linear and not as would be expected based on tagged individuals of this species (Ebert 1968, 1977; Russell 1984).

2.3 Heterocentrotus trigonarius

Weber (1966) made longitudinal sections of 70 spines per individual for 15 H. trigonarius from Teahupoo, Tahiti. He found a straight-line relationship when he plotted the logarithm of test length vs. the maximum number of rings. I have replotted his original data (Fig. 3). The curve bends upward and so is opposite to the direction expected for an individual growth curve.

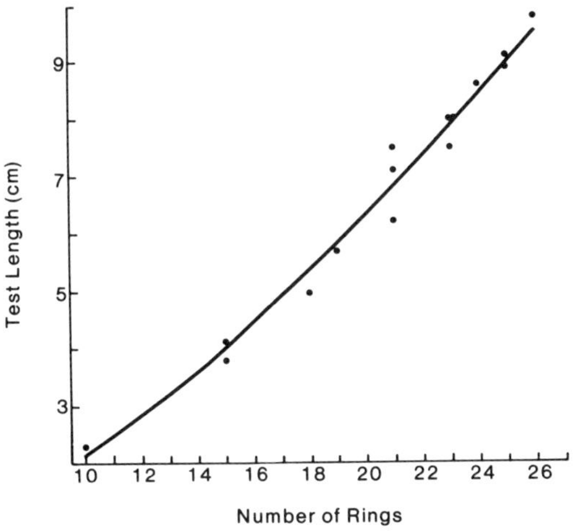

Fig. 3: Heterocentrotus trigonarius, Tahiti, French Polynesia (Weber 1969); each dot is the maximum number of rings in 70 spines; N = 15; D = 1.006exp(0.089R).

2.4 Heterocentrotus mammillatus

Growth of the slate-pencil sea urchin Heterocentrotus mammillatus has been studied at Honaunau Bay on the Kona (leeward) coast of the island of Hawaii and tetracycline tagging was used to determine the growth increment in half-pyramids of Aristotle's lantern (Ebert 1982). These data (Fig. 4) can be used to determine the parameters of a growth curve that relates size to age and can be combined with information on the relationship between half-pyramid length and test length to produce a growth equation that describes test length as a function of age. This relationship for H. mammillatus, is:

$$L_t = 6.768(1 - e^{-0.316t})^{-0.217} \qquad 3)$$

The estimated asymptotic test length is 6.768cm, the growth-rate constant is $0.0316 yr^{-1}$, time (t) is in years, and -0.217 is the shape parameter of the Richards function (Ebert 1982).

The growth curve for H. mammillatus is presented (Fig. 5) together with data on a spine sample from H. mammillatus gathered in August 1976 at Honaunau Bay. As with the data in Figs. 1-3, the relationship between size and ring number does not have the shape of a growth curve. Young individuals have too many growth rings and old individuals have too few. It makes no difference whether average or maximum ring number is used.

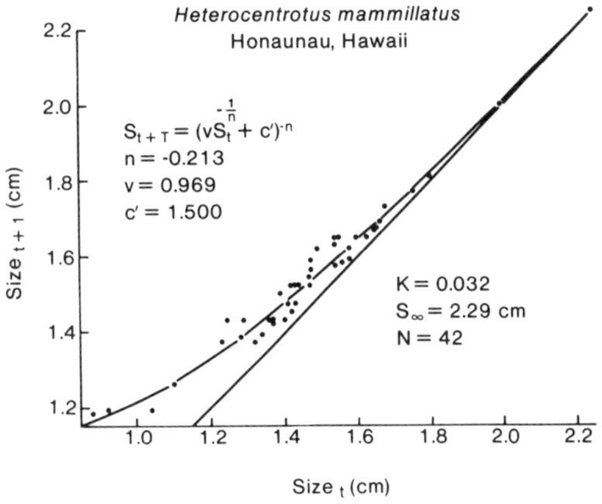

Fig. 4: Growth of half-pyramids of Aristotle's lantern over one year determined by tetracycline marking (Ebert 1982).

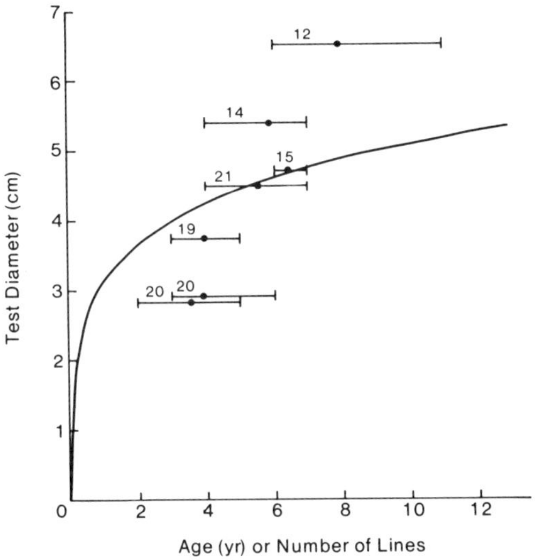

Fig. 5: Growth curve for Heterocentrotus mammillatus (curved line) based on tagged animals compared with number of growth lines in spines (bars); number on each bar is the number of spines examined; dot is mean number and bar is range.

Based on data from the literature and the H. mammillatus spines that were sectioned, I reject the hypothesis that growth lines in the spines are a periodic phenomena and find no reason to believe that these lines can be used to determine age or growth rates of sea urchins.

3 CURRENT WORK ON HETEROCENTROTUS MAMMILLATUS

One hundred H. mammillatus were collected in shallow water (1.5-4m) at Honaunau Bay during November 4-6, 1983. Fourteen spines on each animal were subjected to a variety of treatments to gain insight into the dynamics of calcite deposition, growth line formation, and spine repair. Treatments to primary interambulacral spines were assigned using random-numbers and a map was used to locate spine positions.

The seven treatments, replicated on the two halves of the test, included tagging, total removal, cutting, and drilling small holes. One half of the spines with holes were sanded to simulate severe storm damage and equal numbers of sanded and unsanded spines were tagged with tetracycline. After treatment, sea urchins were returned to a single drop site at a depth of about 3m, which is the same area where animals were studied in 1975-76 (Ebert 1982).

In this paper I will only detail results based on cutting and tetracycline marking; therefore only three treatments will be described.

3.1 Tagging

One spine was tagged on each side of an animal in anticipation of some spine loss during the course of the study. Holes (1.3mm) were drilled using a bit attached to a flexible shaft on a battery powered drill. An attempt was made to make the holes 2cm or closer to the bases of the spines so tags would be protected by other spines and also not likely to catch on coral or rock edges. Nylon monofilament (0.02mm diameter) was threaded through a hole and glass beads added to the line to form a color code. A small piece (ca. 2mm) of spaghetti tubing was added at the end of each set of beads. A square knot was tied in the monofilament to form a small loop close to the spine, the line was threaded back through the spaghetti tubing, and the knot was pulled inside the tubing to protect it and prevent it from untying.

3.2 Partial removal

Spines were sawed-off about 1cm above the milled ring using a carborundum-edged, sabre-saw blade held in file handle with adjustable jaws. Spines were sawed through over half way and then the blade was twisted to fracture the spine. Tips were saved and labeled.

3.3 Tetracycline tagging

A hole (1.3mm) was drilled starting at about mid-shaft. After starting, the drill was tipped so the drill penetrated at a steep angle. Depending on the size of the spine, the holes ranged from 0.5 to 1.5cm deep and with few exceptions, when they went entirely through, were blind holes. I used hydrogenated vegetable oil (cooking shortening) as a carrier for the tetracycline because tetracycline is water soluble and so should remain active longer in an oil carrier than if applied in a water soluble material. One gram of tetracycline-HCl was mixed with 20gm of hydrogenated vegetable oil, the holes were filled with this mixture, and were plugged with bleached spines of Echinometra mathaei.

Seventeen animals that had been tagged in November were collected during March 5-7, 1984, which left tagged individuals in the field for future collection. Additional animals were tagged for collection in June and November.

Using the maps, spines were removed from the sea urchins and, for those spines that had been cut, were paired with their original tips. All spines were treated with 5% sodium hypochlorite to remove organic matter, soaked in fresh water for 24h, and then air-dried. Spines were sectioned by grinding, first on 100 grit and then 150 grit sandpaper, and then were finished with 600 grit carborundum powder on a glass plate with water as the carrier. The spines were cleaned in an ultrasonic cleaner with tap water as the cleaning medium. A UVSL-25 Mineralight^{R} (Ultra-violet Products, Inc., San Grabriel, California) was used to view the tetracycline marks.

4 RESULTS

Spines that were cut-off regenerated new tips during a four month period. Repair was not complete replacement of the lost calcite mass but on average was about 20% of the mass removed. The significant feature of repair is that a new growth line usually was formed during the regeneration process. Fig. 6 is a longitudinal section of a spine that was cut and permitted to regenerate for four months. The regenerated portion is thicker than the original spine shaft and during tip regeneration calcite covered the outside layer of calcite, which now is evident as a "growth line" (Fig. 7). For the particular spine shown in Fig. 6, the new calcite does not extend all the way to the milled-ring and

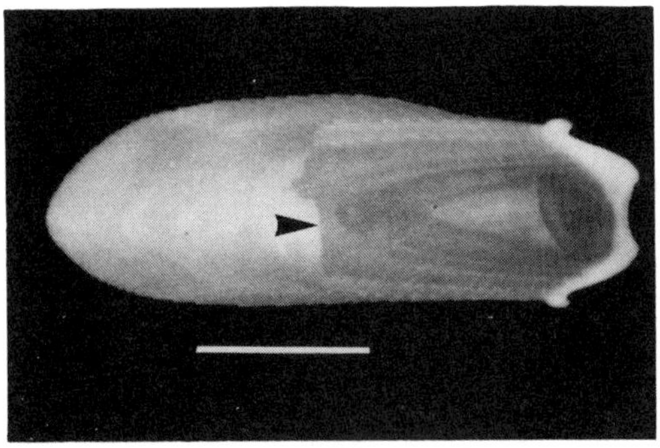

Fig. 6: Midsection of H. mammillatus spine, cut (indicated by arrow) and regenerated for 4 months; scale = 4mm.

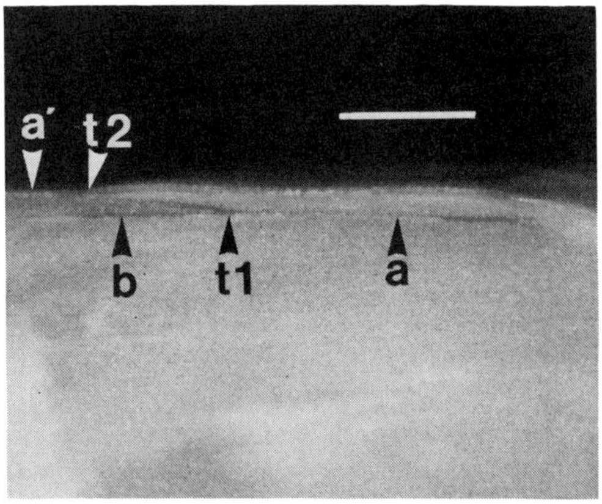

Fig. 8: Longitudinal section of H. mammillatus spine; growth line "a", which is continuous distally with the spine surface at "a'", was added during repair of a hole; growth line "b" ends distally (to the left) at a former break; t1 and t2 are interpreted as extent of trauma to spine surface following damage; scale = 1mm.

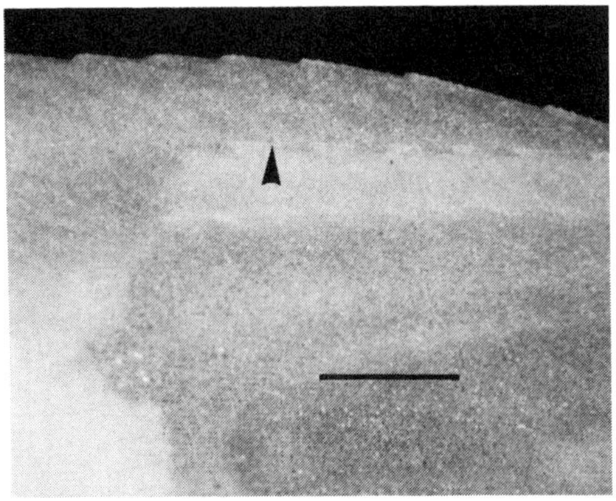

Fig. 7: Detail of Fig. 6 showing "growth line" (arrow) formed during tip regeneration; scale = 1mm.

so only a partial growth line was formed. The formation of a growth line in response to tip damage has been previously shown for Strongylocentrotus purpuratus (Ebert 1967) and is clear in the illustrations given by Weber (1969).

Removal of a spine tip is not needed to initiate at least partial ring formation. Fig. 8 shows the area distal to a hole drilled in a spine, but not tagged with tetracycline, that had regenerated for four months. The growth line labeled "a" was the surface of the spine when the hole was drilled. Regenerated calcite, appearing light gray in the photograph, can be seen sloping into the hole at the right.

Growth line "a" can be followed distally, that is to the left of the figure, and can be seen to bifurcate with one branch leading to the surface of the spine (a') and the other branch remaining inside the spine (b) and terminating at a regenerated break. My interpretation of these lines is that some time in the past several years the spine was broken, which traumatized the spine below the break down to t1. When repair of the tip took place, the surface between the break and t1 was covered with new calcite and so a short segment of new "growth line" was formed. When I drilled a hole in the spine in November 1983, I traumatized the spine from the hole to t2 and as part of damage repair, the surface between the hole and t2 was covered with new calcite.

Tagging spines with tetracycline generally was not successful. Most spines that were treated were shed during a four month period and of the few spines that were recovered, most showed no tetracycline lines possibly because they were not growing when treated. There was, however, one spine that did show a tetracycline mark and the results suggest that lines can form either in response to more subtle trauma or some other causes. Fig. 9 shows the pattern of flourescence at the base of the spine. The internal line is obvious and can be used to determine position in Fig. 10, which is the same spine but with normal illumination to show "growth lines." The tetracycline was incorporated at the spine surface and a growth line was added all the way to the base. During the period from November 1983 to March 1984, another line was added, which appears

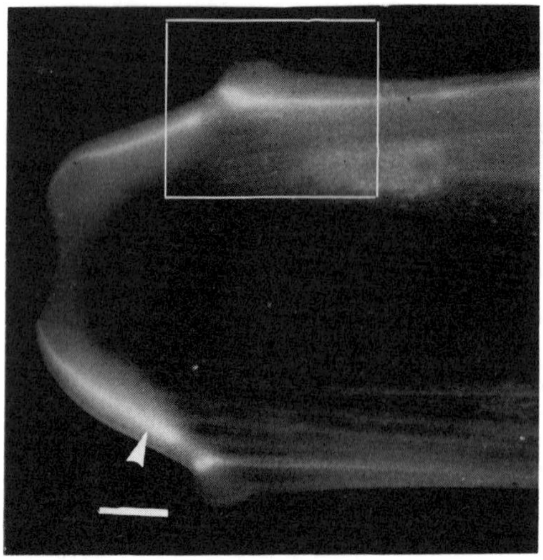

Fig. 9: Midsection of spine tagged with tetracycline in Nov. 1983 and collected in March 1984; glowing line (arrow) is position of spine surface at time of tagging; white rectangle delimits area shown in Fig. 10; scale = 1mm, photo taken with ultraviolet illunimation.

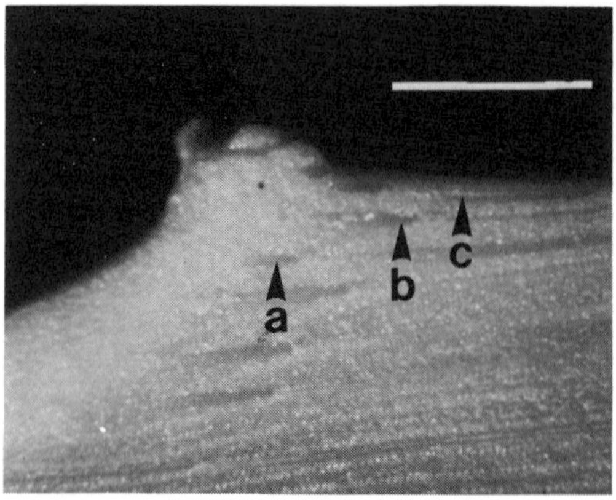

Fig. 10: Detail of spine in Fig. 9 with normal illumination; "a" is a growth line at the position of the tetracycline line; "b" is a second line added during 4 month period; "c" is the external surface of spine; scale = 1mm.

between the line marked with tetracycline and the surface. What caused this growth ring to form is unknown.

5 DISCUSSION

Growth rings in echinoid spines can not be periodic because of the non-correspondence between these rings and age in growh curves determined from tagged individuals or size distributions. The current work on H. mammillatus agrees with the studies on S. purpuratus (Ebert 1967) and shows that lines form in response to trauma, which can be spine breakage but also can be less severe such as a hole drilled in the spine.

My tentative explanation for the linear relationship between test diameter or length and number of growth lines is that small spines are more likely to be damaged than large spines. The susceptibility to damage would have to be determined by some linear function of spine size such as diameter or length and possibly linear position on the test surface.

The additional ring that formed in the spine tagged with tetracycline is very interesting because it was added during the period that the animal was in the field and so probably not in response to any treatment I applied. I doubt that this is a periodic line because no other spines I have as yet examined from this sample show an additional line between the trauma induced line and the outside line; Figs. 6-8 show no such lines yet they were from the same sample as the spine shown in Figs. 9-10.

Although it is unlikely that the growth rings can be used to determine age, they nevertheless are potentially useful in evaluating environmental differences among samples (Ebert 1968). Costs of repair clearly are included in what is considered "maintenance costs" for sea urchins (Ebert 1982).

6 AKNOWLEDGMENTS

This work was supported by a fund from Interpore International, Irvine, CA, and by the US National Science Foundation. J. Harrison assisted with field work in Hawaii and advice from H. Plymale substantially helped in selecting the carrier for tetracycline.

7 REFERENCES

Ebert, T.A. 1967. Growth and repair of spines in the sea urchin Strongylocentrotus purpuratus (Stimpson). Biol. Bull. 133:141-149.

Ebert, T.A. 1968. Growth rates of the sea urchin Strongylocentrotus purpuratus related to food availability and spine abrasion. Ecology 49:1075-1091.

Ebert, T.A. 1975. Growth and mortality in postlarval echinoids. Amer. Zool. 15:755-775.
Ebert, T.A. 1977. An experimental analysis of sea urchin dynamics and community interactions on a rock jetty in California, U.S.A. J. Exp. Mar. Biol. Ecol. 27:1-22.
Ebert, T.A. 1982. Longevity, life history, and relative body wall size in sea urchins. Ecol. Monog. 52:353-394.
Moore, G.P. 1966. The use of trabecular bands as growth indicators in spines of the sea urchin, Heliocidaris erythrogramma. Aust. J. Sci. 29:52-54.
Ricker, W.E. 1975. Computation and interpretation of biological statistics of fish populations. Bull. 191. Fish. Res. Bd. Canada.
Russell, M.P. 1984. Life history traits and resource allocation in the purple sea urchin Strongylocentrotus purpuratus. MS Thesis, San Diego State University.
Weber, J.N. 1969. Origin of concentric banding in the spines of the tropical echinoid Heterocentrotus. Pac. Sci. 23:452-466.

Proceedings of the Fifth International Echinoderm Conference / Galway / 24-29 September 1984

The African sand dollar *Rotula*

JOE GHIOLD
Department of Geology, Louisiana State University, Baton Rouge, USA

ABSTRACT: Rotula orbiculus, a multi-lunate sand dollar, was observed for the first time along near-by islands of Sierra Leone, W. Africa. Rotula is not a suspension-feeder, rather it feeds upon bottom deposits as do most clypeasteroids. A biometric analysis showed that the animal's perimeter is growing faster than the test length as a whole; posterior notches (incomplete lunules) are rarely found in juveniles. The highly "convoluted" posterior margin may serve as sites for food collecting in older individuals.

1 INTRODUCTION

Bizarre life forms have attracted many scientists, if not for their oddity, then for their rare occurrence. The sand dollar Rotula is no exception. Echinologists have pondered about this animal's novel body shape; indeed, the infolding of the posterior test margin is absent in other sand dollars. Functional interpretations of this unique morphological pathway abound in the literature, yet few are based on observations of living specimens.

This paper describes the first observations of living Rotula from the west coast of Africa and explores several aspects of its biology.

Dartevelle (1934, 1940, 1953) gave the most comprehensive treatment of this echinoid, including fossil and extant forms. His work covered systematics, ecology, and proposed a filter-feeding habit by which the posterior notches protruded into the water column in order to sieve passing sea currents. This stationary feeding mode has been mentioned by other workers (Mortensen 1948; Timko 1976; O'Neill 1978; Seilacher 1979). General systematics of the Rotulinae together with descriptions of their external features and evolution were discussed by Koehler (1914), Ikeda (1939), Mortensen (1948), Dartevelle (1953), Durham (1955), Tortoneese (1957) and Seilacher (1979). Meijer and Jangoux (1975) and Seilacher (1979) performed biometric analysis on the test while Sourie (1954), Clark (1955), Collignon (1957), and Longhurst (1958) were concerned with distribution and occurrence.

2 MATERIAL AND METHODS

Rotula orbiculus was collected by dredging off the NW shores of Mut Island, a part of the Turtle Islands, which is located near Sierra Leone, W. Africa (Figure 1). Living populations were observed by skin diving approximately 0.5 km off Mut Island's oceanside. Water characteristics and sea bottom particulars were recorded and samples of host sediment were collected for laboratory analysis.

About 1000 living sand dollars were collected; they were immersed in large buckets of sea water and returned to the laboratory. Echinoids were also dredged in the deep channel of the Sierra Leone River Estuary; individuals were found less frequently there.

2.1 Grain size analysis

Sediment samples were air-dried and mechanically sieved at the University of Tübingen, W. Germany. Dry weight per cent of each fraction was determined and presented in histogram form. Organic content was determined with the LECO carbon analyzer. Sediment composition was examined with the light microscope.

2.2 Laboratory studies

Organisms were placed in large, shallow plastic tubes with sea water and aerated by small piston pumps; sea water was

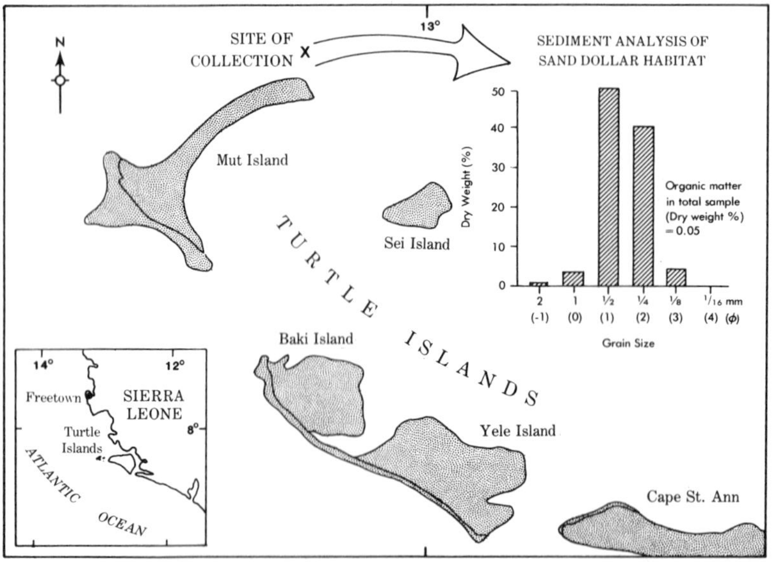

Figure 1. Turtle Islands' area showing location of study site, and histogram showing typical grain size of the host sediment.

changed twice daily. A boom-arm dissecting microscope provided a close-up view of the echinoid's activities. Mirrors, connected to the microscope, allowed a clear view of the oral side of the test. Analysis of gut contents was performed according to Timko (1976). External features were observed with light and scanning electron microscopes.

2.3 Growth analysis

Meijer and Jangoux (1975) and Seilacher (1979) described the general biometrics of Rotula though neither examined the ontogenic growth of the test perimeter. Therefore, a growth analysis was conducted here on a representative population of R. orbiculus to evaluate whether the test margin grows disproportionately to the body as a whole. The size range of the population sampled was 2.5 - 130 mm, and the population totaled 211 individuals. Specimens were photographed at equal distances with a standard SLR camera, and the 35 mm negatives were enlarged and printed to actual size. Using a cursor from a Zeiss Videoplaner, the perimeter (y-axis) of the test was traced together with the test diameter (measured along the anterior - posterior axis) (x-axis) and plotted by computer (Ghiold & Alexander, 1980). A line of isometry was (dashed line) hand-drawn through point "A" and represents the hypothetical case of a population of Rotula that grows perfectly circular, i.e. without infolding of the posterior test margin.

3 RESULTS

3.1 Local setting

Along the coastline of Sierra Leone many rivers discharge into the ocean creating hydraulic conditions that favor the deposition of silt and mud. R. orbiculus is absent intertidally in Sierra Leone. This absence is probably attributable to turbid waters or local deposition of fine sediment. Mut Island located far away from any terrestrial input is surrounded on its oceanside by an extensive sand bank which narrows considerably toward its leeward side. A rippled, bare sand bottom dominates; showing strong tidal currents. At high tide the bank is covered by 2 m of water. Sediments consisted of clean quartz with grain size ranging 2-3ϕ and an organic content of 0.5% (Figure 1).
Sand dollars were lying flat and usually buried under the sediment. At no time, low or high tide - night or day, were sand dollars lying upright with their anterior margin anchored into the sand (c.f. O'Neill 1978, Seilacher 1979). Nor were organisms

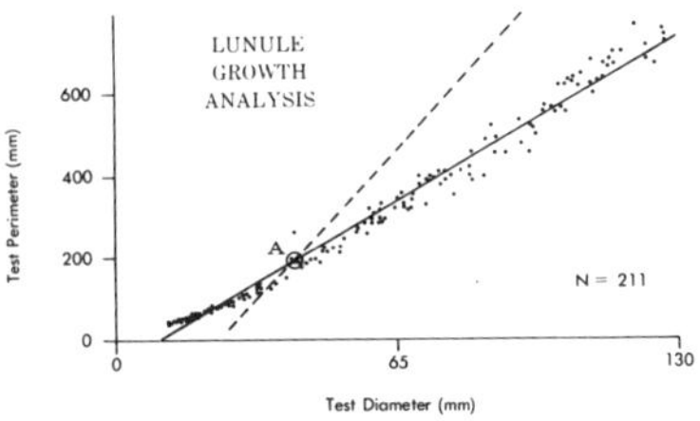

Figure 2. Length of perimeter measured against test length. Solid line is least - squares linear regression line; dashed line is a line of isometry (slope= 1.0) through the point "A", estimating that part of the population with a notched test posterior. N= sample size.

(e.g. barnacles) found attached to live Rotula (c.f. Philippe 1983).

3.2 External appendages

Rotula possesses the major feeding and burrowing spines found in most scutellid sand dollars, i.e. shoe, miliary, test - margin, locomotory and aperture - margin (Ghiold 1984). Shoe spines are bent distally and are highly serrated along the club's edge. Miliary spines lack transparent sac-like covers and are ubiquitous on the test including the food grooves.
Pedicellariae are the least known structures in Rotula. Mortensen (1948) noted one type, bidentate; in addition I observed tridentate and triphylous types. The latter kind have no flexible stalk and are highly concentrated along the posterior notches.
Podia distribution is most striking in Rotula. Podia align in radial, alternating bands, between suture lines of the oral surface. Some extend to the tops of the posterior lunules. They are abundant aborally but not to the extent adorally. Podia within the food grooves were thicker, less tapered and formed a wider terminal knob than other podia.
Ciliary currents flow as in most scutellids except at the posterior notches. Here aboral currents follow the suture lines such that debris is channeled towards the anterior lips of the lunule and never over the "lobes" or "fingers". Thus the food supply route is considerably shortened (Alexander & Ghiold 1980; Smith and Ghiold 1982).

3.3 Growth analysis

The population measured for perimeter growth exhibited an allometric trend for the echinoid's overall test margin (Figure 2). The allometric relationship suggests that the animal's perimeter grows more rapidly than does the body during ontogeny. (Alexander & Ghiold 1980).

3.4 Feeding

Rotula collects and transports food matter similar to most deposit-feeding sand dollars (e.g. mellitids) utilizing both test surfaces (Ghiold 1984). The aboral canopy is not well developed but is capable of grain sieving. Podia share a major role in particle transport and in the gathering of food, especially at the ambitus. The disproportionate lengthening of the posterior notches during the later stages of ontogeny probably increases the periphery's food gathering potential while expediting the transport of food from the aboral surface to the mouth (Smith and Ghiold 1982). Gut contents include diatoms, algal fragments, sand grains, echinoderm spine fragments, sand grains and detrital material.

4 DISCUSSION

Rotula is clearly not a suspension-feeder like Dendraster and direct field observations here support this. There are two important morphologic reasons that confirm Rotula does not feed upright: 1) The petalloids do not exhibit eccentricity in Rotula. For suspension-feeding sand dollars (i.e. Dendraster) eccentricity increases the potential for respiration even while partially buried. 2) Food grooves in Dendraster run short of the anterior margin orally as this portion is normally buried while the animal is feeding upright, supposedly reflecting the grooves' reduced role in feeding near the anterior margins (Figure 3A). Rotula, however, has fully developed anterior food grooves (Figure 3B). One would expect Rotula to show similar food groove patterns anteriorly if it was indeed feeding like Dendraster.
Rotula's alleged upright feeding behavior probably arose by misquotes and misunderstandings from the literature. Dartevelle (1934, 1940) never observed Rotula alive because of the high turbidity in Africa's coastal estuaries. But O'Neill (1978) misread Dartevelle and thus compared Dendraster's mode of feeding to that of Rotula's.

FOOD GROOVE DISTRIBUTION

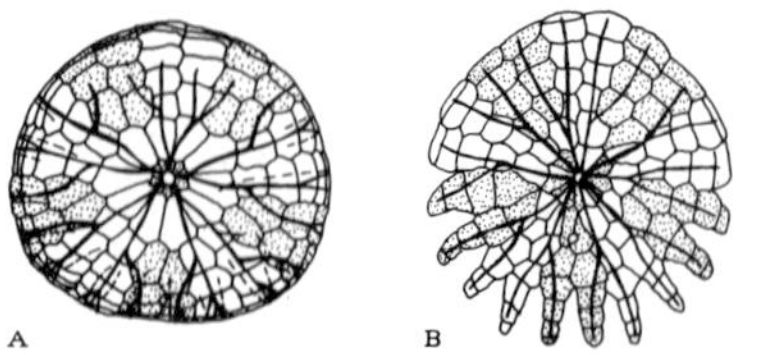

Figure 3. Oral surface of (A) Dendraster excentricus and (B) Rotula orbiculus showing food groove patterns. Organisms about 1/3 actual size.

Philippe (1983) made the same oversight in his interpretation of the fossil echinoid Scutella. Seilacher (1979) never mentions the work of Dartevelle and misread Mortensen's statements. Mortensen (1948: 458) reported the findings of Dartevelle but questioned his conclusions on Rotula's suspension feeding role.

Rotula's uniquely shaped, posterior margin can be explained if a feeding role is invoked (Alexander & Ghiold 1980; Smith & Ghiold 1982; Ghiold 1984). Also, closed anterior lunules in Rotula agusti are necessary because open lunules (notches) at the anterior margin would hinder burrowing (Ghiold 1979; 1984).

Though no populations of Rotula were found on mud, such mud communities have been reported (Dartevelle 1940; Longhurst 1958). It would be interesting to observe how such sand dollars adapt to mud. It would be equally valuable to examine populations of estuary dwelling Rotula and compare their physiology to more stenohaline sand dollars. These and other future studies of the African sand dollar may prove fruitful.

5 ACKNOWLEDGEMENTS

I gratefully acknowledge Mr. & Mrs. U. Beck and family for their kind hospitality and logistical assistance during my stay in Sierra Leone, W. Africa. Thanks are also extended to the crew of the "Tenny" and Sierra Leone University for use of facilities. Special thanks are due to Prof. Adolf Seilacher for his encouragement during my tenure as a SFB-53 Research Fellow at Tübingen. This study was financially supported by the German Science Foundation under the special program SFB-53.

6 REFERENCES

Alexander, D. E. & J. Ghiold. 1980. The functional significance of the lunules in the sand dollar Mellita quinquiesperforata. Biol. Bull. 159: 561-570.

Clark, A. M. 1955. Echinodermata of the Gold Coast. J. West afr. Sci. Assoc. 1: 16-56.

Collignon, J. 1957. Prélimaires à une étude écologiques de la baie de Pointe-Noire. Bull. Inst. Etudes Centrafr. 13-14: 29-37.

Dartevelle, E. 1934. Les Rotulinae des collections du Musée du Congo. Bull. Cercle Zool, Congol. 11: 105-109.

Dartevelle, E. 1940. Les rotules de la côte occidentale d'Afrique. Bull. Inst. roy. colon. belge 11: 175-198.

Dartevelle, E. 1953. Echinides fossiles du Congo et de l'Angola. II. Description systématique des échinides fossiles du Congo et de l'Angola. Ann. Mus. roy. Congo belge 13: 1-240.

Durham, J. W. 1955. Classification of clypeasteroid echinoids. Univ. Calif. Publ. geol. Sci. 31: 73-198.

Ghiold, J. 1979. Spine morphology and its significance in feeding and burrowing in the sand dollar, Mellita quinquiesperforata (Echinodermata: Echinoidea). Bull. Mar. Sc. 29: 481-490.

Ghiold, J. 1984. Adaptive shifts in clypeasteroid evolution: Feeding strategies in the soft bottom realm. N. Jb. Geol. Palaont. Abh. (in press).

Ikeda, H. 1939. Studies on the pseudo-faciole of the scutellids (Echinoidea, Scutellidae). J. Dept. Agric. Kyushyu imp. Univ. 6: 41-93.

Kohler, R. 1914. Echinoderma. I. Asteroidea, Ophiuroidea et Echinoidea. Beit. Kenntn. Meeresfauna West afrikas. 1: 127-309.

Longhurst, A. R. 1958. An ecological survey of the West African marine benthos. Colon. off. fish. Publ., No. 11: 1-102.

Meijer L. & M. Jangoux 1975. Observations biométriques sur Radiorotula orbiculus (Linné) (Echinodermata, Echinoidea, Rotulidae). Rev. Zool. afr. 89: 621-628.

Mortensen, T. 1948. A monograph of the Echinoidea. IV.2. Clypeasteroida Copenhagen: C. A. Reitzel.

O'Neill, P. L. 1978. Hydrodynamic analysis of feeding in sand dollars. Oecologia 34: 157-174.

Phillippe, M. 1983. Déformation d'une Scutelle (Echinoidea, Clypeastroida) Miocène due à la fixation d'une balane: Hypothèse Paléoécologique. Geobios 16: 371-374.

Seilacher, A. 1979. Constructional morphology of sand dollars. Paleobiology 5: 191-221.
Smith, A. B. & J. Ghiold 1982. Roles for holes in sand dollars (Echinoidea): a review of lunile function and evolution. Paleobiology 8: 242-253.
Sourie, R. 1954. Estude Ecologique Sommaire des Fonds Sableux en Baie de Dakar. Ann. l'Ecole Sup. Sci., Dakar. 1: 1-15.
Timko, P. L. 1976. Sand dollars as suspension feeders: a new description of feeding in Dendraster excentricus. Biol. Bull. 151-247-259.
Tortoneese, E. 1957. Introno al clipeastroidi del gen. Rotula. Doriana 2(74): 1-4.

Proceedings of the Fifth International Echinoderm Conference / Galway / 24-29 September 1984

Population dynamics of *Echinocardium cordatum* (Pennant) in the bay of Douarnenez (Brittany)

JACQUES GUILLOU
Université de Bretagne Occidentale, Laboratoire d'Océanographie Biologique, Brest, France

ABSTRACT: An investigation was carried out on a subtidal population of Echinocardium cordatum, over a period of three years, across a bathymetric gradient, from the clean fine sands of the shallower area to the muddy sands of the bay of Douarnenez. Periods of spawning and recruitment are defined, the breeding season being in summer. Large variations in the densities of the recruited juveniles occurred from one year to another. From the length frequency histograms, plurimodal year classes were shown and the life-span of this population was determined. A growth curve is plotted showing a rapid growth process in summer and a slowing down in winter. The spatial distribution analysis revealed some heterogeneity, suggesting migration. A comparison is made with other published data from the geographical range of Echinocardium cordatum.

1 INTRODUCTION

This study of a population of Echinocardium cordatum is part of a more general investigation into the fine sand macrofaunal community of the bay of Douarnenez (Guillou and Robert, 1980).

This species is considered to be the most widely distributed spatangoid in the world, having been recorded from Norway to New Zealand (Hyman, 1955). It is also relatively tolerant towards the nature of the sediment, as it occurs in both fine and coarse sands, and in clean or muddy sediments (Higgins, 1975).

Previous work was carried out in some northern areas, such as the North Sea, by Ursin (1960) and Buchanan (1966), in Scotland by Moore (1936), and in the Mediterranean Sea on the coast of Provençe by Massé (1971). Few results are available concerning the populations of the intermediate area. Most studies emphasize a high variability, both in the spatial distribution and in the biological parameters of the populations. In the bay of Douarnenez, this species occurred in large numbers over several years, which provided ideal conditions for studying its population dynamics.

2 MATERIALS AND METHODS

Echinocardium cordatum is strictly confined to the fine sand of the bay of Douarnenez (Fig. 1), where the environmental conditions are dominated by high sediment instability, due to marked exposure to westerly swells. Periodic sampling was carried out over the period 1977-1979. The sampling stations are distributed along a depth-transect, from the intertidal zone to about 30m, and the samples were collected using a Smith-McIntyre grab and a dredge. They were sieved on a 1mm mesh size screen, fixed in 4% formalin then sorted and preserved in 70% alcohol. The size of individuals was estimated from the length of the anteroposterior axis, measured to the nearest 0.1mm.

3 RESULTS

The results of the biometric study are reproduced in Figure 3. The

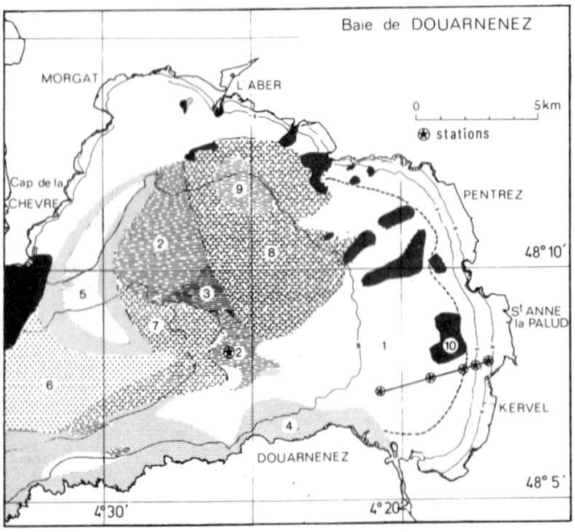

Fig. 1. The sampling stations in the biosedimentary context of the bay of Douarnenez (after Toulement, 1972; Chasse and Glemarec, 1976). (1) clean fine sand; (2) muddy sands.

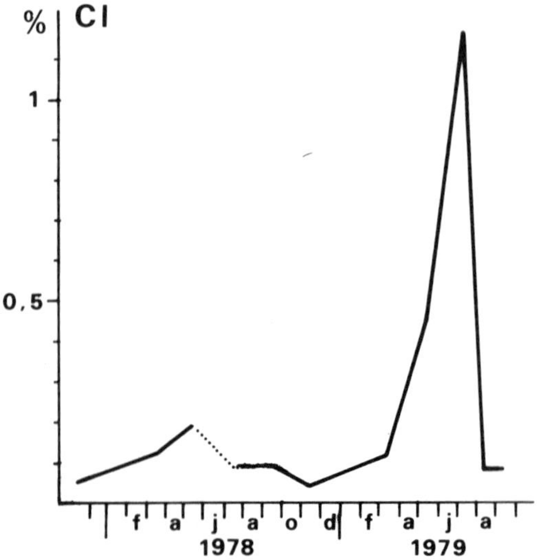

Fig. 2. Variation of the condition index (CI) of Echinocardium cordatum expressed as the ratio of the gonads dry weight to the body dry weight.

analysis of the length frequency histograms for each station identifies three principal zones along the transect:

1. Zone A contains the stations from 0 to 5m, where most of the size classes are represented.
2. Zone B contains the stations from 10 to 20m, where only the medium sizes can be seen.
3. Zone C, where there are no large individuals.

These results clearly show variations in the spatial distribution of the population related to depth.

Recruitment of Echinocardium occurred in zone A, while it was not important or absent in zones B and C. Throughout the sampling period, very few juveniles were observed at a size less than 10mm, which may reflect the sampling procedure. Most of the individuals were recruited between 10 and 20mm.

Variations observed in recruitment:

1. In 1977, recruitment was not important, but the corresponding year-class was regularly observed during the following years (1978 and 1979).
2. In 1978, no recruitment was observed, but this year-class was recorded the following year in low numbers.
3. In 1979, an obvious recruitment occurred in October, which was later than that of 1977.

In order to pin-point the time of recruitment, the spawning period was studied in 1978 and 1979, using a condition index expressed by the ratio of the dry weight of the gonads to the body weight (Fig. 2). The results confirm that spawning in 1978 was weak and occurred between May and July while, in 1979, spawning took place in July and August.

The number of year-classes and hence the life span of the population are defined from these demographic structures (Fig. 3) and show:

1. In zone A: 3 year-classes which correspond to the 1975, 1976 and 1977 cohorts. They can sometimes be divided into two components, probably related to the extended recruitment period.
2. In zone B: only one unimodal structure could be observed at a size which corresponds to the 2nd year-class in the previous zone. Thus, over the years, annual generations must succeed each other through this single mode.
3. In zone C: two year classes are well represented but no individuals larger than 40mm were found.

Figure 4 gives the results of the modal analysis of the histograms,

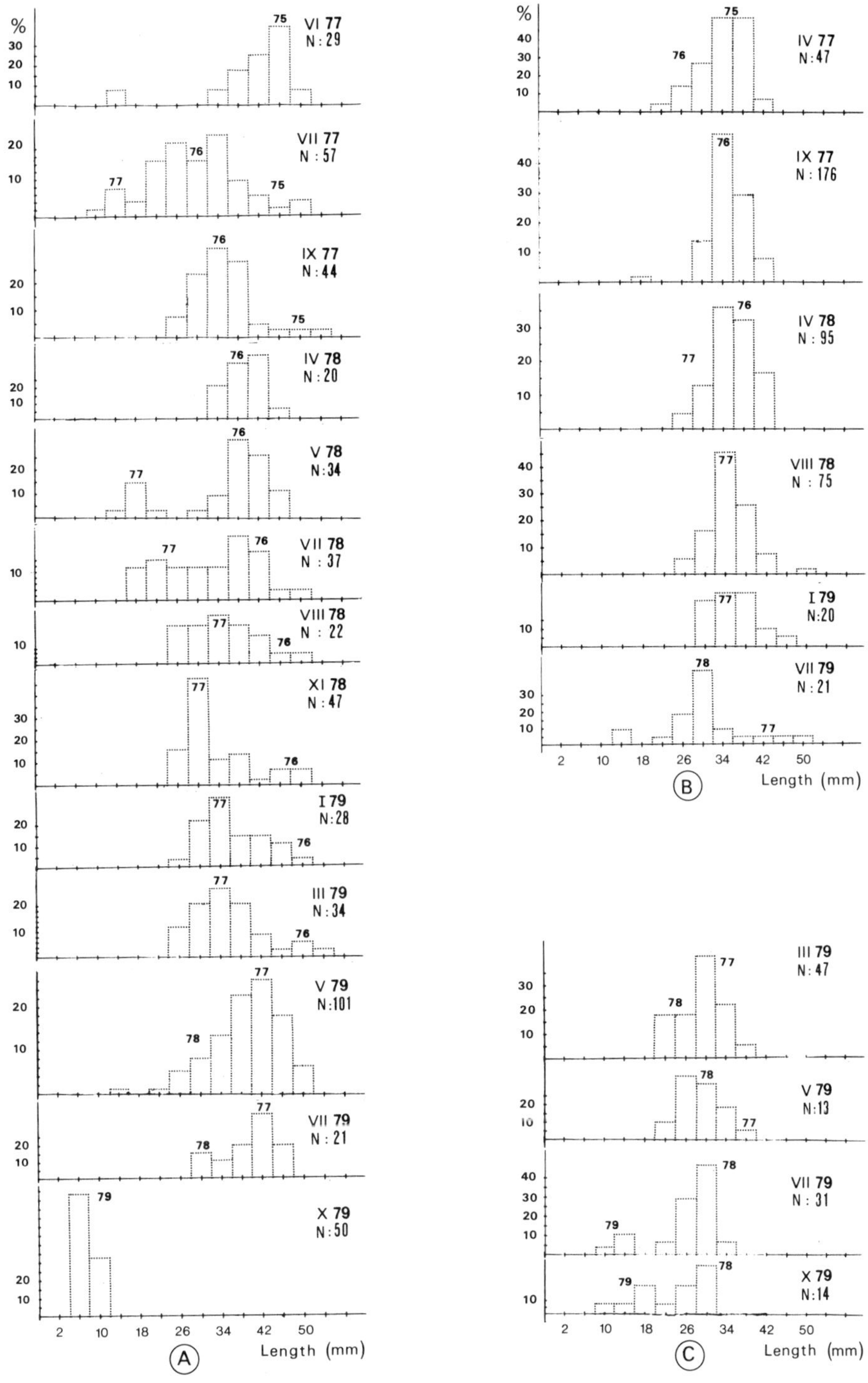

Fig. 3. Size distributions of Echinocardium cordatum in the different zones according to the bathymetric gradient.
Zone A: 0 - 5m; Zone B: 10 - 20m; Zone C: 30m.

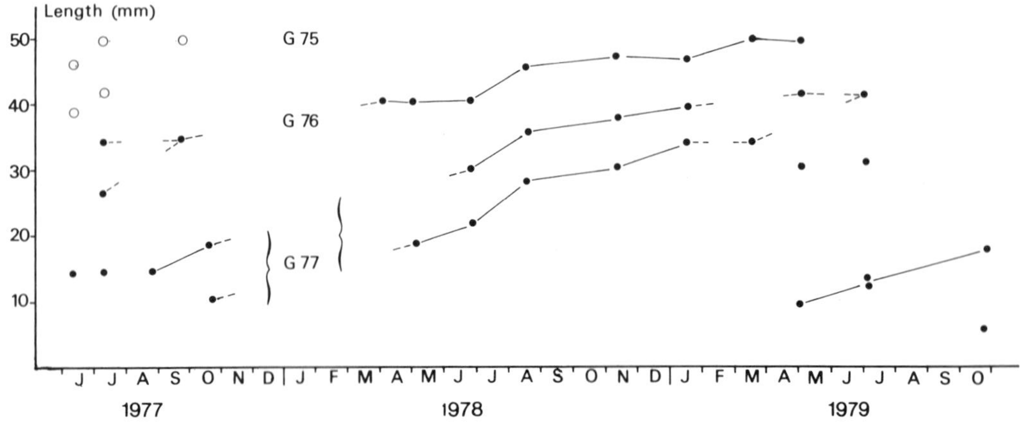

Fig. 4. Growth curves of the successive generations of Echinocardium cordatum followed from June 1977 to October 1979.

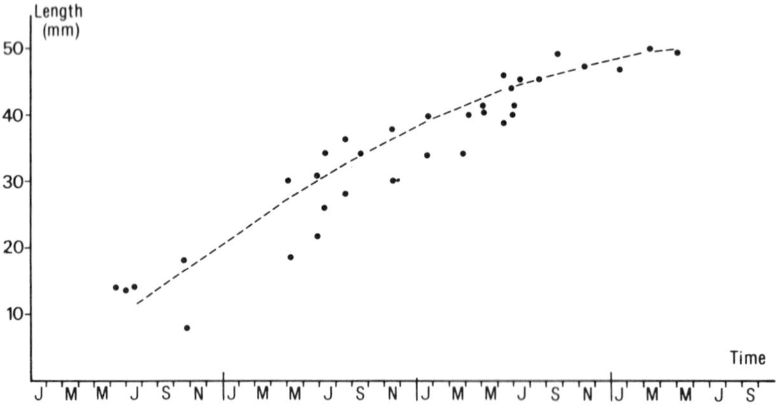

Fig. 5. Average growth curve of Echinocaraıum cordatum estimated from the juxtaposition of the separate curves of the year classes.

in zone A, over the sampling period. It shows seasonal variations in the growth curve of the year classes with a regular growth rate of the juveniles during the first winter, an obvious increase in summer for a short time, followed by a decrease in winter. From these individual growth curves of the year-classes, an average growth curve of the population could be obtained (Fig.5). It shows a self inhibited growth pattern. The mean length attained by the third year was about 50mm, while a theoretical maximal length calculated from the Ford-Walford model was 64mm.

4 DISCUSSION

In the study area of the bay of Douarnenez, Echinocardium cordatum appears to be a summer breeder, as was already evidenced at Morgat, a more sheltered area in the North of

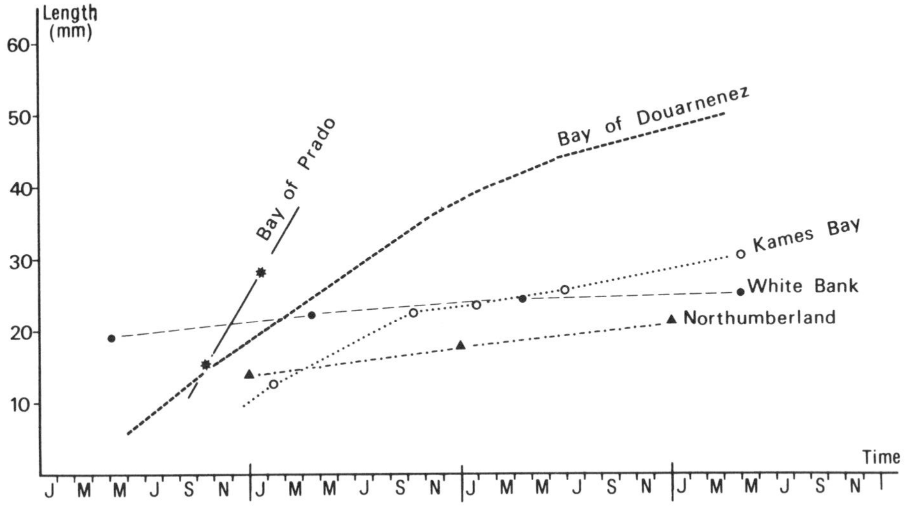

Fig. 6. Growth curves of Echinocardium cordatum from different areas throughout its geographical range: Bay of Prado, Mediterranean sea (Massé, 1971); Bay of Douarnenez; Kames Bay, Clyde estuary (Moore, 1936); White Bank, North sea (Ursin, 1960); Northumberland, North sea (Buchanan, 1966).

the bay (Fameree, 1973). The same observation has been made for most populations of Echinocardium from northern seas: in the North Sea (Ursin, 1960; Buchanan, 1966), in the Irish sea (Moore, 1936), on the French coast of the Channel (Caullery, 1925). On the other hand, in the Mediterranean sea, reproduction is earlier in the year, the individuals being ripe at the end of the winter (Masse, 1971).

It extends over a more or less long period, according to the area. Thus, larvae of Echinocardium were observed by Rees (1954) in the North Sea from May to August. This extended period can explain the plurimodal trend of the year classes, sometimes observed in the size distributions.

An interannual variability can also appear in its reproduction. On the Northumberland Coast, Echinocardium bred every seven years, and only in the intertidal zone (Buchanan, 1966). In the Bay of Douarnenez, this variability is mainly marked at the recruitment level. There was very low recruitment in 1978, and this continued up to 1983, leading to a large decrease in the population of the study area.

Migration is also a significant component in the spatial and temporal variability of the Echinocardium population. This was observed in other areas, notably by Moore (1936), Pequignat (1964), Buchanan (1966), Massé (1971), and mainly consists of movement by adults towards the shore. In the Bay of Douarnenez, this phenomenon can have two aspects: there are migrations of the young individuals which preferentially settle in the shallows and colonize the deeper area, during their second year. But, there is also an opposite migration of adults during their second year towards the shore. This is well illustrated by the demographic structure in the intermediate area (Fig. 3, zone B). These migrations can extend to the infralittoral zone on the beach, which is colonized by adults, while in the more sheltered beach of Morgat, all year classes are well represented. It suggests that hydrodynamism exerts a major influence on the observed distributions.

Biological parameters of the

Echinocardium population, such as growth rate and life span, can be compared with those of populations from the same biotopes in other areas (Fig. 6).

It appears that the growth rate is much higher in southern areas and, as a general trend, lower in the north, an intermediate value being recorded in the Bay of Douarnenez. The life span is one or two years in the Mediterranean (Massé, 1971) while, in the North Sea, 10 to 15 years are suggested by Ursin (1960) and Buchanan (1966). These results show the part played by the latitudinal factor on these biological parameters, as was already noticed for this species by Massé (1971). But these main features can be widely affected by the local conditions, particularly by the texture of the sediment: a reduction in the feeding efficiency of Echinocardium is observed in muddy sediments (Buchanan, 1966) while high values are recorded in clean coarse sands (Higgins, 1974).

REFERENCES

Buchanan, J.B. 1966. The biology of Echinocardium cordatum (Echinodermata: Spatangoidea) from different habitats. J.Mar.biol. Ass.U.K. 46(1): 97-114.

Caullery, M. 1925. Sur la structure et le fonctionnement des gonades chez les Echinides. Trav.Stat. Zool. Wimereux, 9: 21-35.

Chasse, C. & M. Glémarec 1976. Atlas des fonds meubles du Plateau Continental du Golfe de Gascogne. Cartes biosédimentaires, ICA, Brest.

Fameree, D. 1973. Contribution à l'étude du cycle annuel de reproduction de deux espèces d'Echinoides (Echinodermata) des côtes de Bretagne. Mém. Licenc. Sc. Zool. Fac. Sc. Bruxelles: 39p.

Guillou, J. & R. Robert 1980. Principaux aspects de la dynamique d'une population d'Ophiura texturata en Baie de Douarnenez. Proceedings of the European Coll. on Echinoderms. Brussels 1979. In Jangoux Ed. Balkema Publ. Rotterdam: 171-177.

Higgins, R.C. 1974. Specific status of Echinocardium cordatum, E. australe and E. zealandicum (Echinoidea: Spatangoida) around New Zealand, with comments on the relation of morphological variation to environment. J.Zool.Lond. 173, 451-475.

Higgins, R.C. 1975. Observations on the morphology of E. cordatum (Echinoidea: Spatangoida) from diverse geographical areas. J. Zool.Lond. 177: 507-515.

Hyman, L.H. 1955. Echinodermata. The invertebrates, Vol. 4, 763 p. New York.

Massé, H. 1971. Contribution à l' étude quantitative et dynamique de la macrofaune des peuplements des sables fins infralittoraux des Côtes de Provence. Thèse Aix-Marseille: 310p.

Moore, H.B. 1936. The biology of E. cordatum. J.Mar.biol.Ass.U.K., 20(3): 655-671.

Pequignat, E. 1969. Biologie et comportement des E. cordatum (Pennant) en Baie de Seine. Thèse Université Caen, Arch. Orig. C.N.R.S., 3344: 80p.

Rees, C.B. 1954. The distribution of Echinoderm and other larvae in the North Sea, 1947-51. Bull. Mar.Ecol.Edinb. 4: 47-67.

Toulemont, A. 1972. Influence de la nature granulometrique des sédiments sur les structures benthiques. Baie de Douarnenez et d'Audierne (Ouest-Finistère). Cah.Biol.Mar., tome XIII: 91-136.

Ursin, E. 1960. A quantitative investigation of the Echinoderm fauna of the central North Sea. Meddr. Danm. Fisk Og., Havunders. N.S., 2(24): 1-204.

Proceedings of the Fifth International Echinoderm Conference / Galway / 24-29 September 1984

Functional morphology of test, lantern and tube feet ampullae system in flexible and rigid sea urchins (Echinoidea)

MARGIT JENSEN
Zoological Museum, University of Copenhagen, Denmark

ABSTRACT: The sea urchin test is composed of individual plates linked to each other in vertical and horizontal sutures either by connective tissue forming a flexible test or by skeletal processes and collagen fibres forming a rigid test. Musculature attached to the internal stereom of the test is restricted to: 1) ambulacral, peristomial and lantern muscles, e.g. in flexible sea urchins: 2) peristomial and lantern muscles, e.g. in rigid sea urchins with lantern and teeth; and 3) peristomial muscles, e.g. in rigid sea urchins without lantern and teeth. The internal muscles in the flexible test function as a pump. The lantern muscles function as a pump and take part in the feeding mechanism. The respiratory organs in sea urchins are specialized tube feet adapically situated on the test, and ('gills') orally situated on the test. One of the functions of the internal musculature in sea urchins is to pump oxygenated fluid from the respiratory organs to different parts of the body. In addition, the lantern muscles are part of the feeding mechanism.

1 INTRODUCTION

Studies of the morphology, function and evolution of the organs of echinoids have been subject to numerous investigations (Jensen 1972, 1979, 1981; Kier 1982; Märkel 1981; Nichols 1966; Sarasin & Sarasin 1888; Shimizu & Taki 1972; Smith 1978, 1980a,b, 1981; and Taki 1971), but studies of the physiology and evolution of the processes which make the organs function in the living sea urchins are few (Binyon 1972; Farmanfarmaian 1966; Fenner 1973; and Phelan 1977). The present paper discusses the function and evolution of the test, lantern, teeth and coeloms in relation to the respiratory function of the tube feet ampullae system.

2 MATERIAL AND METHODS

All specimens were preserved in 80% ethanol. Specimens for SEM studies were treated as described in Jensen (1979). Specimens for light microscopy were gradually hydrated through a series of ethanols (80%, 60%, 40%), washed in distilled water, and dehydrated through a series of ethanol (40%, 60%, 80%). A small amount of methyl green was added to the 80% ethanol to enhance contrast of the tissue for photography.

3 SKELETAL AND ORGANIC ELEMENTS OF RECENT ECHINOIDS

The echinoid test is composed of individual plates that form five ambulacra and interambulacra, an apical system, a perignathic girdle and a peristome. Ambulacral and interambulacral plates are either bound together by bands of connective tissue, forming a flexible test, or the sutural stereom of the plates interdigitates, forming a rigid test. Plates of flexible sea urchins produce a single-layered sutural stereom, while plates of rigid ones produce a tridimensional sutural stereom.

The interior cavity of the sea urchin test is occupied by the main (perivisceral) coelom, some smaller coeloms, mesenteries, the digestive tract, gonads and ampullae of the tube feet. The mesenteries and the two circuits of the intestine slightly subdivide the perivisceral coelom into compartments. Lantern and teeth of regular echinoids are enclosed in the expanded perioesophageal coelom.

Tube feet are external extensions of the water vascular system, and each tube foot is connected with an internal ampulla through either one or two pore holes in the ambulacral plate. The paired pore holes of the ambulacral plates are either widely separated on the external and internal side of the plate, or the paired pore holes are weakly separated on the external surface, but widely separated on the internal side of the plate. The first type of pore holes bears the thin-walled respiratory tube feet found in clypeasteroids and spatangoids; the latter type bears the extensile tube feet of regular echinoids. Both types of tube feet are connected with leaflike ampullae. Single pore holes of ambulacral plates bear either extensile tube feet with large cylindrical ampullae or non-extensile (slightly extensile) tube feet with diminutive ampullae or without ampullae. These types of tube feet both have strongly developed nerves(Smith 1980b). A basic tube foot consists of a stem which terminates in a sucker or in a pad. The stem consists of: an outer ciliated epithelium, a longitudinal nerve, longitudinal and circular collagen fibres, muscle fibres, ciliated endothelium and an internal collagen septum (Nichols 1959a,b, 1961; Fenner 1973; Smith 1978, 1980b).

3.1 Echinothurioida

The flexible test of echinothurioids is composed of ambulacral and interambulacral plates, which imbricate in all sutures in species of the genera Hygrosoma Mortensen, Kamptosoma Mortensen, Phormosoma Thomson, and Sperosoma Koehler, but only imbricate in the perradial, interradial and adradial sutures of the tests of the species of the genera Araeosoma Mortensen, Asthenosoma Grube, Calveriosoma Mortensen and Hapalosoma Mortensen.

The internal surface of the test of all echinothurioids shows tha the ambulacral plates imbricate adapically, and the interambulacral plates imbricate adorally. In addition, the ambulacral plates overlap the interambulacral plates in the adradial sutures.

This study shows that all plates of the test of the species of the genera Araeosoma, Asthenosoma, Calveriosoma and Hapalosoma are embedded in connective tissue which forms thick bands in the horizontal sutures (Fig. 1). Thick 'pillows' of connective tissue occur between the imbricating plates in the adradial sutures (Figs 3,4).

The tests of the species of Hygrosoma, Kamptosoma, Phormosoma, and Sperosoma form thin layers of connective tissue in the horizontal and vertical sutures (Fig. 5).

The perignathic girdle of echinothurioids is composed of tall auricles and interambulacral ridges (Fig.5). Each auricle consists of the skeletal outgrowth of the two oldest ambulacral plates and a variable number of adradial parts of succeeding primary plates (Jensen in prep.). Echinothurioids enlarge the peristomial area by detaching perradial portions of succeeding primary plates; by detaching succeeding demi plates to the peristome; and by adradial growth of the interambulacral plates. The oldest primary plates divide into an adradial portion without pore holes and a perradial portion with pore holes. The stereom of the adradial portion fuses with that of the basis of the auricle. The connective tissue of the adradial suture is included in the suture between the auricles and the interambulacral ridge. The perradial portion of the subdivided primary plate begins to grow asymmetrically and becomes a crescent-shaped peristomial plate. The pore pair changes to an unipore for the peristomial tube foot. Succeeding demi plates follow the same growth pattern as that of the perradial portions of the primary plates. The interambulacral plates grow in the adradial edges and pull inwards, forming a tall interambulacral ridge.

Plates of single-layered stereom form sheets between the crescent-shaped peristomial plates (Fig.6).

The perioesophageal coelom of echinothurioids encloses lantern and teeth and forms five bulbous sacs (Stewart's organs) that extend from the apical side of the coelom and expand below the compasses, passing through the arcs of the auricles into the ambulacral compartments of the perivisceral coelom. Some echinothurioids bear diminutive external pouches ('gills') of the perioesophageal coelom.

The digestive tract consists of oesophagus, the oral circuit (stomach) and the apical circuit of the intestine with the rectum. The digestive tract of most echinothurioids occupies the central part of the interior cavity of the test, e.g. in H. petersii A. Agassiz and C. hystrix Thomson. The ambital cavity of the ambulacral compartments is occupied by coelomic fluid, and the gonads occupy the main part of the interambulacral compartments of the test. The two circuits of the intestine are horizontally positioned, forming a double ring which encircles the perignathic girdle and lantern. The oral circuit of the intestine forms the peripheral (adambital) festoons, and the apical circuit forms the central festoons of the intestine. Large curvatures of the oral circuit expand adapically in the ambulacra, and large curvatures of the apical circuit expand adambitally in the interambulacra.

This study shows that strong mesenteries fasten the two circuits of the intestine to the internal surface of the test. The mesenteries of the oral circuit (stomach) of the intestine form strong bundles of muscles; these expand from the oral edge of the ocular plate and from the adradial corners of the two series of adapical primary plates to the adapical side of the large curvature of the intestine. A fine collagenous network bridges the two adradial bundles of muscles enclosing the adapical portion of the ambulacrum (Fig. 2). Other bundles of mesenterial muscles expand from the adradial corners of the primary plates of the ambital and adoral part of each ambulacrum and from the adapical side of each auricle; each adradial muscle fan merges into the collagenous septum, which is positioned adambitally to the intestine. The collagenous septa of the two adradial muscle systems of each ambulacrum fasten to the intestine. Thus the adradial muscle fans and the intestine subdivide the perivisceral coelom into five ambulacral and five interambulacral compartments. Sarasin & Sarasin (1888) named the muscle fans 'Längsmuskeln', and since then the mesenterial muscles of echinothurioids have been called adradial muscles.

The mesenteries of the apical circuit of the intestine form loose bundles of collagenous fibres which expand between the ambital curvatures of the intestine and the internal surface of the interambulacra.

Muscle fibres expand from the adoral side of the auricles to the convex edges of the crescent-shaped peristomial plates, forming five tunnels from the auricles to the radial sides of the lantern (Figs 5,6).

Large thin-walled bulbous tube feet without suckers occur on the aboral side of the test, and thick-walled tube feet with small suckers occur on the oral side of the test. Both types of tube feet are connected to bulbous ampullae (Figs 1,3).

3.2. Diadematoida

Diadematoids have a rigid test. The peristomial area is enlarged by adradial growth of the ambulacral and interambulacral plates (Märkel 1981). The movable diadematoid lantern is able to make the flexible peristomial membrane protrude. Most diadematoids bear external pouches ('gills') of the perioesophageal coelom, and some species bear diminutive Stewart's organs.

3.3 Echinoida

Species of Echinoida have a rigid test. The peristomial area becomes enlarged in the same way as in the diadematoids. The peristomial membrane is muscular and extensile. The strongly movable lantern is able to make the peristome protrude. Pouches of the perioesophageal coelom appear in the edge of the peristome.

3.4 Echinoneoida

The plates of the echinoneoid test interdigitate in all sutures, forming a solid test (Fig. 10).

Juvenile Echinoneus cyclostomus H.L. Clark bear lantern, teeth and auricles (Fig. 9). Specimens with a test length of 4 mm have lost all these elements (Jensen in prep.). The lantern of juvenile specimens comprises all skeletal and muscular elements found in the lanterns of regular echinoids (Fig. 7). Teeth of juvenile E. cyclostomus are wedge shaped and composed of triangular tooth plates with large adaxial lappet, like teeth of cassiduloids. The lamellae system is the Fibularia system (Jensen 1981). The peristomial area is slightly increased by asymmetrical resorption in the edge of the peristome (Jensen in prep.), but is small in larger specimens. The peristomial membrane is naked and without tube feet. No external pouches of the perioesophageal coelom are present in the edge of the peristome.

3.5 Clypeasteroida

The rigid test of clypeasteroids bears a small, centrally positioned peristome. The immovable lantern lacks compasses and compass muscles, and the expanded perioesophageal coelom is absent in clypeasteroids.

3.6 Spatangoida

The rigid test of spatangoids bears a small anteriorly positioned peristome. No expanded perioesophageal coelom is developed in spatangoids.

3.7 Cidaroida

Recent cidaroids have a rigid test. The peristomial area becomes enlarged by interradial and adradial growth of the interambulacral plates and by detaching

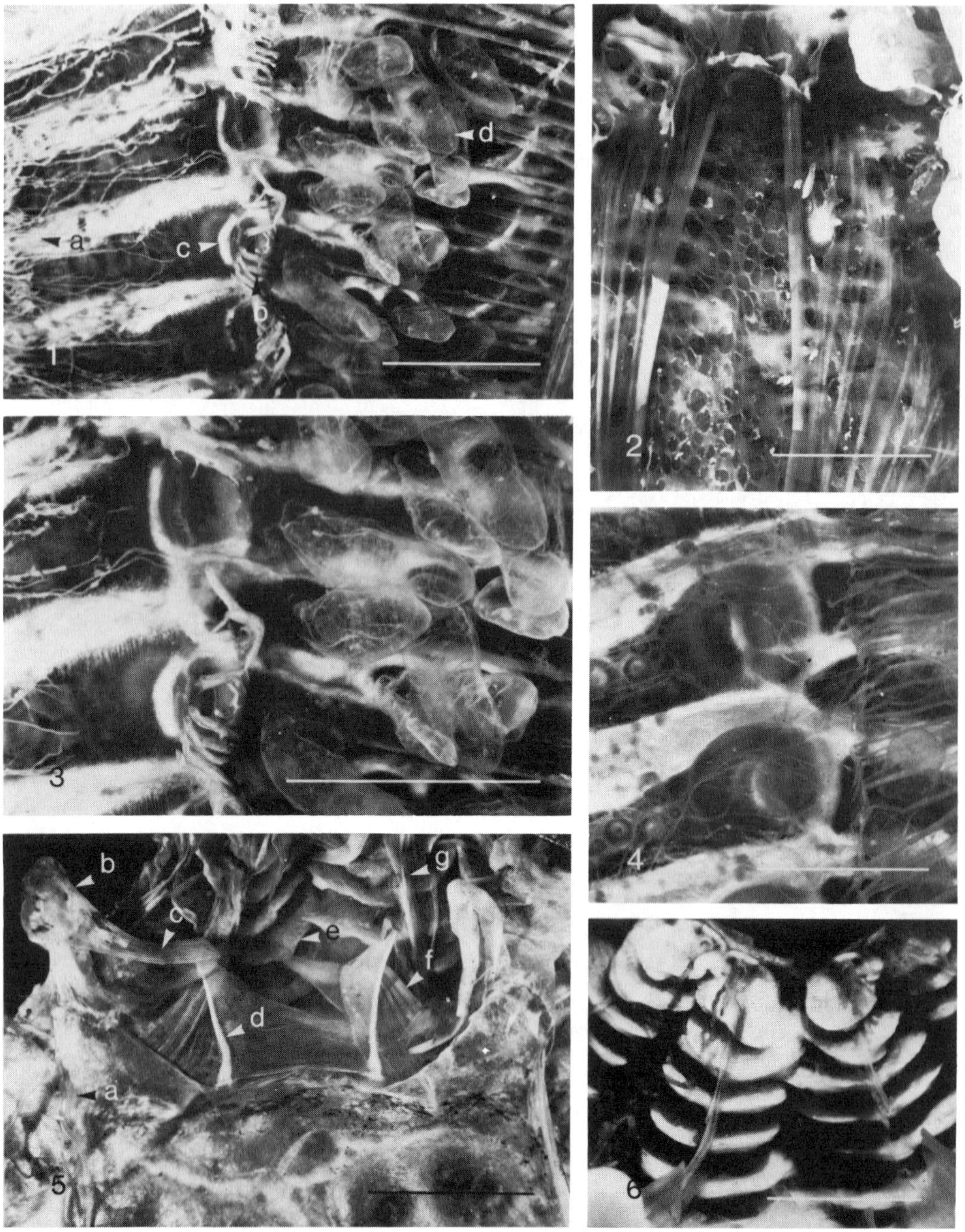

Fig. 1. Calveriosoma hystrix Thomson. Challenger ES 178. Test diameter 11o mm. Internal surface of decalcified aboral part of the test. a) Interambulacral mesenteries; b) adradial muscles; c) collagenous 'pillows'; d) ampullae. Scale = o,5 cm.

Fig. 2. Araeosoma fenestratum (Thomson). Discovery 7993. Test diameter 135 mm. Internal surface of adapical ambulacral plates including the edge of the ocular plate showing the adradial muscles and the collagenous network. Scale = o.5 cm.

ambulacral plates to the peristome (Märkel 1981). The perioesophageal coelom bears large Stewart's organs which extend into the perivisceral coelom.

4 FUNCTION OF TEST, LANTERN AND TUBE FEET AMPULLAE SYSTEM OF RECENT ECHINOIDS

The test of Recent sea urchins is either flexible or rigid.

Echinothurioids are the only Recent echinoids with a flexible test, and these sea urchins are provided with mesenterial muscles which form the adradial muscle fans. The muscle fans produce pulsatory movements and pump the fluid around in the ambulacral and interambulacral compartments of the perivisceral coelom. Peristalsis of the digestive tract and ciliary currents contribute to the flow of the perivisceral fluid. The lanterns of sea urchins are either movable with strong protractor-, retractor- and compass muscles or immovable with weak protractor-, and retractor muscles and without compass and compass muscles. Echinothurioids have movable lanterns which are inclined and fastet via protractor-, retractor-, and compass muscles to the basket-shaped perignathic girdle. The lantern muscles produce rocking and pushing movements of the lantern when the sea urchin is feeding, walking, and possibly all the time. In other words, the lantern forms a stirring block in the centre of the perivisceral coelom. The rocking and pushing movements of the lantern together with the pulsatory movements of the adradial muscle fans circulate the perivisceral fluid. In addition, lantern muscles have an important influence on circulation of the fluid in the perioesophageal coelom and in the Stewart's organs. The coelomic fluid transports oxygen and nutrients to the tissue and transports waste products from the tissues.

Sea urchins with a rigid test lack internal adradial muscle fans.

Do rigid sea urchins get a compensation for the loss of internal muscles and for loss of a flexible body?

I suggest two hypotheses: 1) The function of the lost organ is taken over by another organ. 2) Sea urchins adjust to the new situation, e.g. by subdividing the internal cavity into smaller, narrower compartments.

Species of Echinacea have an almost globular test with a large internal volume. The perivisceral coelom contains a large amount of fluid. The two circuits of the intestine fasten to the internal surface of the test via non-muscular mesenteries forming almost vertically hanging festoons in the interior cavity of the test.

Species of Diadematacea have a rounded test and the two circuits of the intestine are positioned almost horizontally.

Are ciliary currents and peristalsis of the intestinal circuits and the oesophagus sufficient to move the large amount of perivisceral fluid?

I think the large volume of fluid needs a central stirring and pushing organ.

In species of Echinacea and Diadematacea, the tall movable lanterns expand the membrane of the perioesophageal coelom and rise into the central area of the large perivisceral coelom. The movable lantern is able to protrude the elastic membrane of the wide peristome almost to the full height of the lantern. This means that the lantern of species of Echinacea and Diadematacea is a powerful stirring and pushing organ as well as a powerful feeding organ, e.g. in Strongylocentrotus droebachiensis (O. F. Müller).

I suggest that the strong movable lantern with its muscles, and the wide peristome with the elastic membrane are the compensation for the loss of a flexible test and internal muscles in species of Echinacea and Diadematacea.

New investigations by Hanson et al.(1984) show that the perivisceral fluid circulation in Lytechinus variegatus (Lamarck) resembles that of forced stirring and the authors find that the Aristotle's lantern is the driving force.

Ciliary currents, pyramid- and compass muscles produce the flow of the fluid in

---

Fig. 3. Same as fig. 1 showing details of the adradial suture, the horizontal collagenous bands and the ampullae. Scale = o.5 cm.
Fig. 4. Same as fig. 2 showing the adradial suture, thick collagenous bands of the horizontal sutures and the adradial muscles. Scale = o.5 cm.
Fig. 5. Phormosoma placenta Thomson. Challenger AT 192. Test diameter 55 mm. Internal surface of decalcified perignathic girdle and the peristomial area. a) Adradial muscles; b) auricle; c) lantern retractor muscles; d) compass depressor muscles; e) peristomial plate; f) lantern protractor muscles; g) peristomial muscles. Scale = o.5 cm.
Fig. 6. Phormosoma placenta. Challenger AT 192. Test diameter 7o mm. Internal surface of the peristomial area showing the plates, muscles and ampullae. Scale = o.5 cm.

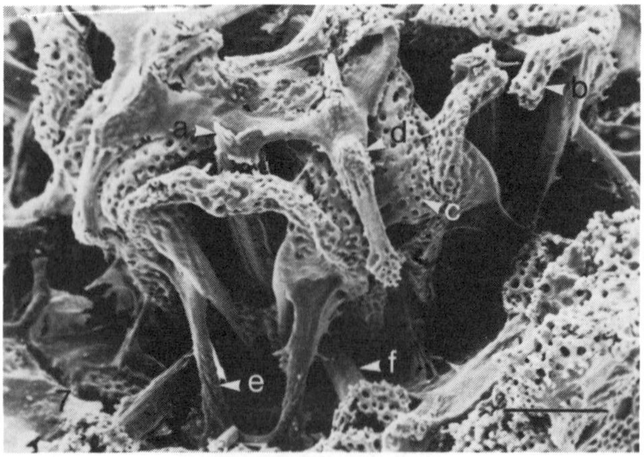

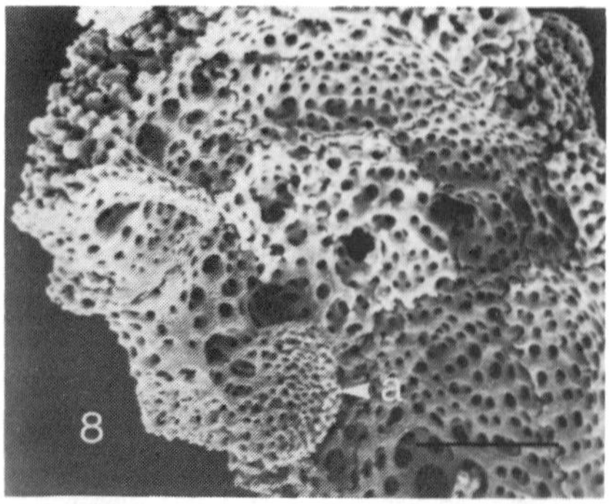

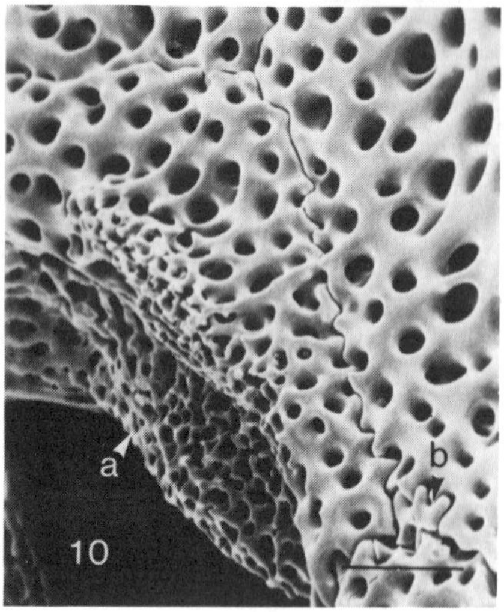

Fig. 7. Psammechinus miliaris (Gmelin). Off Frederikshavn, Kattegat. Test diameter 3 mm. Apical view of lantern. a) Tooth; b) epiphysis; c) rotula; d) compass; e) protractor muscles; f) retractor muscles. Scale = o.1 mm.
Fig. 8. Same as fig. 7. Internal surface of ambulacral and interambulacral plates. Scale = o.1 mm.
Fig. 9. Echinoneus cyclostomus H.L. Clark. Smithsonian Hartford Expedition, 1937. Test length 3.7 mm. Apical view of lantern. a - f as fig. 7. g) Compass elevator muscles; h) auricle. Scale = o.2 mm.
Fig. lo. Same as fig. 9. Internal surface of primordial interambulacral plate and neighbouring interambulacral plates and auricle. a) Edge of the primordial plate. Scale = o.o4 mm.

the perioesophageal coelom.

Recent cidaroids have a rigid test and peristomial membrane with strong plates. The erect lantern has little posibility to protrude the peristomial membrane, but can produce rocking and pushing motions in the internal cavity of the test, and be the driving force of the perivisceral fluid. Circulation of fluid in the perioesophageal coelom and in Stewart's organs is achieved through the action of the lantern muscles and ciliary currents.

Tube feet ampullae system of echinoids is the respiratory, sensory and locomotory organ. This study discuss the tube feet ampullae system as a respiratory organ.

Some of the features which determine the efficiency of a respiratory organ are the size of the respiratory area and the transport medium of the oxygen (Shick 1983).

Echinoids increase the efficiency of the respiratory system either by increasing the number of elements or by specializing some of the elements of the system.

In regular echinoids the number of extensile multifunctioning tube feet is increased by forming compound ambulacral plates, in which elemental plate bears one tube foot and ampulla.

In irregular echinoids the number of tube feet is not increased, but the respiratory system is made more efficient by specialization of the adapically positioned tube feet as leaf-shaped respiratory organs (Shick 1983).

Oxygen diffuses across the respiratory surface of the tube feet and is transported via the fluid of the tube feet ampullae system to the internal ampullae, across the walls of the ampullae to the perivisceral fluid and via the perivisceral fluid to the respiring tissue. Muscular contractions of the tube feet ampullae system contribute to the circulation of the fluid in the system. Regular echinoids are provided with a lantern enclosed in the perioesophageal coelom. Oxygen may be transported via the perivisceral fluid to the perioesophageal coelom and diffuses across the membrane to the fluid of the perioesophageal coelom, which transports the oxygen to the respiring tissue of the lantern. The external outpouches ("gills") of the peristomial membrane may act as a respiratory organ for the tissue in the perioesophageal coelom. This study shows that the Stewart's organs of echinothurioids encroach the ambulacral compartments of the perivisceral coelom. The ambulacral compartments house the internal part of the respiratory tube feet ampullae system and therefore may contain the most highly oxygenated fluid. The position of Stewart's organs indicates that these organs have a respiratory function for the organs in the perioesophageal coelom. Transport of oxygen across the membrane of the perioesophageal coelom may appear in all sea urchins with expanded lantern coelom. Irregular echinoids have a non-expanded perioesophageal coelom. Oxygen is transported from the internal ampullae of the tube feet ampullae system via the perivisceral fluid to the respiring tissue.

Irregular echinoids either have a test with a small internal volume or have a test with a relatively large internal volume. All irregular echinoids have a narrow or small peristome.

Clypeasteroids have a completely flat or a low test with a narrow internal cavity, which usually is reduced by the presence of a system of skeletal pillars. The immovable lantern is not enclosed in a coelom, but is centrally positioned in the perivisceral coelom. This type of lantern may be restricted to a feeding function. The ciliary currents and the peristalsis of the digestive tract may be sufficient to circulate the perivisceral fluid.

Spatangoids have rounded, oval or somewhat flattened test with a relatively large internal volume. The digestive tract subdivides the perivisceral coelom into narrower compartments. Peristalsis of the digestive tract and ciliary currents may be sufficient to circulate the perivisceral fluid. All spatangoids lack Aristotle's lantern.

Recent Echinoneus cyclostomus H.L.Clark and Recent Cassidulus cariboearum Lamarck have rounded almost egg-shaped tests. Juveniles of both species have a lantern. E. cyclostomus has an erect lantern with lantern muscles enclosed in the perioesophageal coelom (Fog. 7). C. cariboearum has an immovable irregular lantern like those of the fibulariids. The stereom of the lantern in juvenile E. cyclostomus resembles an old stereom. The stereom of the two parts of the compasses is solid (Fig. 9). The stereom of the lantern of the juvenile Psammechinus miliaris (Gmelin) resembles as a young stereom with an incomplete suture between the two parts of the compasses (Fig. 7).

Does the lantern tissue in juvenile E. cyclostomus get a sufficient supply of oxygen and nutrients and is it able to get rid of waste products?
Smith (198ob) found non-specialized suckered tube feet with thin ampullae on the aboral and adoral sides of the test of E. cyclostomus. The present study shows that juvenile E. cyclostomus have partitioned isopores with axially positioned neural canal on the adoral side of the test (Jensen in prep.). Smith (op. cit.) found that partitioned isopores with axially positioned neural canal are associated to non-respiratory buccal tube feet of many regular echinoids. The tube feet ampullae system of E. cyclostomus is not very specialized and may be less effective to respiration.

Juvenile E. cyclostomus enclose calcareous particles from the sediment in the aboral circuit of the intestine, whereas the oral circuit of the intestine is empty. This indicates that the juvenile E. cyclostomus has ingested food particles after settling. Juvenile E. cyclostomus stop feeding before lantern and teeth start to be resorbed (Jensen in prep.). This indicates that physiological changes in the oesophagus and intestine begin before skeletal elements of the teeth and lantern start to be resorbed.

Do the juvenile E. cyclostomus lose lantern and teeth as a result of changes in the digestive tract or as a result of insufficient transport of nutrients, oxygen and waste products?

I think the loss of lantern and teeth is

the result of a combination of several components, i.e. physiological, histological and chemical changes in the digestive tract as well as an insufficient amount of oxygen, nutrients and insufficient transport of waste products from the lantern tissue.

5 ACKNOWLEDGEMENTS

I am indebted to Miss A.M. Clark for providing laboratory facilities in the British Museum (Natural History), London and making specimens available for study. Specimens of Echinothurioida, Echinoneoida and Echinoida were generously lent by Dr. D.L. Pawson and Dr. P. M. Kier Smithsonian Institution, Washington, D.C.; Dr. J.D. Gage, Scottish Marine Biological Association, Oban; Mr. D.S.M. Billett, Institute of Oceanographic Sciences, Wormley; and Dr. M.E. Petersen, Zoological Museum, Copenhagen. I thank Dr. K. R. Jensen, Zoological Museum, Copenhagen and Mrs. A. Volsøe for linguistic help. This paper is part of a project supported by the Wellcome-Carlsberg Traveling Research Fellowship no. 33/82 and grants from the Danish Natural Science Research Council nos. 511-15lo7 and 81-4249.

REFERENCES

Binyon, J. 1972. Physiology of echinoderms, Pergamon, Oxford.

Farmanfarmaian, A. 1966. The respiratory physiology of echinoderms. In R.A. Boolootian (ed.), Physiology of Echinodermata, p. 245-265. New York, Interscience Publishers.

Fenner, D.H. 1973. The respiratory adaptions of the podia and ampullae of echinoids (Echinodermata). Biol. Bull. mar. biol. Lab., Woods Hole 145: 323-339.

Hanson, J.L., G. Gust & J.M. Lawrence.1984. Transport dynamics of perivisceral circulation in the sea urchin Lytechinus variegatus. Abstract. Annual meeting of the American Society of Limnology and Oceanographic.

Jensen, M. 1972. The ultrastructure of the echinoid skeleton. Sarsia 48: 39-48.

\- 1979. Primary plates of sea urchin teeth (Echinoidea). Vidensk. Meddr dansk naturh. Foren. 141: 7-27.

\- 1981. Morphology and classification of Euechinoidea Bronn, 186o. A cladistic analysis. Vidensk. Meddr dansk naturh. Foren. 143: 7-99.

Kier, P.M. 1982. Rapid evolution in echinoids. Palaeontology 25: 1-9.

Märkel, K. 1981. Experimental morphology of coronar growth in regular echinoids. Zoomorphology 97: 31-52.

Nichols, D. 1959a. The histology of the tube feet and clavulae of Echinocardium cordatum. Quart. J. microsc. Sci. loo: 73-87.

\- 1959b. The histology and activities of the tube feet of Echinocyamus pusillus. Ibid. loo: 539-55.

\- 1961. A comparative histological study of the tube feet of two regular echinoids. Ibid. lo2: 157-18o.

\- 1966. Functional morphology of the water vascular system. In R.A. Boolootian (ed.) Physiology of Echinodermata, p. 219-244. New York, Interscience Publishers.

Phelan, T.F. 1977. Comments on the water-vascular system, food grooves, and ancestry of the clypeasteroid echinoids. Bull. mar. Sci. 27: 4oo-422.

Sarasin, P & F. Sarasin. 1888. Anatomie der Echinothuriden und die Phylogenie der Echinodermen. Ergebnisse Naturwissenschaftlicher Forschungen auf Ceylon. 1884-1886, 1: 83-153, Pls. lo-17.

Shick, J.M. 1983. Respiratory gas exchange in echinoderms. Echinoderm Studies 1: 67-llo.

Shimizu, M. & J. Taki, 1972. A histological study of the mature test plate and the peristomial membrane in a sea urchin. Bull. Jap. Soc. Sci. Fish. 38: lo21-lo34.

Smith, A.B. 1978. A functional classification of coronal pores of regular echinoids. Palaeontology 21: 759-789, pls. 81-84.

\- 198oa. Stereom microstructure of the echinoid test. Spec. Pap. Palaeontology 25: 1-81.

\- 198ob. The structure and function, and evolution of tube feet and ambulacral pores in irregular echinoids. Palaeontology 23: 39-83, pls. 1-6.

\- 1981. Implication of lantern morphology for the phylogeny of post Palaeozoic echinoids. Palaeontology 24: 779-8ol.

Taki, J. 1971. Tetracycline labelling of test plates in Strongylocentrotus intermedius. Sci. Rep. Hokkaido Fish. Exp. Station. 13: 19-29.

Amoebae in tissues of diseased echinoids (*Strongylocentrotus droebachiensis*) in Nova Scotia

G.M.JONES
Seatech Ltd., Halifax, Nova Scotia, Canada

A.J.HEBDA
Seatech Ltd., Halifax, Nova Scotia, Canada

R.E.SCHEIBLING
Dalhousie University, Halifax, Nova Scotia, Canada

R.J.MILLER
Department of Fisheries and Oceans, Halifax, Nova Scotia, Canada

ABSTRACT: Amoebae (20-30 μm) are identified in tissues of echinoids (Strongylocentrotus droebachiensis) showing symptoms of disease. The amoebae have a parasome and are morphologically similar to the genus Paramoeba. They are distributed widely but in low density in body wall, water vascular system, nerves and gut in individuals in intermediate or late stages of disease. They are less common in individuals in early stages of disease, and have not been found in healthy echinoids. The amoebae ingest cell debris in degenerating tissue, but it is not known whether they are pathogenic or are secondary invaders in diseased echinoids. Affected echinoid tissues are infiltrated by coelomocytes during the course of the disease, but no specific response by coelomocytes to amoebae has been observed.

INTRODUCTION

Mass mortalities of Strongylocentrotus droebachiensis, attributed to disease, have occurred along the Atlantic coast of Nova Scotia between 1980 and 1983 (Miller and Colodey, 1983; Moore and Miller, 1983; Scheibling, this volume and in press). The disease is characterised by progressive loss of oral, tube foot and spine function. Laboratory experiments suggest that a waterborne infective agent is involved (Scheibling and Stephenson, 1984). Histopathological studies have indicated that the symptoms of disease are diffuse and occur mainly in the body wall, water vascular system and adjacent nerve tissue (Jones et al., in press).

Amoebae have consistently been observed in tissues of affected echinoids (Jones et al., in press). The amoebae may be involved in pathogenesis or may be secondary invaders, although no other pathogens are evident (Jones et al., in press).

This paper describes these amoebae and their distribution in tissues of diseased echinoids. Affinities with the genus Paramoeba are discussed.

MATERIALS AND METHODS

Diseased Strongylocentrotus droebachiensis were obtained in the laboratory by exposing healthy individuals to diseased conspecifics, as described by Scheibling and Stephenson (1984), and in nature during an epizootic at Eagle Head, Nova Scotia, in September, 1983. Healthy individuals were obtained from Sandy Cove on the Bay of Fundy, an area that had not been affected by the disease (Scheibling and Stephenson, 1984). Tissues from individuals in early, intermediate and late stages of disease, and uninfected individuals, were prepared for light and electron microscopy as described by Jones et al. (in press).

Six individuals in intermediate or late stages of disease, and three healthy individuals, were rinsed in filtered seawater (0.22 μm membrane filter). A section of body wall (1 cm^2) was excised, rinsed and placed in filtered seawater. The water vascular and nerve tissues were teased apart with forceps and the resultant suspension of cells and fragments was filtered through a Sartorius 0.22 μm cellulose acetate membrane filter. Filters were fixed immediately in Schaudinn's or Stieve's fixatives for 15 minutes or longer, rinsed, and stained in haematoxylin or celestin blue - eosin.

One healthy individual and two individuals in late stages of disease were immersed in separate containers of filtered seawater (100 ml) for 2 hours at room temperature. The bathing medium from each was then decanted, filtered, fixed and stained.

RESULTS

The number of amoebae present in echinoid tissues increases markedly from early to late stages of disease. In early stages (first signs of limp tube feet), isolated amoebae occur in the radial nerve and in nerve tissue at the base of the tube feet. No muscle necrosis is evident in tube feet at this stage. In intermediate and late stages of the disease, amoebae occur throughout the body wall, water vascular system and adjacent tissues (Fig. 1A). Muscle necrosis, sloughing of coelomic lining cells, and connective tissue disruption increase during these stages.

Amoebae are sometimes present in fresh coelomic fluid from late-stage individuals; in one instance numerous amoebae were found. The amoebae are small (20-30 μm) with short, blunt pseudopodia (Fig. 1B). They showed slow locomotion on the slide, with a marked tendency to stay within, or to move into, clumps of amoebocytes. In fixed material the amoebae have a spherical nucleus (4 μm) with a large nucleolus, and an ovoid paranuclear organelle which resembles a parasome. This organelle is 3-4 μm in length, with basophilic polar caps and a central region of low staining affinity (e.g. Figs. 4, 7). Occasionally, two organelles are present, both close to the nucleus.

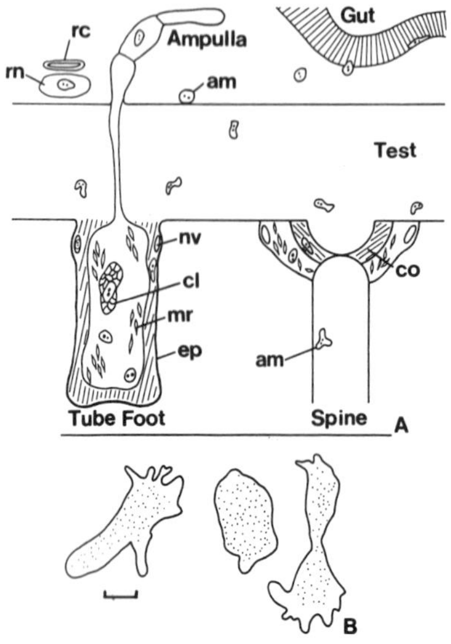

Fig. 1. Amoebae in echinoid tissues. A. General distribution pattern in body wall and adjacent tissues (highly diagrammatic): late stage, with signs of disease in tube feet and spines (muscle necrosis, sloughed coelomic lining cells forming clumps in tube foot lumen). am - amoeba; cl - coelomic lining cells; co - collagen; ep - epithelium; mr - muscle remnant; nv - nervous tissue; rc - radial canal; rn - radial nerve. Not to scale. B. Locomotive form of lining amoebae from coelomic fluid (sketches, bright field illumination). Bar = 5 μm approximately.

In the gut, amoebae appear to emerge from the gut wall into the coelom. Clusters of amoebae occur beneath, within or outside the epithelium (Fig. 2). Little tissue disruption is evident in these regions of emergence.

In the water vascular system, amoebae are common in tube feet, where they occur in lumina or among clumps of sloughed cells, beneath the epithelium and among muscle spindles (Fig. 3,4), and in lumina and walls of ampullae (Fig. 5). They appear to move into ampullae from the coelom across the ampulla wall. Amoebae in lumina of tube feet and ampullae are rounded in shape without prominent pseudopodia (Fig. 3,5).

In the body wall, amoebae occur at all levels in the test, spines and spine bases. Amoebae in calcified tissue are elongated in shape and insinuated between calcite plates, suggesting movement through the organic matrix along an internal-external axis.

In nerve tissue (radial and ampullary nerves, nerve tissue at bases of tube feet and spines), amoebae occur both among cell bodies and at all levels within the neuropile.

In general, amoebae are scattered widely and in low density in echinoid tissues, although they may occur in clusters in the gut wall, water vascular system and radial nerve, particularly in late stages of disease.

Ultrastructurally, there is evidence for ingestion of cellular material by amoebae in late-stage tube feet. Amoebae contain numerous food vacuoles with cell debris, and form pseudopodia (Fig. 6). There is no evidence that intact cells are ingested, or that bacteria or other micro-organisms are present.

There is no histological evidence that echinoid coelomocytes respond to the amoebae, although red and white spherule cells invade the same tissues as the amoebae in increasing numbers as the disease progresses (Jones et al. in press). There is no evidence of phagocytosis or encapsulation of the amoebae by phagocytes.

Stained membrane filters from late stage echinoids contained amoebae, together with muscle spindles, coelomocytes, degenerating cells and other tissue fragments (Fig. 7).

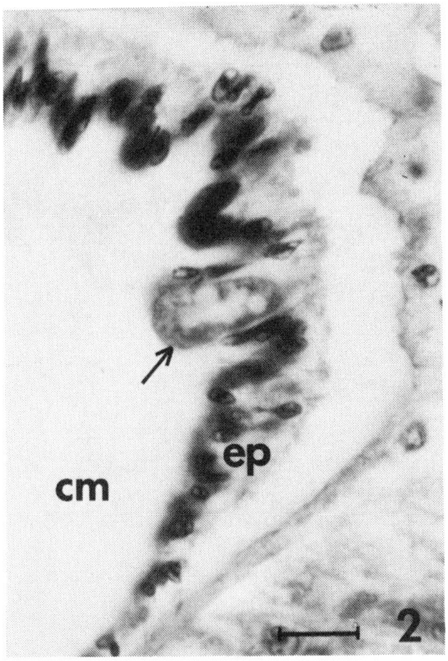

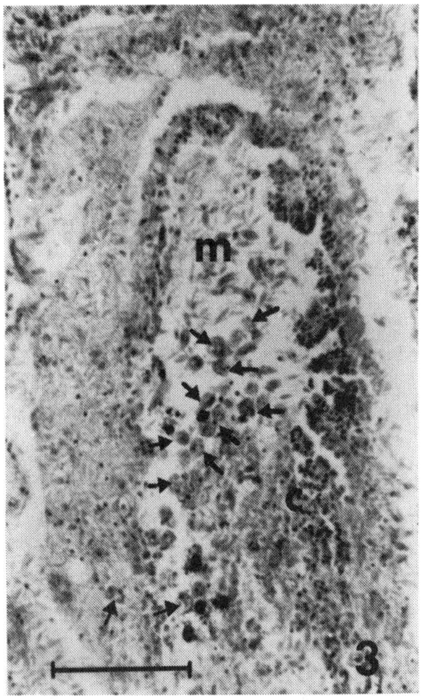

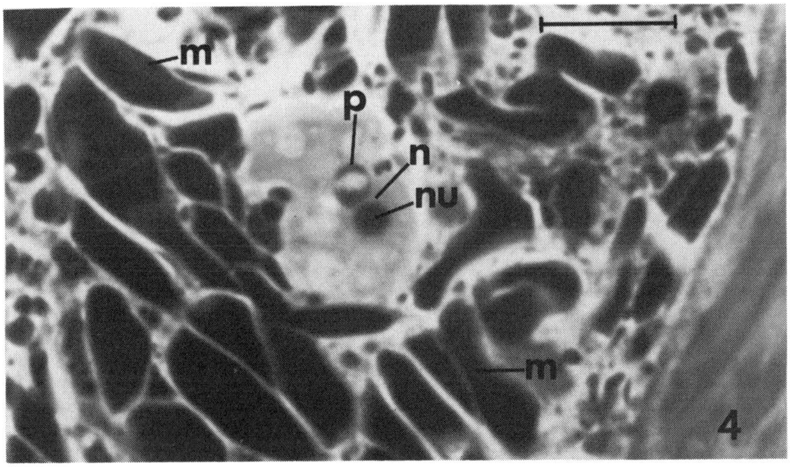

Fig. 2. Amoeba (↑) emerging through the peritoneal epithelium (ep) into the perivisceral coelom (cm). Epithelial cells are distorted by this movement but not extensively damaged. Bar = 10 µm.

Fig. 3. Group of amoebae (arrows) in tube foot lumen. Late stage, with extensive tissue necrosis; lumen contains muscle remnants (m) and clumps of coelomic lining cells (c). Bar = 100 µm.

Fig. 4. Amoeba among muscle remnants (m) in late-stage tube foot, showing closely apposed nucleus (n) with prominent nucleolus (nu) and parasome (p). Field-collected individual. Bar = 10 µm.

Rod-shaped bacteria were occasionally found in clusters in a mucoid substance and are attributed to contamination from gut contents. Amoebae were not recovered from healthy individuals. Sea water surrounding late stage echinoids appeared pink and cloudy after 2 hours. Upon filtration this medium was found to contain numerous coelomocytes of all types, and a few amoebae which tended to appear rounded with dense cytoplasm. The solution bathing one healthy individual remained clear after

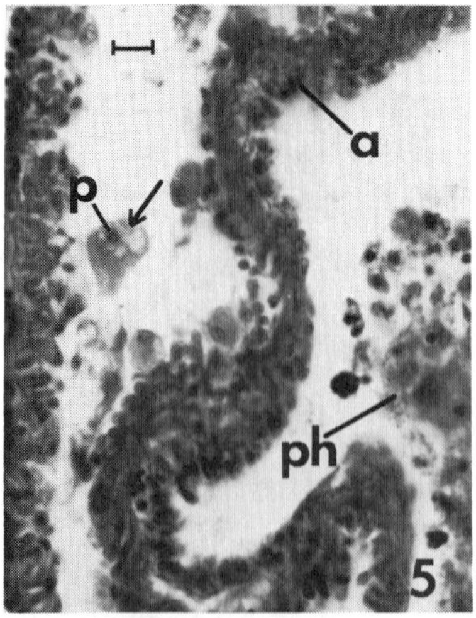

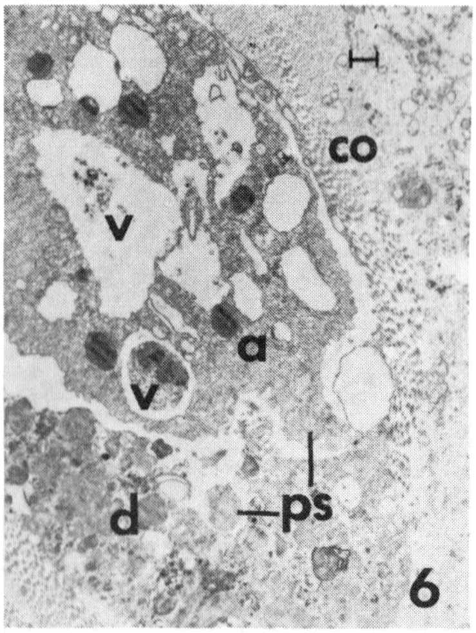

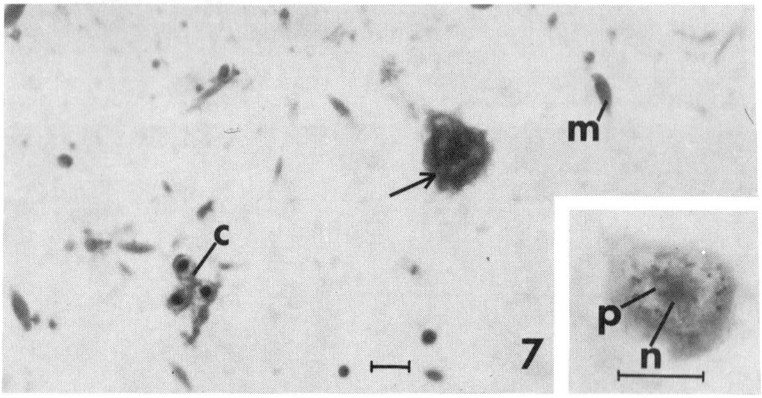

Fig. 5. Amoeba (↑) in lumen of ampulla (a) p-parasome; ph-phagocytes in coelomic fluid. Bar = 10 μm

Fig. 6. EM amoeba (a) within degenerating tube foot connective tissue. The amoeba is surrounded by collagen fibrils (co) and cellular debris (d); pseudopodium (ps) formation appears to be associated with ingestion of debris, and the amoeba contains vacuoles (v) with similar contents. Bar = 1 μm.

Fig. 7. Amoeba (↑), muscle remnants (m) and urchin cells (c) on membrane filters. Bar = 10 μm. Inset: nucleus (n) and parasome (p). Bar = 10 μm.

2 hours and contained very few echinoid cells. No organisms resembling amoebae were observed on the filter.

DISCUSSION

The amoeba found in tissues of Strongylocentrotus droebachiensis is morphologically similar to the genus Paramoeba. The presence of a parasome (nebenkorper) distinguishes this genus from other small mayorellid-like amoebae (Page, 1970, 1973). The parasome is a DNA-containing organelle which is situated near the nucleus and undergoes division, asynchronous with nuclear division (Page, 1970, 1973). It is present in all species of Paramoeba and in the morphologically dissimilar genus Janickina (Grell, 1970; Hollande, 1980). Living amoebae in this study produced short, blunt pseudopodia, resembling those of mayorellids rather than the limax form of Janickina (Page, 1970). Therefore, we suggest that the amoeba is a species of Paramoeba.

Paramoeba is a common marine genus, which has been isolated intertidally from Maine and Europe (Page, 1970, 1973) and in North Atlantic oceanic water from the surface microlayer to at least 300 m depth (Davis et al. 1978). Although most species are freeliving (Page, 1970), P. perniciosa may be an obligate parasite in the blue crab, Callinectes sapidus, off Maryland and Virginia (Sprague et al. 1969; Johnson, 1977), causing mass mortality (grey crab disease) during early summer moulting periods (Sprague and Beckett, 1968; Johnson, 1977; Couch, 1983).

Other species of amoebae described from invertebrate or vertebrate tissues appear to be opportunistic facultative parasites although, as in Acanthamoeba and Naegleria in man, they may be highly pathogenic (John, 1982; Martinez et al. 1975). Cheng (1970) described freshwater Hartmanella associated with mass mortality in oysters, but he considered them to be secondary invaders in debilitated oysters.

Amoebae were consistently present in intermediate or late-stage echinoids, and no other micro-organisms have been found in diseased tissues. However, since the amoebae were not abundant in the tissues sampled, it is not known whether they are pathogenic, or are secondary invaders (e.g. following a viral disease). The distribution of amoebae in tissues coincides with that of the coelomocytes which infiltrate the tissues in response to disease (Jones et al., in press). These coelomocytes may be responding to tissue damage directly or indirectly associated with the amoebae; or both amoebae and coelomocytes may be invading tissue damaged by some other causative agent.

Since this amoeba does invade tissues, and ingest tissue debris, it is behaving as a tissue parasite whether or not it is a pathogen, and its behaviour and effects in echinoids merit further attention.

ACKNOWLEDGEMENTS

This research was supported by D.S.S. contract 10SC.FP101-3-0217 (UP).

REFERENCES

Cheng, T.C. 1970, Hartmanella tahitiensis sp. n, an amoeba associated with mass mortalities of the oyster Crassostrea commercialis in Tahiti, French Polynesia. J. Invertebr. Pathol. 15, 405-419.

Couch, J.A. 1983, Diseases caused by Protozoa. In: Bliss, D.E. (1983) Ed.: The Biology of Crustacea. Vol. 6, pp. 79-11. Academic Press, New York & London.

Davis, P.G., D.A. Caron & J. McN. Sieburth 1978, Oceanic amoebae from the North Atlantic: culture, distribution, and taxonomy. Trans. Amer. Microsc. Soc. 97, 73-88.

Grell, K.G. & G. Benwitz 1970, Ultrastruktur mariner Amoeben. 1. Paramoeba eilhardi. Arch. Protistenk. 112, 119-137

Hollande, A. 1980, Identification du parasome (nebenkern) de Janickina pigmentifera a un symbionte (Perkinsiella amoebae nov. gen. - nov. sp.) apparente aux flagelles kinetoplastidies. Protistologica 16, 613-625.

John, D.T. 1982, Primary amoebic meningoencephalitis and the biology of Naegleria fowleri. Ann. Rev. Microbiol. 36, 101-123.

Johnson, P.T. 1977, Paramoebiasis in the blue crab, Callinectes sapidus. J. Invertebr. Pathol. 29, 308-320.

Jones, G.M., A.J. Hebda, R.E. Scheibling & R.J. Miller (in press), Histopathology of the disease causing mass mortality of sea urchins (Strongylocentrotus droebachiensis) in Nova Scotia. J. Invertebr. Pathol.

Martinez, A.J., S.M. Markowitz & R.J. Duma, 1975, Experimental pneumonitis and encephalitis caused by Acanthamoeba in mice: Pathogenesis and ultrastructural features. J. Infect. Dis. 131, 692-699.

Miller, R.J. & A.G. Colodey 1983, Widespread mass mortalities of the green sea urchin in Nova Scotia, Canada. Mar. Biol. 73: 263-267.

Moore, D.S. & R.J. Miller 1983, Recovery of macroalgae following widespread sea urchin mortality with a description of the nearshore hard-bottom habitat on the Atlantic coast of Nova Scotia. Can. Tech. Rep. Fish. Aquat. Sci. 1230: vii + 94 p.

Scheibling, R.E. (in press), Echinoids, epizootics, and ecological stability in the rocky subtidal off Nova Scotia. Helgolander Meeresunters. 38.

Scheibling, R.E. & R.L. Stephenson 1984, Mass mortality of Strongylocentrotus droebachiensis (Echinodermata: Echinodea) off Nova Scotia, Canada. Mar. Biol. 78, 153-164.

Sprague, V. and R.L. Beckett 1968, The nature of the etiological agent of "gray crab" disease. J. Invertebr. Pathol. 11, 503.

Sprague, V., R.L. Beckett & T.K. Sawyer 1969, A new species of Paramoeba (Amoebida, Paramoebidae) parasitic in the crab Callinectes sapidus J. Invertebr. Pathol. 14, 167-174.

Proceedings of the Fifth International Echinoderm Conference / Galway / 24-29 September 1984

Ecology of juvenile green sea urchins (*Strongylocentrotus droebachiensis*) at an urchin dominated sublittoral site in eastern Newfoundland

D.W.KEATS, G.R.SOUTH & D.H.STEELE
Department of Biology, Memorial University of Newfoundland, St.John's, Canada

ABSTRACT: Urchin population structure was examined over a period of 5 y at an urchin dominated site in eastern Newfoundland, in relation to depth, inter-annual variation, and microhabitat. The distribution of juveniles was the main focus of the study. Juvenile urchin density was low in the two zones with abundant macroalgae (0-2m and > 12m below MLW), and highest in the barrens which occur at intermediate depths. Juvenile urchins were abundant during 1979-1980, but declined over the period from 1981-1983. There is evidence for high predation rates on juveniles. Nevertheless, by inhabiting cryptic micro-habitat beneath undercut corallines and among the uprights of Lithothamnion glaciale, enough survive to maintain the dense population of large urchins, whose grazing maintains the barrens.

1. INTRODUCTION

The green sea urchin (Strongylocentrotus droebachiensis), by selective grazing, controls the biomass, productivity and diversity of benthic hard-substratum communities throughout much of its range (Paine and Vadas 1969; Foreman 1977; Mann 1977; Chapman 1981; Harris 1981; Keats et al. 1982; Hagen 1983; Himmelman et al. 1983a). In Nova Scotia urchin populations expanded during the early 1970's, and decimated kelp (Laminaria) communities along much of the coastline (Mann 1977). Approximately 12 y later urchin populations declined as a result of mass mortality due to disease, and the kelp bed communities have recovered (Miller and Colodey 1983; Moore and Miller 1983). In contrast, much of the Newfoundland coast has been urchin dominated for at least 20 y, and urchin populations show no evidence of decline.

To understand how these dense urchin populations are maintained requires detailed analysis of population structure over a long time period, focusing particularly on the ecology of the juvenile phase. The size structure of green sea urchin populations has been examined in kelp (Laminaria) beds (Miller and Mann 1973; Lang and Mann 1976), following the destruction of kelp beds by urchins (Lang and Mann 1976; Wharton 1980), and in urchin dominated areas of eastern Canada (Himmelman et al. 1983a,b). In areas barren of non-coralline macroalgae, small urchins often predominate; large individuals are rare (Lang and Mann 1976; Himmelman et al. 1983a,b). In the St. Lawrence Estuary growth rates of juvenile urchins were found to be very low (< 2mm/y), and recruitment showed considerable inter-annual variability (Himmelman et al. 1983a,b). This contrasts with established kelp beds, where population density is low, growth rates are much higher, and larger individuals predominate (Lang and Mann 1976; Keats unpublished).

Few studies have presented long-term data on urchin populations at one site within established barrens, and included newly settled individuals in the analysis. The studies in the St. Lawrence Estuary (Himmelman et al. 1983a,b) included the smallest size classes, were fairly long in duration (3 y), but it is not known whether the results apply to the open coast environment of Newfoundland. Here we present urchin size structure data for 5 y at an open coast site in Conception Bay, Newfoundland (52° 51.2'W, 47° 40.0'N) in relation to depth, time, and the removal of larger individuals. These data are intended to determine 1). How urchin population structure changes with depth; 2). If there are fluctuations in recruitment, and if so how they are translated into fluctuations in population structure; 3). If the size range at which predation is important can be identified. These data form part of an on-going long-term study of the relationship between urchins, urchin predators, and the abundance of macro-algae in eastern Newfoundland (Keats et al. 1982).

2. MATERIALS AND METHODS

2.1 Study area

The rationale for, and means of selection of the study area are relevant to another study, and will be presented in more detail elsewhere (Keats et al., in prep.). The general study employed three similar sites, which were located 1 km apart in Conception Bay. The sites are characteristic of the open Atlantic coast of Newfoundland (South 1983), and have a shallow zone with abundant macro-algae, an intermediate zone with abundant urchins and low and variable non-coralline macroalgal biomass, and a deeper zone of high macro-algal biomass.

2.2 Removal of Urchins

Urchins were removed manually from one transect (hereafter referred to as the experimental transect) by smashing them with hammers during June-August, 1980. Subsequently, urchin smashing was conducted as often as was necessary, or weather conditions and logistics permitted (every 1-2 wks in summer, ca. monthly in winter). Urchins were removed from a second transect using surface applied high-calcium quicklime during November of 1981. The area was enlarged by further application of quicklime during July of 1982. A third, control area was maintained for comparative purposes.

2.3 Sampling

From 5 to 10 samples were taken using 0.1 m^2 quadrats in the shallow algae zone (0-2m), at the upper edge of the urchin/coralline zone (2-3m), and well within the urchin/coralline zone (6-9m). Urchin samples from the Agarum/Ptilota zone (12-18m), gathered from the control area during another study (Keats et al. 1982), were also used for depth comparison. Quadrats were placed by blind casting. Large urchin were collected by hand, and smaller urchins by a diver-operated air dredge. The dates on which sampling was conducted are presented in Figs. 1-4.

2.4 Urchin microhabitat

To determine the role of the nature of the coralline algae in governing the abundance of juvenile urchins, urchins of < 5mm test diameter were sampled from the 0.05 m^2 quadrats placed at random in each of the following habitats: 1). on smooth corallines (Phymatolithon laevigatum, P. rugulosum, and Clathromorphum circumscriptum); and 2). branched, undercut corallines (Lithothamnion glaciale).

3. RESULTS

3.1 Analysis of size frequency distributions

Urchin size frequency distributions for the study period, from depths of 0-2m, 2-3m, 6-9m, and 12-18m, are presented in Figs 1-4. Within the shallow algae zone (0-2m) density was low throughout the study, and no single size class was particularly abundant. In general, abundance of the smaller sizes (0-2 and 2.1-4 mm) was greater at 2-3m than at 0-2m, and peak abundance was at 6-9m. At 6-9m there was a marked increase in density, due entirely to an increase in the number of small urchins. The density of larger (> 15mm) urchins was lower at 6-9m, as compared with 2-3m, but was higher than it was at 0-2m. Density was lower within the Agarum/Ptilota zone than at 2-3m and 6-9m and small urchins were rare. It was only within the Agarum/Ptilota-zone that large urchins were found in appreciable quantities.

No temporal trends occur in the size frequency data from 0-2m (Fig. 1). On the control transect at 2-3m population structure was certainly not static, but no strong temporal trends are obvious in these data. The main mode fluctuated from the 3 to the 7 mm size classes. On the experimental transect at 2-3m the mode showed a tendency to shift to the right and became less distinct after removal of larger urchins. This was probably a result of increased growth in response to the increase in macroalgal biomass, but the trend was obscured by the removal of urchins which increased to more than ca 15 mm test diameter.

Fig. 5 shows the abundance of the 1, 3, 5, and 7mm size classes at 6-9m plotted as a percentage of 1979 density (for each size class). The abundance of the 1 mm size class was high during 1979-1980, but was low after 1981 (Figs. 4,5). A decrease in the 3 mm size class showed a lag of ca 1 y behind a decline in the 1 mm size class. There was very little change in the abundance of the 5 mm size class until the summer of 1982, by which time it had decreased considerably. It remained at a similar level during 1983. Very few urchins in the 1 and 3mm size classes were present during the summers of 1982 and 1983. The 5 and 7mm size classes increased

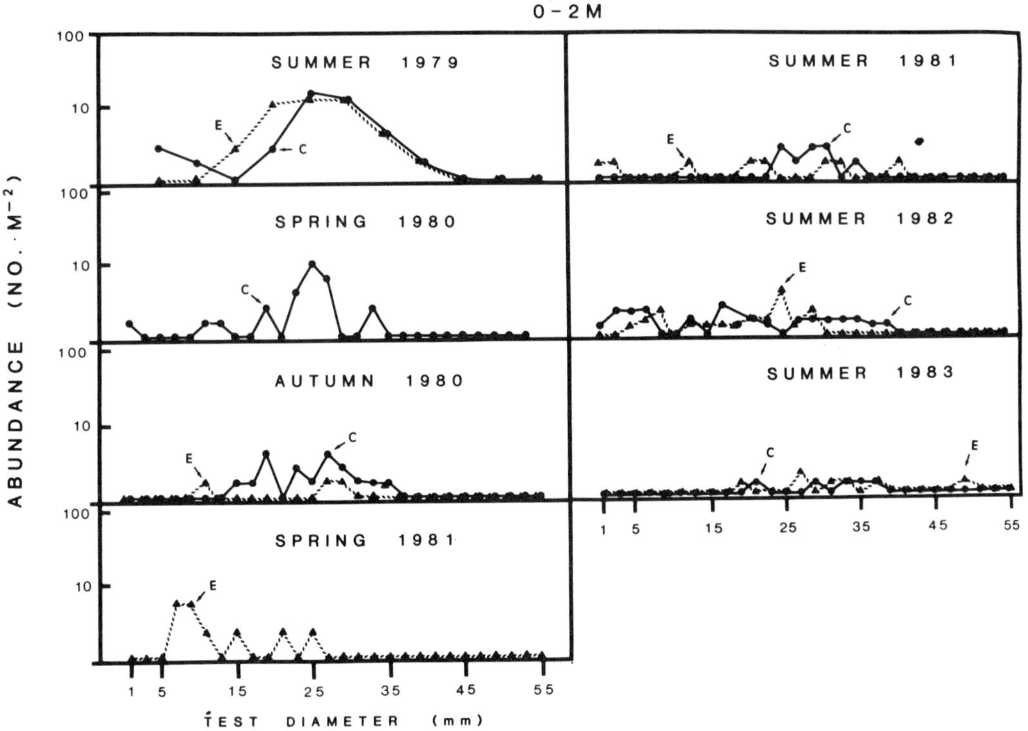

Fig. 1. Urchin size frequency distributions from the shallow algae zone (0-2m). Note: Curves for 1979 are based on 5mm intervals, others on 2mm intervals (C = control transect, E = experimental transect).

in abundance slightly, as the 1 and 3mm classes began to decline, but then decreased again. This suggests that some of the smaller urchins had grown into these larger size classes. The change in the actual density of the 5 and 7 mm size classes could not account for the dramatic decrease in the abundance of 1 and 3mm classes. Nor could migration to other depths account for this change (Figs. 1,2,4). The data suggest that most of the decrease in abundance of smaller urchins was due to mortality, or that mortality at the larger size classes was approximately equal to the number of urchins growing into the larger classes.

A similar trend was evident on the experimental transect (Fig. 3), although the peak remained at 3 mm during the summer of 1982 and 1983. This may indicate reduced mortality of small urchins under the canopy on the experimental transect. The data from the spring of 1982 indicated a tendency for increased growth of smaller urchins after removal of the larger individuals, but this was obscured at other times by the frequent removal of individuals larger than 15mm test diameter.

Both the manual removal and quicklime technique gave poor removal of the smaller (< 15 mm) urchins. This is due largely to the occupancy of cryptic habitats by the smaller urchins (see below).

3.2 Microhabitat of juvenile urchins

Table 1 presents the results of sampling juvenile urchin abundance on smooth (Phymatolithon laevigatum, P. rugulosum, and Clathromorphum circumscriptum) and

Table 1. Estimates of the abundance (no./.05m^2) of juvenile urchins (<10mm) on smooth vs. branched and undercut coralline algae.

	Smooth	Rough
Abundance	0.2	41.7
N	10	10
Std. err.	0.13	6.5

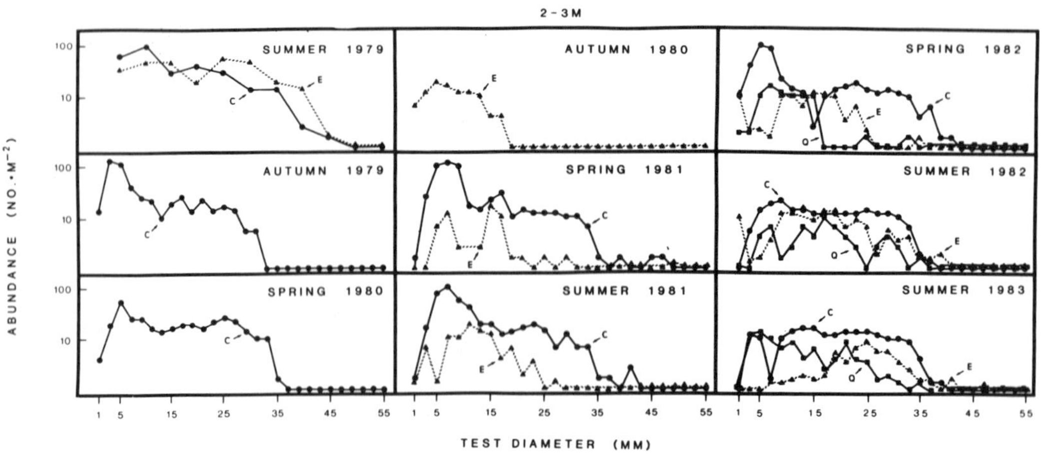

Fig. 2. Urchin size frequency distributions from 2-3m. Note: Curves for summer 1979 are based on 5mm intervals, others on 2mm intervals (C = control transect, E = experimental transect, Q = quicklime transect).

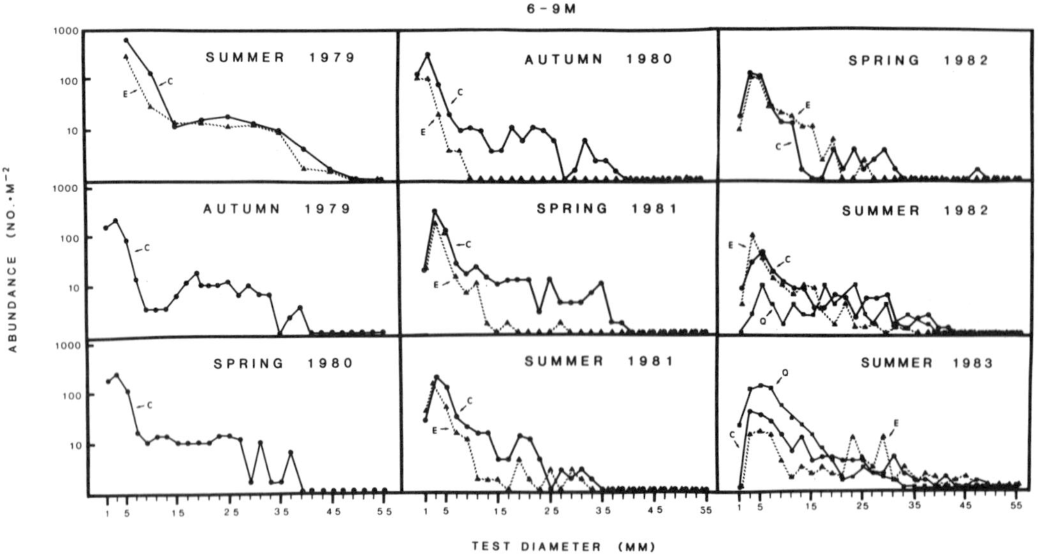

Fig. 3. Urchin size frequency distributions from 6-9m. Note: Curves for summer 1979 are based on 5mm intervals, others on 2mm intervals (C = control transect, E = experimental transect, Q = quicklime transect).

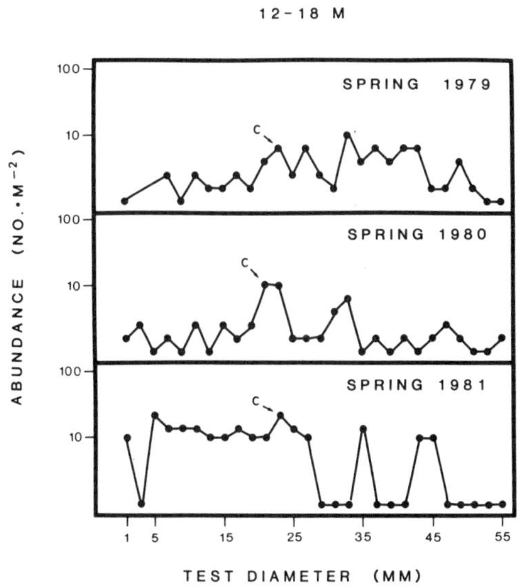

Fig. 4. Urchin size frequency distributions (based on 2mm intervals) for 12-18m on the control transect.

branched, undercut corallines (Lithothamnion glaciale). Juvenile urchins were largely restricted to the branched and undercut corallines.

4. DISCUSSION

4.1 Relation of juvenile abundance and population size-structure with depth, time, and fleshy macro-algae

Changes in population structure with depth were similar to those reported for a population sampled from a loose substratum during 1968 at Portugal Cove, Conception Bay (Himmelman 1969, 1980). Small urchins were uncommon in shallow water, peaked in abundance in the mid to lower barrens, and decreased to a lower density in deeper water. Himmelman (1969) showed that a similar trend was evident, although less pronounced, on bedrock. Air dredge sampling was not employed by Himmelman, however, so it is unlikely that bedrock was sampled as accurately as the loose substratum. Similarly, in Nova Scotia (Wharton 1980) and Quebec (Himmelman et al. 1983a) small urchins were most abundant in barrens.

The low density of small urchins in kelp beds in Nova Scotia (Wharton and Mann 1981), and their low density within the two depth ranges with abundant fleshy macro-algae at the present site suggests that macro-algae may have a negative influence on urchin recruitment. Pearse et al. (1970) hypothesized that large non-calcareous macro-algae reduce urchin recruitment. In support of that hypothesis, Tegner and Dayton (1981) noted a negative correlation between recruitment of Strongylocentrotus franciscanus and Macrocystis pyrifera abundance in California. The hypothesis is worthy of direct testing, especially since there are other complicating factors (including predation, exposure to wave action, the nature of dominant algae with respect to urchin food preferences, and the nature of the substratum). Comparisons between control and urchin removal areas in this study are of little use in this regard, because of the low overall recruitment after 1980.

There are no obvious trends in growth and mortality evident in the data from 0-2m and 2-3m on the control transect, although size-structure was not static. Detailed analysis of the size structure of urchins from the 6-9m samples indicated that growth rates were very low, and inter-annual variation in recruitment and/or mortality was large. Similarly, Himmelman et al. (1983a,b) reported growth rates of 1-1.5mm y^{-1} for urchins in a barren area in the St. Lawrence estuary.

Himmelman et al. (1983a,b) also recorded considerable interannual variation in recruitment; recruitment was high in 1977, but no apparent recruitment occurred during 1978-1981. Propp (1977) also reported large inter-annual variations in population structure of S. droebachiensis, indicative of fluctuations in recruitment, in the Barents Sea. This pattern is common among herbivorous echinoids (Tegner and Dayton 1981, Andrew and Choat 1982), and other benthic invertebrates with pelagic larval phases (Scheltema 1974, Dayton and Tegner 1984).

Himmelman et al. (1983b) recorded a change in the annual growth increment of juvenile urchins from 1-1.5 mm to 6 mm after seaweed biomass increased following adult urchin removal. An increase in the growth rates of juvenile urchins followed regrowth of Macrocystis in California (Pearse et al. 1970). An increase in growth by juvenile urchins after macrophyte increase was evident in the present study, because the mode shifted to the right and became less pronounced. The tracing of modal frequencies through time, however, was not possible. Therefore exact changes in growth cannot be determined. Nevertheless there is ample evidence that the growth of juvenile urchins is food-limited in barren

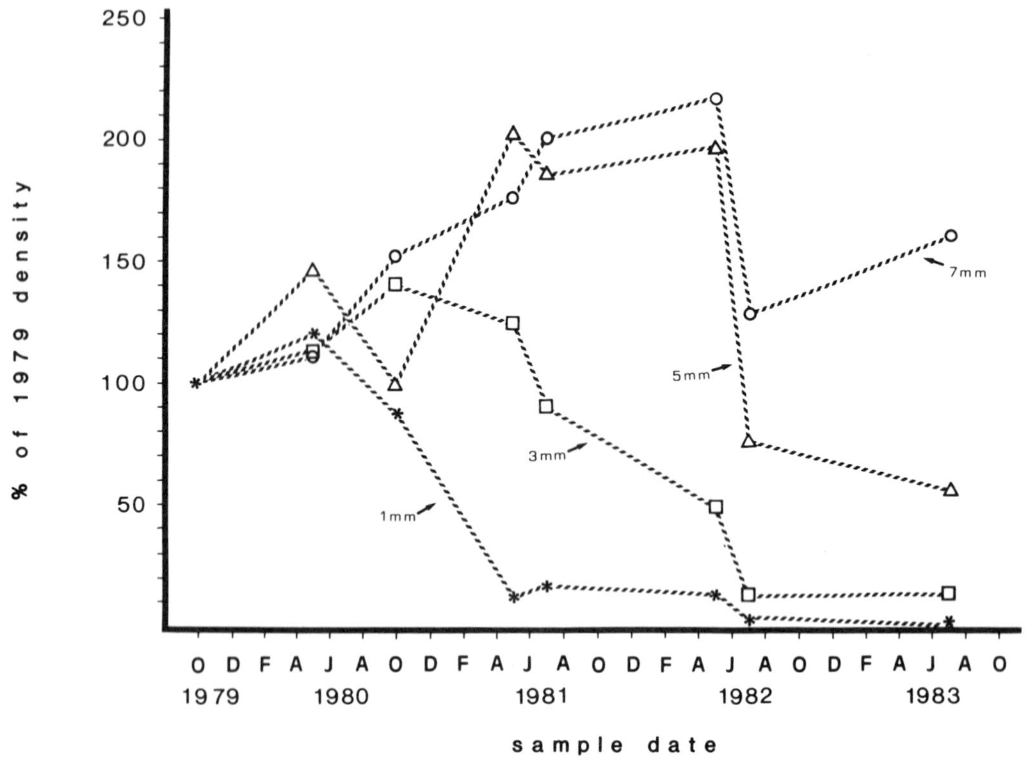

Fig. 5. Abundance of the 1,3,5, and 7mm size classes at 6-9m plotted as percentage of their 1979 density.

conditions (Himmelman et al. 1983a,b). Adult growth and reproduction are also food limited in such habitats (Keats et al. 1984).

4.2 Possible role of predation in determining population structure

A striking feature of the urchin population structure at 6-9m, and to a lesser degree at 2-3m, was the pronounced mode at 3mm and a decrease in abundance in the 3 mm to 10 mm range. This is considered here to be evidence of intense predation on small urchins. Martel (1983) reported that urchins occurred in the alimentary tracts of 100% of 38 territorial male cunners (Tautogolabrus adspersus) examined from St. Philips, a typical urchin/coralline habitat 8 km south of the present study area. Urchins < 5mm occurred in 14%, between 5 and 10 mm in 100%, and larger than 10 mm in 22% of the cunners examined. Martin (1979) reported urchins in the alimentary tracts of 29% of a random sample of 115 cunners from the same site. Urchins made up 14% of the diet of the total sample by weight.

Cunners were the most abundant fish at the study site, and were often seen feeding on the benthos in large aggregations of 20-40 individuals (Keats et al., in prep.). For example, Corallina officinalis supported dense populations of juvenile mussels (Mytilus edulis and Vosella modiolus), polychaetes, and other invertebrates in the spring when the cunners emerged from their over-wintering sites (Keats, unpubl.). Cunners were frequently seen stripping these animals from Corallina, which by late summer was essentially bare. The effects of cunner foraging in other habitats was less obvious to casual observation, but must have been equally severe. These observations and stomach contents analysis indicate that considerable predation on juvenile urchins by cunners occurs in eastern Newfoundland. The predominance of urchins in the 5-10 mm size group in cunner stomachs (Martel 1982), the importance of urchins in the general diet of cunners (Martin 1979), and the marked decrease in abundance of urchins beyond 5 mm suggests that size selective mortality occurs, at least in part due to predation by cunners.

Another predator, the winter flounder (Pseudopleuronectes americanus) was also very abundant in the study area from spring until autumn, and frequently contained urchins in the 5-10 mm range (Keats et al., in prep.).

Further evidence for the importance of predation on juvenile urchins is provided by the observed changes in population structure following a decrease in recruitment. The large mode at 3 mm did not shift significantly to the right; rather it decreased without a compensating increase in the density of larger urchins. Thus size selective predation on juveniles may be an important factor governing the size structure of the urchin population. Given the persistence of high urchin biomass it is evident that despite high levels of predation enough juveniles escape to maintain the adult population. Andrew and Choat (1982) demonstrated experimentally that fish predation on juvenile Evechinus chloroticus in New Zealand was important in controlling population structure, but enough juveniles escaped predation to maintain the urchin/coralline habitat.

4.3 Relation of population structure with the distribution of coralline algae

Propp (1977) observed that juvenile S. droebachiensis were most abundant "under a crust of calcareous algae," and in crevices in the Murmansk area of the Barents Sea. Himmelman (1980) made a similar observation for urchins in eastern Newfoundland. At 6-9m at the study site coverage by the corallines Lithothamnion glaciale and Clathromorphum circumscriptum was 51% (Keats, in prep.). These algae were undercut due to the activities of boring animals, and spaces were present underneath. The erect portions of Lithothamnion uprights also contained spaces among them. It was in these holes and spaces that juvenile urchins were most abundant. Lithothamnion was rare, and Clathromorphum was not undercut in shallower water. Juvenile urchins were rare on the smoother, non-undercut Phymatolithon laevigatum and P. rugulosum, and non-undercut C. circumscriptum. Urchins escape predation in the cryptic habitats, and hence density was greatest in the 2-6mm size range. Undercut Lithothamnion glaciale was also very abundant (73% cover) in the Agarum/Ptilota zone, yet juvenile urchin densities were low. It is therefore evident that factors other than purely the abundance of cryptic microhabitat determines the change in population structure with depth. The presence of the Agarum canopy and Ptilota understory may be important, particularly as both of these species are evidently defended against predation by adult urchins (Himmelman 1980; Larson et al. 1980; Keats et al. 1984). Quantification of the role of the microhabitat as protection from predation, and a determination of whether mortality or settlement factors are responsible for the distribution of juvenile urchins with respect to coralline algae would be valuable.

REFERENCES

Andrew, N.L., & J.H. Choat, 1982. The influence of predation and conspecific adults on the abundance of juvenile Evechinus chloroticus. (Echinoida: Echinometridae). Oecologia (Berl.) 54: 80-87.

Chapman, A.R.O., 1981. Stability of sea urchin dominated barren grounds following the destructive grazing of kelp by sea urchins in St. Margaret's Bay, Eastern Canada. Mar. Biol. (Berl.) 62: 307-311.

Dayton, P.K., & M.J. Tegner, 1984. The importance of scale in community ecology: A kelp forest example with terrestrial analogs. In A New Ecology: Novel Approaches to Interactive Systems. Ed by P.W. Price, C.N. Slobodchikoff, & W.S. Gaud., John Wiley & Sons, Inc., pp. 457-480.

Foreman, R.E. 1977. Benthic community modification and recovery following intensive grazing by Strongylocentrotus droebachensis. Helgolander wiss. Meeresunters. 30: 468-484.

Hagen, N.T., 1983. Destructive grazing of kelp beds by sea urchins in Vestfjorden, northern Norway. Sarsia 68: 177-190.

Harris, L.G., 1981. Studies of community succession in a sea urchin barrens area following sea urchin removal. Am. Zool. 21: 1019.

Himmelman, J.H., 1969. Some aspects of the ecology of Strongylocentrotus droebachiensis in eastern Newfoundland. Thesis (M.Sc.) Mem. Univ. Nfld., 159 pp.

Himmelman, J.H., 1980. The role of the green sea urchin, Strongylocentrotus droebachiensis, in the rocky subtidal region of Newfoundland. Can. Tech. Rept. Fish. Aquat. Sci. 954: 92-119.

Himmelman, J.H., A. Cardinal, & E. Bourget, 1983a. Community development following removal of urchins, Strongylocentrotus droebachiensis, from the subtidal zone of the St. Lawrence Estuary, eastern Canada. Oecologia (Berl.) 59: 27-39.

Himmelman, J.H., Y. Lavergne, F. Axelsen, A. Cardinal, & E. Bourget, 1983b. Sea urchins in The Saint Lawrence estuary: Their abundance, size structure, and suitability for commercial exploitation. Can. J. Fish. Aquat. Sci. 40: 474-486.

Keats, D.W., G.R. South, & D.H. Steele, 1982. The occurrence of Agarum cribrosum (Mert.) Bory. (Phaeophyta, Laminariales) in relation to some of its competitors and predators in Newfoundland. Phycologia 21: 189-191.

Keats, D.W., D.H. Steele and G.R. South. 1984. Depth dependent reproductive output of the green sea urchin Strongylocentrotus droebachiensis (O.F. Mull.) in relation to the nature and availability of food. J. Exp. Mar. Biol. Ecol. 80: in press.

Lang, C. & K.H. Mann, 1976. Changes in sea urchin populations after the destruction of kelp beds. Mar. Biol. (Berl)., 36: 321-326.

Larson, B.R., R.L. Vadas, & M. Keser, 1980. Feeding and nutritional ecology of the sea urchin Strongylocentrotus droebachiensis in Maine, USA. Mar. Biol. (Berl.) 59: 49-62.

Mann, K.H., 1977. Destruction of kelp beds by sea urchins: A cyclical phenomenon or irreversible degradation. Helgolander. wiss. Meeresunters. 30: 455-467.

Martel, G., 1983. Sexual selection and territory size in male cunners (Tautogolabrus adspersus) in Conception Bay, Newfoundland. Thesis (M.Sc.) Mem. Univ. Nfld. 114 pp.

Martin, W.D., 1979. The feeding behavior and food habits of the cunner, Tautogolabrus adspersus (Walbaum) in Conception Bay, Newfoundland. Thesis (B.Sc.) Mem. Univ. Nfld. 42 pp.

Miller, R.J., & A.G. Colodey, 1983. Widespread mass mortalities of the green sea urchin in Nova Scotia, Canada. Mar. Biol. (Berl.), 73: 263-267.

Miller, R.J. & K.H. Mann, 1973. Ecological energetics of the seaweed zone in a marine bay on the Atlantic coast of Canada. III. Energy transformations by sea urchins. Mar. Biol. (Berl.) 18: 99-114.

Moore, D.S., & R.J. Miller, 1983. Recovery of macroalgae following widespread sea urchin mortality with a description of the nearshore hard-bottom habitat on the Atlantic coast of Nova Scotia. Can. Tech. Rept. Fish. Aquat. Sci. 1230: 1-94.

Paine, R.T., & R.L. Vadas, 1969. The effects of grazing by sea urchins, Strongylocentrotus spp., on benthic algal populations. Limnol. Oceanogr. 14: 710-719.

Pearse, J.C., M.E. Clark, D.L. Leighton, C.T. Mitchell, & W.J. North, 1970. Marine waste disposal and sea urchin ecology. Kelp Habitat Improvement Project, A. Rep. Appendix, California Institute of Technology, Pasadena.

Propp, M.V., 1977. Ecology of the sea urchin Strongylocentrotus droebachiensis of the Barents sea: Metabolism and regulation of abundance. Biologiya Morya 1: 39-51.

Scheltema, R.S., 1974. Biological interactions determining larval settlement of marine invertebrates. Thalassia Jugoslavica 10: 263-296.

South, G.R., 1983. Benthic marine algae. In Biogeography and Ecology of the Island of Newfoundland. ed. by G.R. South, Dr. W. Junk Publishers, The Hague.

Tegner, M.J. & P.K. Dayton, 1981. Population structure, recruitment and mortality of two sea urchins (Strongylocentrotus franciscanus and S. purpuratus) in a kelp forest. Mar. Ecol. Prog. Ser. 5: 255-268.

Wharton, W.G., 1980. The extent of destructive grazing by the sea urchin Strongylocentrotus droebachiensis on the eastern and southwestern shores of Nova Scotia. Mimeo. Rep. Dept. Supply Serv. Can. Contract no. 1sc7900058 113 pp.

Wharton, W.G. & K.H. Mann, 1981. Relationship between destructive grazing by the sea urchin (Strongylocentrotus droebachiensis), and the abundance of American lobster, Hommarus americanus, on the Atlantic coast of Nova Scotia. Can. J. Fish. Aquat. Sci. 38: 1339-1349.

Proceedings of the Fifth International Echinoderm Conference / Galway / 24-29 September 1984

Annual reproductive periodicity in eight echinoid species on the Caribbean coast of Panama

H.A.LESSIOS
Smithsonian Tropical Research Institute, Balboa, Panama

ABSTRACT: Monthly samples of eight species of echinoids from the San Blas Islands, Panama were injected with isotonic Potassium Chloride solution to test their readiness to spawn. The percentage of animals that responded to these injections each month was used as an index of the reproductive activity of the populations. The samples revealed annual cycles in the reproduction of Echinometra viridis, Lytechinus williamsi, Clypeaster rosaceus, and Leodia sexiesperforata. These species are reproductively active during Panama's wet season from April to December, and reproductively quiescent during the dry season. Lytechinus variegatus, Tripneustes ventricosus, and Clypeaster subdepressus are ripe throughout the year, while Echinometra lucunter shows non-periodic fluctuations in readiness to spawn. The data do not support the hypothesis (Lessios 1981) that in constant environments rare species should exhibit greater reproductive synchrony than common species.

1. INTRODUCTION

The comparative approach has been employed extensively to investigate possible links between reproductive periodicity of echinoderms and fluctuations in their environment. Comparisons between populations of the same species from different localities have been used to explore the relationship between reproduction and environmental seasonality (Pearse 1968, 1969, 1970, 1974), food availability (Crump 1971; Dix 1970; Gonor 1973), water depth (Moore 1934), and degree of exposure to wave action (Lewis and Storey 1984). Reproductive cycles of different species in the same area have also been compared (Bennett and Giese 1955; Pearse 1968; McPherson 1969; O'Connor et al. 1978; Hendler 1979; Lessios 1981; Tyler et al. 1982). Similarities between reproductive cycles of unrelated species in the same environment often suggest the existence of a strong environmental factor defining a particular time period as optimal for reproduction. For example, the synchronous reproduction of many arctic invertebrate species suggests that they are taking advantage of the spring plankton bloom in this area to release their larvae (Giese 1959; Giese and Pearse 1974). Differences in breeding seasons of various species inhabiting the same area are also instructive; one can ask what aspect of the biology of these species accounts for their different response to the same environment. For instance, dissimilarities in the degree of reproductive periodicity of various species in the environmentally constant deep sea led Rokop (1974) to suggest that population density and mobility are important in determining the presence of reproductive periodicity; rare and sessile species have to maintain reproductive synchrony to increase the probability of fertilization of the gametes they shed into the water. Lessios (1981) also invoked population density to explain why in two abundant species of tropical echinoids reproductive periodicity was lax, while in a third, rare one, breeding cycles were well-defined.

Such correlations seldom constitute elegant demonstrations of the importance of a single environmental factor in the reproduction of the species concerned; they do, however, provide working hypotheses about the relative significance of various parameters, hypotheses which can then be tested either by experiments or by additional comparisons between appropriately chosen localities or species. Thus, Lessios's (1981) post hoc hypothesis that, in a relatively aseasonal tropical environment, population density has a role in determining the presence or absence of reproductive periodicity can be tested in a locality where many echinoid species coexist in different population densities. This paper describes the reproductive cycles of eight species of shallow water echinoids with different population densities at the San Blas Archipelago off the north coast of Panama.

The Caribbean coast of Panama exemplifies the kind of tropical environment envisioned by Orton (1920), when he suggested that in the thermally constant tropics marine animals should breed throughout the year. It is characterized by constant temperatures, small tidal fluctuations, and low but constant primary productivity (Glynn 1972; Meyer & Birkeland 1974; Meyer et. al. 1975; Hendler 1976, 1977). Wind velocity is a variable that does exhibit seasonal changes. During the Panamanian dry season, from mid-December to April, the trade winds blow steadily from the north, creating a seasonal increase in seawater turbulence, reducing cloud cover, and eliminating practically all rainfall. The San Blas Archipelago shares these physical characteristics with the Caribbean coast of Panama. It is also a site of lush coral reefs that support a diverse echinoid fauna. The annual reproductive cycles of eight echinoid species from three different habitats were studied in 1982 and 1983. Tripneustes ventricosus and Lytechinus variegatus inhabit Thalassia beds, L. williamsi and Echinometra viridis are found on live coral reef, E. lucunter and Clypeaster rosaceus live on the shallow reef flat, while C. subdepressus and Leodia sexiesperforata are encountered infaunally in sand bottoms. I collected data on the annual reproductive cycles of these species in an attempt to answer the following questions: 1) Which of these species reproduce cyclically, and which breed throughout the year? 2) If reproductive cycles exist in the various species, do they coincide? 3) If there are differences between the cycles of these species, living in the same environment, what are their possible causes? 4) Is there a negative correlation between population density of the various species and degree of reproductive synchrony?

2. MATERIALS AND METHODS

2.1. Collection sites

The study was conducted near San Blas Point in the San Blas Archipelago, (9° 34' N, 78° 79' W). The infaunal sea urchins Leodia sexiesperforata and Clypeaster subdepressus were collected at an approximate depth of 1 m in a sand patch surrounded by Thalassia beds on Tiantupu reef [see Lessios et al. (in press) for a map of the reefs in the area]. The other species were sampled on or around House Reef, approximately 500 m away. Lytechinus williamsi and Echinometra viridis were collected in areas of live Agaricia and Porites on the reef slope at a depth of 2 to 5 m; E. lucunter and C. rosaceus were gathered at less than a meter of water from the subtidal reef flat; L. variegatus and Tripneustes ventricosus were collected in Thalassia beds (1 to 5 m depth) between House Reef and Nalunega Island.

2.2. Sampling method

Twenty individuals of each species were collected four to two days before new moon each month, for thirteen or fourteen consecutive months. This sampling regime was adopted to avoid possible complications arising from a possible lunar spawning subcycle (Pearse 1975). Each echinoid was placed in a separate container and injected with a 0.5 M solution of Potassium Chloride. KCl induces spawning in ripe echinoids (Tyler 1949; Hinegardner 1975). Echinometra viridis, E. lucunter, Leodia sexiesperforata and Lytechinus williamsi received 5 ml of this solution per animal, while Tripneustes ventricosus, L. variegatus, Clypeaster subdepressus and C. rosaceus received 10 ml because of their larger body cavity. The quantities injected are in excess of what is usually required to induce spawning in ripe individuals. The data consist of the percentage of animals in each sample that spawned in response to the KCl injection. No attempt was made to determine whether shed gametes were fertile. As a means of determining annual reproductive periodicity, this method has some disadvantages when compared with the more widely used gonadal indices (Grant and Tyler 1983a) or oocyte measurements (Grant and Tyler 1983b). It does not distinguish between animals that are at a quiescent stage of their cycle and animals that have recently spawned. It also classifies individuals in one of two stages, and thus makes no allowances for variability in the degree of ripeness. It, therefore, tends to be rather insensitive to subtle changes in reproductive condition. However, the method was chosen because it is a direct measure of readiness to spawn at the time the sample was taken and thus provides information necessary to determine spawning synchrony in each species; it is also rapid, and does not necessarily result in the death of the assayed animals, salient considerations given the total number of animals sampled. The method has proven to be useful in the determination of annual cycles in Lytechinus variegatus and of lunar cycles in two species of Diadema (Lessios 1984).

Population density of each species was determined between July and September 1982 in 1.8 m wide transects, stretching from the shallowest point of each habitat to the habitat end, or (in Thalassia beds) to a distance of 30 m. Numbers of epifaunal echinoids were determined visually; infaunal echinoids were located by searching by feel under the sand to a depth of approximately 5 cm.

A "reproductive synchrony index" had to be devised in order to correlate relative synchrony and population density in each species. For the kind of data used in this study, the best measure of relative reproductive synchrony of a species is the maximum positive deviation from the annual mean percent spawning. This measure attains its highest value if the entire population is reproductively quiescent during some months and spawning in synchrony during others; its lowest value would be attained if each individual spawned at random, so that at any given month the percentage of animals responding to KCl hovered around a mean value. Its chief disadvantage is that it also yields a low value for species in which the majority of individuals is ready to spawn at any given month. However, such year-round breeding may constiute an alternate strategy for avoiding gamete watage. If individuals in a species have the capacity to remain reproductive most of the time, they are under no additional selection pressure to time their reproduction so that it coincides with that of conspecifics. Species that spawn continuously through the year, therefore, are not especially useful in testing the hypothesis that reproductive synchronization in a constant environment helps rare species increase the probability of gamete fertilization. Since each species was sampled for thirteen or fourteen months, the last one or two months of each sample were disregarded in the calculation of annual means.

3. RESULTS

Despite the fact that each species was sampled at the same day of the lunar month, conclusions reached regarding annual reproductive cycles have to be modified with one caveat. The sampling regime assumes that a monthly spawning sub-cycle either does not exist, or, if it does, it follows a lunar period. If the day-to-day variation in readiness to spawn is either haphazard or follows a cycle that is not lunar, it could confound the determination of annual periodicity. That the monthly changes in spawning of seven out of the eight species follow rather regular patterns (see below) suggests that daily variation has not induced spurious results. Specifically, the species can be classified into three functional groups according to their reproductive cycles: a) Species in which the majority of individuals are ripe at any month of the year (year-round breeders); b) species that reproduce during the wet season, but stop reproducing during the dry season (seasonal breeders); c) species in which some individuals are ripe at any given month, but the percentage of ripe individuals fluctuates in a haphazard manner (acyclical breeders).

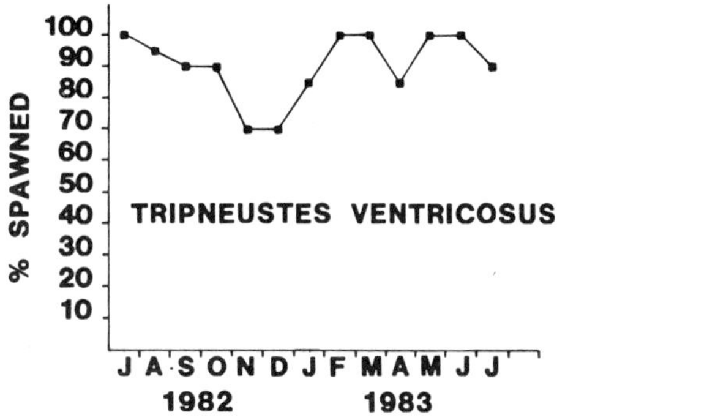

Figure 1. Tripneustes ventricosus: Percent of individuals that spawned when injected with 10 ml of 0.5 M KCl solution. Monthly samples of twenty individuals each were taken 4-2 days before new moon.

3.1. Year-round breeders

Tripneustes ventricosus (Figure 1), Lytechinus variegatus (Figure 2), and Clypeaster subdepressus (Figure 3) spawn through out the year. In these three species the percentage of individuals responding to KCl injections never fell below 60 %. If a more sensitive technique for the assessment of reproductive condition had been employed, an annual peak might have been detected in these three species. The possibility of such a peak does not change the fact that the majority of individuals in these species contain gametes, ready to be released at any month of the year.

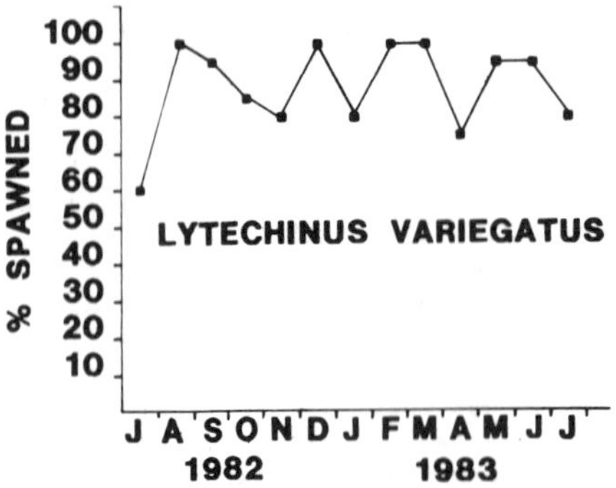

Figure 2. Lytechinus variegatus: Percent of individuals that spawned when injected with 10 ml of 0.5 M KCl solution. Monthly samples of twenty individuals each were taken 4-2 days before new moon.

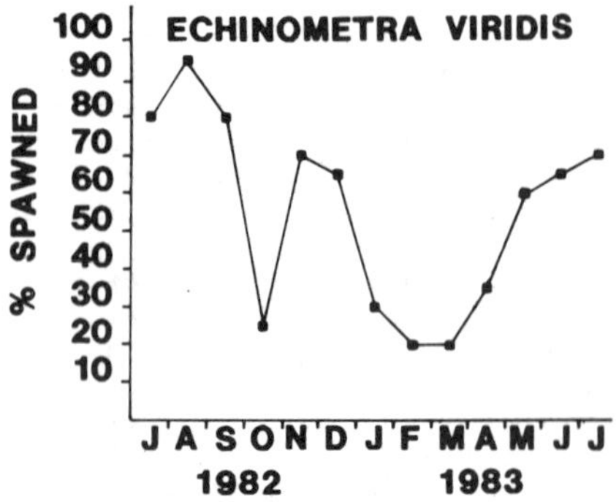

Figure 4. Echinometra viridis: Percent of individuals that spawned when injected with 5 ml of 0.5 M KCl solution. Monthly samples of twenty individuals each were taken 4-2 days before new moon.

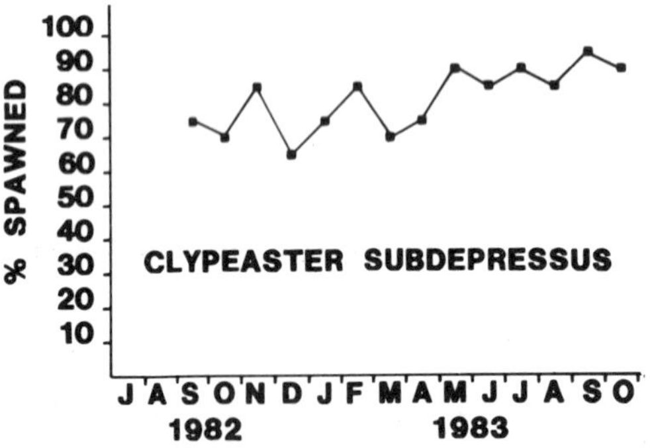

Figure 3. Clypeaster subdepressus: Percent of individuals that spawned when injected with 10 ml of 0.5 M KCl solution. Monthly samples of twenty individuals each were taken 4-2 days before new moon.

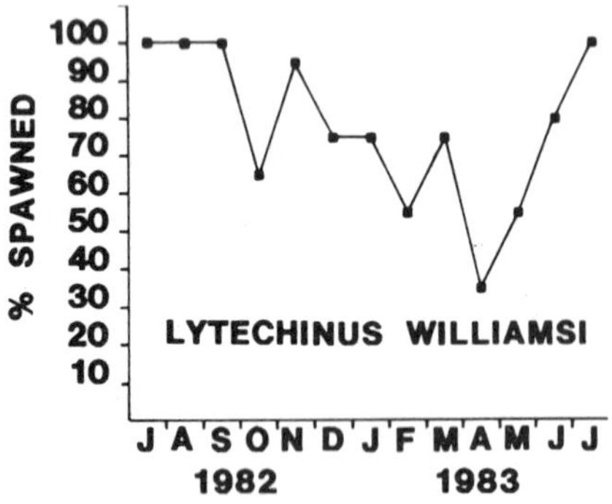

Figure 5. Lytechinus williamsi: Percent of individuals that spawned when injected with 5 ml of 0.5 M KCl solution. Monthly samples of twenty individuals each were taken 4-2 days before new moon.

3.2. Seasonal breeders

Four of the eight species in this study show seasonal variation in their spawning. Echinometra viridis displays a marked reduction in the percentage of animals ready to spawn during the dry season from December to April (Figure 4). Lytechinus williamsi, the other species encountered among live coral, also exhibits lower reproduction during the dry season, but the reduction in ripeness is less pronounced (Figure 5). This species is classified here among the seasonal breeders only because the percent of animals spawning drops below 50 % in April. Reproductive activity in Clypeaster rosaceus peaks in September, and then gradually diminishes, plunging to almost zero during the dry season (Figure 6). Most individuals of Leodia sexiesperforata remain ready to spawn throughout the wet season, but few of them are reproductive in January. There is a gradual increase during the rest of the dry season, until a plateau is reached once again in April (Figure 7). Two of the seasonal breeders, E. viridis and L. williamsi, also show a decline during October in the number of individuals responding to KCl injections. Unlike the dry season reduction, however, which shows a gradual decrease, then a gradual increase from month to month, the

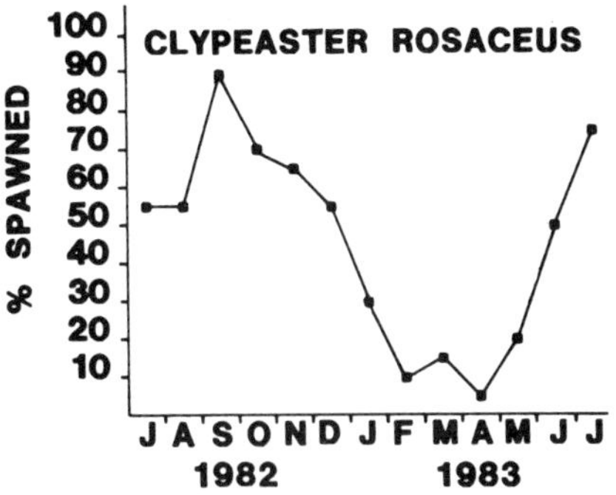

Figure 6. Clypeaster rosaceus: Percent of individuals that spawned when injected with 10 ml of 0.5 M KCl solution. Monthly samples of twenty individuals each were taken 4-2 days before new moon.

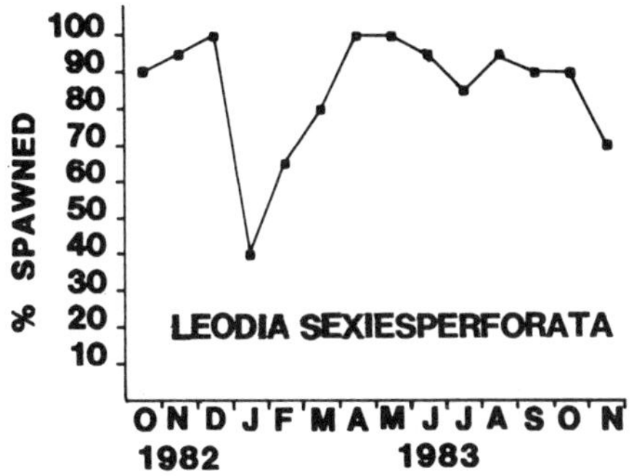

Figure 7. Leodia sexiesperforata: Percent of individuals that spawned when injected with 5 ml of 0.5 M KCl solution. Monthly samples of twenty individuals each were taken 4-2 days before new moon. Note that the origin of the abscissa is different from that of other Figures.

October reduction is limited to this month; the percent of spawning individuals is restored to its high level by November. It is, therefore, likely that this sharp decrease is the result of some environmental anomaly, such as a local increase in temperature, which may have caused spawning in these species (sampled within a few meters of each other) shortly before the time of collection.

3.3. Aseasonal breeders

Spawning in Echinometra lucunter fluctuates between 20 and 75 % throughout the wet

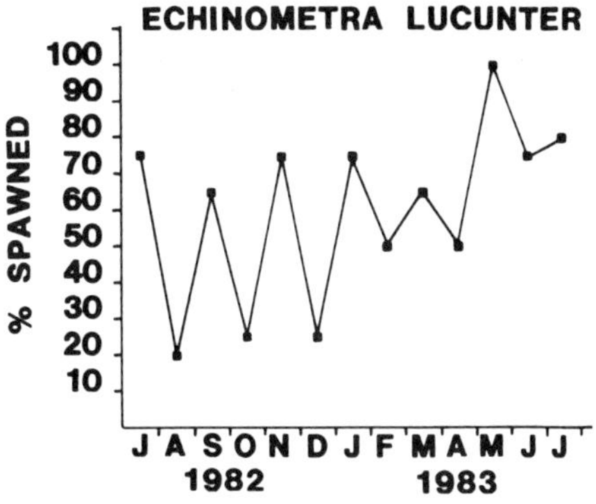

Figure 8. Echinometra lucunter: Percent of individuals that spawned when injected with 5 ml of 0.5 M KCl solution. Monthly samples of twenty individuals each were taken 4-2 days before new moon.

season and between 50 and 90 % in the dry season (Figure 8). No clear annual pattern can be ascribed to these fluctuations; they may be an artifact of daily variation, perhaps induced by unpredictable temperature fluctuations in the shallow habitat the species inhabits.

3.4. Relationship between reproductive synchrony and population density

Population density of every species in the habitat where it was collected is shown in Table 1. Numbers of individuals per square meter vary by three orders of magnitude, presenting an opportunity to test the hypothesis that rare species should be reproductively better synchronized then abundant ones (Rokop 1974; Lessios, 1981). A correlation of the density of each species with the maximum positive deviation from the annual mean of the percent of animals responding to KCl injections yields a Spearman rank coefficient that is not statistically significant ($r = 0.03$, $p > 0.5$). If the three year-round breeders are excluded from the analysis, on the grounds that they may be employing a different strategy to avoid gamete wastage, the correlation coefficient changes towards the expected direction (becomes negative) but remains non-significant ($r = -0.6$, $0.5 > p > 0.02$). Naturally, with a sample size of five species, only a perfect correlation ($r = 1$) would have been significant. It could be argued that comparisons are meaningful only within each habitat, since the mode of gamete dispersal and, therefore, the probability of fertilization

Table 1. Population densities of echinoids on House and Tiantupu reefs, and maximum deviation from the annual mean percentage of animals responding to KCl injections, calculated from data presented in Figures 1 to 8.

Species	Number of Transects	Area Covered (sq. meters)	Mean Density (N/sq. meter)	Maximum Deviation from annual mean %
HOUSE REEF				
Thalassia bed:				
L. variegatus	9	482.66	0.12	11.25
T. ventricosus	9	482.66	0.12	9.58
Reef flat:				
C. rosaceus	9	293.20	0.08	46.67
E. lucunter	9	293.20	0.39	41.67
Reef slope:				
E. viridis	9	93.64	48.77	41.25
L. williamsi	9	93.64	5.66	24.17
TIANTUPU REEF				
Leodia sexiesperforata	10	126.21	1.84	13.75
Clypeaster subdepressus	10	126.21	1.26	14.17

may vary with the living habits of the species and with the degree of clumping of individuals. A habitat by habitat analysis, however, yields mixed results. On the reef flat the expected negative correlation is present; Echinometra lucunter is more abundant and less synchronized then Clypeaster rosaceus. The two infaunal species at Tiantupu, Clypeaster subdepressus and Leodia sexiesperforata, are roughly equal in abundance, and roughly equal in synchrony. The same holds true for the two species from the Thalassia bed, Lytechinus variegatus and Tripneustes ventricosus. However, the results from the reef slope are emphatically different from those expected from the hypothesis. Lytechinus williamsi is an order of magnitude less abundant than Echinometra viridis, yet it is less well synchronized. Thus, none of the ways that one can look at these data provides overwhelming support for the hypothesis that annual reproductive cycles of rare species are more synchronized than abundant ones.

4. DISCUSSION

The simultaneous assessment of reproductive cycles in eight species makes it clear that no single environmental variable can be invoked as an ultimate or proximate factor controlling the mode and timing of reproduction of all echinoids in an area. The hypothesis that in a fairly constant tropical environment rare species maintain better synchronized reproductive cycles to increase the probability of fertilization of their gametes is also not well supported by the data. Though some indications are present that population density may be one of the factors determining degree of synchrony, it seems more prudent to reject the hypothesis until more data are available. Data on monthly spawning cycles of these species are particularly desirable in this context, since reproductive synchrony need not be achieved by annual cycles alone.

In strongly seasonal environments the fluctuation of physical parameters may impose rigid selective regimes, which define the optimal timing of reproduction of most species present. The relatively constant tropical environment of the San Blas may allow enough room for factors other than physical seasonality to become important, so that even species belonging to the same group will follow disparate reproductive cycles. A glimpse of what these factors might be can be gained by trying to determine what characteristics are shared by species that follow a common reproductive pattern.

Three species, Tripneustes ventricosus, Lytechinus variegatus, and Clypeaster subdepressus are ripe for the entire year.

All three have large body sizes; the test diameter of an adult T. ventricosus is about 12 cm, that of L. variegatus about 9 cm, and the anterior-posterior axis of C. subdepressus is about 15 cm. A large body size may allow a species to perceive the environment as more "fine grained" (MacArthur & Wilson 1967), by making small changes in space or time inconsequential for the biological needs of the animal. For example, fluctuations in food availability may be less important for large echinoids, which can store more nutrients in their gonad than small ones. If the rate of utilization of these nutrients is not proportional to body size, the probability that some ripe gametes will be present at any given time in a large animal would be higher. However, a large body cavity alone is not sufficient for the maintenance of year-round breeding cycles. C. rosaceus is also large (ca 14 cm in anterior-posterior axis), yet it shows marked seasonality in its reproduction. The latter species, however, is also rare. By conserving its resources through part of the year, each individual C. rosaceus may be able to produce more gametes during a shorter time in synchrony with its conspecifics. Though the preceding explanation falls in the category of mental gymnastics, it underscores the point that no single factor can explain the variation between species. It may be that a two-factor explanation is necessary; species that are large and common reproduce throughout the year, species that lack one of these attributes have to limit their reproduction to a more discrete time interval.

A seasonal component in the tropical shallow water environment of the San Blas Archipelago is introduced by the changes in wave action, turbidity and salinity caused by the shift in wind pattern during the dry season. Four species seem to respond to these changes by reducing or stopping their reproduction during this period. There is no obvious common characteristic linking these four species. Echinometra viridis and Lytechinus williamsi, are both small and live among live coral. Leodia sexiesperforata has a small body cavity and lives infaunally. Clypeaster rosaceus is large and can live on reef flats, rubble, sand or Thalassia beds. The most interesting question with regard to this group is why all of these species avoid reproducing in the dry season. One can only speculate as to how increased wave action and turbidity in the dry season may affect each of the four species.

Reproductive cycles of some of the species in this study have also been determined in other areas of the Caribbean. Geographic variability in the existence and timing of reproduction seems to be the rule. T. ventricosus and L. variegatus remain ripe throughout the year in the San Blas. The presence of ripe gametes throughout the year in at least some individuals holds true for T. ventricosus and L. variegatus in other localities as well, though the constant ripeness of the majority of the population does not necessarily apply to the entire range of the species. Some ripe individuals of L. variegatus can be found at any time of the year in Florida, but there are clear reproductive peaks in spring and summer (Moore et al. 1963a; Brookbank 1968; Ernest and Blake 1981), with a good deal of variation in timing between local populations (Ernest and Blake 1981) and between years (Moore and Lopez 1972). In Bermuda a single spawning event occurs between mid-April and early June (Moore et. al. 1963a). Tripneustes ventricosus in Florida shows the same pattern as in the San Blas, i.e. constant ripeness of the majority of individuals throughout the year (McPherson 1965), but it also shows peaks in gonadal growth in late spring and early summer (Moore et. al 1963b), or in early winter and early summer (Moore et al. 1963b; McPherson, 1965). In Barbados ripe individuals can be found from January to September, but their percentage decreases drastically after the general spawning in August (Lewis 1958). Echinometra viridis is a seasonal spawner not only at the San Blas but also at two localities on the coast of the Panamanian mainland, where its gonadal content also remains low through the dry season (Lessios 1981). This species also reproduces seasonally in Florida, but there its gonads attain their maximum development during the summer (McPherson 1969). E. lucunter shows no particular annual pattern of reproduction in the San Blas. Development of gonads of this species also fluctuates haphazardly in two localities on the Caribbean coast of the Panamanian mainland (Lessios 1981). In Barbados E. lucunter populations from wave-swept areas spawn once a year, while those from low wave energy habitats spawn twice a year (Lewis and Storey 1984). In Florida this species shows seasonal reproduction, attaining its peak gonadal development during June and July (McPherson 1969). Such variability suggests that local populations have the necessary flexibility to track their respective environments, and that larval exchange does not homogenize the response of populations to the particular environmental challenges they face. Thus, one can hope that ultimate factors controlling echinoid reproduction can be identified in studies limited to a small portion of the species range. In seeking ultimate factors, correlation between reproductive patterns and environmental fluctuations is usually the only available tool. Correlations, however, can be misleading, and the chances that they will mislead increase with decreasing number of species studied. The value of comparing reproductive patterns of many species in the same locality may lie precisely in that, by revealing disparate patterns, they prevent facile explanations and misleading correlations.

ACKNOWLEDGEMENTS

I thank Y. Cerrud, P. Mace, L. Marshall and C. White for technical assistance, and J.H. Christy, H.R. Lasker, and D.R. Robertson for comments on the manuscript. The Kuna Nation and the government of Panama permitted work in the San Blas. Supported by General Research funds from the Smithsonian Tropical Research Institute.

REFERENCES

Bennett, J. & A.C. Giese 1955. The annual reproductive and nutritional cycles in two western sea urchins. Biol. Bull. 109: 226-237.

Brookbank, J.W. 1968. Spawning season and sex ratio of echinoids. Quart. Jour. Florida Acad of Sci. 30: 177-183.

Crump, R.G. 1971. Annual reproductive cycles in three geographically separated populations of Patiriella regularis (Verrill) a common New Zeland asteroid. J. Exp. Mar. Biol. Ecol. 7:137-162.

Dix, T.G. 1970. Biology of Evechinus chloroticus (Echinoidea: Echinometridae) from different localities 3. Reproduction. N. Z. Jour. Mar. Freshw. Res. 4: 385-405.

Ernest, R.G. & N.J. Blake 1981. Reproductive patterns within subpopulations of Lytechinus variegatus (Lamarck). (Echinodermata: Echinoidea). J. Exp. Mar. Biol. Ecol. 55: 25-37.

Giese, A.C. 1959. Comparative physiology: annual reproductive cycles of marine invertebrates. Ann. Rev. Physiol. 21: 547-576.

Giese A.C. & J.S. Pearse 1974. Introduction: General principles. In A.C. Giese & J.S. Pearse (eds.), Reproduction of Marine Invertebrates, v. 1, p. 1-49. New York, Academic Press.

Glynn, P.W. 1972. Observations on the ecology of the Caribbean and Pacific coasts of Panama. Bull. Biol. Soc. Wash. (2): 13-30.

Gonor, J.J. 1973. Reproductive cycles in Oregon populations of the echinoid, Strongylocentrotus purpuratus (Stimpson) I. Annual gonad growth and ovarian gametogenic cycles. J. Exp. Mar. Biol. Ecol. 12: 45-64.

Grant, A. & P.A. Tyler 1983a. The analysis of data in studies of invertebrate reproduction. I. Introduction and statistical analysis of gonad indices and maturity indices. Inter. Jour. Invert. Reproduction 6: 259-269.

Grant, A. & P.A. Tyler 1983b. The analysis of data in studies of invertebrate reproduction. II. The analysis of oocyte size/frequency data, and comparison of different types of data. Inter. Jour. Invert. Reproduction 6: 271-383.

Hendler, G. 1976. Marine Studies. In D.M. Windsor (ed.), Environmental Monitoring and Baseline Data, 1975. p. 131-252. Washington, Smithsonian Institution.

Hendler, G. 1977. Marine Studies. In D.M. Windsor (ed.), Environmental Monitoring and Baseline Data, 1976. p. 139-267. Washington, Smithsonian Institution.

Hendler, G.L. 1979. Reproductive periodicity of ophiuroids (Ophiuroidea: Echinodermata) on the Atlantic and Pacific Coasts of Panama. in S.E. Stancyk (ed.), Reproductive Ecology of Marine Invertebrates. p. 145-156 Belle W. Baruch Library in Marine Science, Columbia. S.C., Univ. South Carolina Press.

Hinegardner, R. 1975. Care and handling of sea urchin eggs, embryos and adults (Principally North American Species). In G. Czihak (ed.). The Sea Urchin Embryo--Biochemistry and Morphogenesis. p. 10-25. Berlin, Springer-Verlag.

Lessios, H.A. 1981. Reproductive periodicity of the echinoids Diadema and Echinometra on the two coasts of Panama. J. Exp. Mar. Biol. Ecol. 50: 47-61.

Lessios, H.A. 1984. Possible prezygotic reproductive isolation in sea urchins separated by the Isthmus of Panama. Evolution 38: 1144-1148.

Lessios, H.A., J.D. Cubit, D.R. Robertson, M.J. Shulman, M.R. Parker, S.D. Garrity & S.C. Levings in press. Mass mortality of Diadema antillarum on the Caribbean coast of Panama. Coral Reefs.

Lewis, J.B. 1958. The biology of the tropical sea urchin Tripneustes esculentus Leske in Barbados, British West Indies. Canad. J. Zool. 36: 607-621.

Lewis, J.B. & G.S. Storey 1984. Differences in morphology and life history traits of the echinoid Echinometra lucunter from different habitats. Mar. Ecol. Prog. Ser. 15: 207-211.

MacArthur, R.H. & E.O. Wilson 1967. The Theory of Island Biogeography. Princeton, Princeton Univ. Press.
McPherson, B.F. 1965. Contributions to the biology of the sea urchin Tripneustes ventricosus. Bull. Mar. Sci. 15: 228-244.
McPherson, B.F. 1969. Studies on the biology of the tropical sea urchins, Echinometra lucunter and E. viridis. Bul. Mar. Sci. 19: 194-213.
Meyer D.L. & C. Birkeland 1974. Marine Studies. In R.W. Rubinoff (ed.), Environmental Monitoring and Baseline Data, 1973. p. 129-254. Washington, Smithsonian Institution.
Meyer D.L., C. Birkeland & G.L. Hendler 1975. Marine Studies. In D.M. Windsor (ed.), Environmental Monitoring and Baseline Data, 1974. p. 226-409. Washington, Smithsonian Institution.
Moore, H.B. 1934. A comparison of the biology of Echinus esculentus in different habitats. Part I. J. Mar. Biol. Ass. U.K. 19: 869-886.
Moore, H.B., T. Jutare, J.C. Bauer & J.A. Jones 1963a. The biology of Lytechinus variegatus. Bull. Mar. Sci. Gulf Carib. 13: 23-53.
Moore, H.B., T. Jutare, J.A. Jones, B.F. McPherson & C.F.E. Roper 1963b. A contribution to the biology of Tripneustes esculentus. Bull. Mar. Sci. Gulf Carib. 13: 267-281.
Moore, H.B. & Lopez, N.N. 1972. Factors controlling variation in the seasonal spawning pattern of Lytechinus variegatus. Mar. Biol. 14: 275-280.
O'Connor, C., G. Riley, S. Lefebvre & D. Bloom 1978. Environmental influences on histological changes in the reproductive cycle of four New South Wales sea urchins. Aquaculture. 15: 1-17.
Orton, J.H. 1920. Sea temperature, breeding and distribution in marine animals. J. Mar. Biol. Ass. U.K. 12: 339-366.
Pearse, J.S. 1968. Patterns of reproduction in four species of Indo-Pacific echinoderms. Proc. Indian Acad. Sci (Sect B) 68: 247-279.
Pearse, J.S. 1969. Reproductive periodicities of Indo-Pacific invertebrates in the Gulf of Suez II. The echinoid Echinometra mathaei (De Blainville). Bull. Mar. Sci. 19: 580-613.
Pearse, J.S. 1970. Reproductive periodicities of Indo-Pacific invertebrates in the Gulf of Suez III. The echinoid Diadema setosum (Leske). Bull. Mar. Sci. 20: 697-720.
Pearse, J.S. 1974. Reproductive patterns of tropical reef animals: three species of sea urchins. Proc. 2nd Int. Symp. Coral Reefs 1: 235-240.
Pearse, J.S. 1975. Lunar reproductive rhythms in sea urchins: A review. J. Interd. Cycle Res. 6: 47-52.
Rokop, F.J. 1974. Reproductive patterns in the deep-sea benthos. Science 186: 743-745.
Tyler, A. 1949. A simple non-injurious method for inducing repeated spawning of sea urchins and sand-dollars. Collecting Net 19. 1: 19-20.
Tyler, P.A., A. Grant, S.L. Pain & J.D. Gage 1982. Is annual reproduction in deep-sea echinoderms a response to variability in their environment? Nature 300: 747-750.

The bald-sea-urchin disease: A bacterial infection

PHILIPPE MAES & MICHEL JANGOUX
Laboratoire de Biologie marine, Université Libre de Bruxelles, Belgium

The causative agent of the bald-sea-urchin disease has been isolated from lesions of diseased individuals. It is a bacterium that would be resistant to the bactericidal substances naturally produced by the echinoids.

INTRODUCTION

The bald-sea-urchin disease is a communicable disease which affects several species of regular echinoids. The disease causes conspicuous body-wall lesions (Johnson 1971, Höbaus et al. 1981, Maes and Jangoux 1984). A lesion generally consists of two different areas, i.e. a greenish colored central part deprived of appendages and a reddish peripheral belt which is markedly swollen. The central part is covered by necrotic tissues. These tissues are invaded by numerous micro-organisms, especially bacteria, while the peripheral swelling (inflamatory tissue) never includes micro-organisms (Maes et al., in press).

MATERIALS AND METHODS

Both diseased and healthy individuals of the echinoid, Paracentrotus lividus were collected at the "Pointe Saint Hernot" (Brittany, France). Experimental infection of echinoids was performed following the method of Maes and Jangoux (1984). (Healthy echinoids were slightly injured, i.e. scarring of the skin and the removal of a few spines, the injured area then being painted using pieces of necrotic tissues taken from lesions of diseased individuals). This method produces typical lesions in healthy echinoids within a few days. The same method was used routinely to test the pathogenic capability of a presumed pathogenic agent (bacterium that had been isolated in vitro.

Bacterial cultures were prepared from necrotic tissues which had been homogenized in sterile sea water. A few drops of the suspension were spread on agar plates containing a culture medium for heterotrophic marine bacteria (Difco marine Broth 2216). Incubation lasted 48 hours at 20°C. Bacteria were then applied to experimentally injured echinoids in order to test their pathogenic capability.

RESULTS AND DISCUSSION

The occurrence of numerous bacteria within the necrotic tissues of diseased echinoids has been emphasized by Maes et al., (in press). These authors suggest that the causative agent of the bald-sea-urchin disease could be a bacterium insensitive to the antibacterial substances naturally produced by the echinoids.

Bacterial cultures from necrotic tissues mostly allowed the isolation of no more than a single strain of bacteria. This bacterial strain is highly infectious for echinoids, producing typical necrosis within a few days of being applied to experimentally injured individuals. In some cases, however, additional strains (i.e. two to three strains) were isolated, but they did not produce any necrosis when tested on echinoids. The bacterial nature of the bald-sea-urchin disease is thus demonstrated. Works are in progress in order to both identify and physiologically characterize the pathogenic bacteria.

It is rather surprising that an almost pure culture of pathogenic bacteria can be directly obtained from necrotic tissues. This suggests that the lesions have a highly selective effect on the invading species of bacteria. The occurrence of presumed antibacterial substances naturally produced by echinoids (i.e. spinochromes) was reported several times (see, e.g., Vevers 1966, Johnson and Chapman 1970, Messer and Wardlaw 1980). Spinochromes are carried to altered body-wall areas through particular migratory cells- viz. the red spherule cells- originating from the general body cavity. Spinochromes were demonstrated in lesions caused by the bald-sea-urchin disease (Maes et al., in press). These authors reported moreover that, in spite of the release of spinochromes, a high number of bacteria occur within the necrotic tissues of the lesions. The present study indicates however that these bacteria presumably belong to a single species. Our results consequently allow us to support the former hypothesis of Maes et al. that the lesion-associated bacterium - which was already presumed to be the causative agent of the disease - would be resistant to the echinoid bactericidal substances.

BIBLIOGRAPHY

Höbaus, E., Fenaux, L., and Hignette, M. 1981. Premières observations sur les lesions provoquées par une maladie affectant le test des oursins en Méditerranée occidentale. Rapp. Comm. intern. Mer Medit., 27: 221-222.

Johnson, P.T. 1971. Studies on diseased urchins from Point Loma. Ann. Rep. Kelp. Habit. Impr. Project (1970-1971), pp. 82-90. Calif. Inst. Technol.: Pasadena.

Johnson, P.T. and Chapman, F.A. 1970. Infection with diatoms and other microorganisms in sea urchin spines (Strongylocentrotus franciscanus). J. invert. Pathol. 16: 268-276.

Maes, P. and Jangoux, M. 1984. The bald-sea-urchin disease: A biopathological approach. Helgoländer Meeresunters. 38: 217-224.

Maes, P., Jangoux, M. and Fenaux, L. in press. La maladie de l'oursin chauve: ultrastructure des lesions et caracterisation de leur pigmentation. Annls. Inst. oceanogr.

Messer, L.I. and Wardlaw, A.C. 1980. Separation of the coelomocytes of Echinus esculentus by density gradient centrifugation. In, Echinoderms: Present and Past, M. Jangoux (ed.), pp. 319-323. Balkema: Rotterdam.

Vevers, H.G. 1966. Pigmentation. In: Physiology of Echinodermata, R.A. Boolootian (ed.), pp. 267-275. Interscience Publ.: N.Y.

Proceedings of the Fifth International Echinoderm Conference / Galway / 24-29 September 1984

An investigation of the distribution of pedicellariae in *Echinus esculentus* (L.)

R.ELIZABETH RAMSAY & ANDREW C.CAMPBELL
School of Biological Sciences, Queen Mary College (University of London), UK

ABSTRACT: Pedicellariae are key characters in echinoid identification. Although the numbers and types possessed by each species is well known there are few accounts of the distribution of these organs over the echinoid test. This project investigates the distribution of pedicellariae in E. esculentus in an attempt to assist our understanding of their role in the life of the animal. Their distribution was determined using a quadrat at three vertical levels on the test. Results showed a complex pattern with pedicellariae varying both in size and numbers up and down as well as round the urchin. It is suggested that environmental factors may influence the development of pedicellariae in various ways. The differences in distribution are related to our knowledge of pedicellarial function.

1 INTRODUCTION

The form and functions of echinoid pedicellariae have recently been reviewed by Campbell (1983). Apart from their roles in defense and grooming, these organs are important as key characters for use in identification and classification (Doederlein 1906), and the varieties and morphology of the different types occurring in each species are well known (Mortensen 1928-1951). There are however very few accounts of the distribution of pedicellariae within a single species. This investigation was carried out on E. esculentus which has four types of pedicellariae, Tridentate, Ophiocephalous, Globiferous and Trifoliate to see how they are positioned on the test and to relate their distribution to what is known of their functions and the animal's way of life.

2 MATERIALS AND METHODS

Sixty one E. esculentus were supplied alive by the University Marine Biological Station, Millport, Isle of Cumbrae in two batches, one of 35 in February 1980 and one of 26 in June 1980. The animals were carefully preserved in 70% IMS on arrival in London and during analysis care was taken to avoid using any specimen which appeared to have shed pedicellariae during transit or preservation and storage.

The number of pedicellariae occurring at different points around and up and down the rigid test of the animal were counted. The ophiocephalous organs on the peristomal membrane were excluded because they are mounted on a flexible part of test where opportunities for resisting sedimentation and fouling are different.

2.1. Sampling

Samples were counted for each ambulacrum and interambulacrum at three different heights: the aboral surface, immediately adjacent to the periproct, the ambitus and the oral surface immediately adjacent to the peristomial membrane. The area sampled in each case was standardised using a paper quadrat 15 x 8 mm inserted into one eye piece of a Wild M5-54065 binocular microscope at fixed magnifications (x 6, x 12, x 25 and x 50). Appropriate compensations were made to ensure standard results and in every case the actual size of each quadrat investigated was checked with calipers. Data were recorded as numbers of pedicellariae/cm^2.

2.2 Data analysis

The data from the two batches were analysed using a Biomedical Data Package Programme BM DP-2PV which was a two-way analysis of variance and co-variance (BMDP, 1977). The programme shows where there are statistically valid differences.

The standard deviations of this programme are based on the actual values of the replicates. The results indicated a significant variation in pedicellarial distribution around the animals.

Because of this the data were checked further using a X^2 test (CHISQ-1ANS.PROG) on the means and variances of the pedicellariae at the 5 symmetry points, keeping ambulacral and interambulacral data separate. If the data points are significantly different then the X^2 value will be larger than the X^2 value obtained from standard statistical tables for a given number of degrees of freedom (the number of degrees of freedom varying according to the number of data points). The results were then sorted using the following reasoning:

i. "✓" means that the X^2 values of the data points was greater than the value given in tables for 4 degrees of freedom at 95% confidence level i.e. the difference between the data points is significance.

ii. "X" means that the X^2 value of the data points was less than the value given in tables for 4 degrees of freedom at 95% confidence level i.e. the difference between data points was small and not significant at the 95% level.

From these data for batch 1 table I was constructed and for batch 2 table IV where "✓" indicates that 5% or more of the X^2 values were significant and "X" indicates that 95% or more of the X^2 values were insignificant.

Table II for batch 1 and table V for batch 2 compare the equivalent symmetry points of all the urchins in the samples making comparisons between urchins rather than within them. To make such a comparison valid the data had to be standardised by finding the mean of the 5 data points and dividing it into each value. These values could then be used for the second X^2 test (MIKES-CHISQ) and the results were subjected to a subsequent X^2 test (MIKES-CHISQ of CHISQ) in order to determine if there was a standard characteristic pattern of pedicellariae round the urchins. The results are shown in tables II and V using the same key as given for tables I and IV.

These two tables were then combined to give tables III and VI using the reasoning

i. Table I gives "✓", Table II gives "✓", then Table III gives "✓" i.e. there is no intrinsic pattern that can be distinguished round the urchin.

ii. Tables I gives "✓" Table II gives "X", then Table III gives "?" i.e. there is a detectable pattern of distribution round the urchin but it is not equal.

iii. Table I gives "X" Table II gives "✓" Table III gives "!" i.e. the pedicellariae are equal all round the urchin but there is no pattern.

iv. Table I gives "X", Table II gives "X" then Table II gives "X" i.e. there is an equal distribution and pattern of pedicellariae around the urchin. The same applies for tables IV, V and VI.

3 RESULTS

3.1 Around the urchins

The data for batches 1 and 2 examined around the urchins is presented in tables I to VI as explained above.

Throughout these the key is:

G. = Globiferous pedicellariae; Td. = Tridentate pedicellariae; Tf. = Trifoliate pedicellariae; O. = Ophiocephalous pedicellariae. In tables I, II, IV, and V A = Ambulacrum; I = Interambulacrum. ✓= significant difference; X = insignificant difference. In tables III and VI

✓ = no pattern distinguished;

? = a pattern is distuinguisable - an uneven distribution;

! = difficult to draw conclusions;

X = a pattern is distinguishable - an even distribution.

Batch 1 Table I

	Height	G.	Td.	Tf.	O.
	(Aboral	✓	✓	✓	✓
A	(Ambitus	✓	X	✓	X
	(Oral	✓	✓	✓	✓
	(Aboral	✓	✓	✓	✓
I	(Ambitus	✓	X	✓	X
	(Oral	✓	✓	✓	✓

Batch 1 Table II

	Height	G.	Td.	Tf.	O.
	(Aboral	✓	X	X	✓
A	(Ambitus	X	X	X	✓
	(Oral	✓	X	X	✓
	(Aboral	✓	✓	X	✓
I	(Ambitus	X	X	X	X
	(Oral	X	X	X	✓

Batch 1 Table III

	Height	G.	Td.	Tf.	O.
A	(Aboral	✓	?	?	✓
	(Ambitus	?	X	?	!
	(Oral	✓	?	?	✓
I	(Aboral	✓	✓	?	✓
	(Ambitus	?	X	?	X
	(Oral	?	?	?	✓

Batch 2 Table IV

		G.	Td.	Tf.	O.
A	(Aboral	✓	✓	✓	✓
	(Ambitus	✓	✓	✓	X
	(Oral	✓	✓	✓	✓
I	(Aboral	✓	✓	✓	✓
	(Ambitus	✓	✓	✓	✓
	(Oral	✓	✓	✓	✓

Batch 2 Table V

		G.	Td.	Tf.	O.
A	(Aboral	✓	X	X	✓
	(Ambitus	X	X	X	✓
	(Oral	✓	X	X	✓
I	(Aboral	X	X	X	✓
	(Ambitus	✓	X	X	✓
	(Oral	✓	X	X	✓

Batch 2 Table VI

		G.	Td.	Tf.	O.
A	(Aboral	✓	?	?	✓
	(Ambitus	?	?	?	!
	(Oral	✓	?	?	✓
I	(Aboral	?	?	?	✓
	(Ambitus	✓	?	?	✓
	(Oral	✓	?	?	✓

In table VII all the standardized results for Batch 1 are presented. These can be compared with table VIII where the standardized result for Batch 2 are given.

The position of symmetry on the test is indicated in brackets, see fig. 1.

Table VII Standardised data

ABORAL		AMBITUS		ORAL	
A	I	A	I	A	I
1.20C	0.99CB	1.04C	1.01CB	1.00C	0.88CB
0.91B	0.98BA	0.95B	0.97BA	0.95B	0.98BA
0.98A	1.01EA	0.99A	1.00EA	0.94A	0.91EA
1.94E	1.03ED	1.03E	0.98ED	0.91E	1.06ED
0.93D	0.99DC	1.00D	1.04DC	0.93D	1.03DC

Table VIII Standardised data

ABORAL		AMBITUS		ORAL	
A	I	A	I	A	I
1.22C	0.92CB	1.08C	0.92CB	1.11C	1.03CB
0.99B	0.98BA	0.97B	0.99BA	0.92B	0.96BA
0.84A	0.97EA	0.91A	1.01EA	0.85A	1.06EA
0.81E	0.93ED	0.98E	1.01ED	0.84E	0.84ED
1.01D	1.1 DC	0.92D	1.01DC	0.85D	0.78DC

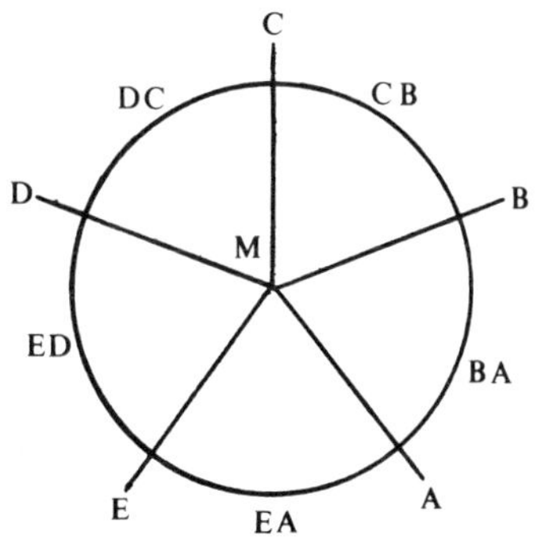

Fig. 1 Diagram to show symmetry of ambulacra A to E and interambulacra BA to EA in relation to madreporite M.

The noticeable feature in these data is that in every case for the ambulacra the largest number of pedicellariae occurs in position C. The picture is less regular for the interambulacra but generally there are more pedicellariae in the segment CB, C, CD, D than opposite.

If the height is disregarded and the data for the ambulacra and intermabulacra for the total number of pedicellariae given at the five symmetry points, then the following table IX is produced. (Position of symmetry is indicated in brackets).

Table IX Standardized data

BATCH 1		BATCH 2	
A	I	A	I
3.24C	2.88CB	3.41C	2.87CB
2.81B	2.93BA	2.88B	2.93BA
2.91A	2.92EA	2.60A	3.04EA
2.88E	3.07ED	2.63E	2.78ED
2.86D	3.06DC	2.78D	2.90DC

The results of the two batches agree very closely. There also seems to be a concentration of pedicellariae in positions B C D.

If the distribution of individual types of pedicellariae is examined we find a range of consistencies. The following results are based on the significance of normalised data and relate to the symmetry of fig. 1.

1. Globiferous pedicellariae
In batch one and batch two these were significantly more abundant in ambulacrum C and interambulacrum CD i.e. at the madreporite. They were significantly less abundant in interambulacra BC and ED i.e. either side of the madreporite.

2. Tridentate pedicellariae
In batch one and two they were significantly more abundant in interambulacra BC, DC and EA in the oral region and significantly less abundant in the interambulacra and ambulacra of the aboral region EA, A, B i.e. away from the madreporite.

3. Trifoliate pedicellariae
In batch one and two they were significantly more abundant in interambulacra in the hemisphere bearing the madreporite and significantly less abundant in ambulacra in the same hemisphere.

4. Ophiocephalous pedicellariae
Appear evenly distributed around the urchin.

3.2 Up and down the urchins
A similar set of tests were then performed on the original data of both batches to investigate the distribution of pedicellariae up and down the individual urchins. The results from the first X^2 test were used to construct tables X and XIII (equivalent to tables I and IV). The same keys apply. The standardized data were then subjected second and third X^2 tests the results being used to construct tables XI and XIV (which are equivalent to tables II and V). The same keys apply. Combining tables X and XI and XIII and XV respectively produced Tables XII and XIV (equivalent to tables III and VI).

Ped = pedicellariae; G = globiferous; Td. = Tridentate; Tf. = Trifoliate; O. = Ophiocephalous.

Batch 1 Table X

Ped. type	Ambulacrum C	B	A	E	D	Intermabulacrum CB	BA	EA	ED	DC
G	✓	✓	✓	✓	✓	✓	✓	✓	✓	✓
Td	✓	✓	✓	✓	✓	✓	✓	✓	✓	✓
Tf	✓	✓	✓	✓	✓	✓	✓	✓	✓	✓
O	✓	✓	✓	✓	✓	✓	✓	✓	✓	✓

Batch 1 Table XI

Ped. type	Ambulacrum C	B	A	E	D	Interambulacrum CB	BA	EA	ED	DC
G	X	X	X	X	X	X	X	X	X	X
Td	X	X	X	X	X	X	X	X	X	X
Tf	X	X	X	X	X	X	X	X	X	X
O	X	X	X	X	X	X	X	X	X	X

Batch 1 Table XII

Ped. type	Ambulacrum C	B	A	E	D	Interambulacrum CB	BA	EA	ED	DC
G	?	?	?	?	?	?	?	?	?	?
Td	?	?	?	?	?	?	?	?	?	?
Tf	?	?	?	?	?	?	?	?	?	?
O	?	?	?	?	?	?	?	?	?	?

Batch 2 Table XIII

Ped. type	Ambulacrum C	B	A	E	D	Interambulacrum CB	BA	EA	ED	DC
G	✓	✓	✓	✓	✓	✓	✓	✓	✓	✓
Td	✓	✓	✓	✓	✓	✓	✓	✓	✓	✓
Tf	✓	✓	✓	✓	✓	✓	✓	✓	✓	✓
O	✓	✓	✓	✓	✓	✓	✓	✓	✓	✓

Batch 2 Table XIV

Ped. type	Ambulacrum C	B	A	E	D	Interambulacrum CB	BA	EA	ED	DC
G	X	X	X	X	X	X	X	X	X	X
Td	X	X	X	X	X	X	X	X	X	X
Tf	X	X	X	X	X	X	X	X	X	X
O	X	X	X	X	X	X	X	X	X	X

Batch 2 Table XV

Ped. type	Ambulacrum C	B	A	E	D	Interambulacrum CB	BA	EA	ED	DC
G	?	?	?	?	?	?	?	?	?	?
Td	?	?	?	?	?	?	?	?	?	?
Tf	?	?	?	?	?	?	?	?	?	?
O	?	?	?	?	?	?	?	?	?	?

The results shown in tables XVI and XVII compare the results of batch 1 and batch 2 and are the ratios of pedicellariae occurring at the three different heights (aboral:ambitus:oral), where the data for the ambital region is taken as 1, and the oral and aboral data is related to the ambital region.

Table XVI Ambulacral Results

	Batch 1	Batch 2
C	0.44:1:0.35	0.76:1:0.62
B	0.42:1:0.35	0.7:1:0.65
A	0.42:1:0.35	0.74:1:0.60
E	0.40:1:0.29	0.56:1:0.41
D	0.36:1:0.42	0.65:1:0.39

Table XVII Interambulacral Results

	Batch 1	Batch 2
CB	0.48:1:0.39	0.65:1:0.51
BA	0.52:1:0.37	0.73:1:0.62
EA	0.48:1:0.37	0.76:1:0.68
ED	0.57:1:0.40	0.75:1:0.55
DC	0.49:1:0.39	0.83:1:0.39

These ratios indicate that the differences caused by height are more marked in Batch 1 (i.e. February data), than Batch 2 (June data).

Further inspection of the data shows that the distribution of different types of pedicellariae with respect to height for both batches is:

Table XVIII

Type of ped.	Area of highest concentration		Area of lowest concentration	
	Batch 1	Batch 2	Batch 1	Batch 2
G	Ambitus	Ambitus	Oral	Oral
Td.	Aboral	Aboral	Ambitus/ Oral	Ambitus
O	Ambitus	Ambitus	Oral	Oral
Tf.	Aboral	Aboral	Aboral/ Oral	Ambitus

In two cases the aboral surface has the largest number of pedicellariae but on no occasion (except for the ophiocephalous pedicellariae where if the peristomal membrane was taken into account the aboral surface might have the smallest number) does the aboral surface have the smallest number of pedicellariae.
This could be indicating that there is a higher concentration of pedicellariae in the aboral region than in other regions. The trends shown here agree largely (but not the absolute numbers) with the results of Davies (1968).

Table XIX showing total numbers of pedicellariae/CM^2 occurring at the three different heights in batch 1 and 2.

	Batch 1	Batch 2
Aboral	49.23	11.05
Ambitus	25.54	10.48
Oral	15.59	7.78

The largest numbers of pedicellariae are on the aboral surface. The lowest concentration of pedicellariae occur on the oral surface, excluding the peristomal ring.

4 DISCUSSION

4.1 Distribution of pedicellariae around the urchin

The interambulacrum DC of the urchin which bears the madreporite carries an abundance of globiferous pedicellariae. There are used in defense against larger animals and when expended are regenerated (Campbell 1983). The presence of so many pedicellariae could mean one of two things, either they are not being used very much, or they are being used a great deal and are being regenerated (Chia 1970). Either of these analyses implies a symmetrical significance and that the urchin is not randomly orientated, but orientated with a purpose relating to the madreporite. Perhaps this is frequently at the functional anterior and the globiferous pedicellariae here are important in protecting their bearer. Alternatively perhaps predators approach the urchin from the direction of the madreporite because of its orientation with respect to currents carrying scents. Again the defensive role would apply. Lewis and Nichols (1980) found that urchins do not appear to migrate in any direction relative to the prevailing current but they did not give details of orientation with respect to currents.

Tridentate pedicellariae have a greater concentration on the side of the urchin bearing the madreporite especially at the aboral level. Tridentates capture small organisms and pick off inactive objects from the test's surface (Campbell and Rainbow 1977). Here as the urchin moves over the substrate a concentration of these organs for dealing with encountered organisms and particles would be useful.

Trifoliate pedicellariae are believed to have a grooming function (Campbell and Rainbow 1977). Again the side of the urchin bearing the madreporite appears better equipped with these organs which would deal either with particles encountered in locomotion or swept towards it by a directional current which the animal "faced" into.

The universally similar distribution of ophiocephalous pedicellariae reflects their use for holding small creeping animals moving over the test (Campbell 1983). Such animals as small crustaceans and polychaetes can move all over the animal, irrespective of its orientation.

4.2 Distribution with respect to height

In both batches there are overall more pedicellariae on the aboral surface of the urchins than on the oral, excluding the peristomal membrane (table XIX). This reflects the likelihood of smothering settlement or colonisation from above rather than from below. This overall trend is reflected in the distribution of tridentate and trifoliate pedicellariae (table XVIII). The globiferous results suggest that these organs may be most usefully deployed at the ambital level which would be the first point of contact with most other benthic organisms. On the oral surface, where they are least abundant, they would be making much contact with the substratum and might suffer loss (they are easily detached) or damage. The ophiocephalous results exclude the peristomal population so the oral population is low only as far as the rigid part of the test is concerned. The vertical distribution of these organs is then in reality more general and consistent with the points made in 4.1.

REFERENCES

Campbell, A.C. 1983. Form and function of pedicellariae. Echinoderm Studies. 1:139-167.

Campbell, A.C. & P.S. Rainbow 1970. The role of pedicellariae in preventing barnacle settlement on the sea-urchin test. Mar. Behav. Physiol. 4:253-260.

Chia, F.S. 1970. Histology of the globiferous pedicellariae of Psammechinus miliaris. J. zool. Soc. Lond. 160:9-16.

Davies, W. 1968. Echinoid pedicellariae unpublished student project. Zoology Dept. Kings College, London University.

Doederlein, L. 1906. Die Echinoiden der deutschen Tiefsee-Expedition. Wiss. Ergebn. Dtsch. Tiefsee-Exped. Valdivia 5(2).

Lewis, G.A. & D. Nichols 1980. Geotactic movements following distribution in the European sea-urchin Echinus esculentus (echinoidea:echinodermata). Prog. Underwater Sci. N.S. ed. H.M. Platt. Pentech Press, Plymouth pp. 171-186.

Mortensen, T. 1928-1951. Monography of the Echinoidea 15 vols. C.A. Reitzel, Copenhagen.

Proceedings of the Fifth International Echinoderm Conference / Galway / 24-29 September 1984

Macro- and microstructure of the primary spines in *Asthenosoma varium* Grube (Echinothuridae: Echinoidea): Affinities with the Diadematidae and Toxopneustidae

M.B.RÉGIS
CERAM, Faculté des Sciences & Techniques de St.-Jerôme, Marseille, France

B.A.THOMASSIN
Centre d'Océanologie de Marseille, France

The microstructure of the primary spines in Asthenosoma varium exhibits two principal forms: i) sharp spines with a poison gland, on the aboral surface; ii) trumpet-shaped spines on the oral surface. The oral spines have four zones along the shaft with variations in the internal and external microstructure. Each zone exhibits microstructural detail which has also been seen in spines of the Diadematidae (particularly Echinothrix) and the Toxopneustidae (Lytechinus, Sphaerechinus and Toxopneustes). These affinities are discussed with respect to the phylogeny and evolution of the regular echinoids.

1 INTRODUCTION

The affinities of the echinoid families Echinothuridae and Diadematidae are disputed by many workers. Whereas Agassiz and Clarke (1909) thought that the Diadematidae stemmed from the Echinothuridae, Mortensen (1935) asserted that the two families represent different and parallel branches proceeding from the Lepidocentridae, which are known as fossils from Ordovician to Carboniferous times. These views were mainly based on test morphology.

The general features and the morphofunctional adaptations of the primary spines of some modern diadematids had recently been studied by Regis (1981) and by Regis and Thomassin (1982). We undertook a similar investigation of echinothurid spines, to allow a comparison of affinities between the two families, and used the shallow-water species Asthenosoma varium Grube which had earlier been described by Mackintosh (1875).

2 MATERIALS AND METHODS

2.1 Sampling location

Specimens of Astenosoma varium were collected in the S.W. New Caledonian lagoon, from the organogenous wrinkles of a coral reef bank, in a muddy area ("Gail Bank"; 30m depth; reference BT/NC 335).

2.2 Types of spines examined

As a first type of primary spine, on the aboral surface, we studied: i) those surrounding the apical system which appeared smallest in size and whose macroscopical structure seemed particularly simple; ii) those borne on the ambulacrum and interambulacrum and whose size progressively increased towards the ambitus.

From the oral surface, as second types of primary spine, we collected: i) the largest and most curved spines from the ambitus, care being taken to ensure that they were unbroken; ii) those from the ambulacral and interambulacral areas, extending down to the peristome.

First, we examined gross spine-morphology and structure for differences between spines of a given type which might reflect their position on the test. In the light of these preliminary observations, we prepared primary spines of both types for SEM examination. After cleaning, transverse sections were made at different levels; some whole spines were also prepared to study both the outer and inner microstructural features, at different magnifications.

3 RESULTS

The presence of the two types of spine described by Mortensen (1935) was confirmed; the primary spines of the aboral surface are totally different in their shape, structure and microstructural organisation from those of the oral surface.

3.1 General morphology

3.1.1 Aboral surface

Observations with an optical lens showed that, except for some slight variations, the primary spines of the aboral surface exhibit the same general pattern. For example, the spines surrounding the apical system, which are the smallest, appear as hollow tubes. They terminate in a fine point and are enclosed in a bag of skin with a large poison gland, "the point of the spine acting like the poison fang of a snake" (Mortensen 1935, fig. 64g., p. 98). The other spines on this surface have the same pattern of organisation, both in the ambulacral and interambulacral areas; those of the former are markedly longer than those of the latter. Moreover, these sheathed spines exhibit some alternating light and dark bands, with two to three annular constrictions, towards the ambitus.

Although Mortensen (1935) mentioned such poison-organs in Asthenosoma varium, he was referring only to the largest spines of the aboral surface. We contend that the smallest spines, with the annular constrictions, also contain venomous cells. This is supported by the study of spine microstructure.

With regard to the venomous spines of A. varium, it is interesting to recall the conflicting views of Sarrasin & Sarrasin (1886, 1888) and Mortensen (1935). The former thought that sheathed spines were essentially embryological characteristics whilst the latter viewed them as a specialisation in the adult.

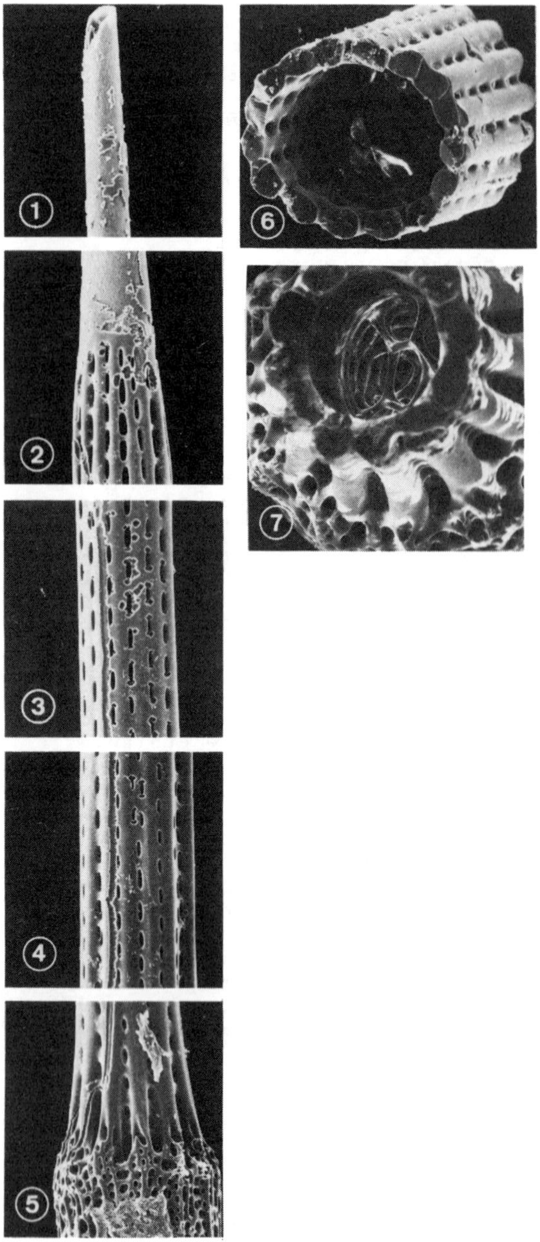

Pl. 1. Asthenosoma varium. Venomous spine of aboral side. (1 - 5), fluted shaft with pores in longitudinal outer view (x 250); (6), transverse section in the third zone (x 375); (7), transverse section above the milled ring (x 600).

3.1.2 Oral Surface

The primary spines display a more complex structure which is very different from that of the aboral surface. They are curved and terminate in a white trumpet-like hoof. This is sharply demarcated from the spine shaft and is finely and densely serrated along almost its entire length. These spines are tallest at the ambitus. The hoof, which is narrow and scarcely wider than the cylindrical spine shaft, tends to be somewhat wider distally; it represents about one quarter of the total spine length. Approaching the peristome, these spines become shorter and less curved. The hoof becomes finer until it eventually disappears, close to the peristome.

As also obtained for the aboral surface, we were unable to differentiate between the spines of the ambulacral and interambulacral zones.

3.2 Microstructural organisation

3.2.1 Aboral surface (Pl. 1)

After cleaning, the calcareous spine-skeleton displays a very homogeneous microarchitecture, irrespective of location or size. It comprises a fluted shaft (65µm in diameter), perforated by equi-sized lenticular pores (12µm) which are disposed in regular longitudinal lines. The shaft terminates in a sharp, hollow point which constitutes one sixth of the total length (Pl. 1, fig. 1-5). This spine-shape is reminiscent of an hypodermic syringe, a comparison which is particularly apt insomuch as we are dealing with a spine which is enclosed in a bag of skin and includes a poison gland. In our opinion, the structural similarity which we noticed between the smaller and bigger spines suggests that all the primary spines may be venomous on this surface. Unfortunately, the state of preservation of our material did not allow histological examination.

Lastly, the transverse sections at different levels of the shaft show that these spines are hollow from base to apex. The wall presents a thin (10µm) cortex, formed of a single layer and traversed throughout by the lenticular pores (Pl. 1, fig. 6). It is only at the foot of the shaft, above the milled ring, that the axial cavity appears to be filled with a loose trabecular meshwork (Pl. 1, fig. 7).

3.2.2 Oral surface (Pl. 2)

The external, microstructural and characteristic organisation of these spines allows us to distinguish four zones which show very different, internal microarchitecture.

i) The distal trumpet-like hoof has a smooth outer face (Pl. fig. 1) but the spine appears to contain a trabecular network. The inner (i.e. cortical) part of the hoof displays a somewhat alveolar (10µm) meshwork of oval spaces disposed in longitudinal, regular series (Pl. 2, fig. 2). This meshwork is delimited close to the central part by a muff, pierced by numerous and regularly distributed pores (Pl. 2, fig. 3). The central part is filled with very fine trabecules which are not joined in the axial part of the hoof (Pl. 2, fig. 3).

ii) The second zone, which is very short, really delimits the hoof itself (Pl. 2, fig. 4); its external wall is fluted and has fine, minute, teeth along each flute. These flutes are separated immediately beneath the hoof by large grooves occupied by an alveolar meshwork, the pores of which are distributed first in three lines, then in two and in quincunces (Pl. 2, fig. 5). The transverse sections show that, at the extreme limit of the hoof which separates the (i) and (ii) zones, the axial cavity is not yet totally filled. On the other hand, in the subapical (ii) zone, the cavity completely disappears, being occupied by a dense trabecular and irregular meshwork (Pl. 2, fig. 6).

iii) The third zone is the most important in relation to the total length of the spine, being at least half its length. It has a microstructure characterised by verticils of scales which arise from the denticules of the (ii) zone as they progressively lengthen. The scales become longer towards the basal part

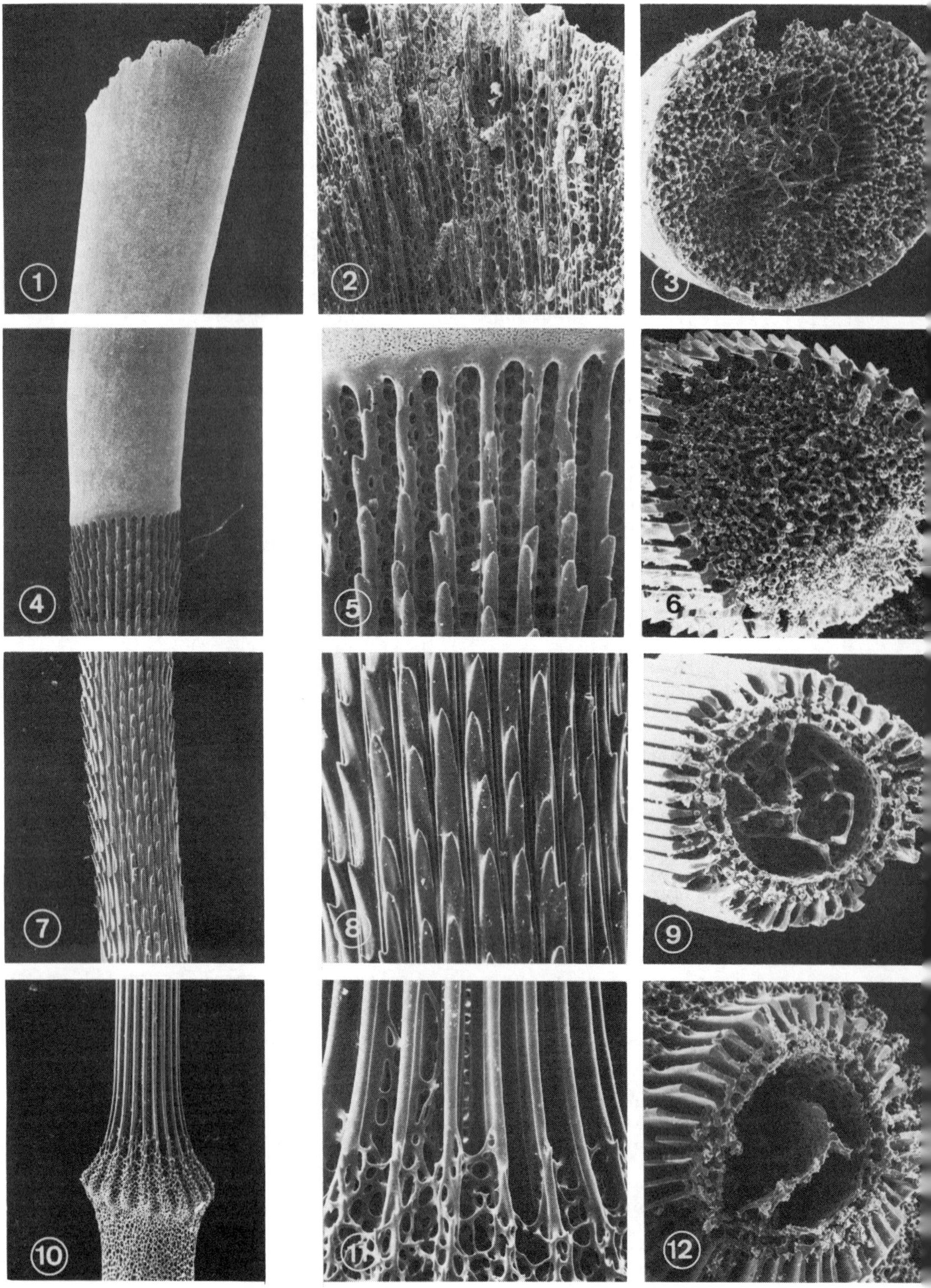
1
2
3
4
5
6
7
8
9
10
11
12

of the shaft. The flutes bearing the scales are themselves pressed closely together (Pl. 2, fig. 7-8). Transverse sections within this zone show that the internal spine microstructure tends to change, and forms an axial and subcircular cavity which occupies half the diameter. This lumen is delimited by a porous muff and is irregularly partitioned (Pl. 2, fig. 9).

iv) In the basal zone of the shaft, the scale verticils disappear and the flutes (16µm) only bear scarce denticules which are no longer in evidence at the spine-foot (Pl. 2, fig. 10-11). Here, the spine is still hollow and has almost the same microstructural organisation as the (iii) zone.

4 MICROSTRUCTURAL AFFINITIES BETWEEN THE PRIMARY SPINES OF A. VARIUM (ECHINOTHURIDAE) AND THOSE OF THE DIADEMATIDAE AND TOXOPNEUSTIDAE

The microstructure of the primary spines of Asthenosoma varium shows analogies with those of the Diadematidae, as described by previous authors (Mackintosh 1875, reviewed in Mortensen 1935), and also, curiously, with those of modern Toxopneustidae.

4.1 The aboral spines

These are characterised by a very peculiar microstructure which seems identical to that of other echinothurid species, e.g. the genus Araeosoma (Mortensen 1935, fig. 64f). By comparison with the hollow spines of the diadematids, the A. varium spines are readily distinguishable by their smooth and fluted external surface which contrasts particularly in regard to the scale verticils, although in other echinothurid species the spines are also ornamented with verticils, e.g. in Phormosoma verticillatum (Mortensen 1935, fig. 64d). They show affinities in the thin cortex, the more important axial cavity and also in the presence of a loose meshwork, at the milled ring level, which looks like some structures observed especially in Echinothrix diadema and Astropyga radiata (Regis & Thomassin 1982, fig. 9c-d).

4.2 The oral spines

The study of the external shaft microstructure of the oral spines allows us to distinguish four zones, each of which exhibits a particular microstructure:

i) The apical zone, with its hoof-shape, is characteristic of most of the Echinothuridae (the genera Phormosoma and Kamptosoma excepted).

ii) The subapical zone is distinguishable from the (i) zone by the presence of flutes, ornamented with minute teeth, alternating with large grooves showing two or three lines of pores. These structures resemble those observed on the spines of some Diadematidae such as Echinothrix diadema (Pl. 3, fig. 7) and also on the distal parts of spines in some Toxopneustidae, e.g. Toxopneustes pileolus (Pl. 4, fig. 7) and, to a lesser degree, Sphaerechinus granularis and Lytechinus variegatus (Pl. 4, fig.1,4).

iii) Here, in the middle and prebasal zone, the microstructure displays verticils (Pl. 3, fig.10) which recall the diadematid spines (Pl. 3, fig. 1, 4)(Mackin-

---

Pl. 2. Asthenosoma varium. Spine of oral side. (1), trumpet-like hoof in longitudinal outer view (x 60); (2), inner part of the hoof in longitudinal view (x 120); (3), transverse section in the hoof (x 180); (4), second zone of the shaft in longitudinal outer view (x 60); (5), part of the second zone of the shaft: flutes with fine minute teeth (x 250); (6), transverse section in second zone of the shaft (x 180); (7), third zone of the shaft in the longitudinal outer view (x 60); (8), part of the third zone of the shaft: the scale verticils (x 250); (9), transverse section in the third zone of the shaft (x 180); (10), fourth zone of the basal shaft in longitudinal outer view (x 60); (11), part of the fourth zone of the basal shaft (x 250); (12), transverse section in the shaft basal zone (x 180).

1
2
3
4
5
6
7
8
9
10
11
12

tosh 1975, Pl. 33, fig. 40).
The flared corollar shape of the verticils resemble that of Diadema setosum.
iv) The basal zone where the microstructure becomes progressively smooth towards the base is similar to that observed at this level on the spines of D. setosum.

In the basal part of the A. varium oral spines, a cylindrical, axial cavity, with a muff perforated by more or less regularly arranged elliptical pores (Pl. 3, fig. 11-12), resembles the diadematid spines (Diadema antillarum, D. setosum, Echinothrix diadema; Pl. 3, fig. 2-3, 5-6, 8-9) described by Mackintosh (1875).

In another connection, an axial cavity of this type, but of lesser diameter, is also found in the foot part of some toxopneustid spines, particularly in Sphaerechinus granularis and Lytechinus variegatus (Pl. 4, fig. 2,5). By contrast, in other toxopneustid species such as Toxopneustes pileolus, the axial cavity begins to be filled with a muff (Pl. 4, fig. 8) and is completely obliterated in Tripneustes gratilla (Pl. 4, fig. 11). The progressive obliteration of this axial cavity is linked to the presence at the spine tip of a central screen surrounded with spurs, as in S. granularis and L. variegatus (Pl. 4, fig.3,6). In T. pileolus, it begins to be obliterated (Pl. 4, fig. 9); in T. gratilla it is completely filled (Pl. 4, fig. 12) and the shaft flutes lack minute teeth (Pl. 4, fig. 10). According to these features, in transverse sections from the apex to the foot, the spine microstructure of S. granularis shows a close relationship to that of the spines of A. varium, except for the apical hoof-zone (Pl. 4, fig. 2-3).

The inner face of the acetabulum of A. varium spines shows a sieve structure, similar to that observed in diadematids, e.g. D. setosum, E. diadema and A. radiata (Pl. 5, fig. 1-2-3), which bears out the relationships between species of these two families. This structure could be observed, at high magnification, on the spines of some toxopneustids, e.g. S. granularis (Pl. 5, fig. 4-5), whereas it is partially obliterated in others, e.g. L. variegatus (Pl. 5, fig. 6-7).

5 CONCLUSION

From Mortensen (1935), the spines of the Echinothuridae recall, in their structure, those of the fossil Lepidocentridae (both families forming the order Lepidocentroida, of the subclass Euechinoidea) more than those of the Diadematidae (classified with the order Aulodonta). By comparison, Mackintosh (1875) saw many affinities between the spines of Asthenosoma varium and those of Centrostephanus and he placed the former among the Diadematidae (Peters). Mortensen (1935), however, saw the spines of these three families as being of the same type. A comparison of the findings of this study on the oral spines (mainly the middle part) of A. varium, with those for modern

Pl. 3. Diadematid and Echinothurid primary spines. (1 - 3), Diadema antillarum: (1), longitudinal outer view of the shaft: the scale verticils (x 60); (2), longitudinal section of the shaft: axial cavity delimited with porous muff (x 60); (3), transverse section of the shaft (x 120); (4 - 6), Diadema setosum: (4), longitudinal outer view of the shaft: the scale verticils (x 120); (5), longitudinal section of the shaft: axial cavity delimited with porous muff (x 1200); (6), transverse section of the shaft (x 90); (7 - 9), Echinothrix diadema: (7), longitudinal outer view of the shaft: flutes ornamented with fine minute teeth (x 120); (8), longitudinal inner view of the shaft (x 120); (9), transverse section of the shaft (x 120); (10 - 12), Asthenosoma varium: (10), longitudinal outer view of the shaft of oral spine: flutes ornamented with fine minute teeth (x 250); (11), longitudinal section of the basal part of oral spine: axial cavity delimited with porous muff (x 180); (12), transverse section at the basal part of the shaft (x 180).

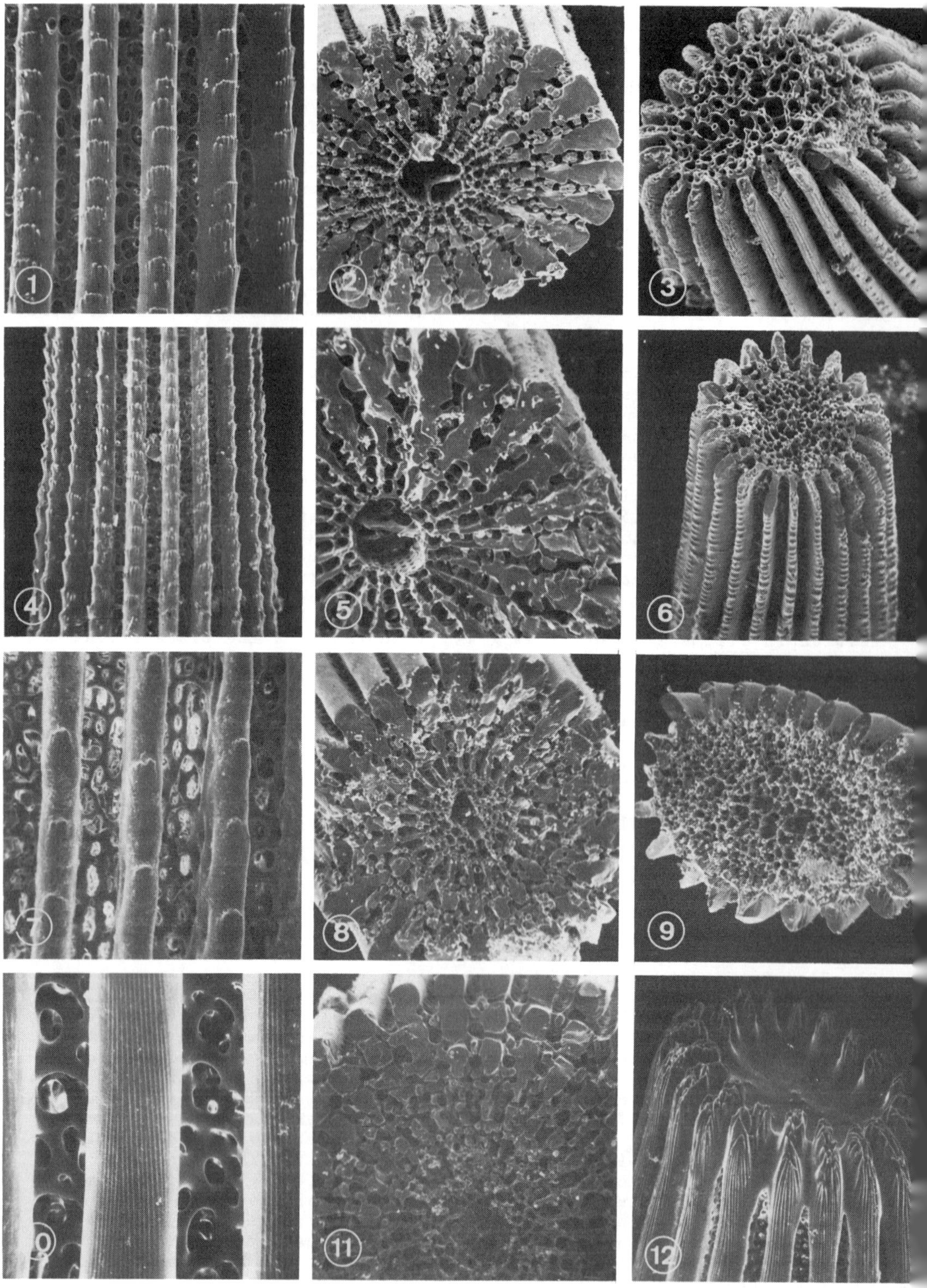

diadematids corroborates this relationship between these families. Spine microstructure is here characterised: i) mainly by an external microstructure with flutes, ornamented by minute teeth or verticils, surrounding a cylindrical axial cavity which is delimited by a muff perforated with elliptical pores; ii) by the inner face of the acetabulum, showing a sieve that probably corresponds to the perforated structure of the sieve tubercles.

According to Hesse (1900), this is the Diadema-type of spine which is found in diadematids, aspidodiadematids and some pedinids (Lissodiadema, Chaetodiadema). All these observations lead to the recent classification of Parker (1982) who includes the orders Echinothuroida and Diadematoida in the superorder Diadematacea (subclass Euechinoidea).

The originality of most of the echinothurid spines lies in the structure of the apical hoof-shaped zone, a structure which also occurs in the aspidodiadematids. Where Mortensen (1940) classified this last family among the Aulodonta, Parker (1982) places it in the order Diadematoida, with the Diadematidae.

So, if from a systematic point of view, the echinothurid A. varium displays true relationships with the diadematids, it also shows other resemblances in its ethology. Most of the modern diadematids colonise bottoms in sheltered areas (either in back reef areas of the littoral zone or in deep areas of the circa- and bathyal zones) and their spine morphology seems to represent a morpho-functional adaptation (Regis & Thomassin 1982). Species with very long spines showing a well developed axial cavity (e.g. D. setosum) behave as an "herbivorous-grazer/seston suspension feeder". A.varium is known to colonise coastal bottoms in sheltered areas, in the deepest zones, as in the New Caledonia lagoon (pers. obs. B.T. and J.L. Menoux) or in inner littoral zones (Mortensen 1935). As with Diadema, it could be detritophagous and is probably sestonophagous, with the ability to assimilate particulate or dissolved organic compounds (presence near W.C. recorded by Mortensen (1940) in the Banda Sea).

The fact that spines of some toxopneustids show morphological resemblances to those of A. varium is more disturbing. Parker (1982) classifies toxopneustids in the order Temnopleuroida, super-order Echinacea, sub-class Euechinoidea. Most authors link this family to the Temnopleuridae, the Phymosomatidae and back to the Cidaridae (Termier & Termier 1953). The similarity between the spine structure of the Diadematacea and that of some toxopneustids is seen in the presence of flutes, surrounding a cylindrical axial cavity which is delimited by a muff with elliptical pores. This similarity is more acceptable as an example of morphological convergence in Lytechinus and Sphaerechinus than as suggesting a phylogenetic link between these families which are systematically far removed. Cuenot (1948), however, saw the cidarids and the diadematids as having evolved separately from a common lepidocentrid or echinothurid ancestor. This is contrary to the

---

Pl. 4. Toxopneustid primary spines. (1 - 3), Sphaerechinus granularis: (1), longitudinal outer view of the shaft (x 250); (2), transverse section of the shaft: axial cavity with muff (x 120); (3), spine tip with central screen (x 250); (4 - 6), Lytechinus variegatus: (4), longitudinal outer view of the shaft (x 150); (5), transverse section of the shaft: axial cavity with muff (x 250); (6), spine tip with central screen (x 150); (7 - 9), Toxopneustes pileolus: (7), longitudinal outer view of the shaft (x 250); (8), transverse section of the shaft: axial cavity partially filled and muff has disappeared; (9), spine tip with central screen partially obliterated (x 120); (10 - 12), Tripneustes gratilla: (10), longitudinal outer view of the shaft (x 600); (11), transverse section of the shaft: axial cavity completely obliterated (x 250); (12), spine tip completely filled (x 250).

Table 1. Affinities and systematic position of Asthenosoma varium (Echinothuridae) with genera of the Diadematidae and Toxopneustidae.

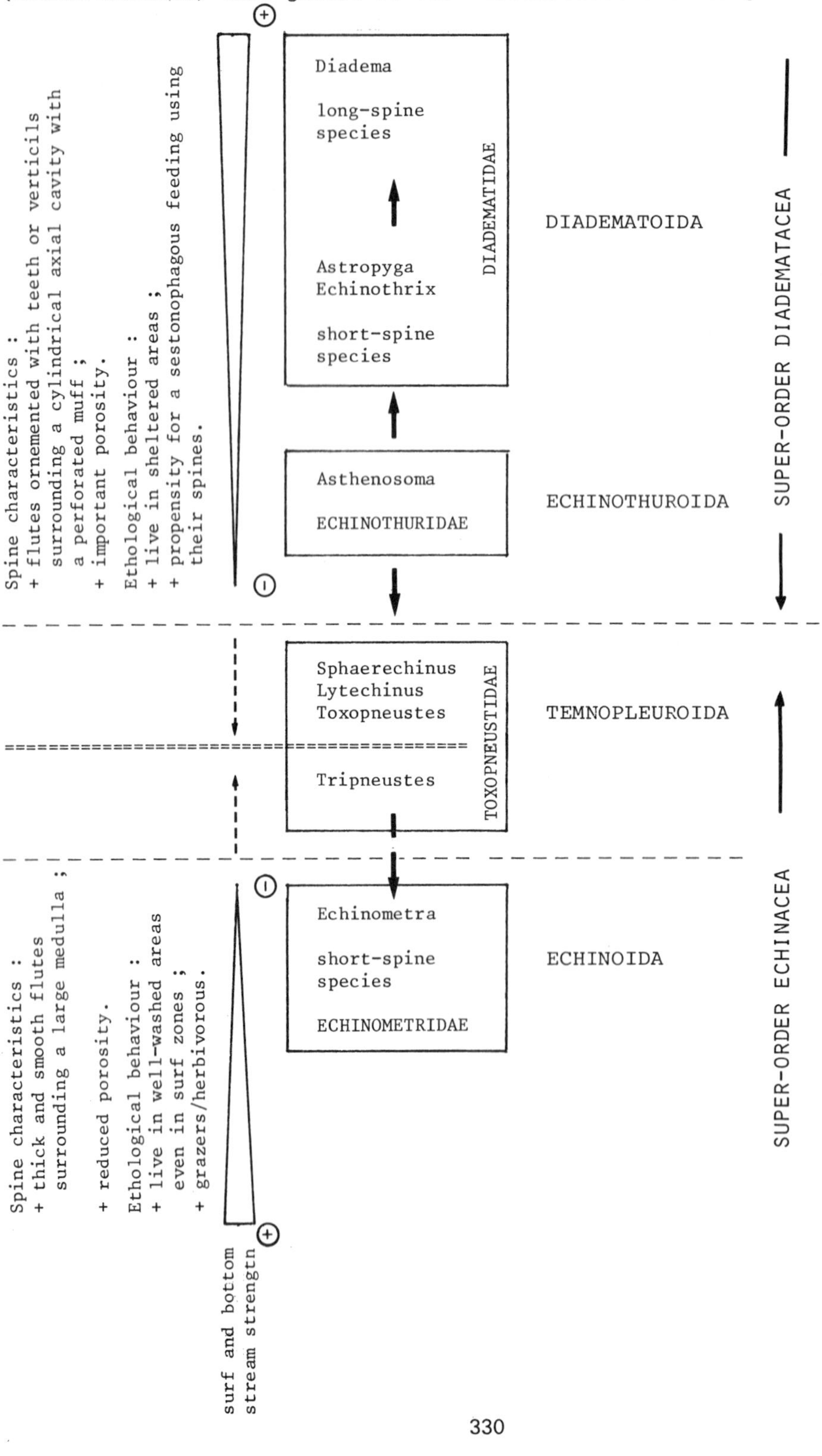

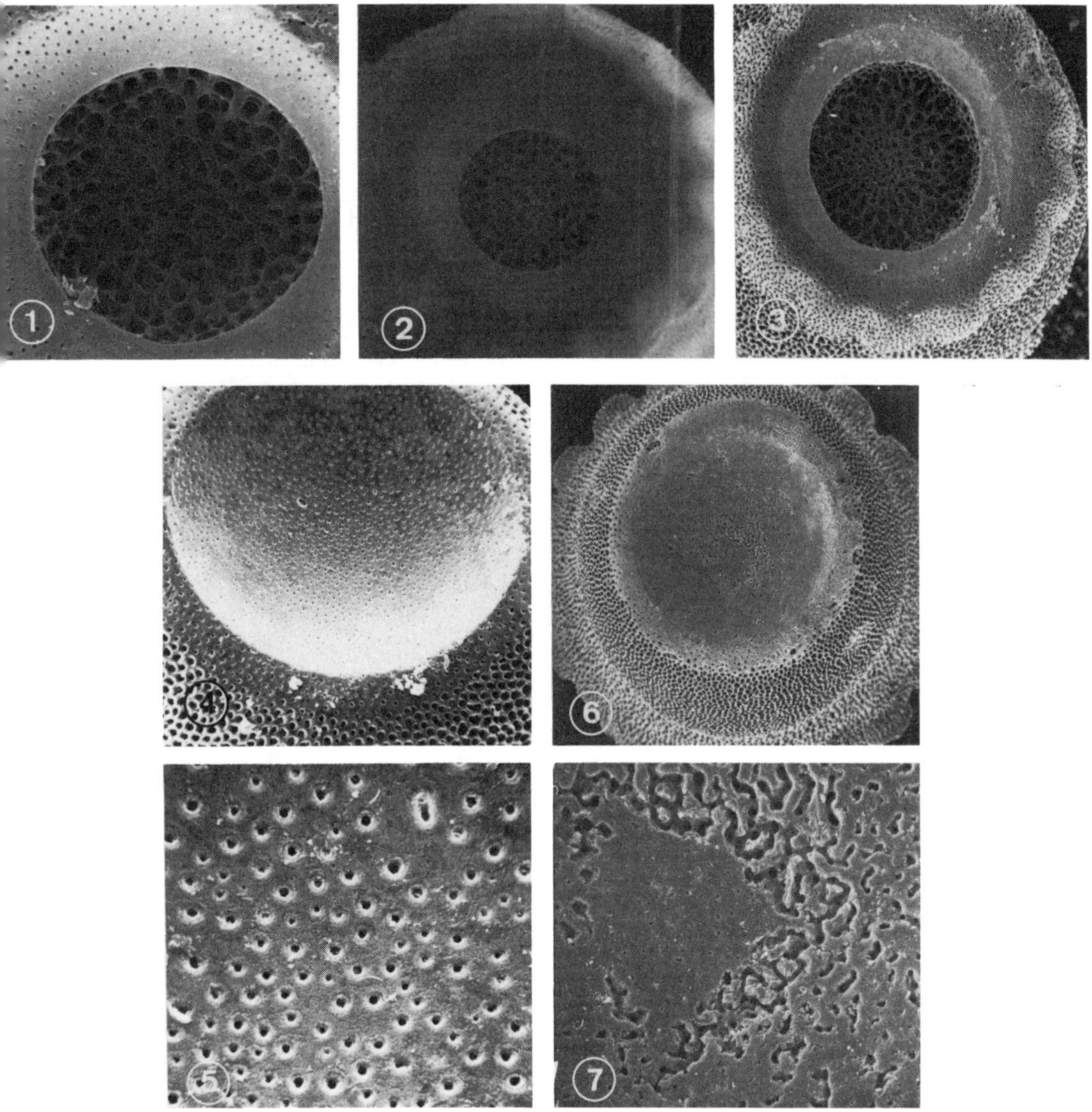

Pl. 5. Inner face of acetabulum. (1), Asthenosoma varium (x 250); (2), Diadema setosum (x 180); (3), Echinothrix diadema (x 120); (4 - 5), Sphaerechinus granularis (x 250; x 1200); (6 - 7), Lytechinus variegatus (x 120; x 600).

view of Gignoux (1933) who saw the Diadema type of spine (Hesse 1900) as having derived from the Cidaris spine-type since the Trias.

In conclusion, Asthenosoma varium (Echinothuridae), on the basis of its spine microstructure, displays close relationships with the Diadematidae, particularly with the short spined species (g. Echinothrix and Astropyga) and shows analogies with some toxopneustid species (Temnopleuroida), e.g. Sphaerechinus granularis, Lytechinus variegatus and, to a lesser degree, Toxopneustes pileolus. In the latter family, which is viewed as something of a turning point, some species show characteristics close to those of the Echinothuridae (Echinothuroida) and the Diadematoida, whilst others (e.g. Tripneustes) show affinities with the Echinometridae (Echinoida). All these opinions are outlined in Table 1.

ACKNOWLEDGEMENTS

Samples were collected in 1983 within the scope of the programme: "Reef and lagoonal ecosystem studies of New Caledonia (CNRS/CPF-ANVAR) SEM pictures were taken with the help of Mrs. Guillaumin (Lab. "Evolution des êtres organisés", Univ. Paris VI.)

REFERENCES

Agassiz, A. & H.L. Clark 1909. Hawaiian Echininoids.Echinothuridae. Mem. Mus. Comp. Zool. Harvard 34, 3.

Cuénot, L. 1948. Anatomie, éthologie et systématique des Echinodermes. In P.P. Grassé (ed.), Traité de Zoologie 11, p. 2-275. Paris: Masson & Cie.

Gignoux, M. 1933. Les oursins réguliers fossiles, évolution et classification. Trav. Lab. géol. Fac. sci. Grenoble 17: 1 - 125.

Hesse, E. 1900. Die Mikrostruktur der fossilen Echinoideenstacheln und deren systematische Bedentung. Neues Jahrb. f. Mineral. 13, Beilage Bd.

Mackintosh, H.W. 1875. Research on the structure of the spines of the Diadematidae (Peters). Trans. roy. Irish Acad, 25 (sci.): 519-558, pl. 31-33.

Mortensen, Th. 1935. A monograph of the Echinoidea. II. Bothriocidaroida, Melonechinoida, Lepidocentroida and Stirodonta). Copenhagen: Reitzel.

Mortensen, Th. 1940. A monograph of the Echinoidea. III. 1. Aulodonta with additions to vol. II (Lepidocentroida and Stirodonta). Copenhagen: Reitzel.

Parker, S.P. (ed.) 1982. Synopsis and classification of living organisms. New York: McGraw-Hill Book Co., 2

Régis , M.B. 1981. Adaptations morphofonctionnelles de la microstructure des radioles d'Echinoides réguliers. Téthys 10: 177-184.

Régis, M.B. & B.A. Thomassin 1982. Ecologie des Echinoides réguliers dans les récifs coralliens de la région de Tuléar (S.W. de Madagascar): adaptation de la microstructure des piquants. Ann. Inst. Oceanogr. Paris (N.S.) 58: 117-158.

Sarasin, 1886. Über einen Lederigel aus dem Hafen vom Trincomalie und seinen Giftapparat. Zool. Anz. 9 p.80.

Sarasin, 1888. Über die Anatomie der Echinothuriden und. Phylogenie der Echinoderm. Ergebn. Naturwis. Forschungen auf Ceylon, 1(3).

Termier, H. & G. Termier 1953. Classe des Echinides. In I. Piveteau (ed.), Traité de paléontologie 3, p. 857-947. Paris: Masson & Cie.

Differential mortality during the embryonic and larval lives of northeast Pacific echinoids

STEVEN S.RUMRILL & FU-SHIANG CHIA
University of Alberta, Edmonton, Canada

ABSTRACT: Rates of predation were determined during the embryonic and larval lives of Dendraster excentricus, Strongylocentrotus franciscanus and S. purpuratus in a series of laboratory feeding experiments. Eggs and early embryonic stages of all three species were most susceptible to predation by brachyuran crab zoea (Hemigrapsus nudus). Swimming blastulae, gastrulae and prism larvae were generally captured at intermediate rates while echinoplutei were captured at relatively low rates. Mean zoea clearance rates decreased from 0.67 to 0.30 liters/zoea/day during the transition of prism larvae into plutei. Reduced susceptibility to predation by H. nudus is related to increasing larval size and to the ontogeny of larval swimming behavior.

1 INTRODUCTION

The life-histories of most temperate echinoids include a pelagic planktotrophic larval stage lasting from 51-141 days in length (Strathmann, 1978). There is little information concerning the survivorship of larval echinoids during the pelagic period. Embryos and larvae risk mortality from several different sources including offshore transport away from suitable habitats (Jackson & Strathmann, 1981), starvation, predation by pelagic and benthic predators (Thorson, 1950; Cowden, et al., 1984) and physiological stress associated with transport into sub-optimal water masses. Interactions between predators and larval prey are particularly noteworthy since their outcomes are directly related to the relative vulnerability of larval stages. Consequently, larval predation may provide a mechanism for the evolution of larval traits which reduce predator-induced mortality.

The pelagic period of planktotrophic echinoids encompasses several changes in larval morphology and behavior. During early embryogenesis the zygote and successive cleavage stages are retained within an extraembryonic vitelline membrane. After hatching the larva emerges as a swimming blastula. Subsequent development following gastrulation gives rise to a swimming prism larva which contains rudiments of spicules and skeletal structures. Eventually an echinopluteus is reached which is characterized by 4, 6 or 8 ciliated larval arms. Within the constraints imposed by this ontogenic pattern there are many interspecific modifications in larval morphology, behavior and rates of development (Mortensen, 1931). The purpose of this paper is to present preliminary findings of stage-specific differences in the susceptibility of embryos and larvae of 3 temperate northeast Pacific echinoids to predation by a zooplanktonic microcarnivore. We also describe interspecific differences in larval susceptibility which are related to larval morphology and rates of pelagic development.

2 MATERIALS AND METHODS

Adult Dendraster excentricus (Eschscholtz, 1831), Strongylocentrotus franciscanus (Agassiz, 1863) and S. purpuratus (Stimpson, 1857) were collected subtidally with SCUBA from Barkley Sound, British Columbia (48°50'N, 125°08'W). Spawning was induced with intracoelomic injections of 0.52 M KCl. Embryos and larvae were cultured in 1 um filtered seawater (12°C) and fed a 2:1 mixture of Dunaliella tercoliecta (Chlorophyta) and Monochrysis leutheri (Chrysophyta). Zoea I and II stage larvae of the grapsid crab Hemigrapsus nudus (Dana, 1851) were used as micropredators. Ovigerous female crabs were maintained in laboratory aquaria and zoeae were collected shortly after hatching. Zoeae were cultured at 12°C and fed brine shrimp nauplii (Artemia salina) prior to experiments with echinoderm prey.

The susceptibility of echinoid embryos and larvae to capture by Hemigrapsus zoea was examined in a series of laboratory feeding experiments. Density-dependent predation rates were determined for each echinoid species by conducting feeding experiments at 7 different prey concentrations (10, 25, 50, 75, 100, 150 and 250 larvae/liter). Stage-specific differences in larval susceptibility were determined at a single prey concentration (100 larvae/liter) for the following development stages: 1) unfertilized ova, 2) pre-hatched embryos, 3) motile blastulae and gastrulae, 4) prism larvae, 5) 4-armed plutei, 6) 6-armed plutei and 7) 8-armed plutei. In each trial of the feeding experiments prey were individually counted into 1 liter glass jars. At the initiation of each trial 1 Hemigrapsus zoea was added to each jar. All feeding experiments consisted of 5 replicate treatments and 3 controls without predators. The jars were strapped horizontally around the long axis of a cylindrical mixing device which was slowly rotated to keep predators and prey suspended. Currents generated within the rotating jars ranged from 0 (center) to 0.97 cm/s (edge). All feeding experiments were run at 10 ± 1.3°C in an incubator with 16L:8D photoperiod. The duration of individual trials varied according to the duration of transient developmental stages. At the end of each trial the number of remaining prey was determined and zoea clearance rates were calculated as

$$C = (V/t)\ (\ln Di - \ln Df)$$

where C is the clearance rate in liters/zoea/day, V is the reciprocal of predator density in liters/zoea, t is the duration of each experiment in hours, and Di and Df are the initial and final densities of prey (Vanderploeg & Scavia, 1979).

3 RESULTS

The density-dependent feeding responses of Hemigrapsus zoea in experiments with embryonic stages of D. excentricus, S. franciscanus and S. purpuratus are shown in Figure 1. For all 3 species of echinoids the mean number of prey ingested increased in a sigmodal manner between densities of 10 and 250 embryos/liter. Rapid increases in predation rate occurred between 50 and 75 embryos/liter. Mean predation rates were consistently greatest for embryos of S. purpuratus ($y = 16.67 + 0.44x$,

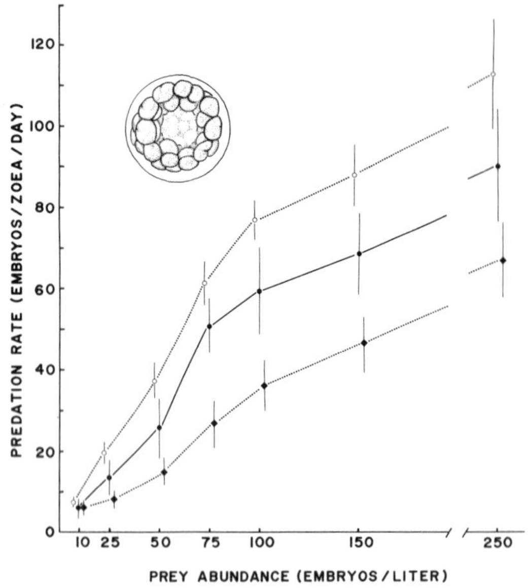

Figure 1. Effects of embryo density on predation rate by H. nudus. (● = D. excentricus; ◆ = S. franciscanus; O = S. purpuratus)

$r^2 = 0.87$), intermediate for D. excentricus ($y = 11.52 + 0.35x$, $r^2 = 0.84$) and lowest for S. franciscanus ($y = 4.29 + 0.26x$, $r^2 = 0.92$). At prey densities greater than 25 embryos/liter differences in predation rates were significant between the 3 species ($p < 0.01$). In all cases the slopes of regression lines were greater than losses in respective controls ($p < 0.001$).

For all species of echinoids, clearance rates of zoeae were highest on non-swimming ova and embryonic stages compared with clearance rates on more advanced larvae (Fig. 2). Clearance rates were intermediate for swimming gastrulae and prism larvae of D. excentricus and S. purpuratus ($p < 0.01$) although the differences were not significant compared with ova and embryos in S. franciscanus ($p > 0.05$). In all species susceptibility to predation decreased sharply during the transition of prism larvae into plutei; pooled clearance rates for plutei having 4-8 arms were significantly lower than combined clearance rates for earlier larval and embryonic stages ($p < 0.001$, Table 1). The duration of the high-susceptibility period lasted about 70 hrs in D. excentricus, 90 hrs in S. franciscanus and 110 hrs in S. purpuratus.

Differences in morphology, body length and swimming behavior are summarized for the development stages of the echinoids

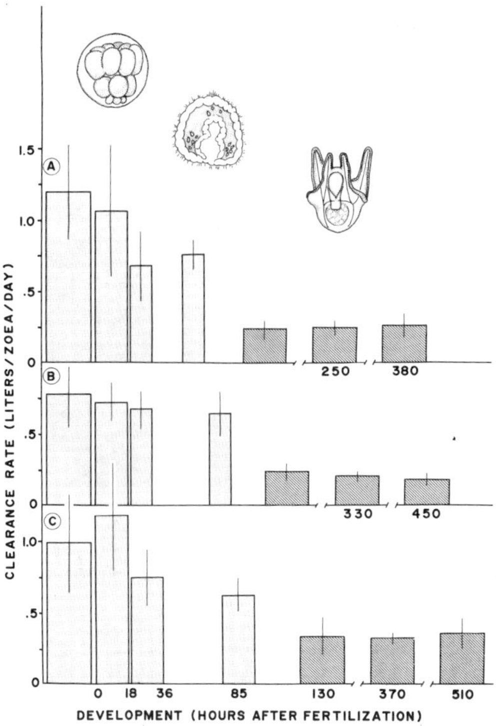

Figure 2. Clearance rates ($\bar{x}$ ±SD) of H. nudus on developing stages of A) D. excentricus, B) S. franciscanus and C) S. purpuratus. Stippled bars = high vulnerability: ova, embryos, blastulae - gastrulae and prism larvae; slashed bars = low vulnerability: 4, 6 and 8-armed echinoplutei.

in Table 2. Larval shape and size are associated with changes in swimming behavior, but all larvae swim at a similar speed (0.94 ±0.13 mm/s). Large body size, elongation of the arms and the reversed swimming ability of pluteus larvae contribute either singly or in combination to the decreased susceptibility of pluteus larvae to predation by zoea (see Fig. 2).

The development of calcite skeletal structures within plutei was not an effective deterrent to predation by Hemigrapsus zoea. Figure 3 shows the internal skeleton of a living Dendraster pluteus. The birefringent remains of the skeleton recovered from zoea gut contents are shown in Figure 4. Similarly crushed skeletons of S. franciscanus and S. purpuratus were also observed in preparations of zoea gut contents.

4 DISCUSSION

Early developmental stages of D. excentricus, S. franciscanus and S. purpuratus were the most susceptible to predation by Hemigrapsus zoea. By comparison, more advanced pluteus larvae of all 3 species were less vulnerable to predation. Differential survivorship of embryos and larvae indicates that features associated with ontogeny function as adaptations for larval defense.

Echinoid embryos and larvae show similar stage-specific patterns of susceptibility in response to predation by other microcarnivorous zooplankton. For example, crab zoea (Cancer, Pugettia) consume ova and embryos of D. excentricus at a high rate compared with pluteus larvae (Rumrill, et al., in prep.). Similar larval mortality patterns have also been observed for D. excentricus and S. purpuratus in response to predation by carnivorous copepods (Pseudocalanus) and euphausids (Thysanoessa) (personal observations). Several parameters including changes in larval size and shape, swimming behavior and physiological development should be considered in assessing the differential susceptibility of embryos

Table 1. Clearance rates ($\bar{x}$ ±SD) and one-way ANOVA comparison of H. nudus predation between high and low vulnerability prey groups (high vulnerability = embryos, blastulae-gastrulae and prism larvae; low vulnerability = 4, 6 and 8-armed plutei; n = 15 trials for each species).

Species	High Vulnerability	Low Vulnerability
D. excentricus	0.92 ±0.32	0.24 ±0.07
S. franciscanus	0.69 ±0.16	0.20 ±0.05
S. purpuratus	0.88 ±0.36	0.37 ±0.13

Source of Variation	d.f.	Sum of Sqrs.	Mean Sqr.	F	p
Between Groups	1	5.64	5.64	136.93	0.001
Within Groups	88	3.63	0.04		
Total	89	9.27			

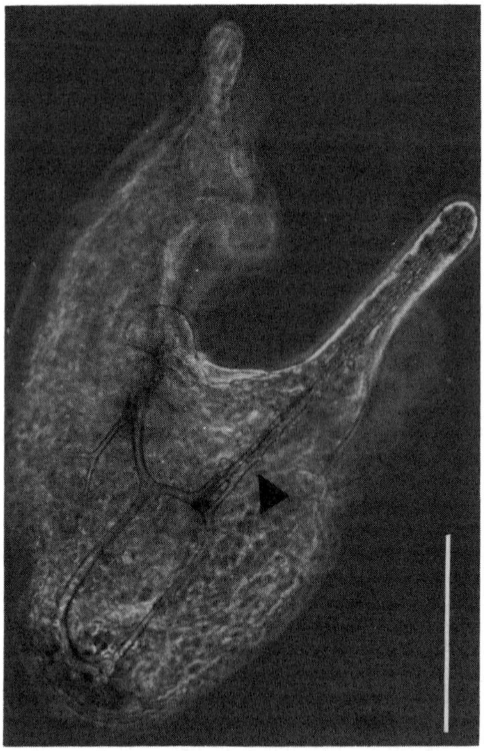

Figure 3. Living pluteus larva of D. excentricus showing internal fenestrated spicule (arrow). Scale bar = 100 um.

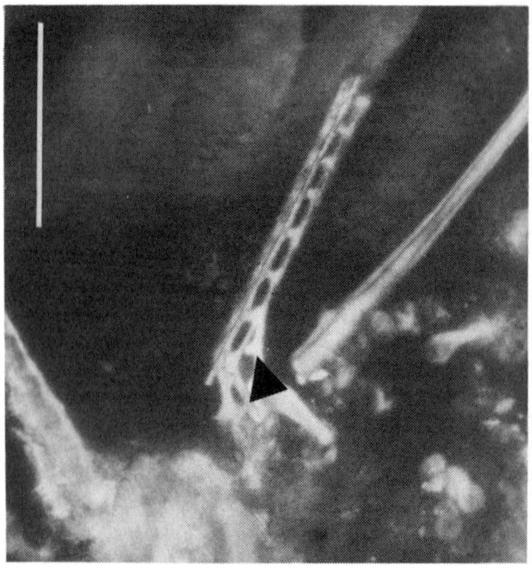

Figure 4. Birefringent remains of D. excentricus spicules (arrow) in gut contents of H. nudus. Scale bar = 50 um.

and larvae to capture by these raptorial predators.

One possible explanation for differential survivorship is that raptorial micropredators such as Hemigrapsus zoeae feed more efficiently on smaller-bodied prey. For example, Costlow and Bookhout (1959) and Sulkin and Epifanio (1975) reported that early zoea stages of the portunid crab Callinectes sapidus continued development only when their diets included small prey such as polychaete larvae, rotifers or echinoid embryos. Furthermore, Levine and Sulkin (1984) showed that newly hatched zoea of Eurypanopeus depressus actively captured non-motile microcapsules in the size range of 40-180 um, although larger capsules were not tested. These results suggest that differential survivorship may be directly dependent upon changes in larval size and that large plutei may attain a refuge from predation simply through growth. However, many crab larvae have been successfully reared through the megalopa stage on diets consisting of large prey such as Artemia (body length > 500 um; Costlow, et al., 1966; Poole, 1966; Trask, 1970; Christiansen & Yang, 1976). Consequently, the low predation rates of echinoplutei reported here are probably dependent upon a combination of ontogenic features including size and swimming behavior.

The development of fenestrated calcite spicules within plutei has been suggested to be a possible deterrent to predation

Table 2. Comparisons of morphology, body length, swimming behavior and speed between development stages of D. excentricus, S. franciscanus and S. purpuratus. Length values = means for all 3 species; swimming speed ($\overline{X}$ ±SD; n = 15/stage) = D. excentricus.

Development Stage	Morphology	Body Length (um)	Swimming Behavior	Swimming Speed (mm/s)
Ova	Sphere	93, 121, 141	Non-Swimming	-
Embryo	Sphere	144	Non-Swimming	-
Gastrula	Prolate Sphere	153	Unidirectional	0.97 ±0.17
Prism	Pyramid	216	Unidirectional	1.03 ±0.12
4-Arm Pluteus	Pyramid	428	Bidirectional	0.84 ±0.13

(Emlet, 1983). However, our experiments with larval strongylocentrotids and D. excentricus showed that predation rates were similar for larvae having fenestrated and non-fenestrated skeletons. The function of fenestrated spicules in echinoid larvae remains to be elucidated.

Cowden, et al. (1984) suggested that D. excentricus plutei utilize chemical defenses to deter benthic predators. However, the presence of larval toxins cannot explain stage-specific predation since it is unlikely that plutei contain toxic substances while earlier stages do not. For example, Lucas, et al. (1979) detected saponin compounds in asteroid larvae at concentrations similar to those found in eggs. Since Hemigrapsus zoea readily consume ova and embryos it is unlikely that chemical defenses are a factor in determining differential susceptibility.

In our experiments the decreased vulnerability of plutei to predation was closely associated with the development of reversed swimming ability. Reversed swimming appears to be directly related to predator avoidance since the response moves larvae away from the immediate capture sphere of the zoea. Echinoid larvae develop the ability to reverse the direction of ciliary beat and swim backwards during the transition from prism larvae into plutei. Mackie, et al. (1969) showed that changes in electrical potential were directly coupled with reversals in ciliary beating when plutei encountered unicellular algae. However, the potential defensive role of ciliary reversal has not been previously identified.

The pattern of differential embryonic and larval survivorship reported here has implications which extend to the ontogeny of other marine invertebrates. High vulnerability of eggs and embryos to predation suggests that planktotrophic species may experience selective pressures to rapidly develop structural, chemical and behavioral defenses. Differential susceptibility may provide an explanation for the short duration of the embryonic period in planktotrophic species as well as for the high incidence of invertebrate life-histories which include the encapsulation or brooding of embryos.

5 ACKNOWLEDGEMENTS

Appreciation is extended to R.E. Foreman, Director of Bamfield Marine Station, for providing research facilities. We also thank J.C. Cureton for preparing the figures and manuscript.

6 REFERENCES

Christiansen, M.E. & W.T. Yang. 1976. Feeding experiments on the larvae of the fiddler crab Uca pugilator (Brachyura, Ocypodidae), reared in the laboratory. Aquaculture 8:91-98.

Costlow, J.D., Jr. & C.G. Bookhout. 1959. The larval development of Callinectes sapidus R. reared in the laboratory. Biol. Bull. 116:373-396.

Costlow, J.D., Jr., C.G. Bookhout & R. Monroe. 1966. Studies on the larval development of the crab, Rithropanopeus harrisii Gould. I. The effect of salinity and temperature on larval development. Physiol. Zool. 39:81-100.

Cowden, C., C.M. Young & F.S. Chia. 1984. Differential predation on marine invertebrate larvae by two benthic predators. Mar. Ecol. Prog. Ser. 14:145-149.

Emlet, R.B. 1983. Locomotion, drag, and the rigid skeleton of larval echinoderms. Biol. Bull. 164:433-445.

Jackson, G.A. & R.R. Strathmann. 1981. Larval mortality from offshore mixing as a link between precompetent and competent periods of development. Am. Nat. 118:16-26.

Levine, D.M. & S.D. Sulkin. 1984. Ingestion and assimilation of microencapsulated diets by brachyuran crab larvae. Mar. Biol. Letters 5:147-153.

Lucas, J.S., R.J. Hart, M.E. Howden & R. Salathe. 1979. Saponins in eggs and larvae of Acanthaster planci (L.) (Asteroidea) as chemical defenses against planktivorous fish. J. exp. mar. Biol. Ecol. 40:155-165.

Mackie, G.O., A.N. Spencer & R.R. Strathmann. 1969. Electrical activity associated with ciliary reversal in echinoderm larvae. Nature 223:1384-1385.

Mortensen, Th. 1931. Contributions to the study of the development and larval forms of echinoderms. I-II. Kgl. Danske Vidensk Selsk. Skr. Nat. Math. Afd. 9:1-39.

Trask, T. 1970. A description of laboratory reared larvae of Cancer productus Randall (Decapoda, Brachyura) and a comparison to larvae of Cancer magister Dana. Crustaceana 18:134-146.

Poole, R.L. 1966. A description of laboratory reared zoeae of Cancer magister Dana and megalopae taken under natural conditions (Decapoda: Brachyura). Crustaceana 11:83-97.

Strathmann, R.R. 1978. Length of pelagic period in echinoderms with feeding larvae from the northeast Pacific. J. exp. mar. Biol. Ecol. 34:23-27.

Sulkin, S.D. & C.E. Epifanio. 1975. Compar-

ison of rotifers and other diets for rearing early larvae of the blue crab, Callinectes sapidus Rathbun. Est. Coastal mar. Sci. 3:109-113.
Thorson, G. 1950. Reproductive and larval ecology of marine bottom invertebrates. Biol. Rev. 25:1-45.
Vanderploeg, H.A. & D. Scavia. 1979. Two electivity indices for feeding with special reference to zooplankton grazing. J. Fish. Res. Board Can. 36:362-365.

Proceedings of the Fifth International Echinoderm Conference / Galway / 24-29 September 1984

L'édification de la face orale au cours du développement direct de *Abatus cordatus,* oursin incubant subantarctique

Ph.SCHATT
Laboratoire de Biologie des Invertébrés marins et de la Malacologie, Muséum national d'Histoire naturelle, Paris, France

ABSTRACT: Direct development of the subantarctic sea-urchin Abatus cordatus: **formation of the oral face.**

At the end of gastrulation, the oral cavity develops within a thickening of the ectoderm. The latter becomes isolated within the mesoderm and extends under the ectoderm of the oral pole of the embryo. At this stage, the oral surface differentiates: the neural plate, water ring, five radial canals and five papillae arise from folds in the ectodermal floor of the cavity. The oral surface is exposed at the beginning of juvenile stage 1 when an opening develops in the ceiling of the cavity.

While the morphology of the oral surface is similar to that described for other echinoids during metamorphosis, no larval stage was observed. This is the first description of an actual direct development in echinoids.

1 INTRODUCTION

Le développement des échinides à larves planctoniques "échinoplutéus" (Agassiz, 1864; Mortensen, 1921,1937; Fell, 1948; Harvey, 1956; Guidice, 1973; Stearns, 1974; Czihak, 1975) et celui d'échinides à larves modifiées dont la vie planctonique est raccourcie, voire absente, (Mortensen, 1921; Williams et Anderson, 1975; Amemiya et Tsuchiya, 1979) sont bien connus. En revanche, celui des échinides incubants n'a encore jamais fait l'objet d'études embryologiques.

Abatus cordatus (Verrill 1876) est un oursin incubant, endémique des îles Kerguelen (Agassiz 1881). Les oeufs lécithotrophiques sont les plus gros connus chez les échinides (Magniez

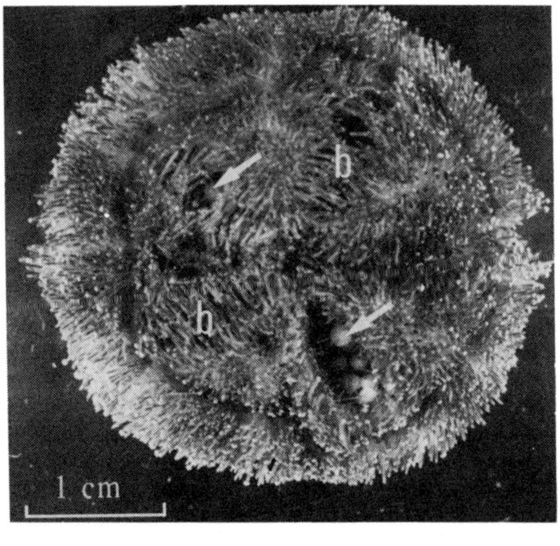

Fig.1. Femelle de **Abatus cordatus.** Les piquants protégeant 2 des marsupia ont été enlevés. (➡): embryons ; (b): marsupia fermés par les piquants imbriqués.

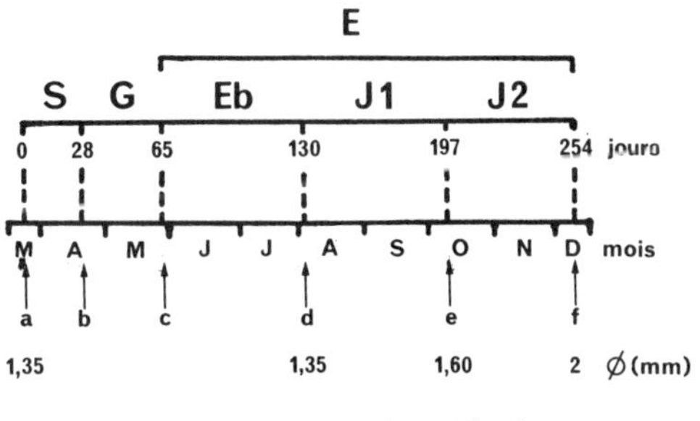

Fig.2. Temps moyen de développement de **Abatus cordatus.** S: segmentation, G: gastrulation, E: embryogenèse, Eb: embryon, J1: juvénile de stade 1, J2: juvénile de stade 2. a: fécondation, b: blastula, c: perte de la membrane de fécondation, d: ouverture de la cavité orale, e: différenciation de la face orale, f: sortie des juvéniles.

1979,1980). La femelle conserve ses embryons et ses juvéniles à l'intérieur de 4 poches incubatrices ou "marsupia" creusées par la déformation du squelette des ambulacres (fig.1). L'incubation dure 8,5 à 9 mois (fig.2).

L'édification de la face orale a lieu durant la période embryonnaire qui dure 2 mois (Schatt 1983) jusqu'à la formation du juvénile de stade 1, défini morphologiquement comme ayant des piquants d'une taille de 0,1 à 0,3 mm (Magniez 1979,1980).

2 MATERIEL ET METHODES

2.1 Récoltes et élevages

Les adultes d'**Abatus cordatus** ont été prélevés de décembre 1981 à mars 1983 à la station intertidale de l'anse du Halage (péninsule Port Jeanne d'Arc). Les oursins ont été élevés au laboratoire de Biomar (Port-aux-Français), dans des aquariums alimentés en eau de mer courante dont la température varie de +2°C, en hiver, à +7°C, en été.

Les embryons et les juvéniles ont été récoltés dans les poches incubatrices et élevés artificiellement dans des coupelles en verre contenant de l'eau de mer filtrée (millipore 0,45 µm), placées à l'obscurité.

2.2 Histologie

Chaque échantillon a été fixé au glutaraldéhyde à 3% tamponné au cacodylate de sodium à pH 7,4. Après déshydratation dans l'éthanol, les échantillons ont été inclus dans l'araldite ACM et coupés à 1 µm d'épaisseur. Ces coupes semi-fines sériées ont été colorées au bleu de toluidine (pH 11,5).

3 RESULTATS

3.1. Fin de la gastrulation

Environ 65 jours après la fécondation, le blastopore est clos et la membrane de fécondation entourant encore la gastrula est rompue. Le diamètre de la gastrula est de 1,35 ± 0,05 mm et reste constant jusqu'au juvénile de stade 1.

Au moment de la fermeture du blastopore, un épaississement de l'ectoderme (fig.3) apparaît à l'opposé du pôle d'invagination (pôle aboral). C'est dans cet épaississement que se développe la cavité orale.

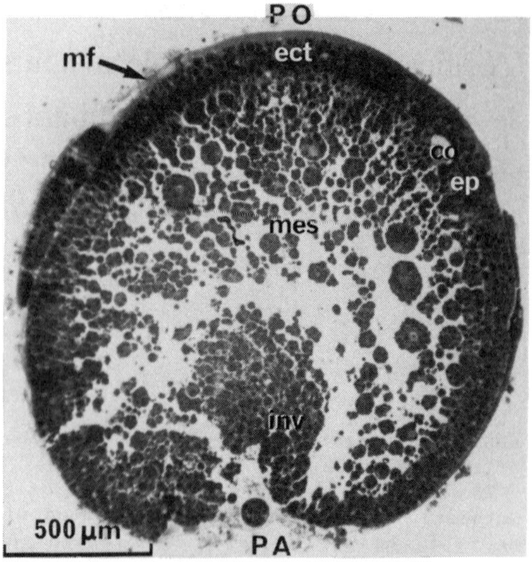

Fig.3. Coupe semi-fine d'une gastrula âgée de **Abatus cordatus**, 65 jours après la fécondation. Formation du bourgeon de la cavité orale. P.O.: pôle oral, P.A.: pôle aboral, ect: ectoderme, ep: épaississement de l'ectoderme, mes: mésoderme, inv: invagination, mf: membrane de fécondation, co: cavité orale.

3.2. La période embryonnaire

Elle s'étende du 65ème jour au 130ème jour après la fécondation.

3.2.1 Organisation générale de l'embryon

L'embryon reste sphérique et présente d'abord une simple dépression aborale figurant la fermeture du blastopore. Au cours de cette étape, une pigmentation rouge apparaît et s'étale sur tout l'embryon ; elle se concentre ensuite au pôle oral. La face orale primordiale, composée des premiers bourgeons de piquants et de pieds ambulacraires, s'édifie dans l'hémisphère oral de l'embryon. Cette formation est isolée du milieu extérieur par un tissu transparent qui est apparemment perforé par les premiers piquants. Les structures larvaires telles que les bras ou des bandes ciliées sont absentes.

L'embryon est entouré d'un épiderme d'une épaisseur de 70 à 80 µm. La cavité coelomique se développe dans le mésoderme (fig.4, 5, 6) au cours de la période embryonnaire. A la fin de cette période, le tube digestif embryonnaire est courbé et suspendu par des mésentères dans la cavité coelomique. Il est constitué de cellules palissadiques riches en vitellus. Aucune nécrose n'est observée. Aucun élément d'un squelette larvaire interne n'apparaît.

3.2.2 La cavité orale

Au début de la période embryonnaire, le bourgeon de la cavité orale s'isole dans le mésoderme sous-ectodermique oral (fig.4). Cette cavité est délimitée par un épithélium supérieur, le plafond, et un épithélium inférieur, le plancher, tous deux d'une épaisseur variant de 70 à 80 µm.

Au cours de la phase embryonnaire, le plafond s'amincit pour n'atteindre plus que 10 µm d'épaisseur au moment de la fusion avec l'ectoderme oral. Le bourgeon s'étend sous l'ectoderme de l'hémisphère oral. La conjonction de ces phénomènes provoque l'augmentation du volume de la cavité orale sans changer le volume de l'embryon (fig.4, 5, 6).

L'épiderme oral fusionne avec le plafond de la cavité orale. Celle-ci s'ouvre à ce niveau, vers l'extérieur. Les premiers piquants et pieds ambulacraires sortent par cette ouverture.

Le plancher conserve une épaisseur constante et participe à la formation de la face orale.

3.2.3 L'édification de la face orale

La face orale est composée de 3 couches tissulaires d'épaisseurs différentes: le plancher de la cavité orale (70-80 µm), le mésoderme (40-50 µm), et l'épithélium coelomique (10 µm).

Au centre de la face orale, le plancher ectodermique présente, pendant la phase embryonnaire, des replis qui vont former la plaque neurale (fig.5). Chez l'embryon âgé, cette plaque neurale est isolée dans le mésoderme et émet 5 nerfs radiaires, issus d'un repli interne de l'ectoderme(fig.6). Une expansion externe de l'ectoderme (ou papille) termine le repli interne formant le nerf radiaire. La longueur de cette papille atteint 170 µm environ (fig.6).

Dans le mésoderme, l'anneau aquifère (fig.5) se différencie ; 5 canaux ambulacraires, localisés sous le nerf radiaire, partent de cet anneau et se terminent dans chacune des 5 papilles (fig.6). Le canal radiaire (nerf radiaire et canal ambulacraire) est formé par la progression de la papille depuis le centre vers le bord de la face orale (fig.6).

Le système nerveux est constitué du côté externe par le cordon nerveux épineural, et du côté interne par le cordon nerveux hyponeural qui délimitent un

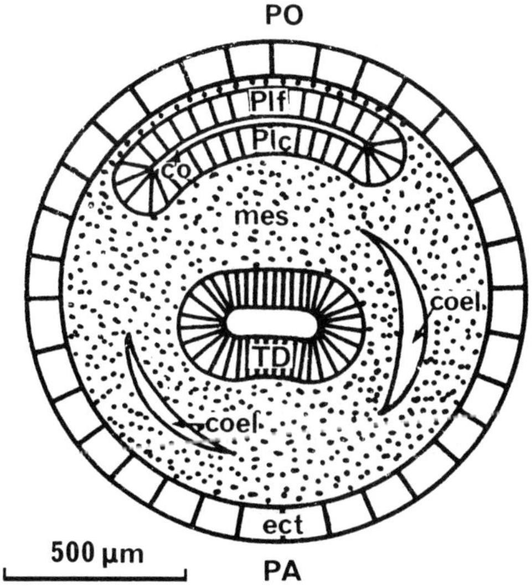

Fig.4. Coupe schématique du début de la période embryonnaire chez **Abatus cordatus**. Le bourgeon de la cavité orale s'isole dans le mésoderme. Abréviations supplémentaires : c.o.: cavité orale, coel: coelome, Plf: plafond, Plc: plancher, TD: tube digestif.

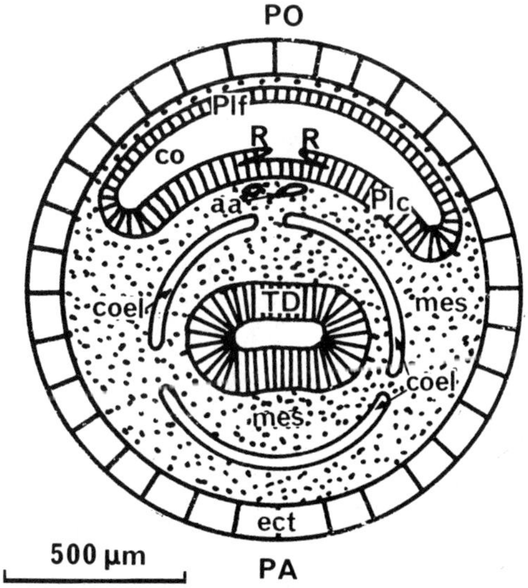

Fig.5. Coupe schématique de l'embryon de **Abatus cordatus** pendant la période embryonnaire. Le bourgeon de la cavité orale s'étend et la face orale débute sa différenciation. Abréviations supplémentaires : R: replis de l'ectoderme, aa: anneau aquifère.

sinus. Le stomodéum se différencie à ce stade mais n'est pas en relation avec le tube digestif ; il est clos du côté de la cavité orale. Les premiers bourgeons de piquants apparaissent et sont formés par le soulèvement de l'ectoderme sous lequel s'isole une petite partie du mésoderme (fig.6).

Le squelette primitif du juvénile et de l'adulte se différencie également chez cet embryon âgé.

3.3 Le juvénile de stade 1

Après l'ouverture de la cavité orale, l'"amnios" (sensu Mortensen, 1921), formé par l'ectoderme et le plafond de la cavité orale, est repoussé vers le pôle aboral (fig.7). La taille du juvénile de stade 1 s'accroît pour mesurer environ 1,6 mm.

L'hémisphère oral du juvénile de stade 1 est constitué de la face orale

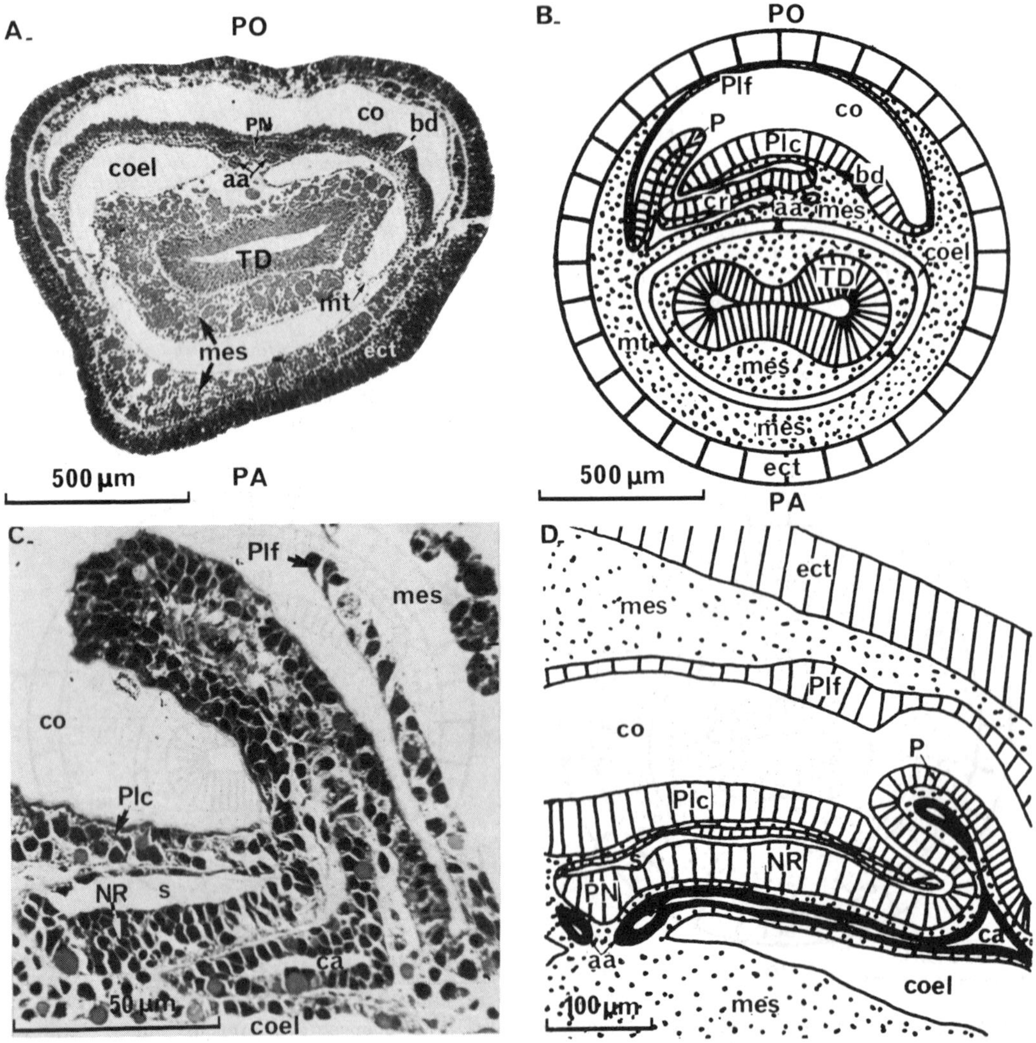

Fig.6. Fin de la période embryonnaire chez **Abatus cordatus**. A: coupe semi-fine de l'embryon, B: coupe schématique de l'embryon, C: coupe semi-fine de la face orale, D: coupe schématique de la face orale. Abréviations supplémentaires : aa: anneau aquifère, bd: bourgeon de piquant, cr: canal radiaire, P: papille, P.N.: plaque neurale, mt: mésentère, NR: nerf radiaire, ca: canal aquifère, s: sinus.

embryonnaire extériorisée (bourgeons de piquants et 5 papilles)(fig.7). L'hémisphère aboral est formé de la partie aborale embryonnaire.

Au centre du pôle oral, la plaque neurale et l'anneau aquifère se prolongent vers les 5 papilles par l'intermédiaire des 5 canaux radiaires. Ces différentes formations ont une structure identique à celle de l'embryon âgé.

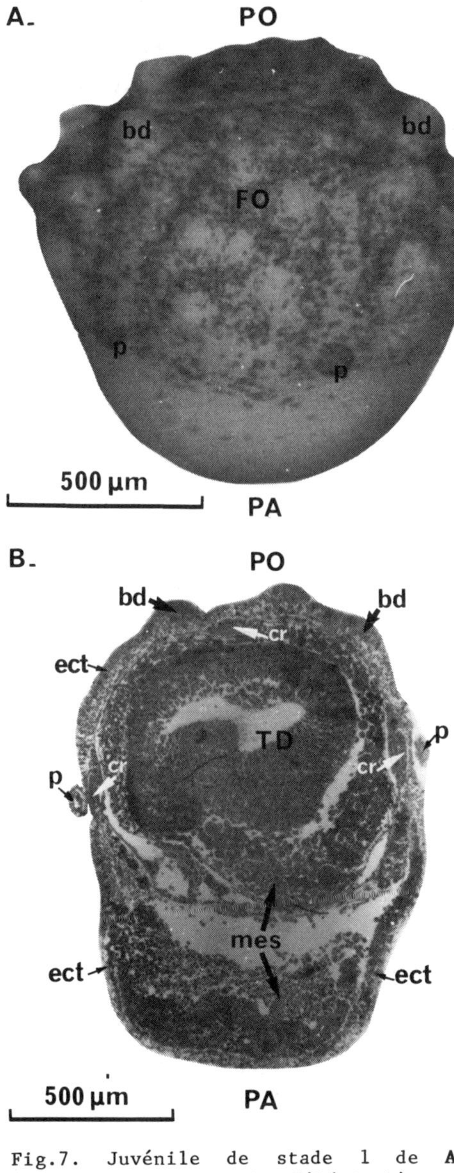

Fig.7. Juvénile de stade 1 de **Abatus cordatus**. A: vue de l'hémisphère oral, F.O.: face orale, B: coupe semi-fine de ce juvénile.

Une grande partie du mésoderme sous-ectodermique est rejetée vers le pôle aboral. Le tube digestif coudé reste embryonnaire chez le juvénile de stade 1, et n'est toujours pas en relation avec le stomodéum.

4 DISCUSSION

Chez les échinides, on distingue classiquement deux types larvaires : l'échinopluteus et les larves modifiées (fig.8).

L'échinopluteus possède en général plusieurs bras plus ou moins développés, supportés par un squelette larvaire (baguettes ou spicules). Les bras et certaines parties du corps possèdent des bandes ciliées délimitant des champs préoral et anal (Mac Bride, 1918 ; Mortensen, 1921,1937 ; Fell, 1948). La métamorphose débute par la formation de la plaque échinienne (épaississement de l'ectoderme) située entre le bras postéro-dorsal et le bras postoral gauche du pluteus. L'invagination de la plaque échinienne et son isolement au dessus de l'hydrocoele donnent naissance à la cavité orale ("cavité amniotique", sensu Mortensen, 1921). Le plancher de cette cavité édifie les 5 premiers pieds ambulacraires et les premiers bourgeons de piquants. A l'ouverture de la cavité orale, ces appendices sont en contact avec le milieu extérieur. Pendant cette étape, les bras du pluteus régressent et se nécrosent.

Les larves modifiées présentent différents aspects :

- un pseudo-pluteus dont les bras courts possèdent ou non un squelette rudimentaire (**Peronella lesueuri**, Mortensen, 1921; **Asthenosoma ijimai**, Amemiya et Tsuchiya, 1979),
- un embryon larviforme (**Heliocidaris erythrogramma**, Williams et Anderson, 1975).

La ciliature externe de ces types larvaires n'est pas disposée en bandes mais répartie uniformément (**P. lesueuri** et **erythrogramma**).

La métamorphose débute par la formation de la plaque échinienne qui s'invagine soit totalement et donne naissance à une cavité orale (**H. erythrogramma**), soit partiellement et reste ouverte sur le milieu extérieur (**P. lesueuri** et **A. ijimai**). Le plancher de l'invagination édifie les 5 premiers pieds ambulacraires et les premiers bourgeons de piquants. L'extrusion de cette face orale donne naissance au juvénile.

Chez **Abatus cordatus**, aucune formation telle que des bras, un squelette larvaire

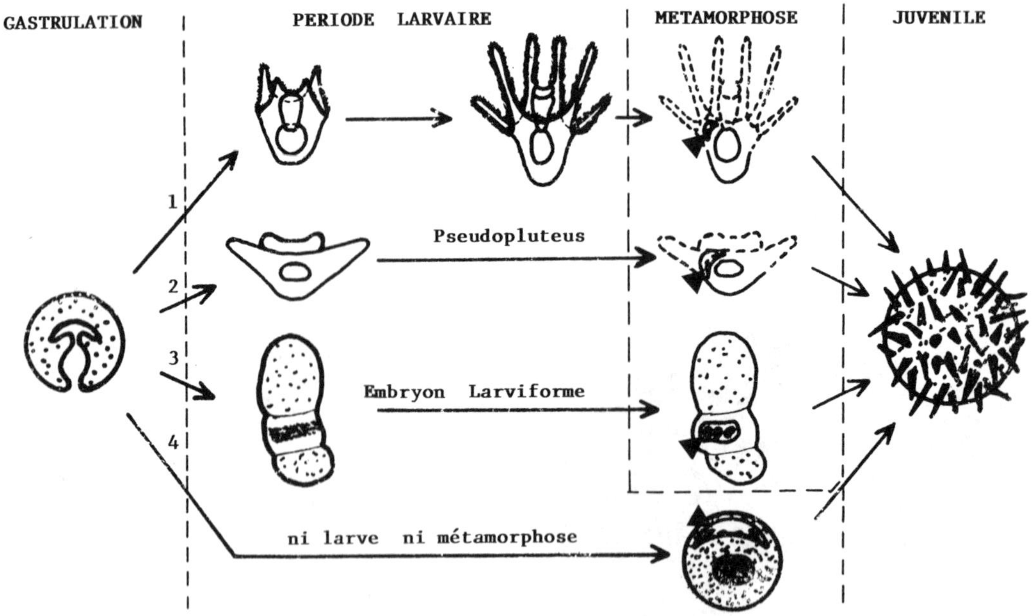

Fig. 8: les différents types de développement chez les échinides.
1: échinopluteus typique (Mortensen 1921-1937, Fell 1948)
2: pseudopluteus: **Asthenosoma ijimai** (Amemiya et Tsuchiya 1979)
Peronella lesueuri (Mortensen 1921)
3: embryon larviforme: **Heliocidaris erythrogramma** (Mortensen 1921, Williams et Anderson 1975)
4: aucune larve: **Abatus cordatus** (Schatt 1983, et présente étude).
Les flèches indiquent la formation de la cavité orale.
Les pointillés représentent les parties des larves qui se nécrosent.

ou des bandes ciliées, n'apparaît au cours de la période embryonnaire. Le développement de **Abatus cordatus** ne présente donc aucun stade larvaire.

En revanche, on constate, comme lors de la métamorphose larvaire des autres échinides, la formation directe d'une cavité orale (fig.8). Elle est issue du creusement de l'épaississement de l'ectoderme qui pénètre dans le mésoderme. Il n'y a donc pas d'invagination réelle. Le plancher de cette cavité édifie une face orale dont la symétrie pentaradiée est marquée par les 5 papilles. Aucune nécrose n'est observée pendant ce processus. **Abatus cordatus** est le premier cas de développement direct décrit chez les échinides :

. pas de phase larvaire même très courte,
. pas de métamorphose avec disparition de la majeure partie des tissus larvaires.

REMERCIEMENTS

Ce travail a été financé par la Mission de Recherche des Terres Australes et Antarctiques Françaises (Programme "Benthos" ; responsable : A.GUILLE).

Je remercie C.BERNARD et P.LEVI de l'aide efficace pour la confection des coupes semi-fines, A.FOUBERT pour la réalisation des photographies, et M.VAN BEVEREN pour la frappe du manuscrit.

J'exprime ma gratitude à N.BOURY-ESNAULT, D.DOUMENC, J.-P. FERAL et M.VAN PRAET pour les fructueuses discussions que j'ai eu avec eux.

REFERENCES

Agassiz, A. 1864. On the embryology of echinoderms. Mem.Amer.Acad.Sci.9: 1-30.

Agassiz,A. 1881. Echinoidea. Rep. Sci. Res. Voy.Challenger. (Zool.). 3: 13-21, 45 pl.

Amemiya, S. and T.Tsuchiya 1979. Development of the echinothurid sea-urchin **Asthenosoma ijimai**. Marine Biology. 52: 93-96.

Czihak, G. 1975. The sea-urchin embryo. Springer Verlag.Berlin. 700p.

Fell, H.B. 1948. Echinoderm embryology and the origin of chordates. Biol. Rev. 23: 81-107.

Guidice, G. 1973. Developmental biology of the sea-urchin embryo. Academic Press, New-York, 469 p.

Harvey, E.B. 1956. The american **Arbacia** and other sea-urchin. Princeton University Press, New Jersey, 298 p.

Mac Bride, E.W. 1918. Development of **Echinocardium cordatum.** Quart. J. Microscop. Sci.59, 63.

Magniez, P. 1979. Modalités de l'incubation chez **Abatus cordatus** (Verrill), oursin endémique des îles Kerguelen. In: Echinoderms. Present and past. Balkema Press, Rotterdam.309-403.

Magniez, P. 1980. Le cycle sexuel d'**Abatus cordatus** : modalités d'incubation et évolution histologique et biochimique des gonades. Thèse de Doctorat de 3° Cycle, Université de Paris 6.

Mortensen, T. 1921. Studies of the development and larval forms of echinoderms. Copenhagen: G.E.C.Gad. 261p., 33pl.

Mortensen, T. 1937. Contributions to the study of the development and larval forms of echinoderms. IV. K. danske Vidensk. Selsk. Skr., ser.9, 7(3): 1-59 + plates I-XIII.

Schatt, Ph. 1983. Embryologie descriptive de l'échinide incubant **Abatus cordatus** (Verrill), oursin endémique des îles Kerguelen. Symbiose.XV,4: 234-235.

Stearns, L.W. 1974. Sea-urchin development. Cellular and molecular aspects. Dowden, Hutchinson, & Ross, Int. Strondburg, Pennsylvania. 339p.

Verrill, A.E. 1876. Contribution to the natural history of Kerguelen Islands. Annelids and echinoderms. Bull. U. S. Nat. Mus. 3 : 64-75.

Williams, D.H.C. and D.T.Anderson 1975. The reproductive system, embryonic development and metamorphosis of the sea-urchin **Heliocidaris erythrogramma** (Val) (Echinoidea : Echinometridae). Aust. J. Zool. 23 : 371-403.

Proceedings of the Fifth International Echinoderm Conference / Galway / 24-29 September 1984

Pedicellariae as a specific character in sea urchin species

HEINZ SPLECHTNA & HELGE HILGERS
Institut für Zoologie der Universität Wien, Abt. für Vergleichende Anatomie und Morphologie, Austria

ABSTRACT: The value of the pedicellariae in classifying within the genus Echinus was assessed by comparing the ophiocephalous pedicellariae of two closely related species. Because of difficulties in distinguishing between Echinus acutus and Echinus melo, using the standard skeletal features, the ophiocephalous pedicellariae are included in the species diagnoses as additional and more clear-cut characteristics. The diagnostic value of this pedicellaria type is discussed; it is shown to be useful not only in respect of the genus Echinus but also the genus Psammechinus.

INTRODUCTION

The problem of distinguishing Echinus acutus Lamarck from Echinus melo Lamarck using such features as the shape and colour of the corona, the arrangement of the primary tubercles on the interambulacral plates and the colour of the primary spines (Moehler 1895, 1921; Mortensen 1903, 1943; Tortonese 1965) is long standing. Based on samples taken over many years during field courses to the Adriatic, these characters were seen to exhibit great variability. The question was posed as to whether it is, in fact, possible to differentiate between the species on morphological grounds. Our work (Hilgers and Splechtna 1976, 1979; Splechtna and Hilgers 1979) on the structure, function and development of pedicellariae indicated a high specificity in the design of these structures. We applied this knowledge in trying to solve the problem outlined above.

Pedicellariae have been used in the classification of orders, families and genera for a long time (Döderlein 1887, 1906; Mortensen 1903, 1943; Jensen 1974, 1982); their value is disputed however where species differentiation is concerned. Of the four main types found in regular echinoids, only the globigerous and tridentate pedicellariae have been used in classification. Ophiocephalous pedicellariae have been regarded as useless in distinguishing the higher systematic categories (Jensen 1974).

To demonstrate the value of ophiocephalous pedicellariae in species diagnosis, the information of Süssbach and Breckner (1911) on Echinus esculentus was used to effect a comparison with our own material. Comparative investigations were also carried out on the related but well separated species Psammechinus miliaris P.L.S. Müller and Psammechinus microtuberculatus Blainville. In addition to material from the Adriatic, we examined specimens from Mediterranean habitats which had been lodged in the Museum of Natural History, Vienna.

METHODS

Some 100 specimens were examined from different habitats in the Mediterranean and Adriatic, together with 10 Echinus esculentus and 30 specimens each of Psammechinus miliaris and P. microtuberculatus. The pedicellariae were removed from the test of the preserved animals, cleared in Eau de Javelle, washed several times in distilled water, dried and prepared for SEM investigation (Cambridge Stereoscan MK IIA).

RESULTS

The diagnoses given in the literature for Echinus acutus and E. melo relate essentially to the following characters: Echinus acutus: shape of test conical with flattened oral side; primary tubercles on all interambulacral plates or variously lacking, particularly above the ambitus; sutures between the plates accompanied by 1 or 2 alternate white and reddish coloured zig-zag bands; colour of spines varied to an extraordinary degree, often green at the base, white distally and reddish to red at the tip. Echinus melo: shape of test globular with less flattened oral side; primary tubercles regularly absent on every second interambulacral plate (fig. 5); sutures regularly accompanied by up to 6 alternate white and brown bands; colour of the primary spines uniformly green.

Problems with identification based on these main characters arise, on the one hand, because of the extremely high variability in the shape of the corona in Echinus acutus (fig. 4a, b, c) recorded by all authors, which gave rise to several subspecies (Mortensen 1943; Tortonese 1965) and, on the other, because the primary tubercles may also be regularly absent on every second interambulacral plate above the ambitus in Echinus acutus. Moreover, the position of the ambitus varies to a high degree as a result of the very different shape of the tests.While the colour of the spines is highly variable and cannot be relied upon as an exclusive specific character, a uniformly green colouration is relatively constant for E. melo. Generally, only a slight deviation from the fixed characters of E. melo leads to E. acutus.

It was necessary, therefore, to find a stable character which would clearly distinguish Echinus acutus and separate it effectively from Echinus melo. Based on our investigations of the ophiocephalous pedicellariae of sea urchins we were able to distinguish clearly the different species by means of the form of these pedicellariae. Tests of this kind were carried out by Süssbach & Breckner (1911) on Echinus esculentus and Echinus acutus, and confirmed by Mortensen (1943). By contrast, Jensen (1974, 1982) questions the value of the ophiocephalous pedicellariae in both the diagnosis of species as well as in higher classification. The ophiocephalous pedicellariae are, in our experience, the first pedicellariae formed by the pluteus and, thus, they are ontogenetically the originals. This conforms to the data given by Gordon (1926). The pedicellaria figured by Burke (1980) could also be an ophiocephalous, based on our knowledge of the development of this pedicellaria.

Each valve of an ophiocephalous pedicellaria is subdivided into 3 parts: a distal spoon-shaped segment provided with gripping teeth at the margin, a median segment bearing the central and peripheral articulations and also the insertions of the adductor muscles, and a proximal handle segment (Fig. 1).

The diagnostic characters in the distal segment are the structures of the margin with the different teeth arrangements, the course of the support from the margin to the central articulation with the median segment, and the shape and size of the articulations. The handle

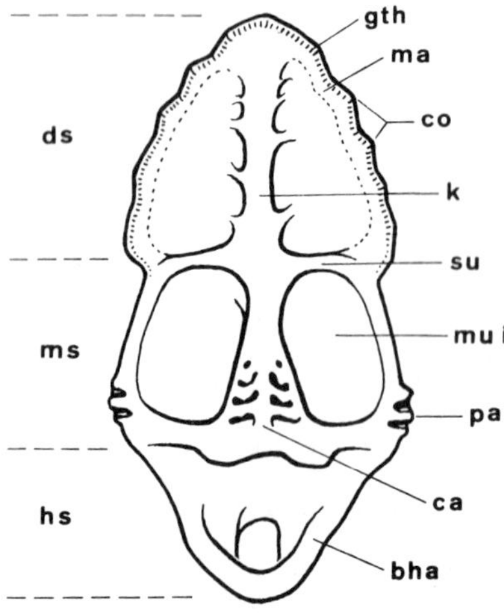

Fig. 1. Scheme of an ophiocephalous valve-skeleton.

segment is not suitable in separating species in the genera considered.

Comparative investigations on the ophiocephalous pedicellariae of Psammechinus miliaris and Psammechinus microtuberculatus gave positive results. The distal segment in Psammechinus miliaris shows a narrow margin which forms relatively poor cogs and continues directly into the margin of the median segment; the fine gripping teeth also continue to this region. Therefore, the margin of both segments is in the same plane and the closed head of the pedicellaria shows a nearly smooth surface in its median part. The supports to the central articulation are isolated, transverse skeletal structures lying above the muscle insertions; its teeth are quite different from the peripheral gripping teeth. In contrast the distal segment of the valve in Psammechinus microtuberculatus has a broad margin with prominent cogs; on both sides, it continues directly to the supports which pass diagonally to the central articulation. The margin of the median segment is clearly separated, so that the valves in this area gape in closed pedicellariae (fig. 2a, b; 6a-d).

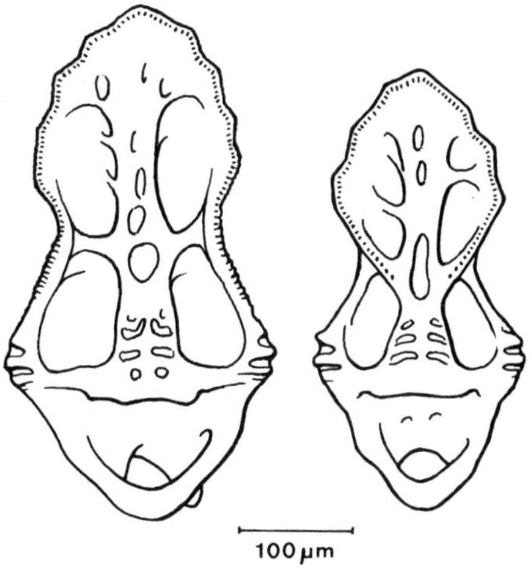

Fig. 2. Ophiocephalous valve-skeleton of a. Psammechinus miliaris and b. Psammechinus microtuberculatus.

Examination of this character in the genus Echinus, in accordance with the Sussbach & Breckner (1911) description of the distal segment, reveals a margin with strong teeth and either lacking or with only slight cogs. The margin of the distal and median segments forms a unit and therefore the outer sur-

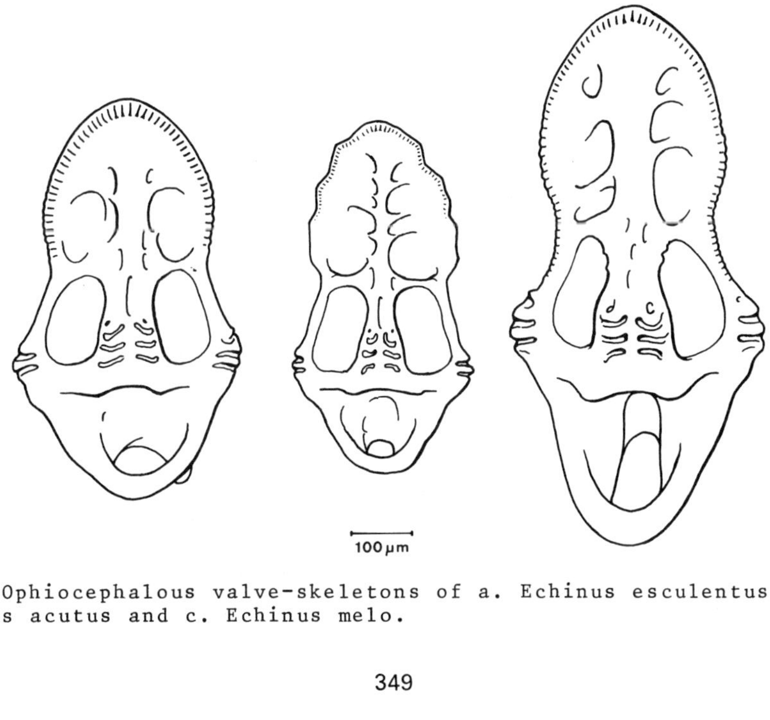

Fig. 3. Ophiocephalous valve-skeletons of a. Echinus esculentus, b. Echinus acutus and c. Echinus melo.

4 a
b
c
5

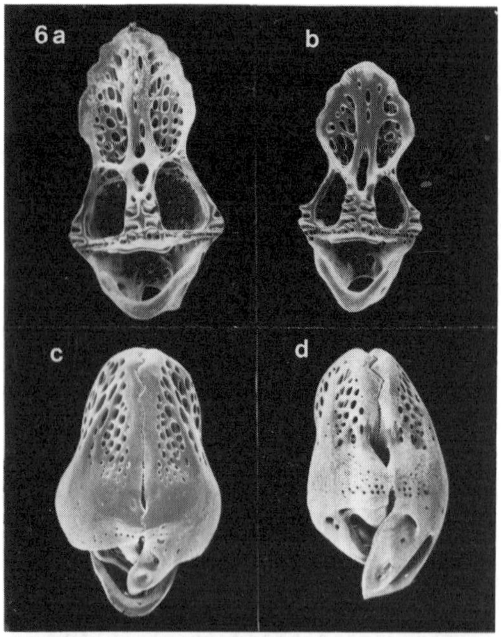
6 a
b
c
d

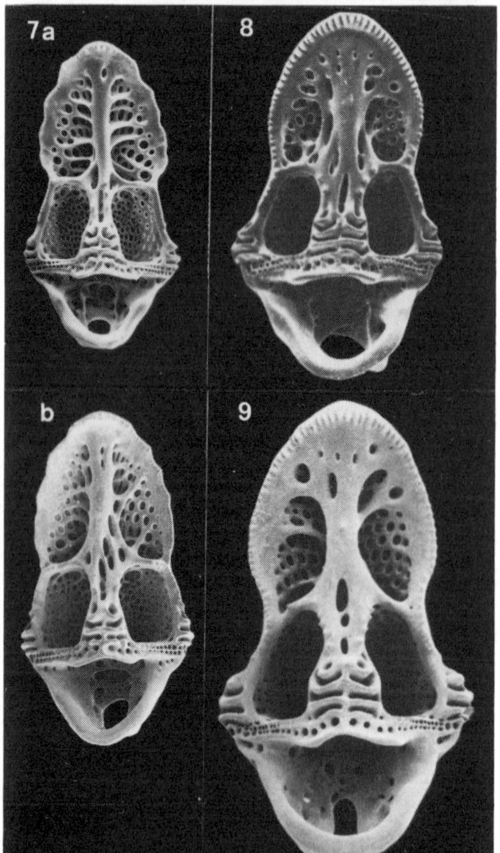
7 a
8
b
9

face of the closed pedicellaria is smooth. The supports of the central articulation are separated and the broad articulation itself continues in a wide median keel. By comparison, the margin of the distal valve segment in Echinus acutus is narrower and provided with prominent cogs which bear fine gripping teeth. The median keel is much narrower and continues in a weakly developed central articulation (fig. 3a, b; 7a, b; 8).

The ophiocephalous pedicellariae and Echinus melo indicates that their use in species diagnosis is only dependable when associated with other characteristics.

ACKNOWLEDGEMENTS

We thank Dr. E. Kritscher, Naturhistorisches Museum Wien, for allowing us to examine the collections under his care. The work was

---

Fig. 4. Cleaned coronae of Echinus acutus, showing variation in the shape of tests and in the arrangement of the primary tubercles on the interambulacral plates: a. normal shape and arrangement, b. almost regular absence of primary tubercles on every second interambulacral plate, and c. shape of the test very low and flattened.

Fig. 5. Cleaned corona of Echinus melo.

Fig. 6. a. Valve-skeleton and c. total head-skeleton of an ophiocephalous pedicellaria of Psammechinus miliaris. b. valve-skeleton and d. total head-skeleton of an ophiocephalous pedicellaria of Psammechinus microtuberculatus.

Fig. 7. c./b. two variations of valve-skeletons of the ophiocephalous pedicellariae of Echinus acutus.

Fig. 8. Valve-skeleton of an ophiocephalous pedicellaria of Echinus esculentus.

Fig. 9. Valve-skeleton of an ophiocephalous pedicellaria of Echinus melo.

supported by Fonds zur Forderung der wissenschaftlichen Förschung, 4354.
of Echinus melo are different to those of Echinus acutus in having i) a broad margin without cogs and strong gripping teeth, ii) a powerful central articulation narrowing the openings to the muscle-insertions, and iii) a strong median keel continuing the broad central articulation. The valves of the ophiocephalous pedicellariae of Echinus melo are clearly distinguishable from those of Echinus acutus. Interestingly, the valves of the ophiocephalous pedicellariae of Echinus esculentus and Echinus melo are quite similar while those of Echinus melo are more slender and more notched at the transition from the distal to the median segment (fig. 3c; 9).

Comparing the 3 species of the genus Echinus, there is a conspicuous difference in the size of the pedicellariae; while those of Echinus acutus and Echinus esculentus are similar, the valves of Echinus melo are about a half as large again (fig. 3, b, c). Variation in the size of the pedicellariae within species is much less than that exhibited by the coronae.

As a key to separating Echinus acutus and Echinus melo, the ophiocephalous pedicellaria of E. acutus has a coarse cogged margin and fine gripping teeth, a poorly developed central articulation which is triangular in shape and continues distally in a small median keel, and broad openings to the muscle-insertions.

In summary, the skeletal valves of the ophiocephalous pedicellariae turned out to be useful characters in distinguishing between Echinus acutus and E. melo and between Psammechinus miliaris and P. microtuberculatus. However, the comparison of Echinus esculentus

ABBREVIATIONS

bha: big handle; ca: central articulation; co: cog; ds: distal segment; gth: gripping teeth; hs: handle segment; k: keel; ma: margin; ms: median segment; mui: muscle-insertion; pa: peripheral articulation; su: support.

REFERENCES

Burke, R.D. 1980. Development of pedicellariae in the pluteus larva of Lytechinus pictus (Echinodermata: Echinoidea). Can.J.Zool. 58: 1674-1682.

Döderlein, L. 1887. Die japanischen Seeigeln. I. Familien Cidaridae und Saleniidae. p. 1-59. Stuttgart.

Döderlein, L. 1906. Die Echinoiden der deutschen Tiefsee-Expedition. V. p. 1-236. Jena.

Gordon, I. 1926. The development of the calcareous test of Echinus miliaris. Phil.Trans.Roy.Soc. Lond.Ser.B. 214: 259-312.

Hilgers, H. & Splechtna, H. 1976. Strukturund Funktionsanalyse ophiocephaler Pedizellarien von Sphaerechinus granularis Lam., Echinus acutus Lam. und Paracentrotus lividus Lam. (Echinodermata, Echinoidea). Zoomorph. 86: 61-80.

Hilgers, H. & Splechtna, H. 1979. Die Entwicklung ophiocephaler Pedizellarien von Sphaerechinus granularis (Echinodermata, Echinoidea). Zoomorph. 93: 265-287.

Jensen, M. 1974. The Strongylocentrotidae (Echinoidea), a morphologic and systematic study. Sarsia 57: 113-148.

Hensen, M. 1982. Pedicellariae in classification of echinoids. In J.M. Lawrence (ed.) Echinoderms: Proceedings of the International Conference, Tampa Bay. p.111-115. Rotterdam, Balkema.

Koehler, R. 1895. Notes echinologiques. Rev.Biol.Nord de la France 7: 20-25.

Koehler, R. 1921. Echinodermes. In Faune de France. p.116-118.

Mortensen, Th. 1903. Echinoidea (part I). In The Danish Ingolf-Expedition. Vol. IV/1. p.152-159. Copenhagen.

Mortensen, Th. 1943. A monograph of the Echinoidea. III.3. Camarodonta II. p.41-57. Copenhagen, Reitzel.

Splechtna, H. & Hilgers, H. 1979. Rasterelektronen- und lichtoptische Untersuchungen an den Pedizellarien von Arbacia lixula L. In M. Jangoux (ed.), Echinoderms: Present and Past, Proceedings of the European Colloquium on Echinoderms, Brussels, 3-8 September 1979. p.79-84. Rotterdam, Balkema.

Süssbach, S. & Breckner, A. 1911. Die Seeigel, Seesterne und Schlangensterne der Nord- und Ostsee. Wiss.Meeresunters. Kiel. N.F. 12. p. 180-184.

Tortonese, E. 1965. Echinodermata. In Fauna d'Italia. p. 328-333. Bologna, Calderini.

Proceedings of the Fifth International Echinoderm Conference / Galway / 24-29 September 1984

Structural analysis of the test of *Echinocyamus pusillus* (O.F.Müller)

MALCOLM TELFORD
Department of Zoology, University of Toronto, Ontario, Canada

ABSTRACT: The arrangement of skeletal elements shows that Echinocyamus pusillus (Clypeasteroida: Fibulariidae) is built to resist a combination of membrane stresses, bending and rotational forces. At the ambitus, horizontal spreading (thrust) is resisted in tension by the oral surface. The test plates increase in thickness from crown to ambitus and continue into the oral surface with almost constant thickness. Internally the interambulacral plates are extended into paired, curved buttresses which extend into the vault of the dome to the expected inflexion point of the bending moments. Thus they are perfectly placed to resist negative bending of the test and vertical displacement of the floor due to rotational forces. The denser stereom of the buttresses increases their stiffness so that they preferentially carry stress to the oral surface. The material between the supports (the ambulacra) contributes significantly to the stiffness and strength of the structure. The ambulacral plates are thickened towards the center to form horizontal components which resist hoop stresses and bending in the ambital region. Where the ambulacral plates join radially, the junction is reinforced by a slight thickening which provides a secondary buttress. A comparison of the morphology of this and other fibulariid species is made. For stress tests, Echinocyamus was loaded at the crown. In almost every case failure occurred around the apical system. The genital and terminal pores of the petaloids concentrate stress so that fracture occurs between them, and the center is punched in. The spines are able to absorb about 15 - 20% of the force before the load is applied directly to the test. These findings are interpreted in light of the known behavior and habitat of Echinocyamus pusillus.

1 INTRODUCTION

Most echinoid skeletons are built like domes of various forms (Telford, 1984). In such structures, radial stresses due to self-weight or applied loads are entirely compressive and, at the margin or ambitus, they are double the magnitude of those at the center or apex (Heyman, 1977; Lin and Stotesbury, 1981). Conversely, circumferentially directed stresses are compressive towards the apex and tensile towards the ambitus, where they are equal in magnitude to the radial stresses already described (Heyman, 1977). Elsewhere (Telford, 1984), I have shown that regular echinoids have greater amounts of collagen binding the sutures in tension, and less, or even none, in sutures transmitting compressive stresses. Radially arranged sutures, especially around the ambitus, may also be sufficiently thickened to constitute distinct ribs through which stresses are carried preferentially. Among the cassiduloids, holectypoids and some spatangoids, there appears to be less reliance on sutural binding with collagen. In these urchins, flexural stiffness and increased strength is attained by very tight-fitting sutures, often with fusion of some trabeculae. The clypeasteroids usually have considerable collagenous reinforcement between the plates (Harold, 1984).

The clypeasteroid family Fibulariidae is composed of several species of tiny echinoids belonging to the genera Echinocyamus and Fibularia. According to Mortensen (1948) the two genera are distinguished by the presence of radiating partitions in Echinocyamus and their absence in Fibularia. These partitions, located in the interambulacra, are not uniformly developed in all species, but are very robust in E. pusillus, the principal subject of this paper. The partitions form part of the system of "supports" that Hyman (1955) described as characteristic of the Clypeasteroida. Although this appears to be widely accepted, I have been unable to find any study that investigates the possible origin or distribution of stresses which would make supports a desirable architectural feature of Echinocyamus and unnecessary in Fibularia.

This paper presents a morphological comparison of the genera, Echinocyamus and Fibularia. Echinocyamus pusillus lives in the interstices of poorly sorted sediments in hydrodynamically active environments (Telford et al, 1983). In the following analysis of E. pusillus, the test has been treated as an internally buttressed dome, subject to external loading by the occasional movement of relatively large pebbles in wave-induced or tidal currents.

2 MATERIALS AND METHODS

2.1 Specimens: Echinocyamus pusillus was collected on the west coast of Scotland, in 10-20 m of water. Specimens were fixed in 10% buffered formalin. A total of 73 individuals was used, ranging from 4.4 to 12.8 mm in length. All other specimens were obtained from the collections of the U. S. National Museum, Smithsonian Institution, Washington. Note: Mortensen (1948) regarded F. craniolaris Koehler and F. ovulum Lamarck as the same. The data below (morphometry and SEM) justify separate treatment of the collections used here, without resolving this taxonomic synonymy.

2.2 Morphometry: Measuremnts were made to the nearest 0.05 mm. Length was the anterior-posterior diameter; width was greatest diameter orthogonal to length; height was the maximum measurement vertical to length and width. Some specimens were embedded in "Bio-plastic", sectioned and test thicknesses were measured by eyepiece micrometer.

2.3 Stress testing: 25 E.pusillus, 5.10 to 9.95 mm length, were crushed to first fracture using an Instron 1122 table testing machine. Specimens were placed oral surface down, and loaded at the apex, with the head advancing at 0.5 mm min^{-1}. Force exerted was recorded in newtons (N) on a standard chart recorder adavancing at 50 mm min^{-1}.

2.4 Statistical treatments: Length, width and height data were analysed by least squares regression. Homogeneity of slopes and significance of differences in intercepts was tested by analysis of covariance. Stress test data were analysed by multiple linear regression. All regressions used log-transformed data. Procedures followed Sokal and Rohlf, 1981.

2.5 Scanning electron microscopy: After removal of spines, specimens were washed and fractured by tapping sharply. The coelomic lining was stripped away to reveal the plate surfaces. Specimens of the unidentified Echinocyamus from Alaska were re-hydrated for 24 hours to soften the lining. Specimens collected dead in dredge hauls or as beach washed material, contained no soft tissue. Selected pieces of test, washed in distilled water and syringed until clean, were mounted on aluminum stubs, freeze dried and ringed with conductive paint to ensure adequate grounding before sputter coating with gold.

3 RESULTS

3.1 Morphometry: Length, width and height data were obtained from 208 Echinocyamus belonging to six species, and from 138 Fibularia belonging to four species. Results of regression analyses are presented in Table 1. Analysis of covariance revealed no significant difference in slopes for four species of Echinocyamus but intercepts were significantly different ($P<0.01$). Similarly, among the four species of Fibularia, differences in slope were not significant but differences in intercepts were ($P<0.01$). From the regression equations the predicted height and width was used to estimate the height/mean diameter and length/width ratios at the maximum observed size. In the genus Echinocyamus, E. pusillus and E. grandiporus were the most alike in these respects and also in development of the internal buttresses (see below). The two most similar Fibularia species were F. ovulum and F. craniolaris. Height and mean diameter data for these four species were compared. There was no significant difference in the means for height but analysis of covariance showed that slopes of the regression lines differed significantly ($P<0.01$). Comparison of the pairs of species revealed no differences in slope within genera, but a real difference between genera ($P<0.01$): these species of Echinocyamus are significantly flatter or lower domed than the Fibularia species.

Among the remaining species it is clear from simple inspection of the data (Table 1) that there is considerable overlap.

3.2 Internal buttresses: In the genus Echinocyamus the internal supports are most strongly developed in E. pusillus (Figure 1). The interambulacral plates are extended towards the center, forming vertical partitions which reach up into the vault of the dome, to the mid-petaloid level. Broad-based, but otherwise similar supports, are found in E. elongatus. In E. grandiporus, E. macrostomus and E. scaber, the buttresses end rather abruptly at the start of the petaloids, without tapering into the vault. The main part of the buttress is a thin sheet but the free edge (towards the coelom) is thickened, forming an almost vertical pillar. Similarly, in E. elegans the buttresses are very thin but lack the thickened free margin. The supports in E. megapetalus are short, reaching little above the ambitus, where they converge slightly. From this level upwards, the buttresses are replaced by thickenings adjacent to the interambulacral sutures, which form shallow ribs tapering into the upper reaches of the dome. The largest species, E. australis, has true supports only in the posterior interambulacrum, flanking the periproct. The interambulacral sutures in the vault of the dome are slightly thickened, forming small ribs between the petaloids and extending down towards the ambitus. The unidentified species from Alaska is a most unusual Echinocyamus with a marginal periproct, enormous gonopores, six hydropores and short petaloids. The interambulacral plates are slightly thickened, forming a more or less smooth pavement internally. The posterior interambulacral plates are extended into very small supports flanking the periproct. In E. pusillus, E. megapetalus and E. australis, the ambulacral plates are thickened in the center and form small circumferential hoops, between which the podia extend. There are no buttresses at all in any species of Fibularia examined in this study. Only in F. volva was there any apparent thickening of the sutures to form shallow ribs, and some thickening of the ambulacral plates forming circumferential hoops.

3.3 Scanning electron microscopy: The test of E. pusillus is extremely thick (Figure 1). The stereom is dense, with small stromal spaces (Figure 2) which almost disappear entirely in the buttresses. The sutures between plates are tight-fitting (Figure 3) with occasional fused trabeculae on the coelomic surface. The sutures are entirely devoid of collagen fibers (Figures 4, 5 and 6). Circumferential sutures include numerous interlocking knobbed and hooked trabeculae (Figures 4 and 5) which form close-fitting compression joints. The radial sutures consist of long, complex spikes of stereom which interdigitate (Figure 6). No collagen was found in the dried Alaskan specimens although other soft structures were complete. The sutural architecture of all Echinocyamus species examined was similar to that of E. pusillus. The test of Fibularia species is not generally as thick as that of Echinocyamus, except in F. cribellum (Figure 7). In this species the stereom is extremely dense and the trabeculae on the coelomic surface are greatly thickened. In F. ovulum (Figure 8) the stereom is almost solid calcite without pore spaces and the sutures are difficult to distinguish because the trabeculae are fused (Figure 9). In other species, such as F. craniolaris (Figure 10), the sutures are quite distinct and very tight-fitting. All of the Fibularia specimens were dead at the time of collection and it was impossible to determine whether or not sutural collagen had originally been present.

3.4 Stress testing: During crushing experiments, traces of applied force (Figure 11) show an initial slow increase as the spines absorb the force. This is followed by a sharp increase as the load is applied directly to the test. Inspection following first fracture showed that almost all specimens (23 of 25) broke around the apical system: the top was punched in. Ten specimens developed radial fractures along suture lines. These were the only fractures in the two in which the apex did not collapse. Three specimens developed circumferential fractures in addition to radial ones, and these were all in the upper half of the dome. Simple linear regression with log-transformed data (Figure 12) showed a significant relationship between breaking load and length ($P<0.001$) and weight ($P<0.001$). The relationship between breaking load and space between gonopores was not significant ($P<0.5$). Multiple linear regression, using breaking load as the dependent variable with the above three independent variables, yielded a highly significant relationship ($P<0.001$) to which the spacing between gonopores contributed a significant amount ($P<0.05$). The coefficient of multiple regression was 0.84.

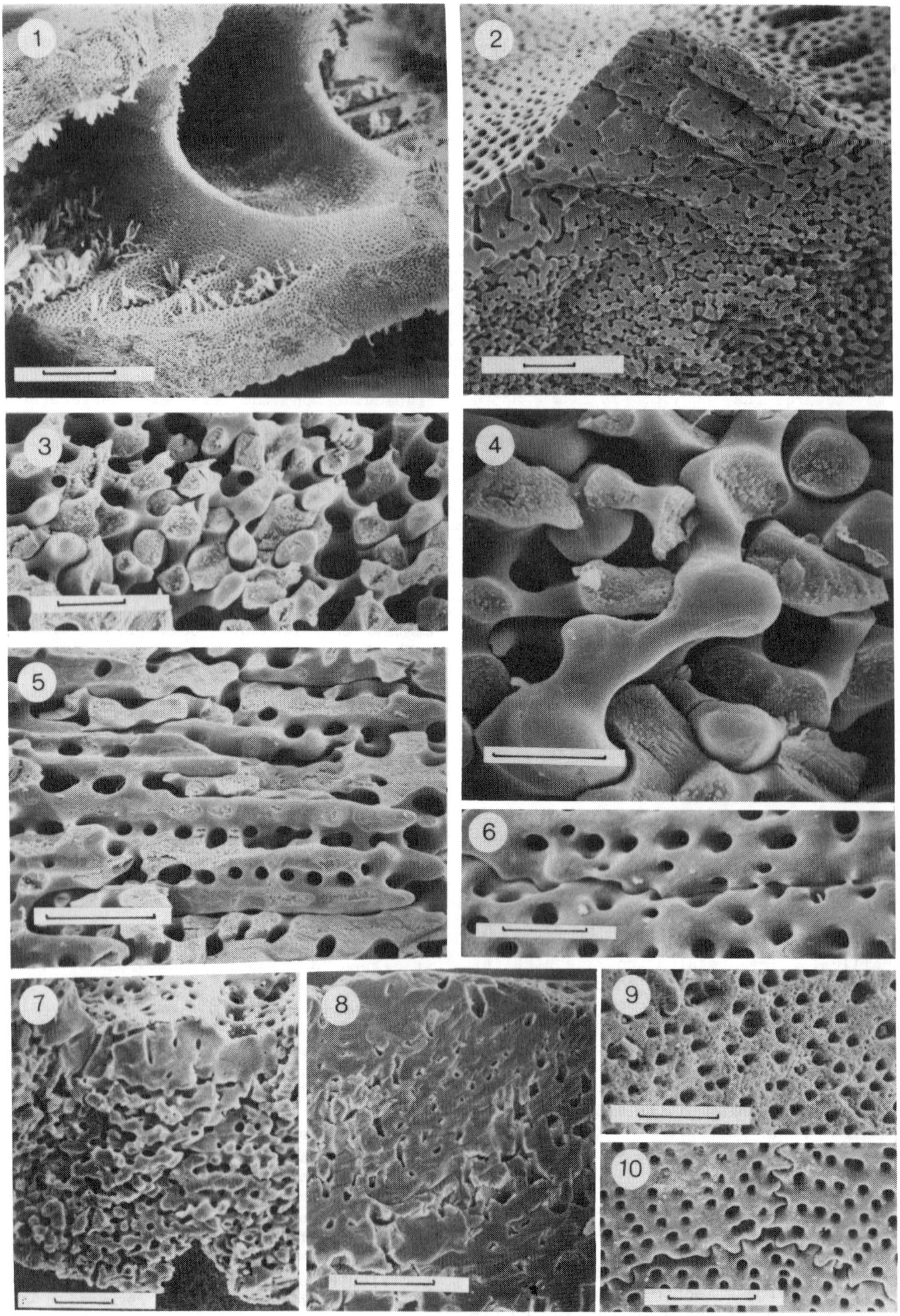
1
2
3
4
5
6
7
8
9
10

Table 1. Regression analyses of height and mean diameter for species of Echinocyamus and Fibularia: a = intercept value, b = slope, r^2 = coefficient of determination. The equation is:

$$height = a. \ mean \ diameter^b$$

Species	a	b	r^2	n
E. pusillus	0.62	0.82	0.92	73
E. grandiporus	0.60	0.83	0.97	30
E. macrostomus	0.70	0.75	0.74	30
E. megapetalus	0.56	0.90	0.95	30
E. australis	0.73	0.85	0.74	30
E. sp. indet.	0.30	1.24	0.74	15
F. ovulum	0.83	0.97	0.92	50
F. craniolaris	0.76	1.03	0.93	40
F. volva	0.51	0.98	0.98	24
F. cribellum	0.65	0.87	0.94	24

Table 2. Comparison of height/mean diameter, length/width and estimated percent total load carried in longitudinal axis of Echinocyamus and Fibularia species at maximum observed size.

Species	H/MD	L/W	%Load
E. pusillus	0.39	1.10	45
E. grandiporus	0.42	1.11	45
E. macrostomus	0.41	1.18	42
E. megapetalus	0.46	1.30	37
E. australis	0.69	1.28	38
E. sp. indet.	0.49	1.16	43
F. ovulum	0.78	1.17	42
F. craniolaris	0.80	1.19	41
F. volva	0.49	1.29	38
F. cribellum	0.52	1.32	36

Figures 1-10. Echinocyamus pusillus (1-6), Fibularia cribellum (7), F. ovulum (8,9) and F. craniolaris (10). 1. Internal buttresses and thickened ambital plates forming hoops (scale = 500 um). 2. Dense stereom of buttress (scale = 100 um). 3. Circumferential compression suture (scale = 25 um). 4. Knobs and hooks in compression suture (scale = 20 um). Scale = 50 um in remaining micrographs. 5. Interdigitating spikes of stereom in radial, tension, suture. 6. Coelomic surface of suture. 7. Dense stereom with some thickened trabeculae. 8. Extremely dense stereom of test, reduced pore space. 9. Sutures obscured by fused trabeculae. 10. More typical fibulariid sutures without trabecular fusion.

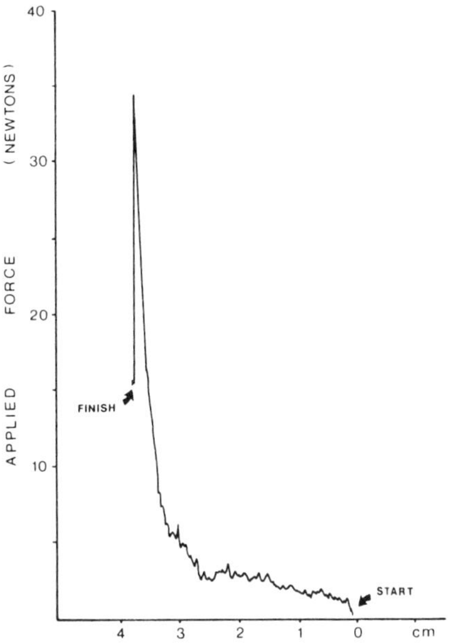

Figure 11. Stress test of Echinocyamus pusillus. The applied force is absorbed by spines at first and then rises sharply as the load is applied to the test. Force drops sharply when the specimen fractures.

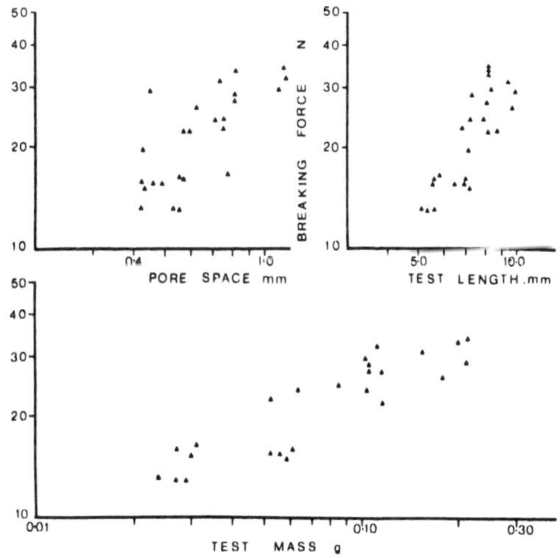

Figure 12. Individual data plots for three independent variables (mass, length and space between pores) from stress testing of Echinocyamus pusillus.

4 DISCUSSION

The fibulariids are all more or less dome-shaped. The domes tend to become flatter as size increases, especially in those species of Echinocyamus in which the internal buttresses are well developed. In engineered domes, the magnitude of circumferential tensile stresses is inversely proportional to the height/diameter ratio: stresses are greater in flat domes than in relatively high domes (Lin and Stotsbury, 1981). The decreasing height/diameter ratios of fibulariids indicates that they will experience circumferential stresses increasing with size. Arches loaded apically show positive, downward bending in the region of the crown, and they bend negatively (outward bowing) toward the edge (Thadani, 1964). The point of inflexion between positive and negative bending occurs about midway between crown and origin of a semicircular arch. Similarly, in domed structures loaded apically, circumferential forces are compressive around the apex, tensile around the margin (Heyman, 1977). Transition between these stresses occurs at about 40° latitude in hemispherical domes. In E. pusillus, the internal buttresses reach upwards to just above the anticipated inflexion point. Domes reinforced with radial ribs are not dependent on pure membrane action (Lin and Stotesbury, 1981). Ribs provide a local increase in flexural stiffness and preferentially carry stress. Directing radial stresses into the ribs effectively decreases the hoop stress around the edge. Thickening of the ambulacral plates in some Echinocyamus species (Figure 1) provides small hoops which resist bending of the shell between ribs. The buttresses are built of denser stereom than other parts of the test and this, too, increases their stiffness. Because of this, the buttresses are able to draw stress and transfer it more directly to the oral surface (and hence the substrate). The oral surface acts somewhat like a rotated tie rod in an arch (Thadani, 1964) or as an extended tension ring (Lin and Stotesbury, 1981). However, in echinoids the edge of the dome is not supported on vertical walls but rather, it merges smoothly into the oral surface following the curve around the ambitus. By analogy with an annular plate supported at the outer edges (Ugural, 1981), it may be inferred that applied loads on the dome surface would exert a rotational force about the ambitus which in turn raises and bends the floor. Another role of the internal buttresses is to resist this rotational force by bracing the oral and aboral surfaces.

The absence of sutural collagen, which is well supported by histological sections (Mooi, unpublished data), might account for the details of sutural architecture. The sutures of E. pusillus are very tight, with long interdigitating processes in tension joints. All of the fibulariids examined have very dense stereom (especially Fibularia) and relatively thick tests (especially Echinocyamus). Both increase flexural stiffness and load carrying capacity (Gordon, 1978) of domed structures. In Fibularia the stereom is especially dense with massively thickened trabeculae on the inner surface. The proportion of the total load which is carried in the longitudinal and transverse axes of an elongate dome is inversely proportional to the ratios of the square of the axes (Lin and Stotesbury, 1981; see also Telford, 1984 for application to echinoids). The length/width ratios for these fibulariids, and percentage of total load carried longitudinally is given in Table 2. Species of Fibularia and those of Echinocyamus without internal supports, show greater test thickness in the transverse than the longitudinal axis. Lateral thickening is less clearly apparent in E. pusillus, probably because length and width become more similar as size increases, and loads are distributed more equally.

The loads required to fracture E. pusillus were regressed with length as a simple measure of size, with weight as a measure of the amount of test material and with space between gonopores because fracture occurred almost exclusively around the apical system. Collectively, these three features, taken as independent variables in multiple regression, explain about 84% of the variance in breaking loads. Other contributing factors would include previous history of fracture, possible damage during collection and the relative size of the gonopores and podial pores. Pores act as centers of stress concentration (Gordon, 1978) and it is probably for this reaon that fractures occur between them so frequently. In both E. megapetalus and E. australis the area between the petaloids is stiffened by slight thickening of the interambulacral sutures, which would decrease radial stresses through the pore rows.

The natural habitat of E. pusillus is poorly sorted sediments in hydrodynamically active environments (Telford et al, 1983; Nichols, 1959). External forces sufficient to fracture the test come from repeated movement of relatively large stones which roll together in wave-induced or tidal currents. The highest load needed to break a specimen was 34 N, the smallest was 13 N. A simple calculation suggests that spherical cobbles of about 15 cm and 10 cm respectively, resting on E. pusillus would be sufficient to fracture the specimens. Strathmann (1981) has noted that echinoid spines may protect the test from structural damage during impact loading. In E. pusillus the spines were able to absorb some 15-20% of the force before it was applied directly to the test. Live E. pusillus usually use their podia to hold pebbles and fragments of shell pressed against the spine tips (Telford et al, 1983; Nichols, 1959). This helps to disperse applied forces over a greater number of spines and a larger area of the test and may thus be interpreted as protective behavior. The distribution of fibulariids is not clear from the literature, but it seems likely that other species occur in much less active environments and in more uniformly sorted sediments.

ACKNOWLEDGMENTS

This work has been supported by the Narural Sciences and Engineering Research Council of Canada through Operating Grant #A4696. I thank E. Lin, Department of Zoology, for technical assistance in SEM. I am greatly indebted to J. Currey, Department of Biology, University of York, for use of stress testing equipment and to A. R. Collins, Department of Engineering, University of Toronto, for critical suggestions as the work progressed. My colleagues, Rich Mooi and Tony Harold provided much stimulating discussion and many useful suggestions.

REFERENCES

Gordon, J. E. 1978. Structures: or why things don't fall down. Penguin Books Ltd., Harmondsworth, England. 395 pp.

Harold, A. S. 1984. Body wall structure in Echinarachnius parma (Lamarck) (Echinoidea: Clypeasteroida). M.Sc. Thesis, Department of Zoology, University of Toronto.

Heyman, J. 1977. Equilibrium of shell structures. Oxford University Press, England. 137 pp.

Hyman, L. H. 1955. The Invertebrates. IV: Echinodermata. McGraw-Hill, New York. 763 pp.

Lin, T. Y., and S. D. Stotesbury. 1981. Structural concepts and systems for architects and engineers. John Wiley and Sons, New York. 590 pp.

Mortensen, Th. 1948. A monograph of the Echinoidea IV:2. Clypeastroida. C. A. Reitzel, Copenhagen. 471 pp.

Nichols, D. 1959. The histology and activities of the tube feet of Echinocyamus pusillus. Q. J. Microsc. Sci. 100: 539-555.

Sokal, R. R., and Rohlf, F. J. 1981. Biometry. W. H. Freeman and Co., San Francisco. 859 pp.

Strathmann, R. R. 1981. The role of spines in preventing structural damage to echinoid tests. Paleobiology, 7: 400-406.

Telford, M. 1984. Domes, arches and urchins: the skeletal architecture of echinoids. Zoomorphology. (In press).

Telford, M., R. Mooi and A. S. Harold. 1983. Feeding structures, behavior, and microhabitat of Echinocyamus pusillus (Echinoidea: Clypeasteroida). Biol. Bull. 165: 745-757.

Thadani, B. N. 1964. Modern methods in structural mechanics. Asia Publishing House, Bombay. 619 pp.

Ugural, A. C. 1981. Stresses in plates and shells. McGraw Hill Book Co., N. Y. 317 pp.

Proceedings of the Fifth International Echinoderm Conference / Galway / 24-29 September 1984

Sea urchins in seagrass communities: Resource management as functional perspective of adaptive strategies

W.P.TERTSCHNIG
Zoologisches Institut der Universität Wien, Abt. Meeresbiologie, Vienna, Austria

ABSTRACT: Feeding behavior and growth of sea urchins was studied with a population of juvenile Tripneustes ventricosus from a seagrass (Thalassia testudinum) stand in Bermuda. Feeding and foraging patterns, preferences and daily consumption rates were determined in field experiments and related to vegetational parameters of the turtle grass community. Although the results confirm that the direct impact of grazing on productive tissue is almost negligible they suggest a revision of the role that sea urchins and other herbivores play in seagrass communities. Their striking inability to efficiently exploit the rich nutritional potential and productivity of seagrasses is interpreted as an evolutionary strategy of maintaining imperfectness on the species level for the benefit of the community. The distinct mode of foraging and selective grazing observed in this study serves as resource management, ensuring persistence and predictibility for both the consumer and the producer. The observed pattern, however, also allows an optimization of feeding and net energy gain through its time-minimizing and energy-maximizing character. Sea urchins grazing at moderate density in seagrass beds function as a community stabilizing factor rather than bearing the potential hazard of exploitation.

1 INTRODUCTION

The importance of perennial algae and seagrass production in shallow coastal waters is widely recognized, and food chains originating from benthic primary producers are receiving increasing interest from ecologists. Sea urchins are among the most important direct consumers of marine macrophytes. Their interactions with algae have been reviewed by Lawrence (1975). Comparatively little is known about sea urchins inhabiting seagrass beds.

Seagrass ecosystem research faces an apparent dilemma: that of explaining why a system of extremely high productivity seems to be relatively unimportant as a source of food (Den Hartog 1980). Sea urchin species living in seagrass beds appear to have lower feeding rates, conversion efficiencies and growth rates than species from algal stands. One might conclude that urchins do suffer from a reduction of fitness (sensu Slobodkin 1964, Schoener 1971, Vadas 1977) when feeding on seagrasses.

Maximum individual fitness is achieved during the life history of a consumer by efficient exploitation of the most valuable prey or prey combination, resulting in increased growth, reproductive capacity and survival. In the evolutionary context, however, the fitness of a population of consumers will greatly depend on the fitness of the producer. Efficiency optimization may carry the risk of reducing the stabilty and persistence of the prey community and consequently involving the negative feedback of a lowered resource predictibility. Assuming that urchins have been associated with seagrasses for a sufficiently long period of time, the evolution of adaptive mechanisms that support their continuous coexistence must be considered a major element of the urchins' strategy of feeding.

Based on 1) the foraging patterns, 2) feeding mode, and 3) growth success in an urchin population as well as on 4) vegetational parameters, 5) nutritional and 6) chemical properties of the seagrass community this study provides evidence that the established opinion of the insignificant role of urchin consumption in seagrasses must be critically reviewed.

When the point of view is widened from the species to the systems level, reduced fitness of individual taxa appears to be balanced by an optimization of community interests.

2 MATERIAL & METHODS

The growth, feeding behavior and foraging pattern of a population of juvenile Tripneustes ventricosus (Echinoidea: Fam. Toxopneustidae) from a seagrass (Thalassia testudinum) stand in Bermuda were investigated during fall, 1982. A small grass bed at the west end of St. George's, NE of Ferry Point was chosen as an experimental area. This location provided a high density of juvenile Tripneustes ($\bar{x}$ = 1,24 / m2) ranging in size from 18 to 55mm, and a well developed Thalassia community with a mean shoot density of 1600/m2. Seagrass biomass amounted to over 90 % of the total community standing crop, the only other important component being the green alga Halimeda incrassata. The seagrass leaves represented the typical fall aspect, with decay exceeding production and maximum epiphytic cover.

Urchin sizes (horizontal diameter) were measured in situ with calipers to the nearest mm.

Field growth rates were determined in 7 mesh wire cages. Each cage covered 725 cm2 of the seagrass bed and contained six Tripneustes ranging in size from 35 to 50 mm. 3 to 5 mm size intervals between individual urchins and a varying coloration allowed individual recognition. Size-specific (5 mm size class intervals) and cage-specific growth rates were determined as increase in mm HD over a period of four weeks.

For comparative purposes, growth rate data were also obtained from the size frequency distribution analysis of field animals. Their sizes were recorded on two occasions with a four weeks interval in 1/4 m2 units along a line transect covering a total of 80 m2; this represents 1/3 of the entire Thalassia stand.

Preliminary observations indicated a distinct mode of feeding in which each sea urchin grazed upon a characteristic area of uniform size ("feeding plot") each day. The selection of cage size was based on the estimate of mean feeding plot size: the 6 enclosed urchins were provided both the approximate area and the leaf biomass equivalent of 2 days of feeding.

The seven cages were placed at visibly different positions in regard to standing crop and degree of epiphytism within the experimental seagrass stand. After confirmation of the estimates on feeding mode and intensity in a pilot study the experment was run for four weeks. The cages were moved to the nearest adjoining area of "ungrazed" Thalassia after the enclosed leaf standing crop had been cleared to sediment level by the six urchins. The mean displacement interval was 48 hours.

Five samples of the above-ground plant standing crop were taken in the vicinity of each of the cages. The sample unit equalled the cage area.

The samples (total N = 35) were rinsed to remove sediment and oven-dried at 90° C to constant weight. Shoot density, leaf standing crop, photosynthetically active leaf biomass (green leaf sections), senescent leaf biomass (epiphyte-covered brownish leaf sections) and Halimeda standing crop were recorded.

Mean Thalassia standing crop values provided an estimate of the consumption rate of each cage population per 48 h interval.

The urchins' activity and foraging patterns were determined by measuring their daily displacement rates within four 1m2 frames positioned in the experimental area. The frames were subdivided in 1/4 m2 units, and a Tripneustes (total N = 16) was initially placed at the center of each subunit. The animals differed in size and coloration to allow individual recognition without applying a direct marking technique.

The positional coordinates of each urchin within the 4 frames were recorded in 24 h intervals, and the magnitude (nearest 5 cm) and direction of displacements determined trigonometrically. Distances from previous positions were considered to be a minimal estimate of the daily movement rate.

After 4 consecutive days, the urchins were replaced with a second set of 16 Tripneustes and allowed to forage for 1 day (day 5). This procedure was repeated once to serve as a control (day 6).

The final part of the experiment again involved freshly collected animals, however, with the frame positions being moved to ungrazed areas (day 7). This experimental design allowed an evaluation of the foraging response towards increasing grazing pressure within a limited area.

Direct observation provided information on the timing and behavioral process of feeding as well as evidence on whether the urchins selected among food items according to their relative availability, or considered a particular prey item only.

3 RESULTS

Growth of juvenile Tripneustes (increase in horizontal diameter over a period of four weeks) was determined directly from individually identifiable cage urchins and from 2 consecutive size frequency distributions of the field population.

Both the field and cage population samples fit the theoretical normal distribution

(χ^2 test for goodness of fit at the 0.05 level).This allowed a comparison of means among the 2 frequency distributions from the field samples and between the field and experimental cage population.

Sample data were
$\bar{x}$ Oct. = 41 mm (confidence interval 1,1), SD 6,4 , N = 136 and
$\bar{x}$ Nov. = 50,1 mm (C.I. 1,7), SD 7,4 , N = 72 for the field population, and
$\bar{x}$ Oct. = 43,3 mm (C.I. 1,6), SD 5,0 and
$\bar{x}$ Nov. = 49,7 mm (C.I. 1,8), SD 5,6 for the cage animals (N = 42).

A mean growth rate of 6,4 mm was determined for the field cage urchins. Considerable variation was observed between the 7 cage populations (range 4,04 mm - 8,41 mm, SD 1,56 , Coeff.var. 24 %), whereas variation between the 4 size classes considered was insignificant (range 6,1 mm - 7,2 mm, SD 0,44 , Coeff.var. 6,7 %, no difference between size class sample means according to t-test at the 0.05 level).

Within each cage, growth varied greatly between the individual urchins. A mean SD of 1,45 and Coeff.var. of 23,5 % indicate that individual fitness and the effects of competition for limited food and space led to differential feeding and growth success.

The natural field conditions supported a significantly higher (t-test at the 0.05 level) growth rate. The mean diameter increase within the field population was 9,1 mm. This amounts to a 40 % surplus for the 4 week period as compared to the cage urchins.

Plant parameters (g d.w./m2) recorded from the sample series at the cage locations I - VII (N = 35) are presented in Table 1.

The corresponding mean feeding rates of the urchins were calculated from cage-specific seagrass standing crop x 6 Tripneustes $^{-1}$ x 2 days $^{-1}$. Due to the relatively uniform distribution of the seagrass st.crop values in the experimental area, consumption rates were very similar between cages: $\bar{x}$ = 1,096 g d.w. per urchin per day, SD 0,10 , Coeff.var. 9,15%.

The high degree of variation observed between cage-specific growth rates indicates that ingestion rates alone are inadequate parameters for sufficiently explaining the variable growth success in the 7 cages. With approx. equal food quantities available per urchin and cage and an assumed equal food processing capability among the experimental animals it appears that qualitative differences in the enclosed seagrass community determined the particular growth-supporting value at the 7 cage locations.

Field observations indicated a strong preference of Tripneustes for the distal, heavily epiphytized Thalassia leaf sections. The basal green parts of the leaves were rarely eaten. Halimeda, the major macroalgal species in the experimental area was completely avoided, even when no other food was available (as observed in the experimental cages prior to displacements).

In an attempt to classify the potential nutritional and growth-determining value within the enclosures, four vegetational parameters were distinguished:
Growth-supporting value was assigned to Thalassia standing crop (A) and the share of epiphytized leaf sections (B). These were considered to be the major attractive and growth-stimulating parameters for the urchins.
Growth-diminishing value was assigned to green seagrass tissue (relative to total biomass, C) and Halimeda biomass (D) which were both considered to be feeding-inhibiting community components.

The four parameters were ranked in accordance to their relative availability (in %) between the 7 sampling sites.

Table 1. Plant parameters (g d.w./m2), feeding rates (g d.w. Thalassia/urchin/day), ranking of food and growth success at the seven cage locations.

Location	Seagrass st. crop	Brown leaf section	Green leaf section	Halimeda st. crop	Feeding rate	Ranking food		Ranking growth
Cage I	180, 0	130, 4	49, 6	61, 2	1, 09	7	:	7
Cage II	209, 6	175, 0	34, 6	39, 6	1, 24	5	:	5
Cage III	188, 8	151, 0	37, 8	3, 7	1, 15	2	:	2
Cage IV	175, 2	146, 5	28, 7	21, 7	1, 05	4	:	4
Cage V	177, 5	146, 7	30, 8	0	1, 07	1	:	1
Cage VI	155, 7	118, 2	37, 5	8, 3	0, 92	6	:	3
Cage VII	178, 2	147, 8	30, 4	19, 8	1, 15	3	:	6
Mean	180, 7	145, 1	35, 6	22, 0	1, 09	Spearman´s rank		
SD	16, 2	17, 7	7, 1	21, 8	0, 10	r_s = 0, 68		

Table 2. Daily movement rates of Tripneustes (expressed as % of experimental population) and decreasing availability of ungrazed area (assuming 1 - 2 feeding plots/day/animal).

Experiment duration	Distances covered daily $x \leq 30$ cm	$30 < x \leq 60$	$60 < x \leq 90$	$x > 90$ cm	$\bar{x}$ /urchin	Ungrazed area during each day
Day 1	75 %	6 %	13 %	6 %	31,3 cm	100 % - 93 %
Day 2	69 %	25 %	6 %	0 %	24,4 cm	93 % - 79 %
Day 3	44 %	44 %	12 %	0 %	>40 cm	86 % - 65 %
Day 4	12 %	38 %	50 %	0 %	>50 cm	79 % - 51 %
Day 5	0 %	6 %	50 %	44 %	>85 cm	72 % - 37 %
Day 6	6 %	6 %	13 %	75 %	>95 cm	65 % - 23 %
Day 7	19 %	44 %	6 %	31 %	>65 cm	100 % - 93 %

The total sum of % A plus % B minus % C minus % D provided a measure of potential attractivity for each cage, the rank of which was compared to the corresponding rank of growth by applying a Spearman's rank correlation (see Table 1).

The correlation between the seven rankings was significantly positive at the 0.05 level, indicating that the varying relative availability of these community parameters did cause the differential growth between the experimental cages. With exception of D (significantly negative), the correlation between A, B or C in single presentation and the ranking of growth was not significant.

A typical pattern of food uptake and activity was maintained in this population of juvenile Tripneustes.
The urchins fed in circular "feeding plots" until they had reduced the original leaf length by 25 - 50 %, which approximately equalled the epiphytized leaf section. The animals then ceased grazing and moved to the next feeding plot, usually in the close vicinity. A relatively constant plot size of 10 - 15 cm diameter was maintained. This represents approx. twice the mean diameter of the urchins and appears to allow maximal ingestion with minimum activity expendure in holding a constant position during feeding periods.

In order to avoid overlapping feeding plots, each urchin had to move a minimum distance between feeding periods; 30 cm (twice the maximum plot diameter) was regarded to be a sufficient distance.

Based on standing crop data, the seagrass biomass removed from a feeding plot could be estimated. It ranged from 0,36 g (10 cm Ø, 25 % consumed) to 1,5 g (15 cm Ø, 50 % consumed). The mean value of 0,83 g d.w. was considered to be a good approximation of actual ingestion during a feeding period.

The movement rates of Tripneustes (N = 16) in the experimental m2 frames are presented in Table 2.
Distances covered daily are pooled in 30 cm - intervals and expressed as % of total experimental population.
The table includes the mean daily displacement rates and the decrease of ungrazed Thalassia available in the experimental frames (in % of total frame area) caused by each frame population assuming 1 - 2 feeding plots per day and animal.

Throughout the experiment no particular frame population or urchin size class showed daily movement rates significantly different from the others.
With increasing duration of the experiment, the increased grazing intensity greatly reduced the availability of the prefered food component in the frame areas. After the second day the urchins showed a steady increase of their daily movement rates. Some of them even increased their rates to distances beyond which they could be recovered. For calculation of mean movement rates minimum estimates for the daily movement of these animals could be considered only, which is indicated by the > - sign prior to $\bar{x}$ values in Table 2.

With up to 77 % of the frame area already covered by feeding plots from the previous days, on the 6 th day the urchins moved for mean distances of >95 cm while foraging for epiphytized leaf tips.
With the final set of urchins the frames were moved to an ungrazed position, the reduced foraging effort resulting in significantly lower rates of >64 cm/day.

64 cm appear to be relatively high when compared with the first two days. I attribute this to the agitation during the preceeding experimental manipulations in the area, with the final set of urchins being most greatly affected.

Directions of daily movements showed an overall random distribution both within and between the individual urchins.

4 DISCUSSION

Much of the current interest on the feeding ecology of sea urchins is related to the impact of food selection and feeding patterns on benthic community structure. Although a general tendency for the diet of sea urchins to reflect the composition of the available flora is apparent, numerous authors have reported preferences for both particular plant species and specific parts of a given plant. There is, however, no clear understanding of the basis for food preferences among echinoids (Lawrence 1975).

The selective feeding pattern of Tripneustes reported in this study has been repeatedly observed in sea urchins inhabiting seagrass communities. Both food choice experiments (Ott & Maurer 1977, Traer 1980) and field observations (Kempf 1962, Rivera 1978, Vadas et al. 1982) indicate a strong preference for senescent and epiphytised leaf tips over young, green seagrass tissue.

The preferred food component is low in energy content per unit of ingestion. The bulk volume is made up of degraded Thalassia (consisting mainly of structural carbohydrates indigestible for sea urchins: Lawrence 1976), calcareous algal crusts and sediment covering the leaf tips. The epiphytic algae, on the other hand, show high levels of soluble carbohydrates, proteins and lipids which are efficiently absorbed by sea urchins. Thus, the epiphytic community alone provides the utilizable organic fraction for Tripneustes.

Green and productive seagrass tissue has a much higher energy content per unit of ingestion (Harrison & Mann1975, Pirc 1984). However, the cell wall structure of seagrasses (cellulose amounting to over 50 % of the organic fraction in Thalassia : Lawrence 1976) makes a major portion inaccessible to net energy gain for sea urchins.

In addition, the green parts of seagrass leaves contain high levels of a tannin compound (chicoric acid : Cariello & Zanetti 1979) which is almost absent in distal leaf sections.

Tannins are known to affect herbivore feeding by altering both taste and digestibility, and may even block the activity of digestive enzymes by complexing with proteins and carbohydrates (Sieburth 1968, Mattson 1980).

Reported daily feeding rates of sea urchins on Thalassia are up to an order of magnitude lower when provided with only green leaf sections (Lowe 1974, Lilly 1975, Greenway 1977) than with the partially epiphytised leaves supplied in the present cage experiments.

Apparently the preferential consumption of leaf tips is due to the avoidance of plant sections maintaining both structural and chemical defense mechanisms. Ingested seagrass tissue is a by-product of little importance, serving merely as a carrier substrate for epiphytic algae.

In order to satisfy their energy demands, young, rapidly growing Tripneustes must therefore ingest a greater amount of their preferred food to compensate for low energy content. Ingestion rates observed in this study (1,09 g d.w./urchin/day) range among the highest recorded for echinoids.

To obtain necessary amounts of the preferred food item the urchins efficiently consume all epiphytised Thalassia leaf tips from the feeding plots. Once a feeding plot is devoid of its distal leaf sections, the urchins change positions and move the minimum distance required to ensure a fresh patch of preferred food. Energy spent in foraging is minimized.

At the same time, energy gain (in food uptake per feeding plot) is maximized. Using the diameter - gut volume ratio provided by Vadas (1982) for similar-sized Lytechinus variegatus, the caged urchins consumed the equivalent of approx. two gut fillings per day. The estimate presented here for the feeding plot - specific consumption of the field animals (0,83 g) corresponds well to the amount of seagrass tissue required for one gut filling and suggests two feeding plots per day and animal.

When the density of Tripneustes within a limited area increases, the probability that feeding plots will overlap is higher, and the distances to be covered between subsequent feeding plots should increase also.

As ungrazed area became less available in the experimental m2 frames, the urchins responded with significantly higher daily movement rates. Thus they increased their foraging effort to ensure consumption of epiphytised leaf tissue, searching for ungrazed seagrass patches rather than increasing their feeding plot - specific consumption by accepting green leaf sections.

Their increased activity and the observed random movement pattern ensures the dispersion of aggregated animals and counteracts overgrazing within limited areas of higher urchin densities.

The effect of food quality rather than quantity on the growth success of Tripneustes is shown by the different cage - specific growth rates.

The adopted feeding strategy maximizes net energy gain, examplified by the higher growth rates of the uncaged field animals. They had unlimited access to preferred food whereas the food of the caged animals consisted of at least 25 % green Thalassia.

Greenway(1977) found that the grazing activity of Lytechinus in areas of previously cropped Thalassia was negligible during the first weeks of blade regrowth and attributed this to insufficient cover (sensu "shelter") for the urchins. In view of my data it could be explained by the lack of attractive leaf tips covered with epiphytes.

However, one must stress that this preferential feeding did probably not evolve from the usual choice between different food items where the "superior" one provided a selective advantage of greater individual fitness. It appears that sea urchins inhabiting seagrass beds only use the epiphytised section because they are not capable of utilizing the nutritional potential of the leaf base.

The question arises as to why natural selection during the history of sea urchin - seagrass interactions did not favor the evolution of better abilities in sea urchins to digest seagrass tissue itself and thus profit from its extremely high productivity.

Why do the urchins adopt the strategy of consuming only the leaf tips, which are depauperated through autolysis, leaching, nutrient relocation and, due to their low organic content, demand very high feeding rates to utilize the energy stored in the epiphytic community ?
The actual selective advantage of this particular feeding strategy becomes evident when the scope of view is extended from the consumer to the community level.

Marine seagrass systems differ from terrestial grasslands in that they lack the feedback loop of remineralisation from herbivores (Ott & Maurer 1977). When photosynthetic tissue is consumed, the energy and nutrients stored in it are transported to neighboring communities and are largely lost to the system. Seagrasses lack all those adaptations known in terrestial grasses that ensure high productivity when grazed but suffer greatly under natural or artificial overgrazing (Greenway 1974, Ott & Maurer 1977, Wittmann & Ott 1982).
From the viewpoint of the marine primary producer it is important to keep the grazing pressure on its photosynthetic tissue as low as possible.
However, the productivity of seagrasses may also be significantly influenced through their epiphytic coverage.

Epiphyte - seagrass relationships and the deleterious effects of epiphytes for the host plant have been reviewed by Orth & Montfrans (1984).
They make up an important contribution to seagrass community production (22 % in Thalassia according to Jones 1969) but excessive epiphytism may have serious consequences for seagrass growth. The impact of overshading by epiphytes reduces the "photosynthetic life" of a seagrass blade by as much as 50 % (Bulthuis & Woekerling 1983) and can be severe enough to cause the complete disappearance of seagrasses (Zostera marina: Sand-Jensen 1977).

In this context, selective grazing on the epiphytised leaf tips is an important beneficial consequence for the host plant as it diverts consumption from photosynthetic tissue to senescent leaf parts. It stabilizes, or even enhances, rather than reduces productivity.

The urchins, on the other hand, maintain this particular feeding pattern only because they are imperfect consumers (sensu Ott 1981). They graze in a "sensible" way in terms of community interests because they were unable to further optimize their food processing capability.
They forage for a food component that allows low absorption efficiency only and ingest a high amount of indigestible plant tissue and inorganic matter together with the epiphytes.

If, however, in the course of evolution the "perfect" sea urchin would have appeared in seagrass beds, the momentary advantage of increasing individual fitness through an efficient utilization of the producer would have led to an overexploitation of seagrass communities.
In the long term, this would have endangered the system's stability and persistence together with the advantages this system provides for the sea urchins : recruitment success, shelter for juveniles, predictible food supply and so on.

The evolutionary concept to maintain an "imperfect" consumer appears to be a characterizing feature of this herbivore - primary producer relationship.
Although caution must be applied when determining the character and evolutionary history of mutual traits in co-occuring species or populations, the sea urchin - seagrass interaction emphasizes the functional perspective of adaptive strategies at the ecosystem level. They ensure that, as a consumer, the sea urchin manages rather than exploits its resource, i.e.
the biter avoids getting bitten.

5 AKNOWLEDGEMENTS

This study was partly funded by a SYDNEY L.WRIGHT and EXXON Corp. fellowship from the Bermuda Biological Station.

I wish to thank the Director and staff of BBS for their notable support and help and Dr. J.Ott for valuable discussions and a critical review of an earlier draft of this paper.

This is Contribution No. 1009 from the Bermuda Biological Station for Research.

6 REFERENCES

Bulthuis, D.A. & W.J.Woekerling 1983. Biomass accumulation and shading effects of epiphytes on the leaves of the seagrass, Heterozostera tasmanica in Victoria, Australia. Aquat.Bot.16: 137-148.

Cariello, L.& L.Zanetti 1979. Distribution of chicoric acid during leaf development of Posidonia oceanica.Bot.Mar.22: 359-360.

Greenway, M. 1974. The effects of cropping on the growth of Thalassia testudinum (König) in Jamaica. Aquacult. 4: 199-206.

Greenway, M. 1977. The production and utilization of Thalassia testudinum König in Kingston Harbor, Jamaica. PhD Dissert., Univ. of the West Indies, Kingston.

Harrison, P.G. & K.H.Mann 1975. Chemical changes during the saisonal cycle of growth and decay in eelgrass (Zostera marina) on the atlantic coast of Canada. J.Fish.Res.Bd.Can. 32: 615-621.

Hartog,C.den 1980. Foreword. In R.C.Phillips & C.P.Mc Roy (eds.), Handbook of seagrass biology: An ecosystem perspective. New York & London: Garland STPM Press.

Jones,A.J. 1969. Primary productivity by the tropical marine grass Thalassia testudinum König and its epiphytes. PhD Dissert., Univ. of Miami.

Kempf, M. 1962. Recherches d' ècologie comparè sur Paracentrotus lividus (Lmk.) et Arbacia lixula L. Rec.Trav.Sta.Mar. Endoume 25: 47-116.

Lawrence, J.M. 1975. On the relationships between marine plants and sea urchins. Oceanogr.mar.Biol.ann.Rev. 13: 213-286.

Lawrence, J.M. 1976. Absorption efficiencies of four species of tropical echinoids fed Thalassia testudinum. Thalassia Yugoslav. 12: 201-205.

Lilly, G.R. 1977. The influence of diet on the growth and bioenergetics of the tropical sea urchin,Tripneustes ventricosus (Lamarck).PhD Dissert., Univ. of British Columbia, Vancouver.

Lowe, E. 1974. Absorption efficiencies , feeding rates and food preferences of Lytechinus variegatus (Echinodermata: Echinoidea) for selected marine plants. MSc Thesis, Univ.of South Florida,Tampa.

Mattson, W.J. 1980. Herbivory in relation to plant nitrogen content. Ann.Rev.Ecol. Syst. 11: 119-161.

Orth, R.J. & J.van Montfrans 1984. Epiphyte - seagrass relationships with an emphasis on the role of micrograzing: A review. Aquat.Bot. 18: 43-69.

Ott, J. 1981. Adaptive strategies at the ecosystem level: Examples from two benthic marine systems. P.S.Z.N.I: Marine Ecology 2(2): 113-158.

Ott, J. & L.Maurer 1977. Strategies of energy transfer from marine macrophytes to consumer levels: The Posidonia oceanica example. In B.F.Keegan, P.O. Ceidigh & J.S.Boaden (eds.), Biology of benthic organisms,p.293-302. Oxford: Pergamon Press.

Pirc,H. 1984. Jahresrhythmische Veränderungen in Bestandesmorphologie, Anatomie und chemischer Zusammensetzung von Posidonia oceanica - Beständen um die Insel Ischia. PhD Dissert.,Univ. of Vienna.

Rivera, J.A. 1978. Aspects of the biology of Lytechinus variegatus(Lamarck) at Jobos Bay, Puerto Rico. MSc Thesis, Univ. of Puerto Rico, Mayagüez.

Sand-Jensen, K. 1977. Effects of epiphytes on eelgrass photosynthesis. Aquat.Bot. 3: 55-63.

Schoener, T.W. 1971. Theory of feeding strategies. Ann.Rev.Ecol.Syst.2: 369-404.

Sieburth, J.McN. 1968. The influence of algal antibiosis on the ecology of marine microorganisms. In M.R.Droop & E.J.F.Wood (eds.), Advances in Microbiology of the sea, p.63-94. London: Academic Press.

Slobodkin, L.B. 1964. The strategy of evolution. Amer.Sci. 52: 342-357.

Traer, K. 1980. The consumption of Posidonia oceanica Delile by echinoids at the isle of Ischia. In M. Jangoux (ed.), Echinoderms: present and past,p.241-244. Rotterdam: Balkema.

Vadas, R.L. 1977. Preferential feeding: an optimization strategy in sea urchins. Ecol. Monogr. 47: 337-371.

Vadas, R.L., T.Fenchel & J.C.Ogden 1982. Ecological studies on the sea urchin Lytechinus variegatus, and the algal - seagrass communities of the Miscito Cays, Nicaragua. Aquat. Bot. 14: 109-125.

Wittmann, K. & J.Ott 1982. Effects of cropping on growth in the mediterranean seagrass Posidonia oceanica (L.) Delile. P.S.Z.N.I: Marine Ecology 3(2): 151-159.

Proceedings of the Fifth International Echinoderm Conference / Galway / 24-29 September 1984

Growth retardation in the regular echinoid *Echinus esculentus* Linnaeus

MARK WHITE, BRENDAN F.KEEGAN & YVONNE LEAHY
Zoology Department, School of Marine Sciences, University College, Galway, Ireland

Morphometrical and age data on the echinoid Echinus esculentus L. are compared for two locations in Irish coastal waters. Ageing was effected by a new acid etching technique. A pronounced differential in growth rate is demonstrated, with apparent retardation at one site being provisionally attributed to curtailment of the algal food supply. Age-associated migratory behaviour is suggested for the same site, with depth ranging being affected by local conditions.

INTRODUCTION

The regular echinoid Echinus esculentus L. is a conspicuous element in the rocky sub-littoral around the Irish coast. In the course of benthic survey work at Kinsale Harbour (South-east Ireland, Map 1a), the Echinus esculentus population in the area was seen to be comprised of smaller individuals than is characteristic of other Irish locations known to the authors.

Studies on Echinus esculentus, including the classic work of Moore (1934, 1935, 1937) and more recent investigations by Larsson (1968), Krumbein & Van der Pers (1974), Edwards (1979) and Nichols (1979a & b) inter alios, demonstrate the echinoid's importance in the bio-economy of the hard bottom sub-littoral communities, and pronounced morphometrical variability over its geographic range. The latter, and our own subjective observations, prompted an investigation as to whether the Echinus population at Kinsale was made up of old urchins whose growth had been stunted or simply comprised many young individuals. If the former was shown to be the case, it was conjectured that a shortage of algal food, due in part to the activities of the urchins themselves, might be res-

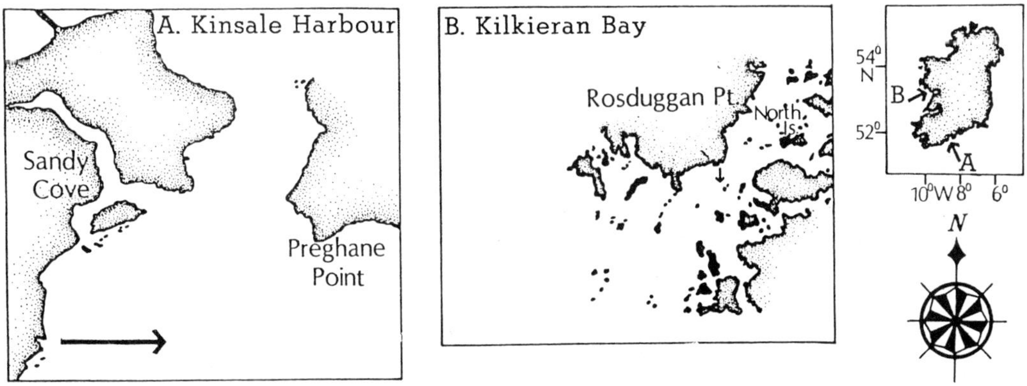

Map 1a & b: Study Areas; arrows indicate sampling sites.

ponsible. A second site at Kilkieran Bay (west coast of Ireland, Map 1b) was chosen for comparative purposes, not on the basis of any presumed environmental similarity but because its population was considered to be more typically constituted (i.e. with larger animals predominating), and food availability was not seen to be a limiting factor. It was assumed at the outset that there was no selection for growth factors between or within the two populations being compared.

STUDY AREAS AND METHODOLOGY

Kinsale Harbour forms the outer part of the Bandon River estuary. The sampling locations, on the western side (Map 1a), are euhaline, with relatively turbid conditions for much of the year. Algal cover on the laminated mudstone bedrock (Van Gelder & Clayton, 1978) decreases rapidly with depth. In fact, the hard bottom below 15m is virtually devoid of macroalgal cover (Plate 1) and, both here and in the shallows, surfaces carry varying amounts of precipitated silt. Euhaline, but markedly less turbid, the Kilkieran locations are mid-way up the bay (Map 1b.), where a granitic boulder field carries a dense algal cover (predominantly Laminaria hyperborea (Gunn)(Foslie) to depths exceeding 20m. The Kilkieran sites are more sheltered than those at

Plate 1. The deep site at Kinsale showing the laminated rock surface and sparse macroalgal cover.

Kinsale which are fully exposed to southerly and south-easterly gales.
Samples were collected using S.C.U.B.A. over the period July 1981 to June 1984, and were taken from two depth levels: shallow-between 5 and 10m, and deep - below 20m.
Urchin tests were cleaned by removing the peristome, Aristotle's lantern, and all the internal organs. Gonads from a sub-sample (228) of these were fixed in 10% formol saline for subsequent sexing using tissue squash preparations. Tests were then soaked in biological detergent and scrubbed to remove surface

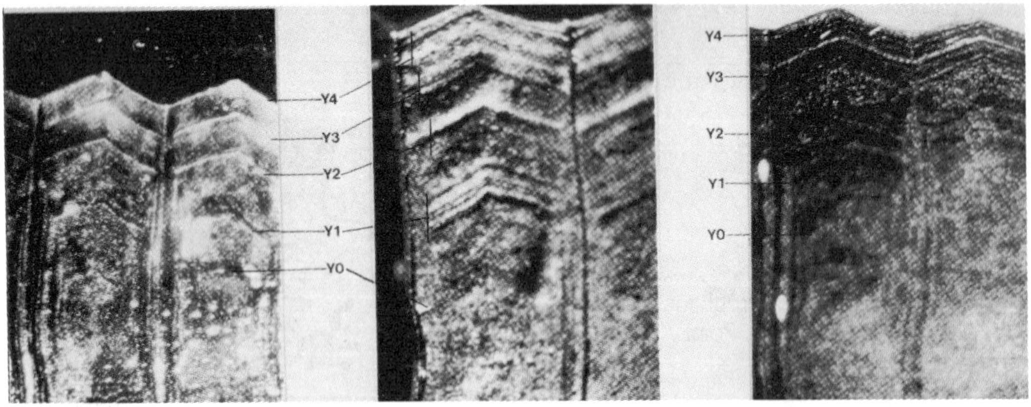

Plate 2: Test plates of Echinus esculentus after treatment to expose growth rings, showing clearly defined annual bands, sub-annual growth checks in discrete groups and poorly defined rings.

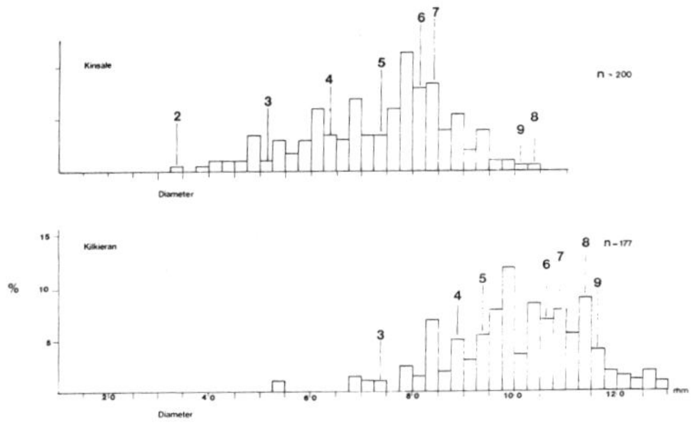

Fig. 1. Diameter frequency histograms for both sites, with mean size for each year class superimposed.

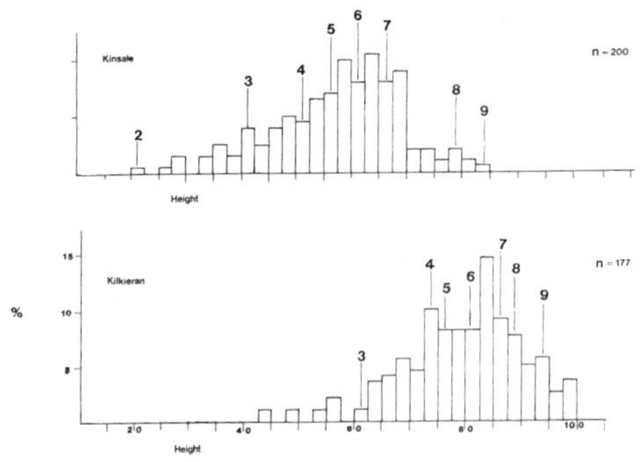

Fig. 2. Height frequency histograms for both sites, with mean size for each year class superimposed.

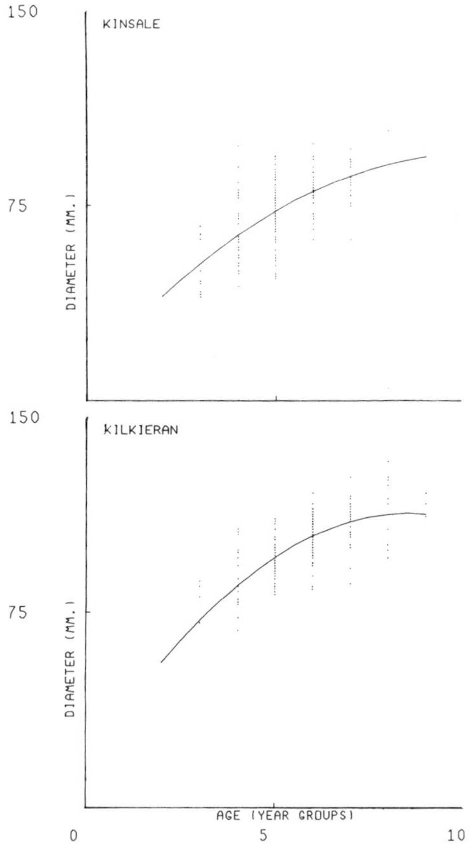

Fig. 3. Diameter/age scatter plots and growth curves for both sites.

structures. They were then air dried before diameter, height, tallness, plate thickness and age were measured. Diameter was defined as the horizontal distance across the ambitus between the mid-point of one ambulacrum and the mid-point of its opposite interambulacrum. Height was taken as the vertical distance between the termini (gonadal plate and perignathic girdle of a single meridian. Both were measured to the nearest 1mm. The ratio of height to diameter (H/D) was used as a measure of tallness. Interambulacral ambital plate thickness was also measured (to ±0.1mm) using the measuring eyepiece in a dissecting microscope.

An ageing technique based on the differential dissolution of the calcite stereom to expose growth rings in the skeletal plates was developed. Dried interambulacral ambital plates were immersed in 14% weight/volume of sodium hypochlorite for 5-10min. to remove the soft organic tissue from the stereom. They were then thoroughly washed in water and placed in 10% hydrochloric acid. Immersion time in the acid varied greatly according to the individual plate and required frequent observation to observe when optimal definition of growth rings appeared to be obtained. Plates were then soaked in toluidine blue histological stain for 2-3 sec and immediately placed on absorbent paper. They were viewed under a dissecting microscope using oblique lighting to count the growth rings. Growth rings were always exposed but to varying degrees (Plate 2). Subannual growth increments or checks were often revealed and tended to

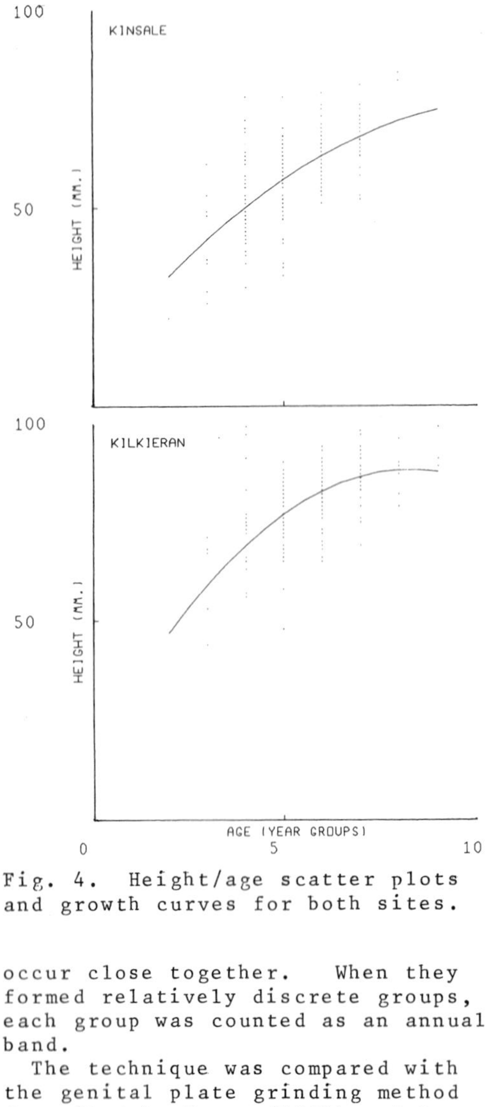

Fig. 4. Height/age scatter plots and growth curves for both sites.

occur close together. When they formed relatively discrete groups, each group was counted as an annual band.

The technique was compared with the genital plate grinding method described by Moore (1935) and later modified by Sime (1982). The comparison was carried out on a total of 228 urchins from both sampling locations, with etched ambital plates being directly compared to ground genital plates.

Data were analysed on a Digital Dec20 mainline computer using the S.P.S.S. and MLab statistical and graphics packages.

RESULTS

Echinus esculentus at Kinsale was found to be significantly smaller in terms of diameter, height and plate thickness than at Kilkieran for both sampling depths. This was found to obtain for individual samples taken in 1981/'82 and 1983/'84. For this reason, and on the basis that the populations had near normal distributions, individual samples for both the shallow and deep locations in both areas are pooled together and treated as such in Fig. 1.

The mean size (diameter and height) was calculated for each year class at each site. Kinsale urchins were found to be significantly smaller than their Kilkieran counterparts of similar age. This was found to be true for both diameter and height (Figs 1 & 2). No significant relationship between sex and test size or shape was found at either site.

The mean age of samples from both sites was calculated and the results compared using the student t-test. Kinsale urchins were found to be significantly younger ($t = 3.71$, $p<0.001$) than those from Kilkieran. This, however, was more a reflection on the larger number of older animals at Kilkieran than a predominance of younger individuals at Kinsale. Accordingly, the Kinsale population can be seen to be made up of smaller individuals (independent of age) than that at Kilkieran.

A comparison of the mean test tallness between sites and between depths within sites proved complex. Although Echinus esculentus from the deep location at Kinsale were taller than those at the same depth in Kilkieran ($t = 4.3$, $p<0.001$), the opposite held for the shallow sites ($t = 5.85$, $p<0.001$). There was no significant difference in mean tallness between the deep and shallow sites in Kilkieran Bay. In Kinsale, the urchins became significantly taller with depth ($t = 9.66$, $p<0.001$).

Interambulacral plates were found to be generally thinner ($t = 7.63$, $p<0.001$) in the Kinsale population. As with tallness, there was no significant difference in plate-thickness between the shallow and deep collections in Kilkieran Bay. Urchins from the shallows in Kinsale Harbour however were found to have thicker plates than those at greater depth ($t = 11.68$, $p<0.001$).

A Marquardt-Levenberg routine was used to fit curves of best fit to diameter/age and height/age data from both sites (Figs. 3 & 4). The curves were polynomial according to the equation $y = ax^2+bx+c$. Juvenile Echinus esculentus are known to recruit into cryptic habitats such as narrow rock crevices (Nichols, pers. comm.) making them extremely difficult to sample. One and two year old animals were not collected during the present study and early growth rates can only be estimated by extrapolating the curves. Growth in the first two years at Kinsale is in the order of 1.75 to 2cm per annum; it then slows gradually with age but does not appear to reach a plateau. Growth in the first two years at Kilkieran is substantially faster and continues at a more rapid rate than at Kinsale until the 7th or 8th year when a definite plateau is attained. The patterns are almost identical for both height and diameter.

In Kinsale a predominance of older urchins in the shallows and young in the deeps was found to be highly significant (Table 1); no such correlation was found at Kilkieran.

Table 1. Percentage occurrence of each age group at the two depth regimes sampled at Kinsale

Age	Shallow	Deep
2	0%	100%
3	10%	90%
4	25.6%	74.4%
5	40.8%	59.2%
6	63.6%	36.4%
7	65.2%	34.8%
8	100%	0%
9	100%	0%

DISCUSSION

It is acknowledged that the ageing technique described here has an inherent element of subjectivity and that in situ experiments with marked individuals are required to provide a basis for more accurate calibration. We feel however that the lack of such calibration does not seriously diminish the value of the technique in comparative studies. The comparison with the genital plate grinding technique of Sime (1982) on 228 urchins showed that there was no significant difference in the results of both methods, although the interambulacral, ambital plates are obviously younger. Etching of the genital plates proved extremely difficult and, on balance, the use of interambulacral ambital plates is preferred for their larger size which 1) gives clearer definition of growth banding and, 2) allows a much larger number to be processed.

At the outset, it was suggested that the Echinus esculentus population at Kinsale Harbour was made up of smaller individuals than characterised populations at other Irish locations known to the authors. This was found to hold in the comparison with the Kilkieran Bay population where urchins were significantly larger than those from Kinsale. Using age data, it was possible to show that the Kinsale urchins were in fact markedly retarded in their growth in comparison to those found at Kilkieran.

The absence of one and two year old urchins from our collections meant that early growth rates could only be inferred by extrapolation from the curves shown in Figs. 3 & 4. Nichols et al.(pers. comm.) were able to age newly settled Echinus esculentus collected from cryptic habitats off Plymouth. Although their data show dramatically different growth rates in the first two years from those inferred here, growth patterns in the latter years appear to be very similar.

Recent studies (Edwards, 1979 and Nichols, 1979a & b) have shown that general statements cannot be made about the relationship between test-size and factors of exposure, and water depth on Echinus esculentus. We would contend that the differential in size with age between our two study areas is due in large measure to the availability of algal food. Turbid conditions in Kinsale will decrease light penetration and in turn limit plant growth. Further, precipitated silt of itself will severely curtail algal settlement and colonization. The generally clear water conditions in Kilkieran are manifested in the luxuriant algal cover within the depths sampled. The aspect of exposure may also have a bearing on test size, the Kinsale sites being much more prone to gale induced turbulence than those at Kilkieran. Our subjective impressions of population densities at the two study locations are that Echinus esculentus is much more numerous in Kilkieran. Ebert (1968) working with Strongylocentrotus purpuratus (Stimpson) stated that food limits biomass rather than numbers and that the ultimate size of the urchins is adjusted to the carrying capacity as well as the physical rigours imposed by the local environment. The carrying capacity of the Kilkieran sites is deemed to exceed the demands of their urchin populations. By comparison, the size and numbers of Echinus esculentus at Kinsale suggest limiting if not unfavourable conditions at the site. This is supported by the presence and absence of a plateau in the growth data (Figs. 3 & 4) from the Kilkieran and Kinsale populations respectively. Nichols et al. (pers. comm.) record a similar plateau in their Plymouth populations but found it to occur in 11 and 12 year old urchins rather than in the 7 and 8 year olds as in Kilkieran. This phase shift is consistent with the contrasting early growth rates between their study and ours.

D'Arcy Thompson (1917) found that Echinus esculentus was flatter in the shallows than at depth and proposed that this was an adaptation to facilitate better adhesion to the substrate. Moore (1935) reported a similar finding, and observed that flatter urchins also had thicker test plates, possibly as a further adaptation to turbulent conditions. In this work, although urchins in the shallows at Kinsale were flatter than those at depth, a comparison between the shallow sites at the two locations showed the Kilkieran population (i.e. the more sheltered site) to be flatter than those at Kinsale. Nichols (1979a) and Edwards (1979) reported conflicting patterns of tallness within and between sites and found that trends could even reverse over consecutive years.

Plate thickness could not be related to exposure at either site. Although the interambulacral, ambital test-plates were found to be significantly thinner at Kinsale, this did not parallel the recorded depth related increase with tallness at the site.

Shoreward migrations in Echinus esculentus associated with increasing age, feeding and reproductive activities have been described by several workers (eg. Elmhirst, 1922) Stott, 1931; Rod house and Tyler, 1978). The uneven distribution of ages at Kinsale would suggest a similar movement. This is interpreted as reflecting the uniformly good feeding conditions across the depths sampled at Kilkieran compared to the stark gradient at Kinsale where the progressive fall-off in algal cover with depth is visibly compounded by the grazing conditions of the urchins themselves.

Work planned and in progress, notably on the slow growing Kinsale population, includes Tetracycline marking experiments to monitor growth and migration, with particular attention to newly settled and juvenile Echinus esculentus. Monthly samples are being used to stage gonadial development and an attempt will be made to relate reproductive processes and events to aspects of the physical and chemical environment.

ACKNOWLEDGEMENTS

The authors wish to thank Dr. C. Maggs, Dr. G. Könnecker and the technical staff of the Benthos Research Group in the Department of Zoology, University College, Galway, for their assistance. The work was carried out within the provisions of the Cost 47 project on Coastal Ecology (financed by the National Board for Science and Technology) and a research contract with Elanco S.A., a subsidiary of Eli Lilly and Company.

Contribution No. 220 from the School of Marine Sciences, University College, Galway.

REFERENCES

Ebert, T.A. 1968. Growth rates of the sea urchin Strongylocentrotus purpuratus related to food availability and spine abrasion, Ecology 49(6): 1075-1091.

Edwards, A.J. 1979. Ecological studies of the edible sea urchin

Echinus esculentus, Bantry Bay, south west Ireland, Prog. Underw. Sci. 4: 197 - 207.
Elmhirst, R. 1922. Habits of Echinus esculentus, Nature 110: 667.
Krumbein, N.E. & Van der Pers, N.C. 1974. Diving investigations on bio-deterioration by sea urchins in the rocky sublittoral of Helgoland, Helgolander. Wiss.Meers.26: 1-17.
Laarson, B.A.S. 1968. Scuba studies on vertical distribution of Swedish rocky-bottom echinoderms. A methodological survey, Ophelia 5: 137-156.
Moore, H.B. 1934. A comparison of the biology of Echinus esculentus in different habitats, Part 1. J.mar.biol.Ass.U.K. 19: 869-955.
Moore, H.B. 1935. A comparison of the biology of Echinus esculentus in different habitats, Part 2. J.mar.biol.Ass.U.K. 20: 109-128.
Moore, H.B. 1937. A comparison of the biology of Echinus esculentus in different habitats, Part 3. J.mar.biol.Ass.U.K. 21: 711-720.
Nichols, D. 1979a. A biometrical study of the British sea urchin Echinus esculentus from diving surveys of three areas over successive years. In: Echinoderms past and present. (Jangoux, ed.) A.A. Balkema, Rotterdam (publ. 1980) 4223pp.
Nichols, D. 1979b. A nationwide survey of the British sea urchin Echinus esculentus. Prog. Underw. Sci. 4: 161-187.
Rodhouse, P.G. & Tyler, I.D. 1978. Distribution and production indices of the sea urchin Echinus esculentus L. in the shallow sublittoral around Lundy. Progr. Underw. Sci. 3: 147-163.
Sime, A.A.T. 1982. Growth ring analysis in regular echinoids. Progr. Underw. Sci. 7: 7-15.
Stott, F.C. 1955. The food canal of the sea urchin Echinus esculentus and its functions. Proc. Zool.Soc.Lond.125: 63-86.
Thompson, D.W. 1917. On growth and form. Cambridge Univ. Press (publ.) 116pp.
Van Gelder, A. & Clayton, G. 1978. New data on early Dinantian (early) Carboniferous) stratigraphy and sedimentation in south Cork, Ireland. Geol.Mijnbouw. 57(1): 25-32.

Proceedings of the Fifth International Echinoderm Conference / Galway / 24-29 September 1984

A respirmetric and direct calorimetric study of ovarian energy metabolism in *Strongylocentrotus droebachiensis*

L.H.BOOKBINDER* & J.M.SHICK
Department of Zoology, University of Maine, Orono, USA
**Present address: Scripps Institution of Oceanography, University of California, San Diego, USA*

Ovaries of the sea urchin S. droebachiensis exhibit changes both in mass and in mass-specific metabolic rate during their annual cycle. Oxygen uptake ($\dot{q}O_2$, $mW \cdot g^{-1}$) by intact ovaries was measured throughout the year at the seasonal sea-water temperature and at the pO_2 prevailing in the sea urchins' perivisceral coelomic fluid (p.v.f.). Coorinated measurements of heat dissipation ($_t\dot{q}$, $mW \cdot g^{-1}$), which includes enthalpy changes associated with both aerobic and anaerobic energy metabolism, indicated that anaerobiosis accounted for 76 to 92% of $_t\dot{q}$ at physiological pO_2. Such a large contribution of anaerobic metabolism in intact ovaries is lively due to their large mass and lack of perfusion by an oxygen-carrying circulatory system. When aerobic metabolism was estimated calorimetrically by subtracting $_t\dot{q}$ measured under anoxia from that measured at physiological pO_2, it consistently agreed with aerobic metabolism measured as oxygen uptake ($\dot{q}O_2$, $mW \cdot g^{-1}$). Ovary oxygen uptake was consistently higher at air saturated pO_2 compared to the lower physiological pO_2, which indicates that aerobic metabolism of this organ was diffusion-limited. Succinic dehydrogenase activity (an index of aerobic capacity), was estimated histochemically and confirmed the low aerobic capacity of intact ovaries, and suggested along with tissue pO_2 measurements that oxygen uptake was restricted to peripheral (< 1mm) tissues.

Seasonal changes in the rate of total internal tissue (ovary plus gut) oxygen uptake ($\dot{Q}O_2$, mW) and total ovary anaerobic heat dissipation ($_t\dot{Q}$anox, mW) (i.e. the product of total tissue mass within an individual sea urchin times the mass-specific $\dot{q}O_2$ or $_t\dot{q}$anox), affect the pO_2 and pH of the perivisceral coelomic fluid. There is an inverse relationship between total $\dot{Q}O_2$ and pO_2 of the p.v.f. Likewise, pH is inversely related both to $_t\dot{Q}$anox and $\dot{Q}O_2$. This pH effect is due to changing amounts of respiratory (CO_2) or metabolic (anaerobic end products) acid added to the coelomic fluid, since pH did not vary seasonally as would be expected from temperature alone. Protein concentration of perivisceral coelomic fluid varied with reproductive condition, reaching maximum values during periods of maximum ovary acidosis, hence increasing the buffering capacity of the p.v.f.

An analysis of acid-base regulation in this sea urchin is not presently possible because neither the actual buffer capacity nor the CO_2 concentration of the p.v.f. is known. The relatively low levels of dissolved lactate (< 20 µM) suggest a minor effect of the release of this anaerobic end product from the tissue into the p.v.f., although it is possible that other acidic end products are present. When the measured $_t\dot{Q}$anox was compared with enthalpy changes calculated from measured ATP and phosphoarginine depletion and lactic acid accumulation in the same ovaries, an average of only 39% of the dissipated heat could be accounted for biochemically. A similar situation exists in other marine invertebrates during anoxia, and the simplest explanation is that there are additional unidentified end products of anaerobiosis. Preliminary HPLC analysis of freshly excised ovaries and ovaries from the same individuals which were exposed to 8 h of anoxia indicate that acetate and propionate are produced. Additional investigation is needed to substantiate this result.

Pheromonal control of metamorphosis in the Pacific sand dollar, *Dendraster excentricus*

ROBERT D.BURKE
Department of Biology, University of Victoria, Canada

Competent larvae are induced to metamorphose by sand from a sand dollar bed or an aqueous extract of the sand. Gel-permeation chromatography and high performance liquid chromatography were used to purify a 980 MW peptide that will induce metamorphosis at concentrations of 10^{-6} - 10^{-5}M. Extracts of whole adults and gonads are also able to induce metamorphosis and adults can condition substrates to induce metamorphosis. It is concluded that the initiation of metamorphosis in Dendraster excentricus is controlled by a pheromone released by adult sand dollars.

Proceedings of the Fifth International Echinoderm Conference / Galway / 24-29 September 1984

The echinoids *Paracentrotus lividus* (Lamarck) and *Psammechinus miliaris* (Gmelin) in the intertidal zone of the French atlantic coast
Introduction to studies on growth and boring activity

C.CUENCA
Natural History Museum, Nantes and Laboratory of Marine Biology, Nantes University, France

This study draws attention to morphological features which are useful in differentiating between Paracentrotus lividus and Psammechinus miliaris. The population dynamics of both species are being investigated in context of their associated biota and biotopes (i.e. intertidal pools) near Douarnenez (South Finistère) and at Lerat (Loire-Atlantique).

At Douarnenez, P. lividus is very abundant in mid-intertidal pools where it bores into siliceous rock.

Both echinoids are known from the coast near Lerat since 1857, when they were first observed by F. Cailliand. Here, over the intervening years and for reasons unknown, the tidal pools have been variously settled by, or empty of, urchins. In 1882, the young of both species were observed under rocks, both at Lerat and at "L'Ile Dumet" (Loire-Atlantique).

In this on-going study, recolonisation and bioerosion by the urchins are being given particular attention.

Skeletal growth systems in echinoid larvae (*Paracentrotus lividus*)

YVES DE GREEF & MICHEL JANGOUX
Laboratoire de Biologie marine, Université Libre de Bruxelles, Belgium

Echinoid planktonic larvae (plutei) typically have slender and elongated arms supported by calcareous rods. Experiments on larvae of the echinoid Paracentrotus lividus (Lamarck) showed that the arms - as well as their supporting rods - can either grow or regress according to environmental conditions. The daily change of the mean length of all skeletal segments has been followed on both normally developed and experimentally regressed (i.e. starved) larvae. A comparison of the growth curves showed that three different systems of skeletal growth co-occur in developing larvae, viz. an endotrophic, an endo-exotrophic and an exotrophic growth system.
The endotrophic growth system concerns rods that develop only during the endotrophic larval period. These rods (i.e. the body rods) reach their final size at the end of that period. Their length does not change whether the exotrophic larvae feeds or not.
The endo-exotrophic growth system concerns rods that grow continuously during the larval life (i.e. the postoral rods and the anterolateral rods).
The exotrophic growth system concerns rods where growth is initiated only in fed larvae (i.e. the posterodorsal rods and the dorsal arch).

Work supported by IRSIA grant 83127 to Yves de Greef, and by FRFC grant 2.4506.83.

Proceedings of the Fifth International Echinoderm Conference / Galway / 24-29 September 1984

Influence of the domestic pollution generated by the discharge of the Marseilles-Cortiou sewage outlet on the distribution and morphometry of *Paracentrotus lividus* (Lmk.)

P.DELMAS & M.B.RÉGIS
CERAM, Faculté des Sciences et Techniques St.-Jerôme, Marseille, France

The outlet of the main-sewer from the Marseilles conurbation takes place on the creek of Cortiou (south of Marseilles), at the sea-surface. On average the discharge is about $9 m^3/sec$ per day. Up to now, the sewage was not pretreated except for a hasty oil-separation and the elimination of large items such as paper, cardboard, plastics, tins, bottles, small pieces of wood, etc... which are retained by a screen system. Four sampling stations were chosen in the investigated area:
(i) station 0 at 500m and station 1 at 1200m from the outlet in rather exposed locations,
(ii) station 2 at 2800m in a more sheltered location, and
(iii) station 3 at 5200m from the outlet and characterized by a permanent current in the shallow and narrow pass between the shore itself and the Maire Island.
The plume chiefly drifts from east to west along the shoreline due to the predominant wind regime and also due to the general cyclonic flow of the Western Basin of the Mediterranean Sea, a part of which enters the funnel between the shore and the chain of small islands.
The average population density of Paracentrotus lividus at 10-12m depth is about 28 indiv./m^2 at site 0 and 42/m^2 at site 1, respectively. At sites 2 and 3, the average density is lower, decreasing from 15(st.2) to 7 (st.3) per square meter at 3-6m depth respectively. The investigation showed that most of the Paracentrotus lividus at stations 0 and 1 are the 25-30 and 30-35 mm size classes, whereas larger specimens are rare here. In comparison, size classes of 35-40, 40-45, 45-50 mm predominate in station 2, while at station 3 the most abundant size classes are 45-50, 50-55, 55-60 mm (some individuals approaching 65 mm). It seems that, at sites 0 and 1, the pelagic larvae which have been carried in from non-polluted areas and are about to settle, metamorphose in small crevices. They then develop rapidly due to the ready availability of organic matter (both particulate and dissolved) but it seems that they cannot, or rarely, develop their gonads, because of the pollutants in the water column. It is probable that they progressively concentrate toxic substances from their food which comprises living and inanimate materials as well as dissolved substances. For example, note the difference in population density between stations 0 and 1.
We suppose that the very marked lengthening of the spines at station 1 in comparison to those of station 3 (where the pollution level is less) corresponds to a local adaptation to exploit the greater amount of dissolved and particulate organic carbon in the polluted area (Pequignat, 1966, 1967, 1972; Regis, 1967, 1978). At the same time, it concentrates some pollutants which disturb the metabolism of this echinoid.

Symbiotic bacteria within the intestinal caecum of the spatangoid echinoid, *Echinocardium cordatum*

CHANTAL DE RIDDER & LOUIS DE VOS
Laboratoire de Biologie marine, animale et cellulaire, Université Libre de Bruxelles, Belgium

The intestinal caecum of most individuals of Echinocardium cordatum harbours peculiar, rounded nodules. Each nodule has a well defined structure and consists of a thick external cortex surrounding a central core. This central core is made of shell or plant fragments, packed diatoms and/or other unidentified debris. The cortex is stratified into three superimposed layers which result from the activity of large sheathed bacteria. The inner cortical layer consists of densely packed empty sheaths. Similar sheaths, inhabited by bacteria, form the middle layer, while the outer layer is made of the same kind of bacteria but unsheathed. Such a cortical structure suggests that the bacteria migrate outside when growing and that they continuously leave the old sheath and build a new one. Iron was detected in the bacterial sheaths by X-ray analysis. Investigations on the nature of the symbiosis are in progress. The symbiotic bacteria are seemingly always present in the intestinal caecum of E. cordatum. Their main characteristic is to attach themselves to debris around which they progressively grow to form irregular to rounded nodules. Bacteria seemingly encounter favourable conditions in the intestinal caecum. These conditions were not investigated but the occurrence of energy-rich material in their immediate environment (e.g. unwanted material of coelomic origin that cross the caecum wall, as well as various ingested debris occurring in the caecum lumen) must be one of them.

Work supported by FRFC grant 2.4525.80.

Proceedings of the Fifth International Echinoderm Conference / Galway / 24-29 September 1984

Oral surface podial feeding in the sand dollar *Echinarachnius parma* (Lamarck)

OLAF ELLERS
Department of Zoology, Duke University, Durham, N.C., USA

Podia on the oral surface and at the ambitus were seen vigorously collecting particles from the substrate. These particles were passed from podium to podium towards the food grooves. In the food grooves, food was loosely aggregated by mucus, passed to the mouth by podia, and ingested. The rate of particle collection and passage along the food grooves was much higher when we added diatoms to the substrate than it was with normal sand substrate. On two occasions pedecellariae were seen chewing on nematodes which were later ingested. Ciliary current flows were marked with dyes and particles. On the aboral surface, flow was radial from the apex to the ambitus. In contrast, currents on the oral surface flow anterior to posterior. On both surfaces, ciliary bands are arranged with the long axis of the cilia perpendicular to the flow. Particles were seen being transported by aboral ciliary currents, but most were lost at the ambitus. Statistical comparison of the substrate and gut contents showed that this sand dollar collects a disproportionately small number of particles larger than 230μm, but does not especially collect particles less than 100μm. This is important since aboral spines are 100μm apart and any particles which were aborally sieved would have to increase the proportion of very small particles in the gut. Thus we were unable to find any evidence in support of the aboral sieve feeding mechanism.

Proceedings of the Fifth International Echinoderm Conference / Galway / 24-29 September 1984

Variations in the rate of ingestion of algal cells with the stage of morphological development of larvae of *Paracentrotus lividus* (Echinodermata)

L.FENAUX, C.CELLARIO & M.ETIENNE
Station Zoologique CERVOV, Villefrance-sur-mer, France

During the exotrophic phase the larvae of echinoids pass through several morphological stages characterized by the development of new appendages near or outside the circumoral area. Formation of new larval arms near the mouth increases the area of the ciliated band which collects the food particles. One may suppose that increasing the length of the ciliated circumoral band would result in a greatest efficiences of particle ingestion.

Variations in the ingestion rate by larvae of the sea urchin Paracentrotus lividus (Lamarck) fed of one of two Haptophyceae Hymenomonas elongata (Droop) Braarud and Monocrysis lutheri (Droop) Braarud are studied. Measurements of ingestion were made on plutei at times corresponding to different stages of the larval development : plutei with two pairs of arms anterolateral and postoral arms, development of posterodorsal arms, preoral arms, epaulettes, pedicellars and the adult rudiment. The development of the larvae fed with a diet of $6x10^5$ to $22x10^5$ um^3 cells of M. lutheri l^{-1}. d^{-1}, was asynchronous with only an increase in the length of larvae without the formation of new appendages. Complete larval development has been achieved with H. elongata using cell concentrations of $8x10^5$ to $55x10^5$ $um^3.l^{-1}.d^{-1}$.

The rates of ingestion of these two algae were compared and related to the differences observed in the morphological development according to the food ingested. For a diet using cell concentrations of H. elongata of $8x10^5$ to $55x10^5$ um^3 $l^{-1}.d^{-1}$, the ingested algal volume varied with the rate of development of the larva. Models of growth has been tested to to illustrate its development. Among mathematical models tested (von Bertalanffy-Brody, Gompertz, logistic function) the logistic function is the one we have used.

Proceedings of the Fifth International Echinoderm Conference / Galway / 24-29 September 1984

Geographic and bathymetric distribution of the clypeasteroids

JOE GHIOLD
Department of Geology, Louisiana State University, La., USA

Maps depicting the geographic and bathymetric ranges of the order Clypeasteroida serve in the interpretation of changes in environmental parameters of this group in space and time. The geographic and bathymetric ranges of 110 extant clypeasteroid species were mapped individually, but were combined here at the generic level for convenience of presentation. The primary data were compiled from Mortensen's monograph and supplemented by other works. In all, 14 genera of the order Clypeasteroida were recorded. Bathymetric ranges of most clypeasteroids are largely confined to the continental shelf areas though the genus Echinocyamus has been found beyond the shelf margin (e.g. E. grandiporus -2500 m, E. macrostomus -2300 m, E. scaber -1886 m). The genera Laganum (700 m), Clypeaster (500 m) and Fibularia (400 m) are the next deepest. All remaining clypeasteroids occur within the range, intertidal -350 m. No clypeasteroid is recorded beyond 3000 m. Cosmopolitan communities appear at first to dominate most biotic provinces; it is, however, necessary to examine the assemblages at the species level to delineate and identify endemic and relict populations. The least specialized groups (with the exception of Echinodiscus) exhibit the greatest geographic distribution, namely, Clypeaster, Echinocyamus, Echinarachnius, Fibularia, Laganum and Peronella. Specialized forms such as Arachnoides, Astriclypeus, Encope, Mellita, Scaphechinus and unusual morphologic types (e.g. Dendraster, Rotula) give smaller geographic ranges. No clypeasteroids are found in Antarctic waters. The particular distributional patterns of these endemic and relict echinoid species can not be explained by vicariance events along, suggesting that other biologic mechanisms are involved. Future quantitative surveys would be more useful if species' locations and depth readings were calculated by ranges rather than as individual data points.

Fine structure of the globiferous pedicellariae of the echinoid, *Sphaerechinus granularis*

MARIANNE GHYOOT & MICHEL JANGOUX
Laboratoire de Biologie marine, Université Libre de Bruxelles, Belgium

Globiferous pedicellariae are defensive, glandular tri-jawed appendages located on the outer body surface of most echinoids. Each jaw is provided with a venom-apparatus and is supported by a calcareous valve. The morphology and fine structure of globiferous pedicellariae in Sphaerechinus granularis were studied using light and electron microscopy, with special attention being given to the venom apparatus. The venom-apparatus comprises a venom gland, a secretory duct and a terminal grooved tooth; the last corresponds to the upper part of the pedicellarial valve. The venom gland is supported by the outer side of the valve and is surrounded by a sheath of smooth circular muscles. It consists of two different regions, i.e. an enlarged basal body and a pair of smaller apical lobes. Cells with enlarged vacuoles, filled with secretory material, occur in both regions. The EM-aspect markedly differs however according to the region considered, from well-defined rounded granules (basal body) to finely granular material (apical lobes). The secretory duct extends to the apical lobes and surrounds the terminal tooth. The duct-lining epithelium consequently lies partly on the tooth itself (i.e. the epithelium of the inner side of the duct) and partly on the pedicellarial dermal tissue (i.e. the epithelium of the outer side of the duct). Cells of the duct's outer epithelium have conspicuous fibrillae (presumably myofibrillae) running all along the cell length. Release of venom by globiferous pedicellariae results from different mechanisms that presumably act together. Stimulation of the sensory hillocks located at the base of the jaw's inner side induce almost simultaneously the closure of the jaws (contraction of jaw's adductor muscles), retraction of the epidermis around the tooth and its insertion into the intruder's integument. Injection of the venom is basically due to the contraction of the muscle sheath surrounding the venom gland. The venom consequently is pushed into the secretory duct (the lumen of which enlarged owing to the contraction of the fibrillae of its outer epithelium) and then into the intruder's integument through the groove of the tooth.

Work supported by IRSIA grant 83327 to M. Ghyoot.

Sea urchin outbreaks and nematode epizootics in Vestfjorden, northern Norway

NILS T.HAGEN
Department of Biology, Dalhousie University, Halifax, Nova Scotia, Canada

Large scale destruction of kelp beds by the sea urchin, Strongylocentrotus droebachiensis, has recently occured off the coast of northern Norway. Barren "Isoyake" bottoms dominated by crustose coralline algae and sea urchins have persisted for at least five years with no signs of macrophyte recovery. A large and previously undescribed nematode endoparasite was found in the coelom of S. droebachiensis from an overgrazed area at Vaerøy Island in 1981, and has been tentatively identified as a species of Echinomermella. In 1983, infection rates of this parasite at eight sites in Vestfjorden ranged from 6 to 56% S. droebachiensis is known to be a host for more than 30 associated species, but this is apparently the first observation of a a nematode associate. This nematode epizootic is apparently triggered by the sea urchin outbreaks, and might be interpreted as a numerical response to increased sea urchin densities. The hatching of juvenile nematodes is probably lethal for the host and preliminary evidence suggests increased sea urchin mortality in areas with a high incidence of parasitism. A related nematode endoparasite, Echinomermella grayi, has been described from the coelom of Echinus esculentus in England. However, no parasites were found in Echinus from Vestfjorden.

Proceedings of the Fifth International Echinoderm Conference / Galway / 24-29 September 1984

Body wall structure of *Echinarachnius parma* (Echinoidea: Clypeasteroida)

ANTONY S.HAROLD
Department of Zoology, University of Toronto, Canada

The tests of sand dollars have relatively low profiles and structural reinforcements that would distribute applied forces. The calcite test consists of 20 columns of coronal plates rediating from the aboral apical system. The surface layer of these plates is a thin layer of irregular labyrinth below which is horizontal galleried stereom. Orientation of the latter is vertical in ambital "pillars" and "partitions". In the plates calcite rods up to 60μm in diameter appear to radiate from plate centers and meet suture faces at right angles. At the plate margin they are produced into long tapering needles that penetrate up to 500μm into the neighbouring plate. The extent of calcification of sutural areas in this manner is most pronounced on the oral surface near the ambitus and on the aboral in the lower petaloid region. All plate sutures are laced with collagenous fibers which are restricted to the skeleton below the surface labyrinth. As plates expand, fibers and galleried stereom appear simultaneously at plate corners. Fiber and gallery rod length therefore increase away from sutural intersections. A similar arrangement of connective tissue exists in regular urchins. The parallel arrangement of the collagenous fiber bundles and the tightness of their winding around skeletal struts indicate a system in tension. These features all make the sand dollar test resistant to deformation and breakage. Allometric analysis shows that the relative amounts of connective tissue and calcite interdigitation in sutures decreases with size.

Proceedings of the Fifth International Echinoderm Conference / Galway / 24-29 September 1984

A comparison of sea urchin recruitment at sites on the Atlantic and Pacific coasts of North America

L.G.HARRIS & JONATHAN D.WITMAN
Zoology Department, University of New Hampshire, Durham, USA

R.ROWLEY
Department of Biological Sciences, University of California, Santa Barbara, USA

A technique was developed for sampling post-metamorphic sea urchins (<1mm test diameter) and other motile fauna from discrete substrates and utilized to look at temporal and spatial patterns of recruitment. Studies were conducted at sites along the coast of New Hampshire in the Gulf of Maine and in the Santa Barbara Channel off of Southern California. In the Gulf of Maine, Strongylocentrotus droebachiensis recruitment was limited to heavy settlement (up to 1 urchin/cm^2) in June, decreasing quickly until no further settlement was observed after the middle of July. Settlement appeared to be random and recruitment was enhanced by prior removal of all motile fauna from substrates. There were no differences in settlement on rocks taken from a kelp bed, a newly established urchin ground and one five years old. Follow up studies showed that survival of urchins was patchy, indicating that populations of S. droebachiensis in the Gulf of Maine are the result of differential survival and not selective settlement. Similar studies conducted at sites in the Santa Barbara Channel indicated that settlement of S. purpuratus occurred over a number of months and in much lower densities. Again, urchin populations appeared to be the result of differential survival.

The *Echinocardium-Lithocystis* Association: Investigation of the host's reactions

MICHEL JANGOUX & CHANTAL DE RIDDER
Laboratoire de Biologie marine, Université Libre de Bruxelles, Belgium

The spatangoid Echinocardium cordatum regularly harbours protozoan parasites belonging to the genus Lithocystis (Sporozoa, Gregarinia). Different developmental stages, from early trophozoites to ripe sporocysts, may occur within the body cavity. The host's coelomocytes react to the presence of motile stages (i.e. trophozoites) of Lithocystis. Trophozoites may be attacked by coelomocytes that progressively surround them. Parasite-associated coelomocytes show a conspicuous transformation, i.e. each cell bears a single spine-like extension so that the whole parasite looks like a minute pin-cushion. It is suggested that the occurrence of sharp-pointed coelomocytes around trophozoites could be a way to prevent gregarine syzygy and, thus, cyst formation. Trophozoites walled off by such coelomocytes will supposedly degenerate. Encysted stages of Lithocystis (viz. gametocysts and sporocysts) are often found embedded in brown bodies. Brown bodies are particular mesothelial, or mesenteral, covered formations occurring in the echinoid body cavity. The presence of cysts within these bodies is supposedly the consequence of their accumulation on mesenteries. Due to the weight of the cysts, the mesenteries are believed to break down eventually and to lead to the formation of the cyst-containing brown bodies.

Proceedings of the Fifth International Echinoderm Conference / Galway / 24-29 September 1984

Anatomy, function and diversity of clypeasteroid non-respiratory podia

R.MOOI
Department of Zoology, University of Toronto, Canada

A typical clypeasteroid non-respiratory podium is composed of three basic systems: 1) a specialized external epidermis, 2) a collagenous connective tissue sheath enclosing a hydrostatic skeleton, 3) a smooth muscle system which operates the tip and stem. The functional anatomy of tissues of each system has now been described with histology and EM. Most podia have both large and small secretory cells in their tip epithelia. Circular fibres form the connective tissue sheath, on the outside of which there are three bundles of longitudinal fibres running the length of the podium. There are five types of non-respiratory podia in the Clypeasteroida. Accessory podia are sensory and manipulatory, barrel-tipped podia collect food and pass it towards the food grooves, where food groove podia aggregate the particles with mucus and move them to the peristome. Large food groove and buccal podia then move the food into the mouth. The diversity of podia accounts for differences in feeding mechanisms of clypeasterine, laganine, rotuline and scutelline forms. Podia of very young sand dollars are different in size and distribution from those of adults. The division of labour amongst podia of the oral surface can be intimately related to the evolution of clypeasteroid feeding mechanisms. Contrary to previous opinion podia, not spines, are the primary feeding organs. In particular, miliary spines cannot now be seriously regarded as the source of feeding mucus. Recent work on clypeasteroid behaviour and podial activity seriously undermines the so-called "aboral rocking sieve" feeding hypothesis and necessitates a reappraisal of outdated accounts of sand dollar feeding mechanisms.

Proceedings of the Fifth International Echinoderm Conference / Galway / 24-29 September 1984

Comparative analyses of free amino-acid composition in eggs and ovaries of sea urchins (the subclass Regularia)

SATOKO NAGAOKI
Department of Biology, Keio University, Kohoku-ku, Yokohama, Japan

It is well established that tissues of marine invertebrates contain a large amount of free amino acirds, and that their composition varies greatly in different animal groups. We, therefore, examined free amino acids in eggs and ovaries of sea urchins, from the viewpoint of comparative biochemistry. Sea urchins of thirty-four species, belonging to five orders, were collected in Japanese waters and in the Mediterranean. Free amino acids were extracted from eggs and ovaries with 80% methanol, and measured in the amino-acid analyzer (Hitachi, Model 835-50; for physiological fluid analysis). According to the kinds of the "major amino acids", which were defined as those contained in over 20% of the total, the sea urchins were divided into the following four groups: (1) Sarcosine type - the Cidaroida (5 spp.), the Diadematoida (1 sp.) and the Arbacioida (2 spp.); (2) alanine-taurine type - the Echinothurioida (3 spp.); (3) taurine type - the Diadematoida (4 spp.); (4) glycine-alanine type - the Arbacioida (2 spp.) and the Echinoida (17 spp.). Thus, biochemical characteristics based on free amino acids generally agreed with the accepted taxonomy. It is interesting that the rather uncommon compounds, taurine and sarcosine, were remarkable only in the primitive orders of sea urchins, and not in the most advanced order, the Echinoida. It is also worth mentioning that the Arbacioida, which is taxonomically placed between the Cidaroida and the Echinoida, showed two different types of amino-acid composition; the family Arbaciidae (2 spp.) belonged to the sarcosine type, and the Phorosomatidae (1 sp.) and the Stomopneustidae (1 sp.) to the glycine-alanine type. This finding suggests that the Arbaciidae are systematically close to the Cidaroida, and that the other two families are closely connected to the Echinoida.

Proceedings of the Fifth International Echinoderm Conference / Galway / 24-29 September 1984

Ecology of the sand dollar *Mellita quinquiesperforata latiambulacra* on the west-central coast of Venezuela (Caribbean Sea)

PABLO E.PENCHASZADEH & MIGUEL E.LAYRISSE
Intecmar, Universidad Simon Bolivar, Caracas, Venezuela

Mellita quinquiesperforata latiambulacra is an important component of sandy benthic communities at 0.5 to 6.0m depth off the west-central coast of Venezuela, on sediments with a high proportion of fines. The maximum recorded density for juveniles was 2570 ind./m^2 while that for the adults was 87 ind./m^2. An almost continuous recruitment through the year was found at Quizandal, with a maximum peak in May and June. The Brady-Bertalanffy growth equation for the three first years of life, on the basis of monthly sampling was obtained: $L_t = 63.0 (1-e^{-0.0943(t-t_o)})$. The instantaneous mortality coefficient was calculated as M = 0.36, with an average life expectancy of 2.8 years and 95% of the theoretical maximum size is reached after 5.6 years of life. The annual mortality rate of the population is calculated to be 30%.

Proceedings of the Fifth International Echinoderm Conference / Galway / 24-29 September 1984

Mass mortality of *Strongylocentrotus droebachiensis* (Echinoidea) and increased macroalgal abundance in the rocky subtidal off Nova Scotia, Canada

ROBERT E.SCHEIBLING
Biology Department, Dalhousie University, Halifax, Nova Scotia, Canada

Between 1980 and 1983, recurrent outbreaks of disease have resulted in mass mortalities of Strongylocentrotus droebachiensis along the Atlantic coast of Nova Scotia. Mortality has been virtually complete (>99%) over an area spanning ≃400 km of coastline and to a depth of at least 15m; partial mortalities have occurred over another ≃150 km. The estimated loss of biomass exceeds 200,000 tons fresh weight. The elimination of echinoids has enabled an extensive colonization of the rocky substratum by kelp and other fleshy macroalgae. Increases in the number and biomass of macroalgal species are reported for areas where mass mortality of echinoids occurred 1, 3 and 4 years previously.

Proceedings of the Fifth International Echinoderm Conference / Galway / 24-29 September 1984

Succession of macroalgae following mass mortalities of *Strongylocentrotus droebachiensis* off Nova Scotia, Canada

ROBERT E.SCHEIBLING
Biology Department, Dalhousie University, Halifax, Nova Scotia, Canada

An outbreak of disease in the fall of 1983 resulted in the elimination of Strongylocentrotus droebachiensis along 108 km (straight-line distance) of the Atlantic coast of Nova Scotia. This extends the range of complete mortality of echinoids (above 15 m depth) to 280 km of coastline between 1980 and 1983. Partial mortality (estimated as 50%) has occurred along another 220 km of coastline during this period. Mortality is estimated as 246 thousand metric tons (wet weight) or 25 billion individuals in the area of complete die-off and 96 thousand metric tons or 10 billion individuals in the area of partial die-off. The elimination of echinoids and concomitant emergence of macroalgae have occurred along different parts of the coast in different years since 1980. Macroalgal communities in areas where echinoids died off one, three and four years previously were compared to existing echinoid-dominated barren grounds and a mature kelp bed where echinoids have been rare or absent for decades. Changes in species' composition, and increases in macroalgal cover, biomass, and density and size of kelp species, characterize the succession from barren grounds to three and four year old kelp (Laminaria) beds. The greatest change occurred between one and three years following echinoid mass mortality. Macroalgal biomass increased from 0.8 kg m^{-2} at one year sites to 7.5 kg m^{-2} at three and four year sites due to the development of a dense canopy of L. longicruris and L. digitata. Macroalgal species diversity (H') was highest at one year sites; diversity was lowest at three and four year sites due to dominance by Laminaria. The estimated increase in macroalgal biomass along 280 km of coastline affected by the echinoid die-off is 2.2 million metric tons. The emergence of kelp beds in these areas reverses another major transition which occurred 1-2 decades earlier, i.e., the destructive grazing of kelp by expanding echinoid populations. Large-scale reciprocal fluctuations in echinoid and macroalgal abundance may characterize the trajectory of a dynamic system which cycles between two alternative community states: kelp beds and echinoid-dominated barren grounds. Disease, by periodically regulating echinoid abundance, may be an important mechanism underlying this cyclicity.

Proceedings of the Fifth International Echinoderm Conference / Galway / 24-29 September 1984

Base sequence complexity of *Clypeaster* and *Echinocardium* DNA

T.YANAGISAWA
Department of Biology, Tokyo Metropolitan University, Japan

DNA was prepared from sperm of Clypeaster japonicus and Echinocardium cordatum: acetone dried spermatozoa were suspended in a solution of 0.1 M EDTA, 4 mM Tris, pH 8.2, and 1% SDS, and treated overnight with Pronase P at 37°C. The digest was shaken several times with a chloroform-octanol mixture and DNA was precipitated by 2 volumes of ethanol. Prepared DNA was sheared in a Virtis 60 homogenizer to short fragments. Fragmented DNA was denatured at 100°C for 5 min and then reassociated in 0.12 M or 0.4 M phosphate buffer, pH 6.8, containing 2 mM EDTA and 0.1% SDS at 60°C or 64°C, respectively. Renatured double-stranded DNA fragments were separated from non-renatured single-stranded fragments by absorption on an hydroxyapatite column. After elution, their concentrations were optically measured. By the analyses of Cot curves obtained, the fold back component of Clypeaster DNA (330 nucleotide long) was found to be 0.08 and those of Echinocardium DNA (370 nucleotide long) was 0.02. Fractions of the fast and slow repetitive, and single copy components were 0.20 (K = 0.66 $M^{-1}.s^{-1}$), 0.20 (0.043) and 0.46 (0.0013) in Clypeaster DNA, and 0.25 (55), 0.25 (0.22) and 0.39 (0.0017) in Echinocardium DNA, respectively. The single copy components (Cot 10^2 - 10^4) were separated, labeled with ^{3}H-thymidine triphosphate by a nick translation reaction and reassociated or hybridized with Clypeaster (Clypeasteroid), Echinocardium (Spatangoida) and Hemicentrotus pulcherrimus (Echinoidea) DNA fragments. The labeled DNA was reassociated more than 80% with the DNA of the same species, but practically no hybridization was observed with DNA of different sea urchin species belonging to different orders.

4. Holothurioidea

Proceedings of the Fifth International Echinoderm Conference / Galway / 24-29 September 1984

Pelagic Holothurioidea (Echinodermata) of the northeast Atlantic

D.S.M.BILLETT
Institute of Oceanographic Sciences (NERC), Wormley, Godalming, Surrey, UK
B.HANSEN
Zoological Museum, Copenhagen, Denmark
Q.J.HUGGETT
Institute of Oceanographic Sciences (NERC), Wormley, Godalming, Surrey, UK

ABSTRACT: Seven species of deep-sea holothurians have been found in pelagic trawls taken in the northeast Atlantic. Four species, Benthodytes lingua, Benthodytes typica, Psychropotes depressa and Psychropotes longicauda, are present only as juvenile forms, but the other three species, Enypniastes diaphana, Peniagone diaphana and Scotothuria herringi are truly pelagic, although they may nevertheless feed at the sediment surface. The taxonomy, geographic distribution, bathymetric range and behaviour are discussed for each species. The pelagic holothurians are capable of linking abyssal regions with the bathypelagic zone and even surface waters, either by producing pelagic juveniles or by direct migration. Their density greater than 500m above the seabed appears to be low although their distribution can be patchy. Dense populations have been observed in continental slope areas.

1 INTRODUCTION

Several holothurian species are known to possess the ability to swim. Most are benthic animals and are capable only of short excursions from the seabed (see Margolin, 1976; Pawson, 1976, 1982; and references therein), while others may have a juvenile pelagic stage (e.g. a juvenile specimen of Psychropotes described as Nectothuria translucida Belyaev and Vinogradov, 1969). A few species, however, can be considered to be truly pelagic although they may nevertheless feed at the sediment surface. These holothurians belong to both the Elasipodida and Aspidochirotida, and are restricted to the open ocean.

The first known specimens of a truly pelagic holothurian, Pelagothuria natatrix (Elasipodida), were procured by the "Albatross" in the Gulf of Panama using a benthic trawl (Ludwig, 1894). Several specimens were observed swimming at the surface, but since the trawl did not possess an opening-closing device it was not possible to assess the depth at which the pelagic holothurians had been caught. Further specimens were obtained by the German Deep-Sea Expedition down to a depth of 1000m in the Indian Ocean using a closing pelagic net (Chun, 1900). This species has subsequently been found as deep as 6776m (Lemche et al. 1976).

Another elasipodid, Peniagone diaphana, was also collected in pelagic nets trawled in the northeast Atlantic (Hérouard, 1923). Initially only post-larval stages were described, but when the material was re-examined the presence of a larger specimen suggested that the ability to lead a pelagic life was not confined to the post-larval stage (Hansen, 1975). However, the benthopelagic lifestyle of this species was not fully appreciated until Barnes et al. (1976) witnessed, from a submersible, dense aggregations of P. diaphana living just above the seabed. In addition to Pelagothuria and Peniagone diaphana another elasipodid, Enypniastes, and two genera of the Synallactidae (Aspidochirotida), Galatheathuria Hansen and Madsen, 1956 and Scotothuria Hansen, 1978, are also truly pelagic.

Although holothurians have been seen swimming at the sea surface it is only in recent years, with the advent of acoustically telemetered pelagic nets, deep-sea photography and submersibles, that the prevalence of this group in the deep-sea pelagic ecosystem has become apparent. Despite the good work from submersibles (Barnes et al. 1976; Pawson, 1976, 1982) little is known of the distribution, abundance and behaviour of this group of enigmatic animals. This paper presents data on holothurians obtained from pelagic and benthic nets and from observations made by deep-sea photography, in the northeast Atlantic. The samples were taken between 10° and 50°N, apart from two records from off the coast of Ghana, just south of the equator.

2 METHODS

Pelagic samples were obtained with a rectangular midwater trawl (RMT 1+8) net system (Baker et al. 1973). This system uses two nets of differing mesh size (0.32mm and 4.5mm mesh) concurrently. The nets are opened and closed by a mechanical release gear which is activated acoustically, and the depth, temperature and speed of the net (from which the volume filtered can be calculated) are continuously telemetered. The net was modified in order to sample within 5m of the seabed at abyssal depths, at first using a system of mercury switches (Boxshall and Roe, 1980) and more recently by mounting a small echo-sounder above the nets. Some of the samples were obtained with a multiple rectangular midwater trawl (RMT 1+8M, Roe and Shale, 1979). Densities per unit volume were calculated only for those stations where replicate hauls were made. Generally the abundance of the holothurians was not sufficient to provide statistically reliable figures.

Pelagic holothurians close to the sediment surface were sampled by a semi-balloon otter trawl and an epibenthic sledge. The otter trawl has a wing-end spread of 8.6m while fishing and is acoustically monitored (Merrett and Marshall, 1981). The epibenthic sledge (Aldred et al. 1976) telemeters the depth, temperature and the distance run, as measured by an odometer, and possesses a time-lapse camera system (Rice et al. 1982). Densities per unit volume were calculated from the photographs.

A wide-area survey camera system taking photographs covering 80m^2 from a height of 10m above the seabed was also used (Huggett, in prep.). The camera takes a photograph every 15 seconds and is towed at approximately 0.5 knots resulting in an overlap of about 40% between adjacent frames. The height of the camera above the seabed is monitored by a near-bottom echo-sounder.

Figure 1 shows the geographic locations of the species reported in this paper. Locality data are tabulated for each species together with details of their density, depth and altitude above the sea floor. The taxonomy, bathymetric range, geographic distribution and behaviour of each species are discussed. Full particulars of stations listed in the text and tables will be found in the Appendix.

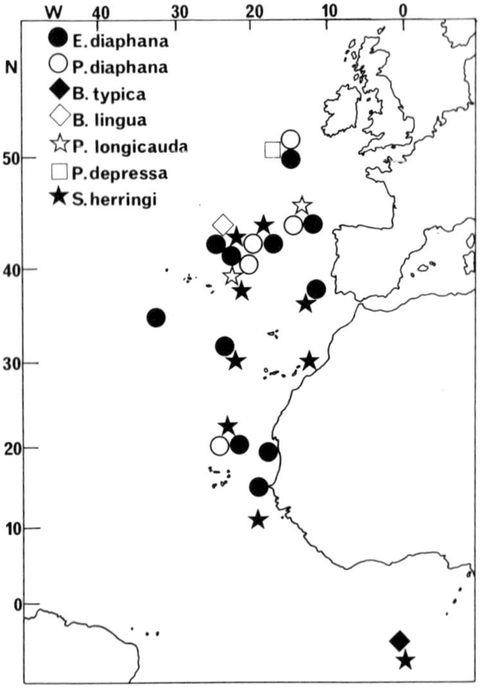

Fig. 1: Locations of samples of pelagic holothurians described in the text

3 RESULTS

3.1 Enypniastes diaphana (Gilchrist, 1920)

Taxonomy: Heding (1950) described a new order, Pelagothurioidea, to include the genera Pelagothuria Ludwig, 1894 and Enypniastes Théel, 1882, which were placed in separate families. In our opinion the separation of the two genera from one another and from the rest of the Holothurioidea should be made at a lower taxonomic level. Enypniastes differs from Pelagothuria in having a number of posterior ventrolateral podia and a thick gelatinous skin, but the genera resemble one another in that the tentacles are connected to the ring canal of the water vascular system directly, rather than to the radial canals. This feature justifies referring the two genera to a separate family, Pelagothuriidae Ludwig, 1894 within the order Elasipodida. The Pelagothuriidae is similar to the other elasipodid families not only in the absence of tentacle ampullae and respiratory trees, but also in the dorsal attachment of the mesentery throughout its length. This feature was demonstrated for Enypniastes by Ekman (1926) who pointed out that the other holothurian orders have a ventral attachment of the third, descending part of the mesentery. Within the Elasipodida, the Pelagothuriidae show affinities to the families Psychropotidae and Elpidiidae which Hansen (1975) united in a separate suborder, Psychropotina.

The species of Enypniastes are soft, fragile animals which are often badly damaged during capture. The taxonomy of the genus is therefore weakly founded. Heding (1950) regarded Planktothuria Gilchrist, 1920 and Euriplastes Koehler and Vaney, 1905 to be synonymous with Enypniastes, and reduced the number of species to two, E. eximia Théel, 1882 and E. ecalcarea Sluiter, 1901. However, Heding apparently overlooked two important taxonomic features which, at present, can be used to separate the Atlantic material, described originally as Planktothuria diaphana, from the Indo-Pacific species. The tentacle of E. diaphana has a characteristic bifid shape, similar to that found in Pelagothuria natatrix (Ludwig, 1894). All the specimens in the present material have this type of tentacle, while only one of the Indo-Pacific species, Euriplastes obscura Koehler and Vaney, 1905 seems to possess them. This species, however, has a straight intestine while in E. diaphana it is looped except in specimens smaller than 10mm. Heding listed both species as synonyms of Enypniastes ecalcarea. The synonymy of the three genera would appear to be correct but with our present knowledge, the three species should be regarded as distinct. All the Atlantic records belong to E. diaphana, including a specimen collected at the sea surface and described as Pelagothuria bouvieri by Hérouard (1906).

The preserved specimens in the present material are 4 to 50mm long. The smallest specimens already possess the adult morphology except that there are only ten tentacles (Fig. 2) and the intestine is not looped. The full complement of twenty tentacles is reached at a body length of 25mm. The anterior brim, or collar, is composed of twelve, occasionally eleven or thirteen, large papillae. Seven to eight pairs of ventrolateral podia form a continuous series around the posterior third of the body. Up to five slender podia may be found more anteriorly on each side. The papillae and podia are almost completely embedded in a thick gelatinous skin. The specimens are translucent and have a light brown pigmentation in alcohol.

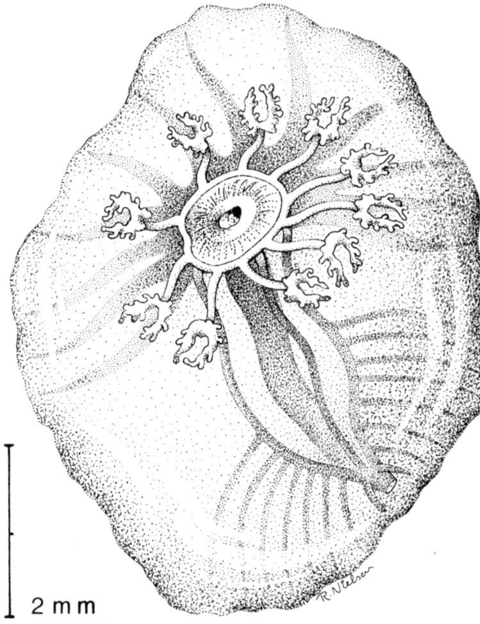

Fig. 2: Enypniastes diaphana. Juvenile specimen from St. 9541#30. Ventral view

There are no dermal deposits, nor is there a calcareous ring.

Distribution: Table 1 lists the material sampled or photographed in the northeast Atlantic. E. diaphana occurs throughout the area considered in this study. Elsewhere, it is known from the type locality off the Cape of Good Hope (Gilchrist, 1920) and may occur in the west Atlantic, where specimens identified as E. eximia were observed from a submersible (Pawson, 1976, 1982). E. diaphana has a wide bathymetric range, 995 to 4980m, but may be found uncommonly also at the sea surface (syn. Pelagothuria bouvieri, Hérouard, 1923). However, the highest incidence of E. diaphana in pelagic nets and the greatest densities occur close to the sediment surface. Densities up to 11 individuals per $10^5 m^3$ were estimated from trawls made on the abyssal plain to the north of the Cape Verde Islands (St. 9541, Table 1). Higher densities, up to 62 per $10^5 m^3$ were calculated from survey-camera photographs for the Madeira Abyssal Plain although densities of only 2 to 7 individuals per $10^5 m^3$ were estimated by the same method for the Azores-Biscay Rise. The density of E. diaphana from the Azores-Biscay Rise was also calculated from

Table 1: Sample details of Enypniastes diaphana collected by pelagic and benthic trawls and photographed close to or on the seabed. For key to gear codes see the Appendix. The number of specimens photographed are printed in brackets. The density from pelagic trawl data has been calculated only where replicate hauls occur at the same depth and in the same area. Estimates based on the presence of a single specimen are printed in brackets.

Station	Net/ Camera	Depth m	Height above seabed m	Number of specimens	Density No. 10^5m^3
8509#27	RMT 8	3500-4000	1000-1500	2	
9131#19	RMT 1	3500-3760	250-500	1	
9541#18	RMT 8	3945-4020	0-20	8	4
9541#19	RMT 1	3970-4040	0-20	1	(6)
9541#19	RMT 8	3970-4040	0-20	19	11
9541#24	RMT 8	3520-3940	100-500	2	
9541#30	RMT 1	995-1500	2500-3000	1	(9)
9541#30	RMT 8	995-1500	2500-3000	4	3
9541#31	RMT 8	1515-2000	2000-2500	1	
9756#10	RMT 8	3000-3500	500-1000	1	
9801#89	RMT 8	3500-3710	1800-2000	1	
9801#90	RMT 8	3300-3500	2000-2200	1	
9801#92	RMT 8	NET FAILED TO CLOSE AT DEPTH		1	
10379#36	RMT 8	2840-2980	0-175	1	(2)
10379#37	RMT 8	2960-2985	0-55	3	7
10651#2	RMT 8	4970-4980	80-100	1	
10656	SC	3737	0-17	(4)	4
10664	SC	3720	0-17	(20)	7
10669	SC	3700	0-17	(5)	2
10678	SC	5422	0-17	(19)	62
10693	SC	5433	0-17	(49)	22
S 126/6	SC	5418	0-17	(36)	15
10874#1	OTSB	1420-1440		132	
10890#2	BN1.5/P	2140-2200		(157)	10^5
11121#10	RMT 8M	3485-3515	575-605	1	
11121#20	RMT 8M	4006-4021	4-19	3	5
11121#21	RMT 8M	3988-4008	17-37	1	(2)
11121#23	RMT 8M	4004-4021	4-21	3	4

pelagic trawls (St. 11121). Only one specimen of E. diaphana was taken in 15 trawls fished at about 3500m depth and 600m above the seabed. However, 7 specimens were taken in 3 out of the 6 trawls fished within 50m of the seabed in the same area, giving an average density of 2 individuals per 10^5m^3. Higher densities would be expected from the survey-camera data than from the net data since many specimens occur within 1 to 3m of the seabed (Pawson, 1976), an area which the pelagic nets did not sample satisfactorily. Only one small specimen was found in the pelagic trawls made at 50°N in the Porcupine Seabight.

A dense patch of E. diaphana was discovered at the sediment surface on the continental slope off northwest Africa (St. 10890#2). A photo-transect over a distance of about 1600m was undertaken using the epibenthic sledge camera system. From 274 frames each covering an area of $2.2m^2$ and a volume of $0.56m^3$, a total of 157 E. diaphana were counted producing an estimated density of one individual per m^3. However, another sledge trawl made in the same area, at a slightly shallower depth, did not capture any specimens (St. 10890#1). A trawl taken on the continental slope at 1430m at a station further south sampled another patch of E. diaphana (St. 10874#1), indicating that E. diaphana is abundant in this highly productive upwelling area, but its distribution is patchy. The factors influencing the distribution of E. diaphana are unknown.

These data together with observations made from submersibles (Pawson, 1976, 1982) and deep-sea photography (Ohta, 1983) indicate that Enypniastes generally lives within a few metres of the sediment surface. However, a few specimens have been taken in the open ocean as high as 3000m above the seabed (Table 1). These specimens are generally quite small, 5 to 10mm long, although one 40mm long E. diaphana was taken at St. 9801#90 at a depth of 3300 to 3500m (2000-2200m above the seabed). In general, large specimens are more common within 500m of the sea floor.

Behaviour: Enypniastes swims vertically in the water column with its anterior end orientated upwards (Pawson, 1982; Ohta, 1983). Apart from the population photographed on the sediment surface at St. 10890#2, Enypniastes is usually found one to three metres above the seabed (Pawson, 1976). One specimen was seen feeding at the sediment surface by Pawson (1982) who estimated that the holothurian could fill its gut within a few minutes before it returned to the water column, where it may "hang" motionless for extended periods (Pawson, 1976). Microscopical analysis of the gut contents of specimens from St. 9541 has shown that E. diaphana does indeed feed on the superficial sediment. The contents include Foraminifera, coccoliths and diatoms, and in some respects resembles the phytodetritus collected from superficial sediments in more northerly latitudes (Billett et al. 1983). The intestines of specimens taken more than 175m above the seabed are completely empty.

Ninety-five per cent of the E. diaphana photographed at St. 10890#2 were facing in the same direction. Brisingid asteroids are normally considered to point their arms into the current (see Tyler et al. 1984), so on the basis of the orientation of brisingid asteroids, the holothurians at this station were facing downstream, as originally observed for Enypniastes by Ohta (1983).

Reproduction: Ohshima (1915) found that E. eximia possessed eggs up to 3.5mm in diameter. The maximum egg diameter in the present material is also 3.5mm, although the majority of the eggs are much smaller. The smallest specimens taken in pelagic nets therefore probably represent recently metamorphosed individuals.

3.2 Peniagone diaphana (Théel, 1882)

Taxonomy: Hansen (1975) considered that the presence of intermediary forms between Peniagone and Scotoanassa warranted the synonymy of the two genera. P. diaphana is the most extreme example of the morphological type ascribed originally to the genus Scotoanassa. It posseses a broad anterior brim, or velum, composed of four anterior papillae, overlying a buccal area which projects postero-ventrally from the body, while the posterior podia are fused into two fan-shaped lobes bordering the hind edge of the body. Observations from a submersible on the swimming activities of P. diaphana have shown that this body form is particularly well suited to a benthopelagic lifestyle (Barnes et al. 1976).

The present material consists of specimens ranging in length between 2 and 110mm. A specimen estimated to be 25cm long was photographed in the northeast Atlantic (Elliott and Thorpe, 1983) but the largest captured specimens are less than half this size. Hansen (1975) reported some large tentacles in a trawl catch from the Tasman Sea which he suggested may have come from an unusually large specimen. The smallest specimens collected are identical to large specimens in external morphology, but juveniles less than 7mm long have smaller and more delicate dermal deposits (Fig. 3). Such a change in the size and robustness of the dermal deposits associated with increasing body size has not been noted in elasipodids before, possibly because very few juvenile specimens have been described.

Distribution: Table 2 lists the material sampled by pelagic and benthic nets. P. diaphana is thought to have a cosmopolitan distribution (Gage et al. in press); it has been sampled in mid-oceanic areas and may reach high densities close to the continental slope (Barnes et al. 1976). It has a wide bathymetric distribution, but unlike Enypniastes it has not been taken shallower than 1900m. P. diaphana has been found up to 1500m above the seabed (Table 2) but still in the abyssopelagic

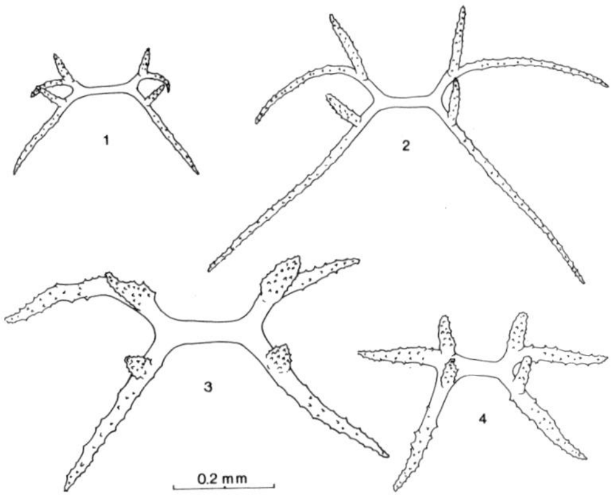

Fig. 3: Peniagone diaphana dermal deposits. 1. Specimen 6mm long (St. 50603#2), 2. Specimen 14mm long (St. 9131#18), 3-4. Dorsal and ventral deposits from a 60mm long specimen (St. 50603#2). The small specimens show no difference between dorsal and ventral deposits

zone. Repeated sampling in the region of the Azores-Biscay Rise (St. 11121) shows that P. diaphana occurs in higher densities close to the seabed. P. diaphana was collected in 5 of the 15 nets fished at about 600m above the seabed (average density: 1 individual per $10^5 m^3$), whereas it occurred in all 6 samples taken close to the sea floor (average density: 19 individuals per $10^5 m^3$). The unusually large catch taken on the Porcupine Abyssal Plain (St. 50515) indicates that it has a patchy distribution, another feature noted by Barnes et al. (1976).

Behaviour: The guts of most of the P. diaphana seen by Barnes et al. (1976) were distended with material probably of sedimentary origin. However, because P. diaphana was never seen to be in contact with the sediment, the authors suggested that it feeds on resuspended detritus. The gut contents of specimens from off northwest Africa (St. 9131#18) and from the Porcupine Seabight (St. 50603) were examined microscopically and were found to contain superficial sediment containing pelagic Foraminifera, gastropods, coccoliths and diatoms. The stomach contents of specimens from the Porcupine Seabight were similar to phytodetrital material collected as a superficial layer overlying the sediment surface (Billett et al. 1983). In most specimens sampled by pelagic nets the anterior part of the gut was devoid of sediment, while the hind gut was often full. Conversely, in

Table 2: Sample details of Peniagone diaphana collected by pelagic and benthic trawls. For key to gear codes see the Appendix. The density in pelagic trawls has been calculated only where replicate hauls have been made at the same depth and in the same area. Estimates based on the presence of a single specimen are placed in brackets.

Station	Net	Depth m	Height above seabed m	Number of specimens	Density No. $10^5 m^3$
8509#27	RMT 8	3500-4000	1000-1500	1	
8509#43	RMT 1	1500-4650	>450	1	
9131#18	RMT 1	3760-3920	100-250	1	(9)
9131#18	RMT 8	3760-3920	100-250	6	5
9541#22	RMT 8	3740-3870	20-100	1	
9801#91	RMT 8M	4300-4520	1000-1200	1	
50603#2	RMT 8	3720-3940	60-260	39	101
50603#3	RMT 8	3700-3900	50-250	12	33
50605#2	RMT 8	2640-2750	5-25	1	
11121#8	RMT 8M	3490-3530	560-600	1	
11121#9	RMT 8M	3485-3580	510-605	1	
11121#11	RMT 8M	3480-3515	575-610	1	
11121#17	RMT 8M	3490-3520	570-600	1	
11121#18	RMT 8M	3480-3520	570-610	6	
11121#20	RMT 8M	4006-4021	4-19	7	12
11121#21	RMT 8M	3988-4008	17-37	11	20
11121#22	RMT 8M	3976-3991	34-39	16	33
11121#23	RMT 8M	4004-4021	4-21	18	26
11121#24	RMT 8M	3991-4007	18-34	4	6
11121#25	RMT 8M	3941-3991	34-84	1	no data
9638#2	OTSB	4043-4104		4	
9640#1	OTSB	3749-3757		4	
9753#7	BN1.5	1942-1942		1	
9756#3	OTSB	4080-4156		7	
9756#5	OTSB	4012-4020		8	
9756#9	BN1.5	4069-4039		1	
50511	OTSB	2420-2460		1	
50515	OTSB	4505-4515		339	
50601#1	BN1.5	4000-4000		12	
50604#1	BN1.5	3490-3550		3	
50711#1	OTSB	4795-4780		6	
10115#1	BN1.5	3900-3950		1	
50812#1	BN1.5	4080-4100		1	
51109#1	BN1.5	3960-3940		1	

specimens taken in benthic trawls, the anterior gut was often full, suggesting that P. diaphana, like Enypniastes, feeds at the sediment surface. The rapid feeding rate of Enypniastes led Pawson (1982) to believe that this holothurian need only remain on the bottom for a few minutes and could explain why only one specimen was seen feeding at the sediment surface. Similarly, the absence of P. diaphana feeding at the sediment surface (Barnes et al. 1976) does not necessarily indicate that P. diaphana feeds on suspended particles.

Reproduction: The reproductive biology of P. diaphana has been described by Tyler et al. (this volume).

3.3 Benthodytes typica Théel, 1882

Distribution: B. typica has a cosmopolitan distribution and is usually found between 1873 and 4700m, although benthic specimens in the western Atlantic have been found between 315 and 1400m (Deichmann, 1954). The only pelagic specimen known is a 13mm long juvenile sampled by the "Michael Sars" North Atlantic Deep Sea Expedition at a depth of 1400m and at a height of about 1600m above the seabed (Grieg, 1921). The present material consists of another juvenile, 7mm long, sampled off the coast of Ghana at a depth of 1205 to 1300m and some 3400m above the sea floor (St. 10523 #18), and eight specimens, 45 to 70mm long, taken in three near-bottom pelagic trawls close to the Mid-Atlantic Ridge (St. 10379). The irregular bottom topography in this area caused problems during the fishing of the trawls and all three samples contained sediment and fauna of benthic origin. Therefore, it is uncertain whether these B. typica specimens were indeed pelagic or not.

Behaviour: Adult B. typica seem unable to swim although other species of Benthodytes, notably B. sanguinolenta, are able to swim for a short time by undulating movements of the body (Pawson, 1976).

Reproduction: B. typica, like Psychropotes, produces large eggs up to 3mm in diameter. Therefore the 7mm specimen caught is a recently metamorphosed individual.

3.4 Benthodytes lingua R. Perrier, 1896

Taxonomy: The material consists of one juvenile specimen, 12mm long. The deposits, which occur solely on the dorsal surface, are slender crosses with bipartite apophyses. There are ten tentacles and three pairs of dorsal papillae.

Distribution: B. lingua occurs in both the North and South Atlantic between 860 and 3192m. The present specimen is the first to be taken in a pelagic net and was sampled at St.11121#10 at a depth of 3485 to 3515m and between 575 to 605m above the seabed.

Behaviour: Like B. typica, adults of this species, which can grow to a considerable size, are incapable of swimming (Pawson, 1982).

Table 3: Sample details of juvenile Psychropotes longicauda collected by pelagic trawls. For key to gear codes see the Appendix.

Station	Net	Depth m	Height above seabed m	Number of Specimens
8509#27	RMT 8	3500-4000	1000-1500	1
11121#1	RMT 8M	3470-3580	510-620	1
11121#2	RMT 8M	3470-3565	525-620	1
11121#8	RMT 8M	3490-3530	560-600	2
11121#9	RMT 8M	3485-3580	510-605	1
11121#18	RMT 8M	3480-3520	570-610	1
11121#21	RMT 8M	3988-4008	17-37	1

3.5 Psychropotes longicauda Théel, 1882

Taxonomy: Belyaev and Vinogradov (1969) described a new genus and species Nectothuria translucida, based on a specimen taken between 4940 and 5930m in the Kurile-Kamchatka Trench, some 2000 to 3000m above the sea floor. The specimen was completely transparent and excluding the length of the large, unpaired dorsal appendage, measured 35mm in length. Hansen (1975) regarded it to be a juvenile P. longicauda but in comparison to the present specimens it is possible that N. translucida is the juvenile form of another species of Psychropotes. The present material contains 8 specimens of juvenile P. longicauda (Table 3) ranging in length between 13 and 23mm. Unlike the adult form, the body wall is thin and transparent and has a reddish hue, particularly around the buccal area. The unpaired, dorsal appendage is transparent and gelatinous and in one case is as large as the body itself. All the juveniles have 10 tentacles, as opposed to the adult complement of 18, and the ventrolateral podia are fused at their base and not free for their whole length as in adult specimens. However, the cross-shaped body deposits measuring 0.8 to 1.0mm across, are the same as those found in small adult specimens.

Similar differences in form were also found between juvenile and adult specimens of a closely related species, Psychropotes semperiana, taken in a benthic trawl on the Seine Abyssal Plain, St. 9035#1. A juvenile measuring 35mm in length has a dorsal appendage about twice as long as the body (Fig. 4). The body and appendage are both gelatinous and the colouration was the same as for juvenile P. longicauda. The identification was based on the form of the body wall deposits which were characteristic of P. semperiana. It is possible that juvenile P. semperiana, like those of P. longicauda, occur pelagically.

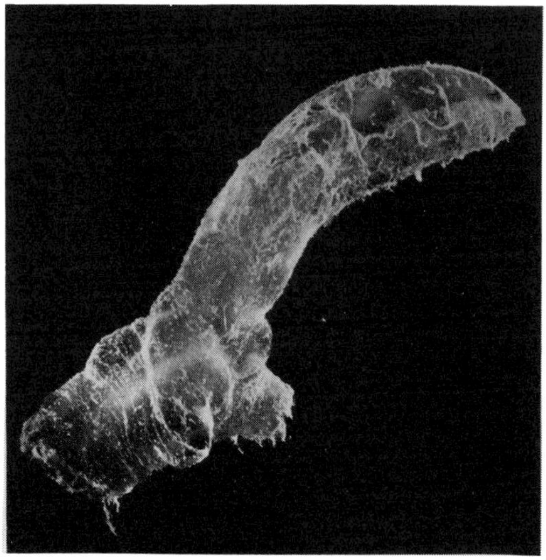

Fig. 4: Psychropotes semperiana. Juvenile specimen from St. 9035#1. The 35mm long body, slightly contracted, has a large unpaired dorsal appendage about 70mm long (right-hand side)

Distribution: P. longicauda is a truly cosmopolitan, abyssal, benthic holothurian. Swimming is only known to occur in juvenile forms. Seven of the eight juvenile specimens sampled pelagically in the north-east Atlantic occurred greater than 500m above the seabed, but still within the abyssopelagic zone (Table 3). From 5 trawls out of 15 sampled at about 600m above the seabed in the region of the Azores-Biscay Rise (St. 11121), 6 specimens of P. longicauda were sampled giving an average density of 1 individual per $10^5 m^3$.

Behaviour: The large gelatinous unpaired dorsal appendage of juvenile P. longicauda may be an adaptation to a pelagic life. The gut in all specimens is empty and the structure of the tentacles suggests that it is unlikely that P. longicauda feeds on particles in the water column.

Reproduction: P. longicauda has the largest eggs known in holothurians, 4.4mm diameter. Hansen (1975) suggested that the large egg size could be an adaptation to a long pelagic development of the holothurian.

3.6 Psychropotes depressa (Théel, 1882)

Distribution: A single, juvenile specimen, 9.6mm long, was taken at St. 9756#2, at a depth of 1000 to 1500m, and 2500 to 3000m above the seabed. Adult P. depressa are the dominant benthic invertebrates in a narrow bathymetric band between 2300 and 2500m in this area.

Behaviour: Adult P. depressa are able to swim for short periods by flexing the whole body (= Euphronoides, Pawson, 1976). The present specimen is the first evidence of a pelagic juvenile stage for this species.

3.7 Scotothuria herringi Hansen, 1978

Taxonomy: Fifteen specimens of this aspidochirotid holothurian have been taken in the northeast Atlantic, including the three specimens upon which the species description was based (Hansen, 1978). Although several of the specimens that have been caught recently are in a poor condition all have cross-shaped dermal deposits typical of S. herringi. The crosses are densely crowded in the superficial layer of the skin but this layer may be almost completely worn off in some specimens. However, other types of spicule frequently occur in the deeper layers of the skin and in the walls of the gonads, intestine and respiratory trees (Fig. 5). These consist of slender, irregular, spinous crosses, tripartite bodies and rods. The ends of the arms and rods are bifurcate or have an enlargement derived from a group of dichotomous divisions. Although these inner spicules are numerous in some specimens they are absent in others, and were not reported in the type material. A re-examination of the paratype held at the Zoological Museum, Copenhagen, failed to reveal their presence. They are not known to occur in the other synallactid genus Galatheathuria Hansen and Madsen, 1956.

Six of the fifteen specimens are intact. The three original specimens used for the species description ranged in length from 48 to 80mm (Hansen, 1978). The three more recent specimens taken at Stns 10536#1, 10651#2 and 10927#1, after preservation in alcohol, are 55mm, 55mm and 70mm long respectively. However, the latter specimen was about 150mm long when fresh (P.J. Herring, personal communication).

The species has hitherto been reported only from the northeast Atlantic. However, two specimens have now been found in the Galathea material from a bottom trawl taken off the coast of Kenya (St. 238, 3960m). The specimens measuring 60 and 80mm in length are in rather poor condition but have both the inner and outer deposits that are typical of the Atlantic specimens. In addition, a specimen and a detached piece of gonad were taken by the Swedish Deep-Sea Expedition at two separate stations in the equatorial Atlantic. Madsen (1953) referred them to Psychropotes minutus Koehler and Vaney, 1905, but expressed reservations. The spicules as illustrated in the paper, and a re-examination of the original deposit preparations confirm that they are identical to the inner spicules of S. herringi. A specimen taken in the Bay of Bengal and described by Koehler and Vaney (1905) as a new genus and species, Dendrothuria similis, had similar inner deposits but none of the outer crosses of the Scotothuria type were found. There were also crosses of the Paelopatides type although it is possible that these were contaminants. Further samples from this area are needed before a synonymy between Dendrothuria and Scotothuria can be considered. Both genera are similar in external morphology and have dendritic tentacles and a large conical pharynx.

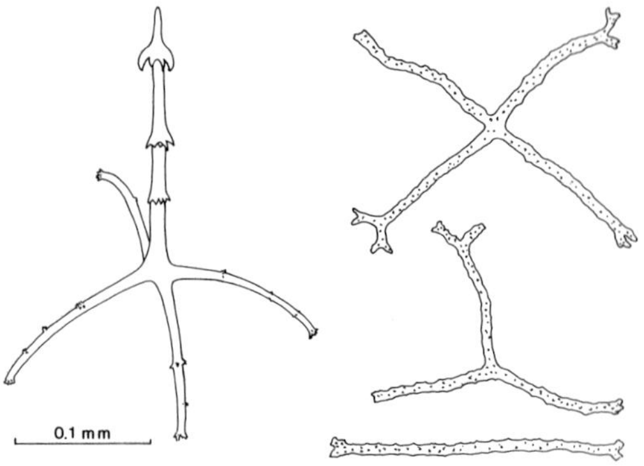

Fig. 5: Scotothuria herringi. Dermal deposits from a fragmented specimen (St. 10651#1). One of the characteristic crosses from the superficial skin and three irregular deposits from a deeper layer

Distribution: The eastern Atlantic and Indian Oceans, 1250 to 4980m. S. herringi was sampled in a bottom trawl in the Indian Ocean, but all the specimens from the northeast Atlantic have been sampled pelagically (Table 4). One of the specimens was taken off the coast of Ghana (St. 10536#1) while the rest were sampled between 10° and 42°N. S. herringi has been found just above the sediment surface (St. 9541#19) but in the same area was also found at least 1500m above the seabed. It has been sampled as shallow as 1250 to 1500m depth and in two instances 3600 to 3900m above the seabed.

Table 4: Sample details of Scotothuria herringi collected by pelagic trawls. For key to gear codes see the Appendix.

Station	Net	Depth m	Height above seabed m	Number of specimens
7824#36	RMT 8	1250-1500	1000-1250	1
7856#52	RMT 8	1250-1500	3650-3900	1
9022#1	RMT 8	2000-2200	0-100	1
9131#21	RMT 8	3000-3500	500-1000	1
9541#19	RMT 1	3970-4040	0-20	1
9541#27	RMT 8	2000-2500	1500-2000	1
9801#90	RMT 8	3300-3500	2000-2200	1
10536#1	RMT 8	2370-3240	1600-2400	1
10651#1	RMT 8M	4550-4980	100-500	1
10651#2	RMT 8M	4970-4980	80-100	2
10927#1	RMT 8	1430-1750	3600-3900	1
11121#22	RMT 8M	3976-3991	34-49	1
11121#23	RMT 8M	4004-4021	4-21	2

Behaviour: Hansen and Madsen (1956) suggested that Galatheathuria swims by undulating the lateral brim in a similar manner to the motion used by the cuttle-fish Sepia. The specimen from St. 10927#1 was sampled by a midwater trawl fitted with a closing cod end which retains the animal in its in situ water. This specimen was kept alive for about an hour during which weak swimming movements, like those proposed for Galatheathuria, were observed (P.J. Herring, personal communication). The guts from three intact specimens have been examined. They are large and are probably capable of great distension. The intestine of a specimen taken 80-100m (St. 10651#2) above the sediment surface, was packed with sediment, but the intestines of the two other specimens sampled greater than 1600m above the seabed (Stns 10536#1 and 10927#1) were completely empty. There was no evidence that these holothurians were feeding on detrital material in the water column. It is known, however, that Pelagothuria natatrix, sampled at a depth of 20 to 1000m and 4000 to 5000m above the seabed, fed on pelagic organisms such as diatoms, Radiolaria and Globigerina (Chun, 1900). Parasitic nematodes were found within the coelom of the specimens of S. herringi from Stns 10536 and 10927.

Reproduction: The ovaries of three specimens of S. herringi were examined in situ microscopically. In the specimen from St. 10927#1, there was a distinct bimodal distribution of egg size with a maximum diameter of 1.8mm. Together with oocytes less than 300µm there were numerous large eggs all of a similar size with an average diameter of 1.6mm. This suggests that S. herringi breeds periodically.

3.8 Paelopatides grisea R. Perrier, 1902

Behaviour: Although P. grisea has never been taken in pelagic nets, it is obvious from photographs taken by the epibenthic sledge camera (St. 9775#3) that it is capable of swimming by undulating the lateral brim (Fig. 6). This type of swimming has been observed from submersibles in the western Atlantic for P. gigantea (Pawson, 1976) which is probably synonymous with P. grisea. The swimming motion is similar to that employed by S. herringi.

Fig. 6: Paelopatides grisea is able to swim by undulating its lateral brim of fused tube feet

4 DISCUSSION

There are three groups of deep-sea holothurians with the ability to swim. Representatives of each group have been found in the northeast Atlantic. First, the benthic holothurian Paelopatides grisea, is able to swim for short periods by undulations of its lateral brim. Second, juveniles of Benthodytes typica, Benthodytes lingua, Psychropotes depressa and Psychropotes longicauda are pelagic although the adults are benthic. These species produce large eggs, but it is not known whether development is direct or through a larval stage. The presence of small juvenile specimens in the water column suggests that there may be a lecithotrophic larva with a long pelagic or "teleplanic" stage. B. typica, P. depressa and P. longicauda are amongst the few deep-sea holothurians that have a world-wide distribution, and so the pelagic dispersal of larvae may result in genetic exchange between oceanic basins. Third, three species that are truly pelagic occur in the northeast Atlantic. All three species have a wide bathymetric range but are generally found close to the sediment surface where they feed on the superficial sediment.

Peniagone diaphana is generally found close to the sediment surface and never more than 1500m above the seabed. In the two instances where it did occur at this height above the sea floor, it was still within the abyssopelagic zone. Enypniastes diaphana, in contrast, although occurring more commonly near the seabed, may also be found in surface waters (Hérouard, 1923), albeit very rarely. Five specimens were found between 995 and 1500m in the present study off the coast of northwest Africa. The bathymetric distribution of Enypniastes encompasses therefore the whole water column. This is similar to the Indo-Pacific species, Pelagothuria natatrix, which has been photographed at 6776m in a deep-sea trench (Lemche et al. 1976) but has been observed also at the sea surface (Ludwig, 1894). It is not known, however, if the records of pelagic holothurians at shallow depths are the result of migrations from the deep-sea. Most of the specimens of Enypniastes sampled greater than 500m above the seabed were small, and not much larger than the maximum egg size, suggesting that these specimens are post-larval stages. The adults do not appear to undergo extensive migrations. However, the presence of large Scotothuria herringi, a holothurian better suited to feeding at the sediment surface, at depths as shallow as 1250m and at least 3600m above the seabed, suggests that active migrations do take place in this species, although it is not clear what purpose they have. It is possible that migration is related to the reproductive cycle which appears to be periodic in this species.

Both Enypniastes diaphana and Peniagone diaphana have patchy distributions and may occur in aggregations. Densities up to one individual per m^3 were estimated for one population of E. diaphana seen feeding at the sediment surface on the continental slope off northwest Africa. Lower densities, 2 to 62 individuals per $10^5 m^3$ were found for abyssal plain areas. Enypniastes diaphana appears to be more abundant close to the continental slope.

The holothurians described here are a faunal group that is capable of linking deep abyssal areas with the bathypelagic zone and even surface waters, either by direct migration or by producing pelagic juveniles. Their density at a height greater than 500m above the seabed appears to be low. However, their distribution can be patchy and dense assemblages have been seen close to the sediment surface in continental slope areas.

Acknowledgements: We would like to thank Dr. M.V. Angel and Dr. A.L. Rice for their criticisms of the manuscript, Mrs P. Talbot for her help in the presentation of the paper, Mr. R. Nielsen for the drawing of E. diaphana and our colleagues at the Institute of Oceanographic Sciences for their help during sampling.

5 REFERENCES

Aldred, R.G., Thurston, M.H., Rice, A.L. and Morley, D.R. 1976, An acoustically monitored opening and closing epibenthic sledge. Deep-Sea Res. 23: 167-174.

Baker, A. de C., Clarke, M.R. and Harris, M.J. 1973, The N.I.O. combination net (RMT 1+8) and further developments of rectangular midwater trawls. J. mar. biol. Ass. U.K. 53: 167-184.

Barnes, A.T., Quetin, L.B., Childress, J.J. and Pawson, D.L. 1976, Deep-sea macroplanktonic sea cucumbers: suspended sediment feeders captured from deep submergence vehicle. Science 194: 1083-1085.

Belyaev, G.M. and Vinogradov, M.E. 1969, A new pelagic holothurian (Elasipoda, Psychropotidae) from the abyssal of the Kurilo-Kamchatka Trench. Zool. Zh. 48: 709-716 (In Russian, English summary).

Billett, D.S.M., Lampitt, R.S., Rice, A.L. and Mantoura, R.F.C. 1983, Seasonal sedimentation of phytoplankton to the deep-sea benthos. Nature, Lond. 302: 520-522.

Boxshall, G.A. and Roe, H.S.J. 1980, The life history and ecology of the aberrant bathypelagic genus Benthomisophria Sars, 1909 (Copepoda: Misphrioida). Bull. Br. Mus. nat. Hist. (Zool.) 38: 9-41.

Chun, C. 1900, Aus den Tiefen des Weltmeeres. Jena: Gustav Fischer Verlag, 549p.

Deichmann, E. 1954, The holothurians of the Gulf of Mexico. Fishery Bull. Fish Wildl. Serv. U.S. 55: 381-410.

Ekman, S. 1926, Systematisch-phylogenetische studien über Elasipoden und Aspidochiroten. Zool. Jb. (Anat.) 47: 429-540.

Elliott, A.J. and Thorpe, S.A. 1983, Benthic observations on the Madeira Abyssal Plain. Oceanologica Acta 6: 463-466.

Gage, J.D., Billett, D.S.M., Jensen, M. and Tyler, P.A. in press, Echinoderms of the Rockall Trough and adjacent areas. 2. Echinoidea and Holothurioidea. Bull. Br. Mus. nat. Hist. (Zool.).

Gilchrist, J.D.F. 1920, Planktothuria diaphana gen. et sp. n. Q. Jl microsc. Sci. 64: 373-382.

Grieg, J.A. 1921, Echinodermata. Rep. scient. Results Michael Sars N. Atlantic deep Sea Exped. 3 (2): 1-47.

Hansen, B. 1975, Systematics and biology of the deep-sea holothurians. Part 1. Elasipoda. Galthea Rep. 13: 1-262.

Hansen, B. 1978, Scotothuria herringi, a new genus and species of bathypelagic holothurians (Holothurioidea, Aspidochirota, Synallactidae). Steenstrupia 5: 33-39.

Hansen, B. and Madsen, F.J. 1956, On two bathypelagic holothurians from the South China Sea. Galathea Rep. 2: 55-59.

Heding, S.G. 1950, Über die Planktothuria der Deutschen Tiefsee-Expedition, nebst einigen Bemerkungen über die Systematik der pelagischen Holothurien. Zool. Anz. 145: 111-118.

Hérouard, E. 1906, Sur Pelagothuria bouvieri (Holothurie pélagique nouvelle) recueillie pendant la campagne du yacht Princesse-Alice en 1905. Bull. Mus. oceanogr. Monaco, 60: 1-6.

Hérouard, E. 1923, Holothuries provenant des campagnes des yachts Princesse-Alice et Hirondelle II (1898-1915). Résult. Camp. scient. Prince-Albert I, 66: 1-163.

Huggett, Q.J. in preparation, The distribution of glacial erratics in the north-east Atlantic: final report.

Koehler, R. and Vaney, C. 1905, Holothuries recueillis par l'Investigator dans l'ocean Indien. I. Les Holothuries de Mer profonde. Calcutta: Indian Museum, 126p.

Lemche, H., Hansen, B., Madsen, F.J., Tendal, O.S. and Wolff, T. 1976, Hadal life as analyzed from photographs. Vidensk. Meddr dansk naturh. Foren. 139: 263-336.

Ludwig, H. 1894, The Holothurioidea. Reports of an exploration off the west coasts of Mexico, Central and South America, and off the Galapagos Islands... "Albatross" 1891. Mem. Mus. comp. Zool. Harv. 17 (3): 1-183.

Madsen, F.J. 1953, Holothurioidea. Rep. Swed. deep Sea Exped. 2: 149-173.

Margolin, A.S. 1976, Swimming of the sea cucumber Parastichopus californicus (Stimpson) in response to sea stars. Ophelia 15: 105-114.

Merrett, N.R. and Marshall, N.B. 1981, Observations on the ecology of deep-sea bottom-living fishes collected off northwest Africa (08°-20°N). Prog. Oceangr. 9: 185-244.

Ohshima, H. 1915, Report on the holothurians collected by the United States Fisheries Steamer "Albatross" in the northwestern Pacific during the summer of 1906. Proc. U.S. natn. Mus. 48: 213-291.

Ohta, S. 1983, Photographic census of large-sized benthic organisms in the bathyal zone of Suruga Bay, Central Japan. Bull. Ocean res. Inst. 15: 1-244.

Pawson, D.L. 1976, Some aspects of the biology of deep-sea echinoderms. Thalassia jugosl. 12: 287-293.

Pawson, D.L. 1982, Deep-sea echinoderms in the Tongue of the Ocean, Bahama Islands: A survey, using the research submersible "Alvin". Mem. Aust. Mus. No. 16, 129-145.

Perrier, R. 1896, Sur les Élasipodes recueillis par le "Travailleur" et le "Talisman". C.r. hebd. Séanc. Acad. Sci. Paris 123: 900-903.

Perrier, R. 1902, Holothuries. Expéd. scient. Travailleur Talisman. 273-554.

Rice, A.L., Aldred, R.G., Darlington, E. and Wild, R.A. 1982, The quantitative estimation of the deep-sea megabenthos: a new approach to an old problem. Oceanologica Acta 5: 63-72.

Roe, H.S.J. and Shale, D.M. 1979, A new multiple rectangular midwater trawl (RMT 1+8M) and some modifications to the Institute of Oceanographic Sciences' RMT 1+8. Mar. Biol. 50: 283-288.

Sluiter, C.Ph. 1901, Die Holothurien der Siboga-Expedition. Siboga-Exped. 44: 1-142.

Théel, Hj 1882, Report on the Holothurioidea dredged by H.M.S. Challenger during the years 1873-1876. Part 1. Rep. scient. Results Voy. Challenger (Zool.) 4 (13): 1-176.

Tyler, P.A., Pain, S.L., Gage, J.D. and Billett, D.S.M. 1984, The reproductive biology of forcipulate seastars (Asteroidea: Echinodermata) from the N.E. Atlantic Ocean. J. mar. biol. Ass. U.K. 64: 587-601.

Tyler, P.A., Muirhead, A., Gage, J.D. and Billett, D.S.M. in press, Gametogenic strategies in deep-sea echinoids and holothurians from the N.E. Atlantic. This volume.

APPENDIX: Station details of all samples listed in Tables and mentioned in the text. The height sampled above the seabed is not included for benthic trawls. Gear codes: RMT 1 (IOS Rectangular Midwater Trawl, mouth area $1m^2$, mesh 320µm); RMT 8 (IOS Rectangular Midwater Trawl, mouth area $8m^2$, mesh 4.5mm). These two nets are trawled concurrently. RMT 1+8 denotes stations where pelagic holothurians occurred in both nets; RMT 8M (IOS Multiple Rectangular Midwater Trawl, mouth area $8m^2$, mesh 4.5mm); BN2.4 (IOS Epibenthic Sledge, mouth area $2.4m^2$, mesh 4.5mm); BN1.5 (IOS Epibenthic Sledge, mouth area $1.5m^2$, mesh 4.5mm); BN1.5/P (IOS Epibenthic Sledge (Photosledge) trawled with camera and without nets); OTSB 14 (Semi-balloon otter trawl, headline length 14m); SC (IOS Survey Camera); HOT (Bottom trawl). See text for gear references.

STATION	POSITION N	W	GEAR	DATE	DEPTH m	HEIGHT ABOVE SEABED m
7824#36	11°03.3'	19°48.3'	RMT 8	9:iii:72	1250-1500	1000-1250
7856#52	29°56.6'	23°05.3'	RMT 8	5:iv:72	1250-1500	3650-3900
8509#27	44°14.4'	12°42.7'	RMT 8	29:iv:74	3500-4000	1000-1500
8509#43	43°33.8'	12°58.4'	RMT 8	1:v:74	1500-4650	>450
9022#1	30°12.3'	11°41.2'	RMT 8	19:viii:76	2000-2200	0-100
9035#1	34°06.0'	11°55.5'	BN2.4	23:viii:76	4453-4457	
9131#18	20°07.7'	21°32.9'	RMT 1+8	21:xi:76	3920-3760	100-250
9131#19	20°14.4'	21°47.0'	RMT 1	21:xi:76	3760-3500	250-500
9131#21	20°16.4'	21°38.1'	RMT 8	22:xi:76	3500-3000	500-1000
9541#18	20°18.5'	21°42.1'	RMT 1+8	18:iv:77	4020-3945	0-20
9541#19	20°19.7'	21°51.3'	RMT 1+8	18:iv:77	4040-3970	0-20
9541#22	20°08.9'	21°25.1'	RMT 8	20:iv:77	3870-3740	20-100
9541#24	20°20.1'	21°45.0'	RMT 8	21:iv:77	3940-3520	100-500
9541#27	20°09.9'	21°26.0'	RMT 1+8	22:iv:77	2000-2500	1500-2000
9541#30	20°01.8'	21°19.8'	RMT 1+8	22:iv:77	995-1500	2500-3000
9541#31	20°00.6'	21°35.8'	RMT 8	23:iv:77	1515-2000	2000-2500
9638#2	49°50.2'	14°07.3'	OTSB 14	9:xi:77	4043-4104	
9640#1	50°03.2'	13°50.6'	OTSB 14	13:xi:77	3749-3757	
9753#7	50°54.4'	12°10.9'	BN1.5	8:iv:78	1942-1942	
9756#2	49°52.8'	14°10.3'	RMT 8	10:iv:78	1000-1500	2500-3000
9756#3	49°48.0'	14°14.8'	OTSB 14	11:iv:78	4080-4156	
9756#5	49°49.3'	14°05.7'	OTSB 14	12:iv:78	4012-4020	
9756#9	49°47.1'	14°01.5'	BN1.5	13:iv:78	4039-4069	
9756#10	49°47.1'	14°04.4'	RMT 8	14:iv:78	3000-3500	500-1000
9756#11	49°49.7'	14°04.7'	RMT 8	14:iv:78	4000-4012	5-20
9775#3	50°56.8'	12°22.4'	BN1.5	22:iv:78	2012-2019	
9801#89	41°52.5'	17°03.6'	RMT 8M	20:v:78	3500-3710	1800-2000
9801#90	41°54.4'	17°08.5'	RMT 8M	20:v:78	3300-3500	2000-2200
9801#91	42°01.5'	17°08.7'	RMT 8M	21:v:78	4300-4520	1000-1200
9801#92	42°04.2'	17°02.4'	RMT 8M	21:v:78	NET FAILED TO	CLOSE AT DEPTH
50511#1	50°32.4'	13°01.4'	OTSB 14	4:vi:79	2405-2435	
50515#1	49°43.9'	15°04.6'	OTSB 14	6:vi:79	4505-4515	
50603#1	49°46.2'	14°01.5'	BN1.5	2:vii:79	4000-4000	
50603#2	49°53.9'	14°08.9'	RMT 8	3:vii:79	3720-3940	60-260
50603#3	49°53.4'	14°04.1'	RMT 8	3:vii:79	3700-3900	50-250
50604#1	50°06.1'	13°53.0'	BN1.5	4:vii:79	3490-3550	
50605#2	50°06.5'	13°12.0'	RMT 8	5:vii:79	2640-2750	5-25
10115#1	49°46.3'	13°56.0'	BN1.5	11:ix:79	3900-3950	
50711#1	49°53.0'	15°36.0'	OTSB 14	18:x:79	4780-4795	
50812#1	49°45.4'	14°10.3'	BN1.5	3:viii:80	4080-4100	
10379#36	35°00.3'	33°01.3'	RMT 8M	14:iv:81	2840-2980	0-175
10379#37	34°58.1'	32°56.5'	RMT 8M	14:iv:81	2960-2985	0-55
10379#38	34°57.0'	32°53.3'	RMT 8M	14:iv:81	2840-2980	0-30
51109#1	49°50.7'	14°02.1'	BN1.5	26:v:81	3940-3960	
10523#18	5°05.1'S	0°29.5'E	RMT 8M	13:v:82	1205-1300	3300-3400
10536#1	7°08.6'S	0°10.4'W	RMT 8M	20:v:82	2370-3240	1600-2400

STATION	POSITION N	W	GEAR	DATE	DEPTH m	HEIGHT ABOVE SEABED m
10651#1	37°36.0'	10°55.0'	RMT 8M	24:ii:83	4550-4980	100-500
10651#2	37°36.0'	10°55.0'	RMT 8M	24:ii:83	4970-4980	80-100
10656	41°52.5'	23°50.0'	SC	5:iii:83	3737	0-17
10664	41°29.0'	23°14.4'	SC	8:iii:83	3720	0-17
10669	42°03.0'	23°53.0'	SC	11:iii:83	3700	0-17
10678	31°23.8'	24°34.6'	SC	17:iii:83	5422	0-17
10693	31°39.9'	24°48.3'	SC	23:iii:83	5433	0-17
S126/6	31°09.9'	23°48.5'	SC	21:iii:82	5418	0-17
10874#1	14°50.8'	17°50.6'	OTSB 14	8:viii:83	1420-1440	
10890#1	19°21.0'	17°41.6'	BN1.5	15:viii:83	2020-2030	
10890#2	19°20.7'	17°48.4'	BN1.5/P	16:viii:83	2140-2200	
10927#1	37°47.4'	21°14.8'	RMT 8	3:ix:83	1430-1750	3600-3900
11121#1	41°28.4'	21°53.5'	RMT 8M	28:v:84	3470-3580	510-620
11121#2	41°33.1'	21°52.7	RMT 8M	28:v:84	3470-3565	525-620
11121#8	41°54.6'	21°44.5'	RMT 8M	30:v:84	3490-3530	560-600
11121#9	41°50.3'	21°44.2'	RMT 8M	30:v:84	3485-3580	510-605
11121#10	41°41.7'	21°45.9'	RMT 8M	30:v:84	3485-3515	575-605
11121#11	41°47.1'	21°46.0'	RMT 8M	30:v:84	3480-3515	575-610
11121#17	41°48.2'	21°40.4'	RMT 8M	3:vi:84	3490-3520	570-600
11121#18	41°52.1'	21°43.2'	RMT 8M	3:vi:84	3480-3520	570-610
11121#20	41°33.0'	21°44.1'	RMT 8M	4:vi:84	4006-4021	4-19
11121#21	41°36.0'	21°43.9'	RMT 8M	4:vi:84	3988-4008	17-37
11121#22	41°38.9'	21°43.8'	RMT 8M	4:vi:84	3976-3991	34-49
11121#23	41°26.1'	21°47.6'	RMT 8M	5:vi:84	4004-4021	4-21
11121#24	41°22.5'	21°44.3'	RMT 8M	5:vi:84	3991-4007	18-34
11121#25	41°18.2'	21°40.9'	RMT 8M	5:vi:84	3941-3991	34-84
GALATHEA 238	3°23'S	44°04'E	HOT	13:iii:51	3960	

Proceedings of the Fifth International Echinoderm Conference / Galway / 24-29 September 1984

Ultrastructural changes in the autotomy tissues of *Eupentacta quinquesemita* (Selenka) (Echinodermata: Holothuroidea) during evisceration

MARIA BYRNE
Smithsonian Marine Station at Link Port, Fort Pierce, Fla., USA

ABSTRACT: Evisceration in the dendrochirote holothurian Eupentacta quinquesemita (Selenka) is associated with the sudden breakdown of three autotomy tissues. These tissues were examined before, during and after evisceration to investigate the presence of autotomy related specializations and to elucidate the morphological events associated with autotomy. The tissues are characterized by a preponderance of connective tissue and also contain muscle and nerve. Axon-like processes containing large electron-dense vesicles (LDVs) are found in the connective tissue and in association with muscle cells. These processes are similar to the neurosecretory-like processes described for other echinoderm autotomy tissues. Autotomy is part of the general phenomenon of variable tensility in echinoderm connective tissues and involves a change in the connective tissue matrix. During autotomy, the matrix loses its structural integrity causing collagen fibril disarray and disorganization of associated cells. Some axon membrane and basal lamina disruption occurs, but the vesicles contained in axons and the LDVs appear to remain in tact. The vesicles contained in locally distributed axons do not appear to be the source of agents that effect connective tissue breakdown; an alternate source is discussed.

1 INTRODUCTION

The autotomy of body parts through a sudden reduction in the tensility of connective tissue structures is characteristic of the Phylum Echinodermata (Emson & Wilkie 1980) and appears to be unique to the group. Echinoderm autotomy differs from autotomy in other invertebrates where body parts are cast off through rupture of muscle tissue specializations (McVean 1975). The sudden breakdown of connective tissue during autotomy is part of the general phenomenon of variable tensility of echinoderm connective tissues (Motokawa 1984; Wilkie 1984). Another aspect of the phenomemon is the reversible stiffening/softening changes demonstrated by echinoderm catch ligaments. The changes associated with variable tensility have attracted attention because they occur in an extracellular tissue and are considered by many workers to be under neutral control (Jordan 1914,1919; Serra-von Buddenbrock 1963; Wilkie 1978,1983, 1984; Holland & Grimmer 1981a,b; Motokawa 1981,1982a,1984; Byrne 1982; Hilgers & Splechtna 1982; Hidaka & Takahashi 1983).

Autotomy is usually associated with anatomical specializations that facilitate ejection of body parts (McVean 1975). This aspect of variable tensility has received relatively little attention, although there are several ultrastructural studies of echinoderm catch ligaments (Holland & Grimmer 1981b; Smith et al 1981; Motokawa 1982b; Hidaka & Takahashi 1983; Wilkie 1983). Thus far, the fine structure of ophiuroid and crinoid arm autotomy tissues has been described (Wilkie 1979; Holland & Grimmer 1981a).

The presence of axon-like processes, filled with large electron-dense vesicles (LDVs), appears to be characteristic of echinoderm connective tissues (Wilkie, 1979,1984; Holland & Grimmer 1981a,b; Smith et al 1981; Byrne 1982; Hilgers & Splechtna 1982; Motokawa 1982b,1984; Hidaka & Takahashi 1983). The LDVs resemble neurosecretory vesicles found in firmly established neurosecretory neurons (Maddrell & Nordmann 1979) and, based on their morphology, LDVs have been suggested to be involved in the control of variable tensility (Wilkie 1979, 1984; Holland & Grimmer 1981 a,b; Smith et al 1981; Motokawa 1982b, 1984; Hidaka & Takahashi 1983).

Holothurian evisceration results in autotomy of the digestive tract and tentacular crown (Kille 1935; Smith & Greenberg 1973; Byrne 1982). In the dendrochirote holothurian Eupentacta quinquesemita (Selenka) evisceration is associated with the irreversible breakdown of three structures, (1) the introvert (the anterior extensible portion of the body wall), (2) the tendon (P-L tendon) that connects the pharyngeal retractor muscle (PRM) to the longitudinal body wall muscle (LBWM), and (3) the intestine-cloacal junction. The fine structure of the intact P-L tendon has previously been reported (Byrne 1982) and will be described here briefly. The ultrastructure of the introvert and intestine-cloacal junction is also described. Particular emphasis is placed on the changes that occur in the three structures during evisceration and for evidence of neural involvement in the connective tissue change.

2 MATERIALS AND METHODS

Specimens of E. quinquesemita were collected using S.C.U.B.A. near Victoria, British Columbia. Evisceration was induced in freshly-collected specimens by squeezing the body transversely with forceps. For transmission electron microscopy, intact and eviscerating specimens were injected with 3% glutaraldehyde in 0.2M cacodylate buffer (pH 7.4). The autotomy tissues were dissected and placed in fresh fixative for one hour at room temperature. The tissues were then rinsed in the same buffer, post-fixed in 1% OsO_4 in cacodylate for one hour at 4°C, decalcified in ascorbic acid (Dietrich & Fontaine, 1975), dehydrated, and embedded in Epon 812. Thick sections were stained with Richardson's stain (Richardson et al 1960) for light microscopic examination. Thin sections were stained with uranyl acetate and lead citrate and viewed with a Philips EM 300.

3 RESULTS

3.1 Structure of intact introvert

The introvert is predominantly dermal connective tissue with the cuticle and epidermis on the outside and the musculature, nerve plexus and peritoneum on the coelomic side (Figs. 1-3). The dermal connective tissue has three layers, the superficial dermis, a dense connective tissue (DCT) layer and a loose connective tissue layer (LCT). The superficial dermis contains ossicles, cross striated collagen fibrils and connective tissue matrix (Fig. 1). The DCT contains abundant collagen fibrils varying in diameter from 20-160nm associated with an interfibrillar matrix, Fig. 4,

Fig. 1. One micrometer cross section of the epidermis (E) and superficial dermis (S). The ossicles (O) are surrounded by collagen fibrils (CF). Scale=1.0μm.

Fig. 2. Cross section of the dense connective tissue (D) and the loose connective tissue (L). M, muscle fibres; Arrow, DCT-LCT interface. Scale=1.0μm.

Fig. 3. Cross section of the 'transparent' LCT (L). M, muscle fibres MC, morula cells; P, peritoneum. Scale=1.0μm.

Fig. 4. LDV-filled process and muscle fibre bundle (M) in the DCT. BL, basal lamina; CF, collagen fibrils. Scale=2.0μm.

Fig. 5. LDV-filled process in LCT (L). The LCT is electron-lucent and contains a few unstriated fibrils (UF). BL, basal lamina. Scale=1.0μm.

Fig. 6. Elongate LDV-filled process in autotomizing introvert near the DCT-LCT interface. CF, collagen fibrils; D, DCT; L, LCT. Scale=2.0μm.

Fig. 7. LDV-filled processes in the gut connective tissue contain axial microtubules (MT) and occasional mitochondria (MI). CF, collagen fibrils. Scale=0.5μm.

Fig. 8. One micrometer cross section of the intestine adjacent to the intestine-cloacal junction. B, brown body; C, connective tissue; EP, intestinal epithelium; G, gut lumen; P, peritoneum. Scale=0.1mm.

Fig. 9. Cross section of a PRM bundle surrounded by P-L tendon collagen fibrils (CF). LDV-filled processes and axons (A) are found with the muscle fibres (M). BL, basal lamina, NT, neurotubules. Scale= 1.0μm.

Fig. 10. Cross section of autotomized P-L tendon. An intact LDV-filled process is alongside a dispersed PRM bundle. The muscle fibres (M) lack thick filaments, and the contractile elements disperse (asterisk). The collagen fibrils (CF) are in disarray, and the sarcolemma (S) has lifted away. An axon-like profile (A) has an irregular membrane profile and contains few vesicles. BL, basal lamina. Scale= 2.0μm.

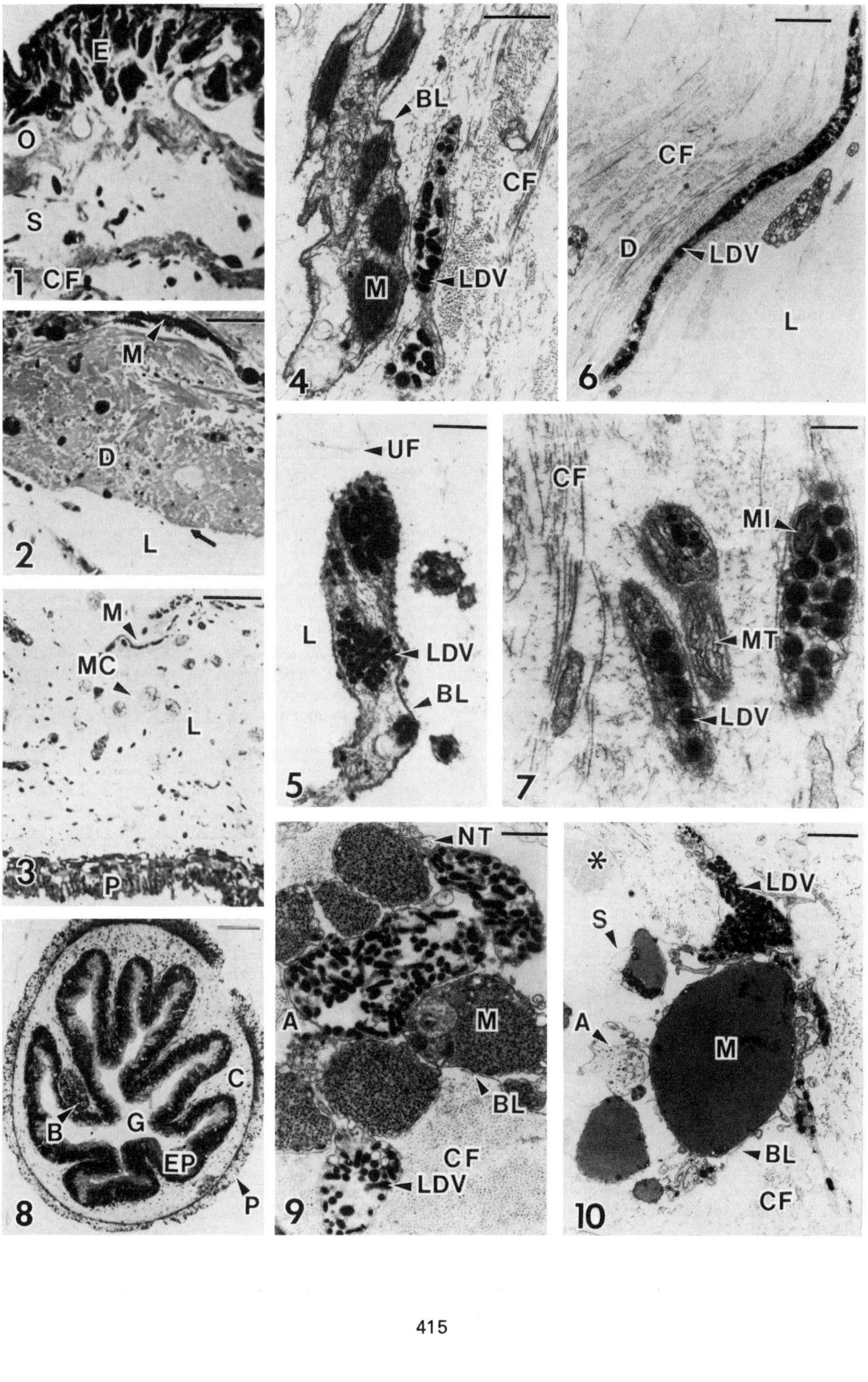
1
E
O
S
CF
2
M
D
L
3
M
MC
L
P
4
BL
CF
M
LDV
5
UF
L
LDV
BL
6
CF
D
LDV
L
7
CF
MI
MT
LDV
8
B
G
EP
C
P
9
NT
A
M
BL
CF
LDV
10
*
LDV
S
A
M
BL
CF

6,11,12). Many of the collagen fibrils have unstriated fibrils (10-14nm in diameter) attached. Bundles of muscle cells surrounded by a basal lamina are scattered through the DCT (Figs. 4,11). Axons are present in the DCT in association with muscle fibres and also surrounded by connective tissue. The axons are ensheathed in a basal lamina and contain neurotubules, occasional mitochondria and two types of vesicles, clear vesicles (70-80nm) and dense-core vesicles (90-140nm). The DCT also contains axon-like processes (Figs. 4, 5) filled with large, electron-dense vesicles (LDVs) that vary in shape from round (150-300nm) through ellipsoidal to sausage-shaped (180-550nm X 125-220nm). Axial microtubules and occasional mitochondria are present in the processes (Fig. 7). These LDV-filled processes are found beside muscle bundles along with axons and are scattered throughout the connective tissue (Figs. 4,6).

The LCT is an amorphous electron-lucent layer of connective tissue matrix with occasional collagen fibrils (40-60nm in diameter) and more common, small diameter (7-12nm), unstriated fibrils (Figs. 3,5,6) Muscle fibres, morula cells, axons and LDV-filled processes are also present in the LCT.

3.2 Structure of intact P-L tendon

The P-L tendon is comprised of a collagenous connective tissue layer surrounding a central muscle region containing the tapered ends of PRM cells and is overlain by peritoneum (Figs. 9,13). Connective tissue ramifies throughout the interior of the tendon so the PRM fibres are completely embedded in connective tissue. The tendon contains abundant small diameter collagen fibrils (30-40nm) and unstriated fibrils (10-15nm) in an interfibrillar matrix. Axons and LDV containing processes similar to those described for the introvert are present in the tendon in association with muscle and connective tissue (Fig. 9).

3.3 Structure of intact intestine cloacal junction

The intestine cloacal junction is the posterior end of the intestine where it joins with the cloaca. It contains a thick connective tissue layer with the gut musculature and peritoneum on the coelomic side and the epithelium on the luminal side (Figs. 8,15). The connective tissue is largely matrix and contains small diameter (4-10nm) unstriated fibrils (Fig. 7). A diffuse layer of collagen fibrils (20-40nm in diameter) is found along the basal lamina near the gut musculature (Fig. 15). Axons are rarely found in the intestinal connective tissue but LDV-filled processes are encountered, especially near the basal lamina (Fig. 7). Morula cells and brown bodies are common in the connective tissue (Figs. 8,15).

3.4 Structure of autotomizing introvert

During evisceration the introvert changes from a solid, opaque structure to one that is soft and translucent. Introvert thickness decreases as it distends, with the viscera and coelomic fluid propelled for-

Figs. 11-16. Similar fields of intact and automized tissues.

Fig. 11. Intact introvert DCT. The muscle fibres (M) are contracted and sarcolemmal extensions (S) lie within basal laminar (BL) folds. CF, collagen fibrils. Scale=2.0µm.

Fig. 12. Autotomizing DCT. The muscle fibres (M) are extended. The sarcolemmal extensions and basal laminar folds have straightened out and the basal lamina (BL) is disrupted (arrows). CF, collagen fibrils. Scale=2.0µm.

Fig. 13. Intact P-L tendon. The peritoneum (P) overlies the relatively compact collagenous connective tissue of the tendon. CF, collagen fibrils; M, muscle; MV, microvilli. Scale=3.0µm.

Fig. 14. Automized P-L tendon. The peritoneum (P) is disrupted (arrow), and the collagen fibrils (CF) are dispersed. M, muscle; MV, microvilli. Scale=2.0µm.

Fig. 15. Intact intestine-cloacal junction. The peritoneum (P), nerve plexus (NP) and muscle (M) layers cover the intestinal connective tissue. Collagen fibrils (CF) are particularly abundant along the basal lamina (BL). B, brown body; MC, morula cell. Scale=5.0µm.

Fig. 16. Autotomized intestine-cloacal junction. The peritoneum (P), nerve plexus and muscle layer are disrupted (arrow), and the collagen fibrils (CF) are dispersed. Scale-5.0µm.

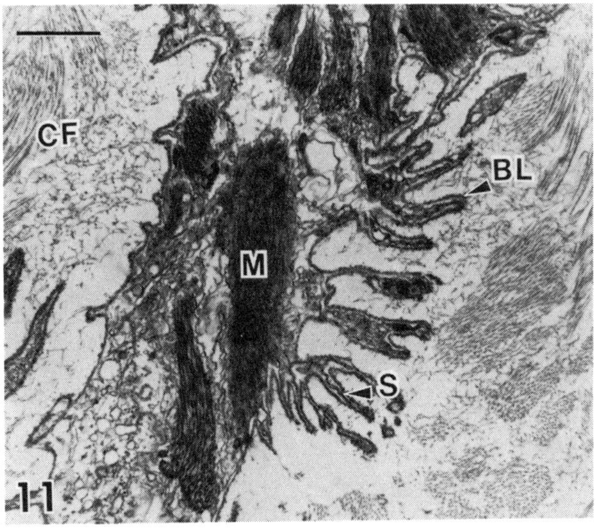
CF
BL
M
S
11

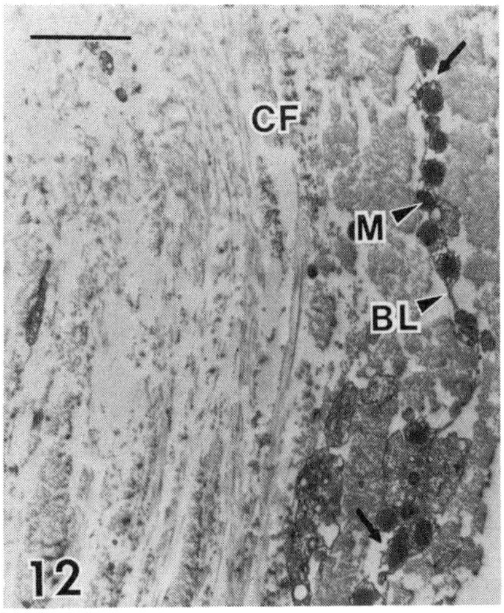
CF
M
BL
12

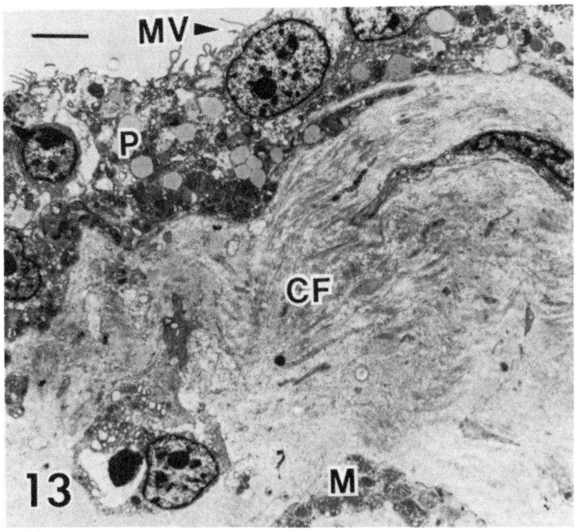
MV
P
CF
M
13

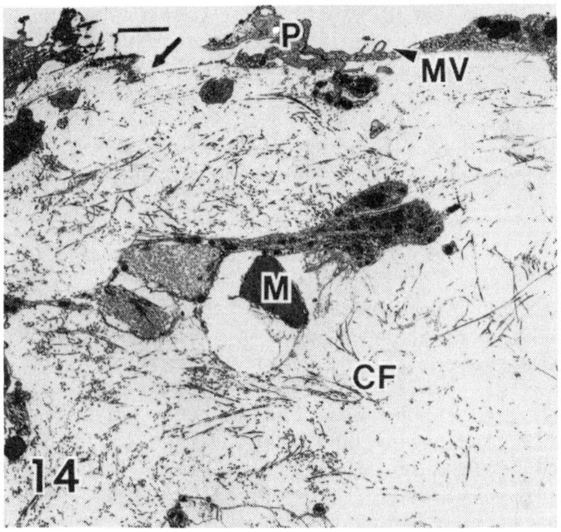
P
MV
M
CF
14

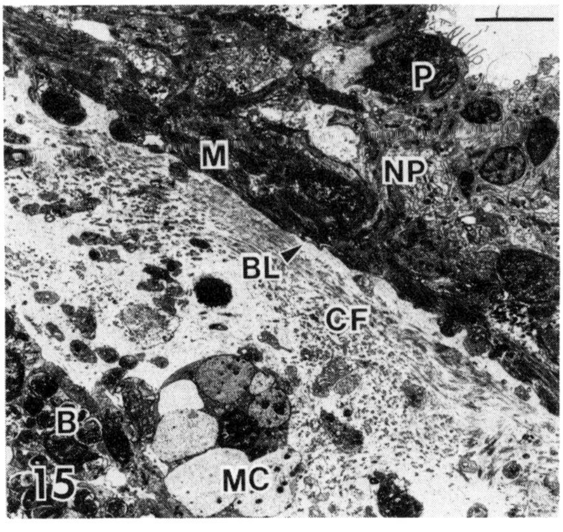
P
M
NP
BL
CF
B
MC
15

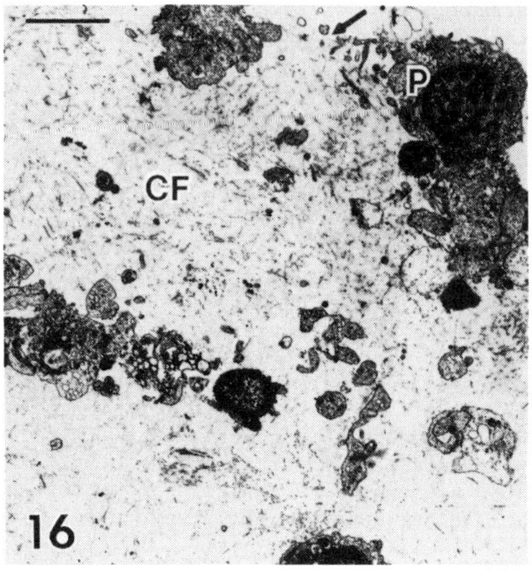
P
CF
16

ward by body wall muscle contraction. As the introvert autotomizes, the superficial dermis and epithelium delaminate from the underlying DCT, and the ossicles embedded within it become visible to the eye. Changes are evident when similar fields of normal and autotomizing DCT are compared (Figs. 11,12). The muscle fibres extend as the introvert distends, and the basal lamina surrounding the muscle fibres is disrupted. Structural examination reveals that the collagen fibrils remain intact and that they appear to slide across and away from each other. This suggests that changes occur in the interfibrilar matrix. Axons and LDV-filled processes also appear to extend, with some damage to axon membranes and the basal lamina, but their vesicular contents appear intact (Fig. 6). The peritoneal cells, muscle fibres and nerve plexus axons dissociate into the coelom, and the LCT is infiltrated by coelomic fluid. Collagen fibrils and unstriated fibrils in the LCT also remain intact, but the matrix loses its structural integrity and the cells embedded in it become disorganized. LDV-filled processes in the LCT and among the dissociated plexus axons retain their vesicles during autotomy.

3.5 Structure of autotomized P-L tendon

During autotomy, the P-L tendon changes from a relatively compact collagenous tissue, to one comprised of disorganized fibrils. Similar fields of normal and autotomized tendon show the increase in interfibril distance and fibril disarray resulting from autotomy (Figs. 13,14). After autotomy, the peritoneum is disrupted, and portions dissociate from the basal lamina. Although the collagen fibrils remain intact, it appears that the associated matrix undergoes a structural change. This results in fibril disorganization and a marked increase in the interfibril distance. The PRM muscle bundles are disrupted, and the contractile filament organization is lost with a disappearance of the thick filaments (Figs. 10,14). Sarcolemmal disruption is evident, and eventually the muscle cell contents are dispersed. Axons and LDV-filled processes also disperse and show membrane irregularity, but for the most part retain their vesicles. Some axon-like profiles devoid of vesicles may represent axons which have lost vesicles during autotomy (Fig. 10). The LDV-filled processes remain surprisingly intact (Fig. 10), although some vesicular loss through mechanical damage might be expected.

3.6 Structure of autotomized intestine-cloacal junction

As for the other autotomy tissues, the intestinal peritoneum, muscle layer and nerve plexus dissociate during autotomy, and the underlying connective tissue is infiltrated with coelomic fluid (compare Figs. 15,16). The collagen fibrils are disorganized, but remain intact. Some axons in the dissociated nerve plexus and connective tissue have an irregular contour, and their basal lamina is disrupted. Vesicles are present in most axons. However, some axons completely surrounded by coelomic fluid contain few or no vesicles, suggesting that the vesicles may have been lost during autotomy.

4 DISCUSSION

Previous examination of echinoderm catch and autotomy tissues revealed that they are collagenous connective tissues (Takahashi 1967; Wilkie 1979; Holland & Grimmer 1981a, b; Smith et al 1981; Motokawa 1982b; Hidaka & Takahashi 1983) and the ultrastructure of the P-L tendon and introvert dense connective tissue is similar. The structure of the introvert loose connective tissue and the intestine differ by having a high matrix content relative to fibril content. The collagen fibrils and unstriated fibrils of the autotomy tissues are characteristically of small diameter and are similar to those observed in other echinoderm connective tissues (Junqueira et al 1980; Holland & Grimmer 1981a,b; Smith et al 1981; Motokawa 1982b; Hidaka & Takahashi 1983). Unlike ophiuroid and crinoid autotomy structures that are comprised entirely of connective tissue (Wilkie 1979; Holland & Grimmer 1981a), the autotomy tissues of E. quinquesemita also contain muscle cells.

Axons and LDV-containing processes are present in the autotomy structures of E. quinquesemita. They are surrounded by connective tissue as well as in association with muscle cells and the nerve plexus. The axons and their vesicular contents appear to be typical of echinoderm neurons. The clear vesicles may be cholinergic, while the dense-core vesicles may be monoaminergic, as suggested elsewhere (Prosser & Mackie 1980; Byrne 1982; Pentreath & Cobb 1982). The LDV-filled processes resemble axons in containing axial microtubules and vesicles, and in their association with muscle fibres. Based on their morphology, the LDV-filled processes of E. quinquesemita are considered to be axons.

The LDVs are similar to the neurosecretory-like vesicles in other echinoderm connective tissues that are typically large, electron-dense and of variable shape. The LDV-containing processes appear characteristic of echinoderm connective tissues and are thought to play a role in connective tissue variable tensility (Wilkie 1979, 1984; Holland & Grimmer 1981a,b; Hilgers & Splechtna 1982; Motokawa 1982b,1984; Hidaka & Takahashi 1983).

The sudden breakdown of the autotomy tissues during evisceration involves a change in the connective tissue matrix, as occurs in arm autotomy in ophiuroids and crinoids (Wilkie 1979; Holland & Grimmer 1981a). During autotomy the fibrillar elements of the autotomy tissues remain intact, but they disperse as the matrix loses its structural integrity. Although it is difficult to describe change in the electron-luscent matrix, the fibril disarray suggests that the matrix has undergone a physical and chemical change, perhaps a sol-gel-like change.

In the autotomy tissues of ophiuroids and crinoids, the LDV-filled processes appear to be the only potential source of autotomy-inducing agents that can be detected microscopically within the connective tissue (Wilkie 1979; Holland & Grimmer 1981a). The contents of the LDVs have been suggested to be proteolytic enzymes or chelating agents that when secreted, effect connective tissue change (Wilkie 1979; Holland & Grimmer 1981a). Axon profiles without vesicles and LDV exocytosis observed in the autotomized crinoid syzygy have been suggested to be part of a massive vesicle exocytosis that plays a role in effecting autotomy (Holland & Grimmer 1981a). However, it is not clear whether these profiles are a cause or a result of autotomy.

During evisceration in E. quinquesemita, the LDVs appear to remain intact. The disruption of axon membranes and basal lamina observed in autotomized tissues may be mechanical rupture occurring during autotomy. Swollen axons devoid of vesicles were observed adjacent to intact vesiculated axons and are likely to be a result rather than a cause of autotomy. Evisceration and autotomy in E. quinquesemita appear to be neurally controlled (Byrne 1983) but, based on morphological evidence, the vesicles contained in locally distributed neurons do not appear to be the source of agents that effect connective tissue breakdown. Numerous neurons and LDV-filled processes are found in association with muscle fibres, especially in the P-L tendon. In ophiuroid and crinoid autotomy tissues, LDV-processes are found only with connective tissue (Wilkie 1979; Holland & Grimmer 1981a); perhaps the LDV-processes of E. quinquesemita are not homologous to those of ophiuroid and crinoid autotomy tissues.

Coelomic fluid factors that alter connective tissue tensility have been isolated from several echinoderms (Motokawa 1981, 1982a,c). There is evidence for the presence of an autotomy-inducing factor in the eviscerated coelomic fluid of E. quinquesemita (Byrne 1983), as found in another dendrochirote holothurian, Sclerodactyla briareus (Smith & Greenberg 1973). The peritoneal disruption, followed by coelomic fluid infiltration into the connective tissues during evisceration, may be important to the autotomy process of E. quinquesemita. The three autotomy tissues are partially or completely bathed in coelomic fluid and in this respect, differ from the arm autotomy tissues of ophiuroids and crinoids. The autotomy tissues of these groups are not associated with a large coelomic fluid medium and so ophiuroid and crinoid arm autotomy may depend on locally distributed cells. In contrast, cells involved in holothurian evisceration may be located at some distance from the autotomy tissues and effect the change through the medium of the coelomic fluid.

Introvert and P-L tendon autotomy was mimicked in vitro by altering the ionic composition of test solutions (Byrne in press). The results suggest that the mechanism of autotomy involves a change in the ionic interations within the proteoglycan matrix. How this is brought about is not known. Although echinoderm autotomy appears to be neurally controlled, the morphological basis for neural mediation has not been established.

ACKNOWLEDGEMENTS

I thank Professor A.R. Fontaine for his enthusiastic supervision of my research. Dr. K.J. Eckelbarger and Mr. J.E. Miller read the manuscript. The work was supported by a University of Victoria Fellowship and this report is Smithsonian Marine Station contribution no. 134.

REFERENCES

Byrne, M. 1982. Functional morphology of a holothurian autotomy plane and its role in evisceration. In International Echinoderms Conference, Tampa Bay, (ed. J.M. Lawrence), pp. 65-68. Rotterdam, Balkema.

Byrne, M. 1983. Evisceration and autotomy in the holothurian Eupentacta quinquesemita (Selenka). Ph.D. thesis, University of Victoria.
Byrne, M. The mechanical properties of the autotomy tissues of the holothruian Eupentacta quinquesemita (Selenka) and the effects of certain physico-chemical agents. J. exp. Biol. (in press).
Dietrich, H.F. & A.R. Fontaine 1975. A decalcification method for ultrastructure of echinoderm tissues. Stain Technol. 50: 351-353.
Emson, R.H. and I.C. Wilkie 1980. Fission and autotomy in echinoderms. Oceanogr. Mar. Biol. Ann. Rev. 18:155-250.
Hidaka, M. & K. Takahashi 1983. Fine structure and mechanical properties of the catch apparatus of the sea-urchin spine, a collagenous connective tissue with muscle-like holding capacity. J. exp. Biol. 103:1-14.
Hilgers H.V. & H. Splechtna 1982. On the control of the detaching of gemmiform pedicellariae in Sphaerechinus granularis (Lam.) and Paracentrotus lividus (Lam.) (Echinodermata, Echinoidea). Zool. Jb. Anat. 107:442-457.
Holland, N.D. & J.C. Grimmer 1981a. Fine structure of syzygial articulations before and after arm autotomy in Florometra serratissima (Echinodermata:Crinoidea). Zoomorphology 98:169-183.
Holland, N.D. & J.C. Grimmer 1981b. Fine structure of the cirri and a possible mechanism for their motility in stalkless crinoids (Echinodermata). Cell Tiss. Res. 214:207-217.
Jordan, H. 1914. Über "reflexarme" Tiere. IV. Die Holothurien. 1 Mitteilung. Die Holothurien als hohlorganige Tiere und die Tonusfunktion ihrer Muskulatur. Zool. Jahrb., Jena, allg. Zool. 34:365-436.
Jordan, H. 1919. Über "reflexarme" IV. Die Holothurien. Zweite Mitteilung. Die Reizbarkeit und der Einflus des zentralen Nervensystems saf die Muskulatur und die muskelähnlichen Fasern der Haut (auf Erregbarkeit und Tonusfunktion). Zool., Jahrb., Jena, allg. Zool. 36:109-156.
Junqueira, L.C.U., G.S. Montes, P.A.S. Mourao, J. Carneiro, L.M.M. Salles & S.S. Bonetti 1980. Collagen-proteoglycans interaction during autonomy in the sea cucumber Stichopus badionotus. Rev. Can. Biol. 39:157-164.
Kille, F.R. 1935. Regeneration in Thyone briareus Lesueur following induced autotomy. Biol. Bull. 69:82-108.
Maddrell, S.H.P. & J.J. Nordmann 1979. Neurosecretion. Blackie and Son Ltd., Glasgow. 173pp.
McVean, A. 1975. Mini-review, autotomy. Comp. Biochem. Physiol. 51A:497-505.
Motokawa, T. 1981. The stiffness change of the holothurian dermis caused by chemical and electrical stimulation. Comp. Biochem. Physiol. 70C:41-48.
Motokawa, T. 1982a. Factors regulating the mechanical properties of holothurian dermis. J. exp. Biol. 99:29-41.
Motokawa, T. 1982b. Fine structure of the dermis of the body wall of the sea cucumber, Stichopus chloronotus, a connective tissue which changes its mechanical properties. Galaxia 1:55-64.
Motokawa, T. 1982c. Rapid change in mechanical properties of echinoderm tissues caused by coelomic fluid. Comp. Biochem. Physiol. 73C:223-229.
Motokawa, T. 1984. Connective tissue catch in echinoderms. Biol. Rev. 59:255-270.
Pentreath, V.W. & J.L.S. Cobb 1982. Electrical conduction and behaviour in "simple" invertebrates. G.A.B. Shelton. ed., Clarendon Press, Oxford. p. 440-472.
Prosser, C.L. & G.C. Mackie 1980. Contractions of holothurian muscles. J. Comp. Physiol. 136:103-112.
Richardson, K.C., L. Jarett & E.H. Fink 1960. Embedding in epoxy resins for ultrathin sectioning in electron microscopy. Stain Technol. 35:313-323.
Serra-von Buddenbrock, E. 1963. Études physiologiques et histologiques sur le tégument des holothuries (Holothuria tubulosa) Vie et Milieu 14:55-70.
Smith, D.S., S.A. Wainwright, J. Baker, & M.L. Cayer 1981. Structural features associated with movement and 'catch' of sea urchin spines. Tissue and Cell 3: 299-320.
Smith, G.N. & M.J. Greenberg 1973. Chemical control of the evisceration process in Thyone briareus. Biol. Bull. 144:421-436.
Takahashi, K. 1967. The catch apparatus of the sea-urchin spine I. gross histology. J. Fac. Sci. Univ. Tokyo IV, 11:109-120.
Wilkie, I.C. 1978. Nervously mediated change in the mechanical properties of a brittlestar ligament. Mar. Behav. Physiol. 5:289-306.
Wilkie, I.C. 1979. The juxtaligamental cells of Ophiocomina nigra (Abildgaard) (Echinodermata: Ophiuroidea) and their possible role in mechano-effector function of collagenous tissue. Cell Tissue Res. 197:515-530.
Wilkie, I.C. 1983. Nervously mediated change in the mechanical properties of the cirral ligaments of a crinoid. Mar. Behav. Physiol. 9:229-248.
Wilkie, I.C. 1984. Variable tensility in echinoderm collagenous tissues: a review. Mar. Behav. Physiol. 11:1-34.

Proceedings of the Fifth International Echinoderm Conference / Galway / 24-29 September 1984

Changes in analyzed organic matter and energetic content of two thin-skinned holothuroids *Leptosynapta galliennei* and *Eumolpadia violacea*, at critical stages of their life cycle

JEAN-PIERRE FÉRAL
Laboratoire de Biologie des Invertébrés marins et Malacologie, Muséum d'Histoire naturelle, Paris, France

ABSTRACT: The biochemical constituents of **Leptosynapta galliennei** (temperate apodid) and of **Eumolpadia violacea** (subantarctic molpadid) have been analyzed relative to reproductive condition. The holothuroids have been reconstructed in terms of wet weight and organic content (analyzed organic matter, AOM) of their body parts and their energetic equivalents were calculated. There is no change in the biochemical levels in all organs except in the gonad, where protein, lipid and carbohydrate accumulations have been observed at full maturity. At this time AOM reaches 30% of the total AOM for testis and 50% for ovary in Leptosynapta galliennei; it is only 15% for the ovary of Eumolpadia violacea. In both cases, nutrients are accumulated directly in the gonad, without any detectable intervention of the body wall constituents. Over the year, the energetic equivalent of a standard L. galliennei is twice that of E. violacea of a same weight and three times at full maturity. This difference is mainly due to the gonad.
It is proposed that the utilization of AOM (or its energetic equivalent) of **all** body parts is the least biased and the most convenient expression for intra- and inter-specific comparisons.

1. INTRODUCTION

Compared to asteroids and echinoids, the biochemical constitution of holothuroids is poorly documented. Since the first studies concerning the nutritive value of the "skin" of Stichopus japonicus* (Tanikawa and Yoshitani 1955, Tanikawa et al. 1955), only two species have been investigated completely, organ by organ - Parastichopus californicus (Giese 1966) and Holothuria (Holothuria) tubulosa (Massin 1978). For the latter, analysis of the main biochemical compounds have been performed at different gonadal stages.

Data are also available

. on the composition of the whole animal (Fish 1967, Turner and Rutherford 1976, Barnes et al. 1976)

. on one type of constituent in all body parts (Allen 1968, Féral and Doumenc 1982, Féral 1983)

. on some constituents in some body parts, mainly the body wall (Lawrence 1972, Prim et al. 1976, Lawrence and Kafri 1979, Smith 1981, Sibuet and Lawrence 1981, Lawrence and Guille 1982).

The only way which can reveal where nutrients are accumulated is to analyse separately all the body parts. If the dosages are done on animals sampled at different seasons, they can indicate what kinds of molecules are the most mobilizable during gametogenesis or starvation (e.g. hibernation).

Many authors have demonstrated a reciprocal relationship between pyloric caeca and ovaries in a number of seastars species (Lawrence and Lane 1982). However such evidence does not clearly exist for holothuroids. Lawrence (1972) on Holothuria (Halodeima) atra and Massin (1978) on H.tubulosa reported that there was no change in the organic levels after starvation or during the life cycle, except for H.tubulosa in the case of the rete mirabile, whose lipid level increased during the gonadal rest. However, it was concluded by Massin (1978) that the quantities accumulated were not important enough to consider the

* *used as trepang*

hemal system as a reserve place. Massin (1978) also pointed out that if experimental starvation occurred during the period of gonadal activity, proteins of the body wall may be mobilized. This mobilization did not happen during the period of gonadal inactivity. Prim et al.(1976) have also indicated that the body wall could be considered as a potential organ of reserve, mainly in protein form.

However, the existence of reserves accumulation in one part of the body of a holothuroid and mobilizable and usable in another part of the organism appears only probable and not generalizable.

The species generally used belong mostly to Aspidochirotida, which are thick-skinned holothuroids. Thus, the body wall appears as the body component which possesses a potential mass of organic material important enough to be a reserve site.

The aim of the present work is to investigate the changes in organic constitution of the body parts of two thin-skinned holothuroid species living in different environments and presenting two types of life cycles : Leptosynapta galliennei, a temperate apodid which breeds once a year and Eumolpadia violacea, a subantarctic molpadid whose reproduction is not strictly seasonal.

2. MATERIAL AND METHODS

2.1. Origin and preparation of the animals

Leptosynapta galliennei (Herapath, 1865) was collected in the sandy intertidal zone in the vicinity of Roscoff (Britanny-France) at four critical stages of its life cycle :

A : beginning of the gonadal growth (November-December).

B : during the gonadal growth and maturation (March-April).

C : at full maturity, just before spawning (late June).

D : just after spawning, at the beginning of the phase of gonadal inactivity (late June-early July).

L.galliennei appears to breed once a year.

Eumolpadia violacea (Studer, 1876) lives in the South Indian Ocean in the vicinity of Kerguelen Islands. The animals were collected by dredge on mud bottoms at 30-60 m depth at two seasons :

July (austral winter) : period of gonadal growth

February (austral summer) : period of full maturity.

The life cycle of E.violacea is not totally known. However, certain characteristics may be drawn : it is remarkable that apparently ripe oocytes (300-350 µm in diameter) occur in the ovary of both samples, as well as smaller ones. The largest oocytes are more numerous in February. This observation might indicate a prolonged or a continuous breeding season. This hypothesis is also supported by the fact that each sample contained mature males with ripe testes and by the presence of reproductively mature and immature individuals of both sexes within a broad size range. Empty gonads in late February and in March were observed. All these premilinary data suggest that although austral midsummer seems to be a main spawning period for E.violacea living in the Golfe du Morbihan in Kerguelen, the reproduction of this species is not completely seasonal.

In this paper L.galliennei of both sexes, and only female E.violacea are considered. Male E.violacea were too heterogeneous in size.

The holothuroids were dissected, weighed and frozen. Reproductive state was estimated by histological examination of a small part of the gonad. Tissues homogenized in distilled water up to 100 mg fresh weight (FW) per ml except for the esophagus (10-20 mg/ml) and the stomach (50 mg/ml) of L.galliennei and the respiratory trees (20-30 mg/ml) of E.violacea.

2.2. Estimation of the main biochemical constituents

All dosages were made on duplicated aliquots of homogenates. Proteins determination is based on a modification of the Folin-phenol reagent method of Hartree (1972) using BSA as a standard. After extraction with a 2 : 1 chloroform-methanol mixture, lipid levels were determined using the sulfophosphovanillin reaction, adapted for marine invertebrates by Barnes and Blackstock (1973), with cholesterol as a standard. Reducing carbohydrate levels were estimated by the phenolsulfuric acid method of Dubois et al. (1956) with glucose as a standard.

The levels of the main biochemical constituents are expressed in µg/mg FW. Quantities of analyzed organic matter (AOM) per organ were calculated and transformed to their energetic equivalents using the conversion coefficients of Brody (1945) and expressed in joules.

Results were tested using nonparametric

Table 1A. Organ indices (%) of Leptosynapta galliennei. A, B, C & D : gonadal stages as previously defined. H : head, BW : body wall, LM : longitudinal muscles, E* : esophagus, S* : stomach, AI* : anterior intestine, PI* : posterior intestine, G : gonad. Symbols (∇,●) : significantly different at α=0.05

		H	BW	LW	E	S	AI	PI	G
A	♂ (1)	*1.8*	*51.1*	*18.6*	*3.0*	*11.3*	*11.3*	*6.8*	*3.0*
	♀ (10)	2.1±0.6	47.1±3.5	20.1±3.5	1.2±0.5	3.7±0.5	11.4±3.2	10.2±1.4	3.9±1.7
B	♂ (3)	*1.3±0.4*	*43.7±0.6*	*17.1±3.2*	*1.1±0.2*	*4.1±1.1*	*8.7±1.1*	*7.8±3.7*	*16.2±2.7*
	♀ (4)	1.3±0.3	42.6±4.0	15.1±1.2	0.9±0.3	3.6±0.5	11.6±3.2	9.6±1.8	15.3±3.9
C	♂ (3)	*1.1±0.2*	*37.9±2.5*∇	*12.4±4.8*	*0.6±0.1*	*4.1±0.7*	*7.7±0.5*	*7.9±3.2*	*28.3±9.7*●
	♀ (3)	1.2±0.2	30.3±1.5∇	7.9±0.4	0.7±0.1	4.5±0.3	8.1±0.6	6.8±1.8	40.5±2.1●
D	♂ (2)	*1.9*	*51.9*	*15.2*	*0.6*	*4.7*	*13.5*	*11.5*	*0.7*
	♀ (2)	1.9	49.2	14.9	1.4	7.9	12.3	11.9	0.5

Table 1B. Organ indices of ♀ Eumolpadia violacea. New abbreviations : J : July, F : February, DT : digestive tract, Ov : ovary, RT : respiratory trees

	H	BW	LM	DT	Ov	RT
J (6)	6.6±0.7	57.5±2.7	9.8±1.4	9.3±1.2	14.1±2.6	2.7±0.4
F (10)	5.4±0.8	58.4±4.9	9.4±2.0	9.7±1.2	15.2±5.7	1.8±0.8

statistical methods which do not suppose the normality of the distribution and can be used for small samples (Siegel 1956, Sokal and Rohlf 1969).

3. RESULTS

3.1. Organ indices, calculated as the ratio of the fresh weight of a body part and the fresh weight of the whole organism (excluding the coelomic fluid and the gut content) multiplied by 100, are given in table 1A for L.galliennei and 1B for E.violacea.

The FW of the nongonadic tissues of L.galliennei is around 1500 mg troughout the year ; the gonad grows from a few mg to ca 1200 mg. The FW of the specimens used of E.violacea was ca 3500-4000 mg.

3.2. Biochemical composition

3.2.1. Organic levels of the main biochemical constituents (μg/mg FW) are given in table 2A for L.galliennei and 2B for E.violacea.

There is no difference between males and females of L.galliennei except in the gonads. There is no significant change in the concentrations of proteins, lipids or carbohydrates in the body wall and in the posterior intestine, some occurs in the anterior intestine. All levels fluctuate in the longitudinal muscles and in the gonads. Ovary and testis are the only organs which are growing. Proteins are the most concentrated in all body parts except in the ovary during the gonadal growth (stage B) in which lipids dominate.

The organic levels of E.violacea are lower than in L.galliennei. There are some differences between the two samples of July and February concerning mainly the ovary and the longitudinal muscles. The levels of proteins, lipids and carbohydrates are higher in the ovary of individuals collected in austral summer. Little changes concerning the carbohydrates occur in the body wall.

3.2.2. Organic contents

The total amounts of AOM in L.galliennei are given in table 3. In the case of E.violacea the total AOM was 140 ± 28 mg in July (n=6) and 231 ± 93 mg in February (n=10). This AOM is distributed as shown in table 4A for L.galliennei and 4B for E.violacea.

In the case of the male L.galliennei, the body wall is the major body component troughout the year. It contains 40 to 50 % of the AOM. The testis reaches 25 % of the AOM at full maturity. Other body parts represent 20 to 25 % and do not variate greatly (tab. 4A).

The ovary represents half of the organic content of the females at spawning time and is the main body component with

* *see Féral & Massin (1982) for definitions*

Table 2A. Organic levels (‰) of Leptosynapta galliennei. Symbols, significantly different at α=0.05 for values taken 2 by 2 in the case of each biochemical constituent

		H	BW	LM	E	S	AI	PI	G
Proteins	A	89±27(10)	84±17(11)	■+▲ 130±16(11)	54±24(11)	70±15(11)	73±14(11)	30±5(11)	♂ ND ♀ 57±16(10)
	B	170±59(7)	67±16(7)	■ 109±20(7)	140±93(7)	81±28(7)	82±19(7)	44±19(7)	♂ ● 82±2(3) ♀ 69±13(4)
	C	57±18(6)	95±33(6)	▲ 96±15(6)	23±18(6)	67±10(6)	69±5(6)	30±5(6)	♂ ● 49±9(3) ♀ 88±26(3)
	D	101±64(4)	82±12(4)	+ 99±14(4)	86±40(4)	65±28(4)	77±10(4)	30±10(4)	♂ ND ♀ ND
Lipids	A	6±6(3)	9±1(11)	■○ 12±3(8)	18±10(8)	7±3(8)	47±11(10)	11±2(11)	♂ ND ♀ 58±19(10)
	B	17±10(7)	10±2(7)	■x 9±1(7)	8±4(6)	10±2(7)	○▲ 52±4(7)	14±3(7)	♂ △ 39±7(3) ♀ ● 95±10(4)
	C	11±3(5)	9±1(6)	○+ 9±1(6)	11(1)	13±1(5)	○ 40±7(6)	10±3(6)	♂ △ 20±3(3) ♀ ● 66±5(3)
	D	7(1)	12±2(4)	x+ 11±2(4)	6(1)	17±4(3)	▲ 44±5(4)	11±5(4)	♂ ND ♀ ND
Carbohydrates	A	17±3(10)	9±1(11)	■ 11±1(11)	14±2(10)	7±1(11)	11±2(11)	6±1(11)	♂ 4(1) ♀ ● 11±1(10)
	B	19±6(7)	8±1(7)	▽ 10±1(7)	10±5(7)	7±1(7)	10±1(7)	6±1(7)	♂ 7±1(3) ♀ 12±3(4)
	C	13±2(6)	8±2(6)	○ 8±1(6)	9±4(6)	7±1(6)	10±1(6)	7±1(6)	♂ 9±3(3) ♀ ● 15±5(3)
	D	22(2)	13±5(4)	▽■○ 15±3(4)	ND	8±6(4)	11±1(4)	7±2(4)	♂ ND ♀ ND

Table 2B. Organic levels (‰) of ♀ Eumolpadia violacea. (*) significantly different at α=0.05

		H	BW	LM	DT	Ov	RT
Proteins	J F	37.1 ± 7.5 (6) 44.4 ± 4.6(10)	30.1 ± 6.1 (6) 37.5 ±10.4(10)	44.5 ± 11.5 (6)* 76.6 ± 15.5(10)*	25.3 ± 5.7 (6) 33.5 ± 8.2(10)	9.2 ± 1.2 (6)* 26.2 ± 8.1(10)*	14.5 ± 4.2 (6)* 36.3 ±11.5(10)*
Lipids	J F	10.1 ± 0.6 (6)* 6.5 ± 2.4(10)*	1.6 ± 0.4 (3) 1.4 ± 0.6(10)	4.3 ± 1.0 (6) 2.9 ± 2.5 (9)	15.4 ± 2.3 (6) 14.8 ± 6.0(10)	13.9 ± 4.6 (6)* 22.6 ± 8.0(10)*	7.0 ± 1.9 (5) 17.5 ±10.5 (4)
Carbohydrates	J F	4.5 ± 1.0 (6) 5.1 ± 0.6(10)	6.7 ± 0.8 (6)* 5.5 ± 0.8(10)*	9.1 ± 1.9 (6)* 11.8 ± 2.4(10)*	4.6 ± 0.8 (6) 4.8 ± 1.1(10)	3.6 ± 0.9 (6)* 6.2 ± 1.5(10)*	3.4 ± 0.8 (6)* 6.3 ± 1.9(10)*

50 % of AOM. The body wall represents not more than 30 % at stage C; the proportions of the other organs are also much smaller compared to the males. An important difference between male and female L.galliennei occurs during the life cycle, due to the greater accumulation of nutrients in the ovary. It should be noticed that the free amino acids, which were not analyzed, may represent a not negligible part of the organic matter in the testis.

Such changes do not appear clearly in the case of E.violacea (tab.4B),

Table 3. Total quantity of AOM (mg) in the whole body of Leptosynapta gallien-nei

	A	B	C	D
♂		*192±40(3)*	*324±171(3)*	*172(2)*
♀	206±44(10)	170±46(4)	424±108(3)	146(2)

Table 4A. Distribution of the AOM (%) in the body parts of Leptosynapta gallien-nei

	H	BW	LM	DT	G
A ♀(10)	2±1	43±6	28±5	22±5	5±1
B♂(3)	*2±1*	*37±4*	*19±6*	*24±2*	*18±1*
♀(4)	2±1	31±8	19±2	22±4	26±6
C♂(3)	*1±1*	*39±7*	*15±4*	*20±2*	*25±10*
♀(4)	1±1	29±5	7±1	13±2	50±1
D♂(2)	*2*	*50*	*18*	*30*	*0*
♀(2)	2	54	18	26	0

Table 4B. Distribution of the AOM (%) in the body parts of ♀ Eumolpadia violacea (*) significantly different at $\alpha=0.05$

	H	BW	LW	DT	Ov	RT
J(6)	9±2	55±4	15±4	10±1	9±2*	2±1
F(10)	6±1	50±5	16±2	10±2	16±5*	2±1

for which the body wall is always the major component (50-55 % AOM) and stays at this level even at spawning time (F). Only the ovary have accumulated nutrients, corresponding to a more than 75 % increase of AOM. The ovary ranks fourth in order of importance in July and becomes third, close behind the longitudinal muscles in February (tab.4B). No difference in the distribution of AOM occurs in the other body parts.

The distribution of AOM in % of proteins, lipids and carbohydrates is given in figures 1A and 1B for the whole body and in tables 5A and 5B for the different body parts, in L.galliennei and E.violacea, respectively.

In the case of the male L.galliennei, proportions of proteins, lipids and carbohydrates stay similar in the 3 samples (fig.1A). In the females, only the carbohydrates remain almost stable. Proteins and lipids do vary at the 4 considered stages (fig.1A).

There is no difference in the gross biochemical composition of female E.viola-cea between the growing phase and the spawning time (fig.1B).

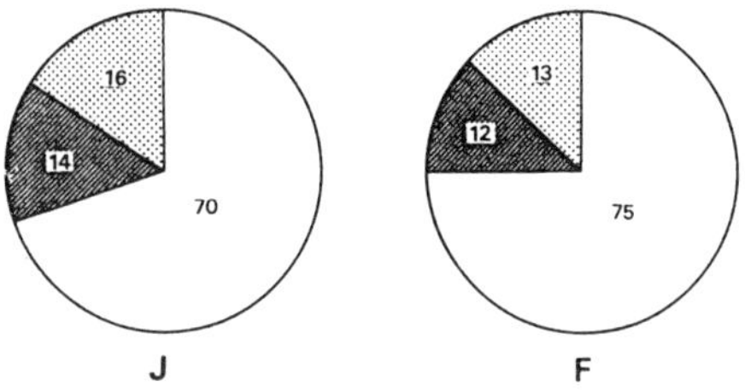

Figure 1B. Distribution of AOM in % of proteins, lipids and carbohydrates in the whole body of ♀ Eumolpadia violacea

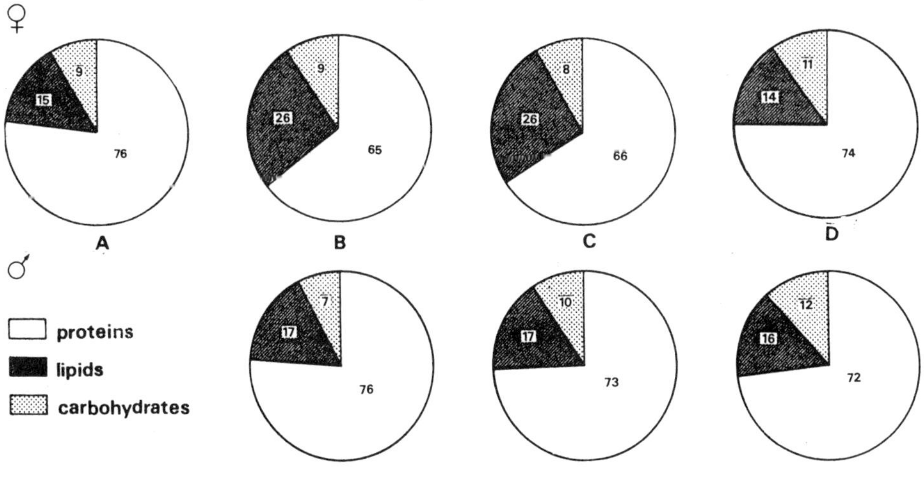

Figure 1A. Distribution of AOM in % of proteins, lipids and carbohydrates in the whole body of Leptosynapta galliennei

Table 5A. Distribution of AOM in % of proteins (P), lipids (L) and carbohydrates (G) in the body parts of Leptosynapta galliennei

			H	BW	LM	E	S	AI	PI	G
P	A	♂								
		♀	78.5±2.4	81.9±2.3	84.6±1.6	59.6±14.1	83.4±3.0	56.7±5.4	64.7±2.3	46.3±6.5
	B	♂	87.5±3.9	79.5±4.5	85.7±2.0	91.2±4.8	85.1±3.8	61.6±1.1	72.6±1.7	64.3±4.0
		♀	75.9±10.4	76.9±4.6	84.5±2.5	78.7±9.4	79.9±3.8	52.4±5.0	62.5±9.5	39.1±3.0
	C	♂	70.1±4.0	81.8±5.2	83.3±1.6	59.6±9.7	75.1±0.8	54.0±1.4	63.2±2.3	63.4±6.5
		♀	69.2±8.4	86.0±3.1	86.3±2.0	46.8±24.3	78.8±2.2	61.5±4.4	66.6±5.4	51.3±9.2
	D	♂	76.9	75.5	76.4	77.4	74.4	60.8	63.5	38.5
		♀	75.8	77.8	81.6	72.3	66.6	55.9	62.7	36.9
L	A	♂								
		♀	5.8±2.3	9.1±1.3	8.3±1.4	21.7±9.9	8.3±2.9	35.0±5.4	22.4±2.6	45.9±7.1
	B	♂	4.6±4.6	11.9±3.1	6.5±1.1	4.8±3.7	9.5±3.3	32.7±1.2	20.3±1.7	30.0±4.2
		♀	12.4±7.0	12.5±3.4	7.8±1.6	5.7±6.3	11.2±2.9	39.9±4.5	25.8±6.4	54.2±3.0
	C	♂	13.0±3.9	9.1±2.2	8.2±1.1	15.9±5.0	16.0±1.4	36.9±0.9	22.6±2.1	25.6±5.7
		♀	14.5±4.7	7.1±1.6	7.3±0.8	25.3±8.5	13.9±1.8	30.0±4.1	18.9±4.0	39.7±7.4
	D	♂	8.4	11.6	10.4	4.0	16.3	31.5	22.5	0.0
		♀	8.2	11.6	7.8	7.8	25.0	35.8	23.5	0.0
G	A	♂								
		♀	15.6±1.7	8.9±1.1	7.1±1.1	18.7±8.3	8.3±1.0	8.3±0.6	12.9±1.6	7.8±1.0
	B	♂	7.9±3.2	8.7±2.1	7.9±0.9	4.0±1.7	5.4±0.8	5.8±0.8	7.1±0.4	5.7±0.2
		♀	11.7±4.2	10.7±1.2	7.8±1.7	15.9±8.9	8.8±1.4	7.6±0.6	11.7±3.3	5.3±3.1
	C	♂	16.9±1.2	9.1±3.0	8.5±0.6	24.5±6.4	8.8±0.8	9.0±0.5	14.2±0.8	11.0±1.2
		♀	16.3±4.3	6.8±1.9	6.4±1.3	27.9±16.7	7.3±0.8	8.4±0.4	14.4±1.5	8.9±1.9
	D	♂	14.8	13.0	13.3	18.6	9.4	7.8	14.1	61.5
		♀	16.1	10.7	10.7	20.0	8.5	8.3	13.9	63.1

Table 5B. Distribution of AOM in % proteins, lipids and carbohydrates in the body parts of ♀ Eumolpadia violacea

	H	BW	LM	DT	Ov	RT
P J	71±3	78±4	76±6	56±3	36±11	58±5
F	80±4	84±3	84±3	64±8	48±8	62±9
L J	20±3	4±1	7±2	34±3	51±11	28±5
F	11±5	3±1	3±2	27±8	41±9	27±7
G J	9±2	18±3	17±5	10±2	13±2	14±3
F	9±1	13±2	13±1	9±1	11±1	11±5

When examining at the level of the different body parts, one can observe that there is no variations of the proportion of proteins, lipids and carbohydrates in the non gonadic tissues of L.galliennei except in the case of longitudinal muscles, where the proportions of carbohydrates are higher, in males and females, at stage D (after spawning). No difference occurs in the testis (stages B and C) where the increase of AOM corresponds to its ponderal growth. This is similar in the ovary only at the beginning of the gonadal growth (stage A) ; then, the increase of the AOM is due to lipids at stage B (more than 25 % of the AOM are dosed in 15 % of the total FW tissues, which represent the ovary at this time) and proteins and carbohydrates at stage C (at full maturity 50 % of the AOM are contained in 40 % of the FW). It seems that the main variation, in amount, of the AOM in L.galliennei concerns the gonad. In the case of E.violacea, the accumulation of AOM in the ovary (9 % AOM in 15 % of FW in July and 16 % AOM in 15 % of FW in February) does

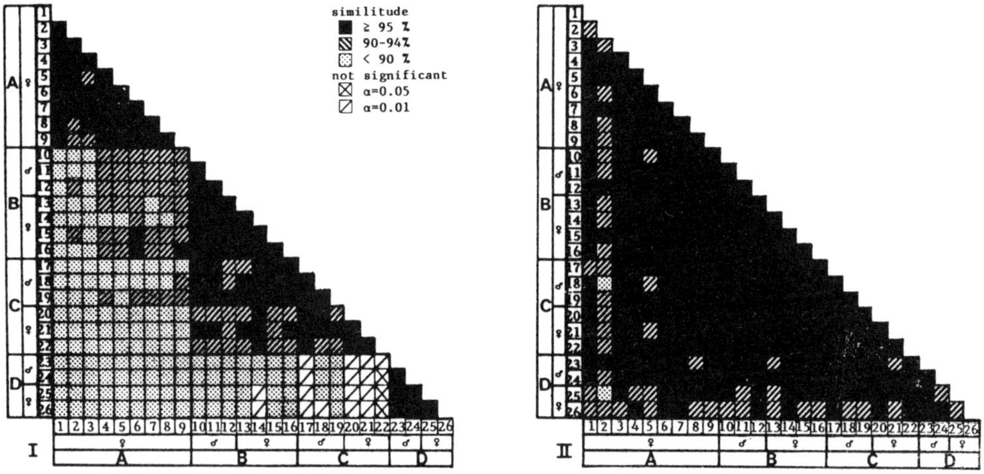

Figure 2. Matrix of similitude representing the degree of similarity (correlation coefficient of Spearmann) existing among all analyzed individuals of Leptosynapta galliennei, in function of the biochemical constitution of all their body parts (I) and without taking into account their gonad (II)

not influence the repartition of the organic content in the same proportion.

3.3. Energetic content

3.3.1. Evidence of the primary role of the gonad on the energetic content

The amounts of proteins, lipids and carbohydrates of L.galliennei were converted to their energetic equivalents. Then, the energetic values of the biochemical constituents of all the body parts of each individual were compared to the energetic values of all the body parts of all the other individuals, using the nonparametric correlation coefficient of Spearmann.

The coefficients are arranged in a matrix of similitude ; the individuals are plotted on the two axes in the chronological order of sampling (A, B, C, D). Half of this square matrix is represented in figure 2I. The value of the correlation coefficient is replaced by conventional colors indicating the level of similarity of the holothuroids taken two by two.

The four groups are quite well individualized at the level of 95 % of similitude except the groups B and C ; this indicates that the maturation rate is not the same for all the individuals of the population. No strong correlation appears between individuals of groups C and D, just before and just after spawning.

When the gonadal biochemical constituents are not taken into account (figure 2II), one can see the existence of an important correlation between all individuals of all groups, whatever the considered season. From the energetic point of view, the nongonadic tissues do not change during the year and do not seem to be involved as an extragonadic reserve place(s).

Table 6. Energetic equivalent (joules/g FW) of the whole body of ♀ Leptosynapta galliennei (L.g.) and ♀ Eumolpadia violacea (E.v.)

	gonadal growth		full maturity	postpawning
	A	B	C	D
L.g.	2880±268	2856±437	3711±468	2757
E.v.	988±79		1284±204	

3.3.2. Energetic content of females of L.galliennei and E.violacea

The greatest differences in the biochemical composition of these holothuroids appear to be in the case of the females.

Examining the energetic content is the most valuable way to compare several species, eliminating organic and inorganic composition variations.

The results for L.galliennei and E.violacea are given in table 6.

L.galliennei is energetically equivalent to two times that of E.violacea and even three times at the full extension of the ovary, before spawning.

4. CONCLUSIONS

- The two studied species have different biochemical composition, which may greatly change during the life cycle and depends on the sex. The variations depend mainly on the importance of the gonadal development, mainly in holothuroids like L.galliennei, whose breeding is seasonal.

- The somatic (nongonadic) tissues do not variate and no body part can be considered as an extragonadic reserve place, even the body wall. This is correlated with the fact that L. galliennei and E. violacea are thin-skinned compared to the other studied species (Lawrence 1972, Prim et al. 1976, Massin 1978). The gonad accumulates nutrients directly (cf. Walker 1982).

- In the case of E. violacea, where reproduction seems to be more or less continuous, the gonad is not so important in terms of energy.

The differences between the two species are related more to the difference of reproductive biology than to the geographical origin.

At the level of a body part, of the organism or of the species, the AOM or its energetic equivalent is the direct expression of the analysis. It is therefor the least biased and the most convenient for intra and interspecific comparisons. The measurement of AOM in all body parts permits a complete reconstruction of mass and energy distribution throughout the whole organism, bringing forth the relative importance of each body part and of each organic constituent in the whole animal. Among echinoderms such data are available only on 3 species (Lawrence 1973, Turner and Rutherford 1976, Turner and Dearborn 1979).

- At the level of population, to estimate energy flow, the use of AOM concerning the whole body could provide a good estimation if done at different seasons and if it takes into account the energy passing through the gonads, especially if this energy may represent 50% of the total energetical equivalent of a species during its life cycle (see Lawrence and Lane 1982, tab. 7).

ACKNOWLEDGMENTS *: Part of this study was supported by the Direction Scientifique des Terres Australes et Antarctiques Françaises (program "benthos", A.Guille, scientific director). I thank M.Van Beveren for her assistance and P.Schatt for his help in Kerguelen. I am also grateful to J.M.Lawrence and P.Magniez for valuable discussions, to J.Douvilliers and C.Leroux (computing department, Roscoff) and to E. von Dorster who improved the English text.*

REFERENCES

Allen,W.V. 1968. Fatty acid synthesis in the echinoderms: Asterias rubens, Echinus esculentus and Holothuria forskali. J.Mar.Biol.Ass.U.K. **48**:521-533.

Barnes,A.T.,L.B.Quetin,J.J.Childress & D.L.Pawson 1976. Deep-sea macroplanktonic sea cucumber : suspended sediment feeders captured from deep submergence vehicle. Science **194**:1083-1085.

Barnes,H. & J.Blackstock 1973. Estimation of lipids in marine animals and tissues : detailled investigation of the sulphophosphovanillin method for "total" lipids. J.exp.Biol.Ecol. **12**:103-118.

Brody,S. 1945. Bioenergetics and growth, 1023pp. New-York,Hafner Publ. C°. Inc..

Dubois,M., K.A.Gilles, J.K.Hamilton, P.A.Rebers & F.Smith 1956. Colorimetric method for determination of sugars and related substances. Analyt.Chem. **28**:350-356.

Féral,J.-P. 1983. Evolution du glycogène et des autres glucides au cours du cycle annuel de l'holothurie apode Leptosynapta galliennei. Symbioses **15**:229-230

Féral,J.-P. & D.A.Doumenc 1982. Glycogen in holothuroids living in subantarctic and temperate waters. In : **International Echinoderms Conference, Tampa Bay,** J.M. Lawrence (ed.), pp.381-385. Rotterdam,Balkema.

Féral,J.-P. & C.Massin 1982. Digsetive systems : Holothuroidea. In : **Echinoderm nutrition.** M. Jangoux and J.M. Lawrence (eds.), pp.192-212. Rotterdam,Balkema.

Fish,J.D. 1967. The biology of Cucumaria elongata (Echinodermata : Holothuroidea). J.Mar.Biol.Ass.U.K. **47**:129-143.

Giese,A.C. 1966. On the biochemical constitution of some echinoderms. In : **Physiology of Echinodermata,** R.A. Boolotian (ed.), pp.757-796. New-York,Interscience Publishers.

Hartree,E.F. 1972. Determination of protein : a modification of the Lowry method that gives a linear photometric reponse. Analyt. Biochem. **48**:422-427.

Lawrence,J.M. 1972. Carbohydrate and lipid levels in the intestine of Holothuria atra (Echinodermata : Holothuroidea). Pacif.Sc. **26**:114-116.

Lawrence,J.M. 1973. Level, content and caloric equivalents of the lipid, carbohydrate and protein in the body components of Luidia clathrata (Echinodermata : Asteroidea : Platyasterida) in Tampa Bay. J.exp.mar.Biol.Ecol. **11**:263-274.

Lawrence,J.M. & A.Guille 1982. Organic composition of tropical, polar and temperate-water echinoderms. Comp.Biochem.Physiol. **72B**:283-287.

Lawrence,J.M. & J.Kafri 1979. Numbers, biomass and caloric content of the echinoderm fauna of the rocky shores of Barbados. Mar.Biol. **52**:87-91.

Lawrence,J.M. & J.M.Lane 1982. The utilization of nutrients by post-metamorphic echinoderms. In : **Echinoderm nutrition,**M. Jangoux and J.M. Lawrence (eds.) pp.331-376. Rotterdam,Balkema.

Massin,C. 1978. Etude de la nutrition chez les holothuries aspidochirotes (échinodermes). Comportement alimentaire, structure et fonctions de l'appareil digestif. **Thèse de doctorat,** université libre de Bruxelles, 240pp. (unpublished).

Prim,P., J.M.Lawrence & R.L.Turner 1976. Protein,carbohydrate and lipid levels of the adult body wall of Actinopyga agassizi, Synaptula hydriformis and Pentacta pygmaea (Echinodermata : Holothuroidea). Comp. Biochem.Physiol. **55B**:307-309.

Sibuet,M. & J.M.Lawrence 1981. Organic content and biomass of abyssal holothuroids (Echinodermata) from the Bay of Biscaye. Mar.Biol. **65**:143-147.

Siegel,S. 1956. Non parametric statistics for the behavioral sciences. 312pp. McGraw-Hill

Smith,T. 1981. Feeding and aspects of the nutritional biology of the dendrochirote holothurian Neopentadactyla mixta (Oestergren, 1898) and the ectoparasitic gastropod Melanella alba Bowdich, 1882. **Ph. D. thesis,** Natn.Univ. Ireland, 117pp. (unpublished).

Sokal,R.R. & F.J.Rohlf 1969. Biometry, the principles and practice of statistics in biological research. 776pp. San Francisco : W.H. Freeman C°. Publ..

Tanikawa,E. & Yoshitani 1955. Studies on the nutritive value of the meat of sea cucumber (Stichopus japonicus Selenka).III.Bull.Fac.Fish.Hokkaido univ.5:346-347.

Tanikawa,E.,M.Akiba & S.Yoshitani 1955. Studies on the nutritive value of the meat of sea cucumber Stichopus japonicus Selenka.II.Bull.Fac.Fish.Hokkaido univ. 5:338-345.

Turner,R.L. & J.H.Dearborn 1979. Organic and inorganic composition of post-metamorphic growth stages of Ophionotus hexactis (E.A.Smith) (Echinodermata : Ophiuroidea) during intraovarian incubation. J.exp.mar.Biol.Ecol. 36:41-51.

Turner,R.L. & J.C.Rutherford 1976. Organic, inorganic and caloric composition of eggs, pentactulae and adults of the brooding sea cucumber Cucumaria curata Cowles (Echinodermata :Holothuroidea). J.exp.mar.Biol.Ecol. 24:49-60.

Walker,C.W. 1982. Nutrition of gametes. In :**Echinoderm nutrition,** M. Jangoux and J.M. Lawrence (eds.), pp.449-468. Rotterdam,Balkema.

Proceedings of the Fifth International Echinoderm Conference / Galway / 24-29 September 1984

Soluble and membrane bound peptide hydrolases of the *Holothuria forskali* Delle Chiaje digestive tract

D.B.JOHNSON, G.V.O'DONOGHUE & J.C.DONLON
Biochemistry Department, University College, Galway, Ireland

ABSTRACT: Five peptide hydrolases from the digestive epithelium of Holothuria forskali have been studied: two were membrane bound (designated M_1 and M_{11}) and three were cytosolic (S_1, S_{11} and S_{111}). Initial studies revealed separate but overlapping substrate specificities. Of seven tripeptides tested with the soluble enzymes only L-Leu-L-Leu-L-Leu and L-Leu-Gly-Gly were hydrolysed, and then only by S_{11}. M_1 and M_{11} also showed weak tripeptidase activity. Activity against dipeptides was considerably higher. In general the enzymes showed a preference for dipeptides having N-terminal hydrophobic aliphatic side chains.

The enzymes have been separated and partially purified, while S_{11} was purified to homogeneity.

1 INTRODUCTION

In earlier reports from this laboratory we have described studies on enzymes of the digestive tracts of some echinoderms. Among the enzymes detected were carbohydrases: the initial characterisation and subcellular localisation of disaccharidases have been described (Johnson et al., 1980; Clifford et al., 1982). While difficulties were encountered in detecting proteolytic activity in some species (Johnson et al., 1980) the presence of peptide hydrolase activity could be demonstrated (McGettigan et al., 1981). The peptide hydrolases of Holothuria forskali digestive tract have now been studied in more detail: they have been separated by centrifugation and by ion-exchange chromatography; their substrate specificities have been studied, and one of them has been purified to homogeneity.

Here we present our findings on the enzymes as a group: the more biochemical characterisation of the purified enzyme will be presented elsewhere.

2 MATERIALS AND METHODS

Amino acids and peptides were from the Sigma Chemical Co. (Poole, England) with the exception of L-Ileu-L-Ileu-L-Ileu, L-Tyr-Gly-Gly, L-Thr-Gly-Gly, L-Leu-L-Tyr-L-Leu, L-Leu-L-Pro, L-Leu-L-TyrNH_2 and L-Met-L-Pro which were from Bachem Feinchimikalien (Switzerland). L-amino acid oxide (type VI, from Crotalus atrox) and peroxidase (type II, from horseradish) were from the Sigma Chemical Co.

Animals, Holothuria (Panningothuria) forskali (Delle Chiaje) were collected off the coast of Galway and brought to the laboratory in sea water. They were kept at 4°C overnight, during which intestinal contents were voided. The anterior intestine (Feral & Massin, 1982) was excised from each animal and washed with ice cold physiological saline. The digestive epithelia were extruded (Johnson et al., 1980) and homogenised in Tris/maleate buffer (5 mM, pH 8.4, buffer A) with a Potter-Elvehjem homogeniser (6 up and down strokes, 900 r.p.m. 1 g wet tissue/5 ml buffer): homogenates were also prepared in Tris/Cl^- buffer (10 mM, pH 8.0; buffer B) under the same conditions. From this homogenate a cytosolic fraction was prepared by centrifugation at 100,000 x g for 90 min.

A brush border membrane fraction was prepared by the method of Schmitz et al. (1973), i.e. by differential centrifugation and by the use of $CaCl_2$ to aggregate microsomes. The brush border pellet obtained following centrifugation at

100,000 x g was washed x 4 with imidazole Cl^-(5 mM)-mannitol (50 mM) buffer (pH 7.0) in order to free the preparation of soluble enzymes. The membrane bound enzymes were solubilized with Triton X-100 (1% v/v) in buffer B.

Starch gels were prepared as described by Smithies (1965) in buffer A which was used also during electrophoresis (150V, 15 h).

Peptidase activity was measured by determination of residues liberated from substrates: that determination was by spectrofluorimetry, following incubation with L-amino acid oxidase, horseradish peroxidase and p-hydroxyphenylacetic acid (Nicholson & Peters, 1978). Following electrophoresis zymograms were developed according to the same principle: an overlay was placed on the starch gel, with L-amino acid oxidase, peroxidase and KI (Donlon & Fottrell, 1972).

Protein was estimated by the method of Hartree (1972).

Ion exchange chromatography was carried out on DEAE-cellulose (DE-52, Whatman Biochemicals), equilibrated in buffer B, and poured to form a column (1.3 x 40 cm) which was then further washed with 3 column volumes of the same buffer. Following sample application a further wash was carried out (2 column volumes of buffer B), and irrigation was continued with a linear gradient of NaCl (0-0.35 M) in buffer B at a flow rate of 17 ml/h. Fractions (1.6 ml) were collected and their A_{280}, peptidase activity (against L-Leu-L-Leu) and conductivity values were determined. When solubilised brush border enzymes were chromatographed the buffers contained Triton X-100 (0.2% v/v).

3 RESULTS

3.1 Detection and designation

When homogenates of H. forskali digestive epithelia were subjected to starch gel electrophoresis, and zymograms were developed, with L-Leu-L-Leu as substrate, four bands of activity were detected, one remaining at the origin (Figure 1A). It was concluded that that band was due to the presence of membrane bound activity. This was shown to be so when cytosolic and solubilised brush border membrane fractions were electrophoresed separately. Thus the cytosolic fraction lacked the band at the origin but contained the three migrating bands. These were designated S_1, S_{11} and S_{111} in order of increasing electrophoretic mobility (Figure 1B). When the solubilised

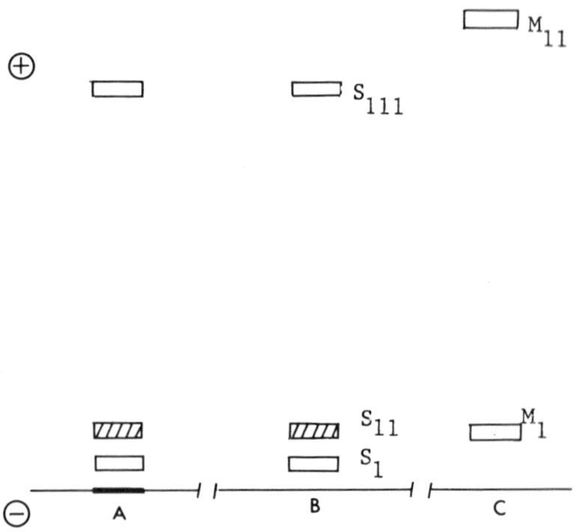

Figure 1. Zymograms of peptidases from A: total homogenate; B: cytosol; C: brush border of H. forskali intestinal mucosa.

brush border membrane fraction was examined two bands of activity were detected, and were designated M_1 and M_{11} (Figure 1C).

3.2 Substrate specificities of cytosolic enzymes

The zymogram development method was used to compare the substrate specificities of the cytoplasmic enzymes. Thus an overlay was applied to an electrophoretogram with the peptide to be tested together with the ancillary enzymes: the appearance of a blue colour indicated enzymic hydrolysis and the electrophoretic mobility of the activity responsible for it. Table 1 shows a range of compounds tested, including only those which gave positive results: the relative colour intensities after 3 h development are indicated by plus and minus symbols. Unless otherwise shown the amino acids residues were of the L configuration. Of seven tripeptides tested only two were acted on and only by S_{11}. Gly-Gly-Leu, Ileu-Ileu-Ileu, Leu-Tyr-Leu, Thr-Gly-Gly and Tyr-Gly-Gly were not hydrolysed, as judged by this technique. It is clear that dipeptides are the preferred substrates and that there is a high degree of overlap in the substrate specificities of the three enzymes. Distinct specificities are clearly indicated, however, by inspection of Table 1. Excluded from the Table are the negative results obtained with Gly-Trp, Gly-Tyr, γ-Glu-Leu, Glu-Trp, Glu-Tyr, Leu-D-Leu, D-Leu-Leu, D-Leu-Tyr, Leu-Pro, Pro-Ilue, Pro-Tyr, Gly-Gly and Leu-β-naphtyhlamide.

Table 1 Peptides hydrolysed by the cytoplasmic enzymes

Substrate	S_1	S_{11}	S_{111}
Ala-Leu	-	-	+++
Gly-Leu	-	-	+++
Pro-Leu	-	-	++
Leu-Phe	-	++	+++
Leu.NH_2	+	-	++
Val-Leu	-	++	++
Leu-Leu	++	++	++
Leu-Ala	++	++	+++
Leu-Gly	+	+++	+
Val-Phe	+	++	+++
Ala-Phe	+	+	++
Phe-Gly	++	++	-
Leu-Gly-Gly	-	+	-
Leu-Leu-Leu	-	+	-

Table 2 Relative rates of hydrolysis

Substrate	M_1	M_{11}	S_{11}
Leu-Gly	100	100	100
Leu-Leu	97	97	58
Leu-Ala	101	93	95
Leu-Tyr	76	67	53
Val-Leu	53	78	12
Ala-Leu	16	107	NT*
Gly-Leu	0	96	0
Gly-Trp	0	68	0
Leu-Pro	0	78	0
Ala-Phe	0	57	1
Pro-Trp	0	21	0
Leu-Gly-Gly	92	24	11
Leu-Leu-Leu	31	62	9
Leu-Leu-Gly	19	32	NT*
Leu-Tyr-Leu	0	0	NT*
Trp-Met-Asp-Phe-NH_2	0	0	NT*

The differences in substrate specificities indicate that the cytosolic activities detected represent distinct enzymes, and that their different electrophoretic mobilities are unlikely to result from simple chemical differences occurring naturally or arising during extraction.

3.3 Substrate specificities of membrane bound enzymes

The substrate specificities of the brush border enzymes were investigated by the quantitative spectrofluorimetric method. To permit that M_1 and M_{11} (10 ml samples) were separated chromatographically on DEAE-cellulose. Figure 2 represents the elution pattern obtained. Starch gel electrophoresis showed that the first peak of activity, which was eluted at 125 mM NaCl, was M_1, while the second peak which was eluted at 280 mM NaCl, was M_{11}. These partially purified samples were incubated with a variety of peptides, and the spectrofluorimetric procedure was used to determine relative rates of hydrolysis of the peptides. Table 2 shows the results so obtained, and indicates both overlap and distinct differences. Considerable hydrolysis of some tripeptides is also indicated.

3.4 Comparison of membrane bound and soluble enzymes

It is clear (Figure 1) that M_1 and S_{11} have identical electrophoretical mobilities, and Tables 1 and 2 suggest a similarity of substrate specificities. While due precautions were taken to ensure that cytosolic activity did not adsorb on the membrane fraction during extraction procedures it was important to rule out this possibility by quantitative study of S_{11}. It was necessary, therefore, to separate S_{11} from other cytosolic peptidases.

Homogenates, prepared in buffer B, were centrifuged at 100,000 x g for 90 min and the supernatant solutions were fractionated with ammonium sulphate. The 45-90% saturation fraction was desalted on Sephadex G-25 and applied (4 ml) to a DEAE-cellulose column. Figure 3 shows that three peaks of peptidase activity were readily separated. S_1 was eluted unretarded, while S_{11} was eluted at a salt concentration of 32 mM. Approximately 8% of the total peptidase activity of the initial homogenate was recovered as S_{11} at this stage.

The relative rates of hydrolysis by S_{11} of some substrates was measured by the spectrofluorimetric method and the results are shown in Table 2. It is clear that M_1 and S_{11} have distinct but overlapping substrate specificities, and are,therefore distinct enzymes.

3.5 Further study of S_{11}

Following its separation on DEAE-cellulose S_{11} has been purified to homogeneity by a

*Not tested.

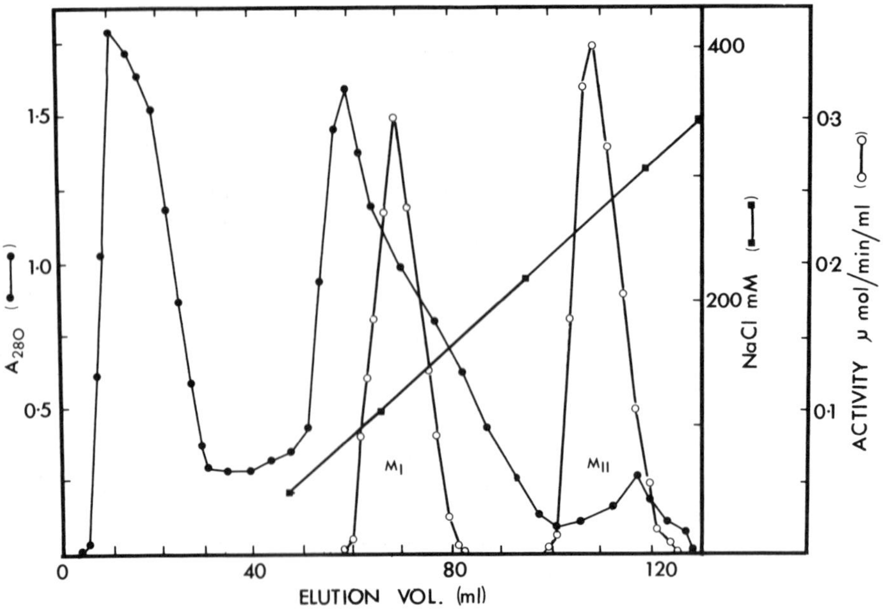

Figure 2. Elution of brush border peptidases from DEAE cellulose

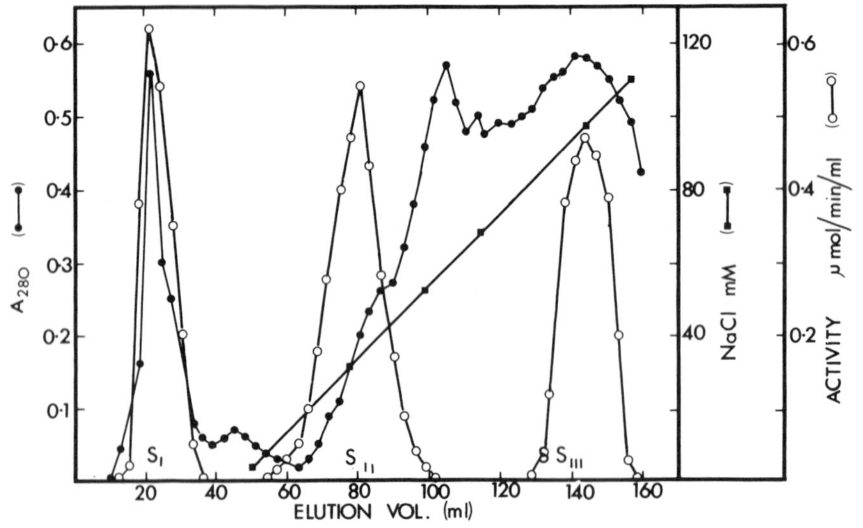

Figure 3. Elution of cytosolic peptidases from DEAE cellulose

combination of chromatofocusing and gel filtration on Sephadex G-200. S_{11} was selected because it was eluted from the ion-exchanger (Figure 2) at a lower concentration of protein than were S_1 or S_{11}, although in total homogenates S_1, S_{11} and S_{111} represented 7%, 38% and 51% of soluble activity. This was determined by extraction of unstained electrophoretograms and assay of the extracts. The apparent molecular weight was found to be 319,000, and a time course study of L-Leu-Gly-Gly hydrolysis (monitored by automated amino acid analysis) showed the enzyme to be an aminopeptidase. Proton induced X-ray emission analysis showed that inactivation of the enzyme by EDTA was accompanied by loss of Zn^{++}.

3.6 Bacterial proteases of H. forskali intestine

The presence of brush border peptidases in H. forskali is consistent with the availability of peptides in the gut. Whether such peptides arise from external sources or from proteolytic activity in the gut has not been established. We have previously reported (Clifford et al., 1982) that PMSF sensitive proteolytic activity could be detected in H. forskali intestine. More recently we have transferred gut contents from that holothurian into sterile bacterial growth media and have cultured the samples at 10°C, 18°C and 37°C. All growth cultures were shown (Walsh & Johnson, unpublished observations) to contain proteolytic activity. Fractionation of the samples on sephadex G-100 revealed a complex pattern of protease activity, not all of which was suppressed by PMSF. Thus it is clear that H. forskali intestine has available a range of proteases of bacterial origin.

When gut contents, without culturing, were centrifuged and applied directly to G-100 columns the elution profiles differed from those obtained with the growth media. This may infer the presence of proteases originating in the animal. However, autodigestion of the proteases may have compliated the profiles in an uncontrolled manner, so that further investigation of this point is essential.

4 DISCUSSION

The digestive epithelia of H. forskali have been shown to contain multiple forms of peptide hydrolase activity and the work of McGettigan et al. (1981) has been extended. On the basis of electrophoretic mobilities and substrate specificities five distinct enzymes have been indicated and they have been successfully separated by subcellular fractionation and by ion-exchange chromatography.

It is clear that the enzymes have broad substrate specificities, although the present study was not fully exhaustive. It should be pointed out that both the zymogram development procedure and the quantitative enzyme analysis depend on the action of amino acid oxidase. Such enzymes have limited substrate specificities, and this in turn limited the range of peptides which could be tested as possible substrates for the peptidases isolated.

Though broad, the substrate specificities differ. S_1 and S_{111} share with leucine aminopeptidase the ability to hydrolyse leucinamide ($leuNH_2$) while S_{11} does not (Smith & Hill, 1960). With the peptides tested S_{11} exhibited tripeptidase activity while S_1 and S_{111} did not.

In general the enzymes showed a preference for dipeptides having N-terminal hydrophobic aliphatic side chains.

Studies of multiple forms of peptidase activity of digestive epithelia generally establish the existence of two groups of enzymes, one bound to the brush border membranes and one soluble or cytosolic. Such studies aim to establish the substrate specificities of the enzymes in order to assign roles (possibly complimentary) to them in the digestive process. Thus brush border enzymes hydrolyse some peptides before uptake into the absorptive cells (Silk, 1981) while smaller peptides, poorly hydrolysed by the brush border enzymes, are absorbed and hydrolysed by the cytosolic enzymes. For a full study of substrate specificities, however, the catalytic constants kcat and Km must be determined (their ratio is the specificity constant) and this can be achieved only with highly purified enzymes. In a continuation of the work presented here the enzyme S_{11} has been purified to homogeneity. Molecular and kinetic properties have been studied and the chemistry and topography of the active centre are under investigation. These results will be presented at a later date.

REFERENCES

Clifford, C., J. Walsh, N. Reidy & D.B. Johnson 1982. Digestive enzymes and subcellular localization of disaccharidases in some echinoderms. Comp. Biochem Physiol. 71B, 105-110.

Donlon, J. & P.F. Fottrell 1972. Studies on substrate specificities and subcellular location of multiple forms of peptide hydrolases in guinea pig intestinal mucosa. Comp. Biochem. Physiol. 41B, 181-193.

Feral, J.-P. & C. Massin 1982. Digestive systems: holothuroidea. In M. Jangoux & J.M. Lawrence (eds) Echinoderm Nutrition, p. 191-212 Rotterdam: Balkema.

Hartree, E.F. 1972. Determination of protein: a modification of the Lowry method that gives a linear photometric response. Analyt. Biochem. 48, 422-427.

McGettigan, S., M. Canning, G. O'Cuinn & D.B. Johnson 1981. Peptide hydrolases in holothurian intestinal mucosa. Comp. Biochem. Physiol. 69C, 169-170.

Nicholson, J.A. & T.J. Peters 1978. Fluorometric assay for intestinal peptidases. Analyt. Biochem. 87, 418-424.
Schmitz, J., H. Preiser, D. Maestracci, B. K. Ghoshi, J.J. Creda & R.K. Crane 1973. Purification of the human intestinal brush border membrane Biochim. Biophys. Acta 323, 98-112.
Silk, D.B.A. 1981. Peptide transport. C in. Sci. 60, 607-615.
Smith, E.L. & R.L. Hill 1960. Leucine-aminopeptidase. In Boyer, P.D., H.A. Lardy & C. Myrbeck (eds). The Enzymes p. 37-62. New York: Academic Press.

Proceedings of the Fifth International Echinoderm Conference / Galway / 24-29 September 1984

Geographic variation of calcareous ossicles and the identification of three species of sea cucumbers (Echinodermata: Holothuroidea) from the eastern Pacific Ocean

PHILIP LAMBERT
British Columbia Provincial Museum, Victoria, Canada

ABSTRACT: Cucumaria curata Cowles, 1907 and C. pseudocurata Deichmann, 1938 are shown to be distinct species, contrary to Rutherford (1977a). C. curata is confined to Pacific Grove, California. C. pseudocurata is common from California to British Columbia. Morphological evidence is also presented which suggests that C. pseudocurata is synonymous with the northern species, C. vegae Theel, 1886.

1 INTRODUCTION

On the west coast of North America three poorly described species of small, intertidal holothurians have been confused and misidentified because of their external similarities. The three species in question are Cucumaria vegae Theel, 1886 described from the Commander Islands, U.S.S.R., C. curata Cowles, 1907 from Pacific Grove, California and C. pseudocurata Deichmann, 1938 from San Mateo Co., California.

The purpose of this paper is to firstly show that Rutherford (1977a) incorrectly synonymized C. pseudocurata and C. curata. Secondly, preliminary morphological evidence is presented to suggest that C. pseudocurata and C. vegae may be synonymous and actually represent two extremes in a geographic cline.

2 MATERIALS AND METHODS

Specimens were obtained from museum and field collections. A list of specimens is available and those collected by the author are housed in the British Columbia Provincial Museum. Field specimens were relaxed with menthol crystals, fixed with neutralized 10 per cent formalin, washed and transferred to 60 per cent isopropyl alcohol.

Ossicles were extracted from pieces of skin (2 mm square) from the mid-dorsal region of specimens between 1 and 3 cm long. The skin was dissolved with fresh household bleach on a glass microscope slide. Bleach was pipetted off and the remaining ossicles were flooded at least three times with distilled water. After drying the ossicles on a slide warmer at 45 C, a permanent slide was prepared with Permount (Fisher Scientific Co.).

Samples of 20 ossicles from each slide were measured to the nearest 1 um. Measurements were obtained with a calibrated scale superimposed on the ossicle through a drawing tube. Transects were made at regular intervals across the slide and only those ossicles that appeared totally in the field were measured. Broken ossicles and those on edge were rejected. The length, width, and number of holes were recorded for each ossicle. Width was taken at a point half way between the two ends at right angles to the long axis.

With the exception of the factor analysis, which was run on the Statistical Package for the Social Sciences (SPSS) (Nie, Hull, Jenkins, Steinbrenner and Bent, 1975) all statistics were run on NWA Statpak (Northwest Analytical, Inc., Portland, Ore.).

3 RESULTS AND DISCUSSION

3.1 Comparison of C. pseudocurata and C. curata

Rutherford (1977a) concluded that C. pseudocurata and C. curata are synonymous. His evidence gives the impression that he never saw a true specimen of C. curata. Rutherford tried to show that the ossicles of C. pseudocurata fit the description for C. curata and made statements which,

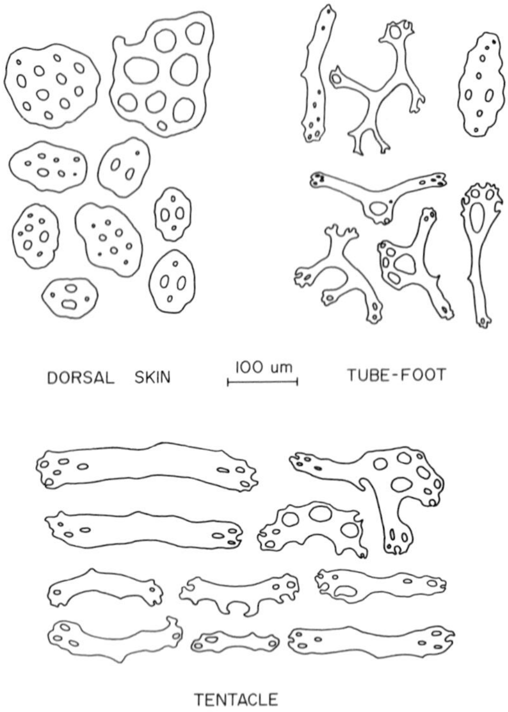

Figure 1. Ossicles of C. curata from Yankee Pt. Pacific Grove. Note 4-holed buttons in dorsal skin.

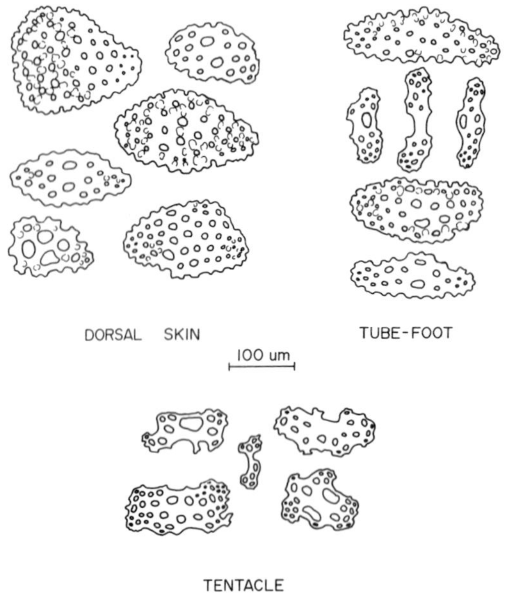

Figure 2. Ossicles of C. pseudocurata from Malpaso Creek, California. Note rod-shaped plates in tube-foot.

upon careful analysis of the original descriptions and specimens of both species, cannot be justified. For example, he stated "...the 4-holed buttons, which correspond with my elongated or rod-shaped plates ...". Having seen the 4-holed buttons in the dorsal skin of the true C. curata (Figure 1) it is difficult to see how they could be confused with the rod-shaped plates in the tube-feet of C. pseudocurata (Figure 2). The original description and illustrations by Cowles (1907) clearly show that the ossicles of the skin are mostly thick, 4-holed buttons, the podia have characteristic 3-legged ossicles, and the tentacles have elongate perforated rods -- 3 distinctive types of ossicles in the 3 regions of the body. Deichmann's description of C. pseudocurata, on the other hand, shows that the skin and tube-feet have oval perforated plates, the tube-foot ossicles being narrower. There are no 3-legged ossicles in the podia of C. pseudocurata. Figures 1 and 2 show the ossicles of these two species from three regions of the body.

One reason for misidentifications in the literature is the external similarity of these two species. Both are 1 to 3 cm long and dark-coloured, but there are several distinctive external characters (see Table 1). I found many museum specimens incorrectly identified as C. curata. These as well as my own field collections, indicate C. curata is only known near Pacific Grove, California in an area of exposed rock with encrusting coralline algae at zero tide (U.S. datum). C. pseudocurata, on the other hand, is common from California to British Columbia, especially within beds of Mytilus californianus at the mid-tide level. At least five references to C. curata (Filice 1950, Smith 1962, Brumbaugh 1965, Rutherford 1977a, 1977b) contain enough descriptive information to identify the specimens as the more common C. pseudocurata rather than C. curata. Other references to C. curata outside the Pacific Grove region are probably incorrect.

In summary, Rutherford's conclusion that C. curata and C. pseudocurata are synonymous is not justified as there is ample evidence that morphologically and ecologically these two species are distinct.

3.2 Synonymy of C. pseudocurata and C. C. vegae

C. pseudocurata has many features in common with C. vegae Theel, 1886, a spe-

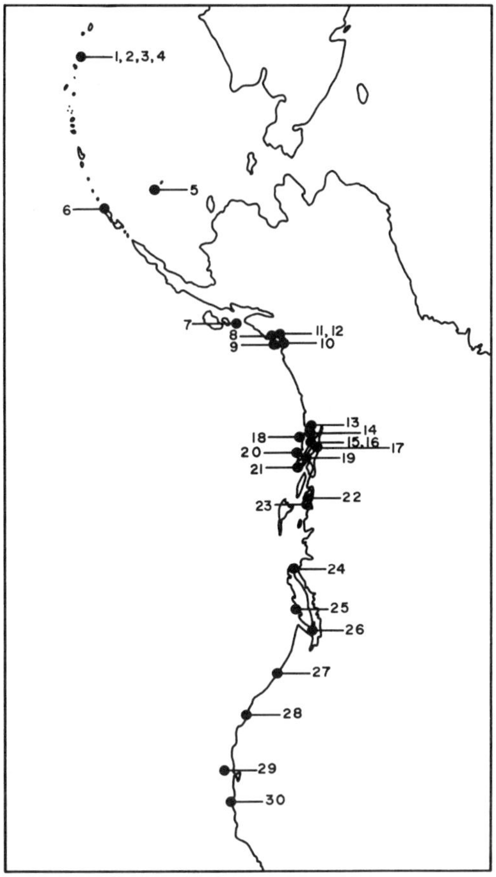

Figure 3. Locations of collection sites on the west coast of North America.

Table 1. Summary of habitat and morphological features.

	C. curata	C. pseudocurata
Habitat	zero tide	mussel zone to zero tide
Tentacles	10 equal	usually 8 equal, 2 smaller
Tube Feet	in 5 rows and dorsal interambulacrum	5 rows
Gonopore	1-20 papillae	1 papilla
Ossicles	See Fig. 1	See Fig. 2

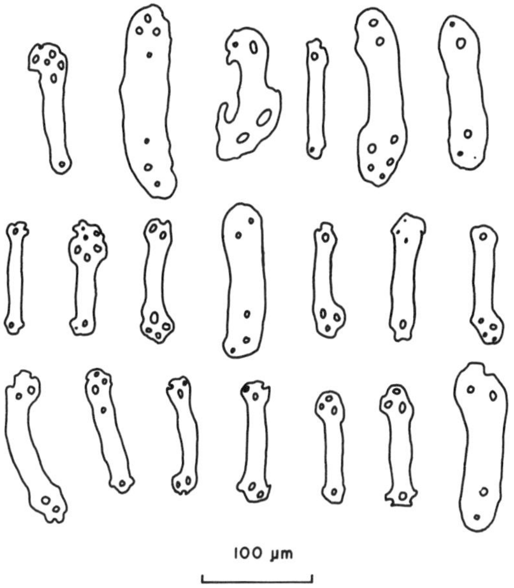

Figure 4. Ossicle sample from site 3 - Buldir Island.

cies described from the Commander Islands at the western end of the Aleutian chain. Externally these two species are indistinguishable. They both have 8 equal and 2 smaller tentacles and 5 rows of podia. However, the skin ossicles of C. vegae from the Aleutians (Figure 4) are made up, primarily, of "spectacle rods" while those of C. pseudocurata from California (Figures 7 and 8) are oval perforated plates. However, samples of ossicles from specimens at a series of 30 sites between these two extremes (Figure 3), show a cline from the northern form to the southern form (Figures 4 to 8). In the area of northern B.C. and S.E. Alaska it is difficult to subjectively assign specimens to either species because of the intermediate nature of the skin ossicles.

To assess the ossicles more objectively, the length, width and number of holes were measured in 20 ossicles from one specimen at each site. Figure 9 shows the mean values $\pm$ 1 standard deviation for length, width, and number of holes at each site. Mean lengths show a wide range of variability with no clear-cut trend related to geographic region. Mean widths are fairly equal from sites 1 to 23 but the means of sites 24 to 30 are substantially higher. The mean number of holes shows a similar trend but sites 26 and 28 are similar to northern spe-

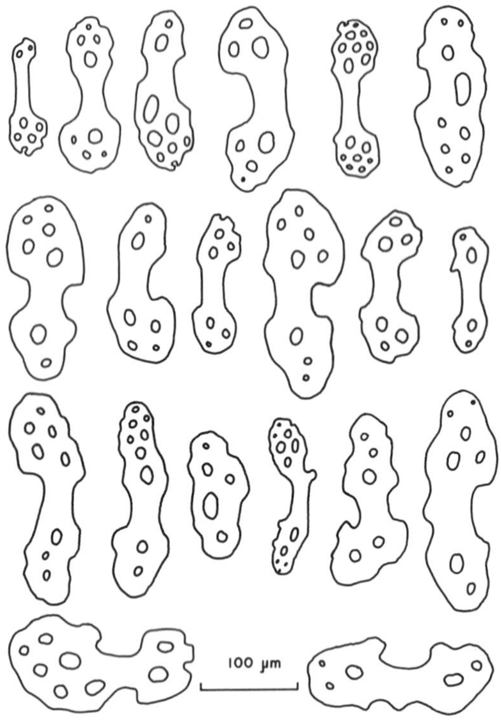

Figure 5. Ossicle sample from site 7 - Sud Island.

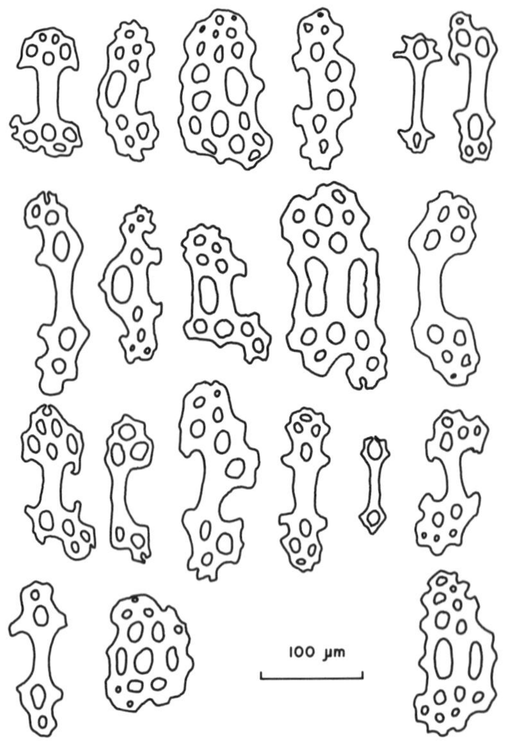

Figure 6. Ossicle sample from site 20 - Tara Island.

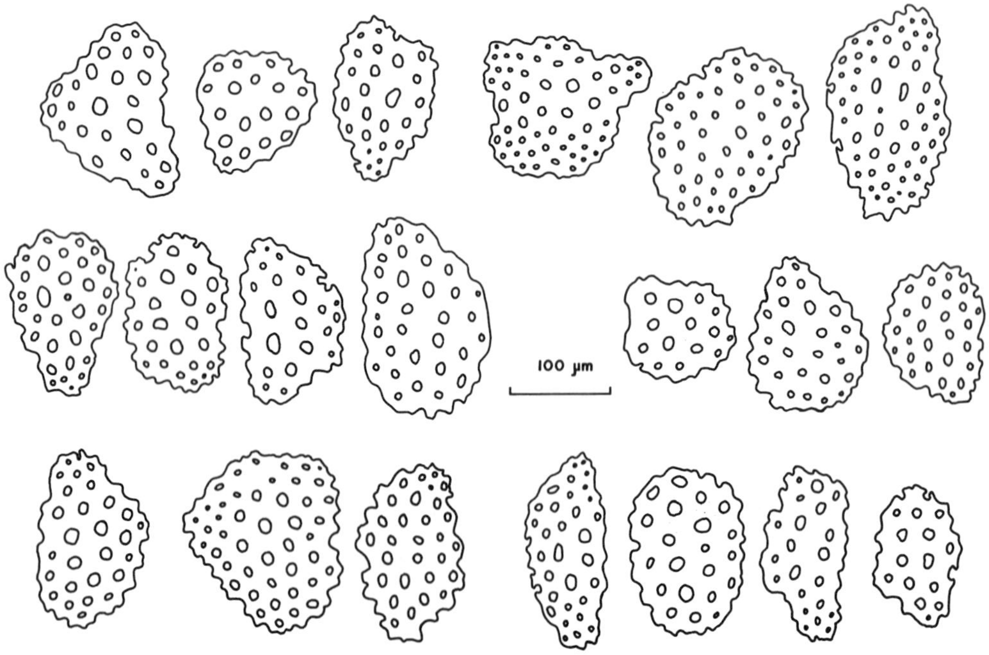

Figure 7. Ossicle sample from site 29 - Farallon Islands.

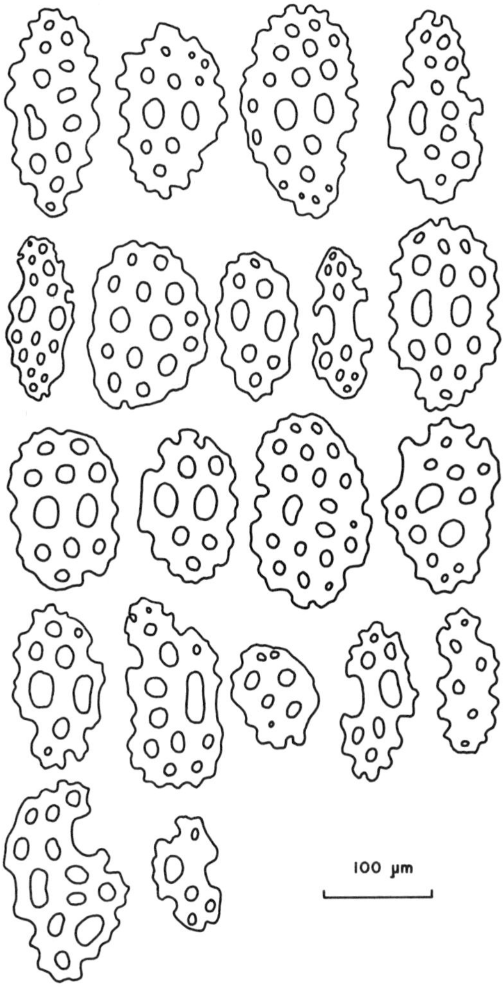

Figure 8. Ossicle sample from site 28 - Trinidad Head.

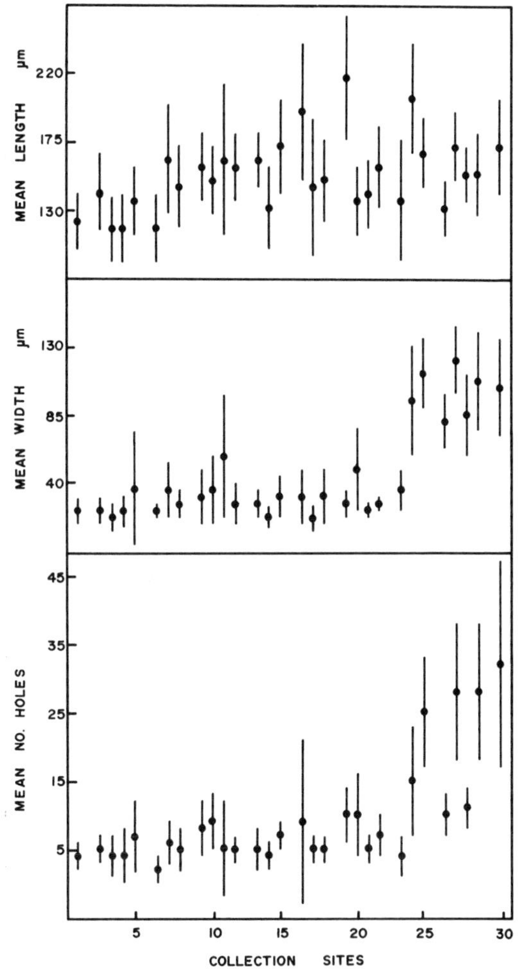

Figure 9. Means (solid circles) ± 1 standard deviation (vertical line) of measurements from 20 ossicles from 1 specimen at each of 30 sites.

cimens. No one geographic region stands out as significantly different in all three measurements. However, the graph of mean width does show an apparent separation of southern sites (24 to 30) from those north of site 24. Morphologically, this seems like a significant break; however, there are 450 km between site 24 and site 25 (Bull Harbour, Vancouver Island north to Dundas I. near the Alaska border). Samples from between these two locations might exhibit transitional ossicles.

Furthermore, Rutherford (1977a) showed that despite obvious changes in the ossicles of C. pseudocurata (he called it C. curata) from oval plates in California to elongate plates in northern B. C., there was no detectable difference in 8 enzyme systems analyzed by starch gel electrophoresis. This result implies that ossicle variations may be due to environmental factors and do not reflect genetic differences. However, this does not rule out the possibility that testing other enzyme systems might reveal some differences.

Figure 10 illustrates ossicle measurements as a histogram of the frequency distributions of length vs. width. The sites were divided into 3 areas: Aleutians (sites 1-6), S.E. Alaska (sites 7-23),

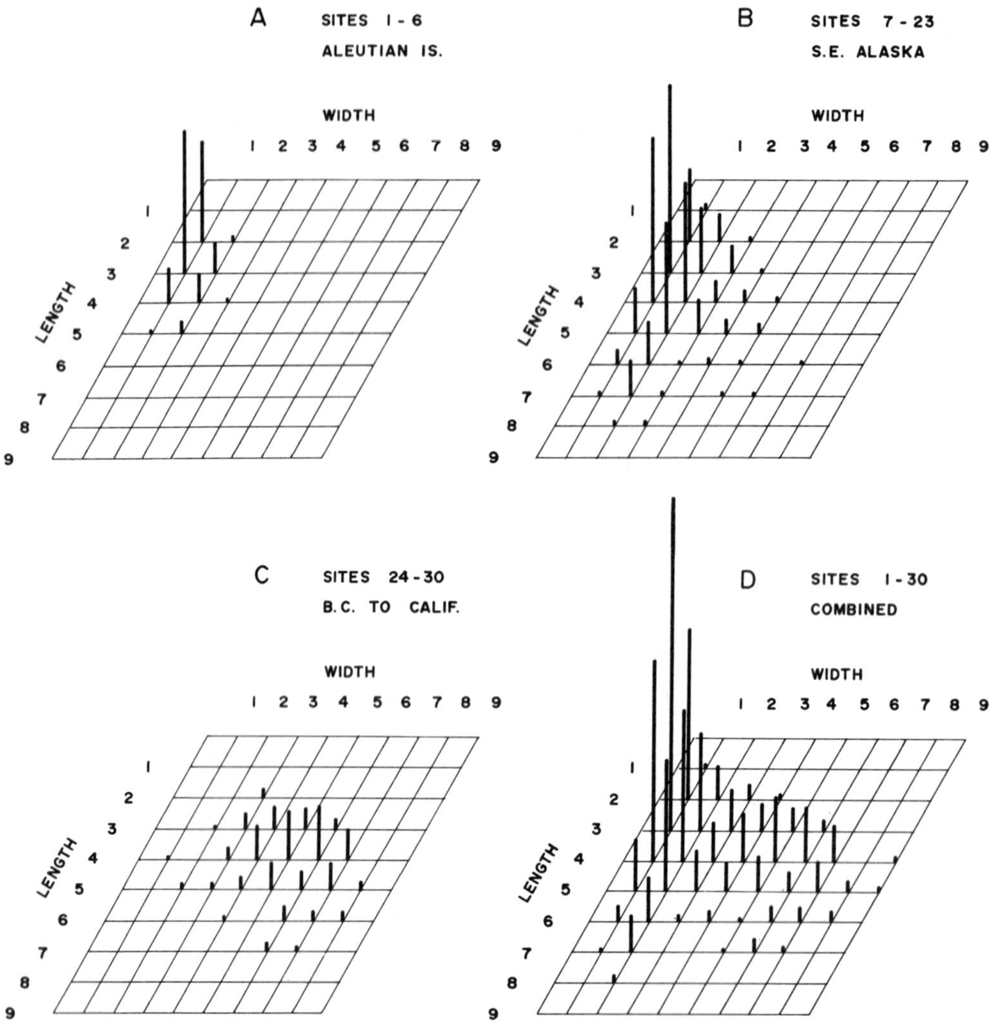

Figure 10. Histograms represent the number of ossicles of a particular length and width. The height of one grid square equals 10 ossicles. Lengths 1 to 9 represent steps of 30 microns between 50 and 350. Widths 1 to 9 represent steps of 20 microns from 5 to 205. (A) size distribution of 120 ossicles from sites 1 to 6 (B) size distribution of 340 ossicles from sites 7 to 23 (C) size distribution of 140 ossicles from sites 24 to 30 (D) size distribution of 600 ossicles from all 30 sites.

and B.C. to California (sites 24 to 30) and plotted in Figure 10A, B, and C to show how ossicles from each area contribute to the combined histogram (Figure 10D). This shows a main peak at the upper left tapering to a smaller peak at the lower right. Plots of length vs. holes and width vs. holes produced similar shaped histograms. The slight valley between the two peaks may also reflect the lack of specimens between Vancouver Island and the Alaska border.

Factor analysis of the ossicle data from 30 sites (Figure 11) did not separate them into discrete geographic groups either. The sites plotted out in a linear fashion with a cluster of northern sites on the right merging into southern sites on the left. Site 24 (Bull Harbour) is intermediate in ossicle shape and geographic location. However, not all sites plot out in their geographic order, site 19 is intermediate geographically but plots out on the far right, while the

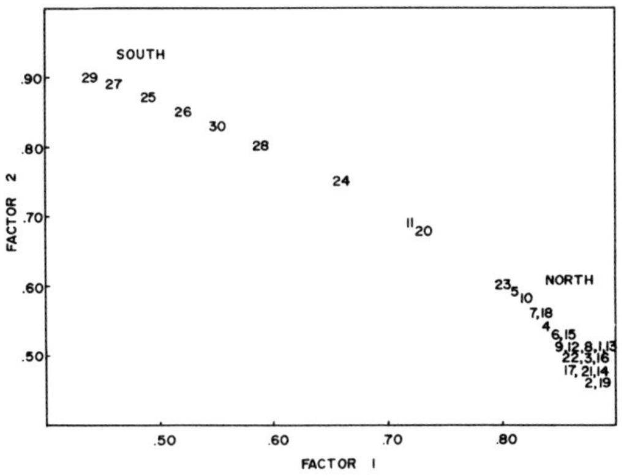

Figure 11. Factor analysis of ossicles from 30 sites (locations as in Figure 3). Northern ossicles are short and narrow. Southern ossicles are large oval plates.

southernmost site (30) is not on the far left as one might expect.

On the basis of morphological characters studied, C. pseudocurata and C. vegae cannot be easily separated. Although externally identical, they appear to represent the two extremes in a geographic cline when the skin ossicles are considered. The data suggest a possible break in the cline in the region between Vancouver Island and S.E. Alaska but a lack of collection sites between these two points and the electrophoretic evidence of Rutherford (1977a) make a species boundary here seem doubtful. Studying ossicle data alone will not answer conclusively the question of synonymizing C. vegae and C. pseudocurata. More information is needed regarding habitat, reproduction, behaviour, and genetic makeup over the entire geographic range. The work reported here is a preliminary study to indicate where efforts might be concentrated to resolve this question. Electrophoretic studies seem to be the next logical step to determine if northern specimens are genetically identical with southern forms.

4 ACKNOWLEDGEMENTS

The Friends of the British Columbia Provincial Museum kindly provided funding for field work in Washington, Oregon and California. I would like to thank the following individuals and institutions for collecting specimens and loaning museum material to me: J. Brumbaugh, M. Roberts, Scott Smiley, Rick Kool, Gordon Green, R. Van Syoc, California Academy of Sciences, Mary Sue Brancato, Museum of Comparative Zoology, Charles O'Clair, Auke Bay Laboratory, Nora Foster, University of Alaska Museum. I would also like to thank Erica Bates and Marilyn Lambert for typing the manuscript.

REFERENCES

Brumbaugh, J. H. 1965. The anatomy, diet, and tentacular feeding mechanism of the dendrochirote holothurian Cucumaria curata Cowles, 1907. Ph.D. Thesis, Stanford University.

Cowles, R. P. 1907. Cucumaria curata sp. nov. Johns Hopkins University Circular. New Series No. 3, 195:180-181.

Deichmann, E. 1938. New holothurians from the western coast of North America and some remarks on the genus Caudina. Proc. New England Zool. Club 16:84-93.

Filice, F. P. 1950. A study of some variations in Cucumaria curata (Holothuroidea). Wasmann Journal of Biology 8(1):39-47.

Nie, N. H., C. H. Hull, J. G. Jenkins, K. Steinbrenner and D. H. Bent. 1975 SPSS - Statistical Package for the Social Sciences, Second Edition. McGraw Hill Inc., 675 pp.

Rutherford, J. C. 1977a. Geographical variation in morphological and electrophoretic characters in the holothurian Cucumaria curata. Marine Biology 43: 165-174.

Rutherford, J. C. 1977b. Variation in egg numbers between populations and between years in the holothurian Cucumaria curata. Marine Biology 43:175-180.

Smith, E. H. 1962. Studies of Cucumaria curata Cowles, 1907. Pacific Naturalist 3(5):233-246.

Theel, H. 1886. Report on the Holothuroidea dredged by H.M.S. Challenger during the years 1873-76. Part II. Challenger Report 14(Part 39):1-290.

Proceedings of the Fifth International Echinoderm Conference / Galway / 24-29 September 1984

The tentacular ultrastructure of dendrochirote holothurians
A comparative SEM study

J.DOUGLAS McKENZIE
Department of Zoology, Queen's University of Belfast, Northern Ireland

ABSTRACT: The tentacles of shallow-water, north European dendrochirote holothurians have been compared at an ultrastructural level using scanning electron microscopy (S.E.M.). Apparent differences between specimens of the same species are mostly ascribable to contraction of the tentacles and the loss of the fuzzy coat and cuticle in some specimens. In Ocnus regeneration of tentacles is seen as a possible source of intraspecific variation in tentacle texture. All the species studied possess papillae and cilia on their tentacles. Differences were found in the fine structure of the tentacles between Thyone species and the other species studied. Structures seen on Duasmodactyla tentacle trunks are considered to have a sensory function. The structural differences and, in some cases, lack of them are discussed in relation to interspecific competition.

INTRODUCTION

In European waters several species of dendrochirote holothurians may be found within the same habitat. Micro-habitat partitioning occurs in some species, supporting Kohn's (1978) hypothesis that habitat rather than food partitioning is the primary mechanism for avoiding competition between suspension feeders. In other cases no such partitioning is obvious (pers. observ.) and interspecific competition is possible. While Massin (1982) states that dendrochirotes all employ the same feeding mechanism and have the same tentacular form, too few studies are available to accept this without further investigation. Indeed suggestions for a radical departure from normal dendrochirote suspension feeding into cryptic deposit feeding have been made for some species (Mortensen 1927, Fankboner 1981) though the evidence for this is weak (McKenzie and Picton 1984).

Large comparative studies at the ultrastructural level have not yet been made for holothurian species though Fankboner (1978, 1981) and Smith (1983) have made valuable starts on this with three dendrochirote species. This paper presents results of a comparative S.E.M. study on north European dendrochirote species. The results of a comparative study using T.E.M. and of other aspects of dendrochirote biology are in preparation for publication elsewhere.

MATERIALS AND METHODS

Specimens were collected by diving or dredging in the sub-littoral zone or from under rocks in the littoral zone. Collection sites include: Brown Rock and Kircubbin, Strangford Lough; St. John's Point, Donegal; Loch Long, Scotland.

Tentacles were dissected from the bodies and fixed in a 3:1 mix of 4% gluteraldehyde (made up in 0.1 M sodium cacodylate buffer with 1% NaCl) and 1% osmium tetroxide (made up in distilled water). Fixation was for one hour followed by a wash in distilled water and dehydration in an acetone series, critical point drying and gold sputter coating. Specimens were examined using a Jeol-35CF. Some specimens were fixed in 4% gluteraldehyde, made up as above, washed and stored in cacodylate prior to post-fixation in 1% OsO_4. Both primary and post-fixation was for one hour.

Specimens of Ocnus brunneus (Forbes), O. lacteus (Forbes and Goodsir), Aslia lefevrei (Barrois), Pawsonia saxicola (Brady and Robertson), Leptopentacta elongata (Düben and Koren), Paracucumaria hyndmani (Thompson), Thyone fusus (O. Fr. Muller), T. roscovita (Hérouard), Duasmodactyla commune (Forbes), Thyonidium pellucidum (Fleming) and Psolus phantapus (Strussenfelt) were examined though no results are presented for Paracucumaria as the tentacles of the only specimen examined were too contracted to provide useful information. The S.E.M. results of Smith (1983) for Neopentadactyla mixta (Östergren) were compared with the findings of the present study.

RESULTS

Considerable differences were often found between specimens of the same species. Most of this variation stemmed either from different states of tentacle and papillae contraction or from loss of the extracellular fuzzy coat and cuticle (Holland and Nealson 1978). Figure 1 shows the effects of cuticle loss on Aslia lefevrei. In 1a and b only slight damage has occurred to the fuzzy coat. In 1b the fuzzy coat can be seen to be just breaking up, exposing papillae secretions. Naked ended cilia (c) are clearly visible. The fuzzy coat and much of the cuticle has been lost in 1c and d. This loss is probably due to physiological trauma rather than a result of fixation as specimens of the same species collected on different days but fixed simultaneously demonstrated such loss in the first specimens captured but not in the one captured the next day. Thyone species were rather more robust than the others with fuzzy coat and cuticular loss seen only rarely. This was surprising as Thyone specimens readily undergo autolysis and evisceration when captured.

Ocnus lacteus and O. brunneus show differences in tentacle texture unrelated to contraction or cuticular loss. In one lacteus specimen three of the ten tentacles had papillae spreading over much of the tentacle rather than being confined to papillae 'discs' as is the norm (Fig. 2a and b). In 2b cilia (c), papillae secretory vesicles (s), microvilli tips (mv) and what are probably bacteria (b) are visible. The needle-like objects in 2a are artefacts caused by buffer electrolytes. The fuzzy coat has been lost though the cuticle is intact.

All the species possessed papillae and cilia on their tentacle buds. Where the fuzzy coat was intact the globular secretory vesicles on the papillae were not visible, the papillae surfaces being slightly pitted and the cilia having naked tips. Cilia were often grouped with up to six cilia in each group. Figure 3 shows such cilia (C) and papillae (P) of Duasmodactyla commune.

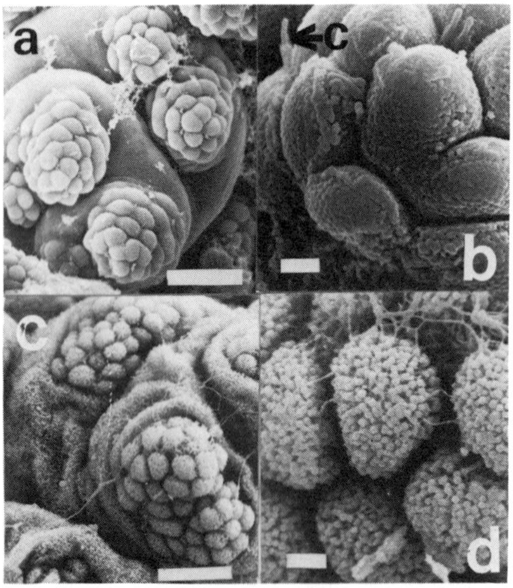

Figure 1. Differences in tentacle appearance due to cuticle loss in Aslia lefevrei. Scale bars: a and c 10 µm b and d 1 µm. c (arrowed) are cilia.

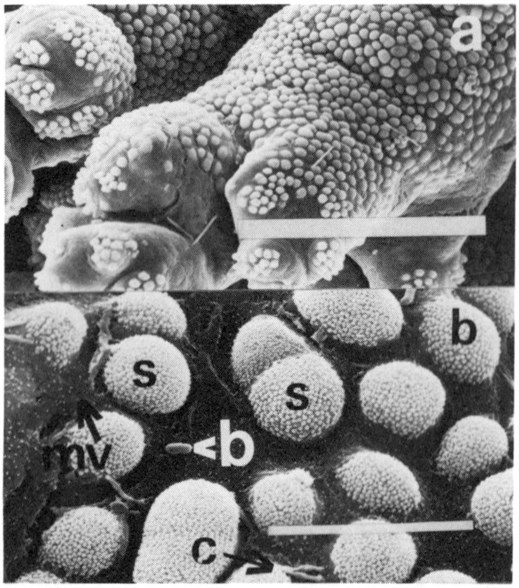

Figure 2. Aberrant tentacles O. lacteus. Scale bar a: 100 µm b: 10 µm. Explanation of symbols in 2b: c (arrowed) are cilia, s indicate clusters of papillae secretions, mv are microvilli tips and b (arrowed) are bacteria.

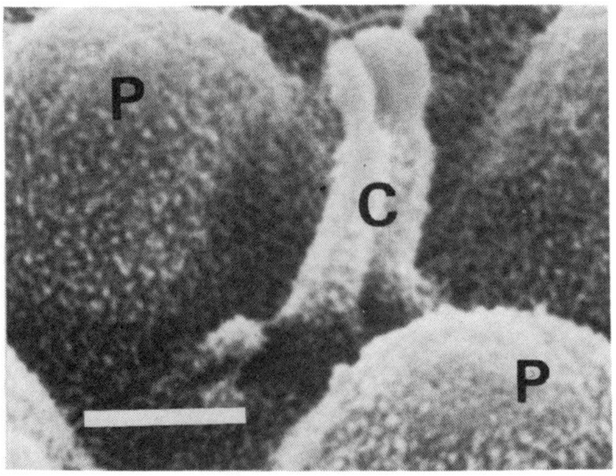

Figure 3. Papillae (P) and naked-ended cilia (C) of Duasmodactyla. Scale bar: 1 μm.

In most species the tentacles had distinctly 'bushy' branches where the side brnaches from the main trunk were short and papillae 'discs' arranged in clusters of up to four 'discs' on each bud (Fig. 4a). The two Thyone species were, however, very different from the others having long, side branches which ended in conical, usually single, papillate areas (Fig. 4b). The arrangement of papillae markedly departed from the norm since papillae extended from the apex to some way down the bud. This is very marked in T. fusus (Fig. 4c) though less so in T. roscovita (Fig. 4d). Cilia have a similar distribution. In roscovita the papillae below the apex are better described as pads. Using 'Sellotape' to remove the fuzzy coat showed that these pads were similar to the apical papillae and have the familiar secretory vesicles underlying the fuzzy coat (Fig. 4d). In Ocnus and Psolus phantapus the papillae are arranged on short evertable projections (Figs. 5a and b).

Globular structures were seen at the bases of the trunk of Duasmodactyla tentacles (Figs. 6a and b). Light microscope examination showed that these bulges covered spicules. These structures were not seen in Thyonidium pellucidum despite suspicions that specimens from the British Isles identified as pellucidum are merely juvenile commune. No significant differences at the S.E.M. level were found between the tentacles of Aslia lefevrei, Pawsonia saxicola, Thyonidium pellucidum, Duasmodactyla commune and Neopentadactyla mixta other than the bulges on Duasmodactyla specimens. (Figs. 1c, 7a, 7b).

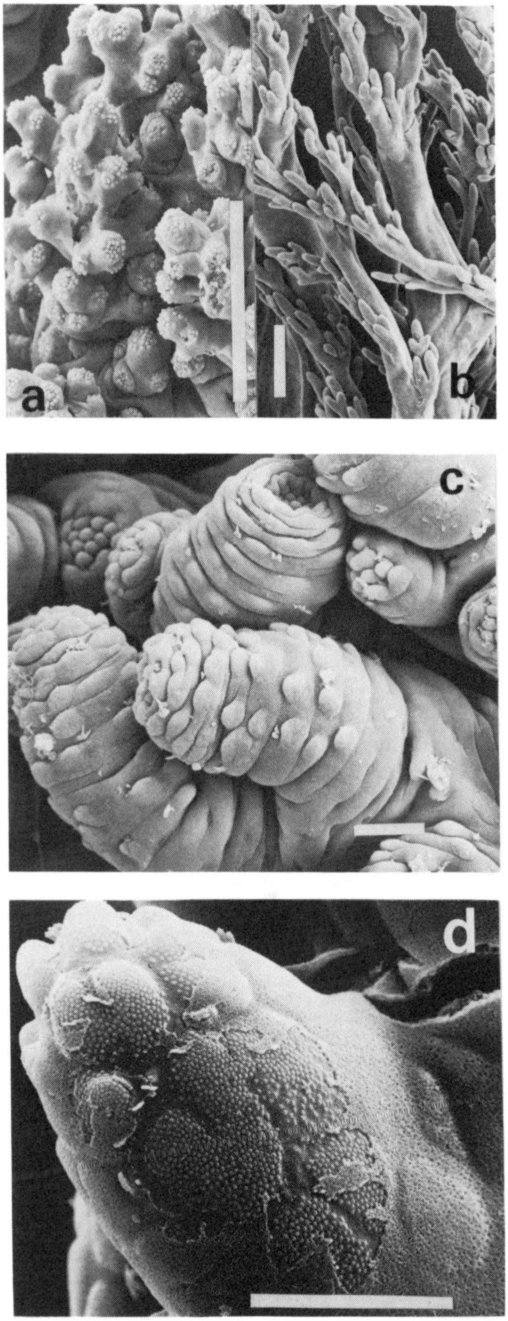

Figure 4a. Tentacles of Thyonidium pellucidum. Scale bar 100 μm. b. Tentacles Thyone fusus. Scale bar 100 μm. c. T. fusus, tentacle buds. Scale bar 100 μm. d. Thyone roscovita tentacle bud. Scale bar 10 μm.

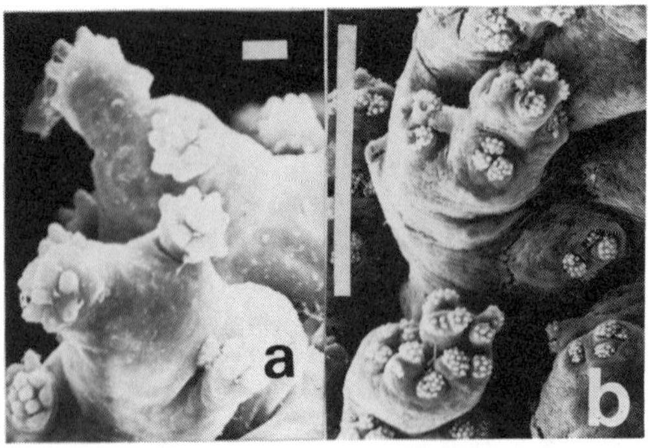

Figure 5a. Tentacle buds of P. phantapus, scale bar 10 µm. 5b tentacles of O. brunneus, scale bar 100 µm.

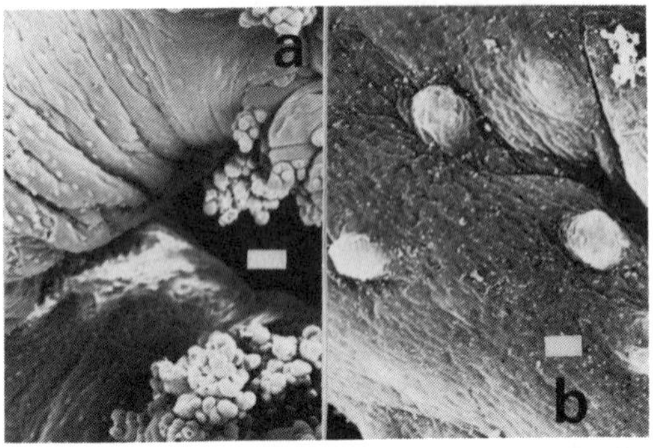

Figure 6. Bulges on trunk of D. commune. Scale bar: 100 µm, b: 10 µm.

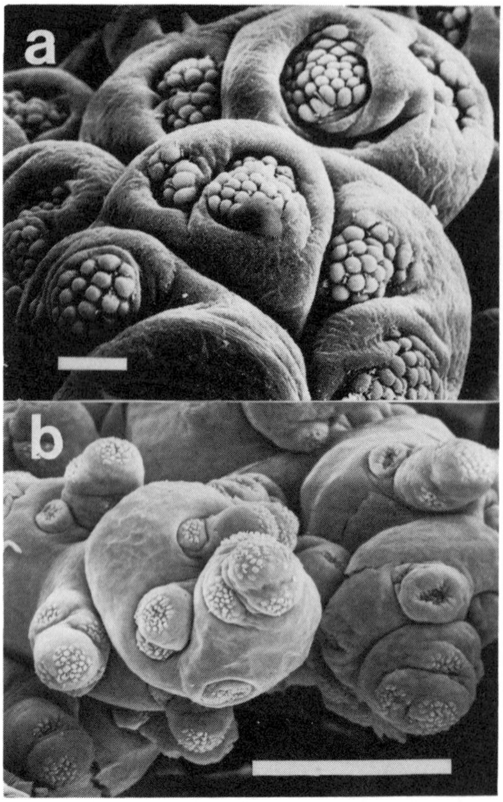

Figure 7a. Tentacle buds of P. saxicola. Scale bar 10 µm. 7b. Tentacle buds of D. commune. Scale bar 100 µm.

DISCUSSION

In Strangford Lough the two Ocnus species are usually found on Modiolus modiolus (L.) shells, sometimes with both species occurring on the same shell. With both species being of a similar size, competition or a way of avoiding it might be expected to occur. While the S.E.M. results on Ocnus showed no significant differences in tentacular structure, there is some evidence of micro-habitat partitioning by the two species (McKenzie 1984) thus supporting Kohn's (1978) hypothesis. No such partitioning can be found between Aslia and Pawsonia. Both are of similar size and may be found together under the same rock or within the same crevice sometimes in very close association with each other. Again no differences were found in the tentacular structure of the two species and with no evidence for either food resource or habitat partitioning it is difficult not to accept that the two species are often in competition.

As the species where competition is a likely possibility show no related structural differences in their tentacles the differences found in the Thyone tentacles are all the more surprising. Both species may occur in mixed communities. T. fusus and T. roscovita are found in Modiolus beds in Strangford Lough along with Pawsonia, both Ocnus species, Paracucumaria, Leptopentacta elongata, Thyonidium and Duasmodactyla (pers. observ.). T. roscovita can also be found in the littoral zone under stones along with Pawsonia as well as in sub-littoral rock crevices. Though striking, the differences between Thyone tentacles and those of other species are really only of degree. The elongation of bud-bearing branches and of the buds themselves would result in a structure similar to that in Thyone. While the structural differences are only those of geometry they may represent considerable functional adaptations. Fankboner (1978) related two forms of food capture for Psolus chitinoides Clark. In the first form of capture the food is held by the (sticky) papillae and in the second food is trapped in the clefts between the tentacle buds. Both he and Smith (1983) noted that the papillae are the only sticky parts of the tentacle. Clearly if more of the tentacle has sticky areas then catches should increase and cleft capture especially would be enhanced by tentacles structured as in Thyone. The sensory cilia will also provide more information on the micro-environment than in species where their distribution is much more restricted. This may allow more efficient 'active' suspension feeding of the type observed by Fankboner (1978) for Psolus. It is even possible that Thyone species employ a different tentacle 'net' shape than other species. Unfortunately comparison of the gut contents of the species studied here are unavailable. Even when such studies are made there may be no differences found as structural variation may influence feeding efficiency rather than the type of food captured. The structural variations may also be only related to the sensory function of the bud epithelium.

The structures on the trunks of Duasmodactyla tentacles are almost certainly sensory in function. Spicules penetrate the structure with the table 'top' in the bulge and the spire in the underlying connective tissue. In adult Duasmodactyla spicules are only found in the introvert but they are common over the whole body surface in small specimens. With nervous connections to the spicules the structures could act as simple mechanoreceptors.

The correct T.E.M. interpretation of the echinoderm cuticle was conclusively illustrated in Holland and Nealson (1978) though its chemical composition, origin and function are still unclear. Only vestiges of the cuticle remained in Fankboner's study (1978) and in the T.E.M. results of Smith (1983) though it was preserved in Smith's S.E.M. specimens. A fuller discussion of the cuticle and its significance in adhesion of objects to holothurian tentacles in inappropriate here but the present study demonstrates that caution is necessary in interpreting S.E.M. results. By far the majority of variation between Fankboner (1978, 1981) and Smith (1983) derive from cuticular loss rather than true differences between the species studied. Preliminary work suggests that such caution should be taken when interpreting the results of S.E.M. and T.E.M. studies on other holothurian orders and that cuticular loss must be considered a serious possibility before attempting to relate surface structure to function.

The variation found in Ocnus specimens may have explanation in the reproductive strategies employed by these species. There is some evidence (McKenzie 1984) for asexual reproduction by fission in Ocnus from Strangford Lough. The different surface textures of Ocnus tentacles may represent varying stages of regeneration. Perhaps the aberrant tentacles seen in O. lacteus (Fig. 2) was, in some way, also the result of regeneration processes.

In conclusion: this study has shown structural variation where it was unexpected and none where it might have been. The formulation of firm statements on whether dendrochirotes do, or do not, employ identical feeding structures and mechanisms are premature but there are sufficient grounds for further investigation. More work is needed to compare gut contents, holothurian community structures allied to studies of feeding in situ and in vivo.

ACKNOWLEDGEMENTS

This work has been funded by a Quota Award from the Science and Engineering Research Council of the U.K. I would like to thank the following: Dr Boaden and the staff of the University Marine Station, Portaferry; Morag MacKinnon and all the other divers who have helped me collect specimens; Mr Jim McCrae and the staff of Q.U.B. science faculty E.M. unit; my supervisor Dr Dai Roberts and all the others who have provided much valuable discussion during this study. I would also like to thank Professor Threadgold for criticism of this manuscript. Lastly I'd like to acknowledge the help of Dr Costello, U.C.G.

REFERENCES

Fankboner, P.V. 1978. Suspension feeding mechanisms of the armoured sea cucumber Psolus chitinoides Clark. J. exp. mar. Biol. Ecol. 31:11-25.

Fankboner, P.V. 1981. A re-examination of mucus feeding by the sea cucumber Leptopentacta (= Cucumaria) elongata. J. mar. biol. Ass. U.K. 61:679-683.

Holland, N.D. & Nealson, K.H. 1978. The fine structure of the echinoderm cuticle and subcuticular bacteria of echinoderms. Acta Zool. (Stockh.) 59:169-185.

Kohn, A.J. 1978. Ecological shift and release in an isolated population: Conus miliaris at Easter Island. Ecol. Monogr. 48:323-336.

McKenzie, J.D. 1984. Description of a neotype for the holothurian Ocnus brunneus (Forbes MS in Thompson, 1840) from Strangford Lough, Northern Ireland (Holothurioidea; Dendrochirotida). Bull. Br. Mus. Nat. Hist. (Zool.) 47(5):265-271.
McKenzie, J.D. & Picton, B.H. 1984. A note on feeding in Leptopentacta (= Cucumaria) elongata. J. mar. biol. Ass. U.K. 64:727.
Massin, C. 1982. Food and feeding mechanisms: Holothuroidea. In M. Jangoux & J.M. Lawrence (eds.), Echinoderm Nutrition, p.43-55. Rotterdam, Balkema.
Mortensen, T. 1927. Handbook of the echinoderms of the British Isles. Oxford: Oxford University Press.
Smith, T.B. 1983. Tentacular ultrastructure and feeding behaviour of Neopentadactyla mixta (Holothuroidea: Dendrochirota). J. mar. biol. Ass. U.K. 63:301-311.

Proceedings of the Fifth International Echinoderm Conference / Galway / 24-29 September 1984

Histology of the autotomy region in the esophagus of *Holothuria scabra* Jaeger

HOWARD SILVER
University of Queensland, Brisbane, Australia
Current address: College of Information Studies, Drexel University, Pa., USA

ABSTRACT: The esophagus of Holothuria scabra begins where the peripharyngeal mesentery joins the gut. At this point, the circular and longitudinal muscle fibers in the gut cross over. Posteriorly, the transition between the esophagus and the stomach occurs at the end of the first constriction of the gut. The autotomy region of the esophagus lies within this constriction. The autotomy region is distinguished by a cell mass of early stage lymphocyte-like cells in the connective tissue layer. The visceral mesentery consists of a portion proximal to the gut, and a portion distal to the gut. The mesentery is also bilaterally differentiated. The autotomy region of the mesentery occurs within the distal mesentery. Structural adaptations which minimise damage or facilitate regeneration appear to be lacking in the autotomy regions of the esophagus and mesentery.

1 INTRODUCTION

In a study of induced evisceration in 37 species of tropical aspidochirote holothurians it was found that autotomy consistantly occurred within restricted zones of weakness (autotomy regions) in the viscera (Silver, 1983). A study was undertaken to describe the histology of the esophagus and visceral mesentery of Holothuria scabra, and to identify structural modifications of the autotomy regions in the esophagus and mesentery.

Histological studies of the digestive tract of aspidochirote holothurians have centered on Holothuria tubulosa (Jourdan, 1883; Hamman, 1884; Cuénot, 1891; Oomen, 1926; Schreiber, 1930; Rosati, 1968; Massin, 1979) and Holothuria atra (Trefz, 1958). I have adopted the terminology used by Rosati (1968), dividing the foregut of H. scabra into a pharynx, esophagus, and stomach.

2 METHODS

2.1 Preparation and fixation

Specimens of H. scabra were collected from Moreton Bay, Queensland and starved for 48 hours. Whole specimens were anaesthetised with 2% diethyl ether and opened by a longitudinal incision along the mid-ventral body wall. Sections of gut and mesentery were removed. In other specimens evisceration was induced by injecting 0.54 M KCl. Material selected for sectioning consisted of:

1. the transition region between the esophagus and pharynx;
2. the first constriction (transition between the esophagus and stomach);
3. the second constriction (stomach);
4. the anterior portion of the expelled viscera;
5. the portion of the esophagus remaining within an eviscerated specimen; and
6. the mesenteries associated with the above portions.

After the material was excised, the portions of gut wall were opened longitudinally. The tissue was pinned out onto parrafin or cardboard and fixed with Bouin's fluid or Gomori's formol sublimate (Culling, 1974). Following fixation, tissue was washed in 50% ethanol, dehydrated and embedded in Tissue Prep (MP 56.5 °C).

2.2 Staining

Sections cut to 6 µm were stained with Mayer's haematoxylin and eosin (Culling, 1974), Mallory's PTAH (Humanson, 1967), and Unna's van Gieson (Drury & Wallington, 1967). To highlight possible neurosecretory, tissue 6 µm and 12 µm sections were stained with Gomori's chrome haematoxylin and eosin (Pearse, 1968), modified aldehyde fuchsin, and Milligan's trichrome (Humanson, 1967).

Photomicrographs were taken with an Olympus BH-2 photoautomat system.

3 RESULTS

3.1 Gross anatomy of the esophagus

The esophagus begins where the gonads originate at a thickening on the left side of the dorsal mesentery (Figure 1). A mesenterial jacket encloses the pharynx, forming a peripharyngeal sinus. The pharyngeal mesentery forms a collar around the gut at the transition between the pharynx and the esophagus.

A reddish pigmented constriction marks the junction between the esophagus and the stomach. Autotomy of the anterior gut occurs within this constriction. Rosati (1968) extended the esophagus to the second constriction of the anterior gut in H. tubulosa but Oomen (1926) and Massin (1979) recognised the first constriction as the beginning of a new zone. Trefz (1958) also recognised the first constriction as a transition between regions of the gut in H. atra.

The esophagus is attached to the dorsal body wall by a thin transparent mesentery. Attached to the ventral side of the esophagus is the ventral haemal sinus which runs the length of the first third of the alimentary canal. The length of the esophagus varies, but for an average sized animal (300 g) the esophagus ranges from 2 - 4 cm in length.

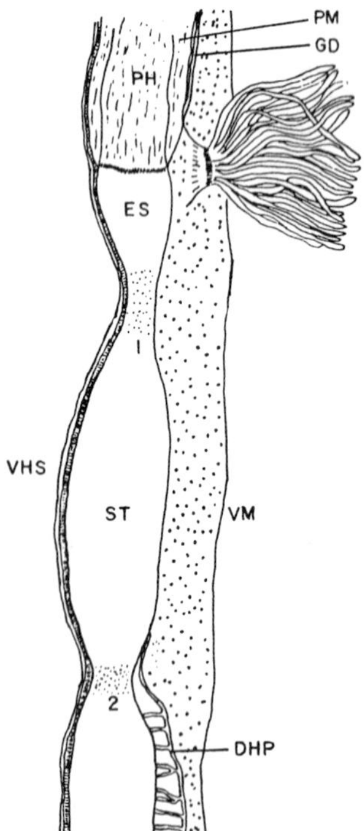

Figure 1. Anatomy of the foregut of Holothuria scabra. (DHP) dorsal haemal plexus (ES), esophagus (G), gonads (GD), gonoduct (PH), pharynx (PM), pharyngeal mesentery (ST), stomach (VHS), ventral haemal sinus (VM), visceral mesentery (1), first constriction of the foregut (2), second constriction of the foregut.

3.2 Histology of the esophagus

The digestive tract of holothurians is composed throughout of four tissue layers: an outer peritoneum (OP), a muscle layer with sublayers of circular and longitudinal muscle, a connective tissue layer (CT), and a lining epithelium (LE). Figure 2 shows a representative transverse section of the esophagus.

Outer peritoneum: The OP consists of a single layer of densely packed cells approximately 9 - 22 µm thick. The epithelial cells are short and cylindrical and contain large nuclei. This layer remains unchanged throughout the esophagus.

Muscle layer: The muscle layer of the esophagus consists throughout of an outer layer of longitudinal fibers and an inner layer of circular fibers. The longitudinal sublayer is poorly developed, being only a few fibers thick and loosely arrayed. The circular layer is organised into bundles of 20 - 30 fibers surrounded by connective tissue and is 18 - 27 µm thick.

Connective tissue: The CT is the most varied tissue layer of the esophagus. It is composed of a loose matrix of collagen fibers. Found within the CT are vascular lacunae, a nerve net, coelomocytes and pigment granules. The esophageal CT is thin, in comparison with that of the pharynx and stomach, varying from 45 - 90 µm.

The CT consists of two indistinctly separated sublayers. The outer sublayer occupies one third of the width of the CT. Projections from the outer CT encapsulate bundles of circular muscle. A nerve net runs longitudinally through the outer CT (Figure 3). The neurons are small cells which were often lodged in the CT projections which encapsulated the circular muscle.

Also found in the outer layer of the esophagus were vascular lacunae. In transverse sections the lacunae appear to be spaces 15 - 25 µm in diameter, adjacent to the circular muscle layer. In longitudinal sections the lacunae are seen to contain fine acidophilic fibers. The lacunae resemble those described in H. tubulosa by Rosati (1968). Another kind of lacunae, found in the stomach, are large spaces (60 - 200 µm) with a thin endothelial lining, situated in the inner CT adjacent to the LE (Figure 4). These have been described in H. tubulosa and H. atra (Oomen, 1926; Trefz, 1958).

The inner sublayer of the CT is a loose network of fibers. Coelomocytes and pigment granules were common in the inner sublayer. The CT layer is separated from the LE by a thin basal membrane.

The autotomy region, within the constriction at the posterior end of the esophagus, is distinguished by a massive proliferation of cells along the inner boundary of the CT (Figure 5). The cell mass appears to form an irregularly thickened collar around the LE, occupying the entire width of the CT in some sections, while in other sections forming a distinct sublayer. The cell mass is composed predominantly of small round cells with large nuclei, 6 - 10 µm across. Cells with similar staining properties occurred all along the inner CT layer of the esophagus but the cell mass was distinguished by its thickness and density of the cells. These cells resembled coelomocytes of the early lymphocyte stage (see Endean, 1958).

Lining epithelium: The esophageal LE contains several layers of cells and is up to 100 µm in thickness. In contrast the LE of the pharynx and stomach are only a single cell layer in thickness. In transverse sections the LE has a rugose appearance. The cells are long, narrow and closely packed. Also present are numerous large mucus secreting goblet cells 20 - 40 µm across. In the esophagus and stomach the lumen had a mucus lining of varying thickness.

There is an abrupt change in the histology of the gut at the junction between the pharynx and esophagus. The most obvious difference between the gut regions is that the disposition of the sublayers of the muscle layer is reversed. In the pharynx the longitudinal muscle lies on the inner side of the muscle layer. The crossover of the circular and longitudinal muscle layers occurs where the pharyngeal mesentery joins the gut (Figure 6).

The histological transition between the stomach and esophagus occurs 4 - 5 mm posterior to the autotomy region. Externally, this corresponds with the end of the first constriction. The transition occurs gradually over a longitudinal distance of approximately 1 mm. The cell mass in the CT layer of intact esophagi is not present in the esophagi of eviscerated H. scabra. Comparison of the length of esophagus attached to the ejected viscera with the distance between the cell mass and the stomach in an intact gut demonstrated that the cell mass occurred within the autotomy region of the esophagus (Figure 7).

3.3 Histology of the mesentery

The dorsal mesentery is composed of three tissue layers: an outer peritoneum, a muscle layer, and an inner connective tissue layer. These tissue layers are continuous with the corresponding outer three layers of the gut. Figure 8 presents a composite view of the dorsal mesentery. The mesentery can be subdivided into a portion proximal to the gut (Figure 9) and a portion distal to the gut (Figure 10). The mesentery is also bilaterally asymetrical, having a left and right side.

The muscle layer was the most varied tissue layer. In the proximal portion it began as a layer of transverse fibers on the right side only. Distally, these fibers gave way to a thick clump of longitudinal fibers, also on the right side. The longitudinal fiber mass ends abruptly, and where it terminates marks the boundary between the proximal and distal portions of the mesentery. In the distal mesentery muscle fibers are found on both sides of the CT layer. In transverse sections muscle fibers alternately appeared to be either transverse, longitudinal, or obliquely oriented. This pattern suggested that the muscle fibers undulate longitudinally through the distal mesentery.

The connective tissue layer is similar in composition to the CT of the gut. Endothelium lined lacunae are found in the proximal mesentery. In the distal mesentery lacunae were lacking. Two kinds of coelomocytes were common: spherule laden, basophilic morula cells, and later stage amebocytes (see Endean, 1958).

Nerve tracts consisted of a single fiber layer situated between the CT and muscle layer. Nerve tracts were found on both sides of the CT layer in the distal mesentery. Short projections from the

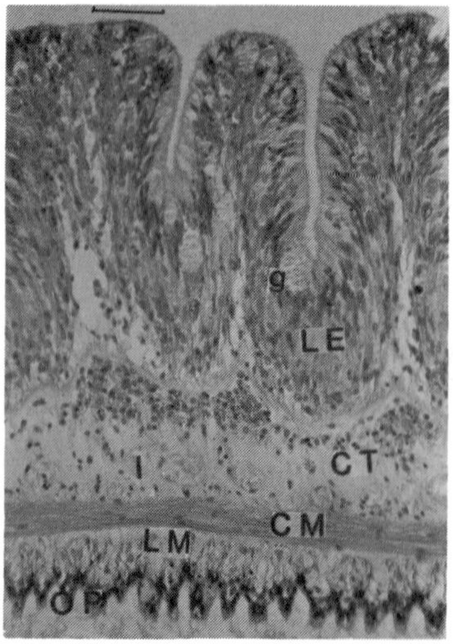

Figure 2. The esophagus of H. scabra showing its four tissue layers (CM), circular muscle (CT), connective tissue (g), goblet cell (l), vascular lacunae (LE), lining epithelium (LM), longitudinal muscle (OP), outer peritoneum. Transverse section, chrome haematoxylin eosin, bar length 35 µm.

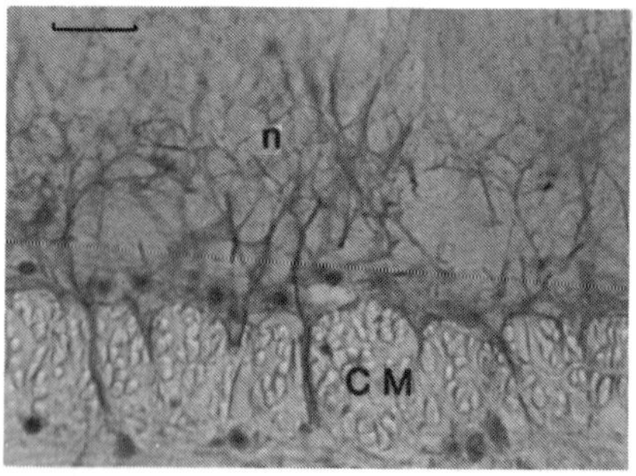

Figure 3. Nerve net in the connective tissue layer of the stomach (n), nerve net. Longitudinal section, chrome haematoxylin eosin, bar length 13 µm.

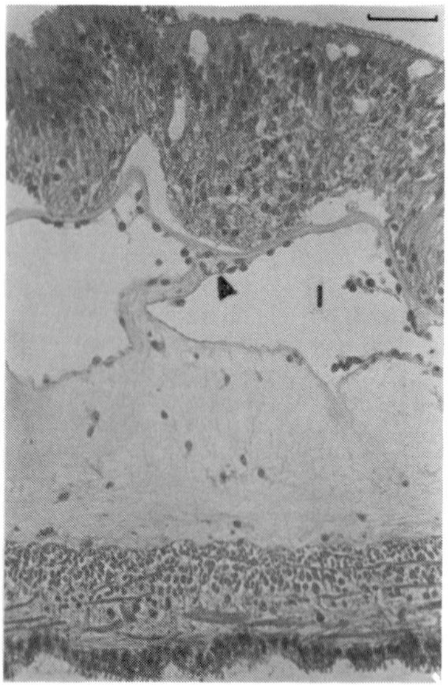

Figure 4. Vascular lacunae in the wall of the stomach (arrow), endothelial lining. Longitudianl section, Mayer's haematoxylin eosin, bar length 25 µm.

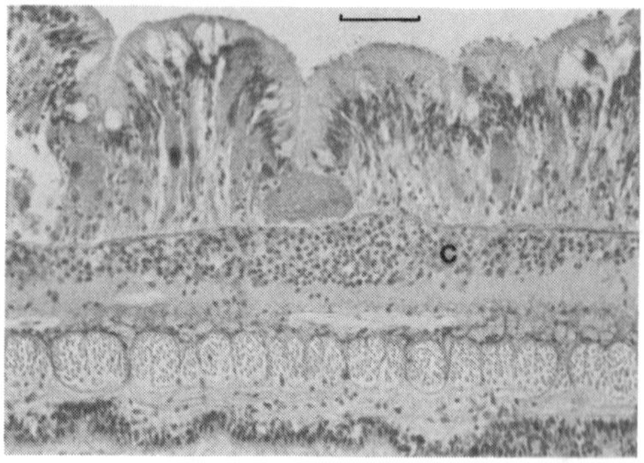

Figure 5. The autotomy region of the esophagus showing the cell mass in the connective tissue layer (c), cell mass. Longitudinal section, chrome haematoxylin eosin, bar length 35 µm.

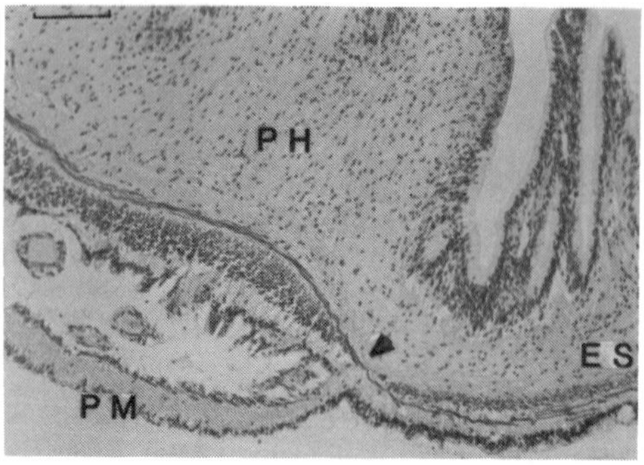

Figure 6. Junction between the esophagus and pharynx showing muscle crossover and the pharyngeal mesentery (arrow), muscle crossover. Longitudinal section, Mallory's PTAH, bar length 50 µm.

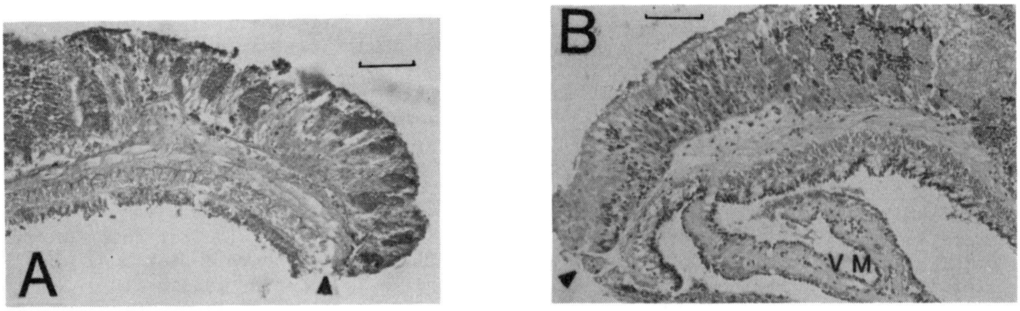

Figure 7. Autotomy region of esophagus following evisceration. A. Longitudinal section of autotomy region anterior to rupture (chrome haematoxylin eosin, bar length 30 µm). B. Longitudinal section of autotomy region posterior to rupture (Mayer's haematoxylin eosin, bar length 30 µm). (arrow), autotomised edge.

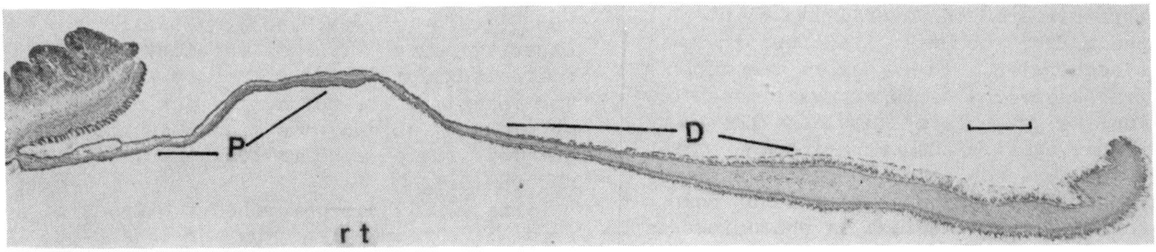

Figure 8. Transverse section of visceral mesentery showing the proximal (P) and distal (D) portions of the mesentery. Unna's van Gieson, bar length .15 mm (rt) right side of mesentery.

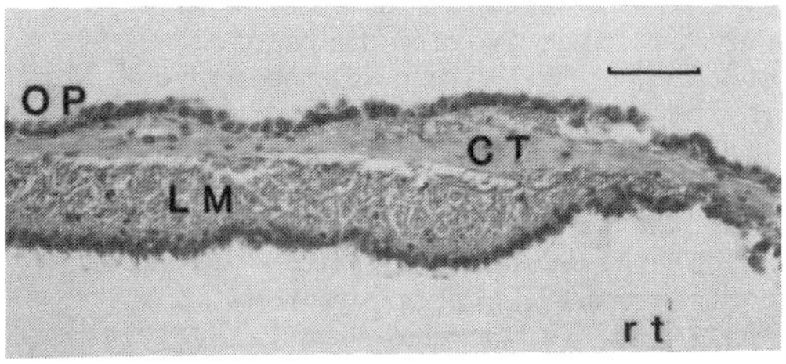

Figure 9. Transverse section of proximal portion of visceral mesentery. Unna's van Gieson, bar length 20 µm.

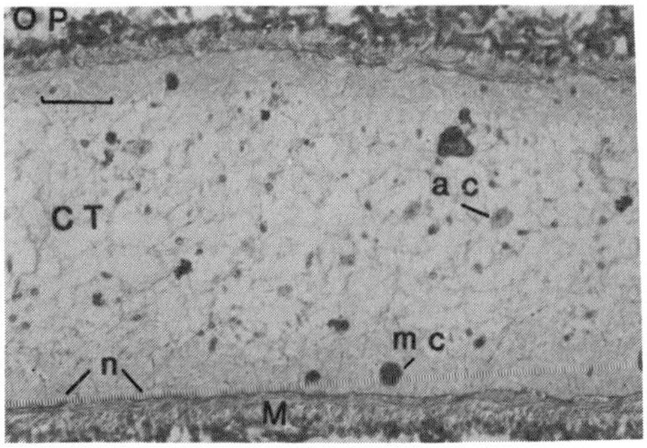

Figure 10. Transverse section of distal portion of visceral mesentery (ac), amebocyte (M), muscle layer (mc), morula cell (n), nerve tract. Chrome haematoxylin eosin, bar length 25 µm.

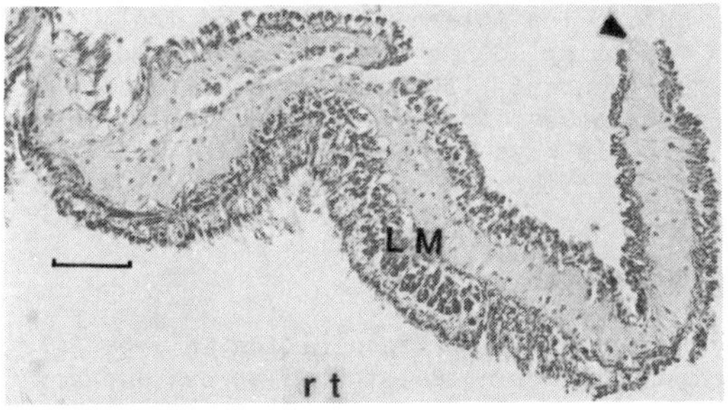

Figure 11. Portion of mesentery remaining attached to the stomach following evisceration (arrow), autotomised edge. Milligan's trichrome, bar length 30 µm.

nerves extended into the muscle layer and CT.

Figure 11 shows the portion of mesentery remaining attached to the posterior portion of the esophagus following evisceration. Autotomy occurred within the distal mesentery. This is indicated by the fact that muscle fibers are seen on both sides of the CT at the autotomised end. The mesentery lacked discontinuities or other anatomical features which could be used to pinpoint the autotomy region. The mesentery, as well, is highly extensible, making it difficult to distinguish specific points within the distal mesentery.

4 DISCUSSION

The histology of the anterior gut of H. scabra differed in several important aspects from accounts given for other aspidochirotes. In H. scabra the esophagus begins where the peripharyngeal mesentery joins the gut. To my knowledge the peripharyngeal sinus has not been previously described in aspidochirote holothurians. Although it has not been described in H. tubulosa, a photomicrograph showing the mesentery is found in Rosati (1968, Figure 22). Visual examination of the guts of H. atra and H. leucospilota showed that they also possess a peripharyngeal mesentery (pers. obs.). A peripharyngeal sinus has been described in the caudate holothurian Molpadida roretzii (Hatanaka, 1939) but the histology of the mesenterial-gut junction was not discussed.

The junction between the the pharynx and esophagus is the site of muscular reversal in H. scabra. In H. tubulosa reversal of the muscle layer has been reported to occur in the autotomy region within the constriction, and the reversal is said to be a feature which predisposes the esophagus to rupture there (Jourdan, 1883, Rosati, 1968). In H. tubulosa and H. atra the constriction has been described as a weak spot in the gut, marked by an attenuation of the tissue layers (Jourdan, 1883; Oomen, 1926; Trefz, 1958). In H. scabra the thickness of the esophageal wall in the autotomy region is comparable with the rest of the esophagus, and there were no features of the autotomy region which would suggest that this site is structurally weaker than the rest of the esophagus.

Little attention has been given in the literature to the histology of the visceral mesenteries other than descriptions of its three layer makeup (eg. Jourdan, 1883; Cunéot, 1891). In H. scabra the CT and muscle layers of the mesentery are transversely differentiated. Autotomy occurred in the distal mesentery. This differs from mesenterial autotomy in the dendrochirotes, Thyone briareus and Eupentacta quinquesemita, for which mesenterial autotomy occurs at the junction with the gut (Kille, 1935; Byrne, pers. comm.).

Rapid, neurally mediated softening of the connective tissue occurs during body wall autotomy in Istichopus bandionotus (Junqueira et al., 1980) and during evisceration in Eupentacta quinquesemita in the introvert, at the PRM-LBWM junction, and at the intestine-cloaca junction (Byrne, 1981, 1982, pers. comm.). In H. scabra visceral autotomy is neurally mediated and can be accomplished without the involvement of the body wall musculature (Silver, 1983). Visual observations of visceral autotomy in H. scabra suggest that the muscles in the gut and mesentery are not active during autotomy. Instead, the tissue in the autotomy region appears to weaken, stretch and rupture.

Byrne (1981) has described two types of nerve-like tissue from the PRM-LBWM junction of E. quinquesemita. The first resembles cholinergic nervous tissue. The second has morphological characteristics of neurosecretory tissue, and resembles the juxtaligamental tissue described from the arm of Ophiocomina nigra (Wilkie, 1979). Further investigation is required concerning the nature of the nervous tissue in the gut of H. scabra, and whether this tissue plays a role in visceral autotomy.

In the arm of O. nigra there are structural adaptations which minimize damage resulting from autotomy and facilitate regeneration (Wilkie, 1978b). Similar modifications are apparently absent in the autotomy regions in the viscera of H. scabra (it is unlikely that the cell mass is involved in regeneration or wound healing because it is no longer present following autotomy). Such adaptations are probably unnecessary because holothurians autotomise internal organs and the resulting wounds are not exposed to the environment. The coelom is an ideal environment for regeneration, which may explain H. scabra's ability to rapidly regenerate its viscera (see Bai, 1971).

REFERENCES

Bai, M.M. 1971. Regeneration in the holothurian, Holothuria scabra Jaeger. Indian J.exp.Biol. 9: 467-491.

Byrne, M. 1981. Functional morphology of a holothurian autotomy plane and its role in evisceration. Proceedings: International Echinoderms Conference, Tampa Bay: 65-68.
Byrne, M. 1982. The functional morphology and mechanical properties of a holothurian autotomy region. Am.Zool. 22: 942.
Cuénot, L. 1891. Etudes morphologiques sur les echinoderms. Archives de Biologie (Paris) 11: 313-680.
Culling, C.F.A. 1974. Handbook of histopathological and histochemical techniques. London: Butterworth.
Dobson, W.E. 1982. Neural mediation of disc autotomy in ophiuroids. Am.Zool. 22: 904.
Dobson, W.E. & R.L.Turner 1983. Morphological and histological basis of disc autotomy in a brittle star. Am.Zool. 23: 1026.
Drury, R.A.B. & E.A.Wallington 1967. Carleton's histological technique 4th ed. Oxford: Oxford University Press.
Endean, R. 1958. The coelomocytes of Holothuria leucospilota. Quart.J.micr.Sci. 99: 47-60.
Hamman, O. 1884. Beitrage zur Histologie der Echinodermen. Heft I. Die Holothurien. Jena G.Fisher, p.309-333.
Hatanaka, M. 1939. A study of the caudate holothurian Molpadia roretzii (v. Marenzeller). Sci.rep.Tohoku Univ. 14: 155-190.
Holland, N.D. & J.C.Grimmer 1981. Fine structure of the syzygial articulations before and after arm autotomy in Florometra serratissima (Echinodermata:Crinoidea). Zoomorphology 98: 169-183.
Humanson, G.L. 1967. Animal tissue techniques. San Francisco: W.H.Freeman.
Jourdan, M.E. 1883. Recherches sur l'histologie des holothuries. Annales Musee d Historie Naturelle de Marsielle v.1(6).
Junqueira, L.C.W., G.S.Montes, P.A.S.Mourao, J.Carniero, L.M.M.Sales, S.S.Bonetti 1980. Collagen-proteoglycans interaction during autotomy in the sea cucumber, Stichopus bandionotus. Rev.Can.Biol. 39: 157-164.
Kille, F. 1935. Regeneration in Thyone briareus Leseur following induced autotomy. Biol.Bull. 69: 82-108.
Massin, C. 1979. Morphologie functionnelle du tube digestif d'Holothuria tubulosa Gmelin (Echinodermata:Holothuroidea). Actes du colloque Europeen sur les echinoderms, Bruxelles.
Oomen, H. 1926. Verdauungspnysiologische studien an holothurien. Pubbl.Staz.zool.Napoli 7: 215-297.
Pearse, A.G.E. 1968. Histochemistry: theoretical and applied. v.I. London: Curchill.
Rosati, F. 1968. The fine structure of the alimentary canal of holothurians. Monitore zool.ital. 2: 49-86.
Schreiber, B. 1930. Studi sull'assorbimento intestinale nelle oloturie. Pubbl.Staz.zool.Napoli 10: 235-277.
Silver, H. 1983. Evisceration in holothurians from the Indo-West Pacific region. Queensland University: MSc thesis.
Trefz, S.M. 1958. The physiology of digestion of Holothuria atra Jaeger with specific reference to its role in the ecology of coral reefs. University of Hawaii: PhD thesis.
Wilkie, I.C. 1978a. Nervously mediated change in the mechanical properties of a brittlestar ligament. Mar.Behav.Physiol. 5: 289-306.
Wilkie, I.C. 1978b. Functional morphology of the autotomy plane of the brittlestar Ophiocomina nigra (Abildgaard). Zoomorphologie 91: 289-305.
Wilkie, I.C. 1979. The juxtaligamental cells of Ophiocomina nigra (Abildgaard) (Echinodermata:Ophiuroidea) and their possible role in mechano-effector function of collagenous tissue. Cell Tissue Res. 197: 515-530.
Wilkie, I.C. 1983. Nervously mediated change in the mechanical properties of the cirral ligaments of a crinoid. Mar.Behav.Physiol. 9: 229-248.

Proceedings of the Fifth International Echinoderm Conference / Galway / 24-29 September 1984

Seasonal torpor in *Neopentadactyla mixta* (Östergren) (Holothuroidea: Dendrochirotida)

TIMOTHY B.SMITH & BRENDAN F.KEEGAN
Department of Zoology, School of Marine Sciences, University College, Galway, Ireland

For six to eight months, bridging Autumn and Spring on the Irish west coast, the holothuroid Neopentadactyla mixta (Östergren) desists from active suspension feeding and retires to depths of 30 - 60cm within coarse sediments. This torpor is marked by a progressive deterioration in general body condition, a reduction in metabolic activity and a decline in the lipid content of the intestine, body wall and gonad. Temperature change may cue this behaviour which is tentatively interpreted as a flight from seasonal impoverishment of the food reservoir and from turbulent water conditions.

INTRODUCTION

Benthic investigations off the west coast of Ireland have shown echinoderms to be numerically prominent in many bottom living communities (Keegan & Könnecker, 1980). An extreme instance involves the sedentary holothuroid Neopentadactyla mixta (Östergren) which has become so numerous in one bay system as to virtually exclude other macroinvertebrates from the loose biogenic gravels ('maerl') which it favours (Keegan, 1974; Cabioch et al., 1978; Smith, 1981). Within a broadly based attempt to ascertain the ecological basis for this phenomenon, the species feeding biology has received particular attention. Könnecker & Keegan (1973) reported on the N. mixta diurnal feeding rhythm, and Smith (1983) described its tentacular feeding behaviour and mode of seston entrapment. These workers were unable to find a population which fed during the winter months. Accordingly, an investigation was mounted to verify if in fact the holothuroid temporarily abstained from macroingestion, as indicated, and to profile certain aspects of the species physiology during this seasonal inactivity.

MATERIALS AND METHODS

Over the period September 1979 to April 1980, monthly collections of Neopentadactyla mixta were taken from within an aggregation in Kilkieran Bay (53°31'N; 9°50'W) where densities of adults and juveniles can exceed 400/m^2 and 15,000/m^2 respectively. The location is relatively exposed to the prevailing winds. An experimental procedure was designed to yield information on the species metabolic rate, its utilization of food reserves, and the affect of its nutritional condition on the gonads.

The entire intestinal tract (flushed clean with sea water) and portion of the body wall were taken from ten animals each month; these were stored in individual containers at -20°C. After subsequent thawing, the body wall material and intestinal tracts from each month were pooled separately, as were the male and female gonadial tissues. A one gram (wet weight) aliquot of each sample, which had been homogenised under liquid nitrogen, was used for biochemical analysis. Following initial extraction with 2:1 chloroform/methanol, the supernatant was assayed for its lipid content

(Marsh and Weinstein, 1966). The residue was homogenised with 20% trichloroacetic acid and gave, after centrifugation, a supernatant and precipitate which were used for glycogen (Brunner, 1964) and DNA assays (Schneider, 1957) respectively. Standards and controls were in accordance with normal procedure and all reagents were Analar grade.

Oxygen consumption was measured for ten animals monthly. Individuals were placed in sealed vessels containing sea water of known dissolved oxygen concentration. After a period of two hours, the dissolved oxygen concentration was again determined using the Winkler method outlined in Strickland and Parsons (1972). The difference recorded was taken as the oxygen consumed and is expressed as ml O_2 per gram dry weight per litre per hour.

Gonad indices were calculated from a monthly sample of fifteen animals. The entire gonad was separated from each individual, four to six gonadial tubules (less than 1% by weight of the gonad) being fixed for histological examination. The remaining tissues were dried to a constant weight and the index was calculated as the gonad weight x 100 divided by the total weight of the animal.

A statistical analysis was carried out on the results obtained for oxygen consumption and the gonad indices. Since the parameters of the distributions around the mean values for each sample are dissimilar, a non-parametric Mann-Whitney U-test (Elliot, 1973) was employed in this regard.

Throughout the sampling period, a log was maintained of the holothuroid's activity/quiescence with respect to feeding and also of its approximate depth ranging within the sediment. The presence or absence of ingested material was noted in the course of removing the intestinal tracts.

RESULTS

General observations.

Neopentadactyla mixta, when feeding, assumes a u-shaped posture with the introvert and oral tentacles projecting into the water column and the terminal anus at or just beneath the surface of the deposit. On full expansion, the tentacle crown can have a spread of c. 140 cm^2 and account for up to one quarter of the animal's total displacement. In strong linear flow, drag on the more or less conically disposed crown forces its 'gape' to be directed obliquely or almost horizontally, rather than vertically, and the individual tentacles will be markedly turbated.

Suspension feeding is achieved by the impingement of seston on the adhesive nodes of the tentacle tips (Smith, 1983), with all particles below a relatively large size being treated as potential food. Whilst the holothuroid tolerates a degree of physical interference with the tentacles and the mouth area, slight vibrations of the immediate substrate provoke total withdrawl. Artificially increased current flow, beyond a calm weather, spring-tide maximum of c. 1.5 m/sec., and bombardment with dislodged sediment particles, had the same effect. Each spring-tide changes the surface topography of the gravel deposits and, even under less than maximum flow, their upper few cm are visibly mobile. The pressures thereby imposed on both the feeding process and position-holding are further exacerbated by wave and swell induced turbulence. A protracted and heavy gale in August 1982 caused the N. mixta population to withdraw and remain deep within the sediment for six to ten days!

Between the September and October sampling sessions of 1979, the whole Neopentadactyla mixta population withdrew into the loose substrate and did not re-emerge until March/April of the following year (fig. 1). Buried to depths of up to 60cm the holothuroid was invariably found with its feeding tentacles retracted within the pharyngeal cavity. The intestinal tracts of all individuals taken during this period were empty of food material, contrasting with the full or partially full condition during times of active feeding. Specimens collected during the winter months could not be induced to resume feeding in the laboratory either by varying the temperature or the photo-period. In many animals taken over January to March, the

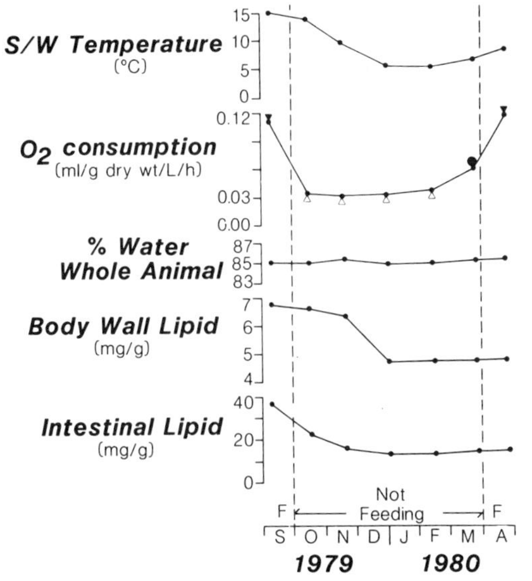

Fig. 1. Metabolic rate and lipid food reserves of the body wall and intestine in Neopentadactyla mixta during the winter of 1979/1980. For metabolic rate, statistically similar values are given the same symbol, i.e.▼, △,●. Other than for O_2 consumption, values are in mg/g wet weight.

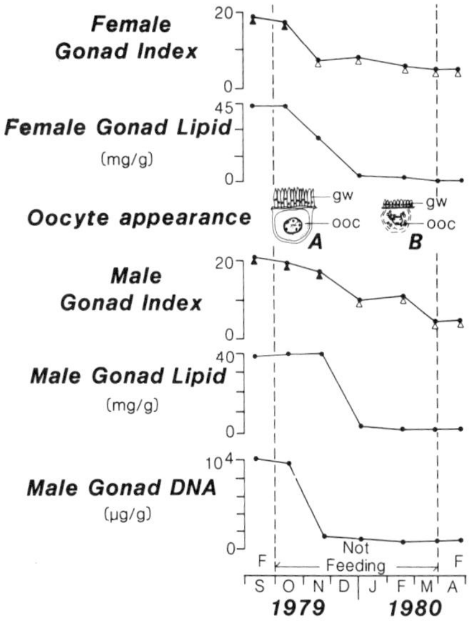

Fig. 2. Effects of torpor on the gonadial tissue of Neopentadactyla mixta during the winter of 1979/1980. For the gonad indices, statistically similar values are given the same symbol, i.e.△,▲. Gw, gonadial wall; ooc, oocyte. Values are for wet wt.

body wall, normally robust and quite thick, was found to be markedly thin and, in some instances, almost transparent.

Physiological changes associated with winter feeding inactivity

On withdrawing into the substrate, Neopentadactyla mixta reduced its oxygen consumption from 0.11ml O_2/gm dry wt./L/hr to 0.03ml O_2/gm dry wt./L/hr. (fig. 1). This rate was maintained until March when there was a significant increase to c. 0.06ml O_2/gm dry wt./L/hr. On re-emerging, oxygen consumption again increased to levels characteristic of the holothuroid whilst actively feeding.

As shown in fig. 1, there was a progressive reduction in the lipid levels of the body wall and intestinal tract throughout the winter and lowest values were recorded from January to April. This reduction in lipid is most marked in respect of the intestine where values fell from c. 38 mg/g to 13 mg/g. Glycogen was found in trace amounts in the September sample only.

Both male and female gonadial tissues were much reduced in weight by April, having undergone a progressive diminution in size over the winter months (fig. 2). The fall-off in the gonadial lipid levels, in both sexes, closely followed this reduction in relative weight (fig. 2). There was also a decrease in the diameter and length of the gonadial tubules as the winter progressed and histological preparations showed that the tubule wall became increasingly more thin over the fasting period. The oocytes, which were large at the onset of winter (up to 300μm in diameter), gradually degenerated and detached from the inner epithelium of the tubule; their cytoplasm became highly vacuolated (fig. 2). In the male gonad, the number of mature sperm and spermatogonia was reduced and this was reflected in the DNA content (fig. 2).

DISCUSSION

Feeding periodicities have been reported for a number of holothuroids, with some exhibiting a daily activity rhythm which Yamanouchi (1956) and Crump (1965) saw as being under the control of external light change. Könnecker and Keegan (1973) inclined towards the same view in respect of an aggregated population of Neopentadactyla mixta which ceased feeding and retracted into the substrate between 1 and 4 hours after sunrise, remained quiescent for 1 to 2 hours and re-emerged over a 4 hour period. On the other hand, Stier (1933) reported a possible endogenous rhythm in Sclerodactyla briareus Lesueur where light did not appear to be the controlling factor. Hunter-Rowe et al. (1976) described the feeding cycle of Thyone fusus (O.F. Müller) as being more likely related to the strength of current flow than to diurnal or tidal rhythms.

Stichopus japonicus Selenka displays a seasonal rhythmicity in its feeding and becomes quiescent when the temperature reaches 24.5°C (Mitsukuri, 1903; Sand, 1962). Engstrom (1982) reported an annual cycle for Cucumaria lubrica (Clark 1901) from Puget Sound, in which there is practically no feeding from late September to February or March. For European waters, Fish's (1967) well documented account of Leptopentacta elongata (Düben and Koren) bears striking similarities to the findings for Neopentadactyla mixta, both in regard to the annual cycle and its effects on the animals.

From the outset, the seasonal torpor of Neopentadactyla mixta is marked by an immediate reduction in metabolic rate which is sustained until macroingestion is resumed. Demands on lipid appear to affect the intestinal tract most immediately, followed by the body wall and gonads. It is generally held that long term nutrient reserves in holothuroids are stored as lipid (Trefz, 1958; Fish, 1967; Krishnan, 1968; Lawrence, 1972). Prim et al. (1976) have suggested that the body-wall also stores large amounts of nutrients in the form of proteinaceous material. The authors did not attempt to ascertain the extent to which such reserves may be used by N. mixta. Feral and Doumenc (1982) reported that in spite of the lack of a specialised storage organ, glycogen is widespread in holothuroids. The fact that it was not detected in other than the September sample of N. mixta could suggest a very early depletion of this carbohydrate during the quiescent period.

As with Leptopentacta elongata (Fish, 1967), the stimulus which initiates, and terminates, the seasonal cycle in Neopentadactyla mixta remains unclear. The evidence points to temperature change as a likely cue for both species. Over the years 1973 - 1981 slight differences in the duration of the fasting phase are retrospectively, and cautiously, attributed to yearly variations in the temperature regime.

The fact remains, however, that feeding could not be induced in a torpid population of either holothuroid by acclimation to higher temperatures. In L. elongata, feeding was suppressed and re-activated by regulating the temperature below and above 8°C respectively (Fish, 1967). Whilst no such experimentation was attempted with N. mixta, it does raise the question of this species ability to accommodate atypical extensions of the fasting phase.

Allowing that a depression from a yearly temperature maximum may be the sensory cue for relative quiesence, what is the significance of this response? The authors see the answer as possibly residing in the ergonomics of the feeding process itself, together with a related refuge-seeking from the effects of seasonal water turbulence.

For all its seeming passivity, the feeding process must be metabolically demanding. Because the experimental technique was so remote from natural conditions, the recorded differentials in oxygen consumption throw little light on the energy required to maintain turgor in the crown and to support the action of tentacles. The possibility has to be entertained that the quality of ambient seston over the winter months may be so poor as to negative the return for feeding effort. This would be compounded by the likelihood that the load-carrying capacity of the tentacles must be considerably lessened when turbulence approaches the level which provokes withdrawl. Such turbulence would typically obtain at the study site for much of the winter period.

One aspect of the overwintering phase suggests that the recorded processes and events may only manifest a 'partial hibernation', a torpor chiefly related to suspension feeding. This is predicated on the fact that, irrespective of the extent of substrate erosion or accretion, the holothuroids maintain a burrowing depth of some 30-50 Clearly, this depth ranging lies within the aerobic zone of the loose deposits and would most likely preclude the possibility of sudden physical dislocation by water movement. It must, however, require active station keeping (i.e. burrowing) with respect to changes in sediment level and these can locally exceed one meter over the September to April period (Keegan, 1972)!

As described by Könnecker and Keegan (1973), Neopentadactyla mixta burrows by advancing the middle portion of the body whose numerous, irregularly arranged, podia shift the sediment particles. Active burrowing, however episodic, would exert a considerable drain on the reserves of a fasting animal, a depletion which could however be lessened by the absorption of dissolved organics directly from the sea-water. Ferguson (1982) ascertained the rates of uptake and release of free amino-acids by three holothuroids and found that those of the thin-skinned Leptosynapta were greater than those of the thicker skinned Tyonella and Thyone. He found that not only does the net uptake of FAA largely maintain the superficial structures of echinoderms but it also facilitates the retention of metabolites. This would be important for animals compelled to desist from active macroingestion for long periods of time. The process would be enhanced by the progressive thinning of the bodywall and the extent to which the interstitial water is apparently renewed in the sediment. Keegan (1972) working on the N. mixta habitat, found that the diminutive bivalve Astarte triangularis could find its nutritional and oxygen requirements at depths exceeding 40cm within the gravel deposit.

ACKNOWLEDGEMENTS

We thank Des Johnson for advice on biochemical procedure and Albert Lawless for technical assistance. The work was partially financed by N.B.S.T. Grant No. 42/79.

This paper is contribution number 229 from the School of Marine Sciences, University College, Galway, Ireland.

REFERENCES

Bruner, R.L. 1964. Determination of reducing value: 3-5 dinitrosalicylic acid method. Meth.Carbohydr. Chem., 4: 67-71.

Cabioch, L., Grainger, J., Keegan, B.F. & Könnecker, G. 1978. Balcis alba (da Costa) a temporary ectoparasite on Neopentadactyla mixta (Östergren). pp.237-243. In D.M. McClusky and A.J. Berry (eds.) 12th European Symposium on Marine Biology. Oxford: Pergamon.

Crump. R.G. 1965. The diurnal activity of holothurians. pp. 43-45. In J.N. Lythgoe and J.D. Woods (eds.) Symposium of the Underwater Association of Malta. London: The Underwater Association.

Elliot, J.M. 1973. Statistical analysis of samples of benthic invertebrates. Freshwater Biological Ass.Science Publ., No. 25.

Engstrom, N.A. 1982. Brooding behaviour and reproductive biology of a subtidal Puget Sound sea cucumber. Cucumaria lubrica (Clark, 1901)(Echinodermata: Holothuroidea). pp. 447-450. In J.M. Lawrence (ed.) International Echinoderms Conference, Tampa Bay. Rotterdam: Balkema.

Feral, J.P. & Doumenc, D.A. 1982. Glycogen in holothuroids living in subantarctic and temperate waters. pp.381-386. In J.M. Lawrence (ed.) International Echinoderms Conference, Tampa Bay, Rotterdam: Balkema.

Fish, J.D. 1967. The biology of Cucumaria elongata (Echinodermata: Holothuroidea). J.mar.biol.Ass.

U.K. 47: 129-143.
Hunter-Rowe, C.K.A., Jeal, F. & West, A.B. 1976. Observations on the behaviour of Thyone fusus (O.F. Müller). Sci.Proc.Roy. Dub.Soc. 5(A): 459-469.
Keegan, B,F. 1972. Benthic studies in Kilkieran Bay and Galway Bay with particular reference to the class Bivalvia. Unpub. Ph.D. Thesis. National University of Ireland.
Keegan, B.F. 1974. The macrofauna of maerl substrates on the west coast of Ireland. Cah.Mar.Biol. 15: 513-530.
Keegan, B.F. & Könnecker, G. 1980. Aggregation in echinoderms on the west coast of Ireland. p.199. In M. Jangoux (ed.) Echinoderms: Present and Past. Rotterdam: Balkema.
Könnecker, G. & Keegan, B.F. 1973. In situ behavioural studies on echinoderm aggregations. 1. Pseudocucumis mixta. Helgolander wiss. Meersunter. 24: 157-162.
Krishnan, S. 1968. Histochemical studies on reproductive and nutritional cycles of the holothurian, Holothuria scabra. Mar.Biol. 2: 54-65.
Lawrence, J.M. 1972. Carbohydrate and lipid levels in the intestine of Holothuria atra (Echinodermata: Holothuroidea). Pac. Sci. 26: 114-116.
Marsh, J.B. & Weinstein, D.B. 1966. Simple charring method for determination of lipid. J.Lipid Biochem. 7: 374-376.
Mitsukuri, K. 1903. Notes on the habits and life history of Stichopus japonicus Selenka. Annot. zool. jap. 5: 1-21.
Prim, P., Lawrence, J.M. & Turner, R.L. 1976. Protein, carbohydrate and lipid levels of the adult body wall of Octinopyga aggasizi, Synaptula hydriformis and Pentacta pygmaea (Echinodermata: Holothuroidea). Comp.Biochm,Physiol.558: 307-309.
Sang, C. 1962. Biology of the Japanese common sea cucumber, Stichopus japonicus Selenka. 226p. Pussan, Korea. (In Korean with English summary).
Schneider, W.C. 1957. Determination of nucleic acids in tissues by pentose analysis. pp. 680-684. In S.P. Colwick and N.O. Kaplan (eds.) Methods in Enzymology Vol. 3. London: Academic Press.
Stier, T.J.B. 1933. Diurnal changes in activities and geotropism in Thyone briareus. Biol.Bull mar.biol.lab.Woods Hole. 15: 259-288.
Strickland, J.D.H. & Parsons, T.R. 1972. A practical handbook of sea water analysis. Bull.Fish. Res.Bd.Can. 125: 1-310.
Smith, T.B. 1981. Feeding and Aspects of the Nutritional Biology of the Holothurian Neopentadactyla mixta (Östergren), 1898 and the Ectoparasitic Gastropod Melanella alba Bowdich, 1822. Unpub. Ph.D. Thesis. National University of Ireland.
Smith, T.B. 1983. Tentacular ultrastructure and feeding behaviour of Neopentadactyla mixta (Holothuroidea: Dendrochirota). J.mar. biol.Ass.U.K. 63: 301-311.
Trefz, S.M. 1958. The Physiology of Digestion of Holothuria atra Jager with special reference to its role in the ecology of Coral Reef. Ph.D. Thesis, University of Hawaii.
Yamanouchi, T. 1956. The daily activity rhythms of the holothurians in the coral reef of the Palao Islands. Publ.Seto mar. biol. Lab. 5: 347-362.

Proceedings of the Fifth International Echinoderm Conference / Galway / 24-29 September 1984

Niche repartition in coexisting sea cucumber species in a kelp bed community

B.VELIMIROV
Institut für Zoologie der Universität Wien, Abt. Meeresbiologie, Vienna, Austria

ABSTRACT: In South African kelp beds two filter feeding sea cucumbers, Pentacta doliolum (Pallas) and Thyone auriata (Q. & G.) form dense patches in which a consistent vertical stratification is apparent. A bottom layer of T. auriata is covered by a dense layer of P. doliolum, so that the former species seems inhibited in maintaining its standard metabolism. The apparent competition for space on the one side, the frequency of the observation of this phenomenon on the other side, and the fact that these patches remain stationary over several years led to the hypothesis that competitive exclusion does not take place, and that this coexistence may be advantageous for one or both species. Differences in morphological and bioenergetic features permit the definition of niche dimensions, enabling the two species to live in close association. The advantage of the association and its zonation pattern is discussed in view of patch stability within the kelp bed ecosystem.

1 INTRODUCTION

A central aspect of niche theory is connected with the amount of resource sharing, which is directly connected to competitive interaction and niche overlap. The idea that there should be an upper limit to how similar the ecologies of two species can be and still allow coexistence has led to concepts like competitive exclusion, character displacement, limiting similarity and maximal tolerable niche overlap (Pianka 1976). Yet, overlap need not necessarily lead to competition unless resources are in short supply, even if one considers that niche overlap is the prerequisite for exploitative competition.

During distribution surveys in South African kelp beds it has been noticed that sea cucumbers, consisting of two species, namely Pentacta doliolum (Pallas) and Thyone auriata (Q. & G.); represented 43% of the total filter feeder biomass in the kelp beds (Velimirov et al. 1977). These species form extensive patches (Velimirov 1983) which remain stationary over several years with maximum patch densities of 1604 g dry weight.m ^{2} for T. auriata and 2017 g dry weight.m ^{2} for P. doliolum.

The peculiarity of the coocurrence of these species is that in patches where both species are found, P. doliolum always forms aggregations on top of T. auriata. It seems, therefore, that P. doliolum competes with T. auriata for space and dominates the latter by forming dense layers which cover the layers formed by T. auriata. Because of the frequency of this observation, it was hypothesized that this obvious niche overlap has no major disadvantage for either species and that competitive exclusion does not take place. Thus attempts were made to detect differences in niche dimension by comparing morphological and bioenergetical features of the two species.

2 MATERIAL & METHODS

In situ measurements and collection of specimens was accomplished using

SCUBA off Oudekraal near Cape Town, in a kelp bed at a depth of 10 to 15 m. Tentacle length and the spanned diameter of extended tentacles was measured to 1 mm with a calibrated glass rod. For each animal, 3 to 5 measurements were taken. Subsequently the animal was transferred into a labelled plastic bag and brought to the laboratory. In an isolated container each animal was relaxed in a seawater/magnesium chloride solution to measure body length. Both, measurements on tentacle length and body length, were repeated on freshly-collected animals of both species which were kept together in tanks for behaviour studies. All tentacle measurements were taken while the water was circulating and with the flow of water directed towards the animals. After relaxation, the animals were fixed in a solution of 10% formaldehyde. Following dissection, intestines which were measured on millimeter-squared paper and gut contents were examinated at 40x using a dissecting microscope.

Respiration rates for both species were determined at 14°C, which is the annual temperature average at the study site (Velimirov et al., 1977) using a Gilson respirometer adapted for large organisms; only freshly-collected animals (within 24 hours of sampling) were used.

3 RESULTS

Observation of cucumbers which were freshly transferred into aquariums showed that the two species reformed patches within the first hour. Similar to the field, a layer of T. auriata was superimposed by a layer of P. doliolum. Random redistribution of the two species within the tank resulted in reformation of patches with the same stratification. During the subsequent dispersion of the animals, it was noted that T. auriata was more sensitive to sudden manipulation stress than P. doliolum. Following manipulation, T. auriata retracted immediately, maintaining a retracted position much longer than disturbed P. doliolum. In some cases, T. auriata reacted to disturbance with evisceration; no such reaction was observed in Pentacta.

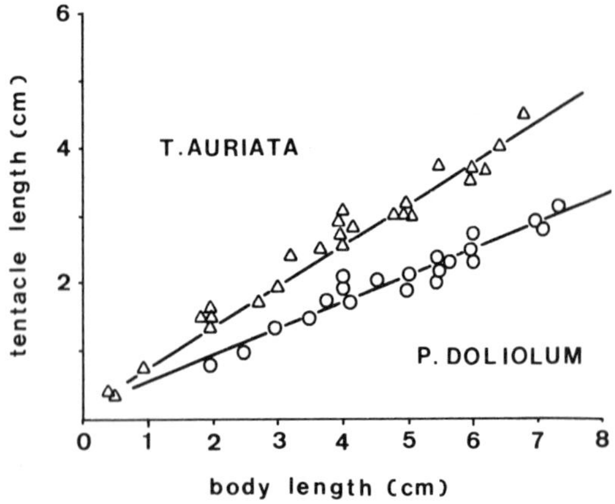

Fig.1. Relationship between tentacle length and body length for Thyone auriaty ($y=0.17 + 0.61x$, $r^2=0.96$) and Pentacta doliolum ($y=0.14 + 0.40x$, $r^2=0.95$).

A plot of tentacle length versus body length for the two species (Fig.1) shows that the deeper occurring T. auriata has a higher reach into the water column as compared to P. doliolum, assuming that the mouth of both species are maintained at equal height above the substrate. The differences between the two species range from 0.7 to 1.5 cm for animals between 2 and 7 cm body

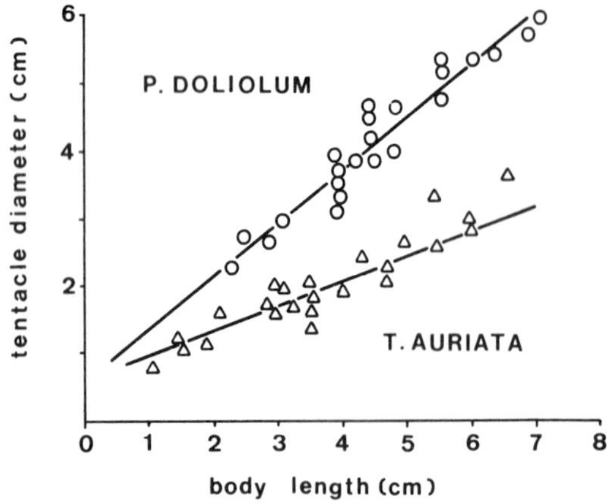

Fig.2. Relationship between spanned tentacle diameter and body length for T. auriata ($y=0.54 + 0.38x$, $r^2=0.79$) and P. doliolum ($y=0.49 + 0.81x$, $r^2=0.91$).

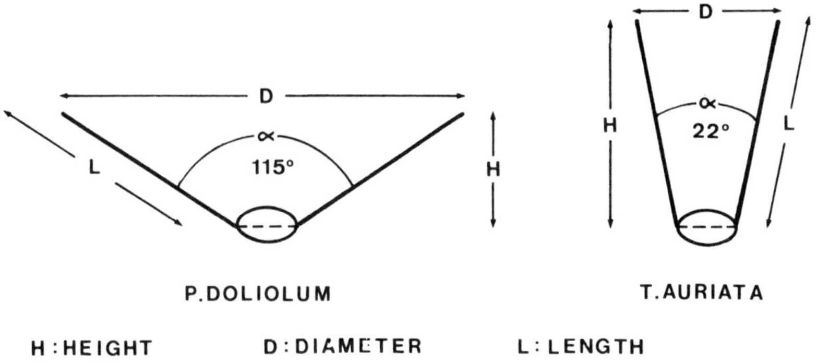

Fig.3. Schematic representation of the filtration funnels for T. auriata and P. doliolum.

length, respectively. An additional difference relates to the mode in which the two species expand their tentacles. Figure 2 demonstrates that the spanned tentacle diameters of the two species differ strikingly, with P. doliolum producing a larger tentacle diameter than T. auriata of the same body length.

Using the spanned diameter and the tentacle length measurements, the angle of the "filtration funnel" formed by expanded tentacles, can be calculated. For these calculations, animals of 5 cm body length were chosen. Average disc diameter for P. doliolum was 1.1 cm (st.dev. = 0.09, n = 16) and for T. auriata 0.98 cm (st.dev. = 0.06, n = 21). For the simplicity of the calculation an oral disc diameter of 1 cm was assumed to be representative for both species.

It can be seen in Figure 3 that P. doliolum forms a flat, plate like filtration funnel, whereas T. auriata forms a long narrow, conical funnel which protrudes twice as high up in the water column as the funnel of P. doliolum.

Another comparison of mean gut length shows another difference between the species. In specimens of comparable size T. auriata clearly has a shorter gut length than P. doliolum.

After quantification of the differences in structural morphologies, the investigation for metabolic differences between the two species led to the determination of respiration rates (Fig.5). Oxygen consumption per gram dry weight is 3 times higher for T. auriata than for P. doliolum; the two regression lines for oxygen consumption being perfectly parallel.

4 DISCUSSION

Hutchinson (1957) assumes that niche overlap cannot be tolerated for any period of time, and that competitive exclusion must occur in the overlapping parts of any two niches. In nature, however, niches often do overlap, without competitive exclusion. At this point, it has to be emphasized that the dense layer of P. doliolum covering T. auriata, prevents the latter species from

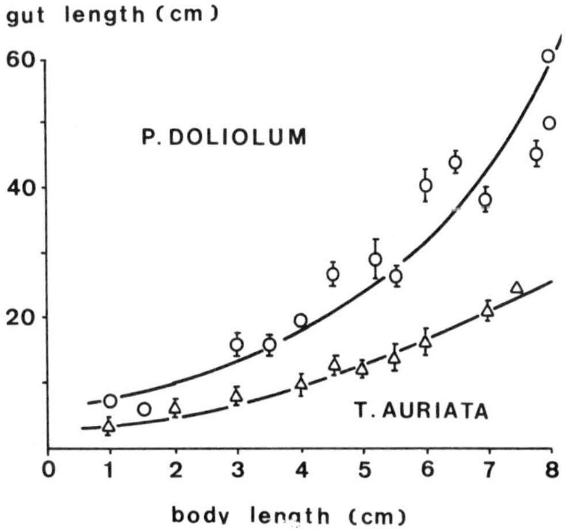

Fig.4. Relationship between gut length and body length for T. auriata ($y = 2.86e^{0.29x}$, $r^2 = 0.95$) and P. doliolum ($y = 5.28e^{0.30x}$, $r^2 = 0.87$).

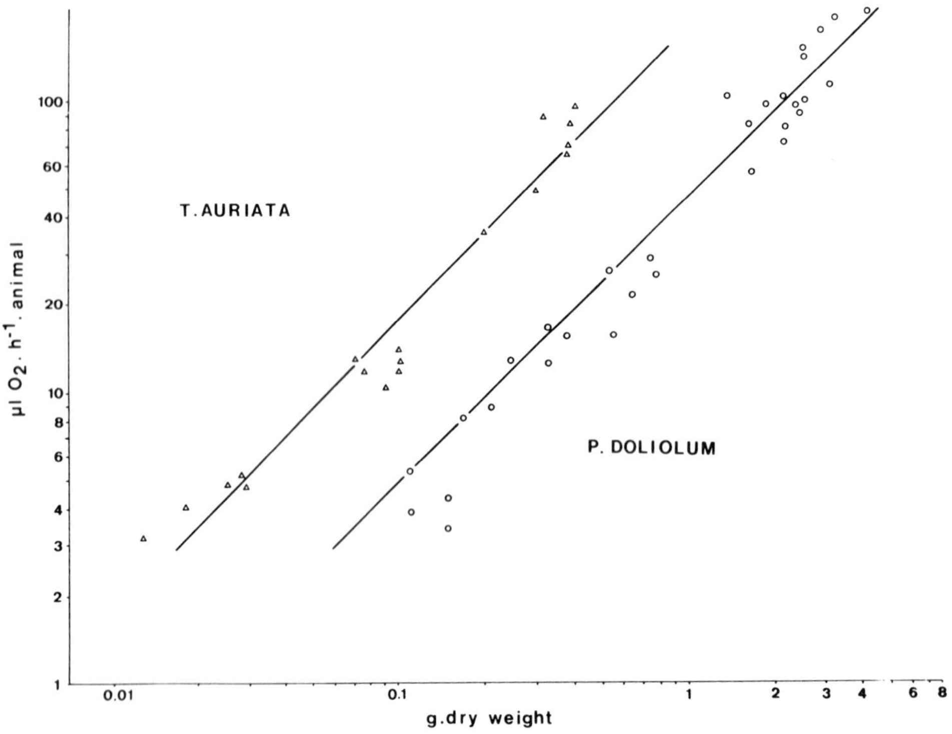

Fig.5. Relationship between respiration rates per weight unit of the two species (T. auriata: $y = 0.01x^{0.98}$, $r^2 = 0.97$; P. doliolum: $y = 0.04x^{0.89}$, $r^2 = 0.96$). Conversion factor dry weight/ash free dry weight for P. doliolum = 0.58, for T. auriata = 0.79.

filter feeding until the former species has commenced filtering activity. During this activity the anterior end of P. doliolum bends upwards, leaving space for individuals of T. auriata to rearrange their position and extend their tentacles. The above measurements and in situ observations on the feeding mode of the cucumbers allow the definition of niche dimensions which permit the coexistence of the two species. The longer extension of the tentacles and the formation of a narrower filtration funnel in T. auriata enables exploitation of the water column at a different level, thus avoiding interference with P. doliolum.

A qualitative investigation of gut contents of P. doliolum, ranging from 4 - 7 cm (n = 26) and T. auriata from 2 - 5 cm (n = 31) showed that both species contained mainly algal debris and detritus, and to a lesser extent small bivalve shells (Anomia sp.), spicules, arthropod remains and all grades of sediment. Although all species were collected from the same site and transferred to formaldehyde at the same time, 47% of the inspected T. auriata specimens had empty guts, in contrast to P. doliolum which always had full intestines. This may well be an indication that T. auriata, with the shorter gut, has a faster gut passage time than P. doliolum or/and that Pentacta feeds over a longer time periode than Thyone.

A further difference is exhibited by respiration rates, indicating differing metabolic rates for the two species. In a recent study, Schiemer (1983) demonstrated that aquatic invertebrates show a strong correlation between the standard metabolism and potential production, thus indicating that T. auriata has a higher intrinsic rate of natural increase than P. doliolum. At the present state of knowledge, it may be too early to speculate on the advantage or disadvantage of a higher standard metabolism for T. auriata, but the significance of a higher potential production, in regard to patch formation, shall be discussed. Considering that the structure and equilibrium of kelp community in which the sea cucumber patches exist are markedly influenced by water movement (Velimirov & Griffiths 1979), the interspecific facilitation of this association becomes apparent. The wave induced movement of kelp fronds (Laminaria pallida) results in a sweeping effect on the sea bottom which clears the substrate from most of the vagile fauna, including herbivores. This sweeping movement promotes kelp bed expansion because sporophytes can grow in the swept area and will take up sweeping as they grow. In monospecific Thyone patches, the specimens on the edges, near the sweeping kelp blades, may be disturbed and forced to regress or be swept away. This scenario could be avoided when the Thyone are covered by a layer of relatively stress-tolerant Pentacta specimens. Because of a potentially higher intrinsic rate of natural increase in T. auriata which implies a faster growth rate (Schiemer 1983) for this species as compared to Pentacta, T. auriata is probably successful in occupying substrate, but may loose patch space easily without the protective layer of Pentacta.

From the available information, one cannot rule out competition, but niche overlap need not necessitate competition,and extensive niche overlap may be correlated with reduced competition (Pianka 1979).

In summary, one may conclude that this sea cucumber aggregation, which may appear as an example of extensive niche overlap and competitive interaction between filter feeders, is a successful association, with clear niche differentiation, interspecific facilitation and patch stabilization in a dynamic system.

5 REFERENCES

Hutchinson, G.E. 1957. Concluding remarks. Cold Spring Harbor Symp. Quant. Biol. 22:415-427.

Pianka, E.R. 1976. Competition and Niche Theory. In R.M. May (ed.), Theoretical Ecology, principles and applications, p.114-141. Blackwell Scientific Publication.

Schiemer, F. 1983. Comparative aspects of food dependence and energetics of free living nematodes. Oikos 41:32-42.

Velimirov, B. 1983. Succession in a kelp bed ecosystem; clearing of primary substrate by wave induced kelp sweeping. Oceanol. Acta Proceedings 17th EMBS, Brest, France, 201-206.

Velimirov, B., J.G. Field, C.L. Griffith & P. Zoutendyk. 1977. The ecology of kelp bed communities in the Benguela upwelling system. Analysis of biomass and spatial distribution. Helgol.wiss.Meeresunters. 30:495-518.

Velimirov, B. & C.L. Griffiths 1979. Wave induced kelp movement and its importance for community structure. Bot.Mar. 22:169-172.

Reproductive patterns among northeast Pacific holothuroids

F.SCOTT McEUEN
Department of Zoology, University of Alberta, Edmonton, Canada

A comparative study of reproductive patterns was conducted using twelve abundant species of holothuroids in the San Juan Archipelago, Washington, USA. Modes of development include viviparity (one sp. of synaptid), external brooding (two dendrochirote spp.), lecithotrophy (seven spp. of dendrochirotes and one molpadid sp.), and planktotrophy (one aspidochirote sp.). Breeding seasons for the viviparous and externally brooding sea cucumbers begin in November and December, the eight lecithotrophic species spawn from December - May, while the single planktotroph has an extended reproductive season of April through September. The unique spawning behaviors exhibited by the different species appear adapted to their particular habitats to aid fertilization and dispersal. Extremes in fecundity and egg size range from 149 eggs of 1015μm diameter of a brooding species to nine million eggs of 206μm size produced by the planktotrophic species. Developmental stages were timed by monitoring cultures maintained at 11°C. Larval types include the auricularia, doliolaria, and a uniformly ciliated larva; the doliolaria may possess two, three, or five ciliary rings. Brooded embryos lack ciliation, metamorphose within the egg membrane, and hatch at the pentactula stage. Larval pelagic periods last 6 to 61 days. Most species' size at metamorphosis is determined by egg size.

Proceedings of the Fifth International Echinoderm Conference / Galway / 24-29 September 1984

Viviparity in a psolid holothurian from the tropical western Atlantic

J.E.MILLER
Harbor Branch Foundation Inc., Fort Pierce, USA

Lissothuria antillensis Pawson is the only psolid holothurian known to brood its young internally. Hermaphroditic individuals produce moderate-size lecithotrophic eggs which develop in the ovotestes up to an average diameter of 400μm. The absence of embryonic stages in the gonad suggests intra-coelomic fertilization, although it has not been determined whether this species is a "selfer" or obligate outcrosser. No block to self-fertilization has been identified, but fully-developed males are frequently found occurring with hermaphrodites. Three developmental stages (embryo, pentactula, post-pentactula) may be incubated simultaneously, with most broods containing all three stages. Brood size and total length (TL) of brooding individuals are positively correlated. Large adults, 13-15mm TL, may brood 80-90 young. The smallest specimen found to be incubating young measured 6mm TL. Parturition apparently occurs through the anus; adults known to have released young show no sign of external rupture and have tentacular crowns intact. In addition, pentactula and post-pentactula stages have been removed from the cloaca of brooding specimens. Newly-released young measuring 1.0-1.5mm TL possess 8 tentacles, 3 pairs of tube feet and a well-developed skeleton of imbricating plates. Incubation may be continuous; brooding adults have been collected during February, May, July, August and October.

Proceedings of the Fifth International Echinoderm Conference / Galway / 24-29 September 1984

An experimental evaluation of the effect of holothurian deposit feeding on infaunal communities

ANN M.MUSCAT
University of Southern California, USA

A manipulative field experiment was designed to test the hypothesis that deposit feeding by an epibenthic holothurian, Parastichopus parvimensis (Clark 1913), influences the distribution and abundance of infaunal populations. The experimental design allowed separation of the effects of P. parvimensis from the other common bottom disturbers in the area, the labrid fishes Semicossyphus pulcher and Halichoeres semicinctus. Five "treatments" were replicated twice: 1) complete exclusion - no fish or sea cucumbers, 2) partial exclusion- fish allowed in, sea cucumbers excluded, 3) sea cucumber enclosure - 1 sea cucumber/m^2, 4) cage control - sea cucumbers and fish allowed access with as much caging material present as possible, and 5) uncaged control - sea cucumbers and fish allowed access with no caging material present. Tests showed no significant cage effects. Infauna were sampled at time 0, 1 year and 2 years, and sorted to major taxonomic group. Predator densities and feeding rates were monitored.
The feeding rates of P. parvimensis varied throughout the year - the greatest amount of sediment being ingested during spring, with a significant decline during late summer through winter. Sea cucumber densities in shallow water (less than 13m) declined significantly during late summer and fall due to a downslope migration. The labrid fish had consistent feeding rates, but their densities also varied throughout the year.
There was no detectable effect of holothurian deposit feeding on any of the infaunal groups. Neither normal, nor ten times normal, sea cucumber densities resulted in a significant change in either the absolute or relative abundances of the infaunal species groups. Release from labrid fish predation, however, did significantly increase the densities of the polychaete, Chaetopterus variopedatus, and the phoronid, Phoronopsis californica.

5. Asteroidea

Pathways of nutrient translocation during vitellogenesis in the sea star, *Asterias rubens*

FRITS B.BEIJNINK & PETER A.VOOGT
Laboratory of Chemical Animal Physiology, State University of Utrecht, Netherlands

ABSTRACT: In vivo determination of turnover rates of coelomic fluid components indicates that translocation through the perivisceral coelomic cavity and coelomocytes contained can not meet the high anabolic demands of developing gonads. Alternatively, various data point to a role of the haemal system in conveying nutritive and vitellogenic substances. Electron microscopy reveals that within early-vitellogenic oocytes, yolk vesicles first appear within the stalk that connects the oocyte with the ovarian haemal sinus. Immunocytochemistry with antiserum against a yolk glycolipoprotein fraction shows a close relationship between ooplasm and haemal ground substance. Whole-body and microautoradiography after ingestion of radio-labelled compounds further corroborates the concept of a translocative function of haemal tissue and, in particular, of the enclosing coelomic sinuses.

1 INTRODUCTION

Defining the pathways and mechanisms of interorgan nutrient transport is a fundamental problem in asteroid physiology which only recently is yielding to investigative efforts. As putative circulatory systems such as the haemal and "perihaemal" sinuses are poorly developed, a long-established consensus exists on a major role of the coelomic fluid in nutrient distribution (review: Ferguson 1982). However, no proof is available that such a mechanism of random diffusion can meet the demands of the developing gonads during gametogenesis. Recent experiments with radiolabelled tracers (Broertjes et al. 1980, 1982) and histological observations on asteroid gonads (review: Walker 1982) have resulted in a revaluation of older concepts (Cuénot 1948) which speculate on a role of the haemal system in gonad nutrition. In this paper, we will present results from a series of studies on the pathways of translocation of nutritive and vitellogenic substances needed for gonad development in the sea star Asterias rubens.

2 RESULTS AND DISCUSSION

2.1 Turnover of coelomic fluid compounds

In an attempt to quantify the amount of nutrients conveyed through the coelomic fluid, we injected tracer doses of radiolabelled oleic acid and leucine into the perivisceral coelomic cavity of A. rubens (Beijnink et al. 1984). By measuring the half time of these compounds through periodic sampling, and by measuring the total concentration of fatty acid and

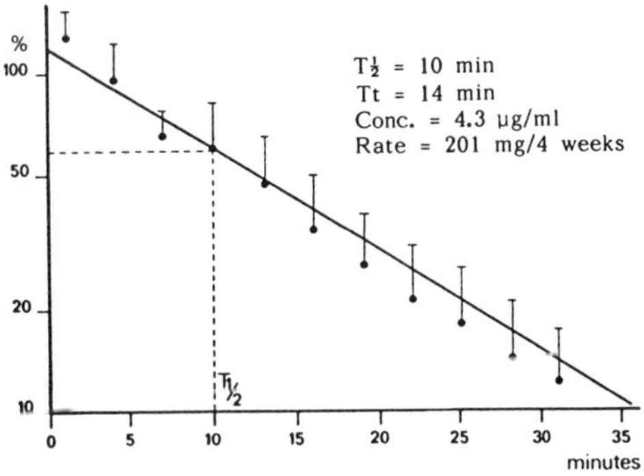

Fig. 1. Assessment of the turnover rate of amino acid in the coelomic fluid of A. rubens by intra-coelomic injection of tritiated leucine. The half time $T\frac{1}{2}$ is estimated from a semi-logarithmic plot of the remaining radioactivity (mean % of the injected dose ± SD, n=3); the turnover time is: Tt = $T\frac{1}{2}/\ln 2$ = 14 min. With a mean amino acid concentration of 4.3±1.9 µg/ml (± SD, n=4), a turnover rate of 201 mg of amino acid/4 weeks can be calculated for a 100 g animal with an average amount of 16.2 ml of coelomic fluid (Modified from Beijnink et al. 1984).

amino acid, turnover rates could be calculated: 74 mg of fatty acid and 201 mg of amino acid (Fig. 1) are conveyed through the coelomic fluid of a 100 g animal in a 4 week period. Since the average increase in lipid and protein content of the ovaries is about three times as high (Oudejans & Van der Sluis 1979), and since only part of the material conveyed through the coelomic fluid is ultimately absorbed by the germinal epithelium, it is clear that nutrient diffusion through the coelomic cavity can not sustain vitellogenesis.

Coelomocytes – derived from the coelomic lining of the digestive system (Fig. 2) – have often been assumed to play a role in lipid transport. A specimen of A. rubens weighing 100 g contains 33 ± 7 mg of coelomocytes (mean dry weight ± SEM, n=10) within the perivisceral coelomic cavity. This amount of cells contains 4 mg of lipid and can accumulate about 1 mg of fatty acid in vitro (Beijnink et al. 1984). Assuming that Asterias is able to renew its population of coelomocytes within three days (Binyon 1982), it is evident that lysis of coelomocytes in target organs can not substantially contribute to lipid transport.

2.2 Morphological and immunological relations between haemal system and oocytes

Employing electron microscopy we have examined the contact sites of vitellogenic primary oocytes with the genital haemal sinus of the ovarian wall (details will be published elsewhere). Early-vitellogenic oocytes are pear-shaped and are attached to the haemal sinus by stalk-like processes. The first ultrastructural indication for vitellogenesis is the appearance of numerous vesicles within the basal region of the oocyte-stalk (Fig. 3). Such vesicles contain flocculent material and often fuse with adjacent lysosome-like vacuoles. Fusion of yolk particles with vacuoles identified cytochemically as lysosomes has recently been reported for Asterina gibbosa by Ferrand (1984). From our study we tentatively concluded that nutritive and vitellogenic substances manufactured extragonadally and stored temporarily in the ovarian wall can pass through the stalk of pear-shaped oocytes.

To verify this concept we applied an immunocytochemical technique – the unlabelled antibody enzym method with peroxidase-antiperoxidase according to Sternberger (1974) – using antiserum raised against one of the yolk

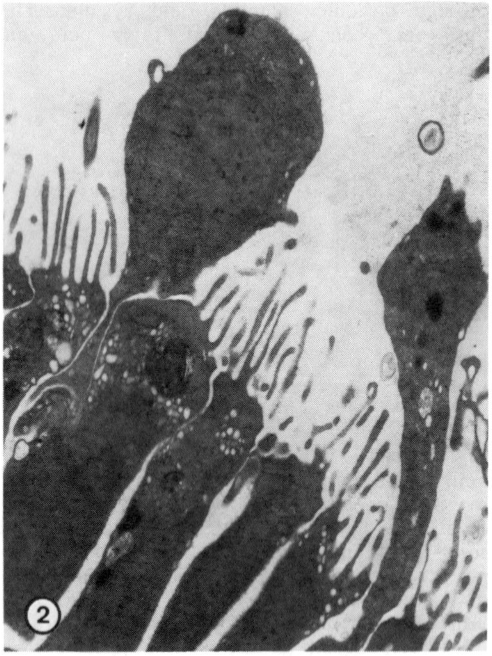

Fig. 2. Protruding cells (putative coelomocytes) within the coelomic epithelium of the pyloric stomach of A. rubens. x7500.

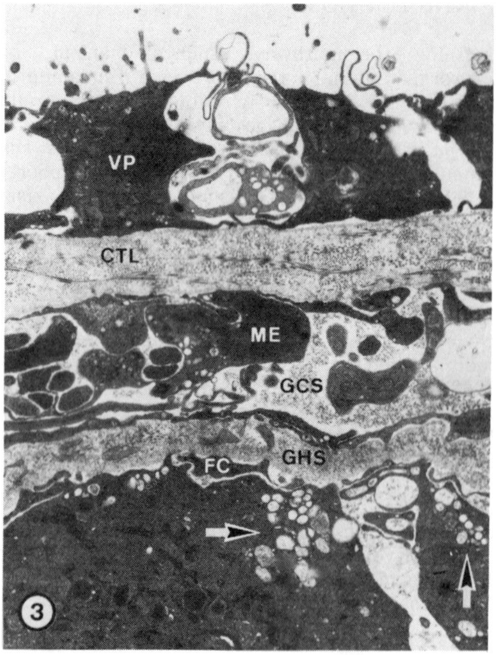

Fig. 3. Vesicles within the basal region of early-vitellogenic oocytes (arrows). FC: follicle cell; GHS: genital haemal sinus; GCS: genital coelomic sinus with myoepithelial lining (ME); CTL: connective tissue layer of outer sac; VP: visceral peritoneum. x5000.

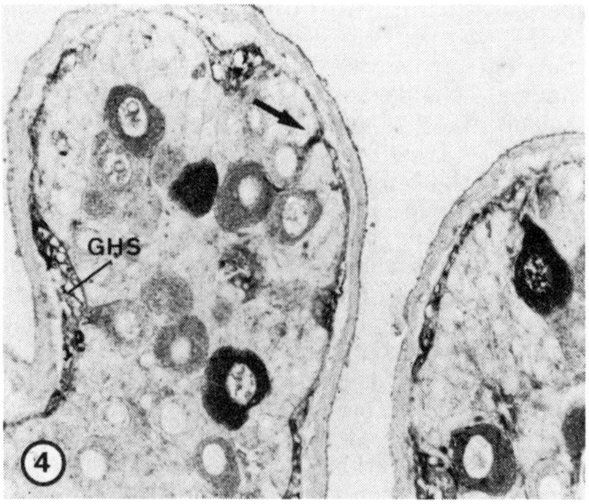

Fig. 4. Immunocytochemical demonstration of yolk glycolipoprotein ("fraction 3", Broertjes et al. 1984b). Note immunoreactivity (dark deposits) of large vitellogenic oocytes, of the stalk of early-vitellogenic oocytes (arrow), and of the genital haemal sinus (GHS). x200.

glycolipoprotein fractions described by Broertjes et al. (1984a,b, these proceedings). Fig. 4 shows immunoreactivity throughout the ooplasm of proper-vitellogenic oocytes, within the stalk of early-vitellogenic oocytes (arrow), and of the ground substance of the genital haemal sinus. These reactions are specific (i.e. suppressed by prior incubation of immune serum with antigen) and are obtained with highly diluted (up to 1:6000) primary antiserum. Specific reactions were also observed within the aboral haemal ring and the gastric haemal tufts. These observations provide strong support for the concept that vitellogenesis in A. rubens depends on the extragonadal synthesis of glycolipoproteins which may be conveyed through – or stored within – the haemal sinuses. However, with the immunocytochemical method used we have not been able to designate a specific site of synthesis of yolk material or precursor; no specific reactions could be observed within the pyloric caeca. Moreover, specific reactions were observed within the genital haemal sinus of the testes (but not in spermatogenic columns) indicating that the studied glycolipoproteins are perhaps not exclusively vitellogenic.

2.3 Whole-body autoradiography

Up to now, autoradiography – a classical approach in studying nutrient translocation – has only been performed on Bouin-fixed paraffin sections of asteroid organs. We employed whole-body autoradiography to follow the fate of ingested 14C-leucine and 14C-palmitic acid (details will be published elsewhere). This technique, which was performed on thick (50 µm) whole-body sections of animals frozen up to 16 days after tracer administration, allows localization throughout the asteroid body of soluble substances that normally are lost in fixation and dehydration procedures (e.g. amino acid not incorporated into protein, and labelled substances within fluid-filled coelomic compartments).

In our experiments which were performed during active gametogenesis (December to February), radioactivity remained localized predominantly in the stomach and pyloric caeca. In sections of animals, frozen one and a half hour after ingestion, labelling of the perivisceral coelomic cavity in regions close to the stomach pointed to initial displacement of tracer through the coelomic fluid and coelomocytes. However, we also observed distinct labeling of the gastric haemal tufts, axial organ, and aboral ring 6 to 24 hours after ingestion of palmitic acid. Gonads became labelled one hour after ingestion of leucine and ultimately displayed levels of radioactivity similar to or higher than those observed in the side lobes of the pyloric caeca (Fig. 5). Administration of palmitic acid resulted in low levels of tracer in the gonad, being present mainly in the periphery of the gonad tissue. Additional microautoradiography on semi-thin (1 µm) sections of glutaraldehyde-fixed gonad tissue revealed substantial

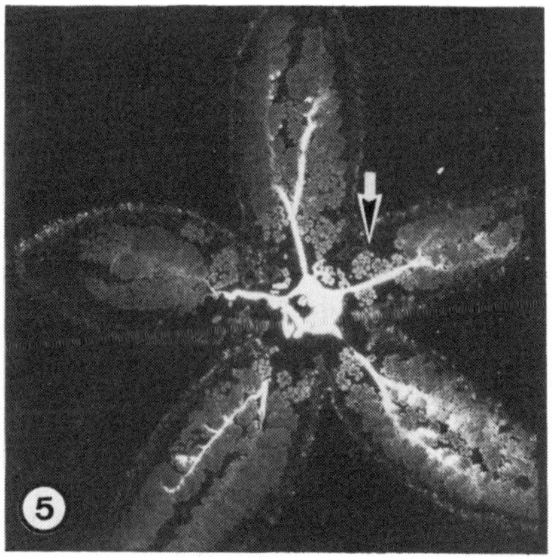

Fig. 5. Contact autoradiograph of a whole-body section of a specimen of A. rubens, frozen in hexane cooled with solid carbon dioxide (- 75°C) one day after ingestion of 10 µCi of 14C-leucine. The pyloric stomach (center), and pyloric and median ducts of the pyloric caeca (visible as broad white bands) are strongly labelled. The gonads are distinctly labelled (arrow). x0.75.

incorporation of label into the ground substance of the genital haemal sinus 24 hours after ingestion of leucine; similar observations were made recently by Ferguson (1984) working with Echinaster.

Whereas these and earlier autoradiographic studies (Ferguson 1970; Broertjes et al. 1980) provide evidence for a role of the haemal system in storage and/or translocation of nutritive materials, it is still far from clear how such processes are effected. Displacement of the mucoid haemal ground substance through the diminuitive haemal sinuses is not likely to occur at rates high enough to account for the prominent levels of tracer observed in the genital haemal sinus shortly after oral administration of labelled compounds. Alternatively, circulation within the fluid-filled "perihaemal" coelomic channels – easily maintained by flagella of the coelomic lining – might be primarily responsible for nutrient translocation; the haemal tissues would then function as storage reservoirs. While it is significant that such a function of haemal tissue has been suggested now for species of three echinoderm classes (Grimmer & Holland 1979; Ferguson 1984; Jackson & Fontaine 1984), it is evident that many problems remain to be solved in order to come to a better understanding of the possible translocative role of the haemal system.

3 ACKNOWLEDGEMENTS

We thank Femmy Brands for performing the immunocytochemical tests and Louis Prinsen and Ientse Van der Sluis for their assistance in tracer experiments. Special thanks are due to Drs. Charles W. Walker, Jan J.S. Broertjes, and Rob C.H.M. Oudejans for advice and interest. The investigations were supported by the Foundation for Fundamental Biological Research (BION) which is subsidized by the Netherlands Organization for the Advancement of Pure Research (ZWO).

4 REFERENCES

Beijnink, F.B., I. Van der Sluis & P.A. Voogt 1984. Turnover rates of fatty acid and amino acid in the coelomic fluid of the sea star Asterias rubens: Implications for the route of nutrient translocation during vitellogenesis. Comp. Biochem. Physiol. 78B: 761-767.

Binyon, J. 1982. The number and origin of the coelomocytes in the perivisceral fluid of the starfish Asterias rubens L. In J.M. Lawrence (ed), Proceedings of the International Conference, Tampa Bay, p. 353. Rotterdam: Balkema.

Broertjes, J.J.S., G. Posthuma, F.B. Beijnink & P.A. Voogt 1980. The admission of nutrients from the digestive system into the haemal channels in the sea-star Asterias rubens (L.). J. mar. biol. Ass. UK 60: 883-890.

Broertjes, J.J.S., J.N. Jens, D. Van Oudheusden, M. De Bruin & P.A. Voogt 1982. Demonstration of nutrient flow in the starfish Asterias rubens (L.) with 125 I labelled proteins. Neth. J. Zool. 32: 472-478.

Broertjes, J.J.S., P. de Waard & P.A. Voogt 1984a. On the presence of vitellogenic substances in the starfish Asterias rubens (L.). J. mar. biol. Ass. UK 64: 261-269.

Broertjes, J.J.S., P. de Waard & P.A. Voogt 1984b. Purification and characterization of vitellogenic substances in the starfish Asterias rubens (L.). Mar. Biol. Lett. 5: 99-104.

Cuénot, L. 1948. Anatomie, éthologie et systématique des Echinodermes. In P.P. Grassé (ed.), Traité de Zoologie, vol. XI, p. 3-363. Paris: Masson.

Ferrand, J.G. 1984. Lysosomes and yolk formation in the oocytes of Asterina gibbosa (Echinodermata, Asteroidea). Int. J. Invert. Reprod. 7: 143-148.

Ferguson, J.C. 1970. An autoradiographic study of the translocation and utilization of amino acids by starfish. Biol. Bull. 138: 14-25.

Ferguson, J.C. 1982. Nutrient translocation. In M. Jangoux & J.M. Lawrence (eds.), Echinoderm nutrition, p. 373-393. Rotterdam: Balkema.

Ferguson, J.C. 1984. Translocative functions of the enigmatic organs of starfish – the axial organ, hemal vessels, Tiedemann's bodies, and rectal caeca: An autoradiographic study. Biol. Bull. 166: 140-155.

Grimmer, J.C. & N.D. Holland 1979. Haemal and coelomic circulatory systems in the arms and pinnules of Florometra serratissima (Echinodermata: Crinoidea). Zoomorphologie 94: 93-109.

Jackson, L.E. & A.R. Fontaine 1984. Nutrient and nitrogenous waste concentrations in the fluid transport systems of a holothurian. Comp. Biochem. Physiol. 78A: 123-128.

Oudejans, R.C.H.M. & I. Van der Sluis 1979. Changes in the biochemical composition of the ovaries of the sea star Asterias rubens during its annual reproductive cycle. Mar. Biol. 50: 255-261.

Sternberger, L.A. 1974. Immunocytochemistry. In A. Oster & L. Weiss (eds.), Foundation of immunologie series, p. 1-246. Englewood Cliffs, New Jersey: Prentice-Hall.

Walker, C.W. 1982. Nutrition of gametes. In M. Jangoux & J.M. Lawrence (eds.), Echinoderm nutrition, p. 449-468. Rotterdam: Balkema.

Proceedings of the Fifth International Echinoderm Conference / Galway / 24-29 September 1984

Fine structure and presumed functions of the gastric hemal tufts of the Asteroid, *Asterias rubens* L.

CATHERINE BOULAND
Laboratoire de Biologie cellulaire et animale, Université d'Orléans, France

CATHERINE YOURASSOWSKY* & MICHEL JANGOUX
Laboratoire de Biologie marine, Université Libre de Bruxelles, Belgium

SUMMARY. Asteroid gastric hemal tufts are paired mesenteric-like organs containing enlarged hemal lacunae. They connect the gut hemal system to that of the gonad through the upper part of the axial organ. Gastric tufts are covered by an heavily folded mesothelium in which microscopic pores occur. Tuft lacunae contains a finely granular ground substance (the hemal fluid) in which two cell types are seen. It is suggested that the gastric tufts are hemal fluid-collector organs and that they could also function as hemal fluid-filtering organs.

INTRODUCTION

The hemal system of echinoderms is made up of more or less developed network of non-epithelialized lacunae running within the supporting connective tissue of the individuals. Hemal lacunae enclose an hemal fluid - that is presumed to be rich energetically - in which single cells may be found. The asteroid hemal system is weakly developed, being conspicuous only at the radial, axial and gonadal levels (Cuénot 1901, Hyman 1955). However hemal lacunae are found associated with the gut. They are confined to the aboral side of the pyloric stomach and of the pyloric caeca, and are connected to the axial organ through paired organs, viz. the gastric hemal tufts (Cuénot 1901).

Gastric hemal tufts occupy a strategic position between the pyloric caeca and the gonads. Their importance in nutrient transfer has been emphasized on several occasions (Broertjes et al. 1980a, 1980b, 1982). The aim of the present paper was chiefly to investigate the gastric tufts topography and fine structure, with special attention to the mesothelial envelope and the types of cells to be found in the tuft hemal lacunae.

*Research assistant of the National Fund for Scientific Research (N.F.S.R., Belgium).

MATERIAL AND METHODS

Individuals of **Asterias rubens** were collected by scuba diving at Scharendijk (Zeeland, Netherlands). They were kept alive in a closed circuit marine aquarium in the laboratory at Brussels.

For light microscopy, the removed hemal tufts were fixed in Bouin's fluid. They were embedded in paraplast and cut in 5 um sections. Masson' trichrome was used as routine stain. Hemal fluid was revealed by the Alcian blue-Periodic acid Schiff staining technique. Fixation and staining procedures were performed according to Ganther and Jollès (1969-1970).

For EM investigations the gastric tufts were fixed in a cooled (4°C) solution of 3 % glutaraldehyde in cacodylate buffer (0.1 M, pH 7.3). They were then washed in the buffer and postfixed with 1 % osmium tetroxide in the same buffer. They were dehydrated by graded ethanol. For TEM-microscopy tufts were embedded in Spurr. Sections were stained with uranyl acetate and lead citrate (Reynolds 1963), and observed with an electron microscope (Philipps EM 300). For SEM-microscopy tufts were dried by the critical point method, using CO_2 as transition fluid. The specimens were mounted on aluminium stubs, coated with gold in a sputter coater and observed with an IST DS-130 scanning electron microsocope.

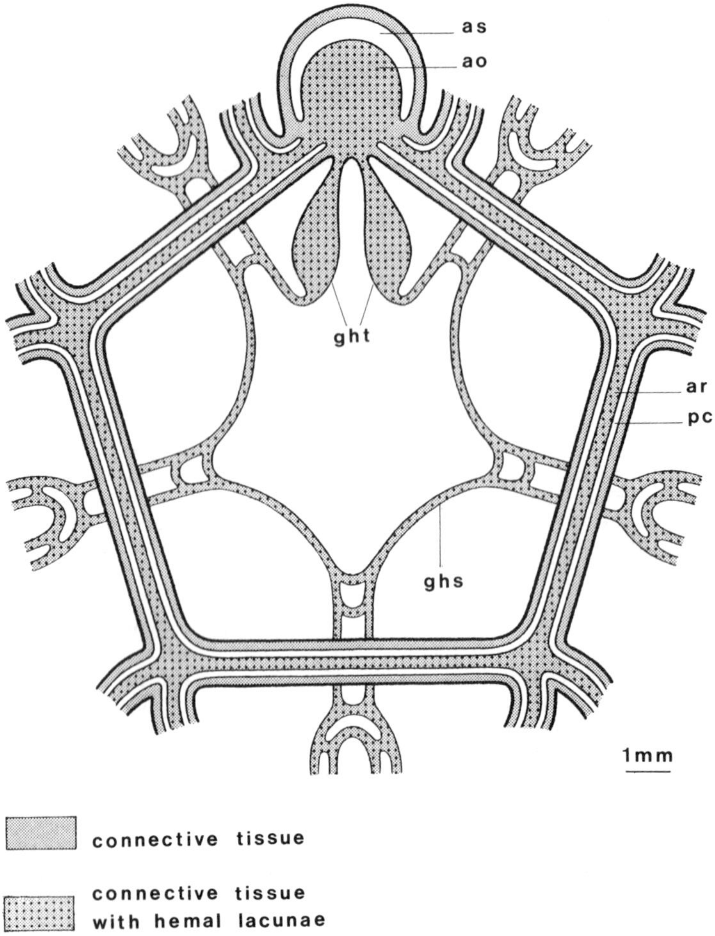

Figure 1. Diagramatic representation of the aboral central hemal system (aboral and pyloric rings). ao axial organ, ar aboral hemal ring, as axial sinus, ghs gut hemal system, ght gastric hemal tufts, pc perihemal coelom.

OBSERVATIONS AND DISCUSSION

The gastric hemal tufts are located in the madreporic interradius (CD interradius). They are paired organs attached to the aboral (coelomic) side of the body-wall through short fenestrated mesenteries. The gastric tufts are connected to the gut hemal system proximally, and to the axial organ distally (Figs 1-3). Both tufts reach the axial organ almost at the same point (Fig. 2). The gastric tuft-axial organ connection takes place at the point where the axial organ contacts the aboral hemal ring (viz. the ring that irrigates the gonads). When going towards the gut the tufts are separated and join the pyloric stomach wall some distance from each other (though they are always in the CD interradius; Figs 1, 3). Most observers reported great difficulties in seeing -either macroscopically or microscopically- the course of the gut hemal system (see, e.g., Ferguson 1982). This system clearly occurs however, but its development is very limited. Actually the gut hemal system of **A. rubens** consists of a peripheral pyloric pentagon and of ten pairs of aboral longitudinal lacunae -sometimes connected- running all along the length of each pyloric caecum. Each tuft is in contact with one half of the gut hemal system (i.e. there is no hemal branch

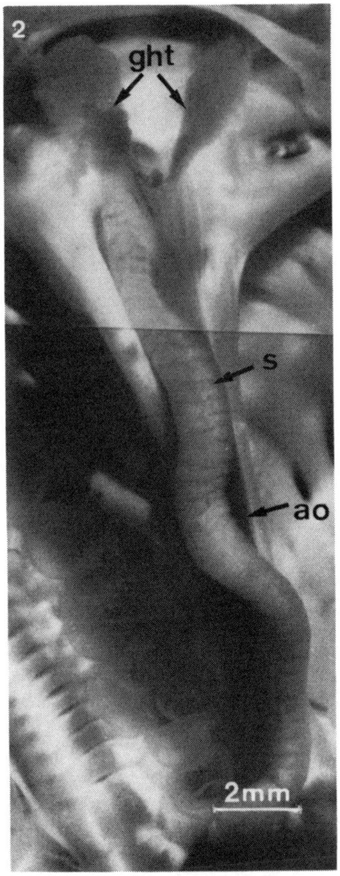

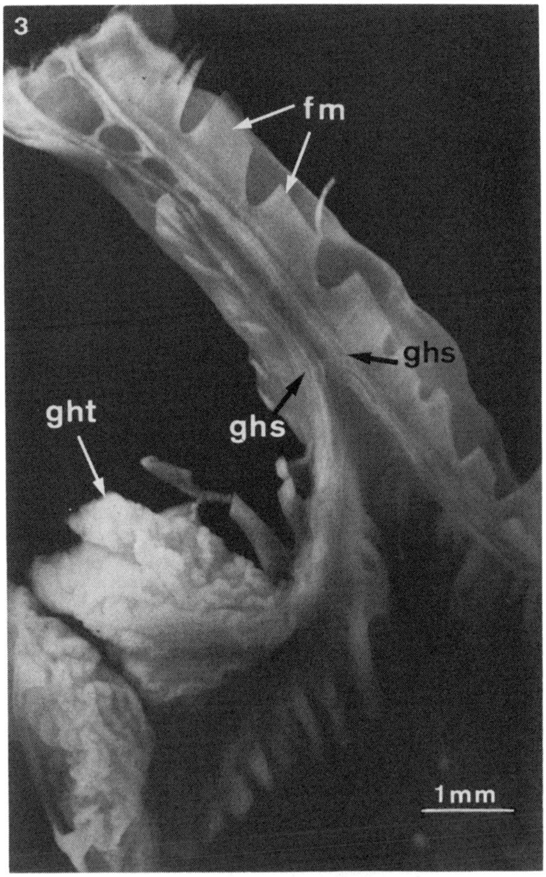

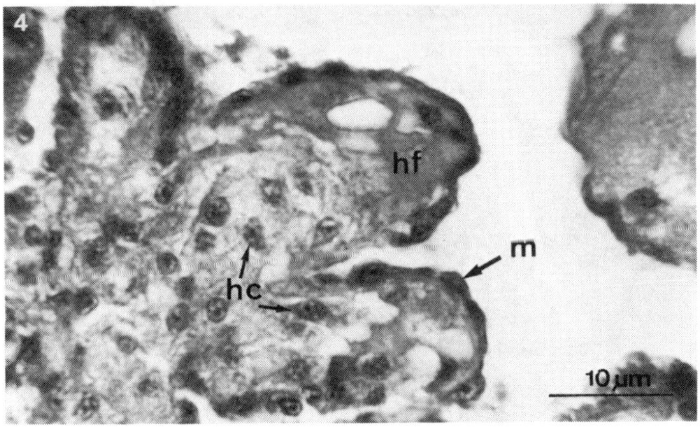

Fig. 2. Macrophotography of the axial complex. ao axial organ, ght gastric hemal tufts, s stone canal.

Fig. 3. Macrophotography of the gastric tuft and of the gut associated hemal system. fm fenestrated mesenteries, ght gastric hemal tuft, ghs gut hemal system.

Fig. 4. Histological section of a gastric tuft. hc hemalocytes, hf hemal fluid, m mesothelium.

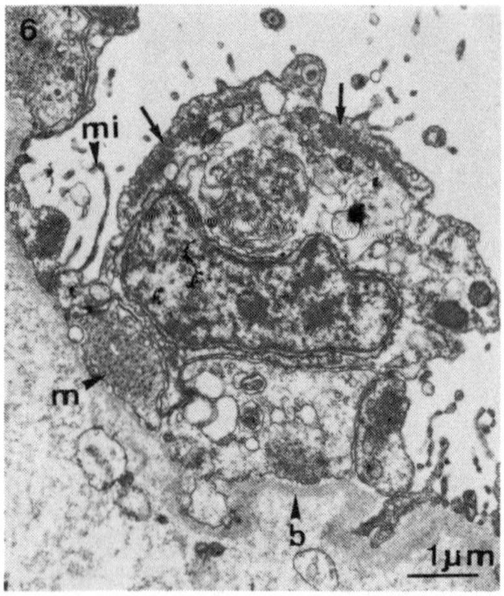

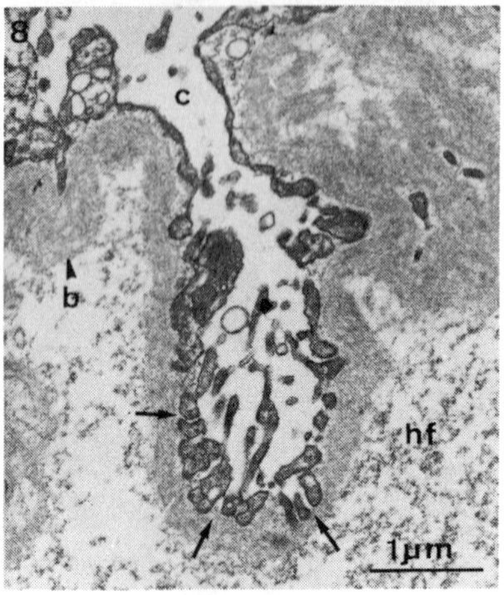

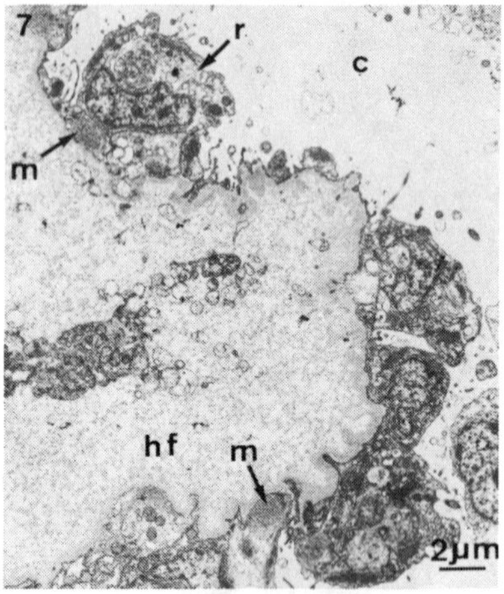

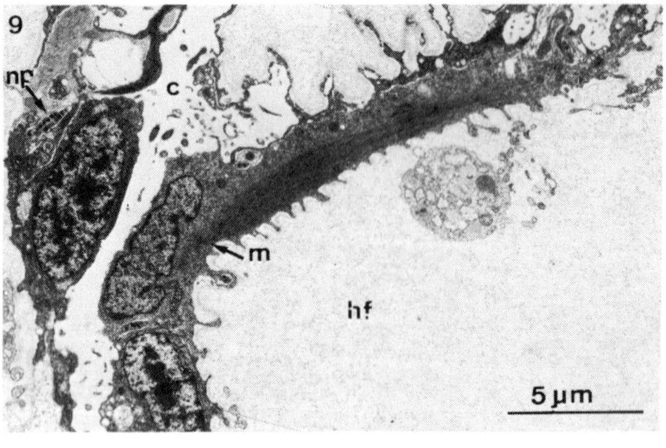

Fig. 5. SEM-view of the coelomic surface of a gastric tuft.

Fig. 6. TEM-view of a revestment mesothelial cell. b basal lamina, m myoepithelial cell, mi microvilli.

connecting the tufts together at their junction with the pyloric hemal pentagon; Fig. 3). Giving their location and anatomical relationships with both the gut and the axial organ, the gastric tufts may be said to resemble hemal collecting organs (Figs 1, 3).

The gastric hemal tufts hang in the general coelomic cavity. They are not surrounded by a perihemal coelom. The gastric tufts basically have a mesenteric structure, viz. a connective tissue layer sandwiched between two mesothelial layers. However the connective layer is considerably distended by conspicuous fluid-filled hemal lacunae (Fig. 4). The hemal fluid is strongly PAS-positive and includes a relative large number of single cells.

The outer (coelomic) side of the gastric tufts is very densely folded (Fig. 5). These folds greatly increase their coelomic surface which is everywhere covered by a flattened mesothelium. The tuft mesothelium consists of two cell types -viz. revestment cells (Figs 6, 7) and myoepithelial cells (Fig. 9)- resting upon a rather thick basal lamina. Revestment cells roughly correspond to the adluminal cells described by Wood and Cavey (1981) in the ambulacral mesothelium of asteroid tube-feet. They are ciliated (a single cilium per cell) and bear scattered apical microvilli. Their most particular feature is that they develop filopod-like cytoplasmic extensions. These extensions sometimes are extremely thin, being markedly thinner than the underlying basal lamina (Fig. 8). Filopodal extensions are not always close to each other, and microscopic mesothelial pores may occur. At the level of these pores the coelomic and hemal fluids are separated only by the basal lamina of the mesothelium (Fig. 8). Such structure strongly recalls the podocyte-like cells decribed by Bargman and von Hehn (1968) in the axial organ of **A. rubens**. There is a very striking similarity between the ultrastructure and hemal relationships of the tuft mesothelium, and the ultrastructure and blood relationships of the vertebrate kidney podocytes. While no physiological investigations have been done, it is tempting to suggest that the hemal fluid could be filtered at the level of the gastric tufts. As for the myoepithelial cells they seem to be less numerous than revestment cells. They could be ciliated when located facing the coelomic cavity (Fig. 9). However they are more often restricted to the basal part of the mesothelium, being covered by revestment cells. Myoepithelial cells always have a bundle of densely packed myofibrillae. They show well-developed basal anchoring structures (Fig. 9). Scattered nervous (hyponeural) processes are also seen near the myoepithelial cells. In situ observation of living tufts allowed to see the contractibility of these organs. The supposed tuft filtration of the hemal fluid would be obviously promoted by their contractions.

The inner part of gastric tufts consists of a very loose connective tissue with scattered collagenic fibrillae. As seen on TEM-pictures the hemal fluid of tufts consists of a finely granular ground substance in which seemingly two types of hemalocytes occur (e.g., Fig. 11), viz. the pseudopodial cells and coelomocyte-like cells. With very few exception those cells are not in contact with the inner side of the mesothelial basal lamina. The pseudopodial cells have more or less elongated cytoplasmic extensions (Figs 12, 13). Their cytoplasm is always very dense. It includes small circular dark-green vesicles the content of which seems to be synthetized by the cell organelles (Figs 11, 12). They also may contain clear irregular-shaped vesicles of various size the content of which has an appearance similar to the hemal ground substances (Fig. 12). Coelomocyte-like cells are irregularly

---

Fig. 7. TEM-view of the tuft mesothelium. c coelom, hf hemal fluid, m myoepithelial cell, r revestment cell.

Fig. 8. TEM-section of a very thin mesothelial area showing microscopic pores (arrows). b basal lamina, c coelom, hf hemal fluid.

Fig. 9. TEM-section showing a myoepithelial cell. c coelom, hf hemal fluid, m myoepithelial cell, np nervous process.

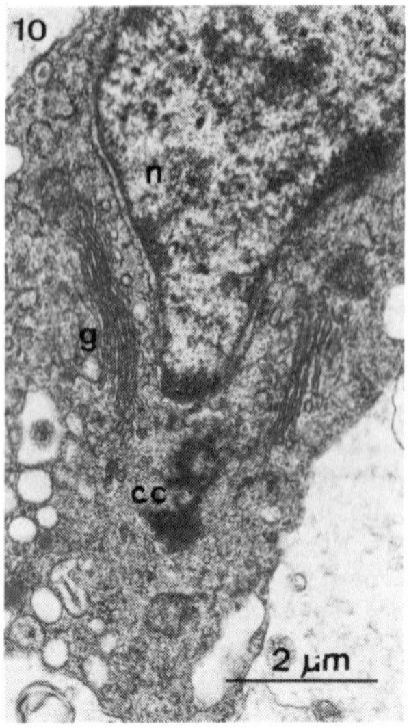

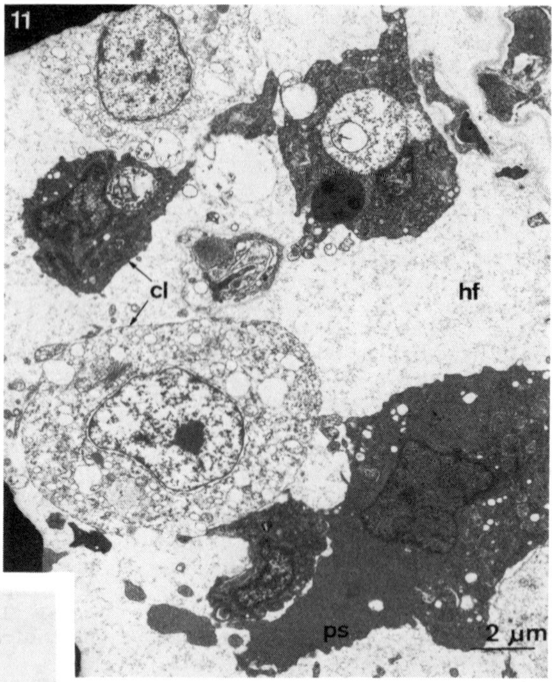

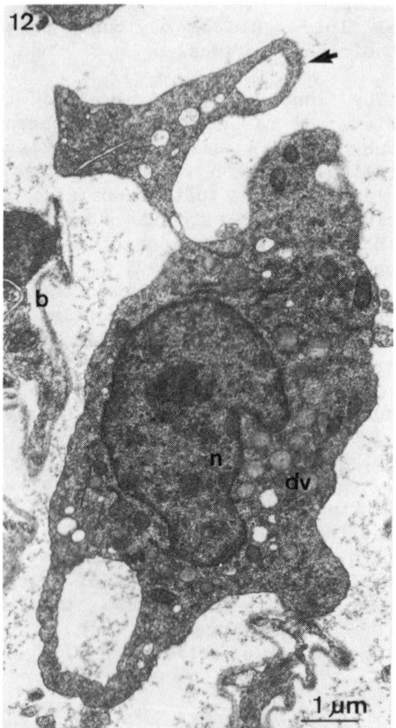

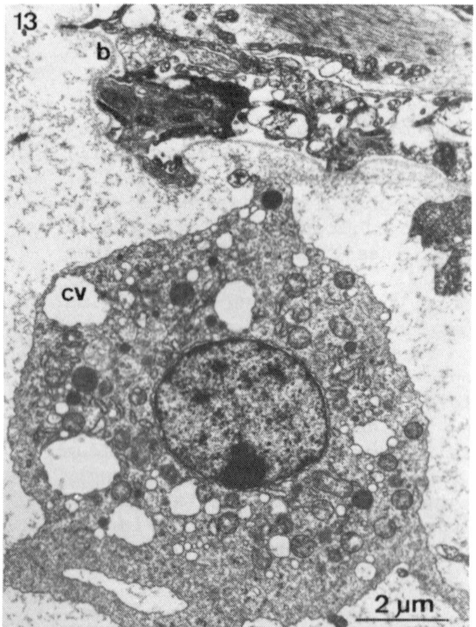

Fig. 10. Enlarged TEM-view of a coelomocyte-like cell. cc centriolar complex, g golgi apparatus, n nucleus.

Fig. 11. General TEM-view of hemalocytes. hf hemal fluid, pm coelomocyte-like cell, ps pseudopodial cell.

shaped though they do not form pseudopodial extensions (Fig. 11). They often show the occurence within their cytoplasm of paired centriolar structures which presumably would indicate a mesothelial origin (see, e.g. Maes and Jangoux 1983). Those cells seem rather active having well-developed Golgi apparatus and ergatoplasmic cisternae, as well as a variable amount of clear vesicles (Fig. 10). Works are in progress to characterize the fine structure, localisation and migratory capabilities of those two cell types more precisely.

Preliminary experiments were done using marine solutions of iron-associated substances -viz. 1 % iron saccharate and 1 ‰ ferritine- pushed into the cardiac stomach cavity. Iron was revealed on histological section using the conventional Perls method (Prussian blue reaction). The occurence of iron-linked particles were seen in cells of the digestive epithelium of pyloric caeca within a few hours, while in the hemal tuft lacunae the presence of iron became obvious only after a 24-hour latent period. (Control experiments -viz. intracoelomic injection of iron saccharate or ferritine- did not show any Perls-positive material within the organs studied). Our results confirm those of Broertjes et al. (1980a) in that they demonstrate that material from the pyloric caeca could be transfered to the gastric tufts. However occurence of iron within the tufts does not means that the injected substances -viz. iron saccharate and ferritine- are translocated. Yet that implies at least that iron particles are translocated. It should be emphasized that iron particles of the tuft hemal lacunae always occured within a cell cytoplasm and that they were never seen in the hemal fluid itself. This allows to suggest that iron particles would be release by the basal side of caecal digestive cells (exocytosis?) and that they would be then carried towards the gastric tufts by coelomocyte-like cells. (Similar cells were observed by Van der Plas et al., 1980, between the caecal digestive cells of **A. rubens**). The migratory capabilities of hemalocytes should be considered. Actually these cells are not necessarily restricted to their hemal lacuna and could penetrate more or less deeply within the connective tissue layer approaching the caecal cells in which the materials to be transfered occur.

Investigations by Broertjes et al. (1980a, b, 1982) demonstrated that elaborated compounds originating from the pyloric caeca are conveyed to the gonads via the hemal lacunae of the gastric tufts. Broertjes et al. (1980a, b) suggested that materials for the gonads that are synthesized in the pyloric caeca would be released in the digestive lumen of the caeca and reabsorbed further on by the digestive cells of the pyloric stomach. Those cells would then transfer the materials to the gastric hemal lacunae which in turn would carry them to the gonads. Broertjes' suggestion seems rather unlikely as these materials would be at least partly destroyed by the extracellular digestive enzymes which are secreted by the digestive epithelium of the pyloric caeca . While the iron-linked substances we used obviously are not similar to the compounds elaborated by and translocated from the pyloric caeca, our results indicate that material occuring in the caecal cells can be transfered to some hemalocytes without passing through the lumen of the pyloric caeca. We would suggest therefore that caecal materials intended for the gonads could be similarly transfered to hemalocytes (or to the hemal fluid?) which then would carry them to the gonad through the gastric hemal tufts and the aboral hemal ring. It can be suggested moreover that the hemal tufts are not one-way organs and that flows of energetic material could also be conveyed from the gonads to the pyloric caeca (e.g., when gonad resorbtion occurs).

Acknowledgments : We thank E. Bricourt, J. Harray and M. Klinkert for technical assistance.

Fig. 12. TEM-view of a pseudopodial cell. b basal lamina, dv dark vesicle, n nucleus (the arrow shows a pseudopodial extension).

Fig. 13. TEM-view of a coelomocyte-like cell. b basal lamina, cv clear vesicle.

BIBLIOGRAPHY

Bargmann, W. & Hehn, G. von 1968, Uber das axialorgan ("mysterious gland") von **Asterias rubens** L., Z. Zellforsch. 88 : 262-277.

Broertjes, J.J.S., Posthuma, G., Den Breejen, P. & Voogt, P.A. 1980a, Evidence for an alternative transport route for the use of vitellogenesis in the sea-star **Asterias rubens** L., J. mar. biol. Ass. U.K. 60 : 157-162.

Broertjes, J.J.S., Posthuma, G., Beijnink, F.B. & Voogt, P.A., 1980b, The admission of nutrients from the digestive system into the haemal channels in the sea-star **Asterias rubens** L., J. mar. biol. Ass. U.K. 60 : 883-890.

Broertjes, J.J.S., Jens, J.N., Van Oudheusden, D., De Bruin, M. & Voogt, P.A. 1982, Demonstration of nutrient flow in the starfish **Asterias rubens** L. with ^{125}I labelled proteins, Neth. J. Zool. 32 (4) : 472-478.

Cuénot, L. 1901, Etudes physiologiques sur les astéries, Archs Zool. exp. gén. 9 : 233-259.

Ferguson, J.C. 1982, Nutrient translocation, in, Echinoderm Nutrition, M. Jangoux and J. Lawrence eds, 373-393, Balkema : Rotterdam.

Ganther, P. & Jollès, G. 1969, Histochimie normale et pathologique I and II, Gauthier-Villars : Paris, 1902 pp.

Hyman, L. H., 1955. The Invertebrates, Vol. IV Echinodermata, The coelomate bilateria, Mc Graw Hill : N.Y., 763 pp.

Maes, P. & Jangoux, M. 1983, Occurence of well-developed pericentriolar complex within the cytoplasm of asteroid coelomocytes (Echinodermata), Develop. comp. Immun. 7 (4) : 691-694.

Reynolds, E. 1963, The use of lead citrate at high ph as an electron opaque stain in electron microscopy, J. cell Biol. 17 : 208-213.

Van der Plas, A.J., Jangoux, M. & Bricourt,E. 1980, The structure of the pyloric caeca of **Asterias rubens** and its changes during the annual reproductive cycle, In Echinoderms : Present and Past, 251-254. Balkema : Rotterdam.

Wood, R.L. & Cavey, M.J. 1981, Ultrastructure of the coelomic lining in the podium of the starfish **Stylasterias forreri**, Cell Tissue Res. 218 : 449-473.

Proceedings of the Fifth International Echinoderm Conference / Galway / 24-29 September 1984

Some characteristics of vitellogenic substances in the starfish *Asterias rubens* L.

JAN J.S.BROERTJES, PIETER DE WAARD & PETER A.VOOGT
Laboratory of Chemical Animal Physiology, State University of Utrecht, Netherlands

ABSTRACT: Homogenates of pyloric caeca, gonads, spermatozoa and eggs from the starfish Asterias rubens were fractionated by gelfiltration chromatography on Sepharose 2B and 4B. Three glycolipoprotein fractions, showing similarities with yolk lipoproteins from a sea urchin, were purified by chromatography on Sephacryl S-400. Electrophoresis, immunological experiments and gelchromatography were used to show the interrelationship between vitellogenic substances. The composition of amino acids, lipids, fatty acids and sugars of different fractions is discussed.

1 INTRODUCTION

Vitellogenesis in oviparous animals is characterized by deposition of yolk in the oocytes. The yolk precursors (vitellogenins) are synthesized in the liver or analogous organs. Untill now, most of the research on vitellogenic substances has been performed on amphibians, birds and insects while data on echinoderms are very scarce.

Oudejans and van der Sluis (1979a) showed that deposition of carbohydrates, lipids and proteins in the ovaries of Asterias rubens starts simultaneously and increases linearly during vitellogenesis, which implies that there are no separate stages of yolk formation for each of these compounds. A similar increase has been reported for asterosaponins in the ovaries of this starfish (Voogt and Huiskamp 1979). The concentrations of carbohydrates, lipids, proteins and asterosaponins remained constant in the ovaries. Such a situation would easily be attained by the production of a macromolecular complex with a constant composition; the pyloric caeca being the most likely production site because it is known that these organs shrink when gonads grow (e.g. Giese 1966).

An indication for such a macromolecular complex comes from Marsh (1965,1968) who detected a very large glycolipoprotein in the eggs of the sea urchin Arbacia punctulata, and from Malkin et al (1965) who isolated a crystalline high molecular weight protein from eggs and embryos of three sea urchin species and gave evidence that this was a yolk component.

From yolk granules of the eggs of the sea urchin Hemicentrotus pulcherrinus, Ii et al (1978) isolated some water-soluble lipoproteins that appeared to be identical with regard to their amino acid composition, but different in lipid content (Deguchi et al 1979). Also from the sand dollar Dendraster excentricus yolk granules were isolated (Ozaki 1979) and three major proteins were detected, the largest being a glycoprotein. Only recently some data have become available on yolk proteins in starfish (Broertjes et al 1984a,b). That so little is known about vitellogenic substances in echinoderms may be due to difficulties in taking samples from the transportroute. To circumvent this problem we decided to study the glycolipoproteins from well defined organs along the supposed transportroute "for gonadal development" (Broertjes and Posthuma 1978, Broertjes et al 1980a,b; 1982).

2 METHODS

Homogenates from pyloric caeca, gonads, gametes and axial complexes were prepared in 10 mM phosphate buffer (pH 6.8) containing 1 mM EDTA, 200 mg NaN_3/l and 150 mg phenylmethylsulphonyl fluoride (PMSF)/l and after centrifuging subjected to gelfiltration on a Sepharose 2B column, connected in series with a Sepharose 4B column. The eluate was monitored for its lipid, carbohydrate and protein content using the vanillin, anthrone and coomassie brilliant blue method respectively. On the ground of these results some fractions could be distinguished, which eventually were purified on Sephacryl S-400. After concentration by ultrafiltration fractions were subjected to horizontal electrophoresis in a 2.75% or a 3% polyacrylamide gel with or without 0.2% sodium dodecyl sulphate (SDS) using a TRIS-glycine buffer (pH 8.9) with or without 0.2% SDS. Fixation, staining and destaining were carried out using standard procedures.

Amino acid analyses were performed after delipidation and hydrolyzation by means of an automatic amino acid analyser using the cationic exchanger Durrum DC 6A with lithium citrate buffers.

Carbohydrate compositions of the fractions in delipidated form were analysed according to the method of Kamerling and Vliegenthart (1982).

Lipids were extracted according to the method of Bligh and Dyer (1959). Total lipids were separated by thin-layer chromatography as described by Björkman et al (1972). They were also separated into neutral lipids, glycolipids and phospholipids by means of columnchromatography according to Kates (1972). Fatty acids were isolated by hydrolyzing the lipid fractions. Unsaponifiable lipids were extracted with hexane, whereafter the waterphase was acidified and later on extracted with hexane. Fatty acids were methylated according to the method of Schlenk and Gellerman (1960). The various fractions of fatty acid methyl esters were analysed by means of gas-liquid chromatography.

An antiserum was raised in rabbit against fraction 3 from ovaries because this fraction appeared to be sufficiently stable. Detection of immunological reactivity was performed by using the double diffusion method after Ouchterlony (Ouchterlony and Nilsson 1978).

3 RESULTS AND DISCUSSION

Figure 1 shows the elution patterns of homogenates of eggs and spermatozoa from Asterias rubens after gelfiltration on Sepharose. Obviously the graphs differ in the fractions III, IV and VI.

Fraction VI shows a low protein content. That of eggs is relatively rich in carbohydrate and lipids and has a strong tendency to foam. Therefore this fraction was examined on a possible reduction of the surface tension by measuring the contents of each tube containing 300 drops after column elution. Figure 1 shows that the carbohydrate-lipid peak of fraction VI from eggs coincides with tubes containing less eluate whereas in that of spermatozoa no reduction of surface tension could be observed. Voogt and van Rheenen (1979) described the presence of asterosaponins in ovaries and eggs of Asterias rubens and noticed that they were absent in testes. Using the method of these authors it was proved unequivocally that asterosaponins are present in fraction VI from ovaries and eggs.

Fractions III and IV seem to be the most likely candidates for possible other vitellogenic substances. Fraction III is only distinctly visible in elution patterns of homogenates from eggs and ovaries. It is characterized by its yellow colour and relatively high lipid content in contrast to the relatively high carbohydrate content of fraction IV. In sperm such fractions could not be detected.

These results made us conclude that the fractions III and IV possibly contain female specific, with other words, vitellogenic substances. Therefore these fractions were purified by repeated gelfiltration on Sephacryl S-400. In this way the peaks III and IV (after purification called 3 and 4) were separated. Moreover an additional peak 2 became visible, which fraction was purified in the same way.

SDS-electrophoresis of fraction 2 showed a band that is also visible in fraction 3 and another band that is the major one of fraction 4. This latter band cannot be due to contamination because it is not found in fraction 3. In electrophoresis without SDS fraction 2 appeared to contain the same band as fraction 3 and also one of the bands of fraction 4, which may imply that fraction 3 and partly fraction 4 are derived from fraction 2. This idea was strengthened when turned out that a purified fraction 2 stored for two months, had partly dissociated into 3 and 4 (Broertjes et al 1984b).

With the techniques used it was not possible to detect the peaks II and III in pyloric caeca and other organs of the supposed transportroute, but this may be caused by too low concentrations in the transportroute, while in the storage organs immediately after biosynthesis and/or assemblage of these female specific proteins release to the transportmedium takes place as is the case in other animals.

To examine the validity of this supposition we raised an antiserum against fraction 3 from ovary.

Figure 2 shows that fraction 2 and fraction 3 from ovary react specifically against the antiserum whereas the reaction of fraction 4 is mainly aspecific.

The reaction against fraction III from

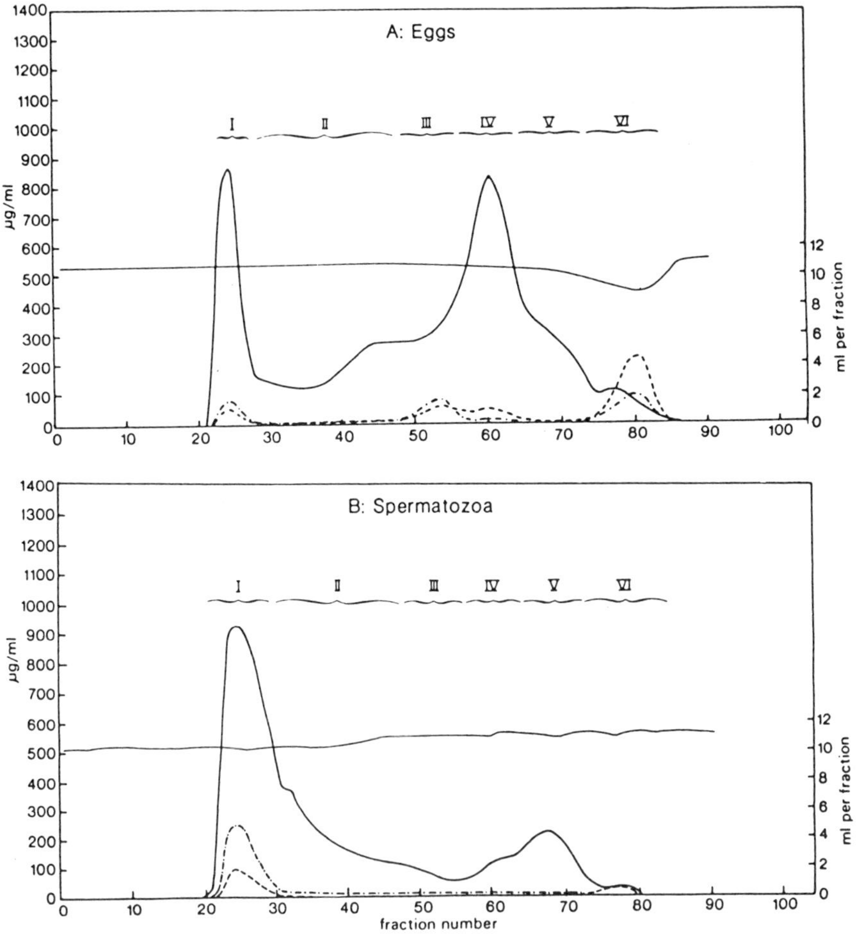

Fig.1: Elution patterns of homogenates from eggs and spermatozoa of Asterias rubens after chromatography on Sepharose. ——ml per tube; ▬▬ proteins; —·—· lipids; — — — carbohydrates.

female pyloric caeca is distinctly specific, which is not the case with the same fraction from male pyloric caeca. No clear reaction could be observed against the relevant fraction from testes and sperm. Very recently we observed a specific reaction against fraction III from female axial complexes.

Analyses of the amino acid compositions of the fractions 3 and 4 showed that they resemble more or less that of a sea urchin (Broertjes et al 1984b).

Lipids from the fractions 3 and 4 turned out to be rather different. They make up about 60% of fraction 3 and 8-10% of fraction 4, whereas the neutral lipids constitute the greater part of the lipids of fraction 3. Among these neutral lipids triacylglycerols and neutral ether lipids - lipids especially important as energydepots (Oudejans and van der Sluis 1979b, Falk-Petersen and Sargent 1982) - are predominant. All lipid classes (neutral-,phospho- and glycolipids) from fraction 3 contain more fatty acids with 20 carbon atoms than those of fraction 4. The percental differences with reference to long chain fatty acids between these fractions are small (Broertjes et al 1984c).

Recently we obtained indications that the carbohydrate moiety of the glycoprotein part of the fractions 2, 3 and 4 belongs to the N-glycosidic oligomannoside type (see Berger et al 1982). Interestingly, glucose is found in a more or less constant ratio to mannose. This is confirmed

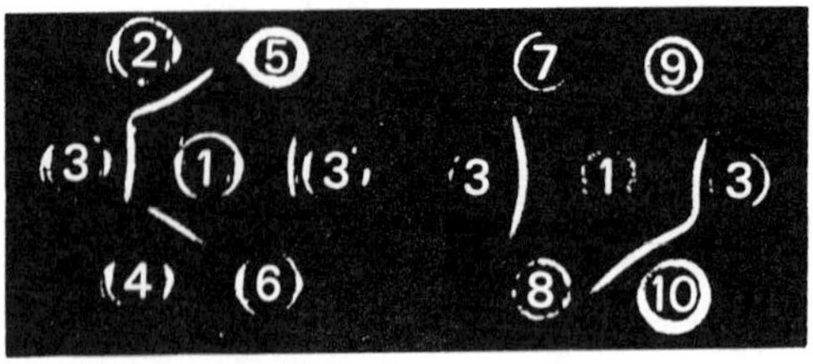

Fig.2: Doublediffusion of antiserum against purified fraction 3 from the ovaries with various other fractions.
In all cases the antiserum is applied in the centre well (1).
2. ovary fraction 2
3. ovary fraction 3
4. ovary fraction 4
5. spermatozoa fraction II
6. spermatozoa fraction III
7. male pyloric caeca fraction II
8. male pyloric caeca fraction IV
9. male pyloric caeca fraction III
10.female pyloric caeca fraction III

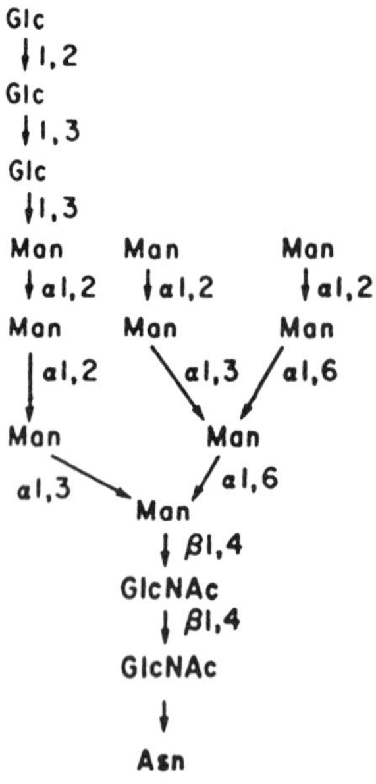

Fig.3: Oligomannoside structure of the carbohydrate moiety of glycoproteins.
Modified after Berger et al (1982).
Man: mannose; Glc: glucose; Asn: asparagin; GlcNAc: N-acetyl-glucosamin.

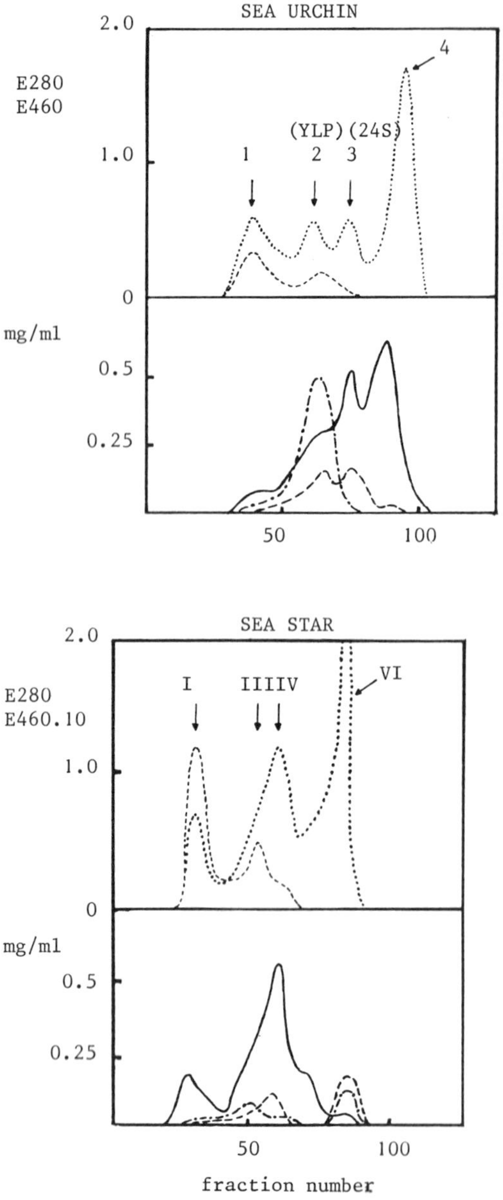

Fig.4: Comparison of elution patterns of a homogenate from yolk platelets out of the sea urchin Hemicentrotus pulcherrinus (after Ii et al 1978) and of a homogenate of ovaries from the starfish Asterias rubens. Both are calculated to the same scale.

——— protein
-.-.-. lipid
— — — carbohydrate
.... E280 (absorbance at 280 nm)
- - - - - E460 (absorbance at 460 nm, for detection of yellow colour.

by structural analyses (Kamerling et al, unpublished results) so that at the moment we propose a structure as given in figure 3.

The strong resemblance in carbohydrate structure of the fractions 2, 3 and 4 gives further support to the above mentioned idea that these fractions are interrelated.

There is a great similarity between the amino acid, lipid and sugar compositions of the vitellogenic substances from Asterias rubens and of the yolk lipoprotein fractions from the sea urchin Hemicentrotus pulcherrinus (Broertjes et al 1984b,c; Deguchi et al 1979).

Figure 4 shows an elution pattern of a homogenate of ovaries of Asterias rubens and of a homogenate from yolk platelets of the sea urchin Hemicentrotus pulcherrinus as given by Ii et al (1978). Despite different experimental conditions (other column, other buffers, other homogenizing method) the similarity is clear. The differences in lipid and carbohydrate content between corresponding peaks of starfish and sea urchin may be the result of the fact that Ii et al (1978) did not use sodium azide as a bacteriostaticum in the elution buffer. Sodium azide causes an underestimation of lipid concentrations when lipids are essayed in its presence (Broertjes et al 1983) as is also the case in carbohydrate estimation (Tong et al 1973).

On the ground of various biochemical features fraction 3 from starfish ovary shows a strong resemblance to the YLP 1, 2 and 3 from sea urchin oocytes (Deguchi et al 1979). Lipid contents and composition, and amino acid composition are very alike, while an oligomannoside structure in sea urchin YLP's is not excluded.

A distinct difference between the elution patterns of figure 4 is the presence in starfish of a fraction in which asterosaponins could be detected, but it is known from literature (Voogt 1982) that in sea urchins saponins are not present.

REFERENCES

Berger,E.G.,E.Buddecke.,J.P.Kamerling.,A. Kobata.,J.C.Paulson.& J.F.G.Vliegenthart 1982. Structure, biosynthesis and functions of glycoprotein glycans. Experientia 38: 1129-1162.

Bligh,E.G. & W.J.Dyer 1959. A rapid method of total lipid extraction and purification. Can.J.Biochem.Physiol. 37: 911-917.

Broertjes,J.J.S. & G.Posthuma 1978. Direct visualization of the haemal system in starfish by a staining procedure. Experientia 34: 1243-1245.

Broertjes,J.J.S.,G.Posthuma.,P.den Breejen. & P.A.Voogt 1980a. Evidence for an alternative transportroute for the use of vitellogenesis in the sea-star Asterias rubens L. J.mar.biol.Ass.U.K. 60: 157-162.

Broertjes,J.J.S.,G.Posthuma.,F.B.Beijnink. & P.A.Voogt 1980b. The admission of nutrients from the digestive system into the haemal channels in the sea-star Asterias rubens L. J.mar.biol.Ass.U.K. 60: 883-890.

Broertjes,J.J.S.,J.N.Jens.,D.van Oudheusden.,M.de Bruin. & P.A.Voogt 1982. Demonstration of nutrient flow in the starfish Asterias rubens L. with ^{125}I labelled proteins. Neth.J.Zool. 32: 472-478.

Broertjes,J.J.S.,P.de Waard. & P.A.Voogt 1983. Interference of azide in the estimation of total lipids by means of the sulphophosphovanillin method. Experientia 39: 594-595.

Broertjes,J.J.S.,P.de Waard. & P.A.Voogt 1984a. On the presence of vitellogenic substances in the starfish Asterias rubens L. J.mar.biol.Ass.U.K. 64: 261-269.

Broertjes,J.J.S.,P.de Waard.& P.A.Voogt 1984b. Purification and characterization of vitellogenic substances in the starfish Asterias rubens L. Mar.Biol.Lett. 5: 99-104.

Broertjes,J.J.S.,P.de Waard.,J.P.Kamerling. & P.A.Voogt 1984c. Some characteristics of the non-protein moiety of vitellogenic substances in the starfish, Asterias rubens L. J.mar.biol.Ass.U.K. 64: in press.

Deguchi,K.,S.Kawashima.,I.Ii. & N.Ueta 1979. Water-soluble lipoproteins from yolk granules in sea urchin eggs. II. Chemical composition. J.Biochem. 85: 1519-1525.

Falk-Petersen,I.B. & J.R.Sargent 1982. Reproduction of asteroids from Balsfjorden, Northern Norway: analyses of lipids in the gonads of Ctenodiscus crispatus, Asterias lincki and Pteraster militaris. Marine Biology 69: 291-298.

Giese,A.C. 1966. On the biochemical constitution of some echinoderms. In: Physiology of Echinodermata (ed. R.A.Boolootian) pp 757-796. Interscience Publishers, New York.

Ii,I.,K.Deguchi.,S.Kawashima.,S.Endo. & N. Ueta 1978. Water-soluble lipoproteins from yolk granules in sea urchin eggs. I. Isolation and general properties. J.Biochem. 84: 737-749.

Kamerling,J.P. & J.F.G.Vliegenthart 1982.

Gas-liquid chromatography and mass spectrometry of sialic acids. In: Sialic Acids, Chemistry, Metabolism and Function (ed. R.Schauer) pp 95-125. Springer Verlag. Wien.
Kates,M. 1972. Techniques of lipidology: isolation, analysis and identification of lipids. In: Laboratory Techniques in Biochemistry and Molecular Biology (eds. T.S.Work & E.Work) vol. 3 II. North Holland Publishing Company. Amsterdam.
Malkin,L.I.,J.Mangan & P.R.Gross 1965. A crystalline protein of high molecular weight from cytoplasmic granules in sea urchin eggs and embryos. Developm.Biol. 12: 520-542.
Marsh,J.B. 1965. Isolation and composition of a water-soluble lipoprotein from Arbacia eggs. Biol.Bull.(Woods Hole,Mass) 129: 415.
Marsh,J.B. 1968. Isolation and composition of a low density lipoprotein from the eggs of Arbacia punctulata. Biol.Bull.(Woods Hole,Mass.) 135:193-199.
Ouchterlony,O. & L.A.Nilsson 1978. Immunodiffusion and immunoelectrophoresis. In: Handbook of Experimental Immunology. Vol. I 3[e]ed. (ed.D.M.Weir) ch 19, pp 19.1 - 19.44. Blackwell Scientific Publications Oxford.
Oudejans,R.C.H.M. & I. van der Sluis 1979a. Changes in the biochemical composition of the ovaries of the sea star Asterias rubens during its annual reproductive cycle. Marine Biology 50: 255-261.
Oudejans,R.C.H.M. & I.van der Sluis 1979b. Storage and depletion of lipid components in the pyloric caeca and ovaries of the sea star Asterias rubens during its annual reproductive cycle. Marine Biology 53: 239-247.
Ozaki,H. 1979. Yolk proteins of the sea-urchin egg. Am.Zool. 19: 998.
Schlenk,H. & J.L.Gellerman 1960. Esterification of fatty acids with diazomethane on a small scale. Anal.Chem. 32: 1412-1414.
Tong,H.K.,K.H.Lee. & H.A.Wong 1973. Interference by azide in the estimation of carbohydrate with anthrone. Anal.Biochem. 51: 390-398.
Voogt,P.A. & J.W.A.van Rheenen 1979. Studies on the possible regulatory function of asterosaponins in oocyte maturation and early embryogenesis of Asterias rubens. Int.J.Invertebr.Reprod. 1: 307-316.
Voogt,P.A. & R.Huiskamp 1979. Sex-dependence and seasonal variation of saponins in the gonads of the starfish Asterias rubens: their relation to reproduction. Comp.Biochem.Physiol.62A: 1049-1055.
Voogt,P.A. 1982. Steroid metabolism. In: Echinoderm Nutrition (eds.M.Jangoux & J.M.Lawrence) ch 18. pp 417-436. A.A. Balkema, Rotterdam.

Proceedings of the Fifth International Echinoderm Conference / Galway / 24-29 September 1984

Immunohistochemical localization of gonad stimulating substance in the seaster *Pycnopodia helianthoides*

G.D.CAINE & R.D.BURKE
University of Victoria, Canada

ABSTRACT: Gonad stimulating substance (GSS) is a peptide extractable from radial nerves that stimulates ovarian follicle cells to release 1-methyladenine, which in turn induces oocyte maturation and spawning in seastars. Although a great deal is known about this neuroendocrine phenomenon, the precise location of GSS in the radial nerve cord, and the pathway by which GSS is transported to the ovary remain unclear. Here we report immunohistochemical localization of GSS to the perihemal epithelium of the radial nerve cord in Pycnopodia helianthoides. The location of GSS appears to correspond to axons containing numerous 100 nm diameter, electron-dense vesicles. We propose that GSS is released into the radial hemal sinus and transported to the ovary within the hemal system.

INTRODUCTION

The neuroendocrine mechanism that initiates spawning in seastars is one of the best studied examples of endogenous control of a reproductive event in an invertebrate. One aspect of the sequence of events that elicits spawning that has been debated is how GSS, produced in the radial nerve cord, is transported to the ovary, its principle target organ. Chaet (1966) suggested that GSS was released into the seawater and acted as a pheromone, inducing spawning in conspecifics. Transport of GSS to the ovaries through nerves was later suggested by Schoenmakers (1981). The prevalent hypothesis (Kanatani, 1973) proposes that GSS enters the water vascular system through radial and transverse hemal sinuses and from there diffuses through the coelom to the ovaries. Attempts to localize GSS histochemically or by dissection suggested that cells in the ependymal layer of the radial nerve were involved (deAngelis, et al., 1972). This complicates the problem, since GSS released from these cells would have to traverse several tissue layers to get to the ovary. This preliminary report presents evidence that GSS could be secreted from axons within the perihemal epithelium of the radial hemal sinus, which is continuous with the ovarian hemal sinus.

MATERIALS AND METHODS

Pure GSS was obtained from 33 g of lyophilized radial nerve cords by two runs of gel filtration on Sephadex G-50, followed by two runs on a Sephadex G-25 column, three ion-exchange runs on DEAE Sephadex A-25, and a final desalting run on Sephadex G-10. Purity was verified by HPLC on a Waters MCH-10 protein column; a single peak was resolved. GSS was conjugated to bovine thyroglobulin (BTG) (Sigma Chemical Co.) with 1-ethyl-3-(3-dimethylaminopropyl) carbodiimide HCl (EDC) (Sigma Chemical Co.) at a molar ratio of 10 EDC: 1 GSS: 0.01 BTG. Female New Zealand white rabbits were bled from the central ear vein to obtain preimmune sera, then each was inoculated with 100 µg of the conjugate in 1 ml Freund's complete adjuvant (Sigma Chemical Co.) by multiple intradermal injections. Subsequent inoculations of 100 µg conjugate in 1 ml Freund's incomplete adjuvant (Sigma Chemical Co.) were carried out at 3 week intervals. Sera were collected every 2 weeks, processed and stored in 0.5 ml aliquots at -20°C, using 0.1% sodium azide as a preservative. GSS antibodies were detected in the sera after 6 weeks and analyzed for titre and specificity using the double diffusion technique on 1.5% agar gels. Specificity was determined by diffusing sera against pure GSS, and tissue extracts of radial nerve cord, tube feet, pyloric caeca, and ovaries. Sera reacted only with pure GSS, radial nerve cord extracts and, to a lesser extent, tube feet extracts.

Tissues for immunohistochemical examination were fixed with 10% formalin/seawater for 10 minutes at 4°C, washed with 3 changes of seawater, and postfixed overnight in acetone at -20°C. Tissues were dehydrated with 3, 1 hr changes of petroleum ether (4°C) then embedded in paraffin. Section (6 μm) were cut and mounted on subbed slides. The sections were cleared in xylene and rehydrated through a graded acetone series, rinsed with distilled water and then washed with phosphate buffered saline (PBS). Tissue sections were incubated with rabbit antiserum (1:100 in PBS) overnight at 4°C, rinsed through 3 changes of PBS, and incubated with goat antirabbit IgG conjugated to horseradish peroxidase (1:50 in PBS) (Sigma Chemical Co.). Staining was developed with 3,3'-diaminobenzedine (Sigma Chemical Co.), 0.4% in 0.14% H_2O_2. Nonspecific binding was blocked by pretreating the tissue with 0.1% bovine serum albumin (Sigma Chemical Co.) in PBS, and by pretreating the primary antiserum with 0.1% Protein A (Sigma Chemical Co.) to bind the F_c fragments. All experiments had two controls: Incubation with preimmune serum; and incubation with the secondary antibody alone. Tissues treated with preimmune serum were slightly darkened, but no discrete staining occured. Tissues treated only with secondary antibody showed similar results.

Tissues for light and electronmicroscopy were fixed in 2.5% glutaraldehyde in Millonig's phosphate buffer (MPB), followed by 1% osmium tetroxide in MPB. The tissues were dehydrated through a graded ethanol series, followed by 2 changes in propylene oxide before embedment in Epon 812. Thick (1 μm) sections were cut with a diamond knife, mounted on glass slides and stained with Richardson's stain. These sections were viewed and photographed on a Zeiss Universal light microscope using Koehler illumination. Thin (55-70 μm) sections for electronmicroscopy were cut with a diamond knife, mounted on parlodion-coated grids, and stained with aqueous solutions of uranyl acetate and lead citrate. Sections were viewed and photographed with a Philips EM-300 transmission electronmicroscope.

RESULTS AND DISCUSSION

The radial nerve cord is made up of three major components (Fig. 1a). The largest component is the ectoneural (radial) nerve, which lies outermost on the oral surface. This is separated from the hyponeural (Lange's) nerve by a connective tissue layer. The third component is the radial hemal sinus, which runs medially and longitudinally through the perihemal space. The radial hemal sinus is bounded by a perihemal epithelium, which also lines the perihemal space.

Immunoreactivity is observed as discrete brown granules, 2 to 3 μm in diameter, within the perihemal epithelium (Fig. 1b). These granules are dispersed throughout the basal portion of the epithelium and appear most concentrated on the medial walls surrounding the hemal sinus (Fic. 1c). No immunoreactivity is observed in the ectoneural or hyponeural components of the radial nerve cord. Ultrastructurally, the perihemal epithelium is a simple, low columnar epithelium. The cells are spatulate in outline, and overlie a mixture of muscle fibers and vesiculated axons. A prominent basal lamina apposes the hemal sinus. The axons are of two general sizes: Large (1-3 μm diameter) and small (300-400 nm diameter) axons. The large axons contain numerous electron-dense, 100 nm diameter vesicles (Fig. 1d). The location, size and distribution of the axons appear to correspond with the location, size and distribution of the immunoreactive granules observed with the light microscope.

Previous attempts to localize GSS in the radial nerve cord have relied on general histochemical techniques that localize neurosecretory substances in tissues (Unger, 1962; Imlay and Chaet, 1967; Atwood & Simon, 1971). Correlation of neurosecretory substances with the presence of GSS assumes that GSS is the only neurosecretory substance in the radial nerve cords. This assumption may not be valid. Efforts to separate the components of the radial nerve cords (deAngelis, et al., 1972) and test for GSS activity have yielded conflicting results, and it is difficult to completely isolate regions of the ectoneural nerve because there are no distinct boundaries between its components. Application of immunohistochemical techniques allow for a more discriminating investigation.

The location of the perihemal axons suggests that GSS could be secreted into the radial hemal sinus, which is continuous with the axial organ and genital hemal sinuses (Fig. 2). Within the ovary, the hemal sinus is directly associated with the germinal epithelium containing oocytes and follicle cells (Schoenmakers et al., 1981); by this route GSS has direct access to its target, the follicle cells. The hemal

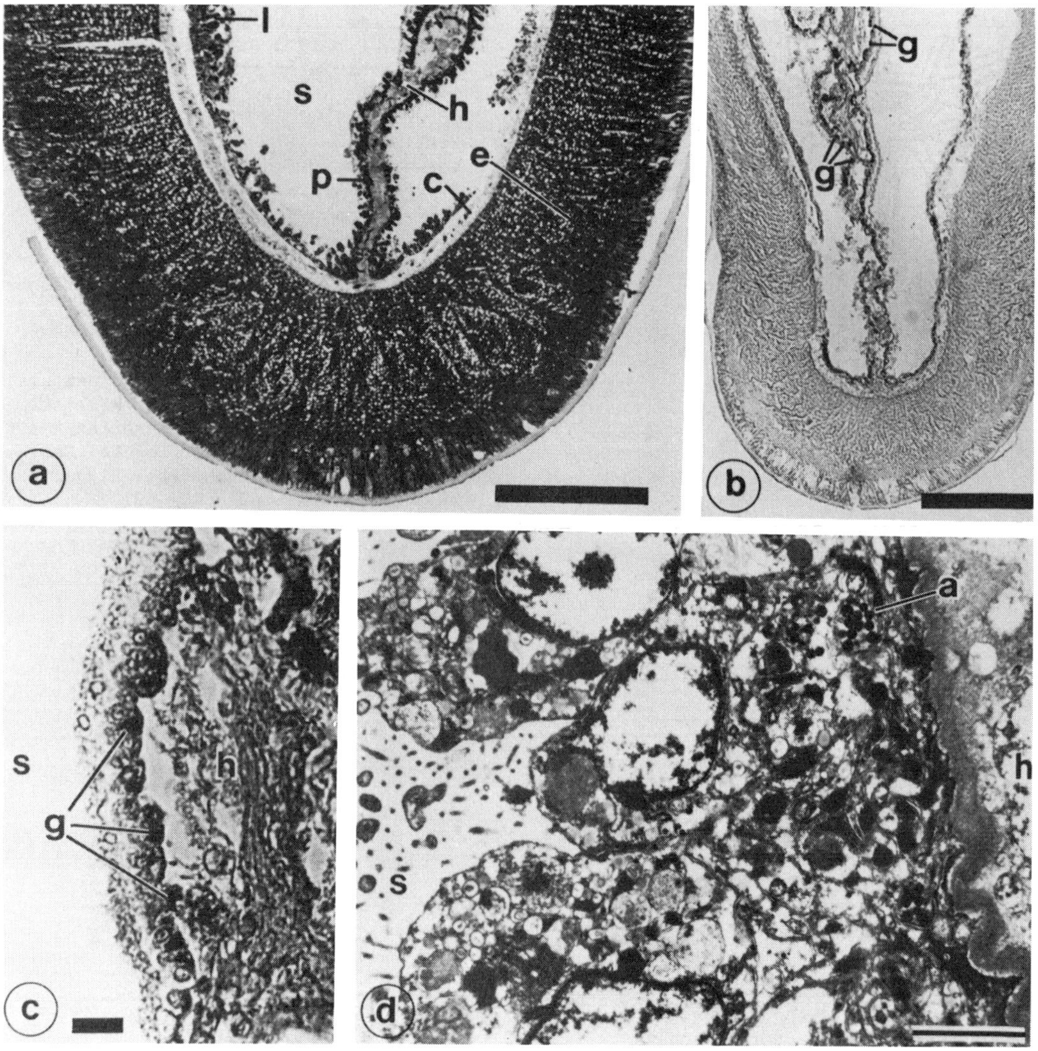

Figure 1: a) Cross section of the radial nerve cord of P. helianthoides. Connective tissue layer, c; ectoneural (radial) nerve, e; radial hemal sinus, h; hyponeural (Lange's) nerve, l; perihemal epithelium, p; and the perihemal space, s. Bar = 100 µm.
b) Low power light micrograph of a cross section of the radial nerve cord stained immunohistochemically. Granules, g. Bar = 100 µm.
c) Higher power light micrograph of a cross section through the perihemal epithelium stained immunohistochemically. Granules, g; hemal sinus, h; and the perihemal space, s. Bar = 5 µm.
d) Transmission electronmicrograph of the perihemal epithelium. Vesiculated axon, a; hemal sinus, h; and the perihemal space, s. Bar = 5 µm.

sinus system has been implicated in nutritive transport during oogenesis, and recent evidence has shown that it does have translocative ability (Ferguson, 1984). It is possible that the hemal sinus system also plays a role in transport of GSS.

ACKNOWLEDGEMENTS

Supported by a Natural Science and Engineering Research Council (Canada) University fellowship and operating grant to RDB, and a University of Victoria Graduate Fellow-

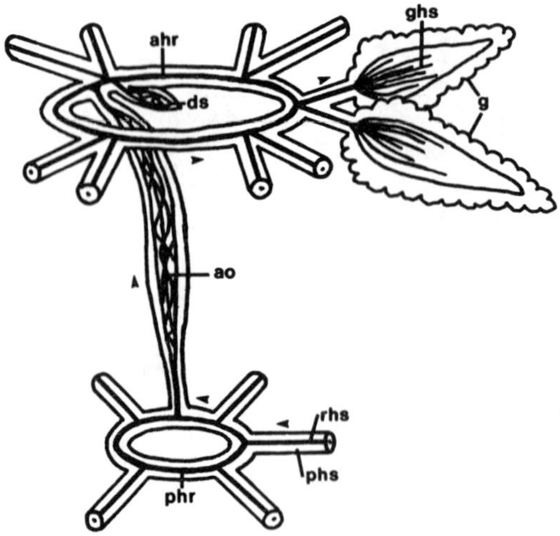

Figure 2: Schematic diagram of the anatomy of the hemal system in seastars. Aboral hemal ring, ahr; axial organ, ao; genital hemal sinus, ghs; gonads, g; periesophageal hemal ring, phr; and the radial hemal sinus, rhs. The gastric hemal tufts have been omitted for clarity. (Redrawn from Ubaghs, 1969).

ship to GDC. We are grateful to Dr. N. Sherwood for valuable technical assistance, to Prof. G.O. Mackie and Dr. M. Paul for their criticism of the manuscript, and to A.W. Gibson for helpful comments and advice.

REFERENCES

Atwood, D.G., & J.L. Simon. 1971. Correlation of gamete-shedding with the presence of neurosecretory granules in Echinaster and Patiria. Amer. Zool. 11: 701-711.

Chaet, A.B. 1966. The gamete-shedding substances of starfish: A physiological biochemical study. Amer. Zool. 6: 263-271.

deAngelis, E., A. Viglia, T. Watanabe, H. Shirai, J. Kubota, & H. Kanatani. 1972. Presence of granules containing gonad-stimulating substance in starfish radial nerve. Annot. Zool. Jap. 45: 16-21.

Ferguson, J.C. 1984. Translocative functions of the enigmatic organs of starfish-the axial organ, hemal vessels, Teidemann bodies and rectal caeca: An autoradiographic study. Biol. Bull. 166: 140-155.

Imlay, M.J., & A.B. Chaet. 1967. Microscopic observation of gamete-shedding substance in starfish radial nerve. Trans. Amer. Microsc. Soc. 86: 120-126.

Kanatani, H. 1973. Maturation-inducing substances in starfish. In J.D. Ebert & T.S. Okada (eds.), International review of cytology, John Wiley & Sons: New York.

Schoenmakers, H.J.N., P.H.J.M. Colebrander, J. Peute, & P.G.W.J. Oordt. 1981. Anatomy of the ovaries of the starfish Asterias rubens (Echinodermata). Cell Tissue Res. 217: 577-597.

Ubaghs, G. 1969. General characteristics of the echinoderms. In M. Florkin & B.J. Scheer (eds.), Echinodermata, Academic Press: New York.

Unger, H. 1962. Experimentelle und histologische untersuchungen uber Wirkfaktoren aus dem Nervensystem von Asterias (Marthasterias) glacialis (Asteroidea; Echinodermata). Zool. Jahrb. Abt. Allgm. Zool. Physiol. Tiere. 69: 481-497.

Proceedings of the Fifth International Echinoderm Conference / Galway / 24-29 September 1984

Distribution, reproductive cycle and morphometric relationships of *Acanthaster planci* (Echinodermata: Asteroidea) in New Caledonia, western tropical Pacific

C.CONAND
ORSTOM, Nouméa, New Caledonia
Currently: Université de Bretagne Occidentale, Brest, France

ABSTRACT : This study of the crown-of-thorns starfish Acanthaster planci in New Caledonia enables us to compare its biological parameters with those from different areas in the Indopacific. Acanthaster has been observed in about fifty locations on the Grande Terre and Loyalty Islands. The density of populations was usually below 0.5 per 100 m^2. However a few locations with higher densities were found which should be monitored. A monthly sampling of the population was conducted for fifteen months at Maître Islet, the station with the highest density (3.3 per 100 m^2). Morphometric relationships and the reproductive cycle were defined. Breeding occurred during the warm season, from November to February. The allometric growth of the gonads showed that the 40 cm diameter specimens may have up to sixty times more oocytes than 25 cm diameter ones and can produce up to sixty million oocytes per spawning season.

1 INTRODUCTION

Acanthaster planci, the crown-of-thorns starfish, a normal constituant of the coral-reefs fauna, generally feeds on coral polyps. When large populations occur on a reef, severe damage is caused by this specialized feeding. Such outbreaks were reported at Green Island on the Great Barrier Reef and in other places of the Indo-Pacific region (Chesher 1969). The question whether these infestations were a new man-induced phenomenon or natural population fluctuations gave rise to scientific research on the species biology, its relations with corals, the distribution of the populations and the recolonization after the infestations has been undertaken. The main results are summarized in the articles of Endean (1973) and Potts (1981). Many hypotheses were proposed to explain the origin and mechanism of the outbreaks.

In New Caledonia the species was mentionned by Catala (1950) but was not observed by Weber and Woodhead (1970) during their 22 day-exploration on the reefs. In the course of the study of the holothurians for the evaluation of the bêche-de-mer fishery potential of New Caledonia, stations were studied in the coral-reef complex and the density of Acanthaster established. Since a large population was observed near Noumea, a study was undertaken to determine its biological parameters and reproductive cycle.

2 DISTRIBUTION AND ABUNDANCE AROUND NEW CALEDONIA

2.1. Methods

Seven cruises were done in 1982 and 1983 around New Caledonia to evaluate the densities and the biomasses of holothurians (Conand 1983). During each cruise in the lagoon, stations were chosen in the different biotopes defined according to the terminology of the morphological features given by Battistini et al. (1975). Recorded at each station were : depth, hydrodynamism, distance to the coastline, percent cover of the different kinds of sediments and dominant species of the flora and fauna. A quadrat was observed during skin-or scuba dives by two divers swimming in a parallel direction. They noted on a mantaboard (Kenchington and Morton 1976), all Acanthaster encountered during two 15 minute periods. The length of the quadrat is given by a low-speed flowmeter fixed under the board, the width is estimated according to the visibility. Densities were calculated for 100 m^2.

Other information on Acanthaster is reported, including personal observations, and communications by the ORSTOM-team of divers and by volunteers.

Table 1 - Distribution and abundance. n = number counted during the dive ; N = numerous ; d = density per 100 m2 ; Or= origin , H = holothurian cruise, P = personnal obs.. O = ORSTOM divers obs., D : diverse ; S = scuba dive, L : snorkel ; juv : juveniles.

ZONE	LOCATION	LONG.E	LAT.S	BIOTOPE	DATE	n	d	Or	OBSERVATIONS
1	Mbo islet	166°13	22°15	sand cay reef, leeward	0583	1	0.05	H	S
	Bancs Ouest reef	166°28	22°27	lagoon reef, windward	0783	15	0.47	H	L
	Maitre islet	166°25	22°20	sand cay reef, windward	0783	131	3.50	H	L
	Nou island	166°23	22°15	fringing reef, slope	0783	4	0.14	H	L
	Abore reef	166°22	22°27	barrier reef, flat	0382	3	0.08	H	L
	Porc-Epic islet	166°34	22°20	fringing reef	0582	34	+0.80	P	L
	Attire islet	166°33	22°31	sand cay reef	0682	5	±0.20	P	L
	Tabu reef	166°27	22°29	lagoon reef	0682	2	-	P	L
	Laregnere islet	166°19	22°20	lagoon reef	0783	3	+0.20	P	S
	Maitre islet	166°25	22°20	sand cay reef, windward	1183	139	-3.31	P	L
	Nakae reef	166°35	22°22	lagoon reef	0182			O	S
	Abore reef	166°17	22°23	barrier, outer slope	0783	1		D	S
2	Champignon islet	166°05	22°06	lagoon bottom and reef	0982	8	0.49	H	S
	Poya reef	165°04	21°27	barrier, inner slope	0982	1	0.01	H	S
	Poe reef	165°20	21°37	barrier inner slope	0883	1		D	L
	Pouembout	164°44	21°14	pass	0183	1		O	S juv.
3	Nienane island	163°42	19°49	fringing reef	0881			O	S juv.
4	Leleizour	164°42	20°25	lagoon bottom, coral heads	0383	1	0.04	H	S
	Napias reef	164°17	20°11	lagoon reef, leeward flat	0383	1	0.08	H	L
	Baaba island	164°00	20°04	fringing reef, slope	0681	3		O	S
	Pam island	164°17	20°13	fringing reef, slope	1183			D	
5	Ana reef	165°56	21°22	lagoon reef	1282	4	0.09	H	S
	Canala bay	165°57	21°25	coral heads on silted bottom	1282	5	0.36	H	S
	Uao reef	165°09	20°43	sand cay reef, windward	1282	1	0.02	H	S
	Sable reef	165°22	20°55	sand cay reef, leeward	1282	3	0.07	H	S
	Sable reef	165°22	20°55	sand cay reef	0182	1		O	S
	Balade reef	164°33	20°13	barrier reef	0383	2		O	S
	Mangalia reef	165°17	20°45	barrier inner slope	1183			D	
6	Unia reef	166°55	21°59	barrier, outer slope	0683	1	0.10	H	L
	Nenou bay	166°23	21°40	coral heads on silted bottom	0683	1	0.17	H	S
	Lavaissiere bay	166°04	21°29	fringing reef, flat	0683	1	0.07	H	S
	Nakety bay	166°05	21°31	fringing reef, slope	1182	1		O	S night
	Port Bouquet bay	166°21	21°39	fringing reef, slope	0483			O	S
	Professor reef	166°13	21°35	lagoonal coral patch	0883	N		D	L
	Thio reef	166°18	21°39	fringing reef slope	1283	4		D	L juv.
	Kouakoue bay	166°39	21°56		0683	1		P	dead juv.
7	Woodin canal	166°49	22°24	fringing reef slope	1083	2	0.28	H	S
	Woodin canal	166°50	22°23	fringing reef flat	1083	13	0.46	H	L
	Prony bay	166°50	22°22	fringing reef slope	1083	2	0.21	H	S
	Prony bay	166°53	22°23	fringing reef slope	1083	3	0.30	H	S
	Ndo islet	166°58	22°41	sand cay reef slope	1083	1	0.12	H	S
	Uime reef	166°40	22°37	lagoon bottom	1083	1	0.03	H	S juv.
	Vua reef	166°45	22°35	sand cay reef, leeward	0582	4	+0.10	P	S
	Kouare reef	166°48	22°47	sand cay reef, leeward	1083	3	-0.15	P	L
	Prony bay	166°50	22°22	fringing reef	0883	N		D	S
	Redika islet	166°37	22°31	sand cay reef slope	1183			D	L
8	Anemata-Ouvea	166°14	20°30	fringing reef	0883	1		O	juv.
	Cape Mande-Lifou	167°03	20°55	fringing reef	0883	1		O	night
	Cape Escarpe-Lifou	167°13	20°40	fringing reef	0883	1		O	juv.
	Pleiades islands	166°28	20°27	fringing reef	1183			D	
	Beautemps-Beaupré reef	166°10	20°25	fringing reef	1183			D	
	Gadgi-Isle of Pines	167°27	22°32	fringing reef	1183			D	

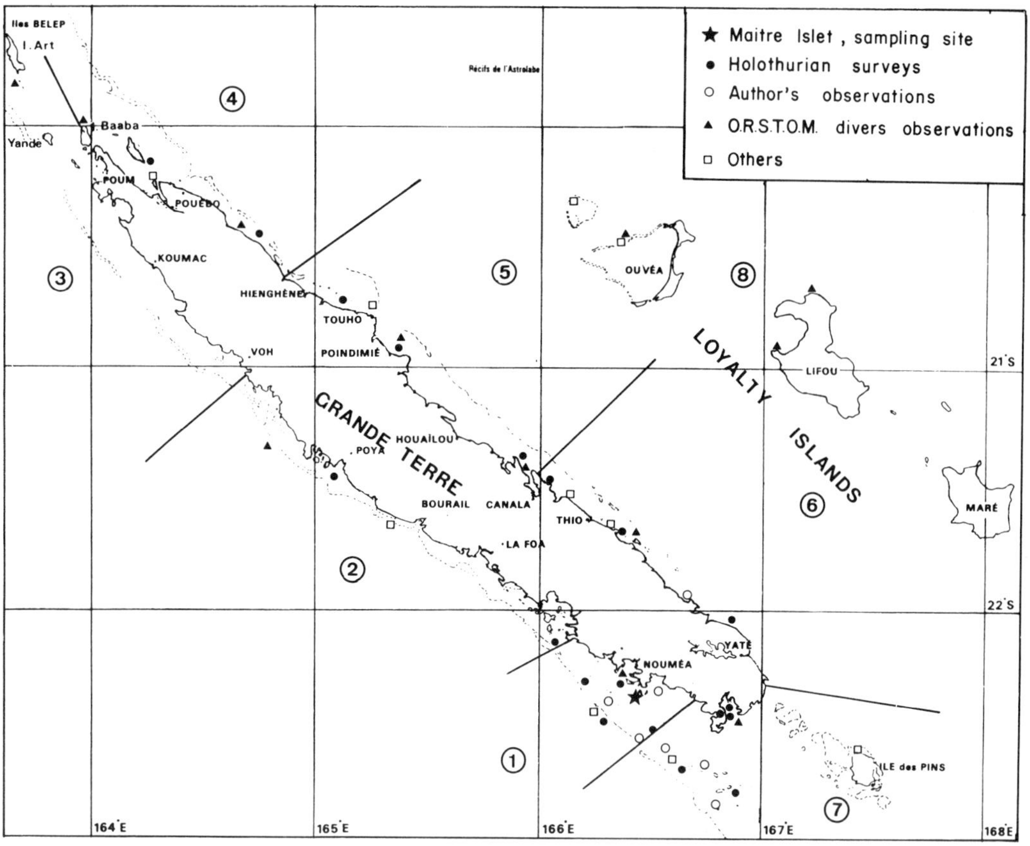

Figure 1. Acanthaster distribution around New Caledonia.

2.2. Results and discussion

Distribution : Acanthaster is distributed all around New Caledonia and the Loyalty Islands (Figure 1 and Table 1). It has been observed in 10 % of the 216 stations of the holothurian surveys. It is more frequent near Noumea in zone 1 and in the South, zone 7, which were more frequently surveyed. It is surprising that Weber and Woodhead (1970) did not observe any in 1967-1968 since hand collecting took place at least in two stations Maitre Islet and Prony Bay in the seventies.

Considering the habitat preferences, the percentages of the stations where Acanthaster occurred among the main reef complexes are : barrier reef 4 %, lagoon floor 4 %, lagoon-reef slopes 14 %, coastal stations 10 %. From the whole set of observations the stations with Acanthaster are distributed : 14 % on the barrier reef, 4 % on the lagoon bottoms, 39 % on the lagoon reefs, 41 % on the fringing and bay reefs. Only one specimen was observed in the shallow waters of the reef flats.

Abundance : densities presented on Table 1 are minimum values since some individuals were probably concealed in coral crevices. At most stations the density was 0 to 0.01 individual per 100 m^2. In a few stations near Noumea and in the South it was between 0.01 and 0.80. The density was the highest (3.30) at Maitre Islet.

Discussion : Potts (1981) summarizes the estimates of normal density populations from different reports, 50 individuals per hectare is probably a maximum. Maitre Islet is the only station of New Caledonia carrying an outbreak population. It is distributed on the windward slope and should not exceed a few thousands of individuals. The other stations with medium densities are located on fringing reefs and lagoon-reef slopes with lush corals. This distribution has also been observed by Pearson and Endean (1969).

The main hypotheses to interpret

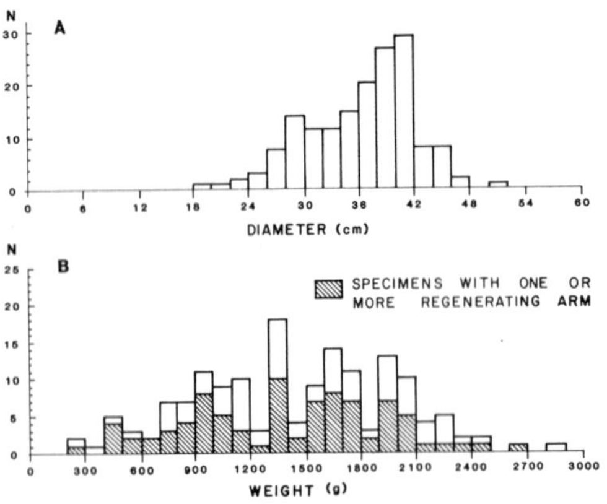

Figure 2. Frequency distribution of diameters (A) and weights (B).

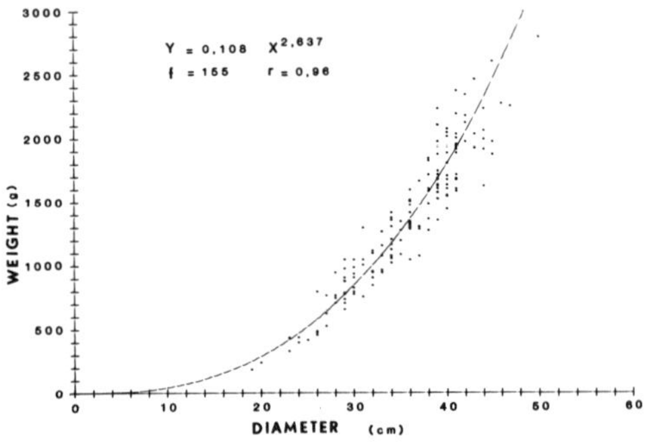

Figure 3. Diameter-weight relation.

outbreaks are generally at variance. On one hand population explosions are new phenomena resulting from man's activities. Primary outbreaks may result from the removal of the predators (Endean 1977) or pollution which weakens coral thus promoting juvenile or adult Acanthaster (Chesher 1969, Fisher 1969). On the other hand outbreaks are natural phenomena occuring in response to environmental perturbations. Adult agregations may follow typhoons (Dana and al. 1972) or larval survival and juvenile recruitment may increase (Birkeland 1982). Thus Lucas (1982) points out that a phytoplankton bloom is a major influence on larval development. Larval life lasts probably one month (Yamaguchi 1973), juveniles are generally cryptic, young individuals become evident on the reefs at 25-35 cm diameter, that is about three years after spawning. Therefore Birkeland shows that outbreaks appear three years after heavy rains. This hypothesis seems rather satisfactory. It is however clear from the scientific litterature that more research is needed to understand the population biology of Acanthaster. This first study in New Caledonia has shown that medium and high density stations are located in the South and near Noumea in places often visited by divers and pleasure boats who anchor on the coral slopes. It is likely that the different theories on the outbreak origins are not really opposed but complementary. Primary outbreaks can result from the interconnected influences of man and environmental conditions. It is therefore important to collect further information in New Caledonia.

3 BIOMETRY AND REPRODUCTION OF AN OUTBREAK POPULATION

3.1. Material and methods

Sampling was conducted at Maitre Islet in the lagoon near Noumea. A large population was observed, feeding on the tabular and branched Acropora association of the windward slope, between 2 and 8 meters in depth. A sample of about fifteen specimens was usually collected, each month, between January 1982 and March 1983. In December 1983 a diameter frequency distribution was established in situ. Samples were brought to the laboratory and the following parameters were recorded : maximum diameter (D) in cm, total wet weight (P) in g, number of arms (N) and regenerating arms (Nr).

Each gonad consists of numerous lobes in two rows near the inner wall of each arm. At first all the gonads were completely removed and their wet weight determined using the method of Yamazoto and Kiyan (1973). Considering the small variation between the arms, the gonad weight (G) in g was then calculated from the weight of three adjacent arms (Ga) and G = Ga/3 x N.

On the fresh gonads the form of the lobes, their colour and consistency were observed then they were preserved in formalin. A fragment was spread onto a hollow plate, sex was determined by microscopic examination and oocyte diameter distribution established for the females. Gonad-index was expressed as a ratio of gonad weight to total weight. The mean value with its confidence interval at the probability level of 0.90 was calculated for each sample.

Fecundity was obtained for ten ripe females using the following method : a fragment of ovary was weighed (p) and put into Gilson's fluid to release the oocytes from

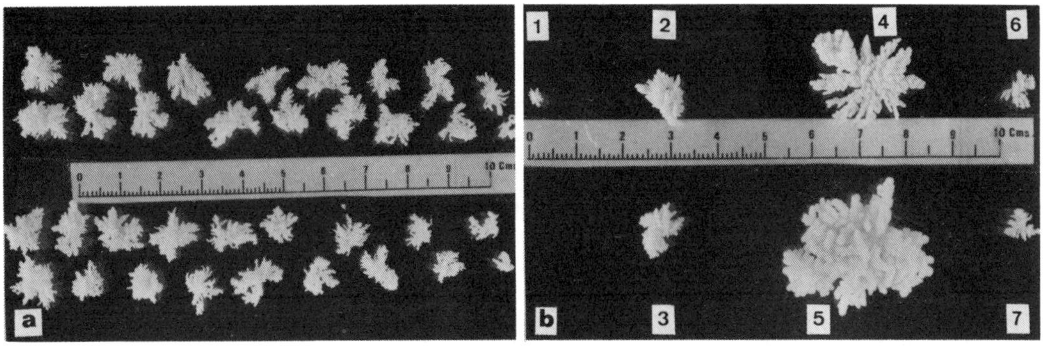

Figure 4. Gonad anotomy. a : gonadal bunches in one arm of a ripe male. b : maturity stages; undetermined : 1; maturation, male : 2, female : 3; ripe, male : 4, female : 5; post-spawning, male : 6, female : 7.

Table 2. Maturity stages scale. n:nb of specimens m:mean, s:standard deviation, G:gonad weight, Ga: gonad weight of 3 adjacent arms, GI:gonad index.

STAGES		G (g/10⁻¹)	Ga/3 (g.10⁻¹)	GI G.100/P
UNDETERMINED				
STAGE 1 : immature	n	5	4	5
STAGE 2 : resting	m	43	3	0.46
	s	44	3	0.18
MALES				
STAGE 3 : maturation	n	6	6	6
	m	327	19	2.24
	s	464	27	2.60
STAGE 4 : ripe	n	40	40	40
	m	1288	79	8.93
	s	737	46	3.85
STAGE 5 : post-spawning	n	23	23	23
	m	219	14	1.42
	s	122	8	1.00
FEMALES				
STAGE 3 : maturation	n	9	9	9
	m	273	17	2.36
	s	274	17	2.20
STAGE 4 : ripe	n	42	41	42
	m	1666	106	11.74
	s	1154	72	6.26
STAGE 5 : post spawning	n	26	26	26
	m	348	21	2.05
	s	249	15	1.20

the ovarian tissues. After washing two successive volumetric subsamples were taken from a homogeneous suspension and the oocytes were counted (n). The fecundity is : F = n x G/p.

3.2 Results and discussion

Acanthaster can reach 60 cm in diameter. The frequency distributions of the diameter and weight measured on the 157 individuals of the entire sample are shown on Figure 2 (A,B). At Maitre Islet the diameter ranged from 19 to 52 cm. It is likely that there are two modes, one at 29 and the other at about 40 cm. No specimen under 19 cm has been found. If the growth curve obtained by Kenchington (1977) on the Great Barrier reef is applied to New Caledonia, then the starfish are from two to more than five years old.

The number of arms was mostly 16 or 17 but ranged from 11 to 21. Among the individuals of the sample 53% presented 1 to 8 regenerating arms, these were found in all size classes.

The diameter-weight relation has been established, $P = 0.108D^{2.637}$ (Fig. 3). It is in accordance with the relation obtained by Nishira and Yamazoto (1972) from a population in Okinawa composed of smaller seastars.

Sexes are separate , males form 45% of the samples, females 50% and undetermined individuals 5 %. . A maturity scale in 5 stages is defined from the macro- and microscopic features of the gonads (Fig. 4 and Table 2). Stage 1 corresponds to true immatures and stage 2 to resting individuals ;they present the same characteristics of very tiny bunches with small lobes, germinal cells are less than 30 µm.

At stage 3 maturation takes place, in both sexes the bunches are larger, whitish in colour and of firm consistency, oocyte-diameters in females range from 30 to 100 µm. Stage 4 corresponds to ripe gonads at their maximum volume, white in males, with sperm running out of a section, yellow in females with oocyte-diameters between 200 and 260 µm. At stage 5, post-spawning, the bunches are deflated and limp, orange in colour, a few lobes contain some residual germinal

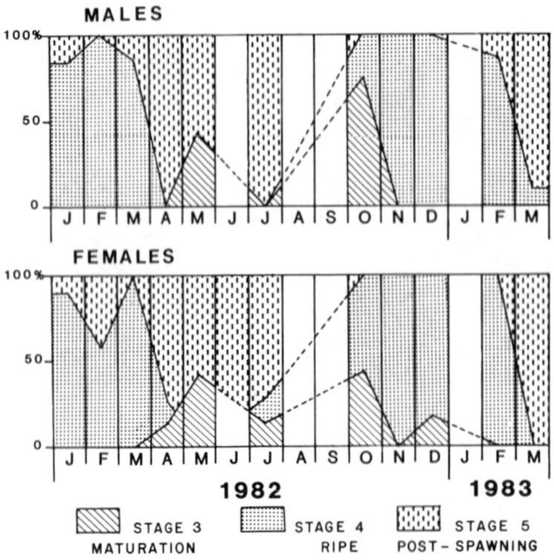

Figure 5. Monthly percentagesof maturity stages for males and females.

cells. Among the entire sample the rare individuals smaller than 22 cm in diameter and 400 g in weight were immature.

Monthly percentages of sexual stages(Fig.5) gives evidence of a similar cycle in males and females. Ripe individuals are predominant from November to March.

The gonad-index cycle (Fig. 6) shows a period of low values from April to July followed by a slow increase until October. The high values from November to February correspond to the spawning season. In females the variance of gonad-index is higher than in males. The decline of the female gonad-index in December could come from a partial spawning.

The reproductive cycle is interpreted from the gonad-index and the sexual stages variations. The long spawning season lasts during the warm waters period when the temperature is from 24 to 28°C. An estimate of the rainfall near Noumea was given by the average of the observations at four stations of the National Weather Service. Heavy rains occurred during the first season in December 1981 with nearly 300 mm and March 1982, but 1983 corresponded to a drought.

Fecundity relations with diameter and weight are curvilinear (Fig. 7). The relation between gonad-weight and fecundity is linear F = 375 + 9019 G, allowing an estimation of relative fecundity of about 90 000 oocytes per ovary gramm. Fecundity is about one million oocytes for a female of 25 cm diameter and goes up to 60 million for a 40 cm one. In a polymodal size population large individuals probably provide a large part of the spawned eggs.

Reproduction is an important factor when

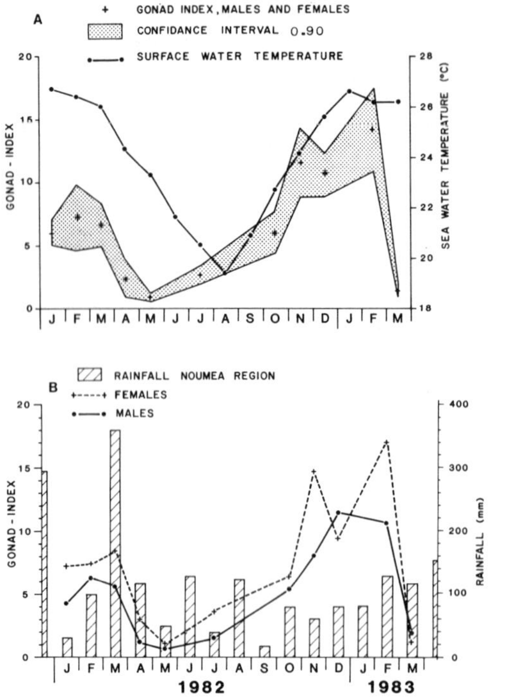

Figure 6. Gonad-index cycle, sea-water surface temperature and rainfall near Noumea.

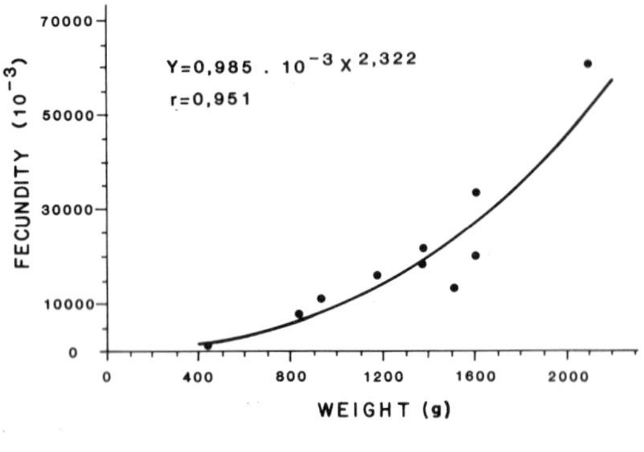

Figure 7. Fecundity - weight relation.

trying to interpret outbreaks. In the hypothesis of the terrestrial runoffs (Birkeland 1982) the duration and the date of spawning are related to environmental factors. It is therefore interesting to compare the reproductive cycles described in different localities. The methods used to establish them were a gonad-index based on the volume of the gonads by Pearson and Endean (1969) and Lucas (1973) on the Great Barrier Reef. This index is not much different from the index with gonad weight used by

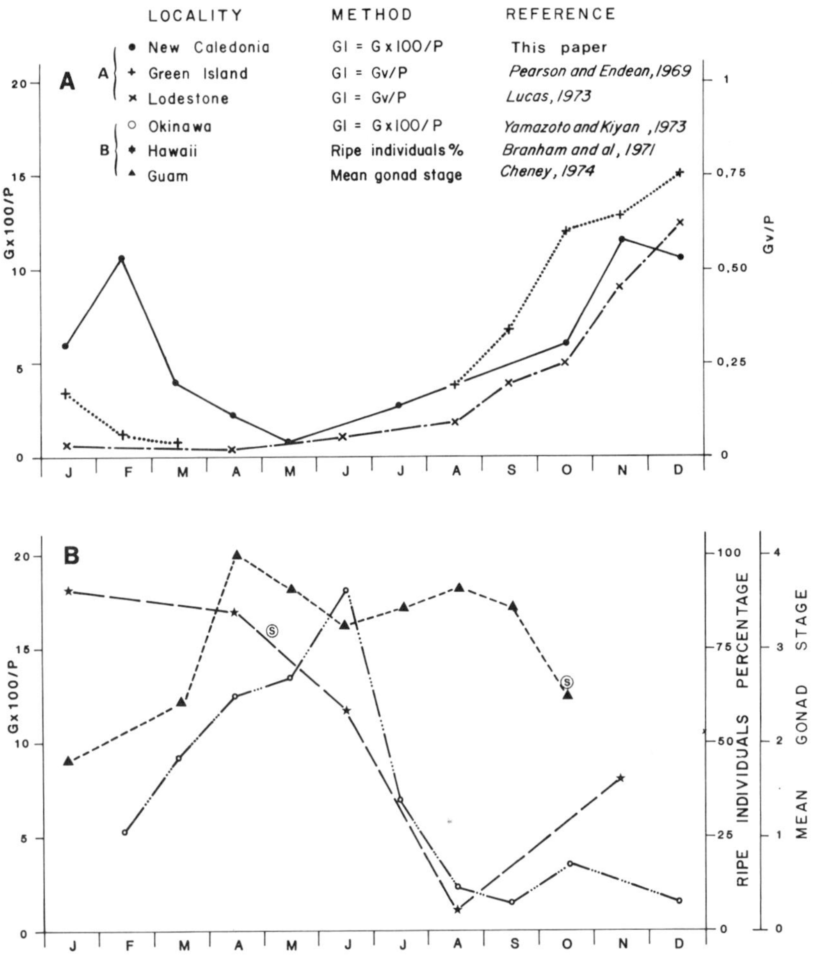

Figure 8. Comparison of reproductive cycles from Southern (A) and Northern (B) Hemispheres.

Yamazoto and Kiyan (1975) and in the present study. Branham et al. (1971) in Hawaii scored subjectively the maturity stages and give their percentage, while Cheney (1974) in Guam used a mean gonad-index based on a subjective four points scale. The figure 8 presents all these data transformed so as to obtain monthly values. In the Southern hemisphere cycles from the different localities are in accordance. In addition Wilson and Marsh (1975) in their study of a population in Western Australia found October and February to be the beginning and the end of the reproductive season with slight interannual variations. In the Northern Hemisphere the cycle in Okinawa corresponds with a six months shift. The main breeding is in June-July and a secondary could occur in October. In Hawaii too, spawning takes place during the warm season but in Guam the variations are less obvious. This can be explained by the data taken at different stations carrying populations with different size distributions.

4 CONCLUSION

The characteristics of an outbreak population in New Caledonia, its reproductive cycle and fecundity have been described. The distribution and density of Acanthaster has been studied in the reef complex of New Caledonia. It was observed in fifty stations mostly at low density populations < 0.5 per 100 m^2. In a few stations with high densities, Maitre Islet and Prony Bay, secondary outbreaks are localized in areas where terrestrial runoff and man-induced disturbance probably act concurrently. Acanthaster population is still monitored at Maitre Islet by establishing biannually in situ a diameter frequency distribution of the individuals and gathering the observations from the divers. In relation to the environmental conditions (weather, hydrology, plankton) studied at the Coastal Station this information may be very useful to understanding the population dynamics and observe if any secundary outbreak appears in early 1985 from the breeding of 1981-82.

ACKNOWLEDGMENTS

I thank R.H. Chesher for his encouragements during this study. I am grateful to the divers of Noumea's Aquarium for collecting the samples, the ORSTOM's divers and the people who communicated their observations, to M. Sibuet for critically reading the manuscript and to C.Hoffschir for his assistance.

REFERENCES

Battistini, R. & al. 1975 - Eléments de terminologie récifale indo-pacifique, Téthys, 7(1) : 1-111.

Birkeland, C. 1982 - Terrestrial Runoff as a cause of Outbreaks of Acanthaster planci (Echinodermata : Asteroïdea). Mar. Biol., 69 : 175-185.

Branham, J.M., S.A. Reed, J.H. Bailey & J. Caperon, 1971 - Coral-eating sea stars Acanthaster planci in Hawaii. Science, 172 : 1155-1157.

Catala, R. 1950 - Contribution à l'étude écologique des îlots coralliens du Pacifique Sud. Bull. biologique, 84, 3 : 233-310.

Cheney, D.P. 1974 - Spawning and aggregation of Acanthaster planci in Micronesia. Proc. 2nd Int. Coral Reef Symp., 1 : 591-594.

Chesher, R.H. 1969 a-Destruction of Pacific corals by the sea star Acanthaster planci. Science, 165 : 280-283.

Chesher, R.H. 1969b- Acanthaster planci, impact on Pacific coral reefs. Final report to U.S. Dep. Interior. Westinghouse Research Laboratories, 151 pp.

Conand, C. 1983 - Distribution et abondance des holothuries du lagon de Nouvelle Calédonie. Rapports de mission - campagnes 1 H à 7 H. Novembre 1983, miméo.

Dana T.F., W.A. Newman & E.W. Fager, 1972- Acanthaster aggregations : interpreted as primarily responses to natural phenomena. Pac. Sci., 26 : 355-372.

Endean, R. 1973 - Population explosions of Acanthaster planci and associated destruction of hermatypic corals in the Indo-West Pacific region. In O.A. Jones & R. Endean (ed.), Biology and Geology of Coral Reefs, p. 389-438. New-York : Academic Press.

Endean, R. 1977 - Acanthaster planci infestations of reefs of the Great Barrier Reef. Proc. 3 rd Int. Coral Reef Symp., 1 : 185-191.

Fisher, J.L. 1969 - Starfish infestation : hypothesis. Science, 165 : 645.

Kenchington, R.A. 1977 - Growth and recruitment of Acanthaster planci (L.) on the Great Barrier Reef. Biol. Conserv., 11 : 103-118.

Kenchington R.A. & B. Morton 1976 - Two surveys of the crown-of-thorns starfish over a section of the Great Barrier Reef. Canberra : A.G.P.S.

Lucas, J.S. 1973 - Reproductive and larval biology of Acanthaster planci (L.) in Great Barrier Reef Waters. Micronesica, 9 : 197-203.

Lucas, J.S. 1982 - Quantitative studies of feeding and nutrition during larval development of the coral reef asteroid Acanthaster planci (L.) J. Exp. Mar. Biol. Ecol., 65 : 173-193.

Nishihira, M. & K. Yamazoto 1972 - Brief survey of Acanthaster planci in Sesoko Island and its vicinity. Okinawa. U. Ryukyus, Sesoko Mar. Sci. Lab. Tech. Rep., 1 : 1-20.

Pearson, R.G. & R. Endean 1969 - A preliminary study of the coral predator Acanthaster planci (L.) (Asteroidea) on the Great Barrier Reef. Fish. Notes Qld., 3 : 27-55.

Potts, D.C. 1981 - Crown-of-thorns starfish-man-induced pest or natural phenomenon. In Ecology of Pests, p. 55-86. C.S.I.R.O. Publication Service.

Weber, J.N. & P.M.J. Woodhead 1970 - Ecological studies of the coral predator Acanthaster planci in the south Pacific. Mar. Biol., 6 : 12-17.

Wilson, B.R. & L.M. Marsh 1975 - Seasonal behaviour of a "normal" population of Acanthaster in Western Australia. Crown-of-thorns starfish Seminar Proceedings, p. 167-179. Canberra : A.G.P.S.

Yamazoto, K. & T. Kiyan 1973 - Reproduction of Acanthaster planci in Okinawa. Micronesica 9 : 185-195.

Yamaguchi, M. 1973 - Early life histories of coral reef asteroids, with special reference to Acanthaster planci (L.). In O.A. Jones & R. Endean (ed.), Biology and Geology of Coral Reefs, p. 369-387. New-york : Academic Press.

Proceedings of the Fifth International Echinoderm Conference / Galway / 24-29 September 1984

The microstructure of the asteroid skeleton (*Asterias rubens*)

PHILIPPE DUBOIS* & MICHEL JANGOUX
Laboratoire de Biologie marine, Université Libre de Bruxelles, Belgium

SUMMARY. A new enzymatic method is proposed to clean whole asteroid ossicles for SEM investigations. Enzymatic digestions were performed on fractured spines - i.e. previously cleaned spines - attempting to demonstrate the occurence of an intratrabecular organic matrix. Particular etching figures were produced (i.e. concentric layers made of tapered prismatic structures). The significance of these figures is discussed.

INTRODUCTION

With few exceptions, echinoderm ossicles are said to behave like calcite monocrystals. That particularity was emphasized by several authors from either light polarization, x-ray diffraction or crystal decoration studies (Nichols and Currey 1968, Donnay and Pawson 1969, Okazaki et al. 1981). The monocrystalline character of ossicles of adult echinoderms is enhanced moreover by the actual lack of a coherent organic matrix within the mineral phase (Currey and Nichols 1967, 1969, Klein and Currey 1970).

While echinoid larval spicules also behave like monocrystals under polarized light, investigations showed that they have a fibrillar organic matrix made of interconnected concentric layers (Benson et al. 1983). This observation recall those of previous authors who noted the occurence of concentric inorganic layers made of small sub-units in echinoid spicules (Millonig 1970, Okazaki and Inoué 1976). Similar inorganic layers were observed also by O'Neill (1981) in the aboral ossicles of an adult asteroid. In this case, the concentric layers seem to be made of calcite microcrystals that are regularly aligned. One may notice that those structures only appear in ossicles that have been loaded in stress relaxation before being fractured.

As already noted by Towe (1972), the occurence of a coherent organic matrix within a mineral phase proves the discontinuity of the crystallographic network and demonstrates the polycrystalline nature of the studied entity. The observation of concentric layers in asteroid ossicles actually could mean that a similarly arranged organic matrix occurs in these ossicles. We therefore attempted to demonstrate the presence of such matrix by trying to etch enzymatically the conchoidal surface of fractured ossicle trabeculae.

MATERIAL AND METHODS

Adult **Asterias rubens** were collected by SCUBA diving at Scharendijk (Zealand, Netherlands). They were kept in a closed-circuit marine aquarium (30 ‰, 10° C) in the laboratory at Brussels.

Investigations were done on adambulacral spines. Spines were sampled on living individuals and preserved in 70° ethanol. Associated tissues - viz. epiderm, derm and ossicle stroma - were digested in incubating the spines for 24 hours at 37° C in a neutral solution of 0.1 % proteinase N (Serva). Ossicles were then washed three times in desionized water (pH 6.2) and preserved in 96° ethanol until investigated. Each set of cleaned spines was SEM-controlled (three spines controlled per set).

* Research assistant of the National Fund for Scientific Research (N.F.S.R., Belgium).

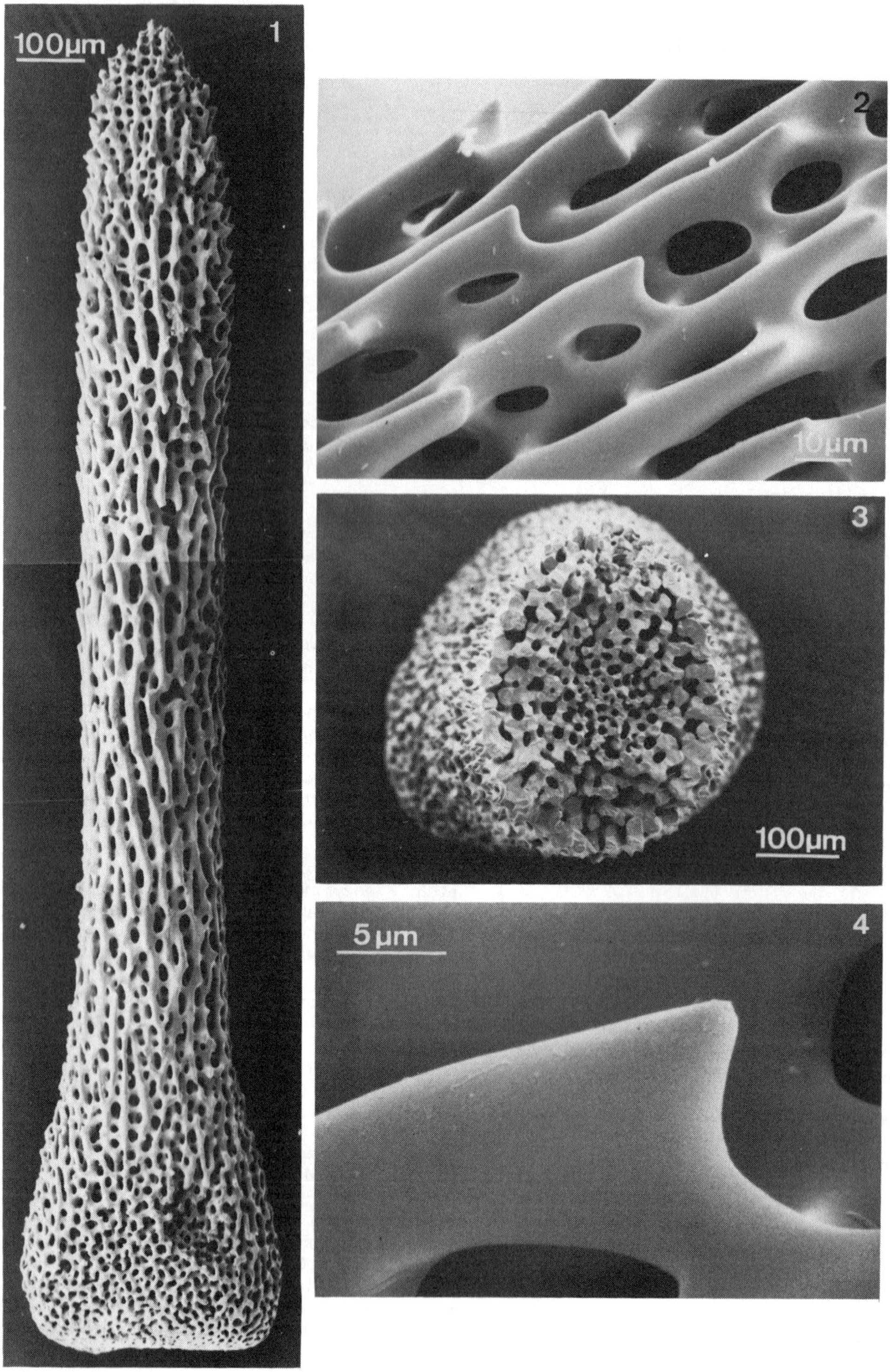
100µm
1
2
10µm
3
100µm
5µm
4

Effect of hydrolysing enzymes on **Asterias** skeleton was investigated as follows. Cleaned spines were washed three times in desionized water before being fractured using a razor blade. Sets of three fractured spines were then transfered into 5 ml of each of the following enzymatic solutions :

- 0.05 % pronase E (Serva) in 0.2 M Tris-HCl buffer (pH 8.0) and 0.05 M $CaCl_2$
- 0.01 % α-chymotrypsin (Serva) in 0.1 M Tris-HCl buffer (pH 7.65) and0.1 M $CaCl_2$
- 0.05 % type IV collagenase (Sigma) in 0.1 M Tris-HCl buffer (pH 7.65) and0.1 M $CaCl_2$
- 0.05 % hyaluronidase (Serva) in 0.02 M Na phosphate buffer (pH 7.0) and 0.45 % NaCl (0.01 % bovine albumine was added to the incubating medium).

Spines were incubated 6 h. either at 37°C (pronase, collagenase and hyaluronidase) or at 25°C (α-chymotrypsin). Control experiments were done with spines incubated in enzyme-free media.

After the incubating period, spines were washed in desionized water and preserved in 96° ethanol. For SEM-observations, spines were air-dried, mounted on aluminium stubs, coated with gold in a sputter coater and observed with a ISI DS 130 scanning electron microscope.

RESULTS AND DISCUSSION

1. Cleaning of intact spines

SEM-investigations on the skeleton of recent echinoderms always need a preliminary digestion of both the poripheral and the intraossicular soft tissues. Most authors prepare the ossicle in digesting the associated soft tissues using sodium hypochlorite. This is a rapid method giving satisfactory results at low SEM magnification. However the progressive loss of activity of sodium hypochlorite solutions makes difficult to control the time necessary to remove the associated tissues. Moreover prolonged exposure of ossicles to sodium hypochlorite may cause local etching of the stereom. This is all-important when only part of the ossicle to be cleaned is embedded in dense connective tissue. For instance sodium hypochlorite will hydrolyse the dense sutural tissue of the basal part of **Asterias rubens** adambulacral spine rather slowly while the tissues covering the spine tip are much more rapidly digested. Alterations of the stereom of the spine tip due to a prolonged stay in the hypochlorite solution will be observed consequently. Such surface alterations of unfractured asteroid ossicles caused by prolonged cleaning in either KOH or sodium hypochlorite have been illustrated by Fontaine and Hall (1981).

In contrast, the use of proteinase to remove ossicle-associated soft tissues showed these tissues are digested completely in a 24 hours period without any alteration of the ossicle stereom (figs 1, 2). Observations of fractured spines (viz. transversally and longitudinally fractured spines) allowed to demonstrate the complete digestion of the intraossicular stroma (fig. 3). Highly magnified spine trabeculae always have a smooth aspect with no sign of etching (fig. 4). Removal of soft tissues using proteinase should be thus recommended in all investigations requiring an important magnification of the stereom.

2. Enzymatic incubations

Non-incubated fractured spine trabeculae showed the classical conchoidal fracture with cleavage steps. After incubation - either in the enzymatic solution or in the control solution - the conchoidal fractures were strongly altered. A differential etching of the fractures was produced, revealing concentric layers made of small individualized structures (figs 5 and 6). This phenomenon - although of variable intensity (compare figs 5a and 6a) - was observed whatever the investigated enzymes. Moreover it occured in the same way and with the same intensity in both the experimental and control tests of each investigated enzyme (compare figs 5a and 5b,and figs

Figures 1-4. SEM-micrographies of adambulacral spines of **A. rubens** after incubation of whole spines in 0.1 % proteinase (24 h., 37° C). Note that the associated tissues are digested everywhere and that there is no sign of etching.

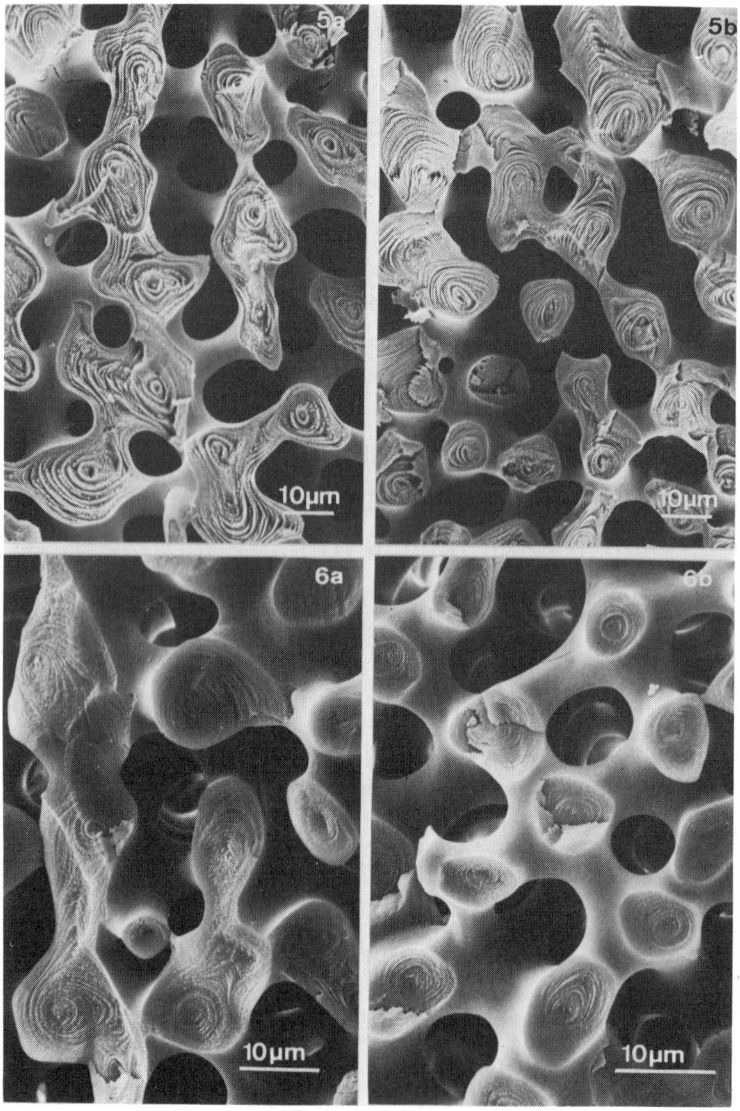

Figures 5-6. SEM-micrographies of trabeculae after enzymatic and control incubations of fractured spines. Fig. 5, experimental (a) and control (b) spines (α-chymotrypsin assay). Fig. 6, experimental (a) and control (b) spines (hyaluronidase assay).

6a and 6b). The etching intensity actually appeared to vary only according to the used buffer solution, whether or not they contain an enzyme component. These results suggest that no organic material has been hydrolysed by the tested enzymes. They seemingly confirm the lack of organic matrix already emphasized by previous authors from biochemical analysis (Currey and Nichols 1969, Klein and Currey 1970). However, one should also consider that the hydrolysis of the eventual intratrabecular organic matrix could have been either concealed by the rate of the dissolution process which reveals the observed concentric layers (figs 5a,b) or uneffected because of the use of an unappropriate enzyme (e.g. hyaluronidase; figs 6a, b).

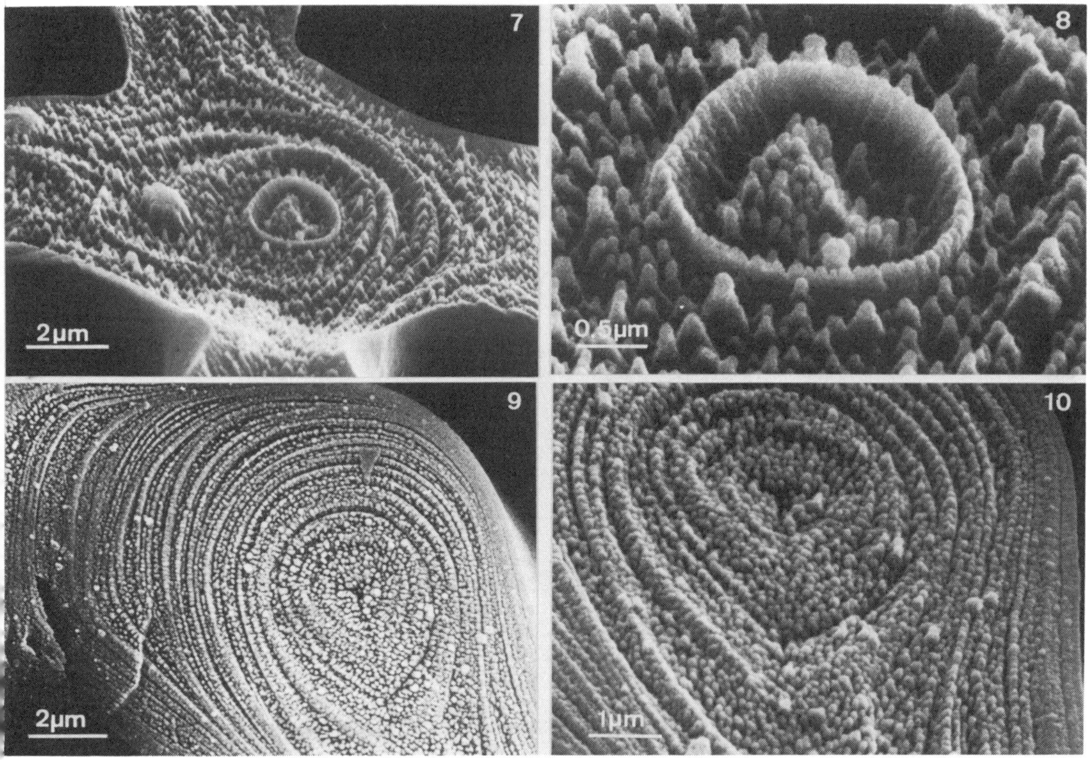

Figures 7-10. Enlarged SEM-micrographies of intratrabecular concentric layers revealed by incubation either in Tris-HCl buffer pH 7.65 (figs 7-8) or in collagenase 0.05 % in Tris-HCl buffer pH 7.65 (Figs 9-10).

3.Structure and significance of the concentric layers

The concentric layers are made of tapered prismatic structures which run parallel to the longitudinal axis of the trabecula (figs 7-10). The limiting faces of these structures are not clearly distinct, a fact that prevents any measurement.

Those structures look very different from the etching figures that Okazaki et al. (1981) produced - using low concentrated acids - in both fractured trabeculae of echinoid ossicles and native calcite monocrystals. Okazaki's figures, which are obviously of physico-chemical origin, are not arranged in concentric layers. As far as the concentric arrangement is concerned those structures also are very different from those obtained with ossicles cleaned using KOH or sodium hypochlorite (see, e.g., Pearse and Pearse 1975, Fontaine and Hall 1981).

The tapered prismatic structures we observed look rather similar to the microcrystals described by O'Neill (1981). O'Neill reported these microcrystals are seen only when the ossicle is fractured after being loaded in stress relaxation. She reported moreover that any other way of fracturing ossicles produced the typical conchoidal fractures of echinoderms. O'Neill noted however that some ossicles - without mentionning which ones - had been stored in 10 % formalin (buffered solution) after being fractured. No information was given on the nature of the buffer. Our results actually showed that buffer solutions alone can cause a differential dissolution which produce structures similar to O'Neill's microcrystals. O'Neill's microcrystalline figures could thus have been produced by etching due either to the formalin itself or to the buffer solution. Actually, they do not definitely prove the polycrystalline nature of asteroid ossicle.

The observed concentric layers are seemingly similar to those "composed of small subunits" described by Okazaki and Inoué (1976) in spicules of echinoid larvae reared in low calcium-magnesium medium. According to them, echinoid spicules would consist of an "accumulation of crystallites" embedded in organic material. This hypothesis was confirmed by Benson et al. (1983) who demonstrated the occurence of a coherent organic matrix within larval spicules.

The significance of the concentric layers revealed on fractured asteroid ossicles remains unclear. Are they only the consequence of a differential dissolution of the skeleton due to particular physico-chemical conditions or do they represent inorganic layers separated by organic laminae (i.e. actual or vestigial laminae) such as those occuring in larval spicules ? Further experiments are needed to answer that question and to eventually reconsider the crystallographic nature (poly- vs. monocrystalline) of the skeleton of adult echinoderms.

Acknowledgements. We thank D. Jehai for revising the English. Work supported by F.R.F.C. grants n^{rs} 2.4506.83 and 2.4525.80.

REFERENCES

Benson, S., E.M.E. Jones, N. Crise-Benson & F. Wilt 1983, Morphology of the organic matrix of the spicule of the sea-urchin **Strongylocentrotus purpuratus** larva, Exp. Cell Res. 148 : 249-253.

Currey, J.D. & D. Nichols 1967, Absence of organic phase in echinoderm calcite, Nature, Lond. 214 : 81-83.

Currey, J.D. & D. Nichols 1969, The question of the organic matrix of echinoderm skeletons, Proc. malac. Soc. Lond. 38 : 546-547.

Donnay, G. & D.L. Pawson 1969, X-ray diffraction studies of echinoderm plates, Science 166 : 1147-1150.

Fontaine, A.R. & B.D. Hall 1981, Biocompatibility of echinoderm skeleton with mammalian cells in vitro : Preliminary evidence, J. biomed. mat. Res. 15 : 61-71.

Klein, L. & J.D. Currey 1970, Echinoid skeleton : absence of a collagenous matrix, Science 169 : 1209-1210.

Millonig, G. 1970, A study on the formation and structure of the sea-urchin spicule, J. submicrosc. Cytol. 2 : 157-165.

Nichols,.D. & J.D. Currey 1968, The secretion, structure and strength of echinoderm calcite. In S.M. Mac Gee-Russel and K.F.A. Ross (eds), Cell structure and its interpretation, pp. 251-261. London, Edward Arnold.

Okazaki, K. & S. Inoué 1976, Crystal property of the larval sea urchin spicule, Develop., Growth and Differ. 18 : 413-434.

Okazaki, K., R.M. Dillaman & K.M. Wilbur 1981, Crystalline axes of the spine and test of the sea-urchin **Strongylocentrotus purpuratus** : determination by crystal etching and decoration, Biol. Bull. 161 : 402-415.

O'Neill, P.L. 1981, Polycrystalline echinoderm calcite and its fracture mechanics, Science 213 : 646-648.

Pearse, J.S. & V.B. Pearse 1975, Growth zones in the echinoid skeleton, Amer. Zool. 15 : 731-753.

Towe, K.M. 1972, Invertebrate shell structure and the organic matrix concept, Biomineralization 4 : 1-14.

Proceedings of the Fifth International Echinoderm Conference / Galway / 24-29 September 1984

Variations in the growth rate of *Asterias rubens* (L.) from west and south Brittany (France)

MONIQUE GUILLOU
Université de Bretagne Occidentale, Laboratoire d'Océanographie Biologique, Brest, France

ANNE GUILLAUMIN
Institut Scientifique et Technique des Pêches Maritimes, Laboratoire Conchylicole, La Trinité-sur-Mer, France

ABSTRACT: The reproductive cycle and growth parameters were studied in populations of Asterias rubens from two bays in Brittany. In the Bay of Douarnenez (Finistère) the population is composed of small individuals. Its growth rate is low with a long life-span (> five years). In the Bay of Quiberon (Morbihan), the growth rate was four times higher with a much shorter life-span (about three years). On average, three year old Asterias were twice as large in the Bay of Quiberon as five year old animals in the Bay of Douarnenez. A discussion of the demographic strategy of this species considers the biological parameters of each population and associated environmental factors.

1 INTRODUCTION

While experiments have shown that the growth and the life-span of Asterias rubens vary according to the geographic area (Bull, 1934, Vevers, 1949, Hancock, 1958), the comparison of demographic structures in the natural environment had never been done. It is indeed very difficult to estimate the age of starfish by measuring their growth.

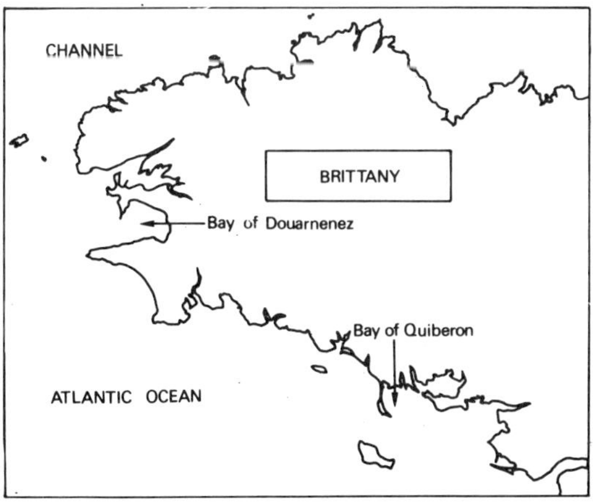

Fig. 1. Location of the study areas

This is only possible when homogeneous feeding and hydrological conditions give rise to regular growth of the individuals (Vevers, 1949, Crump & Emson, 1978, Guillou, 1980, 1983). Two populations in Brittany, in the Bays of Douarnenez and Quiberon, satisfying the above conditions, have been sampled. A comparison of their growth, estimated by modal analysis, showed two very different demographic strategies. Previous studies on this species suggested such variations but did not quantify them. It will thus be interesting to specify their causes by studying each environmental condition.

2 THE STUDY AREAS

2.1 The Bay of Douarnenez

The intertidal population of the "Aber" beach was chosen as a reference. It had a growth rate similar to that of other Asterias rubens populations studied in the bay (Guillou, 1983), but has been sampled more regularly.

The "Aber" beach is situated in the north-west of the bay and is exposed to swell action. The

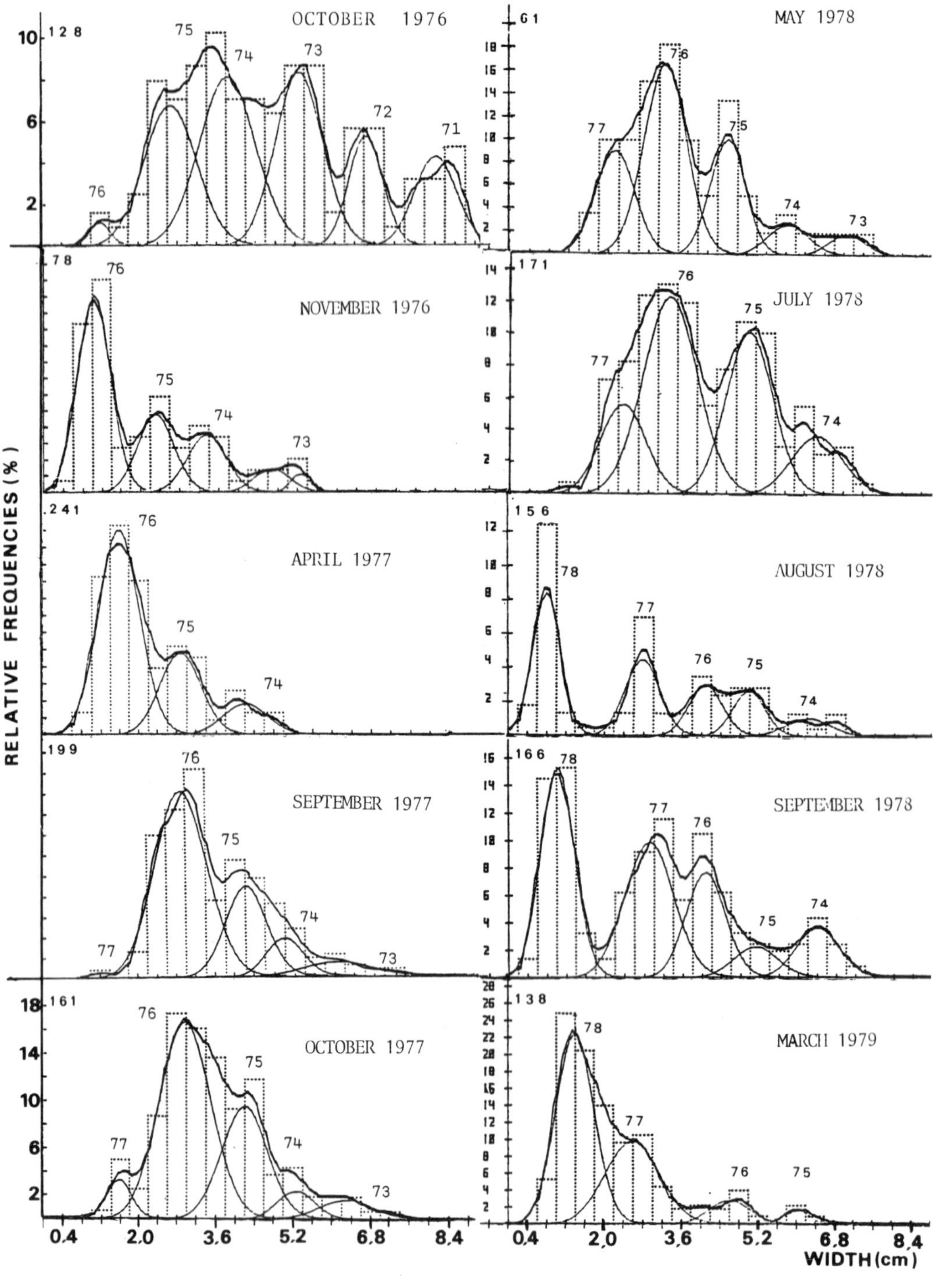

Fig. 2. Size distributions of the population of Asterias rubens in the bay of Douarnenez, with the smoothed curve and the gaussian components superimposed (the effective is situated above to the left).

fauna of the beach material, i.e. a fine sand, is dominated by the bivalve Donax trunculus. Located in the lower part of the beach, the area studied is in the middle of rocks inhabited mainly by mussels, which isolate the starfish population from both the subtidal area and the beach. The sea stars were generally aggregated under the rocks, and could reach ten individuals per tenth of a square metre.

2.2 The Bay of Quiberon

Sampling was carried out in the southern part of the rocky area "La Souris", on a sandy mud bottom, about 15 m deep. The density of sea stars was about 0.2 individuals per square metre. They were associated with high numbers of Crepidula fornicata.

3 METHODS

3.1 Sampling

In the Bay of Douarnenez, the sea stars were sampled monthly at low water during spring tides, from September 1976 to April 1979. They were collected by hand with a view to obtaining all size classes.

In the Bay of Quiberon, sampling (March 1981, September 1983) was achieved by means of an oyster dredge two metres wide. It was equipped with a 4.5 cm mesh net which was reduced to 3.2 cm after March 1983.

3.2 Data analysis

All the sampled individuals were measured either immediately (Quiberon), or after being stored in formalin (Douarnenez). The adopted measurement was the distance from the tip of the longest arm to the opposite interradius (Nauen & Bohm 1979). The data were assigned to size classes of 0.4 cm each at Douarnenez and 1.5 cm at Quiberon. The frequencies were expressed as percentages of the total. The best fit of the observed frequency distributions to a mixture of Gaussian components was established by a semi-automatic method using successively:

1. The logarithmic method of Bhattacharya (1967) which provided an initial estimation for the arithmetic average and standard deviation of each respective Gaussian component.
2. The maximum likelihood method of Hasselblad (1966), using as initial values the preceding estimations. For this purpose, the Normsep (Fortran IV) programme written by Tomlinson (1970) was modified for use with the Hewlett Packard computer by Gros & Cochard (1978).

4 RESULTS

4.1 The population of the Bay of Douarnenez

The consecutive histograms (Fig. 2) show the evolution of the demographic structure during the study period. Previous studies indicated that reproduction is restricted to a limited annual breeding season, i.e. from May to June. Accordingly, the settlement which followed gave rise to a normal size class. It took place in the late Summer or in Autumn (November 1976, August 1978). Thus the second cohort on the same histogramme represents the one year old individuals.

In October 1976, the population is represented by six cohorts. From November, there was high mortality among the oldest specimens and this was probably due to the "Boehlen" oil spill. The progress of each year group can be evaluated by plotting its mean size in relation to time. An adjustment of these curves gives the average growth curve (Fig. 5). It appears that the young Asterias, which settled during the late summer at a size of about 0.8 cm., measured about 2.5 cm one year later. This species lived for at least five years, by which time it reached a mean size of 8 cm. Rarely were individuals larger in size and the observed maximum was 17.5 cm. During these 5 years, growth can be considered as linear, the slow winter growth rate and any loss in size being compensated for by the high summer rate.

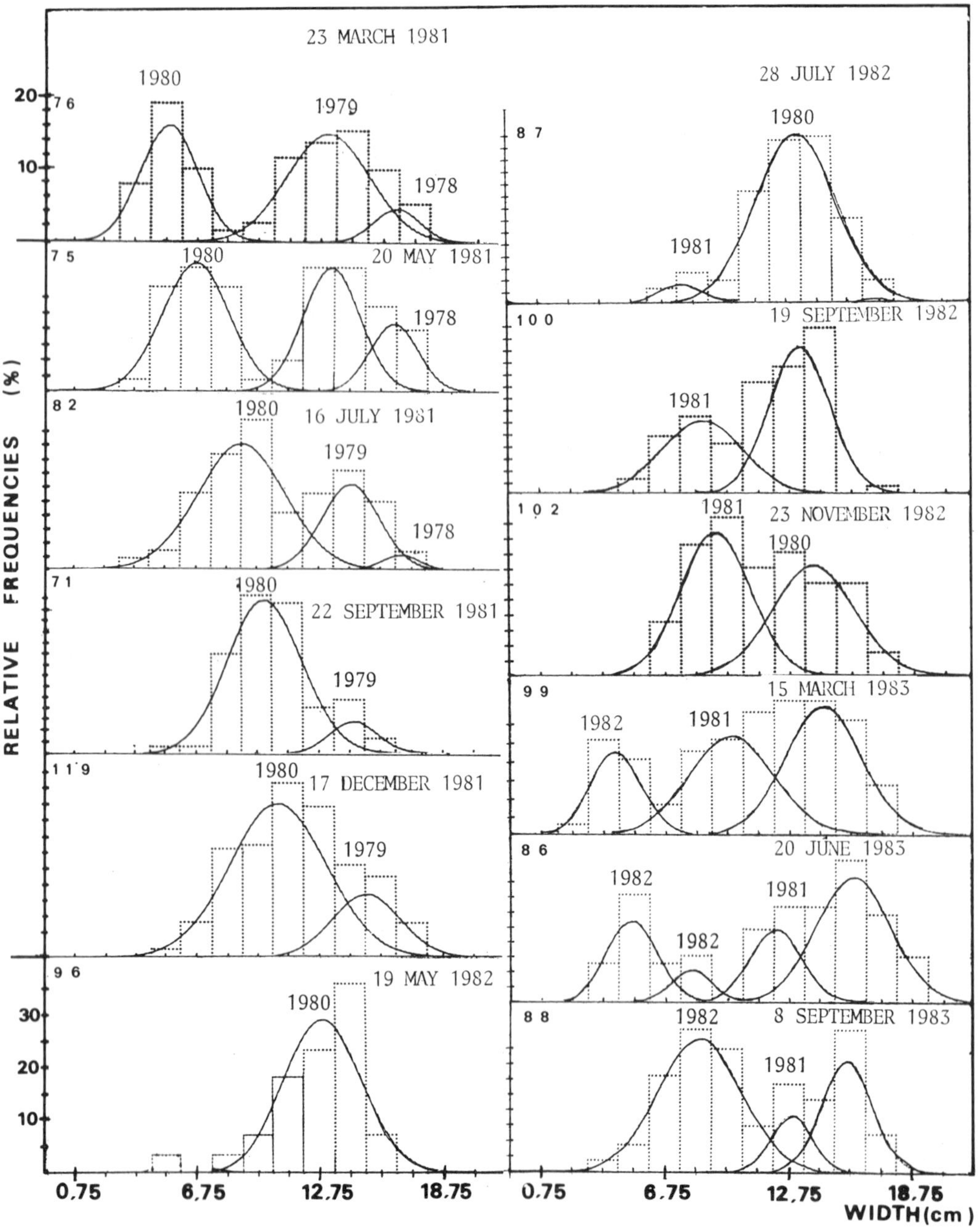

Fig. 3. Size distribution of the population of Asterias rubens in the bay of Quiberon with the normal components.

4.2 The population of the Bay of Quiberon

The demographic evolution of this population is shown in Fig. 3.

With the technique employed in this study, the minimum size of individuals taken was 4 cm. Recruitment observed in March, represented individuals spawned in the previous

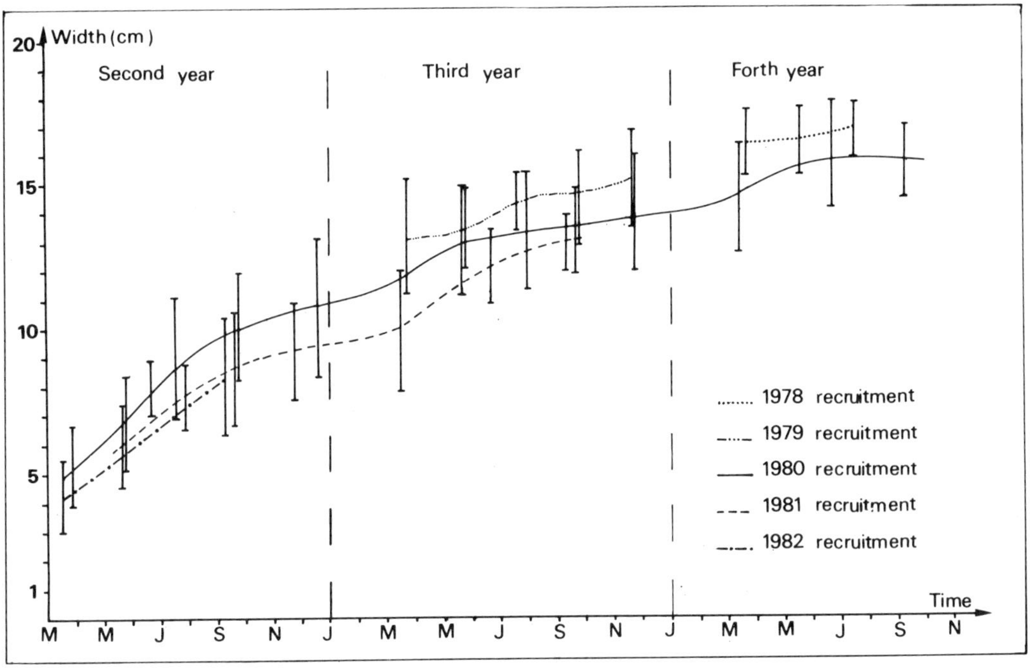

Fig. 4 Average growth curve of each cohort settled in the Bay of Quiberon during the studied period, with the modal length (± 1 dev. std.) of cohorts for each month.

spring (March-April) and settling on the oyster collectors in June or July.

Three cohorts appear on the histogrammes, with the first generally well defined. The growth curve shows winter checks followed by spring increases (Fig. 4). In general, growth was higher during the first year (from 0.5 to about 7 cm), after which it decreased progressively. The oldest specimens reached a mean size of 16 cm after 3½ years.

The intensity of recruitment was variable, i.e. high in 1980 and 1982, and very low in 1981. The cohorts showed growth variations because the 1978 and 1979 cohorts had higher growth rates than those of 1980 and 1981, and the lower the growth rate, the longer the life-span. Knowing that Asterias rubens was not subject to predation and that no larger specimens were found in neighbouring areas, thereby ruling out migration, the length of life was taken as the time from first appearance of the cohort to its complete disappearance by natural mortality. It was approximately three years for the 1978 and slightly less for the 1979 cohorts respectively. Both had a faster growth rate than the 1980 cohort which had a life-span in excess of three years.

5 DISCUSSION

This study shows the great variability of Asterias rubens growth. There is firstly a large spatial variability illustrated by the very different demographies of the two natural populations and a less important temporal variability, with differences in the growth of the settled cohorts from year to year.

What are the causes of these growth variations?

The growth rate of the population in the Bay of Quiberon varies with the density of the adults. The 1980 cohort was the only one in this area after the disappearance of the 1979 cohort in December 1980.

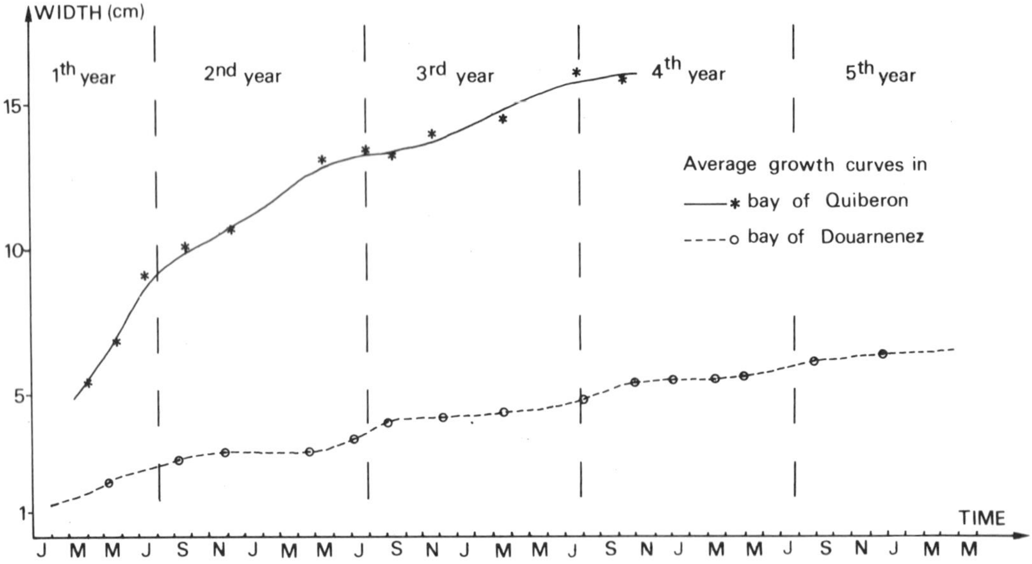

Fig. 5 Growth curves estimated from the averages of the data obtained during many years for each population

Its growth was higher than the 1981 and 1982 cohorts which settled in the same area but amongst a very large (in density and, more especially, in biomass) established population.

The comparison of the two studied populations shows that the mean size of one year old Asterias rubens in the Bay of Quiberon was the same as that of five year old Asterias in the Bay of Douarnenez. In the latter case, almost all the population was less than 8 cm while in Quiberon only 50% was less than this size. The previous observations suggest that the differences in growth rate are attributable to the differences in density of the two stocks. However, though the density was much higher in the Bay of Douarnenez, the mean individual weight was only a tenth of that in the Bay of Quiberon, so it seems we must look elsewhere for the underlying cause of the differential growth rates.

The period of recruitment and, indirectly, the temperature, seem to be important factors. The breeding season at Quiberon can extend from March to June when the temperature increases from 10° to 16°C on average. Settled in early summer in Quiberon, the juveniles grow about 4 cm by the next spring. In Douarnenez Bay, the temperature increase is lower (9.5° to 14°C on average from March to June) and the breeding season begins only in late April or May. The larvae settle in late summer, hide in rock crevices during the winter and growth does not start until the following spring. But later growth is always higher in Quiberon which indicates the effects of other factors.

It is well established that growth depends primarily on the availability of a suitable food source. This can be demonstrated by correlating growth patterns with the quantity and quality of the food supply. This study shows that populations with a fast growth rate reach a greater maximum size than those with a slower growth rate. Thus, the mean size of a population should give an estimate of its growth pattern.

In Table 1, the mean size of some European populations has been correlated with the food, after standardisation of the length measurements. The mean size of Asterias rubens for the Bay of Douarnenez given in the table (4.07 cm $\pm$ 0.11) is the mean of all populations sampled in the bay, in

Tab. 1. Characteristics of some European populations of Asterias rubens (relation between the mean size, the environmental factors and the quality of food).

Author	Area	Depth	Bottom	Prey	Mean width (cm)
Anger et al.	Baltic Sea	15 m	phytal	Mytilus edulis Idotea baltica	2.6
Nauen	Baltica Sea	>15 m	algae, heteregeneous sediments	Abra alba Cyprina	2.6
Vevers	Rame Mud Plymouth-England		mud	poor epifauna buried macrofauna	3
Skaeveland	Borgenfjorden Norway	0 to 40 m	rocky bottom fine sand, Laminaria		3.1
Mash	Millport Scotland	intertidal area			3.6
Guillou	Bay of Douarnenez Brittany	0 to 30 m	fine sands	fine sands bivalves mussels	4.1
Anger et al.	Baltic Sea	15 m	muddy sands	Hydrobia ulvae Macoma baltica Mytilus edulis	4.2
Hancock	River Crouch Essex-England		muddy sands	Barnacles, Crepidula Oysters	4.5
Guillou	Bay of Concarneau Brittany	< 20 m	muddy sands 10 % particles < 63 μ	muddy sands bivalves Cultellus-Abra	4.6
Vevers	Cawsand Bay Plymouth-England				4.8
Guillou	Bay of Concarneau Brittany	20 to 30 m	sandy muds 30-50 % fine particles	sandy muds bivalves Nucula turgida - Abra	5.2
Mash	Kames Bay Scotland	30 m			6.7
Guillou	Bay of Concarneau Brittany	20 to 30 m	muds	Haploops	7.2
Vevers	Plymouth sound England	river mouth		mussels beds suspended material	11.1
Sloan	Morecambe Bay Lancashire-England	river mouth	muddy sands algae	mussels beds	11.2
Guillemin	Bay of Quiberon Brittany	15 m	muddy sands	Crepidula beds	11.4
Guillou	Bay of Brest Brittany	10 to 30 m river mouth	muddy sands	Crepidula Pectinidae	11.8
Vevers	Rame Eddystone Grounds-England	25 to 30 m	mixed substratum (sand and rough ground)	Chlamys and epifauna (worms hydroids, bryozoans)	12.4
Brun	Balsfjord Norway	32 m		Chlamys islandica beds	15.8
Vevers	Outer Ground Eddystone	64 to 70 m	sand with patches of gravel	Chlamys opercularis Cyprina islandica	17.8

one intertidal and three fine sand subtidal areas. The study of Guillou (1983) showed that their growth rates are very similar.
In Table 1, the European populations are divided into two distinct groups: the first has a mean size below 8 cm and the second a mean size above 11 cm. The study of the associated macrofauna shows that all the populations of mean size above 11 cm live on big epibenthic molluscs (i.e. Mytilus, Crepidula and Chlamys). These are generally found in waters rich in suspended material and in river mouths (Vevers 1949, Hancock 1958, Sloan & Aldridge 1981 and personal observations). The other group lives on soft bottoms where the macrofauna may sometimes be important but is buried in the sediment and, thus, harder to obtain. In this group, growth differences can reflect the nature of the bottom and the depth. The largest specimens are generally observed on mud rather than fine sand and in subtidal rather than intertidal areas (Vevers 1949, Mash 1968). This latter observation can be imputed more to migration than differential growth (Guillou 1980, Moore 1984).

The above Table confirms the data of the present study: the population in the Bay of Quiberon belongs to the second group and has a fast growth rate, feeding on the Crepidula beds, whereas the population in the Bay of Douarnenez belongs to the first group and consists of small specimens, for whom portion of the macrofaunal food supply is beyond reach.

CONCLUSION

This study shows that the growth of Asterias rubens does not have a regular pattern. Local factors have a greater effect on the demographic strategy of populations than their position relative to the distribution range of the species. External factors (temperature, hydrodynamic conditions) are less important than the availability of a suitable food supply. The more abundant the biomass of suitable food, the higher the annual growth rate, the shorter the life cycle but paradoxically the higher the maximal size. Thus, in the Bay of Douarnenez, where some macrofaunal food is beyond reach, this species has a slow growth rate and reaches a size of 8 cm after 5 years. On the contrary, in the Bay of Quiberon, it feeds on epibenthic molluscs, and grows rapidly reaching a maximal size of 16 cm after only $3\frac{1}{2}$ years. In both cases, growth is linear up to 8 cm, i.e. during the whole life of the population in the Bay of Douarnenez.

REFERENCES

Anger, K., Rogal, U., Schriever, G. & Valentin, C., 1977. In situ investigations on the echinoderm Asterias rubens as a predator of soft-bottom communities in the Western Baltic Sea. Helg. Wiss. Meeresunters, 29: 439-459.

Bhattacharya, C.G. 1967. A simple method of resolution of a distribution into Gaussian components. Biometrics, 23(1): 115-135.

Brun, E. 1968. Extreme population density of the starfish Asterias rubens L. on a bed of Iceland scallop, Chlamys islandica (O.F. Müller). Astarte 32: 1-4.

Bull, M.O. 1934. Aquarium observations on the rate of growth and enemies of the common starfish, Asterias rubens L. Rep. Dove Mar. Lab. (3) 2: 60-63.

Crump, R.G. & Emson, R.M. 1978. Biologie de Nyctiphanes couchii (Crustacea, Euphausiacea) dans le secteur nord due Golfe de Gascogne. Ann. Inst. Oceanogr. Paris. 54(1): 25 - 46.

Guillou, M. 1980. Données sur la croissance d'Asterias rubens en Bretagne-Sud. In M. Jangoux (ed), Echinoderms Present and Past, p.179-186. Rotterdam: Balkema.

Guillou, M. 1981. Les populations d'Asterias rubens L. (Echinodermata, Asteroidea) en Baie de Douarnenez (Finistère). Thèse de 3e Cycle Fac. Sc. Brest: 122p.

Guillou, M. 1983. La croissance d'Asterias rubens L. (Echinodermata, Asteroidea) en Baie de Douarnenez (Finistère). Ann. Inst. Oceanogr. Paris 59(2):141-154.

Hancock, D.A. 1958. Notes on starfish on an Essex oyster bed, J.

Mar. Biol. Ass. U.K. 37: 313-351.
Hasselblad, V. 1966. Estimation of parameters for a mixture of normal distributions. Technometrics. 8(3): 431-446.
Mash, N. 1968. Arm length and arm damage in Asterias rubens. Nature. 218: 1175-1176.
Moore, R.J. 1984. An investigation into the behavioural and ecological bases for periodic infestations of Asterias rubens. (This volume).
Nauen, C.E. 1978a. The growth of the sea star, Asterias rubens L. and its role as benthic predator in Kiel Bay. Kieler Meeresforsch., Sonderbd. 4: 68-81.
Nauen, C.E. & Bohm, L. 1979. Skeletal growth in the Echinoderm Asterias rubens L. (Asteroidea, Echinodermata) estimated by 45Ca - labelling. J. Exp. Mar. Biol. Ecol. 38: 261-269.
Skjaeveland, S.M. 1973. Ecology of Echinoderms in Borgenfjordeen, North Trondelag. Norway, Miscellaned. 8:51p.
Sloan, N.A. & Aldridge, T.M. 1981. Observations on an aggregation of the starfish Asterias rubens L. in Morecambe Bay, Lancashire. J. Nat. Hist. 15: 407-418.
Tomlinson, P.K. 1970. The resources agency of California, Dept. Fish and Game. Program NORMSEP (FORTRAN IV).
Vevers, H.G. 1949. The biology of Asterias rubens L. I. Growth and reproduction. J.Mar.Biol.Ass. U.K. 28(1): 165-187.

Proceedings of the Fifth International Echinoderm Conference / Galway / 24-29 September 1984

Migration from refuges: A stabilising factor for a sea-star community

P.JOST & K.REIN
Zoologisches Museum der Universität Zürich, Switzerland

ABSTRACT : Factors contributing to the observed stability of a local ecosystem with predator-prey interaction and competition between two sea star species of the genus Astropecten were investigated. Predation by Astropecten aranciacus eliminates about 0.5 - 1 % of the prey population A. bispinosus per day. Recruitment by settlement in this coexistence zone was never observed, but immigration from adjoining patches of the eelgrass Zostera marina was found to at least partially compensate loss due to predation. Other possible refuge zones from which immigration may take place are discussed.

INTRODUCTION

Within a small and locally isolated sea star community at Costa Colostrai, Sardinia (Italy), the two predatory sea star species Astropecten aranciacus L. and Astropecten bispinosus OTTO coexist on a infralitoral sandy bottom; a situation often found in the Mediterranean Sea (Massé, 1970, Tortonese, 1966). At the study site these species compete for food (Ribi et al., 1977). As reported by Massé (1966, 1970 and 1975), A. aranciacus also preys upon other Astropecten species. At the study site in Sardinia predation of A. aranciacus upon A. bispinosus was confirmed first by Ribi and Jost (1978) and later, in more detail, by Schmid and Schärer (1981). In this last study the predation pressure of A. aranciacus upon A. bispinosus in the summer was estimated to be about 1 % of the population of A. bispinosus per day, which equals an average loss of 1.5 individuals of A. bispinosus per day per ha. Since sea star densities remained more or less constant over several years (Ribi, 1978), the losses of A. bispinosus due to predation should be compensated for by either recruitment or immigration or both.

As recruitment by settlement and/or growth of resident individuals cannot compensate for the losses (Schärer, in preparation), immigration mainly by subadult and adult individuals of A. bispinosus seems to be the most important factor contributing to the observed stability of the ecosystem in study. For an estimated amount of immigration of A. bispinosus of 1 % of the population per day, refuges must occur in the surroundings of the ecosystem in study. The important role of spatial and/or temporal refuges for the survival of local populations has been shown by Conell (1970), Birkeland et al. (1971) and Harmelin & Duval (1983).

The observation of juveniles led to the assumption that patches of the eel grass, Zostera marina, could serve as refuges, mainly for Astropecten bispinosus, and also, although less significantly, for A. aranciacus. Considerations of possible refuges led to the presumption that the loss of A. bispinosus due to predation in the coexistence zone (Fig. 1, zone B) was offset mostly by immigration from the eelgrass zone (Fig. 1, zone C) and not from the shallow water region, the other possible refuge zone, (Fig. 1, zone A). The following two sets of hypothesis were postulated :

1. Refuges :
   a) There is a negative relationship between body size of Astropecten bispinosus and eelgrass density
   b) the density of juvenile Astropecten bispinosus is greater in the Zostera zone than in the coexistence zone (zone definitions, see Fig. 1)
2. Migration :
   a) there is a net migration of mainly subadult and adult Astropecten bispinosus from the Zostera zone into the

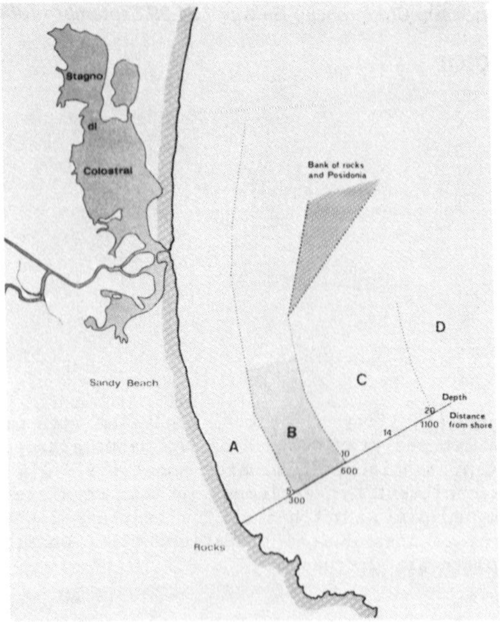

Figure 1. Research area at Costa Colostrai, Sardinia. A = shore zone, sandy bottom, about 75 ha; B = coexistence zone of A. aranciacus and A. bispinosus, sandy bottom, about 75 ha; C = Zostera zone, sandy bottom, covered with Zostera marina, about 150 ha; D = mud zone, sandy to muddy bottom with sparse patches of Posidonia sp. Distances are given in meters.

coexistence zone (i.e. immigration > emigration)

b) this net migration into the coexistence zone compensates for the loss due to predation
c) there is no significant net migration to be observed for Astropecten aranciacus.

The two sets of hypothesis were tested simultaneously by means of two projects.

MATERIAL AND METHODS

These studies were done in July and August 1980 by means of SCUBA diving. Data collection took place about 7:30 a.m. and 6 p.m. during the activity hours of the two Astropecten species (Burla et al., 1972; Ferlin & Ribi, 1978).

The refuge project was done in patches of Zostera marina where the sea stars were counted and the eelgrass density recorded in 202 samples. Sampling was done by means of a grid (5 x 5 m) that was dropped approximately every 10 m along a 50 m line, making sure that the area fully enclosed by the grid was as uniform as possible with regard to the density of Zostera marina.

At each grid position a picture was taken with a NIKONOS underwater camera mounted on a stand, covering an area of 2.5 m^2. The density of Zostera marina was estimated by comparing the negatives with a series of pictures, scoring them to 5 subjectively defined groups, group 0 having no, or hardly any Zostera marina, and group 4 having the greatest density of Zostera. The data were analyzed by a factorial model, using the procedure GLM (General Linear Model) of the Statistical Analysis System package (Ray, 1982). The sea star size (length of longest arm in mm) was used as the dependant and the density of Zostera marina as the independent variable.

Migration rates were estimated along a barrier of 100 m length at the upper edge of the Zostera zone in a depth of 10 m. The barrier separated an area of approximately 2 ha of the coexistence zone from the Zostera zone (Fig. 1). The barrier was made of transparent plastic, imbedded about 10 cm into the bottom, leaving 40 cm above the substratum. Small buoys kept the barrier upright. The barrier was marked into two sections each of 50 m length. All sea stars within 2 m on each side of the barrier were collected, measured and then displaced at least 100 m away from it to prevent their resampling. The species, the radius, the time of day, the part and the side of the barrier where each individual was found were recorded. The migration rates (m_i) were calculated for the two sections of the barrier as

$$m_{ti} = n_{si} - n_{zi}$$

and averaged to

$$\bar{m}_i = (\sum_{j=1}^{k} m_{ti}) / k$$

where m_{ti} is the migration rate per time (morning or evening) of the size class i; n_{si}, the number of individuals at the side to the coexistence zone; n_{zi}, the number of individuals on the other side of the barrier; and k, the number of observations for the size class in consideration. A migration rate $m_i < 0$ means an immigration into, and a $m_i > 0$ an emigration out of, the coexistence zone.

In both zones additional samplings of sea stars were done for size frequency distributions as well as density estimations for both species. For the estimation of the actual predation pressure, prey remains of A. aranciacus were collected by means of a kitchen sieve and analyzed with respect to remains of A. bispinosus. Calculations of the predation pressure were done using the formula

$$p = (e\ S\ Q)$$

(Schmid & Schärer, 1981) where p is the predation pressure; e, the number of eaten individuals of Astropecten bispinosus per 100 samples of prey remains of A. aranciacus; S, the number of impressions produced by A. aranciacus per day (= 1.72, estimated by Ribi & Jost, 1978) and Q, the quotient of the density of A. aranciacus divided by the density of A. bispinosus.

RESULTS

Both species of Astropecten show distinct size frequency distributions, the largest sizes in the coexistence zone, intermediate sizes along the barrier and the smallest ones in the Zostera-zone (Fig. 2 and 3). Kruskal-Wallis χ^2 - Approximations show the

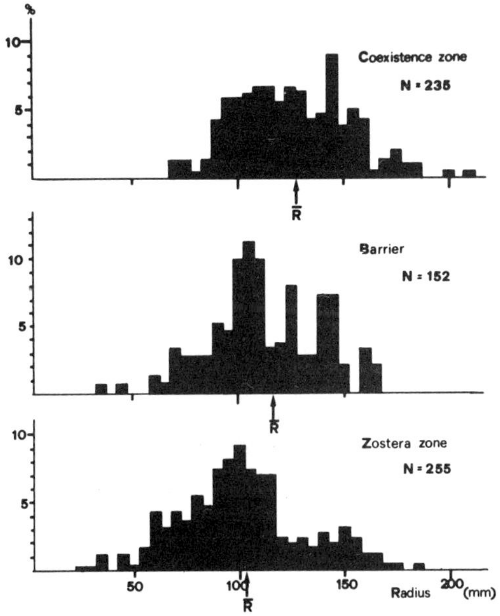

Figure 2. Size frequency distributions of Astropecten aranciacus at the three locations. $\overline{R}$ = mean radius. In the coexistence zone $\overline{R}$ = 125.3 mm; at the barrier $\overline{R}$ = 112.8 mm and in the Zostera zone $\overline{R}$ = 100.5 mm.

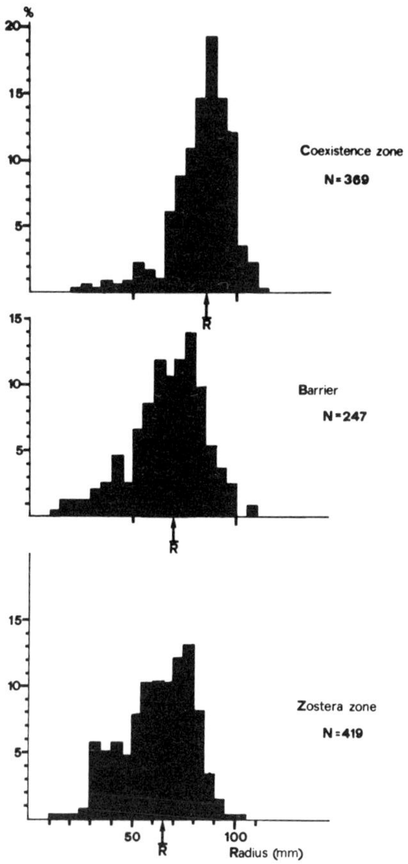

Figure 3. Size frequency distributions of Astropecten bispinosus at the three locations. $\overline{R}$ = mean radius. In the coexistence zone $\overline{R}$ = 85.5 mm; at the barrier $\overline{R}$ = 69.1 mm and in the Zostera zone $\overline{R}$ = 65.1 mm.

following results : $\chi^2 = 86.49$, $P < 0.01$ % and $\chi^2 = 298.9$, $P < 0.01$ % for A. aranciacus and A. bispinosus, respectively).

The analysis of the factorial model shows a negative relationship between sea star size of A. bispinosus and Zostera density ($P = 0.2$ %). For A. aranciacus no significant relationship of Zostera density to sea star size was observed. Figure 4 gives the relations of sea star size to Zostera densities for both species.

The negative influence of Zostera density on the distribution of the different sizes of Astropecten bispinosus clearly demonstrates that small individuals of this species tend to be in dense patches of

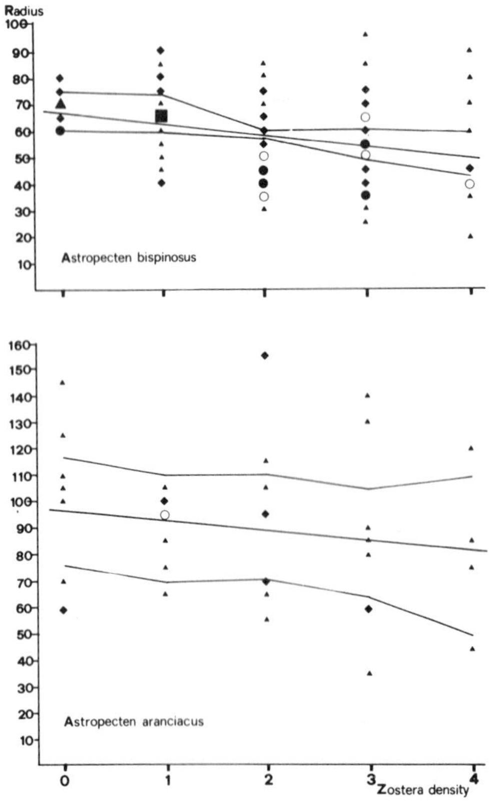

Figure 4: Size of sea stars against Zostera density. Regression lines were calculated as. y = -4.52 X + 66.94 for Astropecten bispinosus and y = -3.57 x + 96.3 for A. aranciacus (the regression being not statistically significantly different from 0). Symbols used : small solid triangle = 1 individual; rhombus = 2 individuals; circle = 3 individuals; solid circle = 4 individuals; big solid triangle = 5 individuals and the big solid square = 6 individuals.

Zostera marina, while the bigger sea stars are found in sparser eelgrass. The same tendency, though not statistically significant, can also be seen for A. aranciacus. Direct observations showed that A. bispinosus with its filigree body structure is much more motile than A. aranciacus with its more massive body structure and that larger sea stars were impeded by the plant growth.

For estimating the predation pressure, 252 samples of impressions with prey remains of Astropecten aranciacus were collected. Two of them contained remains of A. bispinosus,

Table 1: Mean migration rates of the three size classes and the total number of Astropecten aranciacus (with significance levels).

size class (mm)	$\bar{m}$	P (in %)
< 90	+ 0.22	38.5
90 - 130	+ 0.11	50.9
> 90	- 0.35	6.5
Total	+ 0.07	64.7

Table 2: Mean migration rates of four size classes and the total number of Astropecten bispinosus (with significance levels).

size class (mm)	$\bar{m}$	P (in %)
< 40	+ 0.20	56.5
40 - 65	- 0.31	3.4
70 - 90	- 0.16	29.5
> 90	- 0.07	80.1
Total	- 0.20	3.97

that is 0.64 individuals of A. bispinosus eaten per 100 prey samples of A. aranciacus. The density of the two sea stars were estimated as 68.75 and 168.75 individuals/ha for A. aranciacus and A. bispinosus, respectively. This translates to a predation pressure of 0.45 % per day of the population of A. bispinosus or in an average daily loss of 0.75 individuals/ha.

The tables 1 and 2 show the migration rates of the sea stars. There was no significant net migration of Astropecten aranciacus. There was a significant immigration in the subadult size class of A. bispinosus into the coexistence zone (radius = 40-65 mm, m_i = -0.31, signed rank test, P = 3.4 %).

The other three size classes of A. bispinosus (radius < 40 mm, 65-90 mm and > 90 mm) showed no significant net migrations.

An immigration of approximately 0.75 individuals along each 50 m part of the barrier would be expected to compensate for the daily loss of Astropecten bispinosus due to predation by A. aranciacus in the coexistence zone. The mean daily migration rate calculated for the pooled data of A. bispinosus reaches a value for m of only - 0.2.

Although this value is statistically different from zero (signed rank test, P = 4 %) it compensates only for approximately 27 % of the daily loss of A. bispinosus due to predation by A. aranciacus in the coexistence zone.

DISCUSSION

Scheibling (1980a, 1980b) found that the extreme paucity of early juvenile Oreaster reticulatus inhabiting sand bottoms may be due to a development of these sea stars in nursery areas such as beds of Thalassia testudinum where adults are rare. This sea star species is an omnivorous substratum-feeder that also consumes its own larvae and small juveniles. Occasional settlements of juvenile Astropecten occur in the coexistance zone (Schärer, Pabst, Jost, unpublished data). As the juvenile size classes are underrepresented in the coexistence zone (Schärer, in preparation) we assume both species of Astropecten to prey upon their own juveniles. The negative relationship between Zostera density and the size of A. bispinosus indicates that patches of the eelgrass serve as refuges and therefore as nursery areas for the juveniles of this species. This is further supported by the fact that the juveniles (radius < 40 mm) are more frequently found in the Zostera zone than in the coexistence zone (Figure 3).

All size classes of Astropecten bispinosus are active at approximately the same time in both zones. We therefore conclude that the refuges cannot be temporal ones. Analysis of stomach contents from the Zostera zone (Rein, unpublished) showed that all size classes of both species of Astropecten eat the same prey species. The only difference in their diet can be found in the different sizes of the prey items.

The same tendency - smaller sea stars eating smaller prey items - also occurs in the coexistence zone (Ribi, 1978). These differences are due to the anatomy of the sea stars. Thus it is highly improbable that the relationship of small Astropecten bispinosus to dense Zostera patches can be explained by different diets in the two zones. Therefore it can be concluded that patches of Zostera marina function as refuge for juvenile A. bispinosus by physically impeding A. aranciacus.

The observed migration rates of Astropecten aranciacus can be explained by oscillating migration of individual sea stars (Pabst, unpublished data) at the interface of the coexistence zone and the Zostera zone.

Only a part of the hypothesis for A. bispinosus is supported by the results of this study. As expected, mainly the subadults immigrate into the coexistence zone whereas the juveniles show a slight tendency to emigrate from the coexistence zone into the Zostera zone. No net migration of the adult size classes of A. bispinosus was found. This too can be explained by the oscillating migration of individual sea stars.

The difference between the expected and the observed daily migration rate of A. bispinosus can be explained as follows :
- the daily predation pressure of A. aranciacus on A. bispinosus (taken as a function of the availability of other prey organisms) oscillates throughout the year. From about 10 % in spring it decreases to about 1 % in summer and then to zero in autumn and winter (Schärer, in preparation)
- juvenile and subadult A. bispinosus are abundant in the coexistence zone during spring, up to about 50 % of the individuals having arm lengths smaller than 65 mm (Schärer, in preparation).
- seasonal migration towards the shore has been reported by Pabst & Vicentini (1982) for A. jonstoni in spring and by Nojima (1983) for A. latespinosus from spring to autumn.
- the shore zone must be considered as a possible refuge zone too.

We conclude that
- There is no absolute need for a simultaneous compensation of the losses of A. bispinosus in the coexistence zone by immigration.
- the shore zone should play an important role in compensating the losses in the late summer, autumn and winter
- most of the immigration of A. bispinosus (compensating for the annual losses due to predation by A. aranciacus) occurs in late winter and early spring.

REFERENCES

Birkeland, C. Chia, F.-S. & R.R. Strathmann 1971. Development, substratum selection, delay of metamorphosis and growth in the sea star, Mediaster aequalis Stimpson. Biol. Bull. 141:99-108

Burla, H., V. Ferlin, B. Pabst & G. Ribi 1972. Notes on the ecology of Astropecten aranciacus. Mar. Biol. 14:235-241

Connell, J. H. 1970. A predator-prey system in the marine intertidal region. I. Balanus gladula and several predatory species of Thais. Ecol. Monogr. 40:49-78

Ribi, G., H. Burla & P. Ochsner 1976. Beobachtungen über Vorkommen, Abundanz und Körpergrösse von mediterranen Arten der Gattung Astropecten. Helgoländer wiss. Meeresunters., 28:304-317

Harmelin, J. G. & C. Duval 1983. Localisation et dissémination des jeunes de l'oursin Sphaerechinus granularis (Lamarck) en Méditerranée. Rapp. Comm. int. Mer Médit. 28:267-269

Nojima, S. 1983. Ecological studies of the sea star, Astropecten latespinosus MEISSNER. V. Pattern of spatial distribution and seasonal migration, with special reference to spawning aggregation. Publ. Amakusa Mar. Biol. Lab 7:1-16

Pabst, B. & H. Vicentini 1978. Dislocation experiments in the migrating sea star Astropecten jonstoni. Mar. Biol. 48:271-278

Ray, A.A. (ed.) 1982. SAS User's Guide : Statistics, 1982 Edition NC: SAS Institute Inc., 1982, pp. 139. ff.

Schärer, R. Beurteilung von Räubertum und Konkurrenz als regulierende Faktoren eines Ökosystems. PhD Thesis at the University of Zürich, in preparation.

Scheibling, R.E. 1980a. Abundance, spatial distribution and size structure of populations of Oreaster reticulatus (Echinodermata : Asteroidea) on sand bottoms. Mar. Biol. 57:95-106

---- 1980b. Abundance, spatial distribution and size structure of populations of Oreaster reticulatus (Echinodermata : Asteroidea) on sand bottoms. Mar. Biol. 57:107-120

Schmid, P. & R. Schärer, 1981. The predator-prey interaction between two competing sea star species. Mar. Ecol. (P.S.Z.N.I.) 2:207-214

Acknoledgments: This work was partially supported by the Swiss National Science Foundation (Grant No. 3.253-0.77).

Proceedings of the Fifth International Echinoderm Conference / Galway / 24-29 September 1984

Remarks on *Odontaster penicillatus* (Philippi, 1870) (Echinodermata: Asteroidea)

B.LEELING
Zoologisches Institut und Museum, Universität Hamburg, Germany

During the voyage of the fisheries-research vessel "Walther Herwig" of the Budesforschungsanstalt für Fischerei, Hamburg, in the year 1971, a close-meshed station grid was established between Uruguay and Cape Horn. Among others, 25 specimens identified as Odontaster penicillatus were collected in this area. Their taxonomic status and geographical distribution are described and discussed.

1 TAXONOMIC STATUS

The genus Odontaster, established by Verrill in 1880 for the type O. hispidus, is based upon the presence of one large, strong tooth on each jaw, an odd marginal plate, and more or less paxilliform abactinal plates. The status of this genus underwent various changes and expansions, and many synonyms are known from the literature. Within the limits of this report it is not possible to go into full details; for further information see Leipoldt (1895) and Fisher (1940).

The species O. penicillatus was erected by Philippi (1870)(syn. Goniodiscus penicillatus), the main criterion being the shape of the abactinal paxillae. Due to the absence of figures, Fisher (1940) made the questionable suggestion that Sladen's (1889) Gnathaster pilulatus be taken as the type form. Anyway, the main criteria used to distinguish this species from the other members of the genus remained, i.e. the shape of the paxillae, which are tubercle-like in O. penicillatus, and the less dense spinulation of the actinal areas, as compareed with O. meridionalis.

Fisher (1940) proposed to unite the two species, formerly described as O. penicillatus (Philippi) and O. grayi (Bell), under the name of the first one, and to regard them as two formae. In order to distinguish between these two forms, he listed a large number of criteria and, by taking into account certain overlapping effects, he also defined a number of intergrades. Bernasconi (1962) admitted the existence of intergrades, but reduced the distinctive feature between forma penicillatus and forma grayi to different R:r ratios, the shape of f. penicillatus being stellate, ($R > 2r$), and the appearance of f. grayi more or less pentagonal ($R < 2r$).

2 DISTRIBUTION

Thanks to the extensive work of several expeditions over the past hundred years, the distribution of O. penicillatus is quite well known.

Specimens have been found on both sides of South America, around Cape Horn, the Falkland Islands, and on Burdwood Bank. Bernasconi (1962) mentions Puerto Montt, Chile (41° 30' S) as the northern boundary in the Pacific, neglecting Madsen's (1956) report at "La Serena" (about 30° S). In the Atlantic, the northernmost record is from the Buenos Aires province (35°S)(Bernasconi 1962). The southernmost locality is that of Perrier (1891), who mentioned Orange Bay (55°30'S, 68°W). The high abundance of specimens around the Falkland Islands may be attributed to the extensive collections of the R.S.S. "William Scoresby", especially on the Falkland Plateau. On the other hand, only few samples have been recorded from the Patagonian Shelf, as restricted to the coastal area between 40°S and 53°S.

Another important aspect regarding the distribution of this species is the question as to whether the formae penicillatus and grayi might be separated geographically. Ludwig (1905) proposed a differentiation which he based on their bathymetrical distribution, penicillatus being a deeper water species (15-183m), while grayi was found only in depths of 0-35m. However, Fisher (1940), working with a great number of samples, rejected this theory. Ludwig (1905), as well as Bernasconi (1962, 1966), pointed out that the only differences in regional distributions were found in the Cape Horn region, where forma grayi clearly dominated. Additionally, Ludwig (1905) mentions that this form had never been found on the West Coast of Patagonia, but in 1956 Madsen reported the occurrence of one specimen of f.grayi in the "Golfo de Ancud" (about 42°S).

3 RESULTS AND DISCUSSION

The material for the present investigation was collected in 1971 by the German research vessel "Walter Herwig" along the Argentinian Shelf between Uruguay and Cape Horn. A 140 foot ground trawl was dredged for half an hour (0-1,000m depth) for each sample.

All of the 25 specimens collected in depths ranging from 50-105m were clearly identified as Odontaster penicillatus (according to Salden's description, Gnathaster pilulatus = O. penicillatus). Based on the different R:r ratios (Bernasconi 1962), the two formae penicillatus and grayi could be distinguished. Only 1 integrade was found.

It is worth mentioning that a large number of variations in certain features were observed, which Fisher (1940) had used to define different integrades. It was found, for example, that the number of marginal plates was generally less in f. grayi than in f. penicillatus, but since overlapping effects were registered, this character had to be considered as of minor importance.

Another interesting fact was that many of the specimens are pedicillariae, which Bell (1893) used as a criterion even at the species level. Some did not have any at all, while others had only isolated ones in the actino-lateral areas. Sometimes, they occurred on the abactinal surface - either at the base of the paxillae or in the middle of the spinelets on the top. In other cases, pedicellariae were scattered all over the body, including both the supero- and infero-marginalia, something which had not been recorded before. While Koehler (1920) mentions their presence only on the first three ranges of plates adjoining the adambulacral plates, with gaps between them, some of the specimens had their pedicellariae regularly arranged in these rows, and sometimes even in the central actino-lateral area. The occurrence of pedicellariae could not be related to size, shape, regional distribution or bathymetric range. Therefore, it was concluded that probably the most reliable criterion for separating the two formae is the overall body shape, i.e. the R:r ratio. Still, the question remains as to whether the presence or absence of pedicellariae might have taxonomic or even ecological value. Further investigations have to be made in order to solve this problem. Table 1 lists the station numbers of the "Walter Herwig" material and Table 2 presents the R:r ratios.

Our findings fill the gap between older recordings of Odontaster penicillatus on the Atlantic Coast of South America. While it seems that penicillatus is the northern form,and that grayi is clearly restricted to the southern part of Patagonia, it is still not possible to make a definitive statement regarding their geographical distribution. It remains to be seen if they can be separated on the basis of reproduction (temporal races) or behaviour (ethological races).

Table 1: Stations of the "Walther Herwig" material.

Stat.	Latitude S.	Longitude W	Depth (m)	Reg.No.
261	54°00'	67°00'	50	14a
278	52°00'	67°00'	100	16a
283	51°00'	67°00'	105	24
301	49°00'	67°00'	55	32,33,34
303	47°30'	65°00'	78	40
315	45°05'	65°00'	92	1-5
316	44°00'	64°30'	70	6-12
317	44°03'	64°16'	75	13-16 17a
322	43°05'	61°21'	90	50

Table 2: R:r ratio of Odontaster penicillatus and O. grayi

Reg. No.	R	r	R:r	No. Marg. Pl.
		forma grayi		
14a	53	34	1.6:1	17
16a	42	24	1.7:1	16
24	48	30	1.6:1	16
32	52	34	1.5:1	15
33	52	32	1.6:1	16
34	48	30	1.6:1	14
		forma penicillatus		
40	43	21	2:1	16
01	41	20	2:1	19
02	38	18	2.1:1	20
03	36	18	2:1	17
04	30	15	2:1	13
05	28	14	2:1	15
06	36	18	2:1	16
07	36	15	2.4:1	17
08	35	16	2.2:1	17
09	36	17	2.1:1	14
10	33	15	2.2:1	20
11	32	15	2.1:1	17
12	33	15	2.2:1	19
13	43	20	2.1:1	19
14	36	18	2.1	17
15	35	15	2.3:1	20
16	34	17	2:1	16
17a	32	15	2.1:1	18
		Integrade		
50	48	25	1.9:1	20

REFERENCES

Bell, R.J. 1881. Account of the zoological collections made during the survey of H.M.S. "Alert" in the Straits of Magellan and on the Coast of Patagonia. Echinodermata. Proc.Zool.Soc. London: 87-101.

Bell, F.J. 1893. On Odontaster and the allied or synonymous Genera of Asteroid Echinoderms. Proc.Zool. Soc.London: 259-262.

Bernasconi, I. 1962. Asteroideos Argentinos. III. Familia Odontasteridae. Rev.Mus.Arg.Cienc.Nat. Buenos Aires 8(3): 27-51.

Bernasconi, I. 1966. Los Equinoideos y Asteroideos colectados por el buque oceanografico R/V "Vema" frente a last costas Argentinas, Uruguayas y sur de Chile. Rev.Mus. Arg.Cienc.Nat.Buenos Aires 9(7): 147-175.

Bernasconi, I. 1973. Los Equinodermos colectados por el "Walther Herwig" en el Atlantico Sudoeste. Rev.Mus.Arg.Cienc.Nat. Buenos Aires 3(3): 287-334.

Fisher, W.K. 1940. Asteroidea. Discovery Reports 20: 69-306.

Koehler, R. 1908. Asteries, Ophiures et Echinides de l'Expedition antarctique Nationale Ecossaise. Trans.Roy.Soc. Edinburgh 46(3): 529-649.

Koehler, R. 1920. Echinodermata Asteroidea. Australasian Antarctic Expedition 1911-14. Ser.C. 8(1): 308pp.

Koehler, R. 1923. Asteries et Ophiures recuellies par l'Expedition Antarctique Suedoise. Further Zool.Res.Swed.Ant.Exp. 1901-1903 1(1): 145pp.

Leipoldt, F. 1895. Asteroidea der "Vettor-Pisani" Expedition (1882-1885). Z.wiss.Zool.59: 545-654.

Ludwig, H. 1956. Asterien und Ophiuren der schwedischen Expedition nach den Magalhaensländern 1895-1897. Z.wiss.Zool.82:39-79.

Madsen, F.J. 1956. Asteroidea. Reports of the Lund University Chile Expedition 1948-49. Lunds Universitets Arsskrift. N.F.Avd. 2. 52.

Madsen, F.J. 1965. Asteroidea. Scientific results of the "Brategg" Expedition, 1947-48.No.5. Bergen

Meissner, M. 1896. Die von Herrn Dr. L. Plate aus Chile und Feuerland heimgebrachten Seesterne. Arch.Naturgesch:91-108.

Meissner, M. 1904. Asteroideen. Ergebnisse der Hamburger Magalhaensischen Sammelreise 1892/93. Hamburg: 27pp.

Perrier, E. 1891. Echinodermes de la mission scientifique du Cap Horn. 1. Stellerides. Miss.Sci. Cap Horn, Zool. 6: 198pp.

Philippi, R.A. 1870. Neue Seesterne aus Chile. Arch.Naturgesch. 36: 268-275.

Sladen, P.W. 1889. Report on the Asteroidea collected by H.M.S. "Challenger" during the years 1873-76. No. 30. London.

Verrill, A.E. 1880. Notice of the remarkable Marine Fauna occupying the outer banks off the southern coast of New England. Amer.Journ. Sci. 20: 390-403.

Verrill, A.E. 1899. Revision of certain genera and species of starfishes with descriptions of new forms. Trans.Conn.Acad. 10: 145-234.

Proceedings of the Fifth International Echinoderm Conference / Galway / 24-29 September 1984

Size selectivity of prey by *Luidia clathrata* (Say) (Echinodermata: Asteroidea): Effect of nutritive condition and age

JAMES B.McCLINTOCK & JOHN M.LAWRENCE
Department of Biology, University of South Florida, Tampa, USA

ABSTRACT: In the laboratory, adult Luidia clathrata presented with different sizes of buried prey choose the smaller sizes. This selective behavior is probably related to the ease of manipulation and may facilitate maximal intake per unit time. Nutritive condition did not affect size-selectivity as both starved and fed adult L. clathrata choose small prey models. However, starved individuals ingested greater numbers of prey than fed individuals. Small (R<25 mm) L. clathrata showed no significant preference between prey models of different sizes. Examination of stomach contents indicate that populations of Luidia clathrata in the Gulf of Mexico do not ingest different sizes of the sand dollar Mellita quinquiesperforata in direct proportion to their relative abundance, but prefer the smaller individuals.

INTRODUCTION

Feeding behaviors of animals have been increasingly interpreted within the context of optimization theory (for reviews see: Emlen, 1966; Schoener, 1974; Pyke et al., 1977; Hughes, 1980). Selection for efficiency in feeding, which would lead to optimization may act to increase the ability of an animal to choose the best (with respect to fitness) foods (Emlen, 1973). Several studies have found a positive correlation between selection for foods (quantitatively and qualitatively) and somatic and gonadal growth of echinoids (Lilly, 1975; Vadas, 1977). One method to increase maximal energy intake per unit effort is to choose certain sizes of food (Pulliam, 1974; Krebs and Davies, 1978). Size of prey may influence prey handling time (Doi, 1976; Davies, 1977; Krebs et al., 1977) and energy level per individual prey item (Elner and Hughes, 1978; Griffiths and Seiderer, 1982). It is necessary to ascertain the relative effect of handling time and energy level on energy intake.

Few studies have considered the effect of nutritional condition on feeding behaviors in marine invertebrates. Hunger may influence selectivity for different types or sizes of prey, movement patterns, and/or numbers of prey ingested per unit time, but this has not been evaluated. Changes in these feeding behaviors may lead to an increase in somatic and gonadal growth.

Early growth stages in the life cycles of marine invertebrates may be particularily susceptible to predation and changing environmental variables (Kinne, 1970; Birkeland et al., 1971). Consequently, selection for efficiency in feeding behaviors may be important in early developmental stages to insure rapid somatic growth and attainment of reproductive capacity. The biology and ecology of juvenile asteroids is an important, yet neglected, aspect of asteroid life cycles (Sloan, 1980).

The objectives of this study were: 1) to examine the effect of nutritive condition on size-selection of buried foods in adult Luidia clathrata, 2) to examine the effect of age on size-selection for prey and, 3) to extend these laboratory studies to the field with an investigation of size-selection by adult L. clathrata for infaunal individuals of the sand dollar Mellita quinquiesperforata.

MATERIALS AND METHODS

Laboratory investigations

Size-selection for buried prey was investigated in both starved and fed individuals. Specimens of adult Luidia clathrata (mean arm length = 71.6 ± 3.9 mm; n = 40) were collected in May, 1982, from Old Tampa Bay, Florida (approximately 27°55' N, 83°35' W) in 2-3 m water. Starfish were maintained in 40-liter aquariums with artificial seawater (25°C and 20 ppt). Twenty individuals were fed a maintenance diet (1.88 g wet weight/starfish/week) of the bivalve Donax variabilis (Diehl and Lawrence, 1979) for 4-weeks to constitute a fed group, while another twenty individuals received no food for 4-weeks to constitute a starved group.

Selectivity for different sizes of buried prey was investigated using a modification of the methods of Collins (1974) and McClintock and Lawrence (1981). Three sizes of flat, circular bivalve models containing homogenized tissues of Donax variabilis and plaster of Paris (1:15 wt/wt) were prepared. Models measured 0.5, 1.0 and 2.0 cm in diameter and were 0.2 cm deep. Prey models were placed randomly into glass Petri dishes (10 cm diameter) and covered with a 0.5 cm layer of fine sediment. In one experiment, fed and starved individual L. clathrata were offered equal volumes of prey models (5 large, 10 medium, and 20 small). In another experiment, fed and starved individual L. clathrata were simultaneously offered five prey models from each size class. In both experiments individual starfish were permitted to feed for 12 min, and the number and sizes of prey models ingested were recorded.

Juvenile (R<25 mm) Luidia clathrata (mean arm length = 20.2 ± 4.7 mm; n = 45) were collected by bottom trawl (3 m net with 2.5 cm mesh) in May, 1982, in southern Hillsborough Bay in 5-8 m water. Starfish were maintained in 40-liter aquariums with artificial seawater (25°C and 20 ppt). Three sizes of agar food models were prepared using a modification of the methods of Klinger (1982). The agar prey models contained a 1:15 (volume/volume) ratio of Donax variabilis and agar. Food models were square and measured 1 X 1, 2 X 2, or 5 X 5 mm in width and length respectively, and all were 2 mm in depth. Models were placed randomly into glass Petri dishes (6 cm diameter). In one experiment, ten individuals were each offered different numbers of prey models (4 large, 8 medium, and 50 small). In another experiment ten individuals were each simultaneously offered ten prey models from each size class. In both experiments individual starfish were permitted to feed 30 min and the number and sizes of prey ingested were recorded. In both adult and juvenile experiments, a Student's $\underline{t}$ test was used to test for significance (Zar, 1974).

Field investigations

In July 1982, a population of Luidia clathrata was located southeast of Anclote Key, Florida (approximately 28 12' N, 82° 50' W) in 5 m water feeding almost exclusively on the sand dollar Mellita quinquiesperforata. Prey size-selectivity was investigated by collecting twenty individuals of L. clathrata (mean arm length = 95.4 ± 3.7 mm) and measuring the sizes of all intact sand dollars removed from their stomachs. Size-distibutions of M. quinquiesperforata were ascertained by measuring all individuals found within three 0.5 m quadrats placed onto the seafloor. A Kolmogorov-Smirnov goodness-of-fit test was used to test for significance (Zar, 1974).

RESULTS

Laboratory investigations

Fed and starved adult Luidia clathrata showed no significant preferences between small and medium-sized buried prey models (Figures 1 and 2). However, there was a significant (P<0.05) preference for small and medium prey models over large models. Fed and starved starfish offered equal numbers of prey models did not ingest any large prey models, while starfish offered equal amounts of prey models did ingest some of the large prey models. Starved individuals ingested significantly (P<0.05) greater numbers of small and medium sized models than fed individuals over the 12 min feeding period.

Luidia clathrata offered the three sizes of buried prey models depressed the oral-disc into the substratum and brought the prey models into the gut using the tube-feet surrounding the oral-opening. Large prey models were occasionally selected but were typically abandoned

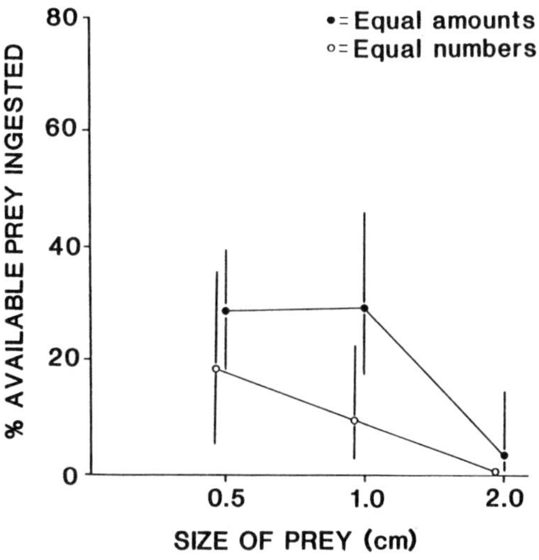

Figure 1. Percent of available prey ingested by fed Luidia clathrata (mean R = 71.6 ± 3.9 mm) simultaneously offered equal amounts (solid circles) and equal numbers (open circles) of three sizes of prey (0.5, 1.0, and 2.0 cm). Means ± 95% confidence limits are given (n = 10). Statistical tests are based on the angular transformation.

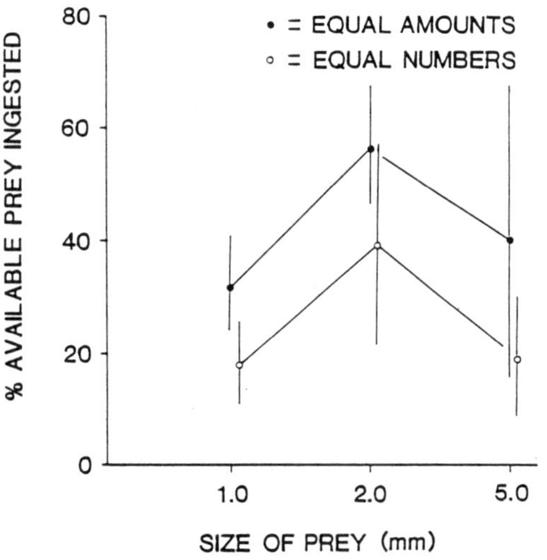

Figure 3. Percent of available prey ingested by juvenile Luidia clathrata (mean R = 20.2 ± 4.7 mm) simultaneously offered different (solid circles) and equal (open circles) numbers of three sizes of prey (1.0, 2.0, and 5.0 mm). Means ± 95% confidence limits are given (n = 10). Statistical tests are based on the angular transformation.

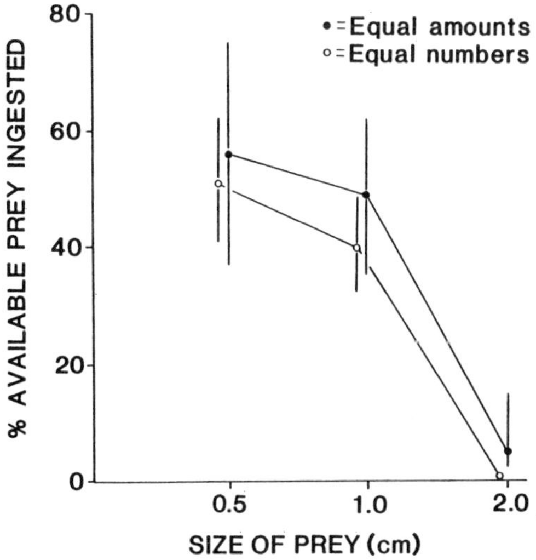

Figure 2. Percent of available prey ingested by starved Luidia clathrata (mean R = 70.4 ± 3.5 mm) simultaneously offered equal amounts (solid circles) and equal numbers (open circles), of three sizes of prey (0.5, 1.0, and 2.0 cm). Means ± 95% confidence limits are given (n = 10). Statistical tests are based on the angular transformation.

following manipulation. In some instances, following the excavation of prey models, starfish assumed a "humped feeding posture" more typical of epifaunal feeding behavior.

Juvenile Luidia clathrata presented with equal and different numbers of agar prey-models showed no significant ($P>0.05$) preference between the three sizes of prey models (Figure 3). However, availability was high and juveniles consistently ingested greater numbers and amounts of medium-sized models. Numbers of models ingested by juveniles were high. In one instance a juvenile (mean radius = 24 mm) ingested 21 small, 16 medium, and 4 large prey models; a total of 512 mm^3. Juveniles all showed humped feeding behavior with non-buried prey, using the tube-feet of the arms to simultaneously move prey models to the oral-opening. Although prey models were smaller, and thus not directly comparable, juvenile L. clathrata ingested greater numbers of prey models per unit time than either fed or starved adult L. clathrata.

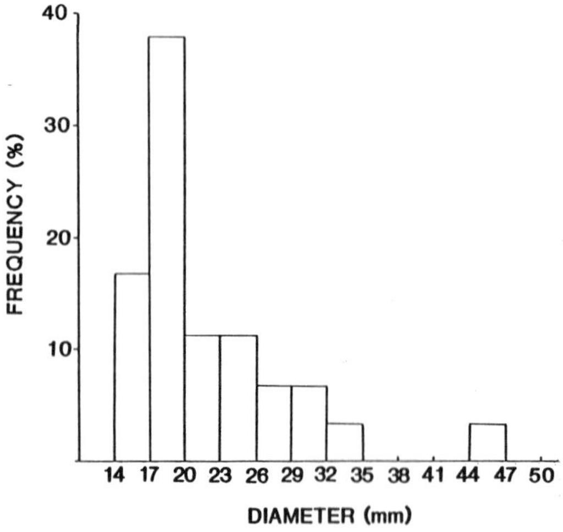

Figure 4. Size-frequency distribution of Mellita quinquiesperforata removed from the guts of twenty Luidia clathrata (mean R = 95.4 ± 3.7 mm) near Anclote Island, Florida (n = 23).

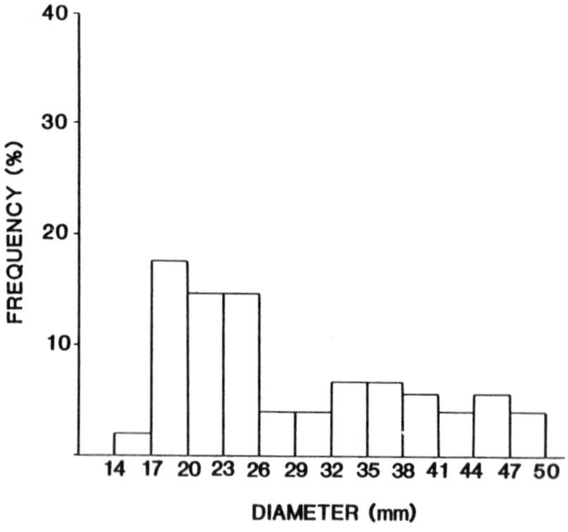

Figure 5. Size-frequency distribution of Mellita quinquiesperforata found co-occuring with Luidia clathrata near Anclote Island, Florida (n = 96).

Field investigations

Luidia clathrata concentrates its feeding effort on the smaller individuals of Mellita quinquiesperforata. Gut contents indicated that the size distribution of ingested M. quinquiesperforata was unimodal and positively skewed with a mean diameter of 22.6 mm (Figure 4). Ninety percent of those L. clathrata collected had sand dollars within their guts. Small sand dollars (14-20 mm diameter) comprised 70 percent of the total number ingested. Densities of M. quinquiesperforata in the study area were 89 ± 4.4 individuals/m^2. There was no significant ($P>0.05$) difference between size distributions of sand dollars from the three 0.5 m quadrats. Pooled size-frequencies of sand dollars within the range of sizes ingested (14-50 mm) by L. clathrata are presented in Figure 5. There were significantly ($P<0.05$) fewer numbers of small (14-20 mm diameter) and greater numbers of large (32-50 mm diameter) sand dollars in field samples than in gut contents of L. clathrata.

DISCUSSION

Adult Luidia clathrata select small prey. Laboratory investigations indicate that medium-sized and small prey models are preferred over larger prey models. In the field, L. clathrata chose the smaller sized specimens of Mellita quinquiesperforata.

McClintock and Lawrence (1981) found that in the laboratory L. clathrata chose small prey models; however these prey models were not buried. The results of the present study indicate that burial of prey does not inhibit the ability of this starfish to "discriminate" between different sizes of prey. Both laboratory and field analyses indicate that L. clathrata excavates prey from the substratum ingesting only the smaller individuals. Several asteroids show similar digging behavior for buried prey. Asterias amurensis and Distolasterias nipon dig for buried bivalves, depressing the oral-disc down into the substratum (Arima et al., 1972). Pisaster brevispinus (R = 10 cm) can extend its tube feet a distance of 8 cm to capture buried prey (VanVeldhuizen and Phillips, 1978; Sloan and Robinson, 1983).

Other asteroids show selectivity for small prey. Luidia foliolata prefers the smaller (< 30 mm) size classes of the cockle Clinocardium nuttallii (Sloan and Robinson, 1983). Astropecten latespinosus and A. scoparius select the smaller individuals of the marine snail Umbonium moniliferum (Doi, 1976). Based on prey-handling times, Doi showed that this behavior results in maximal food intake during limited feeding time. Luidia sarsi selects the smaller specimens among its prey animals (Fenchel, 1965). Fenchel suggested that, by preferring small specimens, L. sarsi may fill the stomach more efficiently than when eating large prey. Astropecten forbesi selects the smaller sizes of the clam Mercenaria mercenaria (Doering, 1980). Doering attributed "preference" for small clams to the inability of small clams to escape via a burrowing response. Christensen (1970) offered equal numbers of small and large bivalves (Spisula subtruncata) to A. irregularis and found a distinct preference for the smaller individuals. He suggested that small bivalves can be ingested more easily, and that large bivalves may escape predation more easily by flight. Tethyaster vestitus ingest the smaller and more easily caught individuals of the asteroid Astropecten riensis (Penchaszadeh and Molinet, 1982). Mellita quinquiesperforata do not show an escape response to Luidia clathrata (McClintock, pers. obs.). Both adults and juvenile sand dollars are within the digging range (1-2 cm) of L. clathrata. Consequently, it is unlikely that preference for smaller sand dollars is the result of size-related differences in burial depth or escape behavior of prey. Luidia clathrata may select small prey due to ease of manipulation, maximal prey intake as related to stomach space, and/or maximal prey intake related to limited feeding time.

Luidia clathrata can ingest relatively large prey. In several instances L. clathrata had Mellita quinquiesperforata measuring 40-50 mm in diameter in their guts. Some asteroids ingest large prey. Odontaster validus attacks and feeds on Acondontaster conspicuus, an asteroid several times its diameter (Dayton et al., 1974). Pisaster ochraceus does not consume small Mytilus edulis in proportion to their abundance, feeding on the larger individuals (Paine, 1976). Similarily, Mauzey (1966) found that P. ochraceus chose large chitons most frequently, while the small Balanus balanoides and M. edulis are of secondary importance. If handling times are relatively uniform between different sizes of prey, then choice of large prey may permit a higher caloric return per unit effort. Asteroids which feed by protruding the cardiac stomach onto macrofaunal prey (low handling times generally) might be expected to select large prey. Luidia clathrata normally ingests its prey (with the exception of extraoral feeding on detritus, McClintock et al., 1983) and selection for smaller prey may be related to mechanical limitations.

Nutritional condition did not influence size selection for prey by adult Luidia clathrata. Holling (1965) and Ivlev (1961) suggested that animals should be most selective when satiated and least selective when starved. Both starved and fed L. clathrata were size-selective, choosing significantly greater numbers of medium and smaller than large prey models. However, starved individuals showed significantly elevated rates of prey ingestion over fed individuals. Numbers of small and medium-sized prey models ingested per unit time approximately doubled following the four-week period. Increased rates of prey ingestion by starved L. clathrata provides a higher energetic return per unit handling time. McClintock and Lawrence (1984) found that starvation in L. clathrata decreases teloreceptive-response times and may influence selectivity for different types of food.

In contrast to adult individuals, juvenile Luidia clathrata showed no selectivity for different sizes of prey models. Juveniles spent significantly less time manipulating large prey items than adults. A decrease in handling time would make large prey more attractive to juvenile starfish and could account for the difference. Juveniles showed similar feeding behavior as adults, generally humping up over non-buried prey and using the tube-feet to move prey to the oral-opening. Although prey models were not buried, laboratory observations indicate that juveniles dig for buried prey, depressing the oral-disc into the substratum. Rates of prey ingestion by juveniles were higher than both starved and fed groups of adult starfish. Only a few studies have investigated feeding behaviors of juvenile starfish. Birkeland et al., (1971) found that juvenile Mediaster aequalis evert their stomachs and insert them over sponges and into individual zooecia of ectoproct Bugula spp. They suggested that, because small asteroids must have a difficult time capturing food, sponges and ectoprocts would seem a suitable prey. Other juvenile M. aequalis fed on a microscopic growth of detritus coating the tubes of the polychaete Phyllochaetopterus prolifica. Small bivalves regularily fall prey to juvenile Crossaster papposus (Birkeland et al., 1971). Juvenile Stichaster australis begin feeding by extruding the cardiac stomach onto the coralline alga Mesophyllum insigna and switch to a carnivorous diet of the bivalve Perna canaliculus after reaching a minimal size (Barker, 1979). Although capable of extraoral feeding (McClintock et al., 1983), juvenile L. clathrata ingested bivalve prey models. Juvenile L. clathrata less than 16 mm R were not available. They may feed on detrital material via extrusion of the cardiac stomach. Juvenile L. clathrata appear to have the manipulative capabilities which would allow them to prey upon small bivalves, ostracods, cumaceans, and other suitable microfauna.

ACKNOWLEDGEMENTS

We thank Brandt Henningson for collecting the juvenile starfish. This research was supported in part by a grant from Sigma Xi to the first author.

REFERENCES

Arima, K., S. Hamaya and Y. Miyakana. 1972. Feeding behavior of the starfish to bivalves. Scient. Rep. Hokkaido Fish. expl. Sta. 14: 63-69.

Barker, M.F. 1979. Breeding and recruitment in a population of the New Zealand starfish Stichaster australis. J. Exp. Mar. Biol. Ecol. 41: 195-211.

Birkeland, C.E., F.S. Chia and R.R. Strathmann. 1971. Development, substratum selection, delay of metamorphosis and growth in the seastar Mediaster aequalis Stimpson. Biol. Bull. 141: 99-108.

Collins, A.R.S. 1974. Biochemical investigations of two responses involved in the feeding behavior of Acanthaster planci (L.). I. Assay methods and preliminary results. J. Exp. Mar. Biol. Ecol. 15: 173-184.

Christensen, A.M. 1970. Feeding biology of the sea-star Astropecten irregularis. Ophelia 8: 1-134.

Davies, N.B. 1977. Prey selection and social behavior in wagtail (Aves: Motocillidae). J. Anim. Ecol. 46: 37-57.

Dayton, P.K., G.A. Robilliard, R.T. Paine, and L.B. Dayton. 1974. Biological accommodation in the benthic community of McMurdo Sound, Antarctica. Ecol. Monogr. 44: 105-128.

Diehl, W.J. and J.M. Lawrence. 1979. Effect of nutrition on the excretion rate of soluble nitrogen products of Luidia clathrata (Say) (Echinodermata: Asteroidea). Comp. Biochem. Physiol. 62A: 801-806.

Doering, P.H. 1980. Defenses of the hard clam Mercenaria mercenaria against predation of the sea star Asterias forbesi. Ph.D. Dissert. Univ. of Rhode Island. 138 pp.

Doi, T. 1976. Some aspects of the feeding ecology of the sea stars genus Astropecten. Publ. Amakusa mar. biol. Lab. 4: 103-116.

Elner, R.W. and R.N. Hughes. 1978. Energy maximization in the diet of the shore crab, Carcinus maenas (L.). J. Anim. Ecol. 47: 103-116.

Emlen, J.M. 1966. The role of time and energy in food preference. Am. Nat. 100: 611-617.

Emlen, J.M. 1973. Ecology: an evolutionary approach. Addison-Wesley Publ. Co. Reading Mass. 185 pp.

Fenchel, T. 1965. Feeding biology of the sea-star Luidia sarsi Duben and Koren. Ophelia 2: 223-236.

Griffiths, C.L. and J.L. Seiderer. 1980. Rock-lobsters and mussels-limitations and preferences in a predator-prey interaction. J. Exp. Mar. Biol. Ecol. 44: 95-109.

Holling, C.S. 1965. The response of predators to prey density and its role in mimicry and population regulation. Mem. Ent. Soc. Can. 45: 1-60.

Hughes, R.N. 1980. Optimal foraging in the marine context. Mar. Biol. Ann. Rev. 18: 423-481.

Ivlev, V.W. 1961. Experimental ecology of the feeding of fishes. Yale Univ. Press. New Haven, Connecticut. 302 pp.

Kinne, O. 1970. Marine Ecology. Wiley-Interscience, London. 681 pp.

Klinger, T.S. 1982. Feeding rates of Lytechinus variegatus Lamarck (Echinodermata: Echinoidea) on differing physiognomies of an artificial food of uniform composition. In: J.M. Lawrence (ed.), Echinoderms: Proceedings of the International Conference, Tampa Bay. pp. 29-32. A.A. Balkema, Rotterdam.

Krebs, J.R., J.T. Ericksen, M.I. Weber and E.L. Charnov. 1977. Optimal prey selection in the Great Tit (Parus major). Anim. Behav. 25: 30-38.

Krebs, J.R. and N.B. Davies. 1978. Behavioral Ecology. Sinauer Associates Inc. Sunderland, Massachusetts. 493 pp.

Lilly, G. 1975. The influence of diet on growth of the tropical sea urchin, Tripneustes ventricosus (Lamarck). Ph.D. Dissert. Univ. of British Columbia. 216 pp.

Mauzey, K.P. 1966. Feeding behavior and reproductive cycles in Pisaster ochraceus. Biol. Bull. 131: 127-144.
McClintock, J.B. and J.M. Lawrence. 1981. An optimization study on the feeding behavior of Luidia clathrata (Say) (Echinodermata: Asteroidea). Mar. Behav. Physiol. 7: 263-275.
McClintock, J.B., T.S. Klinger and J.M. Lawrence. 1983. Extraoral feeding in Luidia clathrata (Say) (Echinodermata: Asteroidea). Bull. Mar. Sci. 33: 171-172.
McClintock, J.B. and J.M. Lawrence. 1984. Ingestive conditioning in Luidia clathrata (Say) (Echinodermata: Asteroidea): effect of nutritional condition on selectivity, teloreception, and rates of ingestion. Mar. Behav. Physiol. 10: 167-181.
Paine, R.T. 1976. Size-limited predation: an observational and experimental approach with the Mytilus-Pisaster interaction. Ecology 57: 858-873.
Penchaszadeh, P.E. and R. Molinet. 1983. Diet of the sea-star Tethyaster vestitus Say (Estropectinidae) in the Gulfo Triste (Venezuela). Bull. Mar. Sci. 33: 187-191.
Pulliam, H.R. 1974. On the theory of optimal diets. Am. Nat. 108: 59-75.
Pyke, G., H.E. Pulliam and E. Charnov. 1977. Optimal foraging: a selective review of theories and tests. Quart. Rev. Biol. 52: 137-154.
Sloan, N.A. 1980. Aspects of feeding biology of asteroids. Oceanogr. Mar. Biol. Ann. Rev. 18: 57-124.
Sloan, N.A. and S.M.C. Robinson. 1983. Winter feeding by asteroids on a subtidal sandbed in British Columbia. Ophelia 22: 125-140.
Vadas, R.L. 1977. Preferential feeding: an optimization strategy in sea urchins. Ecol. Monogr. 47: 337-371.
VanVeldhuizen, H.D. and D.W. Phillips. 1978. Prey capture by Pisaster brevispinus (Asteroidea: Echinodermata) on soft substrata. Mar. Biol. 48: 89-97.
Zar, J.H. 1974. Biostatistical Analysis. Prentice-Hall Inc. Englewood Cliffs, New Jersey. 620 pp.

Proceedings of the Fifth International Echinoderm Conference / Galway / 24-29 September 1984

Role of the body-wall in the nutritional cycle of three populations of English Channel starfish

D.NICHOLS
Department of Biological Sciences, University of Exeter, UK

M.F.BARKER
Portobello Marine Laboratory, Dunedin, New Zealand

ABSTRACT: Annual cycles in the somatic (body-wall) index of three English Channel starfish have been followed over a three-year period. In a subtidal population of Marthasterias glacialis (L.) a significant decline in the mean indices occurred during that period of the reproductive cycle corresponding to gonadial build-up; in an offshore population of Asterias rubens L. there was a significant increase in the mean indices over the corresponding period; and in an intertidal population of A. rubens no significant cycle in this index was revealed. The offshore population of A. rubens was nutritionally stressed and that of M. glacialis at the edge of its range, while that of the intertidal A. rubens had access to abundant food. It is suggested that the different responses of the two populations to stress may be accounted for by different strategies: a less-migratory species like M. glacialis might gain advantage by diverting body-wall material to the gonads prior to spawning, while a strongly migratory species, such as A. rubens, might benefit by maintaining its body-wall, and hence its ability to move away from a nutritionally impoverished area.

1 INTRODUCTION

Starfish, like many temperate shallow-water organisms, show an annual periodicity in reproduction and associated functions. The reciprocal relationship between the relative size of the gonads and that of the food storage tissues of the pyloric caeca underlines the strategy some of these animals adopt in normal circumstances as they prepare for spawning (Barker & Nichols 1983). There is, however, a third major component of the animal, its somatic tissues, constituting principally the body-wall, that has been neglected in studies on the energy pool available for reproduction, maintenance and growth. This paper gives a preliminary account of the involvement of the body-wall in three populations of shallow-water starfish and shows that differences occur between them, which can perhaps be accounted for by differences in the animals' overall behavioural strategies.

Though somewhat unsatisfactory for detailed studies of the annual cycles through which an iteroparous animal passes, the organ index method of assessing relative development of a number of organ-systems across the year is convenient for following dynamic changes (Vevers 1949; Giese 1959; Giese & Pearse 1974; Jangoux & Vloeburgh 1973; Barker 1979) and is particularly useful for assessing the effects of food deprivation (Lawrence 1973; Jangoux & van Impe 1977; Harrold & Pearse 1980). Here, a somatic index is plotted alongside gonadial and pyloric caecal indices over a period which is sufficient to tune out fluctuations in individual years.

2 MATERIALS AND METHODS

Three populations of starfish were studied in this work: an offshore population of Marthasterias glacialis (L.), and an intertidal and an offshore population of Asterias rubens L. Samples (n = 10) of all three populations were taken at roughly monthly intervals from March 1980 until April 1983. The samples of M. glacialis were collected by SCUBA diving from a sub-littoral reef within Plymouth Sound known as Duke Rock (Figure 1) (O S Ref SX 481508) at about 8 m depth. The intertidal A. rubens site was at Hollicombe Point, Torbay (O S Ref SX 898618), a biologically rich reef extending some 150 m to seaward at low-water; mature starfish were taken from beneath rocks and within crevices at E L W S tide-level. The offshore A. rubens

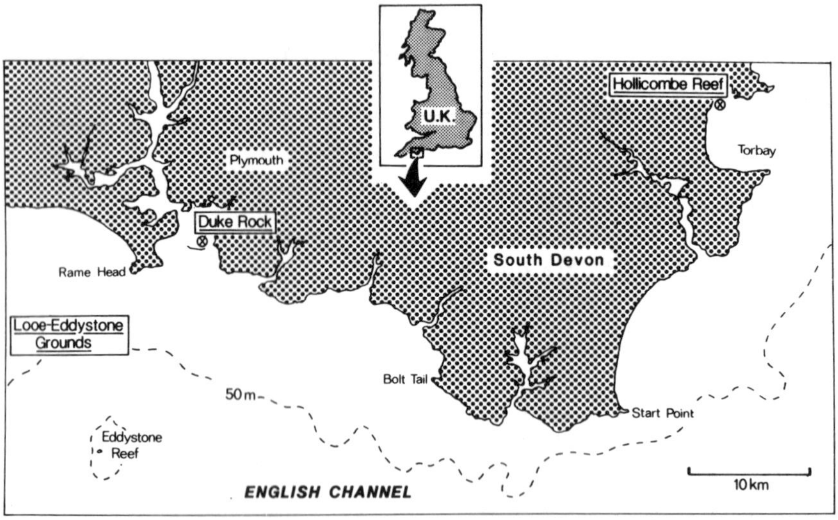

Figure 1 Map showing the sampling sites used in this study. An intertidal population of Asterias rubens was sampled at Hollicombe Reef and a subtidal one from the Looe-Eddystone Grounds, while a subtidal population of Marthasterias glacialis was sampled from Duke Rock, within Plymouth Sound.

were collected by the Marine Biological Association's R V 'Squilla' using an Agassiz trawl from the shelly Looe-Eddystone Grounds about 15 km south-south-west of Rame Head, Cornwall. The crew of the vessel describe these Grounds as nutritionally poor. All samples were maintained in circulating sea-water until processed, generally within 24 h of collection.

Individuals of each sample were first measured (diameter from one arm-tip to a line joining the tips of the two opposite arms), then opened by a median dorsal incision along each arm. The pyloric caeca and gonads (where present) were removed, blotted dry and their volume determined by displacement. The eviscerated body was blotted dry and weighed. Gonadial and pyloric caecal indices were determined by the method of Pearse (1965), i.e.,

$$\text{organ index} = \frac{\text{organ volume}}{\text{eviscerated wet weight}} \%$$

Somatic indices were calculated using the formula:

$$\text{somatic index} = \frac{\text{eviscerated wet weight}}{\text{diameter}}$$

3 RESULTS

The annual cycles of the somatic index (S I) of M. glacialis for the period March 1980 to April 1983 are plotted on coincident axes in Figure 2. From the immediate post-spawning period each year (about July) to the commencement of the next season's spawning (March) the S I shows a gradual decline. Compounding the three years, Spearman's rank shows a positive correlation (r_s significant at < 0.01), though one of the years studied, that for 1982-83, does not show the significant negative slope shown by the other two (1980-81, r_s significant at < 0.05; 1981-82, r_s significant at $\simeq 0.01$; 1982-83, r_s not significant at > 0.05). The 1982-83 season was somewhat anomalous, however, in that there was no real peak to the S I cycle: the sample for September 1982, when the peak might be expected to have occurred, contained at least four starfish with regenerating arms, and this may have affected the relative biomass of the body-wall, and hence the S I for that month and the trend for the whole year. From about March/April each year, that is, corresponding to the peak of relative gonad size, the S I shows a general rise until gonadial build-up recommences in about September.

The annual cycle of the S I for A. rubens from the Looe-Eddystone Grounds for the

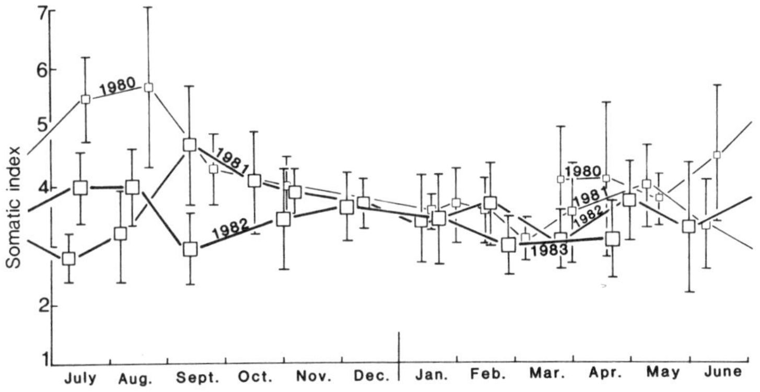

<u>Figure 2</u> Three annual cycles of the somatic index of samples (n = 10; vertical bar-lines are 95% confidence limits) of the starfish Marthasterias glacialis from a subtidal population from Duke Rock, Plymouth Sound, plotted on coincident axes. Size of symbols increases for later years.

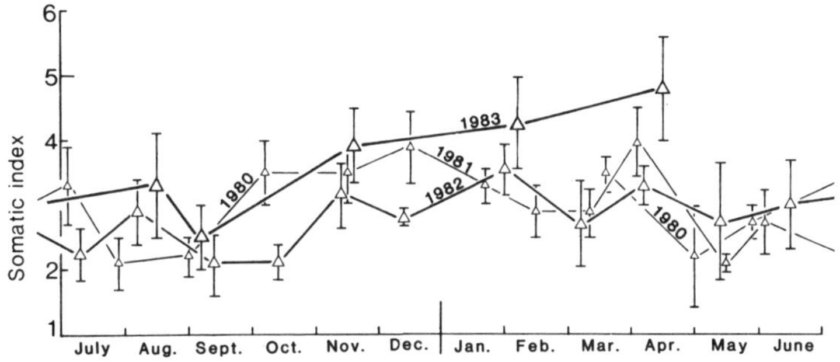

<u>Figure 3</u> Three annual cycles of the somatic index for samples (n = 10; 95% confidence limits) of the starfish Asterias rubens from a population at 50 m depth on the Looe-Eddystone Grounds, English Channel, plotted on coincident axes. Size of symbols increases for later years.

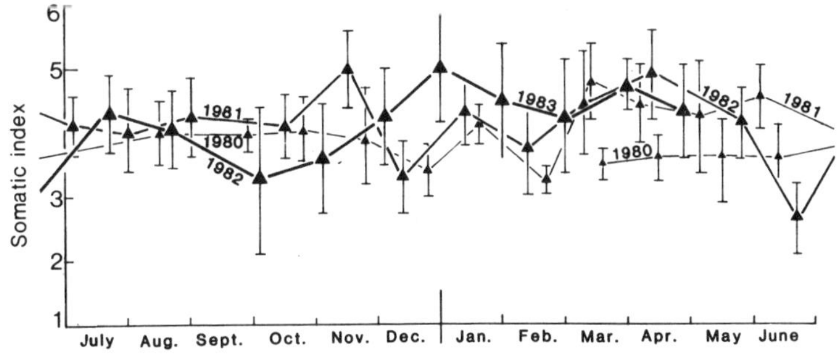

<u>Figure 4</u> Three annual cycles of the somatic index of samples (n = 10; 95% confidence limits) of the starfish Asterias rubens from an intertidal population at Hollicombe Reef, Torbay, English Channel, plotted on coincident axes. Size of symbols increases for later years.

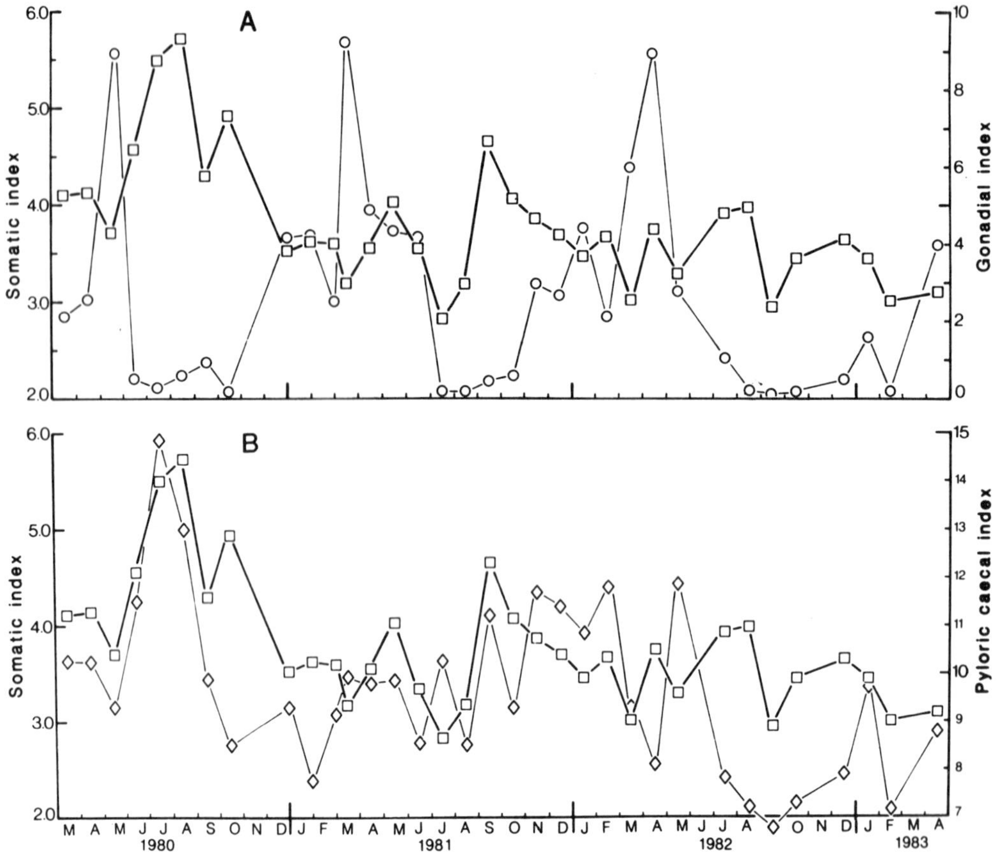

Figure 5 Comparisons of annual cycles for means of samples (n = 10) of Marthasterias glacialis from Duke Rock, Plymouth Sound. A, somatic (squares, left-hand axis) and gonadial (circles, right-hand axis) indices. B, somatic (squares, left-hand axis) and pyloric caecal (diamonds, right-hand axis) indices.

period March 1980 to April 1983 are plotted on coincident axes in Figure 3. From the immediate post-spawning period each year (about July) to the start of the next spawning period (about March) there is an increase in the S I. Compounding the three years studied, the correlation revealed by Spearman's rank is significant at < 0.01. The samples from late 1982 until April 1983, when sampling ceased, were subjectively judged as of better 'quality' than preceding samples from the same time of year, in that the bodies of the starfish were more turgid and the animals more active in the aquarium upon capture. This may be reflected in the higher values for the S I for this series of samples. However, even if the three samples for November 1982, February 1983 and April 1983 are omitted, the slope is still significant ($r_s > 0.01$). From the end of March to about May, that is, corresponding to the spawning season, there is a rapid decline in the S I in all years studied.

The annual cycle of the S I for samples of A. rubens from an intertidal site at Hollicombe Reef, Torbay, for the period March 1980 to April 1983 are plotted on coincident axes in Figure 4. In contrast to the other populations studied in this work, the period from post-spawning to gonadial build-up in the Hollicombe population (July to January) shows no significant slope; further, there is no obvious rise or fall coincident with the spawning period itself. Hollicombe Reef is a comparatively rich reef, with many patches of the mussel Mytilus edulis, a preferred food of A. rubens (Sloan 1980; Sloan & Aldridge 1981), and for most months the first ten individuals found, which

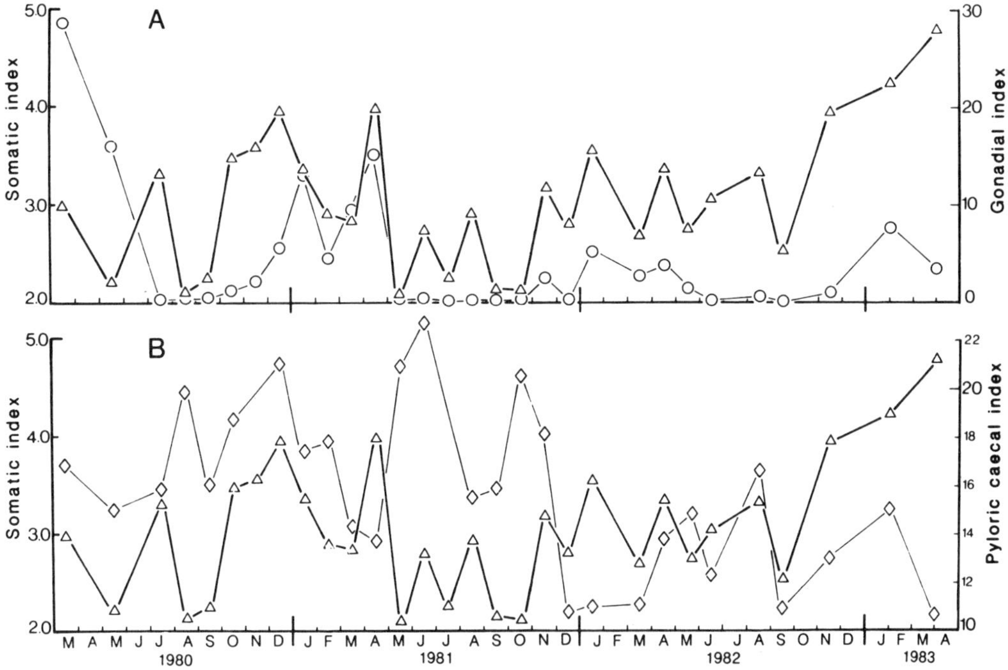

Figure 6 Comparisons of annual cycles for means of samples (n = 10) of Asterias rubens from the Looe-Eddystone Grounds, English Channel. A, somatic (triangles, left-hand axis) and gonadial (circles, right-hand axis) indices. B, somatic (triangles, left-hand axis) and pyloric caecal (diamonds, right-hand axis) indices.

constituted the sample, were all good quality, large, mature individuals. The one exception to this over the three-year sampling period was June 1982, when an onshore wind and low pressure caused a worse-than-usual tide, and in consequence the sample collected that month was of poor quality and rather flaccid starfish that may have been washed on to the reef by the heavy weather. This sample produced a lower-than-expected S I for June 1982.

4 DISCUSSION

In 1930 R.A. Fisher wrote: "It would be instructive to know ... what circumstances in the life history and environment (of an organism) would render profitable the diversion of a greater or lesser share of the available resources towards reproduction." More recently, Calow (1979) has commented on the 'investment priorities' of various invertebrates when faced with the need to allocate reduced resources and has remarked that reproduction is most likely to have an adverse effect on a parent when it uses materials and energy that are otherwise needed for its metabolic well-being; if reproduction competes more successfully than other vital processes of the organism for the limited supply of energy, then both the reproductive performance and the future survival of the organism are put at risk.

This phenomenon has been only cursorily studied in starfish. For instance, Harrold & Pearse (1980) remark that most previous workers dealing with the relation between food storage and gonad build-up have not addressed the question of an allocation strategy by starvation experiments, and they conclude that keeping up a high reproductive output may sometimes jeopardise survival by channelling resources away from the body maintenance, which might be risky in times of food shortage or other stress situations.

In this paper an attempt is made to investigate resource allocation with regard to the body-wall in three starfish populations; two of these are ecologically

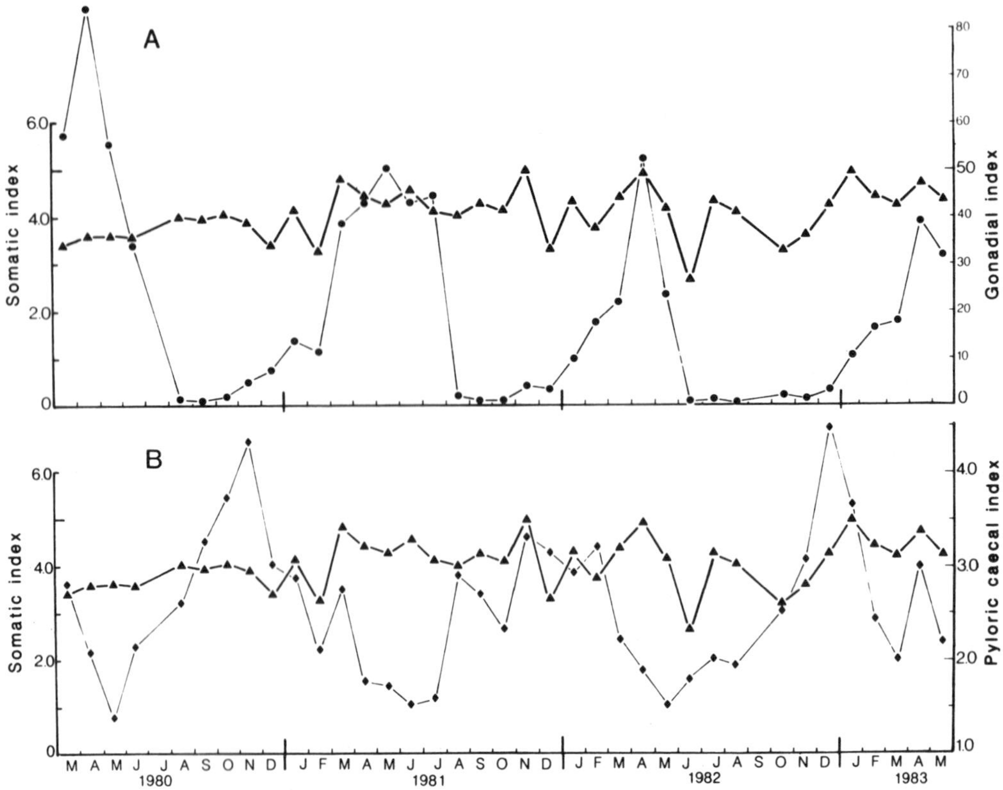

Figure 7 Comparison of annual cycles for means of samples (n = 10) of Asterias rubens from an intertidal site at Hollicombe Reef, Torbay, English Channel. A, somatic (triangles, left-hand axis) and gonadial (circles, right-hand axis) indices. B, somatic (triangles, left-hand axis) and pyloric caecal (diamonds, right-hand axis) indices.

different populations of one species and the third is a population of another species with a very different mode of life. The results show that two of the three populations, thought to be nutritionally impoverished or otherwise deprived (A. rubens from Looe-Eddystone and M. glacialis), show a fluctuating somatic (body-wall) index over the annual cycle, while the third, from a nutritionally rich site (A. rubens from Hollicombe) shows no such cycle. Interestingly, preliminary results of samples from another starfish, Sclerasterias mollis, which, like the Hollicombe A. rubens, lives in a food-rich area, also shows no cycle for the S I, at least so far as one year's results show (Barker, in prep.).

The question arises: how does the S I cycle relate to cycles of other organ indices which indicate resource allocation in starfish? Figure 5A shows the relationship between the S I and the gonadial index (G I) for M. glacialis. An inverse relationship between these cycles is indicated. For instance, gonadial build-up begins in October 1980 and peaks in March 1981, while over the same period the S I declines. Similarly, in the following season the gonadial build-up begins in September 1981 and peaks in April 1982, and this period also shows a decline in the S I. For the last season studied, October 1982 to April 1983, an inverse relationship exists, but it is less marked, and this may be linked to the poor sample in September 1982, which included a high proportion of regenerating individuals.

An inverse relationship between the gonadial index and the pyloric caecal index (P I) for M. glacialis has been shown to exist (Nichols & Barker, 1984 a), so it might be expected that the S I would show a direct relationship with the P I, that is, the converse of its relationship with the G I. Figure 5B shows that in two out of the three years studied there is remarkable conformity between these two indices. In particular, at pyloric caecal peaks (July 1980, May and September 1981) the S I also shows peaks; the anomalous year of 1982-83 does not show such an obvious relationship, however, and it appears that such a lack of conformity cannot be explained by a single anomalous sample in September 1982, since there is a trough in the P I over an eight-month period (May 1982 to January 1983) and the anomalous sample is only one of five taken during this time.

It has been pointed out elsewhere (Nichols & Barker 1984 a) that the population of M. glacialis sampled for this work lies at about the eastern up-Channel limit of this species' range, and may for this or some other physiological reason be under nutritional stress. If so, it might be an advantageous strategy for resources to be directed from body-wall tissues to those of the developing gonads, in much the same way as occurs in corixid bugs (Calow 1979). Some starfish, including A. rubens, do not apparently behave in this way: their strategy appears to be to allocate preferentially to non-gonadial tissue under conditions of stress (Nimitz 1971; Jangoux & van Impe 1977; Harrold & Pearse 1980), so that the general condition of the non-reproductive parts of the body is not seriously impaired. This can be seen in the Looe-Eddystone population investigated in the present study. There is no inverse relationship between the somatic index and the gonadial index: indeed, the two are roughly in phase (Figure 6A). Neither do the somatic and pyloric caecal indices show any obvious relationship (Figure 6B). In A. rubens populations in food rich areas (e.g. the Hollicombe population of this study) there is little seasonal variation in the somatic index, while the gonadial and pyloric caecal indices show their characteristic inverse relationship, and this is shown in Figure 7, A and B. We suggest that as A. rubens is a migratory species (Dare 1982), it is an advantage for this starfish to maintain the tissues of the body wall in such a condition that it could move away from a food-impoverished locality to one offering better conditions.

This is probably of considerable advantage to an iteroparous animal (Cole 1954), since it enables such animals to tide themselves over a period of reduced resources, in anticipation of better times to come. There are important energy costs involved in reproduction. Natural selection determines a particular strategy for each species of animal that will maximise its chances of producing offspring: costs must not exceed benefits, since to do so will reduce fitness. In the case of the two starfish species studied in this work, the ability to migrate possessed by one (Asterias) appears to favour somatic maintenance under stress conditions, while the more stationary habit of the other (Marthasterias) favours an investment in reproduction at the expense of the body-wall.

REFERENCES

Barker, M.F. 1979. Breeding and recruitment in a population of the New Zealand starfish Stichaster australis (Verrill) (Echinodermata: Asteroidea). J.exp.mar. Biol.Ecol. 41:195-211.

Barker, M.F. & D.Nichols 1983. Reproduction, recruitment and juvenile ecology of the starfish Asterias rubens and Marthasterias glacialis. J.mar.biol.Ass. U.K. 63:745-765.

Calow, P. 1979. The cost of reproduction - a physiological approach. Biol.Rev. 54:23-40.

Cole, L.C. 1954. The population consequences of life-history phenomena. Q.Rev.Biol. 29:103-137.

Dare, P.J. 1982. Notes on the swarming behaviour and population density of Asterias rubens L. (Echinodermata: Asteroidea) feeding on the mussel, Mytilus edulis L. J.Cons.perm.int.Explor. Mer 40:112-118.

Fisher, R.A. 1930. The genetical theory of natural selection. O.U.P.

Giese, A.C. 1959. Comparative physiology: annual reproductive cycles of marine invertebrates. A.Rev.Physiol. 21:547-576.

Giese, A.C. & J.S.Pearse 1974. In Reproduction of marine invertebrates. Academic Press.

Harrold, C. & J.S.Pearse 1980. Allocation of pyloric caecum reserves in fed and starved sea stars, Pisaster giganteus (Stimpson): somatic maintenance comes before reproduction. J.exp.mar.Biol.Ecol. 48(2):169-184.

Jangoux, M. & E.van Impe 1977. The annual cycle pyloric of Asterias rubens L. (Echinodermata: Asteroidea). J.exp.mar.Biol.Ecol. 30:165-184.

Jangoux, M. & M.Vloebergh 1973. Contribution à l'étude du cycle annuel de reproduction d'une population d'Asterias rubens (Echinodermata, Asteroidea) du littoral Belge. Netherlands J.Sea Res. 6:389-408.

Lawrence, J.M. 1973. Level, content and calorific equivalents of the lipid, carbohydrate and protein in the body components of Luidia clathrata (Echinodermata: Asteroidea: Platyasterida) in Tampa Bay. J.exp.mar.Biol.Ecol. 11(3):263-274.

Nichols, D. & M.F.Barker 1984 a. Reproductive and nutritional periodicities in the starfish Marthasterias glacialis from Plymouth Sound. J.mar.biol.Ass.U.K. 64:461-470.

Nichols, D. & M.F.Barker 1984 b. A comparative study of reproductive and nutritional periodicities in two populations of Asterias rubens (Echinodermata: Asteroidea) from the English Channel. J.mar.biol.Ass.U.K. 64:471-484.

Nimitz, Sister M.A. 1971. Histochemical study of gut nutrient reserves in relation to reproduction and nutrition in the sea-stars, Pisaster ochraceus and Patiria miniata. Biol.Bull.mar.biol.Lab., Woods Hole 140(3):461-481.

Pearse, J.S. 1965. Reproductive periodicities in several contrasting populations of Odontaster validus Koehler, a common Antarctic asteroid. Antarct.Res. Ser. 5:39-85.

Sloan, N.A. 1980. Aspects of the feeding biology of asteroids. Oceanog.mar.biol: An.Rev. 18:57-124.

Sloan, N.A. & T.H. Aldridge 1981. Observations on an aggregation of the starfish Asterias rubens L. in Morecambe Bay, Lancashire, England. J.Nat.Hist 15:407-418.

Vevers, H.G. 1949. The biology of Asterias rubens: growth and reproduction. J.mar.biol.Ass.U.K. 28:165-187.

Histochemical localization of NADPH producing enzymes by the tetrazolium salt method and of acetyl CoA carboxylase by a novel peroxydase-labelled avidin method in the storage organs and gonads of the sea star, *Asterias rubens*

ROB C.H.M.OUDEJANS, PIERRE N.M.KONINGS & PETER A.VOOGT
Laboratory of Chemical Animal Physiology, State University of Utrecht, Netherlands

ABSTRACT: By using histochemical methods the localization of the NADPH producing enzymes glucose-6-P dehydrogenase, 6-P-gluconate dehydrogenase and $NADP^+$-dependent isocitrate dehydrogenase has been determined in pyloric caeca, ovaries and testes of the sea star, Asterias rubens. Highest activities of all three dehydrogenases occur in pyloric caeca, especially in the median duct and in the follicle cells of the ovaries. Acetyl CoA carboxylase was localized via a novel method in pyloric caeca and ovaries. Activity of dehydrogenases and the carboxylase coincide in the apical part of the caeca pylorica storage cells and in the follicle cells, suggesting that fatty acid biosynthesis is situated in that part of the pyloric caeca and in the follicle cells.

1 INTRODUCTION

During the annual reproductive cycle of Asterias rubens reserve materials are synthesized and stored in the pyloric caeca and used for yolk production during the vitellogenic stage of ovarian development (Voogt et al. 1984). Proteins and lipids are the predominant reserve substances, whereas carbohydrates occur only in relatively small amounts. The lipids are supposed to be synthesized from dietary carbohydrates (Oudejans & Van der Sluis 1979).
In previous studies we have focussed on the de novo synthesis of fatty acids by studying the enzymes involved in the production of the substrates acetyl CoA and NADPH. (Oudejans et al. 1983) and the activities of acetyl CoA carboxylase and fatty acid synthetase utilizing these substrates in the complex pathway of fatty acid synthesis (Oudejans et al. 1984).
All data obtained so far are based on experiments in which homogenates of total organs and subcellular fractions obtained by differential pelleting have been used. Since one may anticipate that particular cells or cell compartments are involved in specific biosynthetic tasks rather than the total organ we have tried to localize enzyme activities by histochemical procedures.
In this way information can be gathered on which cells are charged with fatty acid synthesis. To demonstrate activities of $NAD(P)^+$-dependent dehydrogenases several methods using various tetrazolium salts are currently available (Lojda et al. 1976; Seidler 1984). For this purpose both paraffin and cryostat sections can be used.
Apart from very laborious immunological techniques, no method for the histochemical localization of acetyl CoA carboxylase and fatty acid synthetase is available.
In this paper we present data on the localization of glucose-6-P dehydrogenase, 6-P-gluconate dehydrogenase and $NADP^+$-dependent isocitrate dehydrogenase using the tetrazolium salt method.
Moreover, a novel method is described to detect activity of acetyl CoA carboxylase using peroxydase-labelled avidin, which due to its very specific binding capacity for the biotin group of acetyl CoA carboxylase, can be visualized via diaminobenzidine resulting in a brown precipitate.
Results will be discussed with reference to our current knowledge on vitellogenesis in sea star.

2 MATERIALS AND METHODS

2.1 Chemicals

3,3'-Diaminobenzidine tetrahydrochloride (DAB) was purchased from BDH Chemicals (Poole, UK). Polyvinyl-pyrrolidone (PVP) (mean MW 360,000), DL-isocitrate and peroxydase-labelled avidin (type VI) were from Sigma (St Louis, USA). According to the maufacturer 50% of the avidin is active with a peroxydase activity of 45 units mg^{-1}. Nitroblue-tetrazolium (NBT) was obtained from Calbiochem (San Diego, USA), $NADP^+$, glucose-6-P and 6-P-gluconate were from Boehringer (Mannheim, FRG). All other chemicals of the best quality available were from local suppliers.

2.2 Animals

Mature sea stars (Asterias rubens) were collected in the Wadden Sea (The Netherlands) and kept in well-aerated running sea water of 12°C. Animals were weighed, sexed and of each individual the pyloric caeca index and gonad index (proportional weight of total fresh weight) were determined and also the stage of ovarian development according to Schoenmakers et al. (1981, 1984).

2.3 Histological procedure

Pieces of pyloric caeca and gonads were frozen directly after dissection in liquid nitrogen and sectioned at 8-10 µm on a cryostat. Unfixed cryostat sections were air-dried (5 min). Alternatively, pieces of tissue were fixed in Bouin (24 hr) and embedded in paraffin and sectioned at 5 µm.

2.4 Localization of dehydrogenases

Unfixed, air-dried sections were incubated in a mixture consisting of 0.2M TRIS buffer supplemented with 0.15 g PVP, 0.05 mg $NADP^+$, 0.67 mg of the appropriate substrate and 0.2 mg NBT ml^{-1} (pH 7.4). Incubation lasted for 45 min at 37°C. Subsequently the sections were rinsed in aqua dest (2 min) and embedded in water-mounted medium. Controls were run without substrate and/or $NADP^+$.

2.5 Localization of acetyl CoA carboxylase

A dilution range (1:1000 to 1:100,000) of the peroxydase-labelled avidin was prepared in 0.05M TRIS buffer (pH 7.6) and stored in portions of 10 µl at -20°C until use to avoid repeatedly thawing and freezing. Paraffin sections were washed twice in xylene (5 min) and in methanol (25 min) to inhibit pseudoperoxydase, dehydrated via an alcohol range, washed in TRIS (pH 7.6) with 0.9% NaCl (5 min), in the same buffer with 0.01% H_2O_2 and 0.02% Triton X-100 (to inhibit pseudoperoxydase) and in TRIS without additions (5 min). Subsequently sections were incubated overnight in peroxydase-labelled avidin (10 µl of the appropriate dilution dropped onto each section). Then the sections were washed in TRIS with 0.9% NaCl (5 min) and TRIS (15 min.). Complexed peroxydase-labelled avidin was visualized by incubating with DAB (5 mg DAB in 10 ml TRIS + 100 µl 1% H_2O_2) for a variable time. Then the sections were washed in aqua dest (10 min), dehydrated and embedded in glycerol-gelatin.
The procedure for cryostat sections was identical to that for paraffin sections, except for the xylene washing and dehydration by alcohol, which were omitted. Embedding was directly in water-mounted medium. The methods described above are mainly based on Van Noorden & Polak (1983).
In order to modulate the activity of the acetyl CoA carboxylase, cryostat sections were preincubated (30 min) in 0.05M TRIS (pH 6.9) to which either 20 mM $MgCl_2$ and 1.25 mM ATP were added or 20 mM citrate.
Moreover, the following controls were run: a) without incubation with peroxydase-labelled avidin, b) without suppression of pseudoperoxydase activity, c) incubation at pH > 9.6. Preincubations were at the given pH 6.9, d) incubation of rat liver sections (both paraffin and cryostat sections).

3 RESULTS

3.1 $NADP^+$-dependent dehydrogenases

Since the investigated dehydrogenases are localized in the cytoplasm, care has to be taken to prevent diffusion of the reaction products. In order to obtain reliable results viscous incubation media with PVP have been used. Crucial to the procedure for cryostat sections is to section them directly from pieces of tissue frozen immediately after dissection. Otherwise degeneration of tissue results in bad pictures. In lipid-rich tissue (e.g. pyloric caeca) reaction products tend to accumulate in lipid droplets.
Localization in pyloric caeca.
The three dehydrogenases appear to be present particularly in the cells of the median duct. In the side lobes activity is absent juxta-proximal of the nuclei of the storage cells and in the rest of the cytoplasm the distribution is nearly homogenous (Fig 1b,1f) In the transition zone between median duct and side lobes activities are increasing towards the median duct. Highest activities were observed for glucose-6-P dehydrogenase, medium activities for isocitrate dehydrogenase and slightly less for 6-P-gluconate dehydrogenase. No differences were observed between pyloric caeca of female and male animal
Localization in ovaries.
Highest activities of the dehydrogenases are found in the cytoplasm of follicle cells (Fig 1a) and in the cytoplasm of small oocytes. Also the haemal system is -although

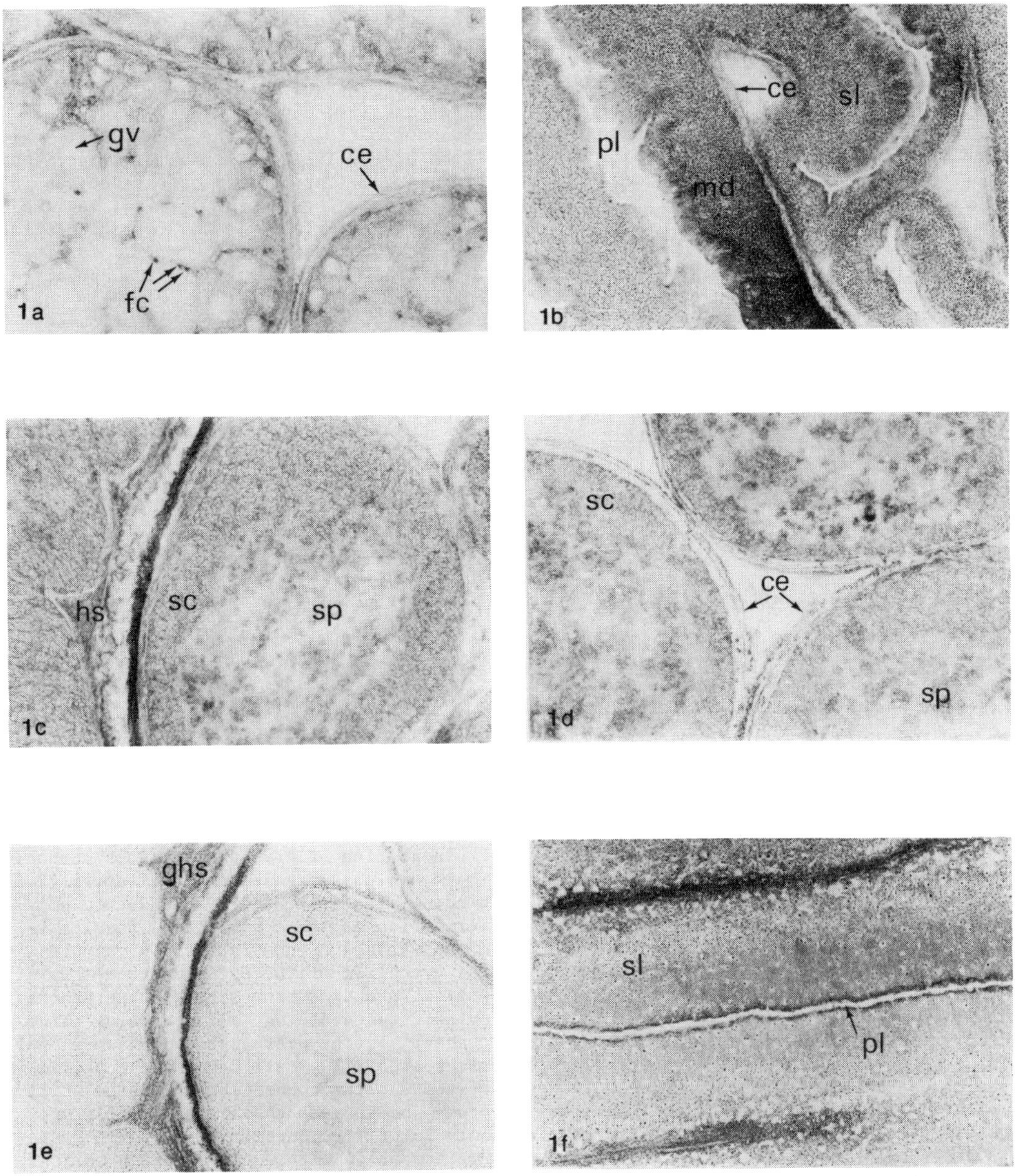

Fig. 1. Histochemical localization of $NADP^+$-dependent dehydrogenases in pyloric caeca, ovaries and testes of the sea star Asterias rubens.
Gonad index 2.4 for a-b and 1.9 for d-f; pyloric caeca index 24.0 for a-b and 24.9 for d-f.
a-c: Glucose-6-P dehydrogenase in a) ovaries, b) pyloric caeca and c) testes
d : 6-P-gluconate dehydrogenase in testes
e-f: $NADP^+$-dependent isocitrate dehydrogenase in e) testes and f) pyloric caeca.
Abbreviations used: cd cell debris, ce coelomic epithelium, fc follicle cell, ghs genital haemal sinus, gv germinal vesicle, hs haemal system, md median duct, pl pyloric caeca lumen, sc spermatogenic column, sl side lobe, sp spermatozoa.

faintly- coloured, whereas the cytoplasm of large oocytes and the coelomic epithelium is only slightly coloured. No activity was observed in germinal vesicles and in the nuclei of follicle cells. The distribution of all three dehydrogenases was rather identical; the activity of 6-P-gluconate dehydrogenase was like that for glucose-6-P dehydrogenase and that of isocitrate dehydrogenase was rather low.

Localization in testes.

In general, in the spermatogenic columns small granules of reaction products are found. The transition zone between the columns and the lumen of the testis is clearly visible (Fig 1c-e) due to the absence of granules, whereas in the lumen ripe spermazoa, phagocytes and cell debris do occur. Also the coelomic epithelium contains small granules. The ground substance of the haemal system is homogenously coloured with small granules, whereas the genital haemal sinus is fully transparent. The wall of the testis lobes contain high activities.

3.2 Acetyl CoA carboxylase

Sectioning of frozen pyloric caeca tissue appeared to be difficult due to its high lipid content and consequently less good pictures were obtained than with paraffin-embedded tissue. However, enzyme activity in tissue frozen directly can be manipulated by stimulating or inhibiting substances. Incubation of sections for too long time resulted in dark brown pictures, probably due to non-specific binding of the avidin. Ultimately we have chosen short incubation times (1-2 min). Incubating the sections at $pH > 9.6$ permitted longer incubation times, since then the non-specific binding of avidin is drastically decreased.

Localization in pyloric caeca.

In paraffin sections (Fig 2b) the median duct, side lobes and mesenteries of the pyloric caeca contain reaction products. In the median duct a few cells appear dark-coloured (Fig 2c). In successive sections the same cells are coloured. We suppose these cells to be mucocytes,which occur frequently in the median duct and only sporadically in the side lobes (Van der Plas et al. 1980). Both in the median duct and in the side lobes the coelomic epithelium contains dark brown granules. Also the connective tissue is coloured, whereas small light brown granules occur throughout the cytoplasm of pyloric caeca storage cells. Using more diluted avidin or shorter incubation times with DAB only the coelomic epithelium is coloured.

Incubation of cryostat sections resulted in a dark brown zone on the apical side of the pyloric caeca cells. The cytoplasm is light brown coloured with basally transparent vacuoles. The coelomic epithelium contains dark brown granules.

In cryostat sections of the ovaries, the connective tissue and the haemal system are light brown. The genital haemal sinus remains transparent. The ooplasm of the oocytes is light brown, whereas the germinal vesicles are transparent. Only a few follicle cells are dark brown coloured.

In paraffin sections the ovaries appear to be rather transparent (Fig 2a). Only the coelomic epithelium and the haemal system are distinctly coloured as are the follicle cells. With increasing avidin concentration and increasing DAB incubation time the whole ovary appears to be coloured.

3.3 Control incubations for acetyl CoA carboxylase

In order to evaluate the new method several control incubations have been performed. Not any colour was visible after incubations without the peroxydase-labelled avidin. In paraffin and cryostat sections incubated without suppression of pseudoperoxydase activity collagenic tissues appeared to be dark brown and also, in cryostat sections only, the haemal system.

After incubation at $pH > 9.6$, longer incubations with DAB (5-15 min) are permitted without colouring of the background. The specific binding of the peroxydase-labelled avidin remained unchanged.

Preincubation of cryostat sections with citrate, which decreases the availability of binding places for the avidin due to polymerization of the carboxylase, results in disappearance of the dark brown granules in the cytoplasm of the pyloric caeca (Fig 2f), whereas the dark brown apical zone of the pyloric caeca storage cells remains being present. In the ovaries the ooplasm is fully transparent. The cell debris, the haemal system and the connective tissue is still brownish, whereas the follicle cells are only faintly coloured.

Preincubation with $MgATP^{2-}$, which shifts the carboxylase from the polymeric to the monomeric form, increased the brown colour of the apical zone of the pyloric caeca cells (Fig 2e) and in the cytoplasm more dark brown granules are found as compared to sections which are not preincubated. In the ovaries (Fig 2d) the follicle cells are dark brown. In the cytoplasm no differences were observed.

In order to compare the results obtained with sea star tissues, paraffin and cryostat sections of rat liver were incubated in the same way. Effects of stimulating and/or in-

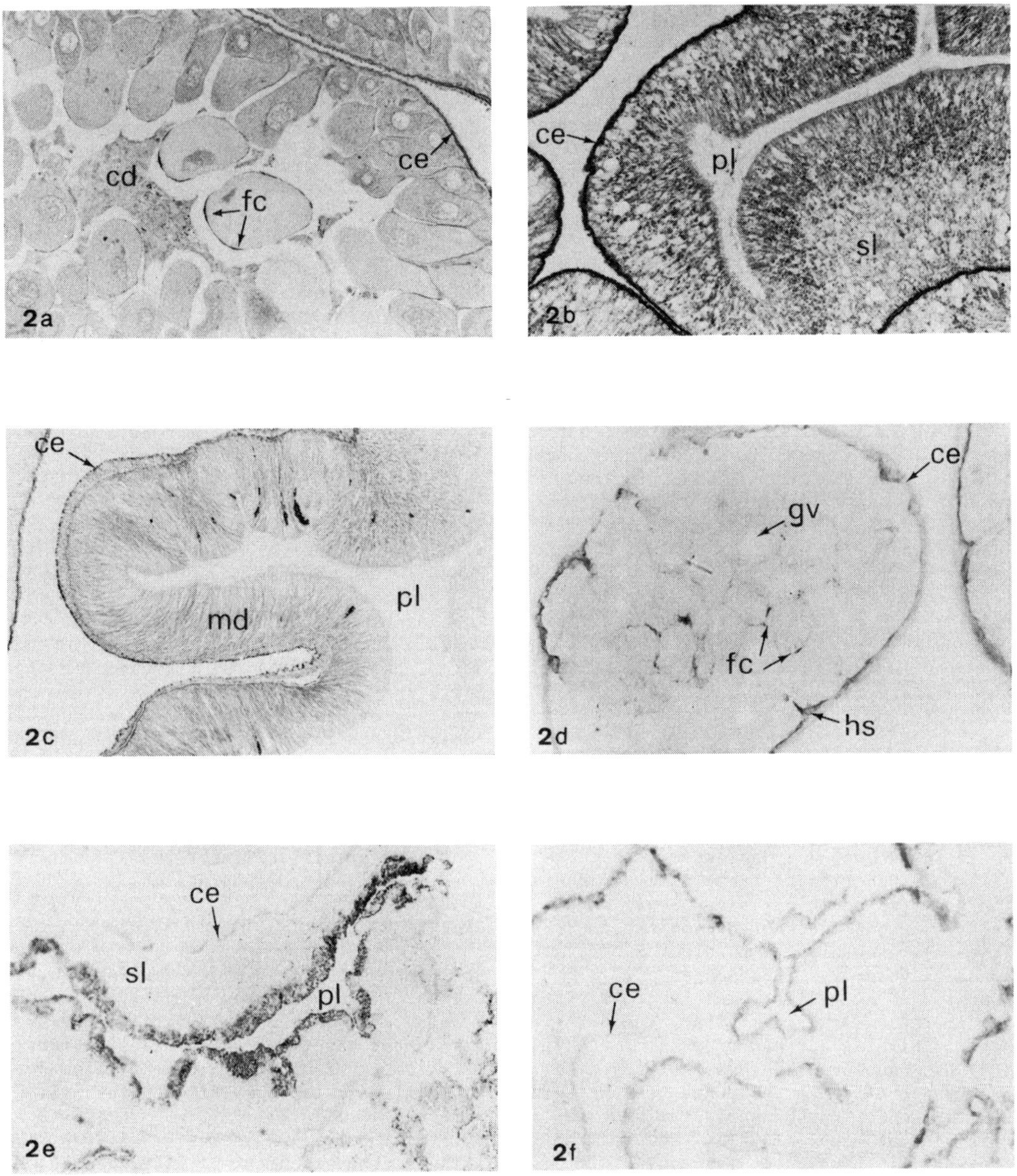

Fig. 2. Histochemical localization of acetyl CoA carboxylase in pyloric caeca and ovaries of the sea star Asterias rubens.
Gonad index 2.2 for a-f and pyloric caeca index 26.4 for a-f.

a-c: Paraffin sections of a) ovaries, b) pyloric caeca and c) median duct of pyloric caeca

d-f: Cryostat sections of d) ovaries, preincubated with $MgATP^{2-}$, e) pyloric caeca, preincubated with $MgATP^{2-}$ and c) pyloric caeca, preincubated with citrate.

For abbreviations used see Fig. 1.

hibiting substances were comparable to those obtained with sea star.

4 DISCUSSION

The substrates and products of the acetyl CoA carboxylase reaction are unsuited for a histochemical localization of the enzyme. So, we have tried to develop a novel method for detection by using the highly specific binding capacity of avidin for biotin, which is an integral component of acetyl CoA carboxylase. By labelling the avidin with peroxydase an efficient detection of small amounts of enzyme is possible after incubation with DAB, which results in a brown precipitate at the binding place of the avidin. However, biotin is not exclusively occurring in acetyl CoA carboxylase, but also in several other biotin containing carboxylases. From the literature it is known (Moss & Lane 1972) that acetyl CoA carboxylase is the only biotin-containing enzyme which cannot bind avidin when it is in the active polymeric form. In this way we have tried to increase the binding of avidin to the enzyme by inactivating the carboxylase with $MgATP^{2-}$ to the monomeric form or to decrease it by activating the carboxylase with citrate to the polymeric form (Moss & Lane 1971; McNeillie et al. 1981) (Fig 2d-f).
Application of these stimulating or inhibiting substances shows that particularly the apical part of the pyloric caeca storage cells and the follicle cells of the ovaries react in the way one may expect, so these cells may function in fatty acid synthesis. Non-specific reactions also occur, particularly in mucocytes of the median duct (Fig 2c), probably due to binding of avidin to negative groups (Guesdon et al. 1979; Wood & Warncke 1981). This effect is strongly suppressed when incubating at a $pH > 9.6$. After finishing our experiments, Amersham (UK) has announced the availability of a new range of high performance streptavidin systems, which may be very promising in this kind of experiments in reducing the non-specific binding of avidin.
The distribution of all three dehydrogenases over the pyloric caeca shows strong similarities. They are found all over the cytoplasm of the storage cells, particularly juxta-proximal and in the infranuclear zone. However, highest activities occur in the median duct, which can be explained by its high activity of proliferation (Van der Plas et al. 1983).for which nucleic acids and membrane lipids will be needed. In the ovaries, all three dehydrogenases appear to be present particularly in the follicle cells and in the unripe oocytes which are found close to the germinal epithelium. Primary oocytes contain only very low enzyme activities. It is difficult to conclude in what sense follicle cells are concerned with synthesis of materials for oocyte growth. Their high capacity to produce NADPH may be necessary for other purposes, particularly since in Asterias, the follicle cells are not distinctly differentiated into secretory cells (Beijnink et al. 1984a).
In the testes activity of the dehydrogenases also is restricted to the young stages of spermatogenic development, since in the ripe spermatozoa no enzyme activity can be demonstrated. Probably the enzyme activities are associated with the axial somatic cells, which are supposed to be the suppliers of materials needed for spermatocyte growth (Walker 1980).
In conclusion, in Asterias rubens NADPH is produced in pyloric caeca, ovaries and testes, but quantitatively the pyloric caeca predominate. However, to what extent this NADPH is used for de novo fatty acid synthesis remains to be investigated.
The coincidence of NADPH producing dehydrogenases and acetyl CoA carboxylase in the apical region of the pyloric caeca storage cells and in the follicle cells of the ovaries suggests that fatty acid synthesis is localized there. However, based on current knowledge of vitellogenesis in Asterias rubens (Beijnink et al. 1984b; Broertjes et al. 1984; Voogt et al. 1984), pyloric caeca cells are most likely the site for fatty acid synthesis specifically for the production of vitellogenic glycolipoproteins.

REFERENCES

Beijnink, F.B., I. Van der Sluis & P.A.Voogt 1984b. Turnover rates of fatty acid and amino acid in the coelomic fluid of the sea star Asterias rubens: Implications for the route of nutrient translocation during vitellogenesis. Comp.Biochem.Physiol. 78B: 761-767.

Beijnink, F.B., C.W.Walker & P.A.Voogt 1984a The microenvironment of vitellogenic oocytes in the sea star, Asterias rubens; interactions between the ovarian haemal system, follicle cells and primary oocytes. Cell Tissue Res. (in press).

Broertjes, J.J.S., P. De Waard & P.A.Voogt 1984. On the presence of vitellogenic substances in the starfish Asterias rubens (L.). J.mar.biol.Ass.U.K. 64: 261-269.

Guesdon, J.L., T.Ternynck & S.Arrameas 1979. The use of avidin-biotin interactions in immunoenzymatic techniques. J.Histochem. Cytochem. 27: 1131-1139.

McNeillie, E.M., R.A.Clegg & V.A.Zammit 1981

Regulation of acetyl CoA carboxylase in rat mammary gland. Biochem.J. 200: 639-644.

Lojda, Z., R.Gossrau & T.H.Schiebler 1976. Enzym-histochemische Methoden. Berlin: Springer-Verlag.

Moss, J. & M.D.Lane 1971. The biotin-dependent enzymes. Adv.Enzymol. 35: 321-422.

Moss, J. & M.D.Lane 1972. Acetyl coenzyme A carboxylase. III. Further studies on the relation of catalytic activities to polymeric state. J.biol.Chem. 247: 4944-4961.

Oudejans, R.C.H.M., J.H.Boot & P.A.Voogt 1984. De novo synthesis of fatty acids in the sea star Asterias rubens during its annual reproductive cycle. Adv.Invertebr. Reprod. 3: 621.

Oudejans, R.C.H.M., W.G.E.J.Schoonen & P.A. Voogt 1983. Production of NADPH and acetyl CoA for lipid biosynthesis in the pyloric caeca of the sea star, Asterias rubens. Comp.Biochem.Physiol. 74B: 351-355.

Oudejans, R.C.H.M. & I. Van der Sluis 1979. Storage and depletion of lipid components in the pyloric caeca and ovaries of the seastar Asterias rubens during its annual reproductive cycle. Mar.Biol. 53: 239-247.

Schoenmakers, H.J.N., P.H.J.M.Colenbrander, J.Peute & P.G.W.J. Van Oordt 1981. Anatomy of the ovaries of the starfish Asterias rubens (Echinodermata). A histological and ultrastructural study. Cell Tissue Res. 217: 577-597.

Schoenmakers, H.J.N., M.J.Goedhart & P.A. Voogt 1984. Biometrical and histological aspects of the reproductive cycle of the ovaries of Asterias rubens (Echinodermata). Biol.Bull.,Woods Hole 166: 328-348.

Seidler, E. 1984. Zum Aussagewert histochemischer Tetrazoliumtechniken. Acta histochem.Suppl. 30: 145-152.

Van der Plas, A.J., F.Brands & P.A.Voogt 1983. An autoradiographic investigation of the proliferative activity of pyloric caeca and gonads of Asterias rubens. J.Morphol. 177: 51-68.

Van der Plas, A.J., M.Jangoux & E.Bricourt 1980. The structure of the pyloric caeca of Asterias rubens and its changes during the annual reproductive cycle. In M.Jangoux (ed.), Echinoderms: Present and past, p. 251-254. Rotterdam: Balkema.

Van Noorden, S. & J.M.Polak 1983. Immunochemistry today. In J.M.Polak & S. Van Noorden (eds), Immunochemistry. Practical applications in pathology and biology, p. 10-42. Bristol: J.Wright & Sons.

Voogt, P.A., J.J.S.Broertjes & R.C.H.M.Oudejans 1984. Vitellogenesis in sea star: Physiological and metabolic implications. Comp.Biochem.Physiol. (in press).

Walker, C.W. 1980. Spermatogenic columns, somatic cells and the structural environment of germinal cells in the testes of sea-star. J.Morphol. 166: 81-107.

Wood, G.S. & R.Warncke 1981. Suppression of endogenous avidin-binding activity in tissue and its relevance to biotin-avidin detection systems. J.Histochem.Cytochem. 29: 1196-1204.

Proceedings of the Fifth International Echinoderm Conference / Galway / 24-29 September 1984

Patterns of nitrogen excretion in seven species of asteroids

WILLIAM B.STICKLE
Department of Zoology and Physiology, Louisiana State University, Baton Rouge, USA

ABSTRACT: The percentage of total nitrogen excreted as ammonia, urea and primary amines was determined in the asteroids Crossaster papposus, Dermasterias imbricata, Evasterias troschelii, Henricia leviscula, Orthasterias koehleri, Pisaster ochraceus and Pteraster tesselatus. The percentage of nitrogen excreted as ammonia varied among the species from 21-87%, urea accounted for 0-66% and primary amines accounted for 0-21%. Urea accounted for a significant percentage of total nitrogen excreted by four species of carnivorous asteroids which had been starved for several weeks, including: D. imbricata (40% of total nitrogen excreted), O. koehleri (66%), P. ochraceus (53%) and P. tesselatus (58%). Consequently, urea must be synthesized endogenously in these species and is not a catabolic product of ingested arginine. Significant activities of three urea cycle enzymes from the pyloric ceca of P. ochraceus have been determined including; arginase, glutamine synthetase and carbamyl phosphate synthetase.

1. INTRODUCTION

Urea comprises a significant percentage of total nitrogen excreted by a number of echinoderm species (Diehl and Lawrence, 1979; Stickle et al., 1982). The urea excretion rate of the asteroid Luidia clathrada was not affected by diet, (Diehl and Lawrence, 1979) but that of Leptasterias hexactis was directly related to its ingestion rate of Mytilus edulis (Shirley and Stickle, 1982). The later observation suggests that urea excretion in L. hexactis occurs as a result of arginine ingestion coupled with a 5:1 arginase:urease activity ratio in the pyloric ceca (Stickle et al., 1982).

The objectives of the present study are: (1) to determine patterns of ammonia, urea and primary amine excretion in seven species of asteroids; (2) to determine the role which nutritional history plays in altering the excretory pattern of asteroids. Of fundamental importance to the second objective is whether the source of metabolic substrate for urea synthesis is endogenous or must it be ingested, and (3) to determine the activity of three Krebs Ornithine Cycle enzymes in order to partially characterize this metabolic pathway.

2. MATERIALS AND METHODS

Specimens of Crossaster papposus, Dermasterias imbricata, Henricia leviscula, Orthasterias koehleri, Pisaster ochraceus and Pteraster tesselatus were collected in the vicinity of the Friday Harbor Laboratories, Washington, U.S.A. and placed in a flow through sea water system. They were held for several weeks at 12-13°C and 30°/ooS prior to the determination of excretion rates. Evasterias troschelii were collected near Juneau, Alaska, U.S.A. and flown to Baton Rouge, Louisiana where they were maintained at 13°C and 30°/ooS prior to the determination of excretion rates. Data for Leptasterias hexactis (Shirley and Stickle, 1982; Stickle et al. 1982) are included in the tables for the sake of comparison with data for the other seven species.

Nitrogen excretion rates were measured by placing sea stars into 50-7,000 mls of Millipore-filtered (0.45 µM pore size) and oxygenated sea water, adjusted to 30°/ooS and maintained at 12-13°C. Blank containers, filled with Millipore filtered and oxygenated sea water only, were included as controls. Sea stars were removed from the excretion containers after two to six hours depending on their size and the volume of incubation water.

Table 1. Sample size (N) nutritional history (C= carnivore, S= suspension feeder; F= fed prior to experiment and S= starved prior to experiment) and body composition (%) data are given for water (H_2O), dry weight (DW), skeletal weight (SKW) and soft tissue weight (STW) for the seven species of sea stars studied as $\bar{x}\pm SE$, along with data for Leptasterias hexactis (a) reported from Shirley and Stickle (1982). The W:D:T Ratio is the ratio of wet to dry to soft tissue weight for each species.

Species	N	NUTRITIONAL HISTORY	BODY COMPOSITION H_2O	DW	SKW	STW	W:D:T RATIO
ORDER FORCIPULATA							
Evasterias troschelii	10	C-F	78±4	22±4	11±2	11±3	9.1:2.1:1
Leptasterias hexactis[a]	18	C-F	70	30	14	16	7.4:2.1:1
Orthasterias koehleri	1	C-S	72	28	20	8	12.4:3.4:1
Pisaster ochraceus	5	C-F	78±1	22±1	14±1	8±1	12.4:2.2:1
ORDER SPINULOSICA							
Crossaster papposus	1	C-S	82	18	10	8	12.0:2.2:1
Dermasterias imbricata	2	C-S	82±0	18±0	10±1	8±1	11.4:2.1:1
Henricia leviscula	7	S-F	59±1	41±1	26±2	15±1	6.8:2.8:1
Pteraster papposus	1	C-S	86	14	8	6	15.9:2.2:1

Duplicate water samples were taken for ammonia, urea and primary amine analyses and excretion rates calculated as µM/g tissue dry weight/h. Ammonia content was measured by the phenol-hypochlorite method of Solorzano (1969). Urea was measured by a modified urease technique (Sigma Technical Bulletin #640). Primary amines were measured by the fluorescamine method of North (1975) using leucine as a standard.

Total nitrogen excretion was calculated as ug/g dry tissue weight/h by multiplying the rates of ammonia, urea and primary amines excreted by the atomic weight of nitrogen excreted in each form, 14.01, 28.02 and 16.00 respectively, for each sea star and summing the µg/h excreted in each form. Percentage nitrogen excreted in each form was calculated after converting all rates to µg.

The number of individuals of each species assayed, their feeding mode, as given in Lambert (1982), nutritional history and body composition are given in Table 1. P. ochraceus were starved for several weeks before excretion rates were determined, then they were provided several bivalve species as prey for two weeks before excretion rates were again determined.

The wet weight of each sea star was determined after excretion experiments were completed then each sea star, with the exception of starved P. ochraceus, was dried to constant weight at 90°C. Each sea star was then digested in 3X its volume of 10% KOH. The remaining skeletal material was filtered, rinsed, dried and weighed for calculation of tissue dry weight.

Activities of arginase, carbamyl phosphate synthetase and glutamine synthetase were determined in the pyloric ceca of P. ochraceus by Drs. Paul M. Anderson and Carol A. Casey according to methods given in Casey and Anderson (1982). Activities were calculated on a µM/g wet weight/h basis.

All excretion rate and percentage composition data are given as $\bar{x}\pm SE$ if the excretion rate of more than one animal was determined. Excretion rates and the percentage composition of ammonia, urea and primary amines of starved and fed P. ochraceus were compared by use of the Students-T test.

Table 2. Nitrogen excretion. Rates (R) are given as µM/g soft tissue/h, $\bar{x}\pm SE$, except for total N which is given as ug/g soft tissue/h, N= sample size. F= fed and S= starved. Percent composition (P) of excreted nitrogen is also given. [a] from regressions given in Shirley and Stickle (Mar. Biol. 69:155-163. 1982).

Species		VAR	N	NH_4^+	UREA	PA	TOTAL
ORDER FORCIPULATA							
Evasterias troschelii		R	10	2.9±0.6	0.1±0.0	0.1±0.0	46.6±9.9
		P		87±1	9±1	4±0	
Leptasterias hexactis[a]		R	18	0.7	0.4	0.0	23.0
		P		61	33	6	
Orthasterias koehleri		R	1	0.6	0.6	0.0	24.8
		P		34	66	0	
Pisaster ochraceus	F	R	5	2.1±0.44	0.4±0.2	0.1±0.0	42.0±12.2
		P		76±7	20±7	5±1	
	S	R	6	1.1±0.2	0.9±0.2	0.1±0.0	40.8±8.9
		P		45±8	53±8	2±1	
ORDER SPINULOSIDA							
Crossaster papposus		R	1	0.8	0.0	0.9	13.0
		P		84	0	17	
Dermasterias imbricata		R	2	0.6±0.2	0[illegible]±0.3	0.0±0.0	19.1±12.5
		P		57±22	[illegible]0±22	3±5	
Henricia leviscula		R	7	1.0±0.7	0.0±0.0	0.0±0.0	14.8±9.8
		P		76±6	20±5	4±1	
Pteraster tesselatus		R	1	0.2	0.3	0.2	15.7
		P		21	58	21	

3. RESULTS

Total nitrogen excretion rates of four species of sea stars belonging to the order Forcipulata are considerably higher than those of the four species of sea stars belonging to the order Spinulosida (Table 2). In a classical sense, Pisaster ochraceus (fed), Evasterias troschelii, Leptasterias hexactis, Henricia leviscula, Crossaster papposus and Dermasterias imbricata are ammonotelic whereas P. ochraceus (starved), Orthasterias koehleri and Pteraster tesselatus are ureotelic. However, all but E. troschelii and C. papposus excrete significant quantities (>10% of total nitrogen) of urea.

The source of excreted urea in P. ochraceus, O. koehleri, P. tesselatus and D. imbricata must be considered endogenous because these sea stars had been starved for several weeks before excretory patterns were determined. Urea comprised a significantly greater ($p<0.05$) percentage of total nitrogen excreted when P. ochraceus was starved for two weeks than when the same individuals were allowed to prey ad libitum on several species of bivalves prior to the determination of excretory rates. Ammonia excretion was significantly higher when sea stars were fed than starved. Since the total nitrogen excretion rate was not significantly different in fed and starved sea stars starvation resulted in a quantitative shift from ammonia to urea excretion.

Significant activity of two enzymes associated with the Krebs Ornithine Cycle have been determined in the pyloric ceca of P. ochraceus and trace activity of a third enzyme detected. Glutamine synthetase activity of 180 µM/h/g wet tissue, arginase activity of 5 µM/h/g wet tissue and trace activities of carbamyl phosphate were observed in the pyloric ceca. To date, no attempt has been made to determine the activity of other enzymes involved in urea synthesis.

4. DISCUSSION

Surprisingly, endogenously synthesized urea is excreted in such quantities by starved Pisaster ochraceus, Orthasterias koehleri and Pteraster tesselatus that they must be classified as ureotelic and Leptasterias hexactis, Henricia leviscula and Dermasterias imbricata excrete an appreciable percentage (>10%) of the total nitrogen excreted as urea. Luidia clathrata has also been shown to excrete from 8-17 percent of the sum of ammonia and urea nitrogen as urea whether starved or fed a diet of moderate or high protein content (Diehl and Lawrence, 1979). These findings are at odds with the generalization of Baldwin (1967) that the major nitrogenous end product of aquatic animals [with the exception of elasmobranchs] is ammonia.

The well known paradigm of ammonotelism in aquatic animals, ureotelism in amphibious animals with a restricted water supply and uricotelism in terrestrial animals and in cleidoic eggs (Needham, 1931; Baldwin, 1967; Hoar, 1984) has recently been challenged by observations of nitrogen excretion patterns from three phylogenetically distinct groups of deuterostomic animals. The results of this study indicate urea to be an extremely important excretory product of several species of asteroid echinoderms. Likewise, Markus and Lambert (1983) found urea to comprise 40 to 50 percent of the soluble nitrogen excreted by the ascidians Styela clava, S. partia and S. plicata. Finally the embryonic softshell turtle, Trionyx spiniferus, has been found to excrete 70 percent of total nitrogen as urea, 24 percent as ammonia, and only 6 percent as the anticipated urate into the cleidoic egg at the time of hatching (Packard and Packard, 1983). In light of these findings the proposed relationship between water availability, toxicity and the form in which nitrogen is excreted must be carefully reexamined. A knowledge of the selective pressures which have led to these exceptions to the Baldwin-Needham hypothesis will enhance our understanding of nitrogen metabolism in the asteroids.

Neither the selective forces which have lead to the production of significant rates of urea excretion in five of the seven species of sea stars studied nor the endogenous mechanism(s) by which urea is produced is known. Significant activities of the only three Krebs Ornithine Cycle enzymes assayed to date in the pyloric caeca of P. ochraceus have been identified. Urea could be synthesized by the Krebs Ornithine cycle if it is found to be intact in the asteroids. Urea could also be produced from purine metabolism but less than 10 percent of the total nitrogen excreted by animals usually originates in this way (Hoar, 1984). Another possible source of endogenously produced urea is, as suggested by Diehl and Lawrence (1979) for Luidia clathrata, the de novo synthesis of arginine coupled with a high arginase:urease activity as shown for Leptasterias hexactis by Stickle et al. (1982). Certainly, it will be impossible to identify selective forces which have led to the excretion of significant quantities of urea by the asteroids without first identifying the metabolic path ways by which urea is synthesized.

One possible explanation for the excretion of urea by the asteroids may be the high degree of anaerobic metabolism in certain organs, including the pyloric ceca of the sea star Asterias rubens where heat production under anoxia was 68-69% of heat production under normoxic perivisceral fluid PO_2 levels (Shick, 1983). Lack of perfusion in this thick organ is believed to be responsible for the evolution of a high degree of anaerobic metabolism. Lack of perfusion of deep tissues may also have been the selective force which has led to the synthesis of urea in the asteroids because it is less toxic than ammonia.

A distinct gradient in total nitrogen and ammonia excretion rate of the sea star species exists with respect to their activity pattern and vertical distribution. Higher excretion rates exist in the ecologically dominant intertidal carnivores, Evasterias troschelii and Pisaster ochraceus than in the intertidal Leptasterias hexactis, Dermasterias imbricata and Henricia leviscula with the lowest rates occurring in the least active subtidal carnivores O. koehleri, C. papposus and P. tesselatus.

With two exceptions, primary amine loss from the sea stars studied is of little significance to their nitrogen budgets. Primary amines accounted for 17% of the nitrogen lost by the soft bodied Crossaster papposus and contributed 21% of nitrogen lost by the mucous producing Pteraster tesselatus. These results were obtained with environmentally realistic ambient total primary amine concentrations of 0.5-3.0 µM. Similar results have been obtained for Luidia clathrata in which negligible amino nitrogen was excreted (Diehl and Lawrence, 1979).

Ammonia and total nitrogen excretion rates of the sea stars investigated in this study are of the same magnitude as in a number of other echinoderm (Diehl and Lawrence, 1979; Ellington and Lawrence, 1974; Emerson, 1969 and Stickle et al., 1982) and molluscan species (Emerson, 1969; Bayne and Scullard, 1977 and Stickle et al., 1985) but are significantly lower than those of the polychaete Nephytys caeca (Emerson, 1969) and the crustaceans Macrobrachium rosenbergii (Nelson et al., 1977) and Pagurus sp. (Emerson, 1969). Echinoderms have classically been found to have low metabolic rates (Binyon, 1972; Shick, 1983) but, as discussed by Stickle et al (1982), that statement is largely due to the expression of echinoderm metabolic data on a wet or dry (including skeletal components) weight basis. The ratio of wet to dry organic tissue weight varied between 6.8 and 15.9:1 for the species studied herein while the total dry weight to tissue dry weight ratio varied for 2.1-3.4:1. Consequently metabolic rates of the sea stars presently being studied will be underestimated by these ratios if reported on a wet or total dry weight basis.

Now that the excretion of significant quantities (>10% of total N) of urea has been documented for six species of asteroids, including Leptasterias hexactis, the metabolic pathways by which urea is synthesized must be characterized before any attempt is made to identify selective forces which have led to this apparent anomaly of the Baldwin-Needham hypothesis.

5. ACKNOWLEDGEMENTS

Most of the research on this project was carried out at the University of Washington's Marine Laboratory at Friday Harbor. I am grateful to the Director, Dr. Dennis Willows, for maintaining the laboratory in an atmosphere which is technically supportative, intellectually exciting and esthetically pleasing. Dr. Stanley Rice of the NMFS laboratory in Auke Bay, Alaska kindly sent me the Evasterias troschelii used in this project. Thanks go to Martin Kapper for reviewing this manuscript and to Phuong Khanh Nguyen for typing it. This project would have been impossible without a Summer Fellowship provided by the Louisiana State University Council on Research.

REFERENCES

Baldwin, E. 1967. Dynamic aspects of biochemistry. 5th ed. Cambridge: Cambridge Univ. Press.

Bayne, B.L. & C. Scullard. 1977. Rates of nitrogen excretion by species of Mytilus (Bivalvia: Mollusca). J. mar. biol. Ass. U.K. 57:355-369.

Binyon, J. 1972. Physiology of echinoderms. Oxford: Pergamon Press Ltd.

Casey, C. & P.M. Anderson. 1982. Subcellular localization of glutamine synthetase and urea cycle enzymes in liver of spiney dogfish (Squalus acanthias). J. Biol. Chem. 257:8449-8453.

Diehl, W.J. III & J.M. Lawrence. 1979. Effect of nutrition on the excretion rate of soluble nitrogenous products of Luidia clathrata (Say) (Echinodermata: Asteroidea). Comp. Biochem. Physiol. 62A:801-806.

Ellington, W.R. & J.M. Lawrence. 1974. Coelomic fluid volume regulation and isosmotic intracellular regulation by Luidia clathrata (Echinodermata: Asteroidea) in response to hyposmotic stress. Biol. Bull. mar. biol. Lab., Woods Hole 146:20-31.

Emerson, D.N. 1969. Influence of salinity on ammonia excretion rates and tissue constituents of euryhaline invertebrates. Comp. Biochem. Physiol. 29:1115-1133.

Hoar, W.S. 1984. General and comparative physiology, 3rd ed. Englewood Cliffs, New Jersey: Prentice-Hall.

Lambert, P. 1981. The sea stars of British Columbia, Handbook 39. Victoria, British Columbia: British Columbia. Provincial Museum.

Markus, J.A. & C.C. Lambert. 1983. Urea and ammonia excretion by solitary ascidians. J.Exp. Mar. Biol. Ecol. 66:1-10.

Needham, J. 1931. Chemical Embryology. Cambridge, Cambridge Univ. Press.

Nelson, S.G., A.W. Knight & H.W. Li. 1977. The metabolic cost of food utilization and ammonia production by ju-

venile Macrobrachium rosenbergii (Crustacea: Palaemonidae). Comp. Biochem. Physiol. 57A: 67-72.
North, B.B. 1975. Primary amines in California coastal waters: utilization by phytoplankton. Limnol. Oceanogr. 20:20-27.
Packard. G.C. & M.J. Packard. 1983. Patterns of nitrogen excretion by embryonic softshell turtles (Trionyx spiniferus) developing in cleidoic eggs. Science 221:1049-1050.
Shick, J.M. 1983. Respiratory gas exchange in echinoderms. In M. Janoux & J.M. Lawrence (eds) Echinoderm Studies, Vol. 1, p. 67-110. Rotterdam: Balkema.
Shirley, T.C. & W.B. Stickle. 1982. Responses of Leptasterias hexactis (Echinodermata: Asteroidea) to low salinity II. Nitrogen metabolism, respiration and energy budget. Mar. Biol. 69:155-163.
Solorzano, L. 1969. Determination of ammonia in natural waters by the phenolhypochlorite method. Limnol. Oceanogr. 14:799-801.
Stickle, W.B., T.C. Shirley & T.D. Sabourin. 1982. Patterns of nitrogen excretion in four species of echinoderms as a function of salinity. In J.M. Lawrence (ed.), Proceedings of the international echinoderm conference, Tampa, Florida, p. 371-377. Rotterdam: Balkema.
Stickle, W.B., M.A. Kapper, E. Blakeney & B.L. Bayne. 1985. Effects of salinity on nitrogen metabolism of the muricid gastropod Thais (Nucella) lapillus. J. Exp. Mar. Biol. Ecol. (submitted).

Proceedings of the Fifth International Echinoderm Conference / Galway / 24-29 September 1984

Comparison of chemical senses in the sea star *Marthasterias glacialis* with chemical senses in some fishes

T.VALENTINČIČ
Department of Biology, Ljubljana, Yugoslavia

D.OTA
Institute of Biology, Ljubljana, Yugoslavia

ABSTRACT: Low molecular weight compounds: lactic acid, some amino acids, and acetylcholine iodide release a feeding response in hungry sea stars (Marthasterias glacialis). Single amino acids known from electrophysiological studies to stimulate fish chemoreceptors do not release feeding response when presented one at a time to naive brown-bullheads (Ictalurus nebulosus). Conditioning procedures established a single amino acid-food search association in the fish. Either single amino acids or mixtures of them can make-up a searching image in the fish. The second aim of this work is a comparison of behavioural data on chemoreception in Marthasterias glacialis with electrophysiological data from structure-activity studies of chemoreceptors in the fishes Ictalurus nebulosus and Salmo gairdneri from the literature.

1 INTRODUCTION

Acetylcholine iodide induces opening of the mouth , whereas adrenaline causes its closure in Asterias forbesi (Anderson 1954). The following substances release behavioural responses of the whole animals: lactic acid and urea (Zafiriou 1972); L-glutamic acid (Castilla, 1972); L-cysteine, L-proline, several neutral amino acids, lactic acid, and acetylcholine iodide (Valentinčič 1973, 1975, 1980 a and b); betaine and some amino acids (Moore and Huxley 1976). Behavioral sensitivity to lactic acid (3x 10^{-11} M) found in Marthasterias glacialis (Valentinčič, in press) and high sensitivity to the same chemical (10^{-11}M) recorded electrophysiogically in Ophiura texturata (Moore, personal communication) are among the highest in aquatic organisms.

Naive fishes are behaviourally not responsive, or seldom responsive, to single chemicals. To consitute full stimulatory activity, mixtures of low molecular compounds are required (Carr 1982). Conditioning establishes single chemical-behaviour association in fishes.

Electrophysiological investigations of fish chemoreceptors are probably the most extensive among aquatic organisms. Close relationships between marine and freshwater fish chemoreceptors have been demonstrated (Caprio 1980; Goh and Tamura 1980). Standard electrophysiological techniques enbled studies of sensitivity and structure-activity relationships of amino acids in gustatory and in olfactory sense organs. Examples of gustatory and olfactory chemosensivity as found in channel catfish and rainbow trout will be cited (Hara 1976, 1977, 1982; Caprio, 1975, 1978, 1982). Our results on chemoreceptors in Marthasterias glacialis will be compared with structure-activity data cited from literature.

2 METHODS

2.1 Experiments with the sea star

The sea stars Marthasterias glacialis were collected at different localities in the Gulf of Rijeka from 1978 to 1983. They were raised in closed system aquaria. Large biological filters lowered concentration of organic substances added into sea water in less than three hours. The sea stars were starved from 15 to 30 days to increse their behavioural responsiveness to feeding stimuli. Responsiveness of the sea stars to feeding stimuli depend on hunger, which can be determined using live mussles or fish meat. L-cysteine and L-leucine were used as hunger controls later. All animals responsive to meat extract would respond to L-cysteine also. The L-cysteine controls were repeated 44 times with a group of 19 animals. They were used to demonstrate constancy and reproducibility of the response. 1 cm^3 of sea water and 20 minutes later 1 cm^3 of test substance at concentration 3×10^{-3} M, were introduced with an automatic Oxford micropipette. A strong flow of water provided fast distribution of the substances. The chemicals reached remote corners of the aquaria in less than 1 minute. Final dilution of the substances was obtained after a maximum of 2 minutes: 3×10^{-7}M. Behaviour of animals (orienting responses, movements of single arms, and crawling), were recorded every five minutes of the control and of the test. Tests and controls were compared using a modified Wilcoxon test. Crawling behaviour was scored more than movements of single arms and movements of single arms were scored more than orienting responses. Later tests were compared with the corresponding L-cysteine responsiveness controls using the same statistics.

2.2 Experiments with fish

Brown bullheads (Ictalurus nebulosus), were obtained from the fish farm, Crna Mlaka, near Karlovac. They were adapted to individual aquaria (100 dm^3) during a period of 1 month. The fish stayed in plastic tubes used for shelters during 16 hours of illumination daily. The responsiveness to chemicals was tested when fish were inactive. Air bubbles mixed aquarium water, and the test stimuli reached the animals in shelters in less than 1 minute. The behaviour of animals was observed 2 minutes after the injection of 1 cm^3 of aquarium water and during 2 minutes after the addition of 1_2cm^3 of test solution containing 10^{-2}M of L-alanine, L-arginine, L-cysteine, acetylcholine iodide respectively. A final concetration of 10^{-7} M was attained in test aquaria in less than 2 minutes. Swimming consists of numerous turns and short straight swims within a limited space in the aquarium. The number of more than 90 degree turns was counted as a measure of the amount of swimming. Conditioning procedure consisted of the test, with a food reward at the end of the test.

2.3 Comparison of the behavioural responses of Marthasterias glacialis with data on fish chemosensitivity from literature

The test concetration (3×10^{-7} M) was the lowest possible to induce full behavioural response to hunger control stimuli, L-cysteine and L-leucine.Substances less effective at either receptor or behavioural level stimulated smaller numbers of animals and to a lesser degree. Substances with behavioural threshold values below 3×10^{-5} M were probably not detected in our experiments. Structure-activity studies in fishes used higher concetration (10^{-4} M) thus detecting more and less sensireceptor responses. The amplitude measured in a nerve response served as a quantitative measure of stimulus-receptor binding.

The data on the sea star sensitivity to chemicals are presented as: full stimulatory effectiveness = more than 80% of the sea stars that responded to L-cysteine responded to the test chemical; high s. e.= 50 - 80%; medium s. e. = 30 - 50%; low s. e. = less than 30 % and statistically different from the sea water control.

Relative stimulatory effectiveness of a chemical close to 100% described in fish chemoreceptors probably correspond to full behavioural effectiveness in the sea star. Values of the relative effectiveness lower than 100% would either correspond to estimates high, medium, and low effectiveness or the substance is not an adequate stimulus for feeding. Values of relative stimulatory effectiveness close to 0 mean that the substance was not detected by fish chemoreceptors studied. In Marthasterias glacialis the estimate "no effectiveness" means that the substance was not detected by chemoreceptors related to feeding at test concentration $3x10^{-7}$ M. With this limitations in mind results of our studies and results of studies in the two fishes Ictalurus punctatus and Salmo gairdneri cited from Caprio 1975, 1978, and 1982 and Hara 1976, 1977, and 1982 can be compared.

3 RESULTS

3.1 Innate response of Marthasterias glacialis and conditioned response of Ictalurus nebulosus to chemicals

L-cysteine provoked food search in hungry sea stars in 44 tests during 44 weeks. No improvement with experience was observed even if the same feeding stimulus was repeated several times in succesion.

Brown bullheads (Ictalurus nebulosus) brought from nature do not respond to single chemicals such as L-arginine (Fig. 2, first trials), L-alanine, L-cysteine, and acetylcholine iodide. If a food reward was offered to the fish 2 minutes after the injection of the signal chemical, all the fish became conditioned after 30 to 50 trials. Nine brown bullheads were offered L-arginine in 2 to 3 trials daily. (Fig. 2). Most animals would respond to the stimulus after the conditioning procedure was accomplished. Different groups of brown bullheads were conditioned to L-alanine, L-cysteine, and acetylcholine iodide.

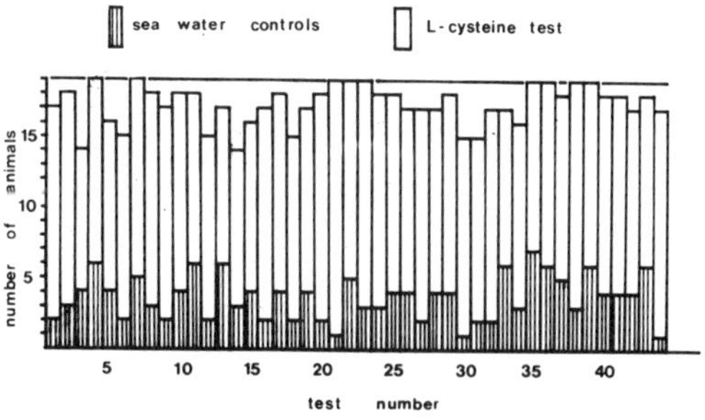

Fig. 1: Responsiveness of 19 Marthasterias glacialis to sea water controls and to L-cysteine at $3x10^{-7}$ M in 44 tests. The response was considered positive if the animals crawled for more than 5 minutes.

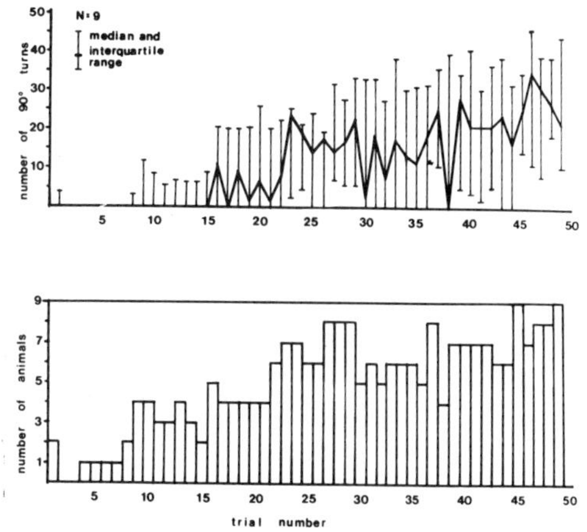

Fig. 2: Responsiveness of 9 brown bullheads (Ictalurus nebulosus) to L-arginine during 49 conditioning trials. A: Median number and interquartile range of turns above 90 degrees. B: Number of animals leaving shelters and starting the search for food reward.

3.2 Comparison of chemical senses

Controlled hunger brought all the test sea stars into an almost perfect responsive state to feeding stimuli: both isomers of lactic acid, both isomers of cysteine, L-proline, betaine, acetylcholine iodide, and several analogues (Table 1). Ideally, if a perfect manipulation of a physiological state of an animal would be possible, all the sea stars would respond to a feeding stimulus above its feeding behaviour threshold. The differences in threshold in different animals were : from 10^{-11} to 10^{-9} M in L-lactic acid and from $3x10^{-9}$ to $3x10^{-7}$ M in L-cysteine (Fig. 3). The behavioural threshold studies of effective neutral amino acids yielded results similar to those in L-cysteine.

L-isomers of naturally occurring amino acids release feeding behaviour in the sea star Marthasterias glacialis and stimulate chemoreceptors of the two fishes (Table 1). D-isomers of neutral amino acids are not stimulatory or cause little stimulation in the fish chemoreceptors. On the contrary L- and D- isomers of cysteine, alanine, and lactic acid release feeding effec[illegible] in Marthasterias glacialis. Th

TABLE 1

COMPARISON OF STIMULATORY EFFECTIVENESS OF NEUTRAL AMINO ACIDS IN MARTHASTERIAS GLACIALIS AND IN SOME FISHES

(Results of statistic comparison of tests with sea water controls are given as: NS, +, ++ for not significant and significant differences at 5% and 1% levels respectively; the comparisons of tests with the corresponding L-cysteine controls are given: ns, x, xx for not significant and significant differences at 5% and 1% levels.

Techniques	behaviour	electrophysiology		
Species	Marthasterias glacialis	Ictalurus punctatus	Ictalurus punctatus	Salmo gairdneri
Chemoreceptors	related to feeding	olfaction (neural)	extra oral taste	olfactory bulb
Standard	L-cysteine L-leucine	L-alanine	L-alanine	L-serine
Function of standard	responsiveness control	normalization of data..........		
Concentration	3×10^{-7} M	10^{-4} M	10^{-4} M	10^{-4} M
Reference	Valentinčič, in press	Caprio, 1978, 1982	Caprio 1975, 1978 1982	Hara, 1976, 1977 1982

Substance	stim. act.	Statistics SW	Cys (P)	Relative stimulatory effectiveness		
GLYCINE	full	++	ns	53.5	53.4	88.0
L-ALANINE	high	++	ns,x	100	100	112.3
D-ALANINE	high	++	ns	--	57.2	30
L-VALINE	med.	++	ns,x	--	46.7	65.9
D-VALINE	no	NS	xx	--	--	11
L-NORVALINE	high	++	ns	101.3	44.6	124.5
L-LEUCINE	full	++	ns	--	32.6	92.7
D-LEUCINE	no	NS	xx	--	--	16
L-ISOLEUCINE	full	++	ns	--	34.8	77.1
L-NORLEUCINE	high	++	ns	93.9	42.8	135.3
L-SERINE	high	++	ns,xx	91.7	69.5	100
L-THREONINE	no	NS	xx	--	50.0	53.8
L-CYSTEINE	full	++	ns	118.7	51.7	100.2
D-CYSTEINE	full	++	ns	--	--	52
L-ARGININE	high	++	ns	75.0	86.6	49
L-ASPARTIC ACID	low	+	xx	78.0	42.9	14.8-50.1
L-ASPARAGINE	no	NS	xx	90.1	37.2	117.3
L-GLUTAMIC ACID	low	+	xx	66.9	27.0	6.8-32.1
L-GLUTAMINE	no	NS	xx	104.6	60.6	119.3
ß-ALANINE	no	NS	xx	63.2	44.2	7.1
α-METHYL-ALANINE	high	++	ns,x	50.9	27.0	63.0
α-AMINOBUTYRIC ACID	--			107.2	64.0	131.0
N-ACETYL-GLYCINE	med.	++	xx	--	34.8	--
N-ACETYL-L-ALANINE	--			49.5	44.0	26.8
N-ACETYL-L-LEUCINE	no	NS	xx	--	--	--
N-ACETYL-L-CYSTEINE	full	++	ns	--	--	8.8
N-METHYL-GLYCINE	low	+	xx	--	25.3	1.1
BETAINE	full	++	ns	--	--	2.6
GLYCINE METHYL ESTER	no	NS	xx	--	43.9	50.1
L-ALANINE METHYL ESTER	no	NS	xx	98.0	98.8	112.7
L-ALANINE ETHYL ESTER	no	NS	xx	103.0	97.6	108.6
L-CYSTEINE METHYL ESTER	full	++	ns	--	--	--
L-CYSTEIC ACID	no	NS	xx	51.3	--	10.1
GLYCYL-GLYCINE	no	NS	xx	--	--	19.9
L-ALANYL-GLYCINE	no	NS	xx	--	72.3	--
GLUTATHIONE	high	++	ns	--	27.3	48.6
LACTIC ACID	full	++	ns	--	--	--
GLYCOLIC ACID	high	++	xx	--	--	--
ACETYLCHOLINE IODIDE	full	++	ns	--	--	--

sponse to amino acids with long fourth α-moiety is stereoselective in the sea star. Amino acids L-leucine, L-isoleucine, and L-norvaline are very effective substances while their D-isomers do not influence behaviour of the sea stars. The basic amino acid, L-arginine stimulates the asteroid and taste in the channel catfish. Acidic amino acids have little effect as feeding stimuli in Marthasterias glacialis. Their relative stimulatory effectiveness in the fishes has medium values. L-asparagine and L-glutamine do not influence behaviour of the sea star. Both substances are highly effective in fish olfaction. β-alanine is effective only in the catfish taste.

α-methyl-alanine has the same stimulatory effectiveness as L-alanine in Marthasterias glacialis while this substance stimulates the fish receptors a little. The effectiveness of α-aminobutyric acid was high in all fish species cited. Acetylated products are little effective in fish chemoreceptors. N-acetyl-L-cysteine release feeding behaviour of the sea star regularly. Other acetylated products of neutral amino acids triggered the sea star behaviour on some occasions. Betaine was found to be one of the most effective substances in asteroid behaviour. This substance is a poor stimulant in rainbow trout olfaction.

One of the big differences between the sea star and the fishes was found for amino acid esters. The only effective ester in Marthasterias glacialis feeding was L-cysteine methyl ester which had the same effectiveness as the parent L-cysteine. On the contrary the two fishes can detect neutral amino acid esters by olfaction and by taste. The L-alanine esters were most effective. Di- and tri-peptides of effective amino acids were found ineffective in both groups studied with the exception of catfish taste. Reduced glutathione was the only tripeptide which induced vigorous search and confusion in the sea star ability to discriminate food items from other small objects.

Three important key stimuli for feeding are listed in the bottom of table 1. Lactic acid was found to be the most effective in Marthasterias glacialis (Fig. 3).

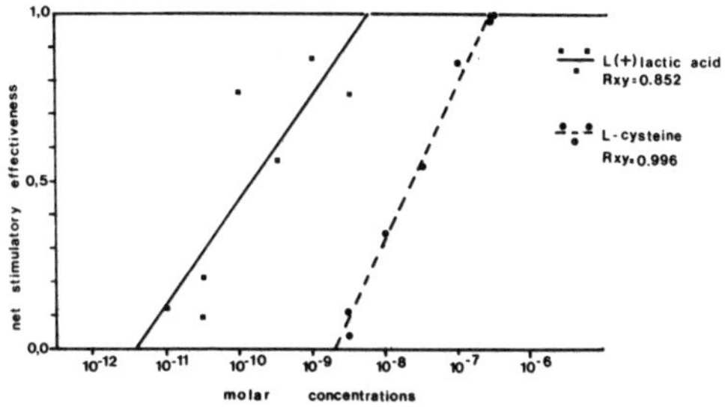

Fig. 3: Behavioural threshold values for L(+)lactic acid and for L-cysteine in 19 hungry Marthasterias glacialis. The net stimulatory effectiveness was calculated according to the formula: (p(test)-p(control))/(1-p(control)) (Abbot's correction, Bliss, 1967).

4 DISCUSSION

Marthasterias glacialis does not change responsiveness to feeding stimuli if the stimuli are repeated. On the contrary naive brown bullheads brought from nature did not respond to stimulation with low molecular substances: L-arginine, L-cysteine, L-alanine, and acetylcholine iodide. All brown bullheads were conditioned to the same substances after 30 to 50 trials. Searching image in fishes can be produced either with natural mixtures released from dead or living prey or with single chemicals applied in a laboratory.

According to McClintock and Lawrence (1984) the sea star Luidia clathrata can be ingestively conditioned to living prey and their juices. It seems that this type of conditioning can not operate with the single chemicals in Marthasterias glacialis. These substances are innate feeding stimuli and no further increase in responsiveness can be expected. No habituation to feeding stimuli was observed in the course of our experiments.

Echinoderms have normal chemosensory abilities. They can be compared to chemosensory abilities of some fishes irrespective of their marine or freshwater life. There are " receptors " typical for each animal species studied. The following differences can be summarized for chemical senses in Marthasterias glacialis and in the two fish species cited:

- The fishes do not detect D-isomers of amino acids while the sea star is not stereoselective for lactic acid, cysteine, and short side chain neutral amino acid alanine.
- Amino acid esters effective in the fish chemoreceptors are not stimulatory to the sea star. An exception to this role is the ester of cysteine which stimulates activity in the sea star.
- Acetylated products of amino acids are not effective in the fishes. The only effective acetylated amino acid in the asteroid chemosensitivity was N-acetyl-L-cysteine.
- Neutral amino acids not containing the sulphydryl group should interact with receptors different from the cysteine receptor in Marthasterias glacialis. This appears not to be the case in the fishes.
- Basic amino acid L-arginine is readily detected both by the two fishes and by the echinoderm.
- The tripeptide reduced glutathione contains sulphydryl group. It has some of the cysteine stimulating properties in Marthasterias glacialis. Other peptides were not effective in both groups compared.
- It would be worth searching lactic acid and acetylcholine sensitivity in fishes. This is supported by our conditioning experiments with acetylcholine iodide in the brown bullhead (Ictalurus nebulosus).

REFERENCES

Anderson, J.M. 1954. Studies on the cardiac stomach of the starfish, Asterias forbesi. Biol. Bull. mar. biol. Lab. Woods Hole, 107(2):157-173.

Bliss, C.I. 1967. Statistics for biology. London, McGraw-Hill.

Caprio, J. 1975. High sensitivity of catfish taste receptors to amino acids. Comp. Biochem. Physiol., 52A:247-251.

Caprio, J. 1978. Olfaction and taste in the channel catfish: an electrophysiological study of the responses to amino acids and derivatives. J. Comp. Physiol., 123: 357-371.

Caprio, J. 1980. Similarity of the olfactory receptor responses (EOG) of freshwater and marine catfish to amino acids. Can. J. Zool., 58: 1778-1784.

Caprio, J. 1982. High sensitivity and specificity of olfactory and gustatory receptors of catfish to amino acids. In Hara, T.J. (ed.), Chemoreception in fishes, p. 109-134. Amsterdam, Elsevier.

Carr, W.E.S. 1982. Chemical stimulation of feeding behaviour. In Hara, T.J. (ed.), Chemoreception in fishes, p. 259-273. Amsterdam, Elsevier.

Castilla, J.C. 1972. Avoidance behaviour of Asterias rubens to extracts of Mytilus edulis, solutions of bacteriological peptone, and selected amino acids. Mar. Biol. 15: 236-245.

Goh, Y. & T. Tamura 1980. Olfactory and gustatory responses to amino acids in two marine teleosts - red sea bream and mullet. Comp. Biochem. Physiol. 66C:217-224.

Hara, T.J. 1976. Structure-activity relationships of amino acids in fish olfaction. Comp. Biochem. Physiol. 54A:31-36.

Hara, T.J. 1977. Further studies on structure-activity relationships of amino acids in fish olfaction. Comp. Biochem. Physiol. 56A:559-565.

Hara, T.J. 1982. Structure-activity relationships of amino acids as olfactory stimuli. In Hara, T.J. (ed.), Chemoreception in fishes, p. 135-157. Amsterdam, Elsevier.

McClintock, J.B. & J.M. Lawrence 1984. Ingestive conditioning in Luidia clathrata (Say) (Echinodermata: Asteroidea): Effect of nutritional condition on selectivity, teloreception, and rates of ingestion. Mar. Behav. Physiol. 10: 167-181.

Moore, R.J. & C.J. Huxley 1976. Aversive behaviour of crown-of-thorns starfish to coral evoked by food related chemicals. Nature, Lond., 263:407-409.

Valentinčič, T. 1973. Food finding and stimuli to feeding in the sea star Marthasterias glacialis L. Neth. J. Sea Res. 7:191-199.

Valentinčič, T. 1975. Amino acid chemoreception and other releasign factors in the feeding response of the sea star Marthasterias glacialis (L.). Proc. 9th European mar. Biol. Symp. p. 693-705, Aberdeen University Press

Valentinčič, T. 1980. Structure-activity relationships of amino acids in Marthasterias glacialis chemo-

reception. In Jangoux, M. (ed.) Echinoderms: Present and Past, p. 331-335. Rotterdam, Balkema.
Valentinčič, T. 1980. Structure-activity studies of amino acids and acetylcholine in chemoreception of the starfish Marthasterias glacialis. In Brzin, M., Sket, D. & H. Bachelard (eds.) Synaptic constituents in health and disease, p. 683. Ljubljanam Oxford, Mladinska knjiga, Pergamon Press.
Zafiriou, O. 1972. Responses of Asterias vulgaris to chemical stimuli. Mar. Biol. 17:100-107.

Proceedings of the Fifth International Echinoderm Conference / Galway / 24-29 September 1984

The effect of feeding and starvation on the level and content of nucleic acids in the pyloric caeca of *Luidia clathrata* (Say)

STEPHEN A.WATTS & J.M.LAWRENCE
University of South Florida, Tampa, USA

ABSTRACT: The concentration of DNA increased in the pyloric caeca with starvation although the total amount decreased. Assuming the loss of equal sized cells, approximately 30% of the decrease in the size of the pyloric caeca could be attributed to a decrease in cell number during the initial stages of starvation. Approximately 80-100% of the decrease in the size of the pyloric caeca could be attributed to a decrease in cell number with prolonged starvation. It seems that initial utilization of nutrients in the pyloric caeca with starvation involves a decrease in cell size as well as a decrease in cell number. After the potential for decrease in cell size has been exhausted, further nutrient demands are met primarily by cell destruction. The concentration of RNA decreased significantly only after prolonged starvation. RNA:DNA ratios were 5.4-6.4 in fed individuals and 3.8-4.2 in starved individuals. The greatest decrease in the ratio occured in the first month of starvation. The minimal (maintenance) RNA:DNA ratio is 3-4.

1 INTRODUCTION

Changes in size and biochemical composition of the pyloric caeca is a common occurrence in asteroids (see Lawrence and Lane 1982). These changes can be the result of seasonal changes in the physiology and reproductive condition of the organism, or related to the nutritional condition of the organism. Although many studies have examined the changes in size and biochemical composition of the pyloric caeca, most investigators have used wet or dry weight of the organ or ratios of biochemical components as an index of changes in biochemical composition. Giese (1966) indicated that determination of DNA was necessary for correct interpretation of changes in the biochemical composition of the organ, as the quantity of DNA would serve as a better index of the number of cells present in a tissue.

Greenfield (1959) found no distinct seasonal trend in the concentration of nucleic acids from the pyloric caeca of Pisaster ochraceus. This suggests that seasonal changes in the size of the pyloric caeca are the result of changes in cell number. Van der Plas and Voogt (1982) found that DNA content in the pyloric caeca of Asterias rubens roughly followed the pyloric caeca index, indicating that the number of cells follows the size of the pyloric caeca. However, they also found the level of DNA varied inversely with the pyloric caeca index. From this they concluded that both changes in cell number and cell content are involved in changes in the size of the pyloric caeca.

Relative changes in the RNA content of cells are indicative of changes in the cell's protein-synthesizing capacity. However, since cell size varies, the RNA:DNA ratio may give a more accurate interpretation of the relative protein-synthesizing capacity of the tissue. Greenfield (1959) indicated that there was no seasonal variation in the level of RNA with changes in the size of the pyloric caeca. However, the RNA:DNA ratios (calculated by the present authors) during September-November (high PCI) was 4.6, while the ratio dropped to 2.6 during March-May (low PCI). Assuming that the variable RNA is involved in protein synthesis, these data indicate an approximate doubling of the protein synthesizing capacity of the pyloric caeca during periods of maximal growth. Van der Plas and Voogt (1982) indicated that an increase in the size in the pyloric caeca of Asterias rubens, was associated with an increase in the RNA levels concommitant with a decrease in DNA levels. These data imply that there is an increase in the RNA:DNA ratio with an increase in the PCI. The RNA:DNA ratios (calculated by the present authors) ranged from 1.5 (during the periods of high PCI) to Ø.7 (low PCI), and again indicate an increase in the protein-synthesizing capacity with an increase in the size of the pyloric caeca.

Although seasonal variations in the size of the pyloric caeca occur in many asteroids, the nutritional condition of the organism can also affect the size of the pyloric caeca. The purpose of this present study was to examine the level and content of DNA and RNA in fed and starved individuals of Luidia clathrata, and to determine the effects of nutritional condition on the mechanisms and time course of pyloric caeca growth. We have also examined the influence of nutritional condition and pyloric caeca sizeontheRNA:DNA ratios.

2 MATERIALS AND METHODS

Specimens of Luidia clathrata (Say)(6.5-7.5 cm radius) were collected in June 1983 in Old Tampa Bay, Tampa, Florida on the south side of the Courtney Campbell Causeway in 3-4 m water. Field temperature and salinity were 27oC and 19o/oo. Individuals were placed in aerated aquaria in a constant environment chamber (25oC, 22-24o/oo S) for 3 weeks prior to the beginning of the experiment. The starfish were fed a maintenance diet of the bivalve Donax variabilis(Dehn 1981). After the three weeks, 1Ø individuals were dissected and the pyloric caeca analysed for the level and content of nucleic acids (TØ). Forty additional animals were divided into two groups of twenty individuals. One group was fed the bivalve Donax variabilis ad libidum; the other group was starved. Ten individuals from each group were dissected and analysed after one (T1) and two (TØ) months.

The pyloric caeca were removed from all individuals, weighed, and the pyloric caeca index (PCI) calculated by wet weight pyloric caeca/wet weight individual x 1ØØ. Samples of the pyloric caeca were removed and placed on ice for nucleic acid determinations.

The TCA nucleic acid extraction procedures of Schneider (1945) are not appropriate for the lipid rich pyloric caeca. The following procedure was developed to extract intact, biologically active DNA and RNA from the pyloric caeca. These procedures work equally well on other echinoderm tissues.

Approximately Ø.25 g pyloric caeca was homogenized on ice in NET buffer (Ø.Ø5 M Tris-HCl pH 8.Ø, Ø.1 M NaCl, Ø.Ø2 M EDTA). Tissues were homogenized (5% w/v) in the presence of proteinase K (Scientific Products, 1ØØ ug/ml homogenate). The homogenate was made to 2% SDS, mixed and incubated for 1.5-2 hours at 45-5ØoC. After incubation, two volumes of prewarmed chloroform:isoamyl alcohol (24:1) were added, vortexed and incubated for 1Ø min. The samples were then centrifuged at high speed in an IEC clinical centrifuge for 1Ø min and the aqueous layer removed. One volume of phenol (saturated with NET buffer) and one volume of chloroform were added to the aqueous layer, mixed, incubated for 1Ø min, and re-centrifuged. An additional extraction with phenol:chloroform or phenol may be necessary for tissues rich in lipids or pigments. The clarified aqueous phase was removed without disrupting the interface of the phases. The aqueous phase was then precipitated with two volumes of 1ØØ% ethanol. After overnight storage at -1ØoC, the nucleic acids were recovered by centrifugation for 1Ø min and resuspended in 3 ml cold Ø.Ø1 M Tris-HCl pH 8.Ø, Ø.ØØ1 M EDTA buffer. The DNA recovered by this procedure is protein free, intact, restrictable, and almost exclusively nuclear in origin (Herrin et al., in press).

DNA levels were measured by the diphenylamine colorimetric analysis of Dische (1955) using herring sperm DNA as the standard. RNA levels were measured by the orcinol colorimetric analysis of Ogue and Rosen (195Ø) using yeast RNA as the standard. For those tissues which have low yields of nucleic acids, DNA can be determined by the method of Burton (1956) and RNA by the method of Almog and Shirey (1978). The statistical significance of all data was determined by Student's t-test.

3 RESULTS

The effect of two months' starvation or feeding on the pyloric caeca index (PCI), concentrations of nucleic acids, and the RNA:DNA ratio in Luidia clathrata is shown in Table 1. The PCI increased during the first month in those individuals which were fed ad libidum. No further increase in the PCI occurred during the second month of feeding. The PCI in individuals after one month's starvation was approximately one half of those fed ad libidum. No further decrease in the PCI occured during the second month of starvation. The concentration of DNA decreased with feeding and increased with starvation. The greatest change in the concentration of DNA occured during the first month.

Table 1. Pyloric caeca index (PCI), concentrations of DNA and RNA, and the RNA:DNA ratio in Luidia clathrata at time 0 (T0) and after one month (T1) and two months (T2) of starvation or feeding. Numbers indicate mean with SEM in parentheses. n=8-10.

	Time (months)		
	T0	T1	T2
PCI			
Fed	11.9(0.9)	15.2(1.1)	14.8(1.2)
Starved		8.4(0.7)	8.8(0.7)
DNA(ug/gFW)			
Fed	813(57)	664(38)	735(35)
Starved		1007(47)	963(60)
RNA(ug/gFW)			
Fed	3964(88)	4230(205)	3914(163)
Starved		4099(215)	3431(162)
RNA:DNA			
Fed	5.0(0.3)	6.4(0.3)	5.4(0.4)
Starved		4.2(0.3)	3.8(0.2)

The concentration of RNA was significantly different ($P<0.05$) between fed and starved individuals only after two months. However, the RNA:DNA ratio decreased after one month of starvation.

Figure 1 shows the relation of the RNA:DNA ratio to the PCI. With an increase in the PCI there is a curvilinear increase in the ratio, with the maximal change in the ratio occuring between PCI values of 8-12.

Individual variability in the pyloric caeca weight and the level and content of DNA within an experimental group was high. For this reason, the total DNA vs the PC wet weight for all individuals and exposure times are presented in Figure 2A. Line (I) indicates the theoretical changes in total DNA which would have occured if there had been a change in cell number of equal-sized cells. The horizontal line (II) indicates the theoretical changes in total DNA which would have occured if there had been a change in cell size but not in cell number. In general, PC values ranging from 0-3 g represent starved individuals, while values from 2-5 g represent fed individuals. If we assume the pyloric caeca weight indicates nutritional condition, several conclusions are apparent. During the initial stages of starvation there is a decrease in the amount of DNA but an increase in concentration of DNA (Figure 3A and B). Assuming loss of equal-sized cells, approximately 30% of the decrease in the size of the pyloric caeca could be attributed to a decrease in cell number during the initial stages of starvation. With prolonged starvation, approximately 80-100% of the decrease in the size of the pyloric caeca could be attributed to a decrease in cell number.

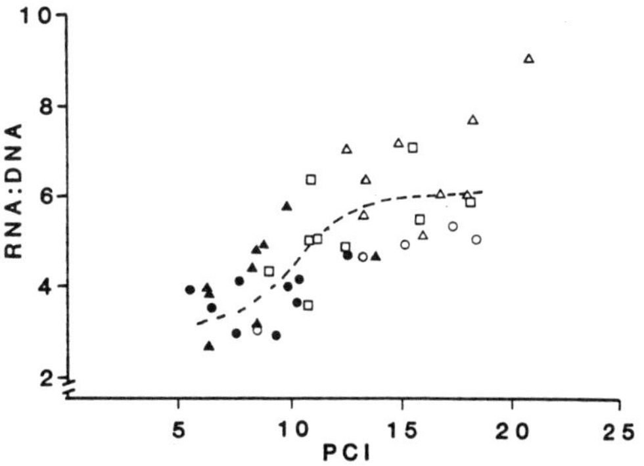

Figure 1: The relation of the RNA:DNA ratio to the pyloric caeca index (PCI) in Luidia clathrata at time 0 (T0) and after one month (T1) and two months (T2) of starvation or feeding (T0 □ ; fed T1 △ ,T2 ○ ; starved T1 ▲ ,T2 ●).

The relation of total DNA to the PCI is shown in Figure 2B. This figure is comparable to Figure 2A, although variability is higher when using the PCI. However, using the PCI instead of the PC weight allows comparison of DNA levels in individuls of different sizes, as the PCI is based on the ratio of the PC weight to the total body weight.

There is an inverse relation between DNA concentrations and the PC weight and the PCI (Figures 3A and B, respectively). Concentrations of DNA appear to stabilize at high PC weights and PCI. At low PC weights and PCI there is a rapid increase in concentrations of DNA.

4 DISCUSSION

Lawrence (1973) examined the effects of one month's starvation on the size and biochemical composition of the pyloric caeca of Luidia clathrata. The percent decrease in the size of the pyloric caeca in this study parallels the results found in his earlier study. The greatest change in the size of the pyloric caeca occured during the first month in both starved and fed individuals. Luidia clathrata appears to reach a maximum PCI of 20 under optimal laboratory conditions. Lawrence (1973) reported a maximal PCI of 12 and Dehn (1981) reported a maximal PCI of 14 in field populations of L. clathrata. The greatest decrease in the PCI during the early stages of starvation can be correlated with metabolic studies in

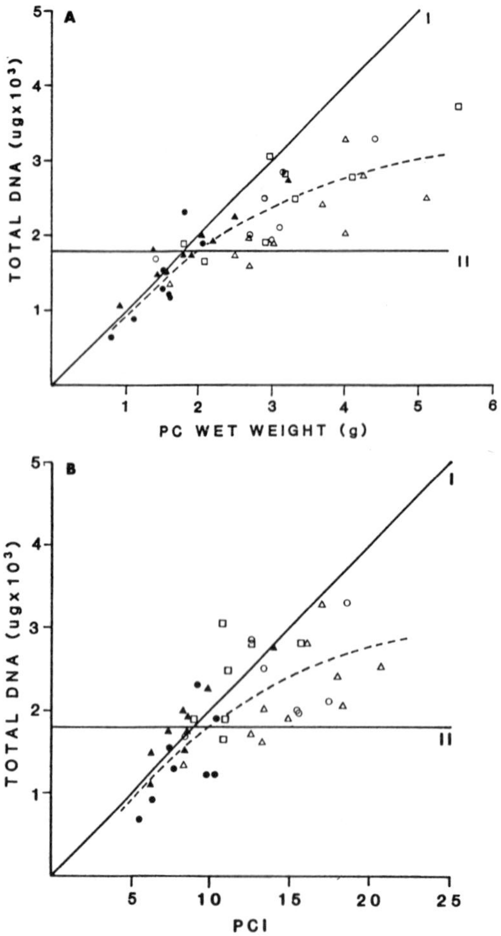

Figure 2: (A) Total DNA vs pyloric caeca wet weight in Luidia clathrata at time Ø and after one month (T1) and two months (T2) of starvation or feeding. The dashed line indicates the curve of best fit for all individuals. Line I represents the theoretical changes in the total amount of DNA if there is a change in cell number of equal-sized cells. Line II represents the theoretical changes in total DNA if the cell size changes and the cell number remains constant (TØ □ ;fed T1 △ ,T2 ○ ;starved T1 ▲ T2 ●).(B) Total DNA vs the pyloric caeca index (PCI).

other echinoderms. Respiratory rates of echinoids decrease quickly to a relatively stable, lower rate with starvation (see Lawrence and Lane 1982). The activities of several enzymes of intermediary metabolism decrease during starvation in the asteroid Echinaster spinulosus (Watts and Lawrence 1982). These lower rates of metabolism with starvation are probably involved in

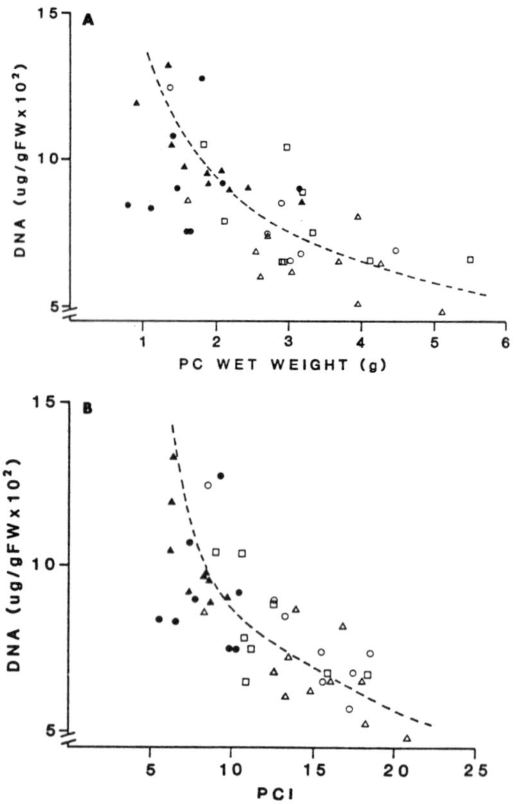

Figure 3: (A) Level of DNA vs the pyloric caeca wet weight in Luidia clathrata at time Ø and after one month (T1) and two months (T2) of starvation and feeding. Dashed line represents the curve of best fit (TØ □; fed T1 △,T2 ○; starved T1 ▲ , T2 ●). (B) Level of DNA vs the pyloric caeca index (PCI).

maintenance of the organisms.

Level and content of DNA varied directly with the pyloric caeca weight in similar-sized individuals. If we assume that all individuals are from the same population and differ only in nutritional condition, we see that an increase in the weight of the pyloric caeca is associated with an increase in the amount of DNA and a decrease in the concentration of DNA in the pyloric caeca. These data indicate that there are changes in the size and number of cells during growth of the pyloric caeca. Van der Plas and Voogt (1982) found similar results in Asterias rubens. Although there is an apparent change in the size and number of cells, the relative contribution of each mechanism controlling the size of the pyloric caeca depends on the relative size of the pyloric caeca. Assuming the loss of equal-sized cells, approximately 3Ø-5Ø% of the decrease in the size of

A. Decrease in the PC (decrease in cell size)

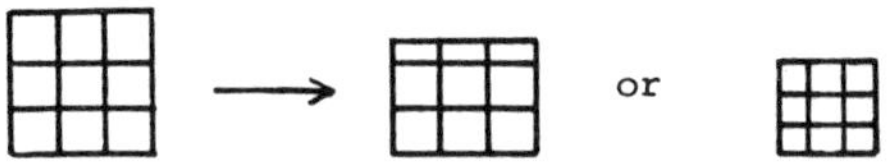

decrease in size of some cells — decrease in size of all cells

No change in the amount of DNA, but there is an increase in the concentration

B. Decrease in the PC (decrease in cell number)

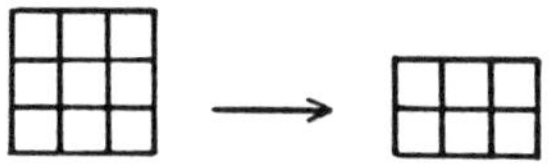

decrease in the number no decrease in size

Decrease in the amount of DNA, but not in the concentration.

C. Initial decrease in size of the pyloric caeca primarily by a decreae in cell size followed by a decrease in cell number.

Figure 4. Diagram of possible mechanisms of size changes of the pyloric caeca (PC).

the pyloric caeca could be attributed to a decrease in cell number during the initial stages of starvation in those individuals which are at or near maximum pyloric caeca size (PCImax). During the latter stages of starvation, when pyloric caeca sizes are less than 1/2 the PCImax, approximately 80-100% of the decrease in the size of the pyloric caeca could be attributed to a decrease in cell number.

Although many investigators have postulated that changes in the size of the pyloric caeca may be the result of changes in cell size, number, or both, none have considered the relative size of the pyloric caeca with respect to the maximal size under optimal conditions (PCImax). In L. clathrata changes in cell size occurs in the pyloric caeca only in those individuals with a PCI greater than 1/2PCImax. In those individuals with a PCI less than 1/2PCImax, changes in cell number is the major mechanism in controlling the size of the pyloric caeca. Lawrence (1973) suggested that the changes in the size of the pyloric caeca of L. clathrata were the result of cell destruction or production. Based on the low PCI reported in his study, his conclusions are comparable to the present study. However, when the PCI of L. clathrata exceeds 10, changes in cell size becomes increasingly important as a mechanism of pyloric caeca growth.

Based on these data, the growth of the pyloric caeca in L. clathrata can be described in three phases. First, assuming the animals are in nutritionally poor condition (low PCI), growth in the pyloric caeca is due to a rapid increase in cell production, as indicated by an increase in DNA content. Second, as the nutritional condition of the organisms improves (PCI > 1/2PCImax), growth of the pyloric caeca is the result of changes in cell size as well as cell number. Finally, further growth of the pyloric caeca is the result of an increase in cell size and cell number with a concommitant increase in the perivisceral cavity due to growth of the body wall.

Since the pyloric caeca contain a heterogeneous cell population, it is not clear which cell types are changing with changes in the size of the pyloric caeca. Van der Plas et al. (1981) found that only the storage cells in the digestive epithelium changed seasonally in the pyloric caeca of Asterias rubens. Although seasonal changes in the histology of the pyloric caeca of L. clathrata have not been examined, Klinger and Dehn (1980) found that the cells in the pyloric caeca of L. clathrata are similar to those cells in other asteroids. It is assumed that mechanisms of changes in cell types among asteroids are similar. A decrease in the number of these large storage cells would result in a decrease in the total amount of DNA present in the tissue (Figure 4). However, a decrease in the size of these cells may result in an apparent increase in the concentration of DNA in the tissue, as indicated in these data. Alternatively, large increases in DNA concentrations may be due to increases in ploidy, but this has not been demonstrated. There are seasonal changes in the ploidy of the asteroid stomach epithelium (Vorobjev 1982). It is difficult to estimate the relative contribution of either mechanism of organ growth without knowing the changes in the type and quantity of cells in the pyloric caeca.

The PCI can be used to determine changes in the number and size of cell populations with changes in the pyloric caeca size (Figures 2B,3B). The advantage of using the PCI is that individuals of different body sizes may be compared. However, reproductive condition may influence the PCI, and care should be taken when interpreting data using the PCI in individuals differing in reproductive states.

The concentration of RNA differed between fed and starved individuals only after two months. There was only a 13% decrease in RNA production in starved individuals after two months. However, the change in RNA concentration is not an accurate interpretation of protein-synthesizing capacity in a tissue where there is a heterogeneous population of both cell sizes and types. In this case it is more appropriate to use the RNA:DNA ratio to determine the protein-synthesizing capacity of the tissue. This ratio gives the average amount of RNA per cell, and is independent of tissue weight. RNA measured by the techniques presented in this paper represent predominantly ribosomal RNA, with minor contributions of messenger and transfer RNA. The RNA:DNA ratio in the pyloric caeca of L. clathrata decreased with one month's starvation, although there was no decrease in the concentration of RNA. This indicates a large decrease in the protein-synthesizing capacity of the pyloric caeca with starvation. The greatest increase in the RNA:DNA ratio occurs near 1/2PCImax, indicating that the greatest increase in synthetic activity occurs during the periods of cell growth. The minimal (maintenance) ratio is approximately 3-4 in L. clathrata, and represents the level of RNA production necessary for maintenance of the organisms.

These data indicate the importance of knowing the nutritional and physiological state of the organism when considering the changes in the biochemical composition of the pyloric caeca. The PCI relative to the PCImax can be used as a approximate estimate of the nutritional state of the organisms. In addition, the RNA:DNA ratio can indicate the physiological well-being of the organisms.

REFERENCES

Almog, R. & T.L. Shirey 1978. A modified orcinol test for the specific determination of RNA. Anal. Biochem. 91: 130-137.

Burton, K. 1956. A study of the condition and mechanisms of the diphenylamine reaction for the colorimetric estimation of deoxyribonucleic acid. Biochem. J. 62: 315-322.

Dehn, P.F. 1981. Growth and reproduction in Luidia clathrata (Say) (Echinodermata: Asteroidea). PhD dissertation, Univ. of South Florida.

De Mik-Van der Plas, A.J. 1981. Pyloric caeca of the starfish Asterias rubens L., during the reproductive cycle. PhD dissertation, Utrecht, The Netherlands. Dische, Z. 1955. Color reactions of nucleic acid components. In E. Chargaff & J. Davidson (eds.), The Nucleic Acids, Volume 1. p.285. New York, Academic Press.

Giese, A.C. 1966. On the biochemical constituents of some echinoderms. In R.A. Boolootian (ed.), Physiology of Echinodermata, p.757-796. New York, Interscience Publ.

Greenfield, L. 1959. Biochemical and environmental factors involved in the reproductive cycle of the sea star Pisaster ochraceus (Brandt). PhD. dissertation, Stanford University.

Herrin, D., C. Helling & S.A. Watts 1985. Relative levels of actin and actin gene sequences in tube feet and pyloric caeca of the adult starfish Luidia clathrata (Echinodermata: Asteroidea). Comp. Biochem. Physiol. in press.

Klinger, T.S. & P.F. Dehn 1980. Histological structure of the pyloric caeca of Luidia clathrata (Asteroidia). Florida Sci. 43 Suppl.1.:16.

Lawrence, J.M. 1973. Level, content, and caloric equivalents of the lipid, carbohydrate, and protein in the body components of Luidia clathrata (Echinodermata: Asteroidea) in Tampa Bay. J. exp. mar. Biol. Ecol. 11: 263- 274.

Lawrence, J.M. & J.M. Lane 1982. The utilization of nutrients by post-metamorphic echinoderms. In M. Jangoux & J.M. Lawrence (eds.), Echinoderm Nutrition, p. 331-371. Rotterdam, A.A. Balkema.

Ogue, M. & G. Rosen 1950. The nucleic acids of plant tissues. I. The extraction and estimation of desoxypentose nucleic acid and pentose nucleic acid. Arch. Biochem. 25:262-276.

Schneider, W.C. 1945. Phosphorus compounds in animal tissues. Extraction and estimation of desoxypentose nucleic acid and pentose nucleic acid. J. Biol. Chem. 161: 293-303.

Van der Plas, A.J. & P.A. Voogt 1982. The biochemical composition of the pyloric caeca and its relation to DNA levels in the female starfish Asterias rubens during the reproductive cycle. J. Comp. Physiol. 148: 49-58.

Vorobjev, V.A. 1982. The morphology of mitoses and seasonal changes in proliferation with the starfish Asterias amurensis stomach epithelium. Histologia 24 No.5: 580-585.

Watts, S.A. & J.M. Lawrence 1982. Effect of starvation on enzyme activities and nutrient stores in the pyloric caeca of Echinaster sp. (Echinodermata: Asteroidea). Comp. Physiol. Biochem. 72B: 461-463.

Proceedings of the Fifth International Echinoderm Conference / Galway / 24-29 September 1984

Consequences of predation by the asteroid *Evasterias troschelii* Stimpson in a soft-sediment ascidian community

CRAIG M. YOUNG
Florida State University, Tallahassee, USA

ABSTRACT: Individuals of the forcipulate asteroid Evasterias troschelii collected from rocky habitats eat mostly molluscs and barnacles, generally rejecting solitary ascidians as prey, whereas individuals from soft bottoms eat ascidians readily. Feeding biology was studied in detail in a muddy subtidal community to determine how predation might affect the population dynamics and abundance patterns of ascidian prey. Food preferences are predictable and correlate negatively with the toughness of the prey tunic. Chelyosoma productum, an ascidian with soft tunic, is readily pulled apart by the starfish tube feet, whereas the species that have leathery tunic, including Styela gibbsii and Pyura haustor are opened only by digesting a hole in the tunic. Differential predation on ascidian species resulted in a shifting of relative abundances under experimental field conditions. Starfish did not selectively remove individuals of particular sizes from ascidian aggregations, so population structure did not shift as density was declining. Clumping reduced vulnerability to predation in Styela gibbsii, a species that exhibits gregarious settlement during the larval stage, but not in Chelyosoma productum, larvae of which exhibit similar behavior.

1 INTRODUCTION

Subtidal ascidians are consumed by a variety of asteroids, most of which do not specialize on ascidian prey, but rather switch to ascidians in circumstances where preferred prey, often molluscs, are not available (Gulliksen and Skjaeveland, 1973; Keough and Butler, 1979; Mauzey et al., 1968). In this study I consider the implications of one such dietary shift that occurs when an asteroid characteristic of rocky bottoms finds itself in a soft-sediment community dominated by epifaunal ascidians. Specifically, I discuss how asteroid predation influences relative abundance, population structure and possibly evolution of larval behavior in the ascidian assemblage.

Evasterias troschelii, a starfish common in the northeastern pacific, occurs on subtidal rock, cobble, sand, and mud. On rock, about 96% of their diet consists of sedentary animals with hard shells, including molluscs, barnacles and brachiopods, whereas on sand, a relatively large proportion consume ascidians (Mauzey et al., 1968). E. troschelii feed on bivalves extraorally by arching over the prey, and either inserting the cardiac stomach in a crack between the valves or pulling the shells apart with tube feet (Christensen 1957). A large individual can generate a sustained force as high as 5700 g.

2 MATERIALS AND METHODS

2.1 Field Sites

Assemblages of ascidians and asteroids on soft bottom were examined by diving at Broken Pt., Shaw Island, Washington (26-30 m depth), at Dyes Inlet, Bremerton Wa. (10-18 m) and in front of the Friday Harbor Laboratories, San Juan Island, Wa. (15-20 m). Deeper sites sampled by dredging with the research vessel Hydah included Hankin Pt., Shaw Island, the "Potato Patch" between Orcas and Shaw Islands, and off the Northwestern shore of Blakely Island, Washington. Observations of feeding Evasterias troschelii were made at numerous rocky subtidal sites in the San Juan Islands as well as the three shallow muddy sites mentioned above. Information on species composition, population density, size distributions, spatial pattern and clump organization of the ascidian populations was recorded routinely; most of these data will be reported in greater detail elsewhere.

2.2 Laboratory Feeding Experiments

Feeding experiments were run in circular, running seawater aquaria of plexiglass, 1 m in diameter and 25 cm deep. Because no more than two such tanks were available at any given time, replicate runs were made on consecutive days. In the first series of experiments, which consisted of 8 runs, 4 starfish collected from a rocky site (Reuben Tarte Pt., San Juan Island) were maintained in one tank, and four from a muddy site (Broken Pt., Shaw Island) were kept in the other. Four each of the three most common ascidians occurring on soft sediment were placed in each tank. On subsequent mornings the consumed ascidians were counted then were replaced with fresh individuals. In the second series of experiments (7 days), run 6 months later, 4 newly-collected asteroids from a muddy bottom were used, and 5 ascidians per species were offered to the starfish daily instead of 4.

2.2 Field Feeding Experiments

Feeding experiments were done in field enclosures to evaluate how predation by Evasterias troschelii might modify the composition of an existing ascidian assemblage under natural conditions. An initial attempt to conduct the experiments in situ at Broken Point failed because of difficulties in relocating specific sites and short working times at 26-30 m depth. For convenience and safety, the experiments were therefore set up on a shallower mud bottom about 50 m south of the Friday Harbor Laboratories Breakwater. This site differed from the Broken Pt. site in three main particulars: 1) shallower depth (17 m), 2) more drifting algae, 3) lower densities of ascidians and the polychaete Phyllochaetopterus prolifica, which is a major substratum for ascidians. The ascidian assemblage at Broken Pt. was reconstructed at the experimental site within 2 large enclosures of galvanized "chicken wire" mesh. Each cage consisted of a 30 cm high strip of mesh that surrounded a 2 m by 2 m square plot. Corners of the cages were held by steel stakes that were painted and wrapped in rubber in order to reduce electrolysis. Large wire staples (25 cm long) spaced about every 0.5 m secured the lower edges of the cages to the bottom.

All ascidian individuals and clumps

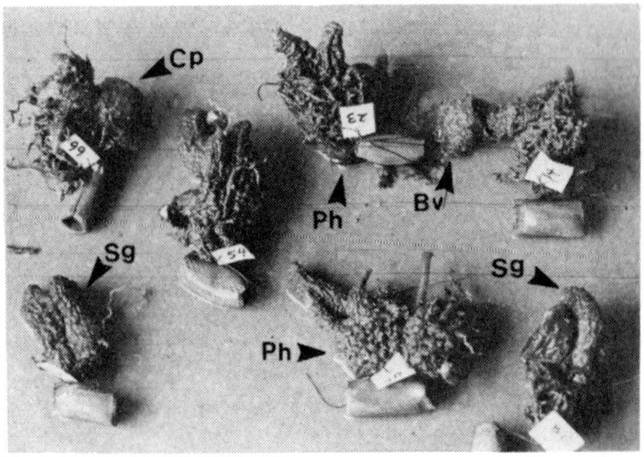

Figure 1. Multiple species ascidian clumps labelled and weighted for field feeding experiment. Four ascidian species are represented: Pyura haustor (labelled Ph), Styela gibbsii (Sg), Boltenia villosa (Bv) and Chelyosoma productum (Cp).

in a randomly-chosen area of approximately 8 m^2 at Broken Point were collected and brought into the laboratory, where they were characterized and labelled. Each animal was measured with calipers along whatever axis gave the best correlation with wet weight for that species. A piece of braided nylon thread was tied through the basal portion of the clump or threaded with a needle through a posterior lobe of tunic. A small label of waterproof paper and a 2 cm length of 1 cm diameter lead pipe were tied to the thread (Fig. 1).

After all clumps were prepared, they were held in running seawater aquaria for several days in order to assure that the labelling procedure had not damaged any individuals. The clumps were classified according to the number of individuals present (i.e. individual animals, pairs of animals, or more than 2 animals), and an equal number of each clump type was randomly assigned to each cage. The ascidians were carried underwater in mesh bags and distributed haphazardly in the two cages by scattering them while swimming several meters above. Two Evasterias troschelii were added to each cage. The cages were only partly effective in retaining the seastars, however, so missing stars had to be replaced every week or two.

After the experiments ran for 30 days, the substratum within each cage was searched carefully for ascidians, labels and weights. These remains were examined in the laboratory in order to determine the identities and sizes of missing animals and to seek direct evidence of predation by starfish.

3 RESULTS

3.1. Field Observations

On muddy and sandy bottoms, ascidians sometimes occur as isolated individuals attached to shells or to parchment tubes of the polychaete Phyllochaetopterus prolifica. More commonly they occur in single-species or multiple-species aggregations in which most individuals are attached to one or more ascidians. Although as many as 8 species of solitary ascidian may occur in these assemblages, over 90% of the individuals are invariably from 4 species, Pyura haustor, Styela gibbsii, Boltenia villosa and Chelyosoma productum. The first three of these are stolidobranchs with tough, leathery tunic; C. productum is a phlebobranch with relatively soft, gelatinous tunic.

Evasterias troschelii individuals living in rocky subtidal habitats are often more than twice the diameter of the largest individuals I have observed on muddy or sandy bottoms (about 35 mm). Moreover, the ratio of basal arm width to arm length appears to be much larger in hard-bottom individuals than in their soft-sediment counterparts.

Half of the 20 individual starfish surveyed on hard bottom were feeding; of these, 7 were eating molluscs (Pododesmus cepio, Humilaria kennerlyi, Crepidula sp.) and barnacles (Balanus nubilus). Three individuals at a rocky site where ascidians were particularly common (Reuben Tarte Pt.) were eating Pyura haustor and Chelyosoma productum. By contrast, 13 of 14 feeding individuals (92.8%) encountered in muddy sites were eating ascidians; only one individual was eating a clam. The ascidians consumed were Styela gibbsii (6 attacks), Pyura haustor (4 attacks) and Chelyosoma productum (3 attacks). These findings are in general agreement with those of Mauzey et al. (1968), which included observations on 448 starfish.

3.2 Laboratory Feeding Experiments

In 8 consecutive 1-day laboratory feeding experiments, 4 Evasterias collected from the rocky subtidal zone at Reuben Tarte Beach consumed only a single Chelysoma productum. After the 8 day experiment was terminated, the starfish were left with ascidians for an additional 30 days. The starfish did not attack any ascidians during this period even though no alternative foods were available. By contrast, asteroids collected from muddy subtidal sites (Broken Point) consumed ascidians at an average rate of 1.23 per day. Table 1 gives the log-likelihood analysis of these data. The results were virtually identical in both experiments. All interactions involving the factor "day"´ were non-significant, indicating that the behavior of the asteroids did not differ significantly among temporal replicates. The interaction between species and consumption was significant in both sets of experiments, showing that asteroids do not consume all ascidian

Table 1. Three-factor log-likelihood analyses of the association among replicates (R), prey species (SP), and consumption (C) in 2 laboratory feeding experiments.

Expt. No.	Source of Variation	df	G	p
1	R X SP	14	2.1469	ns
	R X C	7	9.6448	ns
	SP X C	2	29.1662	<0.001
	R X SP X C	14	8.2373	ns
2	R X SP	12	0.4810	ns
	R X C	6	4.6052	ns
	SP X C	2	12.7669	<0.01
	R X SP X C	12	6.5259	ns

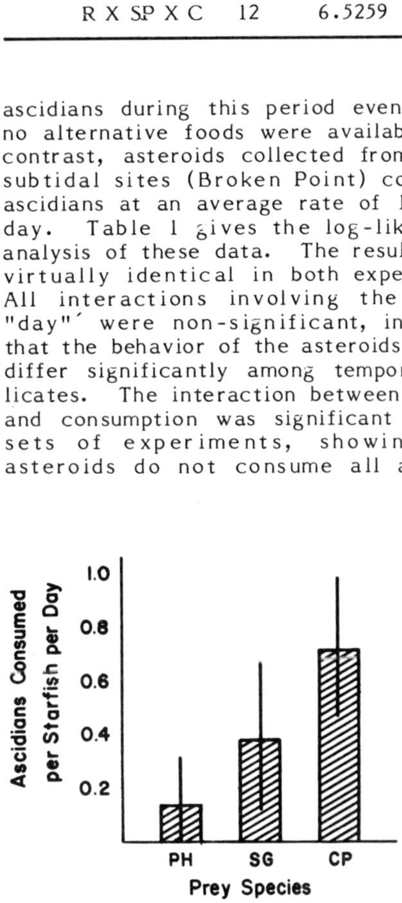

Figure 2. Mean + 1 S.D. daily consumption by species in laboratory experiments with Evasterias troschelii collected from soft-sediment. PH: Pyura haustor; SG: Styela gibbsii; CP: Chelyosoma productum.

species tested according to the proportions available.

Evasterias consumes only the soft internal tissues of ascidians; the tunic is always left more-or-less intact. Table 2 shows the attack points on the common species of prey. The attack points vary significantly with species; Pyura haustor may be attacked on either end, Styela gibbsii is nearly always attacked posteriorly and Chelyosoma productum

Table 2. Two-factor contingency table showing the relationship between attack site and prey species. The association is significant at $p < 0.001$ ($G = 26.4773$; 4 df).

Species	Attack Site			
	Top	Side	Bottom	Total
C. productum	6	19	6	31
S. gibbsii	3	0	13	16
P. haustor	4	0	4	8
(Total)	13	19	23	55

is attacked laterally. These differences in site of attack reflect two different feeding methods, both of which may be employed by the same individual starfish for attacking different tunicate species. The stolidobranch ascidians with hard tunic are entered by digesting a small hole in the tunic, either posteriorly or anteriorly. In the case of Styela gibbsii, which is cylindrical and leathery,

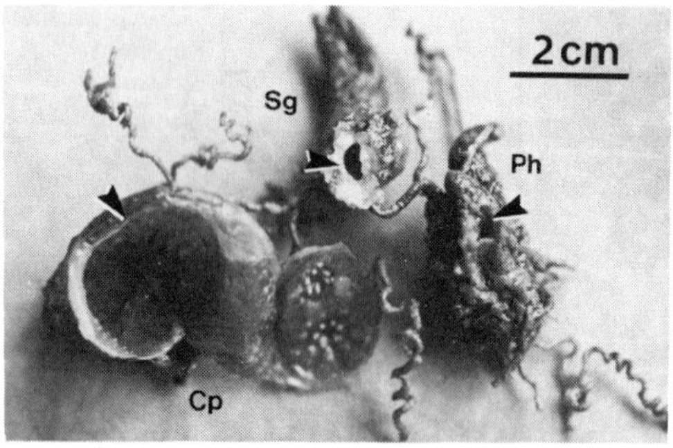

Figure 3. Remains of the ascidians Pyura haustor (Ph), Styela gibbsii (Sg), and Chelyosoma productum (Cp) following attack by Evasterias troschelii. Holes created by the starfish are indicated by arrows.

Figure 4. Oral view of Evasterias trochelii feeding on Chelyosoma productum. Note tube foot attachment.

the thinnest portion of the tunic is at the posterior attachment surface. Chelyosoma productum is opened without everting the stomach (Fig. 4). The starfish hunches over the prey, attaches tube feet to all sides of the ascidian, then pulls laterally until the animal is ripped apart. In a few instances, the anterior plates of C. productum were removed intact, suggesting that the ascidian had been pulled apart along the anterior-posterior axis.

3.3 Field Feeding Experiments

During the 30 day field feeding experiment, starfish left the experimental enclosures on occasion and had to be replaced. Thus, the experimental data do not give true feeding rates under field conditions. Moreover, other sources of mortality (e.g., suffocation in mud) probably played some role. At the conclusion of the experiment, 66 animals were missing completely while 16 tests of dead individuals were still in place. Of the 16 tests remaining, 12 showed clear evidence of predation by Evasterias troschelli, suggesting that predation was probably a major sources of mortality during the experiment. Analysis for the three most abundant species revealed a significant three-way interaction between clump size, species and consumption in cage A but not in cage B (Table 3). Thus, a significant difference in mortality processes existed between the two replicates. In cage B, the species/clump size association was non-significant but consumption interacted significantly with both clump size and species. An examination of data (Table 3) reveals that the species/consumption interaction results largely from much lower than expected consumption of Pyura haustor and higher

Table 3. Three-way contingency tables of data from field feeding experiment, showing the relationship among prey species, clump size and consumption. Analysis is presented in Table 4.

Cage No.	Species	Clump Size	Consumption Eaten	Not Eaten
1	C. productum	1	7	4
		2	3	7
		3+	15	7
	S. gibbsii	1	8	0
		2	1	7
		3+	2	23
	P. haustor	1	0	5
		2	1	13
		3+	0	17
2	C. productum	1	7	4
		2	5	0
		3+	11	18
	S. gibbsii	1	4	3
		2	4	5
		3+	4	27
	P. haustor	1	2	4
		2	0	9
		3+	1	33

Table 4. Three-factor log-likelihood analyses of the associations among prey species (SP), clump size (S) and consumption (C) in field feeding experiment (data in Table 3).

Cage No.	Source of Variation	df	G	p
1	SP X S	4	3.2573	<0.01
	SP X C	2	23.4078	<0.001
	S X C	2	10.4711	<0.001
	SP X S X C	4	19.4881	<0.001
2	SP X S	4	3.1781	ns
	SP X C	2	23.4869	<0.001
	S X C	2	15.0281	<0.001
	SP X S X C	4	4.9852	ns

than expected consumption of Chelyosoma productum. This result agrees with the laboratory data, and again suggests that predation was a major source of mortality in the cages. The clump size/ consumption interaction occurs primarily because fewer individuals in large clumps are consumed

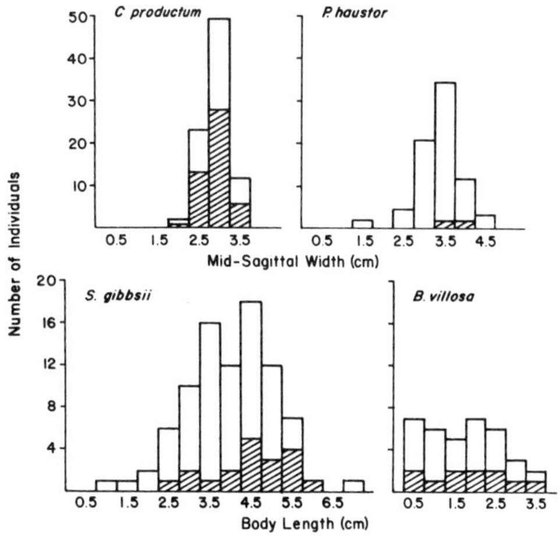

Figure 5. Size-frequency distributions of ascidians available at the beginning of field experiment (entire histogram bars) and ascidians consumed (shaded portions of bars). Numbers on the abscissa represent upper limits of size classes. All differences between available and consumed were non-significant by the Kolmogorov-Smirnov goodness of fit test.

than expected, and more individuals occurring singly are consumed than expected. In both cages, this latter phenomenon is most striking for Styela gibbsii; the majority of isolated individuals are consumed, but only a very small percentage of the clumped individuals are eaten. The same phenomenon occurred to a lesser extent with Chelyosoma productum in cage B, but in cage A relatively large numbers of Chelyosoma individuals occurring in large clumps were attacked and consumed. Boltenia villosa were not included in the analysis because they generally occurred only in large, mixed-species clumps and not as individuals or pairs. B. villosa in large clumps were attacked readily; 23% were consumed in Cage A and 40% were consumed in cage B.

To determine if Evasterias troschelii preferentially consumes animals of certain sizes, I compared the size-frequency distributions of animals consumed and available in the field feeding experiment (Fig. 5). The difference was not significant for any of the four species.

4 DISCUSSION

The results of this study suggest that the asteroid Evasterias troschelii can alter the relative abundances of ascidian species in soft-sediment communities by preying selectively. Available evidence does not support the hypothesis that size-selective predation modifies the population structure of any of the prey species investigated. Ascidians of various sizes are taken in approximately the same frequency at which they occur. Consumption rates of ascidian prey are related inversely to tunic toughness. In other words, starfish eat the prey that are easiest to open first.

The rank order of preferences was quite consistent for all of the laboratory and field experiments. However, in both the present study and that of Mauzey et al. (1968) Evasterias troschelii were found eating Pyura haustor much more often than Chelyosoma productum in the field; based on the feeding experiments, one would predict the opposite result. Two explanations for this apparent discrepancy are possible. First, encounter probabilities may not be the same for the two prey species. Indeed, heavy predation on C. productum by starfish might result in a dietary shift to a less preferred prey species such as Pyura haustor. Second, differences in handling time for the two prey species may bias the data because the diver is more likely to observe predation on the species that takes longest to consume (Fairweather and Underwood, 1983). This bias may be very important in the present example, since P. haustor must be opened by digestion, while C. productum can be torn open very quickly.

Christensen (1957) in a classic study of asteroid feeding showed that Evasterias troschelii could feed on clams either by pulling them open with the tube feet or by inserting the stomach through small gapes between the valves. E. troschelli use variations of these methods for opening ascidians. Chelyosoma productum is ripped apart by pulling laterally with opposing tube feet, while stolidobranch species with hard, leathery tunic are entered by digesting a hole through which the stomach may be inserted. I observed only one instance where the starfish had apparently inserted its stomach down the siphon of an ascidian (Pyura haustor). In this case, the internal siphon epithelium had been partially disgested so that the siphonal aperture was larger than normal. Even though the starfish were obviously capable of digesting holes in the tunic, they fed only on the internal tissues; tunic was always left behind more-or-less intact.

Three of the four major ascidian species in soft-sediment communities, Chelyosoma productum, Styela gibbsii and Pyura haustor, occur most commonly as members of single-species or multiple- species aggregations. All three are known to exhibit gregarious settlement behavior as larvae (Young, 1982; Young and Braithwaite, 1980). Tadpoles settle preferentially on the tunic of established conspecifics or on individuals of closely related species. Does this tendency to aggregate enhance the likelihood that individual ascidians will avoid predation by Evasterias troschelii? Pyura haustor, the species with the hardest tunic, seems to suffer little predation either in clumps or as individuals. Chelyosoma productum, which Evasterias can rip apart with the tube feet, is preyed upon readily in clumps. Clumping improves the chances of surviving a starfish attack only for Styela gibbsii. S. gibbsii are cylindrical ascidians, typically abvout 1 cm in diameter and 4-5 cm long. They attach to the substratum (or to conspecifics) only by the posterior end. In 81% of the attacks recorded in this study, E. troschelii attacked S. gibbsii on the posterior (attachment) end. Starfish readily consume ascidians that are attached to 2mm diameter polychaete tubes, but not individuals attached securely to other ascidians in a clump. For Styela gibbsii, therefore, it is possible that predation by asteroids is a selective pressure favoring evolution of gregarious larval behavior.

Starfish collected from rocky sites where molluscs and barnacles were abundant did not feed on ascidians in the laboratory, even when ascidians were the only prey offered for over a month. This suggests that like other starfish, E. troschelii may exhibit ingestive conditioning (reviewed by Sloan, 1980). In the San Juan Islands, where rocky and muddy habitats are very large and have very different potential prey populations, it seems likely that an individual starfish may be forced to specialize on a narrow group of prey items, and and therefore undergo "conditioning" not just on the short term, but for most or all of its life. Further experiments on prey selection, dietary reversal and nutrition in E. troschelii may help clear up some of the present controvery on the subject of conditioning (Sloan, 1980). Preliminary observations on habitat-specific differences in starfish size and allometry suggest that starfish may do much better on hard bottoms where non-ascidian prey are abundant.

5 ACKNOWLEDGEMENTS

I thank Ronald Shimek and Michael Aley for underwater assistance, Robyn Young for labelling the ascidian clumps, and Michael Keough for writing the G-test program. The field work was completed while I was a graduate student at the University of Alberta, in residence at Friday Harbor Laboratories. Portions of this work were supported by N.S.F. grant OCE-8400406.

6 REFERENCES

Christensen, A.M. 1957. The feeding behavior of the seastar Evasterias troschelii Stimpson. Limn. Oceanogr. 2:180-197.

Fairweather, P.G. & A.J. Underwood 1983. The apparent diet of predators and biases due to handling times of their prey. Oecologia 56:169-179.

Gulliksen, B. & S.H. Skjaeveland 1973. The sea-star, Asterias rubens L., as a predator on the ascidian, Ciona intestinalis (L.), in Borgenfjorden, North-Trondelag, Norway. Sarsia 52:15-20.

Keough, M.J. & A.J. Butler 1979. The role of asteroid predators in the organization of a sessile community on pier pilings. Marine Biology 51:167-177.

Mauzey, K.P., C. Birkeland & P.K. Dayton 1968. Feeding behavior of asteroids and defense responses of their prey in the Puget Sound Region. Ecology 49:603-619.

Sloan, N.A. 1980. Aspects of the feeding biology of asteroids. Oceanogr. Mar. Biol. Ann. Rev. 18:57-124.

Young, C.M. 1982. Larval behavior, predation and early post-settlement mortality as determinants of spatial distribution in subtidal solitary ascidians of the San Juan Islands, Washington. PhD Dissertation, University of Alberta, 260 pp.

Young, C.M. & L.F. Braithwaite 1980. Larval behavior and post-settling morphology in the ascidian Chelyosoma productum Stimpson. J. Exp. Mar. Biol. Ecol. 42:157-169.

Proceedings of the Fifth International Echinoderm Conference / Galway / 24-29 September 1984

On the status and affinities of the goniasterid genus *Calliaster* Gray, 1840 (Asteroidea)

AZNAM AZIZ & MICHEL JANGOUX
Laboratoire de Biologie marine, Université Libre de Bruxelles, Belgium

The goniasterid genus Calliaster (type-species: Calliaster childreni Gray) is distributed in the whole Indo-West Pacific Ocean, ranging mostly between 50 and 450m depth. It is a well defined genus, whose validity was never questioned by subsequent authors.

The genus Calliaster may be diagnosed as follows:
A genus of goniasterid with five elongated, tapering to blunt arms. Disk relatively large. Aboral surface slightly convex on disk and inflated along the midradial line of arms; oral surface flat. Abactinal plates generally smooth, flat or slightly convex, and roundish to elliptical in shape; each plate surrounded by a series of small granules; most carinal plates of arms and disk and some abactino-lateral plates of disk bearing an enlarged spine or tubercle; occurrence of small, secondary irregularly arranged abactinal plates on the central part of the disk. Marginal plates large and massive, generally smooth, bearing spines or tubercles and surrounded by a series of small granules. Actinal intermediate plates almost flat, rectangular to polygonal in shape, smooth or with scattered granules; most plates surrounded by a series of small granules and bearing a prominent spine or tubercle. Adambulacral plates with 3 to 17 elongate, blunt-tipped, cylindrical to slightly compressed furrow spines; furrow spines in a palmate series or forming a close-comb; 1 or 2 (rarely 3) prominent subambulacral spines forming a transverse series. Mouth plates triangularly elongated with 4 to 6 oral (marginal) spines similar to the furrow spines, and 1 to 2 (exceptionally up to 6) slender prominent suboral spines. Papulae single, distributed all over the abactinal surface except on the lateral interradial areas of the disk and on the most distal part of arms; no marginal or actinal papulae. Excavate pedicellariae with broad, flattened, round-tipped or spatulate jaws may be present; pedicellariae located usually on abactinal and/or actinal intermediate plates, more rarely on marginal plates.

As already noted by previous authors, the genus Calliaster is related to the genus Hippasteria Gray, 1840. Their most marked differences concern both the number of furrow spines (usually 2 to 3 in Hippasteria; up to 17 in Calliaster) and the structure of the abactinal skeleton. In Hippasteria, secondary plates occur all over the abactinal skeleton, forming regular reticulated meshes, while in Calliaster these plates develop only in the center of the disk, forming a rather irregular network. As for the abactinal armature, it should be emphasized that the carinal and centrodiscal plates of Calliaster always have an enlarged spine or tubercle, while those of Hippasteria usually support a large, slit-like, bivalved pedicellaria.

Many striking affinities apparently occur between Calliaster and three other goniasterid genera, viz. Milteliphaster Alcock, 1893, Astrothauma Fisher, 1913, and Mabahissaster Macan, 1938. Works are in progress to define the precise generic status of each of these genera in relation to the genus Calliaster.

ACKNOWLEDGEMENTS

Work supported by AGCD grant to A. Aziz (Zoologist of the National Institute of Oceanology, Indonesia).

Proceedings of the Fifth International Echinoderm Conference / Galway / 24-29 September 1984

Activity throughout the year and translocation of the gonad stimulating substance (GSS) in the asteroid, *Asterias rubens*

CATHERINE BOULAND
Laboratoire de Biologie cellulaire et animale, Université d'Orléans, France
& Laboratoire de Biologie marine, Université Libre de Bruxelles, Belgium

A quantitative biological test was routinely used to determine the GSS activity of the radial nerve cord throughout the year, as well as to investigate the possible occurrence of GSS in selected "hemal" organs (i.e. axial organ and gastric hemal tufts). The test used was that of Bouland and Jangoux (1983). It facilitates the establishment of a "spawning index" for pieces of mature ovaries in the presence of crude extracts of tissues or organs that are supposed to have GSS activity. The results indicate that the radial nerve cord has GSS activity, whatever the period of the year. However, this activity varies, being lower during the summer months. It regularly increases during ovarian growth, reaching a maximum at the spawning period. Preliminary assays with hemal organs and coelomic fluid indicate that the GSS is conveyed either through the hemal system itself or through hemal-associated tissues. They show, moreover, that the GSS seems to be continuously translocated throughout the year, whatever its concentration in the nerve cord.

Proceedings of the Fifth International Echinoderm Conference / Galway / 24-29 September 1984

Asexual reproduction in the sea star *Stephanasterias albula*

SALLY F.CARSON & P.V.MLADENOV
Mount Allison University, Sackville, Canada

A population of the fissiparous sea star Stephanasterias albula was studied from May, 1983 to May, 1984. Fission involves a splitting of the disc of the sea star into two approximately equal halves. Each half then regenerates the missing body parts to form a complete individual. S. albula is thus frequently characterized by an asymmetrical body consisting of a set of small and a set of larger arms. Arm number in S. albula is variable, although individuals with 6 arms predominate. Size-frequency analysis has shown that most S. albula range from 4-18 mm in length of longest arm. The complete absence of very small individuals suggests that larval recruitment is absent or minimal. No individuals have yet been found to contain gonads. This population appears, therefore, to be maintained solely by asexual propagation. The incidence of fission is highest in the summer months and may be correlated with seasonal changes in photoperiod and/or water temperature. S. albula can regenerate new arms to 60% of the length of the long arms in roughly 9 months and individuals probably split once a year. It is hoped that further study of this population will provide a better understanding of the evolution and ecological significance of asexual reproduction in echinoderms.

Proceedings of the Fifth International Echinoderm Conference / Galway / 24-29 September 1984

Sexual and asexual reproduction in geographically separated populations of the fissiparous asteroid *Coscinasterias calamaria* Gray

R.G.CRUMP
Orielton Field Centre, Pembroke, Dyfed, UK

M.F.BARKER
Portobello Marine Laboratory, Dunedin, New Zealand

Sexual reproduction and asexual reproduction by fission were studied in four populations of Coscinasterias calamaria (Gray), two in Otago Harbour in the south island of New Zealand and two in the north island near Auckland. The annual reproductive cycle in both islands of New Zealand is clearly defined with a spawning season between November and January. In both sites the pyloric caeca index was approximately inverse to the gonad index cycle as found in other forcipulate asteroids. Substantial differences in the sex ratios of mature starfish occur at each site studied, with 1:1 ratios in two populations, one population heavily biased towards females and a fourth consisting almost entirely of males. Morphometric variation in arm number due to splitting was studied and the frequency of splitting varied considerably between the four populations. Generally speaking sublittoral C. calamaria divide less frequently than intertidal starfish and populations in which food is less abundant or of poorer quality are more fissiparous and put less energy into sexual reproduction, than populations with plentiful readily available food in the form of mussels. The significance of the different reproductive strategies in C. calamaria is discussed.

Proceedings of the Fifth International Echinoderm Conference / Galway / 24-29 September 1984

The significance of the larval coelom in characterising the bipinnaria and brachiolaria in asteroid ontogeny

W.ERBER
Institut für Zoologie der Universität Wien, Austria

Within asteroid embryology there are two larval types, the bipinnaria and the brachiolaria. Depending on the presence or absence of one of these there are said to be three modes of development, the indirect, the direct and the so-called non-brachiolarian mode. All the bipinnariae of the different species show typical common morphological features. The bipinnaria stage is suppressed in the ontogeny of some asteroid species. However, the external features of the later larval stages at the onset of metamorphosis are much more variable. A comparison of the different asteroid developmental types shows that each of these larvae has one common external feature and therefore will be called brachiolaria (Strenger, A., Erber, W. 1983: Z. zool.Syst.Evolut.-Forsch.2, 235-239). This is the bipartition of the larval body into an anterior portion designated as a larval stalk and a posterior portion which transforms into the adult. This definition will be confirmed by the metamorphogenesis of the larval coelom. The bipinnaria has always paired coelomic rudiments on both sides of a functional tripartite gut. Leaving aside some variations in external morphology and the larval gut the brachiolaria stage is always characterized by an unpaired anterior preoral coelom. This preoral coelom is a pure larval organ and it is the essential prerequisite for the development of a functional or vestigal larval stalk within metamorphosis. In the proper sense of this definition, the term non-brachiolarian development is not correct. The brachiolaria is an essential stage in the embryology of every known asteroid and is at least traceable in the vestigial larval stalk and the preoral coelom.

Proceedings of the Fifth International Echinoderm Conference / Galway / 24-29 September 1984

Acid phosphatase during yolk formation and early development in *Asterina gibbosa* (Echinodermata: Asteroidea)

J.G.FERRAND
Biologie Cellulaire et Animale, UER Sciences, Université d'Orléans, France

Previous histochemical and biochemical works (Delavault and Ferrand 1972, Ferrand 1980, 1982) have shown the presence of different lysosomal enzymes, specially acid phosphatase, in the oocytes of Asterina gibbosa and other Asteroids.

This abstract reports the results of a complete cytochemical study using Gomori's reaction, both in the course of the vitellogenesis and during early development of the egg of Asterina gibbosa.

During oogenesis, at the end of the previtellogenic stage, dictyosomes secrete acid phosphatase and we observe true lysosomes which participate in yolk formation by fusion with endoplasmic reticulum cisternae (Ferrand 1984 a). So, when vitellogenesis begins, all yolk globules contain this enzyme. However, the intensity of histochemical reaction decreases with the onset of oocyte maturity (Ferrand 1984 b).

After spawning, acid phosphatase is not detectable again until the blastula has been formed. Activity increases during the formation of the lecitotrophic larva, at the yolk globules level. No acid phosphatase is found in the Golgi apparatus of embryonic cells, so it is presumed that the enzyme previously observed in the vitellogenic oocyte, inactivated in the late oocyte and during the early cleavage stages, reappears when yolk consumption occurs, i.e after blastulation.

Experiments using trypan blue during embryogenesis inhibit yolk consumption. This seems to prove the participation of lysosomal hydrolases in normal development (Ferrand 1984 b).

REFERENCES

Delavault, R. & J.G. Ferrand 1972. Etude de la phosphatase acide au cours du cycle sexuel annuel, dans les ovocytes d'Asterina gibbosa (Echinoderme, Astéride).

Ferrand, J.G. 1980. Hydrolases lysosomiques et vitellus des ovocytes d'Asterina gibbosa et d'Asterias rubens (Echinodermes, Astérides). In M. Jangoux (ed.), Echinoderms, Present and Past, p. 393-398. Rotterdam, Balkema.

Ferrand, J.G. 1982. The activity of hydrolases in ova and sperm of Starfishes. In J.M. Lawrence (ed.), Int. Echinoderms Conference, Tampa Bay, p. 485-488, Rotterdam, Balkema.

Ferrand, J.G. 1984 a. Lysosomes and yolk formation in the oocytes of Asterina gibbosa (Echinodermata, Asteroidea). Int. J. Invert. Reprod. Develop. 7 : 143-147.

Ferrand, J.G. 1984 b. La phosphatase acide au cours de la vitellogenèse et de l'embryogenèse précoce chez Asterina gibbosa Pennant (Echinoderme, Astéride). Arch. Biol. (Bruxelles) 95 : in press.

Proceedings of the Fifth International Echinoderm Conference / Galway / 24-29 September 1984

Adaptive aspects of the annual caeca/gonad reproductive cycle in *Asterias forbesi*

D.R.FRANZ
Brooklyn College, City University of New York, USA

In the North Atlantic, the seastars Asterias forbesi and A. rubens are ecological analogs, and probably share a common ancestor. The continental shelf environments of the NW Atlantic (A. forbesi) and NE Atlantic (A. rubens) differ principally in marine climate. Therefore, each species probably is physiologically adapted to its unique climatic environment. Interspecific differences in the annual caeca/gonad cycle may represent such physiological adaptations. In a coastal sublittoral population of A. forbesi from New York studied from 1980 to the present, spawning occurred from late June to mid-July at bottom temperatures of 16 - 18°C. Feeding is minimal from June through August, the period of highest temperatures (16 - 20°C). Caeca and gonads increase in September, and continue to grow during Fall and Winter, although a pause may occur in feeding as well as caeca and gonad growth in mid-winter (<4°C). Caecum weights peak in April and decline precipitously until spawning in July. Gametogenesis begins in September and continues through the Fall and Winter, during which time large numbers of medium-sized oocytes as well as some large oocytes of spawnable size accumulate. Between April and May, a proportion of accumulated oocytes increase to spawning size, possibly reflecting translocation of nutrients from caeca to ovaries. For at least three months after spawning, unspawned oocytes are resorbed in the gonad. The long-term storage function of the caeca in A. forbesi is de-emphasized relative to A. rubens, with about 65% of gametogenic activity occurring before the cyclical drop in caecum weights, as compared to about 10% in A. rubens. The period during which caeca and gonad indices are inverse does not exceed two months in A. forbesi as compared to five months in A. rubens. Unique aspects of the gonad/caeca cycle in A. forbesi are probably adaptations to the relatively long periods when animals are unable to feed due to extreme temperatures, as well as unpredictably severe climatic conditions. In particular, during the Fall/Winter feeding period, energy is partitioned simultaneously for growth, storage (in caeca) and immediate gametogenesis.

Coelomic microcanaliculae within the asteroid body-wall

MICHEL JANGOUX, PHILIPPE DUBOIS*, ANNE LAMBERT & CATHERINE YOURASSOWSKY*
Laboratoire de Biologie marine, Université Libre de Bruxelles, Belgium

In the course of a study of the fine structure of the asteroid body-wall (i.e. Asterias rubens and Marthasterias glacialis), unexpected microcanaliculae were discovered in the stroma of spines and pedicellariae (H.D. = 2,5 to 6,5 µm) as well as in the uncalcified dermal tissue (H.D. = 9 to 12 µm). Such microcanaliculae have not been reported before for any echinoderm group. Each microcanalicula is lined by an epithelium. Epithelial (adluminal) cells are flattened and attached by short intermediate junctions. Each cell bears a cilium and numerous microvilli, filling the microcanalicular lumen. The cell cytoplasm contains small clear vesicles (presumably endocytic) and a few mitochondria. A well-developed Golgi apparatus generally occurs. Myoepithelial cells and nervous processes are often located under the adluminal cells, close to the epithelial basal lamina. Body-wall reconstructions from E.M. pictures suggest that microcanaliculae form a highly convoluted system throughout the stroma and, less densely, throughout the uncalcified dermal tissue. It is obvious from its fine structure that the canalicular epithelium is a mesothelium and, therefore, that microcanaliculae are of coelomic nature. It was seen, moreover, that enlarged canaliculae occur between the coelomic side of the body-wall and the inner surface of the plate ossicles. One may presume that ossicle microcanaliculae are involved in the calcification/resorption process of skeletal ossicles. Similarly, they must take a prominent part in the general turnover of the body-wall. On the other hand, the structure of the microcanalicular system itself accounts for the uniformity of the asteroid peripheral muscles (e.g. spine and pedicellarial muscles) which always appear to be made of mesothelial cells. Finally, it can be said that the course of the microcanalicular network within the body-wall reflects that of the accompanying nervous system (i.e. the hyponeural nervous system). This implies that the hyponeural system is everywhere located very close to the epineural system.

* Research assistants of the NFSR (Belgium). Work supported by IRSIA grant 81382 to A. Lambert, and by FRFC grant 2.4506.83.

Proceedings of the Fifth International Echinoderm Conference / Galway / 24-29 September 1984

Investigations on the movement and excitability of the rosettes of crossed pedicellariae in the asteroid, *Marthasterias glacialis*

ANNE LAMBERT & MICHEL JANGOUX
Laborntoire de Biologie marine, Université Libre de Bruxelles, Belgium

The rosettes of crossed pedicellariae of Marthasterias glacialis show a definite excitability towards different stimulations. The reaction of pedicellariae to mechanical and chemical stimulation were studied, both qualitatively and quantitatively, in order to propose a functional interpretation. Stimulation experiments were all performed under a stereoscopic microscope, coupled with a photographic camera. The rosettes were cut off and put into a petri dish containing filtered sea water. A platinum wire (30 µm of diameter) was used to perform mechanical stimulations of either the spine, the rosette, the outer or the inner side of the pedicellaria. Chemical stimulations were effected with a peristaltic pump delivering uniformly weak test solutions. (The test solutions were rough extracts of either the mussel Mytilus edulis, the crab Carcinus maenas or the hermit-crab Eupagurus bernhardus). The study of the rosettes' excitability was carried out through a separate analysis of each movement contributing to their overall activity. For each stimulation considered, 5 rosettes from 15 asteroids were cut off (total effective number = 75 rosettes). The results were analysed using the Wilcoxon statistical test.
Our results showed that pedicellariae are undoubtely very sensitive and typically react when mechanically and chemically stimulated. The range of reactions induced by chemical stimulations appears to be larger than that induced by tactile stimulations. For instance, the repeated opening and closing movements of pedicellarial valves or the retracting responses of the pedicellariae were not observed when mechanically stimulated. A statistical correlation test (Spearman test) showed that the different reactions recorded occur independently of each other. As a rule, it seems that the reactions of the rosettes of crossed pedicellariae occur randomly and lack coordination. This seemingly aberrant behaviour could reflect a preventive but relatively poorly organized reaction to the presence of potential predators. Collective movements of all the rosettes of a given body area can either deter living organisms, including potential predators, or simply prevent their access to the asteroid body surface. The pedicellariae thus form a kind of armour that responds to some perturbations of the environment.

Proceedings of the Fifth International Echinoderm Conference / Galway / 24-29 September 1984

Changing properties of somatic accessory and germinal cells during the amitotic/mitotic and premeiotic/meiotic transitions of spermatogenesis in *Asterias vulgaris*

D.A.LAROCHELLE & C.W.WALKER
University of New Hampshire, Durham, USA

Ultrastructural, autoradiographic and fluorescence microscopic observations have been made on the spermatogenic epithelium during the amitotic/mitotic transition of spermatogonia and premeiotic/meiotic transition of primary spermatocytes. At each of these critical steps in spermatogenesis, the somatic accessory and germinal cells are characterized by specific and predictable features of their cytoplasm and membranes. In the summer (June - mid-September), amitotic spermatogonia are enclosed by somatic accessory cells which form follicle-like compartments of the spermatogenic epithelium. Throughout the amitotic phase, spermatogonia remain isolated in this fashion from contact with cells in other follicles, from the lumen of the testis where phagocytosis and later secretion occur and also from the accumulating potentially nutritive contents of the underlying genital haemal sinus (GHS) of the testicular wall. Until late September-early October, such amitotic spermatogonia are apparently suspended in G1 and do not incorporate significant amounts of ^{3}H-thymidine or amino acids nor do they fluoresce brightly following treatment with sulfoflavin or FITC-ConA. Following the intial mitotic origin of primary spermatocytes (October), further intimate interactions with somatic accessory cells occur during the formation of spermatogenic columns (Walker, J. Morph., 166 (1980)). Although the flagellated primary spermatocytes forming columns are not coupled cytoplasmically, they regularly become increasingly competent tc undergo meiosis 1 as they ascend along the axes of columns from base to tip. Results from concurrent autoradiography and sulfoflavin staining demonstrate the relative height within columns at which primary spermatocytes undergo DNA and basic nucleoprotein synthesis and suggest that these two processes are uncoupled. Deaminated proteins are lysine-rich. FITC-ConA is bound to the cytoplasm of SA cells within columns and within the testicular lumen and also to the nuclear membranes of primary spermatocytes; such binding is blocked following pretreatment with a-methyl-d-mannoside. Based on these preliminary observations, we discuss mechanisms possibly involved in promoting resumption of the cell cycle in amitotic spermatogonia and cytokinetic events in competent primary spermatocytes suspended in prophase of meiosis 1.

An investigation into the behavioural and ecological bases for periodic infestations of *Asterias rubens*

RICHARD J. MOORE & ANDREW C. CAMPBELL
School of Biological Sciences, Queen Mary College (University of London), UK

Two populations of Asterias rubens on the north Kent coast are being monitored by marking studies over a 12-month period. One, a shallow sub-tidal site at Hampton supports a moderately dense, but dispersed, population of large Asterias; the other is an intertidal mussel bed off Margate.

Densities of starfish on transects at the two sites in March and April were generally low, but in May, an influx of starfish from deeper water was apparent. By June, an aggregated band of small Asterias at densities >50 m^{-2} had invaded the Mytilus bed at Margate. In contrast, the Hampton population, although showing a general density increase into the intertidal zone, remained dispersed, feeding predominantly on Crepidula fornicata.

Complementary laboratory studies are investigating the importance of interactions between individual starfishes effected by the effluent from feeding activities. Experiments with a Y-maze have shown that prey being consumed by a starfish exerts a greater chemical stimulation than unattacked prey, but the response is flexible.

It is suggested that mutual attraction through feeding activities is an important component of the mechanism of aggregation of Asterias, and our experiments support this view: Starfish sampled from the field populations and tested in the Y-maze showed that the Margate starfish were more greatly attracted to starfish feeding on prey than were Hampton starfish. The Hampton starfish were also less likely to respond to a current-borne chemical stimulus, and appeared to detect prey primarily by contact chemoreception.

The authors are currently investigating the possible influence of environmental factors such as prey availability (ingestive conditioning), temperature and photoperiod on the transition between the two types of response to effluent from feeding activities.

The possible importance of photoperiod is suggested by observations on starfish held under solitary and grouped regimes. Starfish maintained in groups with superabundant prey (Mytilus edulis) grew significantly faster - and consumed greater quantities of prey - than starfish maintained solitarily. This appeared to be a consequence of greater daytime feeding activity by grouped starfish compared to those held solitarily. An advantage of aggregation in Asterias may be to reduce the pressure by visually-guided predators. The high incidence of arm injuries to Asterias in the intertidal mussel beds at Margate suggests that the large number of gulls (Larus spp.) there are an important cause of mortality.

This research is supported by the Natural Environment Research Council of the U.K.

Regulation of spermatogenic proliferation in *Asterias vulgaris* by polyamines

FRANK F.SMITH
University of New Hampshire, Durham, USA

Polyamines have been shown to be essential for nucleic acid synthesis and proliferation in a wide variety of cell systems. Ornithine decarboxylase (ODC), the rate-limiting enzyme in polyamine synthesis, generally attains a transient peak in activity shortly before the initiation of DNA synthesis. In Asterias vulgaris, ODC activity increases greatly during the proliferative phase of spermatogenesis. Intact testes near the beginning of the proliferative phase were supplied with exogenous polyamines (a mixture of 50μM putrescine, 10μM spermidine, and 10μM spermine) and subsequently showed a significant increase in incorporation of ^{3}H-thymidine into DNA. However, prior intracoelomic injection of polyamines showed no sustained effect on thymidine incorporation; after 24 hours incubation in polyamine-free medium, the rate of thymidine incorporation was statistically indistinguishable from that in control specimens. These results suggest a direct role for polyamines in the regulation of spermatogenic proliferation in A. vulgaris. Rather than inducing a fundamental shift in proliferative program, the polyamines appear to have satisfied a regulatory requirement in the existing program of otherwise mitotically competent cells.

Supported by NSF PCM 80-03670 to C. W. Walker

Proceedings of the Fifth International Echinoderm Conference / Galway / 24-29 September 1984

The *Acanthaster* infestations: A step in the damaging-evolution of the coral reef ecosystem

BERNARD A.THOMASSIN
Centre d'Océanologie de Marseille, Station marine d'Endoume, Université d'Aix-Marseille II, France

The starfish Acanthaster planci (L) is a common inhabitant of the indo-pacific coral reef ecosystems (in the sensu stricto meaning defined by Harmelin-Vivien et al., 1982). In undisturbed environments, it occurs normally at a density of 1-2 ind./ hectare in the coral built biotopes (reef flats, few outer barrier reef slopes, and lagoonal slopes). Over the last 25-year period, Acanthaster infestations occuring in various coral reef areas have been studied and several hypotheses advanced to explain this phenomenon. These population blooms occured in natural environments without anthropogenic influences as well as in those exposed to various stresses linked to human environmental modifications ("pollution"). From personal observations on Acanthaster density increases or true explosions, occuring in coral reef areas exposed or otherwise to human interferences (Madagascar, Comoro Is., Red Sea and New Caledonia), to analyses of the literature, it can be interpreted that the population increases are associated with the first regressive evolutionary steps of a coral reef ecosystem, i.e. a tipping over of community equilibrium (characteristic of the "mature stage" of the coral reef ecosystem) towards the degrading processes (characteristic of the "senescence stage") that lead to the death of the reef. The first stages of this increase in the population are linked to an alteration of coral health (as a response to various kinds of stress) This is generally followed by an eutrophication of the marine environment beneficial to Acanthaster larvae and juveniles by development of phytoplankton and the benthic algal communities (in the space-time competition between corals and benthic algae). The Acanthaster infestation has a "snow-ball" progression with increasing feed-back effects which lead to the complete destruction of the coral communities. The coral polyps are preyed upon and algal communities grow upon dead coral colonies; this leads to major disruption of the trophic relationships in the coral reef ecosystem. Infestation disappears when coral food becomes insufficient and when the sessile photophilous assemblages (dominated by corals), characteristic of the coral reef ecosystem s.s., are replaced by algal substitution facies.

Proceedings of the Fifth International Echinoderm Conference / Galway / 24-29 September 1984

Comparison of steroid levels in the gonads and pyloric caeca of male and female *Asterias rubens:* Relation to reproduction

PETER A.VOOGT
Laboratory of Chemical Animal Physiology, State University of Utrecht, Netherlands

Recently the presence and metabolism of steroids in female sea stars have been studied rather extensively. The presence of progesterone and oestrone in ovaries and pyloric caeca has been established beyond doubt (Dieleman and Schoenmakers, 1979) and the main pathways in steroid metabolism have been outlined (Voogt and Schoenmakers, 1980). Further the levels of progesterone and oestrone in the ovaries and pyloric caeca have been estimated throughout the year (Schoenmakers and Dieleman, 1981) and a physiological function of these steroids in reproduction has been discussed (Voogt et al., 1984). These and older studies are all concerned with female animals and as a result nothing is known about the presence and the levels of steroids in male animals. In this study we showed the presence of progesterone and oestrone in the gonads and pyloric caeca of male sea stars and estimated the levels throughout the reproductive cycle. The patterns obtained turned out to be sex-specific as well as organ-specific for both steroids The progesterone levels in the gonads showed an inverse course: the onset of gonadal growth is marked in the female by a sharp decrease of the progesterone level in this organ, whereas in the male it is marked by a peak value of progesterone. It is likely that both steroids in the testes are involved in the regulation of gametogenesis. The onset of a new reproductive cycle, taking the abrupt decrease of the pyloric caeca-index as a marker, coincides with a strong decrease of the progesterone level and an increase of the eostrone level in the pyloric caeca of the male sea star. It is submitted that progesterone and oestrone in the pyloric caeca act in concert to provide the materials required for gametogenesis.

REFERENCES

Dieleman, S.J. and Schoenmakers, H.J.N. (1979) Gen. comp. Endocrinol. 39, 534-542
Schoenmakers, H.J.N. and Dieleman, S.J. (1980) Gen. comp. Endocrinol. 43, 63-70
Voogt, P.A. and Schoenmakers, H.J.N. (1980) In: Echinoderms - present and past, (M. Jangoux, ed.), Balkema, Rotterdam, pp. 349-353
Voogt, P.A., Oudejans, R.C.H.M. and Broertjes, J.J.S. (1984) Advances in Invertebrate Reproduction, Elsevier 12pp. (in press)

Proceedings of the Fifth International Echinoderm Conference / Galway / 24-29 September 1984

Development of the larval coelom in the asteroid, *Asterias rubens*

CATHERINE YOURASSOWSKY*
Laboratoire de Biologie marine, Université Libre de Bruxelles, Belgium

The morphology of coelomic pouches in developing larvae of Asterias rubens was studied on laboratory reared larvae. Movements of the coelomic pouches were investigated both in vivo (facilitated by the transparency of the larvae) and in SEM microscopy (by microdissection of various developmental larval stages). The observed development of the larval coelom generally agrees with that described by Gemmill (1914). Some differences occur however. They mostly concern the relative movements of the posterior part of the right and left somatocoelic cavities, as well as the moment when the left somatocoel and the right axohydrocoel fuse together. (A new diagramatic representation of the development of the larval coelom is proposed), SEM-investigations showed, moreover, that local concentrations of mesenchyme cells occur on parts of the coelomic pouches' outer surfaces. These cells could be involved in the morphogenetic movements of the coelomic pouches. Special attention was given to the coelomic structures associated with the brachiolar arms. Each brachiolar arm is topped by several adhesive papillae and includes an expansion of the preoral coelom. Differentiation of adhesive papillae appears to occur only when both the mesothelium and the ectoderm of the brachiolar arm tip are very close to each other.

* Research assistant of the NFSR (Belgium). Work supported by FRFC grant 2.4525.80.

6. Ophiuroidea

Proceedings of the Fifth International Echinoderm Conference / Galway / 24-29 September 1984

The brittle star, *Ophiactis savignyi* (Müller & Troschel), an inhabitant of a Pacific sponge, *Damiriana hawaiiana* de Laubenfels

HUBERT CASPERS
Institut für Hydrobiologie und Fischereiwissenschaft, Universität Hamburg, Germany

ABSTRACT: The interior of the marine sponge, Damiriana hawaiiana, which is common on the tidal flats of Coconut Island, Kaneohe Bay, Hawaii, contains a variety of animals, most of which are commensal. The most important inhabitant is the six-armed ophiuroid, Ophiactis savignyi. Often, more than 20 specimens of different sizes fill the canal system. They obtain enough oxygen and nutrients from the water, that circulates through the sponge. Many egg-bearing specimens were found. Usually, the brittle stars form a tangled mass, limiting the movement of the individuals, and it seems that they remain for their entire lives in the host. They are part of a well protected benthic community adapted to the structure and activity of a large sessil animal. The relationship between the brittle star and its host is discussed.

1 INTRODUCTION

Sponges, in both freshwater and marine habitats, are known to be inhabited by a variety of invertebrates, mainly polychaetes and crustaceans, and also some fish. Pearse (1934) examined five species of sponge near the Dry Tortugas and reported on three species of brittle stars, Ophiocoma echinata, Ophiothrix oerstedii, and Ophiactis savignyi, the last of which seemed to favor Spongia officinalis as a host. A member of the Demospongia, Damiriana hawaiiana, is found on the tidal flats of the coral platform in Kaneohe Bay, Oahu, Hawaii. Specimens sometimes attain a diameter of over 10 cm. A general external examination of the living sponge reveals a few branches, about the size and shape of fingers, arising from an amorphous basal mass (Figure 1). The osculum can open as wide as 6 mm, and it can be closed by a membrane that maintains a circular form as the aperture shrinks. Breaking open a sponge reveals what Pearse (1934) called a "living hotel". Many invertebrates of very different taxa can be washed out. Peculiar are the "nests" of accumulated brittle stars of different sizes (Figure 2). The whole habitat is characterized by the slimy secretions of the host, which restrict the movement of the inhabitants, not only in the open canals but also in the vascular system of the sponge.

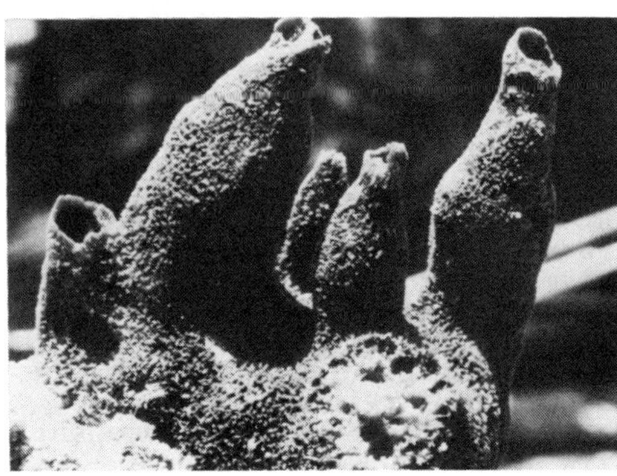

Figure 1. A colony of Damiriana hawaiiana

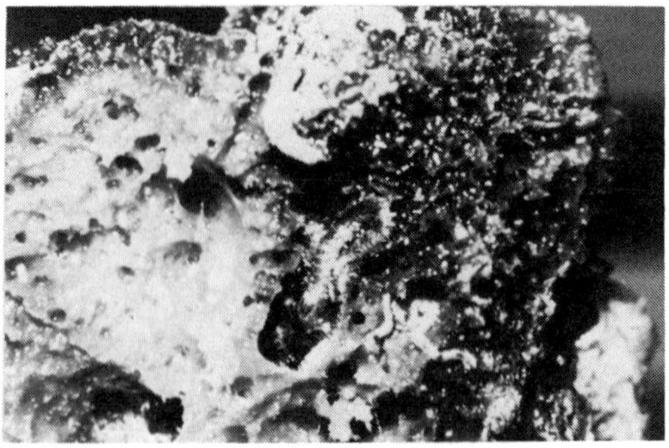

Figure 2. Accumulation of brittle stars in the sponge

2 THE BRITTLE STAR, OPHIACTIS SAVIGNYI (MÜLLER & TROSCHEL)

Ophiactis savignyi is a common species in Hawaii, ranging widely throughout the Indo-West Pacific, the East Pacific and the warmer waters along the West and East Coasts of the Americas (Edmondson 1946, Clark 1976). Normally, these brittle stars live in the sediment and between branches of corals (McCloskey 1970, Uebelacker 1977, Sloan 1979, 1982, Sloan et al. 1979). Clark (1976) designated the species as "primarily epizoic in sponges but also hides in crevices of any kind". In Hawaii, some have been reported to inhabit sponges, but such a mass occurrence as that in Damiriana has not been known before.

Usually Ophiactis has six arms (exceptionally five or seven). It is remarkable that the brittle stars in the sponges are often damaged, with two or three arms broken and in different stages of regeneration. The slimy habitat seems to promote the breaking of the arm segments. However, broken arms were never found in the sponges; apparently they break into fragments and are removed by the sponge.

Hyman (1955, p.669) mentioned that the reproduction by fission is responsible for the success of O. savignyi in populating the tropics. This kind of reproduction is a perfect adaptation to a permanent life in sponges.

At least eleven species of Ophiactis have been reported to be epizoic on sponges or fissiparous. As mentioned, about half of the brittle stars in Daminiana had arms in different stages of regeneration. These observations correspond to reports of many authors about free-living specimens. O. savignyi is the best known fissiparous ophiuroid species.

Clark (1976) emphasized the great regenerative powers of many species of echinoderms, correlated with autotomy and generally limited to single arms or parts of them. Fissiparity is wide-spread among ophiuroids, occurring in nine families out of seventeen. It can be regarded as a form of asexual reproduction, also called schizogony. The most obvious characteristic of fissiparous species is their supernumerary arms. The majority of them have six or more arms and an asymmetrical appearance due to incomplete regeneration. Clark found that specimens of Ophiactis versicolor from the Indian Ocean had the capability of coiling the parts of the arms downward.

Hyman (1955) reported that ophiuroids very commonly live in groups. Allee (1927) and Allee & Fowler (1932) described such aggregations of Ophioderma brevispina as essentially an expression of negative phototaxis combined with positive thigmotaxis. This may explain the behavior of Ophiactis in the sponge.

As mentioned above, a movement of the arms in the slimy environment of the sponge may increase the possibility of autotomy, but this is probably the same in open habitats and could be "naturally planned". According to Clark (1967), a dorsoventral flexure, the epizoic habits, and fissiparity are combined in the genus Ophiothela. Checking the literature, however, it has not been definitely established that autotomized arms result from asexual reproduction. No suppression of sexual reproduction has been observed. Clark (1967) described fissiparity as a phenomen "in which division takes place across the disk and results in the separation of two approximately equal parts, both of which are capable of regenerating to form a complete specimen. It can therefore be regarded as a form of asexual reproduction". This contrasts with the observations of many authors, who found arms split off but no division of the disk. This is the case with the individuals in Damiriana. Therefore, it is uncertain if the autotomy of the arms in the sponges really results in asexual reproduction.

3 OPHIACTIS IN THE SPONGE

Sometimes up to 20 specimens of this brittle star live together in clumps or "nests", usually in the broad base of the sponge colony. The discs of the largest brittle stars measure 5 mm in diameter, and the arms have a length up to 32 mm. All size groups, from tiny individuals to adults with sperm and eggs, have been found together. Often, the long arms extend through the lumen of the water passages, and some were observed in the excurrent channel of the sponge. The cavities with the clumped Ophiactis were fully filled with the slimy secretion of the sponge, and the brittle stars are scarcely able to move their arms. In the spaces crowded with brittle stars, the walls were observed to have a blackish coating, apparently consisting of detritus, possibly formed by faecal material from the echinoderms.

As mentioned above, many fertile specimens were found. Clumps of eggs adhered to their discs, surrounded by a kind of membrane. When a single brittle star is taken out of the sponge and kept in sea water, the egg clumps break up. An egg is about 0.08 mm in diamter. Attempts to rear the eggs were unsuccessful.

Apparently, no fertilized eggs or larvae leave the sponge. On the other hand, it is remarkable that in the larger sponges, young brittle stars were often completely absent. Therefore, the maintenance of the brittle star population might be solely by asexual reproduction after larvae first enter the water-vascular system of the sponge in its first stage of growth. This kind of development by asexual reproduction in such hosts as sponges could permit the brittle stars to remain in a safe refuge for several years. Young growing specimens can also remain in the sponge for a long period of time, even if they cannot reproduce asexually. Perhaps larvae of young brittle stars have difficulties entering sponges already crowded with older echinoderms. Only when the sponges are destroyed do the brittle stars and other inhabitants have to transfer to another habitat in order to survive. If the brittle stars produce sperm and eggs in the sponge, the fertilization of the eggs within the host seems to be possible. The question then raised is how the fertilized eggs leave the sponge. Many such ecological questions have still to be clarified. The most simple explanation is that young brittle stars or their larvae enter the sponge and begin a protected existence in the host. It is remarkable that in another Hawaiian coastal region, a tiny sponge was found to be inhabited by only a single Ophiactis.

4 NATURE OF THE ASSOCIATION

The problem of classifying the relationship between the brittle star and the sponge basically depends on what the brittle stars feed on in the sponge. In a comprehensive monograph on "Echinoderm Nutrition", edited by Jangoux & Lawrence in 1982, Warner described food and feeding mechanisms of Ophiuroidea. Hunt (1925) reported that the stomachs of Ophiactis balli collected off Plymouth contained detritus and small invertebrates, and he observed deposit feeding. Pentreath (1970) noted suspension feeding (mucus spine) by O. resiliens in New Zealand waters. This species maintains a feeding posture with arms extended stiffly upward in currents stronger than 15 cm/s. The food is then concentrated at the proximal ends of the arms. This species, like Ophiothrix fragilis, appears to be a specialized suspension feeder with filtering tube-feet (see also Stephens & Virker 1966, Stewart 1979). Pentreath also reported that large suspended food particles act as chemical stimuli in the arm podia.

None of these feeding processes seem to be employed by brittle stars in sponges. Apparently nothing is known about their diet and feeding methods.

Semantic problems arise when attempting to describe the mode of life of the brittle stars in the sponge. This is due to a lack of standardization in the ecological terminology (Odum 1971). To some authors, symbiosis is used to describe all relationships among different organisms, including parasitism. To others, symbiosis is synonymous with mutualism, the state in which both animals benefit from the relationship. The benefits to the brittle stars are not matched by advantages to the sponge, so the relationship is not mutualism, but it might be called symbiosis in a broad sense.

The term parasitism is reserved by some authors for a relationship where one organism lives on or in another and feeds on it. To others, the term means a relationship where one organism obtains a benefit from a host, while the host suffers some disadvantage. The blocking of the water circulation in the sponge might reduce the amount of food it can take in, so the term parasitism applies in one sense but not the other. A purely commensal relationship is also not the case because the brittle stars obtain benefits from the sponge. It is therefore not yet possible to exactly define this relationship. I will therefore simply declare that this is a case of a brittle star using a sponge as its protector and life-support system, whereby the sponge receives no benefits and may lose some efficiency in its water circulation system.

As mentioned, Clark (1976) described Ophiactis savignyi as being primarily "epizoic" in sponges. It might be better to call this "endozoic". This life style could be adapted by species that usually hide in any available crevice and find a particularly good hiding place in sponges, which provide a regular water flow, supplies of oxygen and food, and the protection against predators. This permits the brittle stars to live together in an entangled mass. If the sponge can survive, the situation remains ideal for the commensals.

In a paper just published on the association of several Ophiothrix species and Callyspongia vaginalis in the Belize Barrier Reef, G. Hendler (1984) discusses potential interaction and stresses the nocturnal behavior of O. lineata, "moving its arms like a pendulum, very slowly sweeping across the surface of the sponge". The moving arms thus cleared a swath through the film of sediment clinging to the sponge. Such feeding activity was not observed in the Hawaiian Ophiactis savignyi, but it is possible and could explain some of the damage to the arms, if predators feed on them when they move outside the sponge. Hendler, in his experiments, did not observe attacks of predators on brittle stars in the lumen of the sponge, but many marauding fish attacked brittle stars on the sand and on the exterior of sponges. He wrote, "fish did not molest the individuals in the sponge, even when the arms of injured brittlestars protruded from the osculum". Hendler thus designates sponges as a refuge from predation, especially since many sponges are toxic to fish. It may also be mentioned that the Hawaiian Ophiactis live in a sponge with cavities that are fully filled with slimy secretions that could prevent them from moving their arms.

Hendler reported striking differences in the mean size of Ophiothrix lineata inside and outside the sponge. The largest specimens were inside. This is contrary to the observations of Ophiactis in Hawaii, where in Damiriana, numerous young individuals coexisted with adult specimens.

Hendler suggests that sponges are a source of food for brittle stars: "In the association between Callyspongia vaginalis and Ophiothrix lineata, the sponge provides the brittlestar with detrital material and a suitable surface from which to feed.... clear evidence for a commensal relationship." The brittle stars do not collect material that can be utilized by sponges. Nevertheless, "the question of brittlestars as parasites of sponges is, at present, unresolved." There is a discussion about whether the feeding activity of Ophiothrix may benefit the sponge: "As the brittlestars feed, the incurrent filtering surface of the sponge is cleaned and this may promote greater pumping efficiency."

5 ACKNOWLEDGEMENTS

These observations were made during a study program on marine chironomids at the Hawaii Institute of Marine Biology on Coconut Island under a National Science Foundation grant. I thank all my colleagues at this institute for their hospitality.

REFERENCES

Allee, W.C. 1927. Studies in animal aggregations: Some physiological effects of aggregation on the brittle starfish Ophioderma brevispina. J.Exp.Zool. 48: 475-495.

Allee, W.C. & J.Fowler 1932. Studies in animal aggregations: Further studies on oxygen consumption and autotomy of the brittle star Ophioderma. J.Exp.Zool. 64: 33-50.

Clark, A.M. 1967. Variable symmetrie in fissiparous Asterozoa. In N. Millot (ed.), Echinoderm Biology, p.143-157. Zool.Soc. London, Academic Press.

Clark, A.M. 1976. Echinoderms of coral reefs. In O.A.Jones & R. Endean (eds.), Biology and Geology of Coral Reefs III, Vol.2 p.95-123. New York and London, Academic Press.

Edmondson, C.H. 1946. Reef and Shore Fauna of Hawaii. Bernice P.Bishop Mus. Spec.Publ.22.

Hendler, G. 1984. The association of Ophiothrix lineata and Calyspongia vaginalis: A brittle star-sponge cleaning symbiosis? Marine Ecology 5, p. 9-27.

Hunt, O.D. 1925. The food of the bottom fauna of the Plymouth fishing grounds. J.mar.biol. Ass.U.K. 13: 560-599.

Hyman, L.H. 1940. The Invertebrates, Vol.1. Protozoa through Ctenophora. New York and London, McGraw Hill.

McCloskey, L.R. 1970. The dynamics of the community associated with a marine scleractinian coral. Int.Revue ges.Hydrobiol. 55: 13-81.

Odum, E.P. 1971. Fundamentals of Ecology. Third Edition. Philadelphia and London, W.B.Saunders.

Pearse, A.S. 1934. Inhabitants of certain sponges at Dry Tortugas. Carnegie Inst.Wash., Tortugas Lab.Publ.28: 117-124.

Pentreath, R.J. 1970. Feeding mechanisms and the functional morphology of podia and spines in some New Zealand ophiuroids (Echinodermata). J.Zool.Lond. 161: 395-429.

Sloan, N.A. 1979. Microhabitat and resource utilization in cryptic rocky intertidal echinoderms at Aldabra Atoll, Seychelles. Mar. Biol. 54: 269-279.

Sloan, N.A. 1982. Size and structure of echinoderm populations associated with different coexisting coral species at Aldabra Atoll, Seychelles. Mar.Biol. 66: 67-75.

Sloan, N.A., A.M.Clark & J.D.Taylor 1979. The echinoderms of Aldabra and their habitats. Bull.Br.Mus. nat.hist.(Zool.) 37: 81-128.

Stephens, G.C. & R.A.Virkar 1966. Uptake of organic material by aquatic invertebrates. IV. The influence of salinity on the uptake of amino acids by the brittle star, Ophiactis arenosa. Biol.Bull. 131: 172-185.

Stewart, M.G. 1979. Absorption of dissolved organic nutrients by marine invertebrates. In M.Barnes (ed.), Oceanogr.Mar.Biol.Ann.Rev. 17, p. 163-192. Aberdeen Univ. Press.

Uebelacker, J.M. 1977. Cryptofauna species/area relationship in the coral reef sponge, Gelliodes digitalis. In D.L.Taylor (ed.), Proc.Third Internat.Coral Reef Symp. Vol.1, p. 69-73. Miami: Univ.of Miami Press.

Warner, G. 1982. Food and feeding mechanisms: Ophiuroidea. In M. Jangoux & J.M.Lawrence (eds.), Echinoderm Nutrition, p.161-184. Rotterdam, Balkema.

Proceedings of the Fifth International Echinoderm Conference / Galway / 24-29 September 1984

Post-autotomy feeding behavior of *Micropholis gracillima* (Stimpson): Implications for regeneration

LEE ANN JEROME CLEMENTS
Belle W. Baruch Institute for Marine Biology and Coastal Research and Marine Science Program, University of South Carolina, Columbia, USA

Abstract: The activity of recently autotomized Micropholis gracillima individuals was monitored for eight days. Burrowing time immediately following autotomy was not significantly different for autotomized and non-autotomized individuals. Differences in behavior were observed over time ($P < 0.001$). Animals that lost one arm were less active than non-autotomized individuals. Individuals that autotomized both the disc and one arm and those that autotomized only the disc had depressed activity for the first 84 hrs. No differences among treatments were observed after the initial layer of tissue had formed over the oral frame (108 hrs). No material was observed in the regenerated guts of animals that had autotomized their discs (sampled 180 hrs after autotomy). The most frequent behaviors among discless animals were probing the sediment surface with extended arms and waving arms in the water column. Although these activities are usually associated with feeding, they may not always be associated with particle ingestion. Changes in behavior following autotomy may alter energy intake and the time required for regeneration.

1 INTRODUCTION

Partial predation on soft-bodied organisms is common in benthic habitats (Edwards & Steele, 1968; DeVlas, 1979a; Peterson & Quammen, 1982). The most frequently cropped and regenerated portions of benthic infauna include feeding appendages (tentacles, palps and siphons: Peterson & Quammen, 1982; Woodin, 1982) and posterior segments (DeVlas, 1979b; Woodin, 1984); however, these losses usually represent only small losses of biomass. Echinoderms, notably holothurians and ophiuroids, may lose large portions of their individual biomass (viscera and appendages) when disturbed (Pawson, 1966; Emson & Wilkie, 1980). Ophiuroids of the family Amphiuridae frequently autotomize their discs, including the gut and gonads, as well as arms. Estimates of the percentage of regeneration in natural populations of amphiurids are sparse, but several studies show variability within habitats that may be related to predation intensity (Singletary, 1970; Sides, 1981; Turner et al., 1982; Bowmer & Keegan, 1983; personal observation). Turner et al. (1982) have suggested that disc autotomy enables Ophiophragmus filograneus to burrow quickly, thus escaping from its predators. Autotomy of an easily regenerated appendage may also provide a diversion to the predator, and may be a strategy evolved to avoid more extensive predation.

The energetic and ecological costs of autotomy have not been evaluated. In addition to tissue loss, loss of arms may influence overall activity, feeding ability and energy intake. Disc autotomy involves loss of the gut and gonads, so feeding activity, energy absorption and reproduction are likely to be altered during regeneration. The purpose of this study was to examine the burrowing time and feeding activity of an amphiurid ophiuroid, Micropholis gracillima (Stimpson), following disc and arm autotomy.

2 MATERIALS AND METHODS

Micropholis gracillima individuals were collected from an intertidal mudflat in North Inlet Estuary near Georgetown, SC, USA (33° 20'N 79° 10'W). Individuals missing discs and/or arms at the time of collection were not used in the experiment.

Twenty-four hours prior to collection of the animals, 40 cores of natural sediment

(11 cm diam; 5 cm depth) were taken from the mudflat described above. The cores were transported to the laboratory, set up in a running seawater tank (0.5 x 1.5 x 0.1 m) in a 4 x 10 core rectangular array, and allowed to acclimate for 24 hrs. To preserve the natural sediment topography and heterogeneity, cores were disturbed as little as possible and macrofauna present at the time of collection were not removed from cores. Treatments were randomly assigned to cores, and experiments were run at ambient temperature and salinity (T: 24°C; S: 33 ppt).

Four treatments were used to distinguish between the effects of arm and disc autotomy: 1) "Arm"--induced to autotomize one arm 5-9 segments distal to the disc, 2) "Disc"--induced to autotomize the disc, 3) "Disc/Arm"--induced to autotomize the disc and one arm, and 4) "Control"--intact arms and complete disc. Each treatment was replicated 10 times. Arm autotomy was induced by grasping the arm with a pair of sharply pointed forceps (Wilkie, 1978). Arms were neither cut nor pulled, and autotomy always occured between the grasping point and the disc several segments from the position of the forceps. Disc autotomy was induced by breaking the attachment of the disc to the oral frame at one arm, inducing the ophiuroid to detach the disc at the remaining four arms (Dobson, 1984). Complete disc autotomy usually occurred within 30 seconds.

Each individual was randomly assigned to a treatment and induced to autotomize appropriately. As soon as autotomy was complete, the animal was placed in a core, oral side down, in contact with the sediment surface. The length of time required for the individual to burrow was recorded. Burrowing was defined as any process which resulted in obscuring the disc from view, and was considered complete when the disc/aboral surface was not visible above the sediment surface.

Observations of arm activity of all individuals were made every 12 hours, for eight days. During each 1-hour observation period the animals were censused every 5 min, number of arms present on the surface and type of behavior of each arm were recorded. Individual activity was expressed as the sum of the arms observed during each census for each 1-hour observation period.

Three types of behavior were observed and recorded for each individual: 1) "probing", during which an arm is extended from the burrow and probes the sediment surface, 2) "waving", when an arm is extended into the water column and moved back and forth, generally with the tube feet extended, and 3) "irrigating", which consists of an undulating motion of the arm while partially extended from the burrow (see Woodley, 1975). These behaviors are the only arm movements visible at the sediment surface. Waving and probing are usually associated with feeding, while irrigating is thought to be a means of transporting oxygenated water through the burrow (Woodley, 1975).

Burrowing times were analyzed using a Kruskal-Wallis test (Daniel, 1978). Separate analyses were performed on waving, probing and irrigating activities and on the sum of these (total activity). Data were analyzed by two-way repeated measures Analysis of Variance (ANOVA) (Winer, 1971). When significant effects ($P < 0.05$) were indicated between treatments or among observation periods all pairwise comparisons of activity were made using Bonferroni t-tests (Kirk, 1982).

3 RESULTS

3.1 Burrowing

Burrowing time did not differ significantly between treatments ($P > 0.10$) although there was a trend toward shorter burrowing time as degree of autotomy increased (Table 1). Upon contact with the sediment all of the animals actively moved about on the surface of the core for a few seconds, probing the surface with their arms. Burrowing was almost always accomplished by traveling down into an existing hole or crevice rather than pushing sediment aside with the podia (Woodley, 1975). One arm would probe the hole and then the disc and remaining arms would follow it.

Table 1. Mean time (min) and 95% confidence intervals for burrowing after autotomy. N = number of replicates

TREATMENT	N	MEAN $\pm$ 95% C.I.
Control	10	3.116 $\pm$ 2.136
Arm	10	2.217 $\pm$ 2.706
Disc	10	1.676 $\pm$ 2.010
Disc/Arm	10	1.147 $\pm$ 0.772

3.2 Activity

Activity was significantly different among treatments and observation periods (Fig. 1, Table 2). The interaction of treatment and observation period was significant (Table

2) so pairwise comparisons of observation periods were made separately for each treatment.

Table 2. Two-way repeated measures ANOVA: activity.

SOURCE	DF	MS	F
Treatment	3	1363.195	8.32 ***
Error 1	34	163.854	
Observation period	13	478.461	17.34 ***
Interaction	39	53.699	1.95 ***
Error 2	442	27.586	

*=0.05, **=0.01, ***=0.001

Control animals were the most active throughout the experiment (Fig. 1), with three peaks of activity at 24, 96 and 180 hrs (Fig. 1A). These peaks are significantly different from activity levels at all other observation periods (Table 3). No observations were made at 120 hrs due to a heavy rainstorm that increased the turbidity of the water.

Individuals missing one arm had similar activity patterns to the controls; however, activity was generally lower (Fig. 1B), and there were no significant differences among activity levels at any observation period (Table 3).

Activity in the disc and disc/arm treatments was lower than that in both the arm and control treatments for the first 108 hrs (Fig. 1 C&D). Activity was not significantly different for the first 108 hrs in the disc treatment and the first 132 hrs in the disc/arm treatment (Table 3). Likewise, activity was not significantly different among observation periods 96 to 180 hrs for either treatment (Table 3).

3.2.1 Irrigating

Significant differences in irrigating activity were detected among observation periods (two-way repeated measures ANOVA, observation period, F=20.11, df=13, 442, P <0.001). Treatments were not significantly different (F=0.82, df=3, 34, P=0.489). Irrigating activity increased significantly at 132 hrs in all treatments (Fig. 1) This corresponded to a heavy rainstorm and accompanying decreased salinity and increased turbidity.

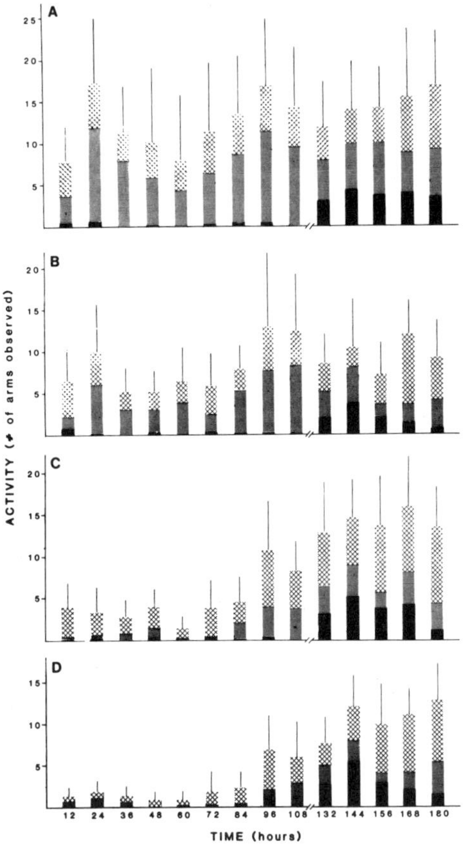

Figure 1. Post-autotomy activity (arms observed hr^{-1}) at each observation period: A. Control; B. Arm; C. Disc; D. Disc/Arm; ( ) probing; ( ) waving; ( ) irrigating; error bars represent 95% confidence intervals.

3.2.2 Waving and Probing

Waving was the most frequently observed behavior in all treatments (Fig. 1), and there were no significant differences in waving activity among treatments (two-way repeated measures ANOVA, treatments, F=1.78, df=3, 34, P=0.169). Significant differences were detected among observation periods (F=9.60, df=13, 442, P <0.001), and the interaction between treatment and observation period was not significant (F=1.01, df=39, 442, P=0.451). Activity at 12, 24 and 72-156 hr observation periods was significantly different from high activity periods (168 and 180 hrs) and low activity periods (36-60 hrs) (Table 4).

Table 3. Results of pairwise comparisons of mean activity (arms observed hr^{-1}) among observation periods. Within each treatment means with the same letter are not significantly different at the 0.05 level.

TREATMENT	HOURS													
	12	24	36	48	60	72	84	96	108	132	144	156	168	180
Control	7.8 a	17.8 b	11.2 a	10.2 a	7.9 a	11.4 a	13.4 a	16.8 b	14.2 a	11.8 a	14.0 a	14.2 a	15.5 a	16.8 b
Arm	6.5 a	10.0 a	5.2 a	5.2 a	6.3 a	5.7 a	7.9 a	12.8 a	12.4 a	8.6 a	14.4 a	7.1 a	12.0 a	9.1 a
Disc	4.0 a	3.4 a	2.9 a	4.0 a	1.6 a	3.9 a	4.7 a	10.8 ab	8.4 ab	13.0 b	14.9 b	13.9 b	16.1 b	13.7 b
Disc/Arm	1.3 a	1.9 a	1.3 a	0.8 a	0.8 a	1.8 a	2.4 a	6.8 ab	6.0 ab	7.6 ab	12.1 b	9.9 b	11.1 b	12.8 b

Table 4. Results of pairwise comparisons of mean waving activity (arms observed waving hr^{-1}) among observation periods for all treatments combined. Means with the same letter are not significantly different at the 0.05 level.

HOURS													
12	24	36	48	60	72	84	96	108	132	144	156	168	180
3.2 ab	3.3 ab	2.1 a	2.4 a	2.0 a	3.3 ab	3.0 ab	5.4 bc	4.2 ab	3.7 ab	5.3 bc	5.1 bc	7.4 c	7.3 c

Probing was the second most frequently observed behavior (Fig. 1). Differences in probing activity were detected among treatments and observation periods with significant interaction between the two factors (Table 5). For control individuals, probing activity was significantly more frequent at 24, 84, 96 and 108 hr observation periods (Table 6). However, individuals in the arm treatment were more active only at 96 and 108 hr observation periods (Table 6). No significant differences were detected among observation periods for the disc and disc/arm treatments (Table 6).

Table 5. Two-way repeated measures ANOVA: probing.

SOURCE	DF	MS	F
Treatment	3	806.387	13.29 ***
Error 1	34	60.674	
Observation period	13	74.106	7.99 ***
Interaction	39	20.347	2.19 ***
Error 2	442	9.277	

*=0.05, **=0.01, ***=0.001

4 DISCUSSION

Overall activity, the combination of maintenance activity (irrigating) and feeding activity (waving and probing) was influenced by autotomy. After autotomy an organism would be expected to show reduced activity to conserve energy and avoid further injury while regenerating lost tissue. Intact M. gracillima individuals were more active than animals that had lost one arm, the disc or both the disc and one arm (Fig. 1). Activity tends to decrease with increasing tissue loss.

Peaks of activity among control individuals do not correspond to any identified

Table 6. Results of pairwise comparisons of mean probing activity (arms observed probing hr^{-1}) among observation periods. Within each treatment, means with the same letter are not significantly different at the 0.05 level.

TREATMENT	HOURS													
	12	24	36	48	60	72	84	96	108	132	144	156	168	180
Control	3.2 a	11.3 b	7.9 ab	5.7 a	4.3 a	6.1 a	8.4 b	11.0 b	9.5 b	5.0 a	5.6 a	6.3 a	4.8 a	5.7 a
Arm	1.4 a	5.9 a	3.0 a	2.9 a	3.8 a	2.2 a	5.2 ab	7.4 b	8.1 b	3.2 a	4.4 a	1.7 a	2.2 a	3.6 a
Disc	0.2 a	0.4 a	0.8 a	1.2 a	0.2 a	0.2 a	2.0 a	3.7 a	3.8 a	3.3 a	3.8 a	1.9 a	3.8 a	3.2 a
Disc/Arm	0.5 a	0.8 a	0.6 a	0.0 a	0.1 a	0.1 a	0.3 a	2.1 a	2.8 a	2.4 a	2.5 a	1.2 a	2.1 a	3.8 a

changes in the laboratory or field environment, and cannot presently be explained. Significant increases in activity with time in the disc and disc/arm treatments seem to be indications of recovery from disc loss similar to the resumption of defecation after tissue loss by the polychaete Abarenicola pacifica (Woodin, 1984). However, for M. gracillima, total activity may not be an accurate measure of recovery since it includes irrigating as well as waving and probing. Irrigating activity is not different among treatments and increases for all treatments at 132 hrs, indicating that it was not influenced by tissue loss.

Waving and probing activities have been associated with feeding (Woodley, 1975), and are probably better measures of recovery after disc autotomy. Presumably autotomized individuals cannot feed, so feeding behavior would be futile and expend energy needlessly prior to regeneration of a functional gut. A layer of disc integument is usually regenerated over the aboral surface of the oral frame by the fifth day following autotomy (Singletary, 1970; personal observation), which corresponds to increased overall activity observed at 96-108 hrs. Although no ingested material was observed when the animals were examined at 180 hours, the newly regenerated disc integument may function as an absorptive surface for nutrients. However, increased feeding activity was not soley due to partial disc regeneration since waving activity increased at 72 hrs and probing was never significantly different in the disc and disc/arm treatments (Fig. 1; Tables 4&6).

Probing and waving were the most frequently observed behaviors in the disc and disc/arm treatments and may increase epithelial contact with oxygenated water, facilitating respiration, or with dissolved organic matter (DOM) in the water column and sediment pore water. DOM represents a food source used by many invertebrates (Stephens, 1981; Wright, 1982; Nell et al., 1983), and may be a source of nutrients for regeneration after autotomy. Numerous studies have indicated that organisms lacking functional guts or feeding organs actively absorb amino acids from the surrounding water (e.g. Crowe et al., 1982; Manahan & Crisp, 1982; Manahan, 1983). Smith et al. (1981) indicated that DOM is actively taken up by recently eviscerated crinoids, and other echinoderms seem to be able to utilize DOM as a food source (Ferguson, 1971; Ferguson, 1980).

As a predator avoidance mechanism, autotomy would work only if it provided a diversion, and/or increased the ability of the animal to escape. Autotomized arms probably provide a diversion as they continue to move after autotomy (Hendler, 1983; personal observation), and Buchanan (1964) reported that Amphiura filiformis arms emit a phosphorescent flash after autotomy. Escape via burrowing was not significantly quicker among autotomized M. gracillima, although there was a trend towards shorter burrowing times (Table 1). These results are supported by studies of burrowing by Ophiophragmus filograneus following disc autotomy (Turner et al., 1982) which showed that burrowing was not significantly different shortly after autotomy. The trend towards shorter burrowing time may have been induced by the increased handling necessary to induce autotomy and not because of any advantage of disc loss since all the animals pulled themselves down crevices and burrow holes regardless of treatment. However, during the experiment, I observed that autotomized individuals usually moved about the sediment surface more quickly than non-autotomized animals.

Disc autotomy among amphiurid ophiuroids is common. Stancyk (1974) reported that in over two years of monthly samples the percentage of Ophiophragmus filograneus individuals regenerating discs ranged from 2.5% to 56% with a mean of 26.7 + 15.1%. Amphiura filiformis in Galway has a disc regeneration frequency of 3.6 + 1.7% (Bowmer & Keegan, 1983). Singletary (1970) reported disc regeneration frequencies of 5.5%, 3.3% and 1.5% for Ophionephthys limicola, Amphioplus coniortodes and Micropholis gracillima respectively. Micropholis gracillima in N. Inlet has a disc regeneration frequency of approximately 50%, with lower frequencies during the late summer months (personal observation). The variability in disc regeneration frequencies within a habitat are thought to be related to variations in predation intensity throughout the year (Turner et al., 1982; Bowmer & Keegan, 1983), and Turner et al. (1982) reported that disc autotomy occurred in response to sediment disturbance similar to that produced by stingrays, a known predator of ophiuroids.

Regeneration of autotomized discs requires time and energy. Disc-autotomy as a means to avoid complete predation is unlikely unless regeneration of the gut and gonads is energetically inexpensive or can occur quickly enough to prevent loss of reproductive potential. Turner and Murdoch (1976) implied that regeneration is accomplished by using stored material since disc regeneration was accompanied by loss of arm material. However, weight loss in the arms could be due to factors other than the relocation of energy into the regenerating disc, and more than one source of energy is probably involved in regeneration. The time and energy required for regeneration also may change with the reproductive state of the animal, as would feeding behavior and frequency. In any case, post-autotomy behavioral changes like those observed in this study could significantly alter energy availability and utilization by regenerating animals.

6 ACKNOWLEDGEMENTS

I am grateful to Dr. S.A. Woodin, Dr. S.E. Stancyk, and M. Luckenbach for their helpful comments and reviews of this manuscript. The Belle W. Baruch Institute for Marine Biology and Coastal Research provided laboratory space, and I am grateful to the staff for all their help and encouragement. Financial support for this research was provided by the Piedmont Chapter of the Explorers Club. Contribution #xxxxx of the Belle W. Baruch Institute for Marine Biology and Coastal Research.

7 REFERENCES

Bowmer, T. & B.F. Keegan, 1983. Field survey of the occurrence and significance of regeneration in Amphiura filiformis (Echinodermata: Ophiuroidea) from Galway Bay, west coast of Ireland. Mar. Biol. 74: 65-71.

Buchanan, J.B., 1964. A comparative study of some features of the biology of Amphiura filiformis and Amphiura chiajei (Ophiuroidea) considered in relation to their distribution. J. Mar. Biol. Ass. U.K. 44: 565-576.

Crowe, J.H., L.M. Crowe, P. Roe & D. Wickham, 1982. Uptake of dissolved organic material by nemertean worms: association of worms with arthrodial membranes. Amer. Zool. 22: 671-682.

Daniel, W.W., 1978. Applied nonparametric statistics. Boston, Houghton Mifflin Company, 503 pp.

DeVlas, J., 1979a. Annual food intake by plaice and flounder in a tidal flat area in the Dutch Wadden Sea, with special reference to consumption of regenerating parts of macrobenthic prey. Neth. J. Sea Res. 13: 117-153.

__________, 1979b. Secondary production by tail regeneration in a tidal flat population of lugworms (Arenicola marina) cropped by flatfish. Neth. J. Sea Res. 13: 362-393.

Dobson, W.E., 1984. Nervously mediated histolysis of connective tissue during disc autotomy in Ophiophragmus filograneus (Echinodermata Ophiuroidea). M.S. Thesis, Florida Institute of Technology, Melbourne, 68 pp. [unpublished].

Edwards, R.R.C. & J.H. Steel, 1968. The ecology of 0-group plaice and common dabs at Loch Ewe. I. Population and food. J. Exp. Mar. Biol. Ecol. 2: 215-238.

Emson, R.H. & I.C. Wilkie, 1980. Fission and autotomy in echinoderms. Oceanogr. Mar. Biol. Ann. Rev. 18: 155-250.

Hendler, G., 1983. The association of Ophiothrix lineata and Callyspongia vaginalis: a brittlestar-sponge cleaning symbiosis? Mar. Ecol. 5: 9-27.

Kirk, R.E., 1982. Experimental design, second edition: procedures for the behavioral sciences. Belmont, California, Brooks/Cole Publishing, 911 pp.

Manahan, D.T., 1983. The uptake and metabolism of dissolved amino acids by bivalve larvae. Biol. Bull. 164: 236-250.

Manahan, D.T. & D.J. Crisp, 1982. The role of dissolved organic material in the nutrition of pelagic larvae: amino acid uptake by bivalve veligers. Amer. Zool. 22: 635-646.

Nell, J.A., M.E. Skeel & P. Dunkley, 1983. Uptake of some dissolved organic nutrients by the sydney rock oyster Saccostrea commercialis. Mar. Biol. 74: 313-318.

Pawson, D.L., 1966. Ecology of holothurians. In, R.A. Boolootian (ed.), Physiology of Echinodermata, New York, Interscience Publishers.

Sides, E.M., 1981. Aspects of space utilization in shallow-water brittlestars (Echinodermata: Ophiuroidea) of Discovery Bay, Jamaica. Ph.D. Dissertation, University of the West Indies, Mona, Jamaica.

Singletary, R.L., 1970. The biology and ecology of Amphioplus coniortodes, Ophionephthys limicola and Micropholis gracillima (Ophiuroidea: Amphiuridea). PhD thesis, Univ. of Miami, Florida, USA, 137 pp.

Smith, D.F., D.L. Meyer & S.M.J.Horner, 1981. Amino acid uptake by the comatulid crinoid Cenometra bella (Echinodermata) following eviceration. Mar. Biol. 61: 207-213.

Stancyk, S.E., 1974. Life history patterns of three estuarine brittle-stars (Ophiuroidea) at Cedar Key, Florida. PhD dissertation, University of Florida, Gainsville, Florida.

Turner, R.L. & J.D. Murdoch, 1976. Potential of arms as a nutrient source for disc regeneration in brittlestars. Amer Zool. 16: 288 [abstract].

Turner, R.L., D.W. Heatwole & S.E. Stancyk, 1982. Ophiuroid discs in stingray stomachs: evasive autotomy or partial consumption of prey? In, J.M. Lawrence (ed.), The International Echinoderms Conference/Tampa Bay/14-17 September 1981: Proceedings, Rotterdam, Balkema pp. 331-335.

Wilkie, I.C., 1978. Arm autotomy in brittle-stars (Echinodermata: Ophiuroidea). J. Zool. Lond. 186: 311-330.

Woodley, J.D., 1975. The behavior of some amphiurid brittle-stars. J. Exp. Mar. Biol. Ecol. 18:29-46.

Winer, B.J., 1971. Statistical principles in experimental design. New York, McGraw Hill Book Co., 907 pp.

Woodin, S.A., 1982. Browsing: importan in marine sedimentary environments Spionid polychaete examples. J. Exp. Mar. Biol. Ecol. 60: 35-45.

__________, 1984. Effects of browsing predators: changes in infauna following tissue loss. Biol. Bull. 166: 558-573.

Wright, S.H., 1982. A nutritional role for amino acid transport in filter-feeding marine invertebrates. Amer. Zool. 22:621-634.

Proceedings of the Fifth International Echinoderm Conference / Galway / 24-29 September 1984

Intracellular studies on the nervous system of an echinoderm

JAMES L.S.COBB
Gatty Marine Laboratory, St. Andrews, Scotland, UK

ABSTRACT: Intracellular techniques have been used to study the chemical synapse across the basement membrane that separates the ectoneural nervous system from the hyponeural. Lucifer Yellow dye fills have shown that all the neurones break up into a fine plexus of neurites where they make synaptic contact. The hyponeural motor cells show inhibitory and excitatory junction potentials concomitant with activity in the ectoneural system. These junction potentials summate to produce spike potentials in the axon hillock region of the motor axons. The motor nerves are inhibited on one side and excited on the other during rapid arm flexures.

1 INTRODUCTION

The last two decades have seen great progress in neurobiology, initially on cellular function and at present much investigation is at the molecular level. Although some progress has been made on the neurobiology of echinoderms much of it has been speculative since it has been based on ultrastructural studies alone. The small size of most neurones has made them relatively intractable to study using modern neurophysiological techniques. The brittlestars however have neurones an order of magnitude larger than other echinoderms and perform rapid behavioural responses to stimulus. There is however no evidence to suggest that the layout, and hence presumably function of the nervous system, is different from the other Eleutherozoan classes.

There is some confusion in the echinoderm literature recently about the nervous system, and it is necessary to clarify matters particularly with regard to the hyponeural system. This system is purely motor, and although developmentally difficult to prove unequivocally, it is clearly mesodermal in origin. It is associated with large 'skeletal' type muscles. These are the lantern in echinoids, the body wall muscle/ossicle system in asteroids, the ossicle muscles in brittlestars and the large body wall muscles in holothurians. The hyponeural system is cholinergic and ultrastructurally nerve endings always contain only small agranular vesicles. The ectoneural system is sensory, interneuronal and motor. The main motor innervation by the ectoneural is of the tubefoot/ampulla system but this is indirect across a basal lamina (see Cobb, 1970, Florey and Cahill 1980). The only directly innervated muscles may be those of the pedicellaria and spines. It is possible even in these cases that there may be hyponeural motor nerves that have been overlooked and these systems should be re-examined.

The gut endothelium also contains a plexus of the ectoneural system and at least part of the function of this is to innervate secretory cells (Cobb unpublished). There is a separate plexus within the mesodermal muscle layer of the gut but it has not been proved to be connected with the hyponeural nerves of the circumoral ring in spite of substantial effort (Cobb unpublished). Its origin is likely to be similarly mesodermal but it would be better to call it the visceral motor plexus until its relationship to the other nervous system is clarified.

The presence of separate nervous systems in each germ layer and the lack of direct connection between them is one of the major enigmas of echinoderm biology. This report shows using intracellular

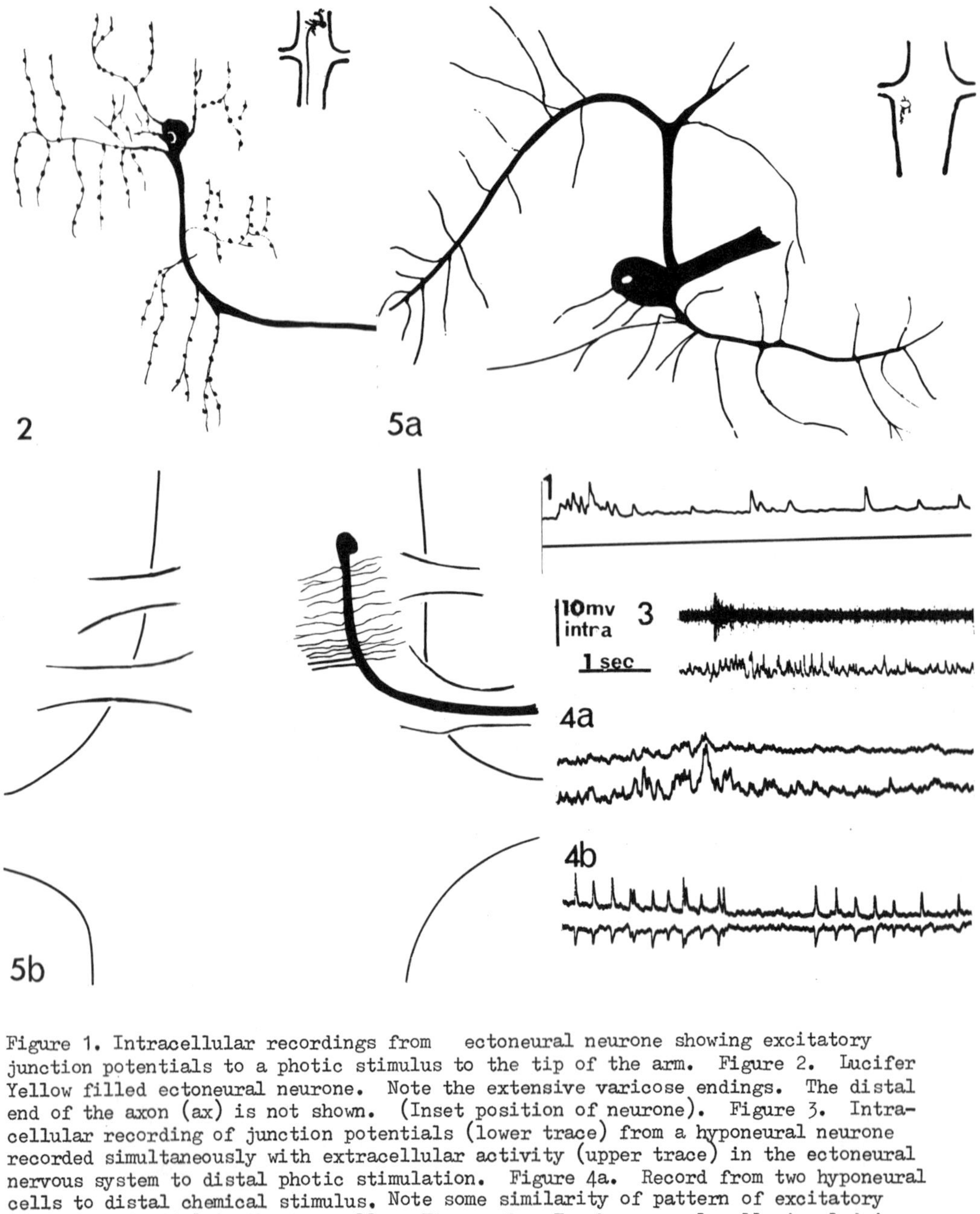

Figure 1. Intracellular recordings from ectoneural neurone showing excitatory junction potentials to a photic stimulus to the tip of the arm. Figure 2. Lucifer Yellow filled ectoneural neurone. Note the extensive varicose endings. The distal end of the axon (ax) is not shown. (Inset position of neurone). Figure 3. Intracellular recording of junction potentials (lower trace) from a hyponeural neurone recorded simultaneously with extracellular activity (upper trace) in the ectoneural nervous system to distal photic stimulation. Figure 4a. Record from two hyponeural cells to distal chemical stimulus. Note some similarity of pattern of excitatory junction potential in the two cells. Figure 4b. Two hyponeural cells impaled in mirror image position on either side of the mid-line in a single ganglion. One cell shows inhibitory junction potentials and the other excitatory to a distal electrical stimulus. The similarity is remarkable and striking. Figure 5a. Hyponeural dye filled cell showing extensive non-varicose dendrites. The axon (ax) from this cell innervated the oral region of the inter-vertebral muscles and was broken off short. Figure 5b. Hyponeural dye filled cell showing non-varicose dendrites. The axon (ax) joins one of the two main nerve branches that ascend to the vertebral muscles.

recording and dye marking techniques that there is a conventional physiological connection between them.

2 MATERIALS AND METHODS

Large specimens of Ophiura ophiura were obtained from Millport, Isle of Cumbrae on the West coast of Scotland. Single arm preparations were used involving removal of the oral plates to study the ectoneural nerves tissue and the fracturing and dissection of the vertebral ossicles to reach the hyponeural ganglia. Conventional glass micro electrodes of about 60 mΩ resistance were used for recording and were produced on a Campden horizontal electrode puller. Conventional amplification and recording systems were used for the intracellular studies. Lucifer Yellow dye was iontophoretically injected in 3M LiCl using 500msec pulses of 10n.amp negative current at 1Hz. Preparations of dye filled axons were normally examined fresh but some were fixed, dehydrated and cleared in methyl salicylate.

3 RESULTS

3.1 Ectoneural nervous system, physiology

The neurones of this tissue are more difficult to penetrate than the hyponeural and the results obtained so far are preliminary. It is clear that some impalements are made in the cell bodies of the nerve cells which lie in a layer on the surface but in many cases the electrodes are penetrating the layer of axons lying aboral to the cell bodies. All cells in which reasonably stable resting potentials were obtained showed action potentials when the cells were depolarized by injected current and also on rebound after injection of hyperpolarizing current. The resting potential varied between -30mv to -75mv and action potentials in the latter were up to 90mv with overshoot. It is clear however that many impalements were of a relatively poor quality and it is likely that the lower resting potentials are a consequence of this. A small number of cells impaled showed some spontaneous activity and these usually appeared as excitatory junction potentials, Figure 1. Many impalements caused substantial initial bursts of injury spike potentials. Junction potentials and spike potentials were obtained to a photic or chemical stimulus to the distal part of the nerve cord in some single arm preparations.

3.2 Single cell morphology

About forty individual cells have been filled with Lucifer Yellow but in many cases the entire cell failed to fill. It has also not proved possible yet to characterize electrophysiologically each filled cell. All cells filled showed characteristic fine varicose branched endings, Figure 2, and two basic cell types were established. One of these cell types has a basic longitudinal orientation (IG cells) and the other a transverse one (T cells). Cells that were completely filled invariably showed the varicose plexus of endings at each end whether they were IG or T cells. Each varicose plexus normally consisted of very fine branches arising from large trunks and covered quite large areas relative to the size of the nerve cord. Many fills in IG cells faded before the distant terminal region and the maximum distance they run has not been determined. One cell could be traced through three segments. It is interesting to note that the general morphology of these cells resembles those of vertebrates rather than those in other invertebrate phyla.

3.3 Hyponeural tissue, physiological studies

These cells are relatively easy to impale and preliminary characterization is complete. The cell body region is non-excitable but nearly all impalements show some spontaneous activity. Junction potentials, which may be excitatory or inhibitory, can be recorded up to 15mv and although the spontaneous level of these is usually low, stimulation of the nerve cord distally with photic, chemical or electrical parameters produces large bursts of junction potentials concomitant with typical electrical activity recordable from the ectoneural system, Figure 3. Junction potentials summate to either depolarize or hyperpolarize the resting potential and the former, if above threshold, give rise to action potentials in the axon hillock region. Neuronal activity within the ectoneural nervous system thus excites or inhibits motor activity within the hyponeural nervous system, involving a chemical synapse across the basement membrane. Experiments using two micro electrodes show that distal stimuli may produce ectoneural activity that inhibits motor nerves on one side of the arm while exciting those on the other, Figure 4.

3.4 Single cell morphology

Approximately 100 Lucifer fills of hyponeural cells were obtained and all these showed a non varicose but extensive dendritic tree of fine processes against the basal membrane, Figure 5. One or two axons arise from the cell bodies but axons to the oral region of the vertebral muscles cannot be dissected free and would become damaged when the nerve cord is removed for examination under the fluorescence microscope. There is no evidence from these fills that branches of hyponeural cells cross the midline or pass longitudinally from one ganglion to another.

4 DISCUSSION

These intracellular studies have the long-term goal of describing the cellular basis for the integration of behaviour by the nervous system. They have however already clarified a number of problems identified by previous electron microscope work. The Lucifer Yellow dye fills have shown unequivocally that both the ectoneural and hyponeural 'giant' fibres break up into a fine plexus to make synaptic contact with other neurones. This explains the failure to demonstrate large chemical or electrical synapses using the electron microscope. It has shown unequivocally there is a physiological connection between the ectoneural and hyponeural nervous systems.

It has provided the first clues as to integrating mechanisms in showing that motor neurones on one side of the arm are inhibited while the contra-lateral ones are excited and this would correlate behaviourally with rapid arm flexures. In the long term an understanding of function at the cellular level has to examine three basic processes. The first is the way sensory input into a small number of peripheral ganglia can produce a burst of activity within the ectoneural system which dictates and controls a complex piece of subsequent whole animal behaviour. The second process to be understood is how the bursts of propagated activity carried from ganglion to ganglion along the radial nerve cords and through the circumoral connectives are interpreted and integrated within each individual ganglion. This is clearly highly complex functionally since an individual ganglion is requires to interpret the input burst of activity, integrate this with levels of activity derived from other sources (some of which may be internal). It must then produce a co-ordinated output onto the hyponeural system as well as possibly an output via its own motor nerves. It also possibly modifies the original input burst of activity which is propagated on within the nervous system. Finally the way in which the output is integrated by the motor system requires describing.

There is as yet no intracellular information on the sensory system of echinoderms but there is accumulating a substantial amount of information using extracellular techniques (see Moore, in press). The complex integration within ganglion of ectoneural systems has only been superficially probed to-date but there is some evidence that there are at least two classes of neurones, one longitudinal associated with the onward propagation of bursts of nervous activity and one transverse which may be involved with integration of activity within a single ganglion. The single most pressing question to be answered is whether the longitudinal class of interneurones transmit information about all modalities of sensory stimulus or whether there are separate pathways. Present studies provide some evidence that there are not separate pathways and that conflicting or unnatural stimuli produce activity in the nervous system that is meaningless to the animal and can in some circumstances produce a catatonic state of paralysis in the animal. It may well be that the non-centralized nature of the nervous system and the physiological mechanism of encoding co-ordinating information have been highly restrictive in the evolution of complex behaviour and indeed perhaps of the phylum as a whole.

The present study has described in detail the function of the hyponeural motor system and shown that at least at the cellular level the physiology of the nervous system is conventional. Further studies should show how the unique classes of interneurones involved in the segmental ganglia of the radial nerve cords co-ordinate behaviour in a non-centralized nervous system. This will be of great comparative value in the overall study of the function and evolution of nervous systems.

REFERENCES

Cobb, J.L.S. 1970. The significance of the radial nerve cords. Cell Tiss.Res. 108: 457-474.

Florey, E. & M.A. Cahill. 1980. Cholinergic motor control of sea urchin. Cell Tiss. Res. 88: 281-292.

Moore, A. (in press). Proc. V Int. Echinoderm Conf. B. Keegan (ed.) Balkema, Rotterdam.

Proceedings of the Fifth International Echinoderm Conference / Galway / 24-29 September 1984

Hemal transport of ingested nutrients by the ophiuroid, *Ophioderma brevispinum*

JOHN C.FERGUSON
Eckert College, St. Petersburg, USA

ABSTRACT: Specimens of Ophioderma were fed small clams labeled with C^{14} amino acids, and after 8 to 24 hours sacrificed, sectioned, and studied as stripping film autoradiographs to observe translocation of the nutrient material. While some radioactivity was detected in the stomach tissues of the brittlestars, more notable concentrations were found in the hemal structures, especially the radial hemal canals. By 12 hours, this activity extended even to the arm tips. Label, at generally lower levels, was also seen in the oral hemal ring, axial organ (especially the aboral end), genital hemal connectives, transverse hemal canals, and in the connective tissue layer of the tube feet. Individual cells within the oral papillae also became highly labeled. Very little activity was found in the gonads, muscles, or most other parts not mentioned above. Additional experiments were conducted in which the brittlestars themselves were placed directly in the C^{14} nutrient medium before being processed into autoradiographs. These animals presented a similar picture of labeled nutrient distribution within the hemal tissues. Further, their epidermis, genital bursae, and mouth areas became heavily labeled. A sharp demarcation was noted between the oral epidermis and the much less labeled stomach tissue.

1 INTRODUCTION

The hemal systems of echinderms have confused investigators for centuries (See review of Ferguson, 1982). Their potential as true circulatory entities has been discounted in recent years, primarily because their histological structure is often more "sponge-like" than vascular and obvious propulsive mechanisms are generally absent. Furthermore, work by Irving (1924), Budington (1942), Farmanfarmaian and Phillips (1962), and Ferguson (1964a,b, and 1968) demonstrated the importance of coelomic circulation in internal transport. In light of this background, it was most surprising that autoradiographs published by Ferguson (1970) of Echinaster injected with C^{14}-labeled amino acids showed remarkable affinity of hemal elements for the tracer material. Additional autoradiographs and caloric evaluation of Asterias led Broertjes et al. (1979 and 1980) to the conclusion that nutrients were admitted from the digestive system into hemal channels, and that this translocation was important to vitellogenesis. Most recently, an extensive autoradiographic study on Echinaster by Ferguson (1984) clearly showed distribution of ingested nutrients within hemal channels to the gonads, tube feet, and distal body parts. While this latter investigation did confirm the importance of the hemal structures as a distributary mechanism of metabolic products, it could not demonstrate prominent direct connections between the digestive organs and the major hemal elements. It appeared more likely that the hemal tissues assisted in the distribution of derived metabolic products removed from circulation in the coelomic fluids rather than those taken directly from the digestive organs.

The present study was undertaken to provide additional confirmation and extension of these findings. It has used similar autoradiographic methods, this time applied to the ophiuroid, Ophioderma brevispinum.

2 MATERIALS AND METHODS

A nutrient medium was prepared from Millipore filtered (0.45μ) sea water containing

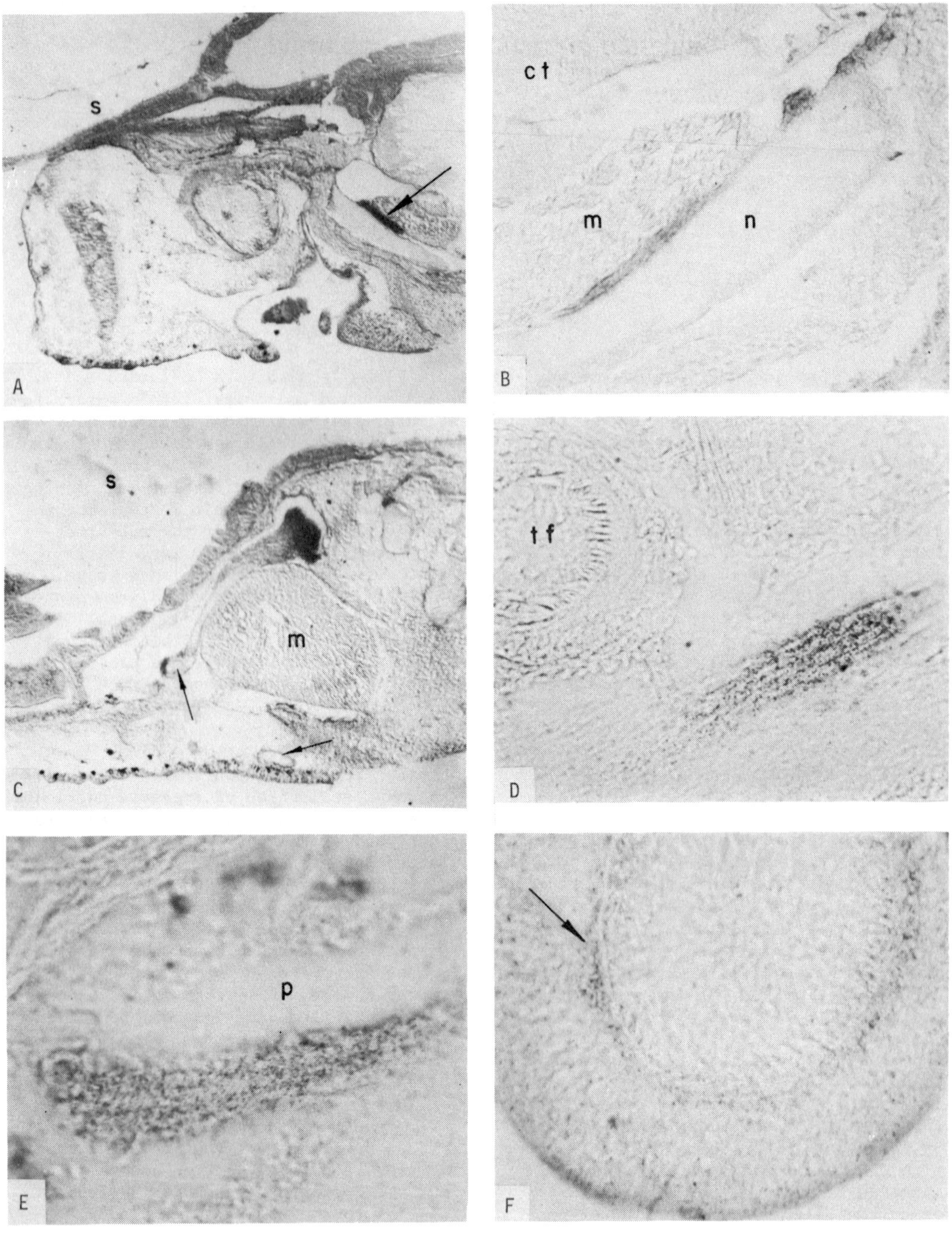

Figure 1. Unstained autoradiographic sections of Ophioderma 12 hours after start of feeding. A. Adjacent to mouth (left). Label is heavily concentrated in radial hemal canal (arrow). Moderate label is found in stomach (top) and in scattered epidermal cells (bottom). Other darkening is natural pigment. (35X). B. Detail of a radial hemal canal connecting to oral hemal ring (top right). (125X). C. Axial organ. Most activity is concentrated at aboral end (upper); a smaller amount lower, near stone canal (arrows). (40X). D. Longitudinal view of conspicuously labeled hemal substance in radial hemal

100 μCi C^{14}-labeled synthetic algal protein hydrolysate, 375 mg glycine, 990 mg glucose, and 10 mg chloramphenicol per liter. Surf clams (Donax variabilis) were then emersed in this and allowed to assimilate the nutrient material for about 24 hours. They were then removed, rinsed, broken open, and fed to specimens of Ophioderma brevispinum in a shallow container, which was being continually flushed with a rapid flow of fresh sea water. Most of the animals completed feeding within 5 to 6 hours. Specimens were sacrificed at 8, 12, and 24 hours from the initiation of feeding. Some additional studies were made on specimens of Ophioderma placed directly in the labeled nutrient medium for up to 12 hours. These were sacrificed at 4.5, 8, 22, and 31 hours after being put in the medium.

Pieces of tissues were fixed and decalcified in Bouin's, cut as 8 μ thick paraffin sections, and prepared as Kodak AR-10 autoradiographs with exposure times ranging from 3 to 6 months.

3 RESULTS

The autoradiographs of the fed specimens revealed labeled elements within tissues, all localized to very specific sites (See Fig. 1). Activity was detected rather uniformly throughout the stomach (Fig. 1,A), but only in a moderate intensity. Considerably greater concentrations of tracer were found in all the major hemal tissues, including the oral hemal ring (Fig. 1,B), the axial organ (Fig. 1,C), the radial hemal vessels (Fig. 1,A&D), and small scattered aboral vessels, perhaps genital connectives.

The observed distributions of labeled material in the hemal tissues was to some extent progressive. While intensity fluctuated from point to point along the structures, highest incorporation was within the radial hemal vessels of the disk (Fig. 1,A). The retention there was even more pronounced than most parts of the axial gland, in which the greatest uptake was found at the aboral end, and secondarily, the oral end (Fig. 1,C). Although incorporation within the radial hemal canal was strongest in the disk area, some radioactivity could be detected even in the most distal parts of these vessels at 8 hours. By 12 hours, radioactivity in the hemal canals of the arm tips was very prominent (Fig. 1,D). Also, by 12 hours label could be clearly detected in the transverse connections of the radial hemal canals (Fig. 1,E), and in the middle connective tissue layer of the tube feet (Fig. 1,F). The distribution of label in the 24-hour specimen was not materially different from that of the 12-hour one, except that the overall intensity was not as great. This could have been due to either ingestion of a smaller quantity of food or metabolic utilization of the material.

In all the specimens, strongly labeled cells were detected within the oral papillae, along with a small amount in the oral epidermis. Except for these areas, all of the other parts of the animals--epidermis, connective tissues, muscles, gonads, etc.--remained essentially free of label (generally less than 2X background in grain density).

The specimens of Ophioderma that had been placed directly in the labeled liquid nutrient medium displayed a very similar pattern of tracer distribution within their hemal tissues. In addition, these animals also showed a great deal of uptake by all the external epidermal cells, and even stronger uptake by the linings of the genital bursae. Very little evidence of movement of absorbed materials from these sites to deeper parts of the body was evident, even after 31 hours. A little activity was seen in the distal portions of the stone canal and ampulla, but not in the radial water vessels themselves.

Again, uptake into the buccal epithelium and cells within the oral papillae was conspicuous. Most surprising was a sharp demarcation between uptake in the oral epithelium and the stomach, noted in all specimens. The level of label retained by the stomach was only a small fraction of that retained by the oral epithelium and other exposed surface parts.

4 DISCUSSION

The observed rapid uptake, distribution, and retention of labeled nutrient materials by the hemal system is most remarkable. The

---

canal near arm tip. (330X). E. Labeled transverse hemal canal extending to an out of view tube foot on left. (330X). F. Transverse section of tube foot. Most label is concentrated in middle connective tissue layer (arrow). (330X). (ct, connective tissue; m, muscle; n, nerve cord; p, perihemal coelom; s, stomach; tf, tube foot).

methods used, of course, only visualize insoluble metabolic products, and thus do not give a complete picture of nutrient translocation. It is obvious, however, that within a few hours of ingestion, the hemal tissues become the major reservoirs in the body of derived insoluble nutrient products, and they remain so for some time. Direct movement of material within the hemal vessels could not be detected, but as noted, retention in the radial hemal canal was initially more pronounced in its proximal portion than in the more distal parts. Likewise, label did not appear in the connective tissue layers of the tube feet until after it was found in the hemal vessels.

Thus, these results provide further confirmation that the hemal tissues are very active entities within the echinoderm body, and that they probably assist in the distribution of derived metabolic materials. The hemal system does not, however, appear to function in the manner of a normal vascular system. The minor retention by the stomach tissues, the lack of conspicuous connections between the stomach and the major hemal vessels, and the rapid appearance of label within the oral papillae and the distal parts of the radial hemal vessels all lend credence to the belief that dissolved products are distributed by the more rapidly circulating coelomic and interstitial fluids, and that the insoluble hemal substance is derived from soluble elements contained in these fluids. The hemal system appears to deliver its products to the tube feet, the gonads, and possible other specific parts of the body.

It is at this time still difficult to understand the physical and biochemical nature of the hemal system and how it succeeds in accumulating and transporting metabolic materials. The hemal substance is probably rich in glycoprotein glycans, and it may have remarkable abilities to bind and perhaps enhance the diffusion of amino acids and other nutritional products. The echinoderm hemal system thus may be emerging as the model of choice for study of the complex biochemical interactions associated with extravascular tissue nutrition. Similar physical-chemical mechanisms may provide nutritional support of human connective tissues and operate in many other places within the body.

REFERENCES

Budington, R.A. 1942. The ciliary transport-system of Asterias forbesi. Biol. Bull. 83:438-450.

Broertjes, J.J., G. Posthuma, F.B. Beijnink, & P.A. Voogt 1979. The admission of nutrients from the digestive system into the hemal channels in the sea-star Asterias forbesi. J. Mar. Biol. Assoc. U.K. 60:883-890.

Broertjes, J.J., G. Posthuma, P. denBreejen, & P.A. Voogt 1980. Evidence for an alternative transport route for the use of vitellogenesis in the seastar Asterias rubens(L.). J. Mar. Biol. Assoc. U.K. 60:157-162.

Farmamfarmaian, R.A. & J.H. Phillips 1962. Digestion, storage and translocation of nutrients in the purple sea urchin (Strongylocentrotus purpuratus). Biol. Bull. 123:105-120.

Ferguson, J.C. 1964a. Nutrient transport in starfish. I. Properties of the coelomic fluid. Biol. Bull. 126:33-53.

Ferguson, J.C. 1964b. Nutrient transport in starfish. II. Uptake of nutrients by isolated organs. Biol. Bull. 126:391-406.

Ferguson, J.C. 1968. Transport of amino acids by starfish digestive glands. Comp. Biochem. Physiol. 24:921-931.

Ferguson, J.C. 1970. An autoradiographic study of the translocation and utilization of amino acids by starfish. Biol. Bull. 138:14-25.

Ferguson, J.C. 1982. Nutrient translocation. In M. Jangoux & J.M. Lawrence (eds.), Echinoderm Nutrition, p. 373-393. Balkema, Rotterdam.

Ferguson, J.C. 1984. Translocative functions of the enigmatic organs of starfish--the axial organ, hemal vessels, Tiedemann's bodies, and rectal caeca: an autoradiographic study. Biol. Bull. 166:140-155.

Irving, L. 1924. Ciliary currents in starfish. J. Exp. Zool. 41:115-124.

Proceedings of the Fifth International Echinoderm Conference / Galway / 24-29 September 1984

Neurophysiological studies on the perception of environmental stimuli in *Ophiura ophiura* (L.) (Echinodermata, Ophiuroidea)

A.MOORE
The University of St. Andrews, Scotland, UK

ABSTRACT: The neurophysiological responses of an ophiuroid, recorded extracellularly from the radial nerve cord, to chemical, photic and vibration stimuli are presented and discussed in relation to the behaviour of the whole animal in its natural environment. The responses to shadow and water movement in the aquatic medium are related to the detection of predators. An interoreceptor located within the arm spines is suggested as the receptor detecting water movement. Neurophysiological evidence of the ophiuroids ability to detect chemicals is presented, and the disadvantages of using behavioural changes to monitor the sensory discriminatory powers of echinoderms to perceive their environment are discussed.

1 INTRODUCTION

Neurophysiological studies on the sensory discriminatory powers of ophiuroids to perceive their environment are now possible. The technique involves recording extracellularly from the radial nerve cord of a brittlestar using simple suction electrodes. Brehm (1977) showed that such recordings were possible on ophiuroids and Stubbs (1982) subsequently published a preliminary paper on the neurophysiology of the photic response in Ophiura ophiura.

This paper describes the responses of the common ophiuroid Ophiura ophiura, to photic, vibration and chemical stimulation using these techniques. It shows that the brittlestar is particularly sensitive to the onset of shadow and movement in the aquatic environment and correlates these findings with whole animal behaviour. It also shows that this species is able to detect consistently very low levels of amino acids in its environment and provides the first electrophysiological evidence of chemoreception in an echinoderm.

The present report is an overview of progress achieved and does not aim to present a detailed analysis of all the results obtained from studies of the three sensory parameters.

2 MATERIALS AND METHODS

Ophiura ophiura is a small, subtidal ophiuroid commonly found around the coasts of Britain at depths of up to 200 metres. Specimens for this study were collected by beam trawl in the St. Andrews Bay area from a depth of 6 metres. The animals were maintained in sand bottomed circulation tanks with a water temperature of 5°C and in total darkness.

2.1 Electrophysiology

Single arm preparations of the experimental animal were initially dissected to reveal the radial nerve cord (R.N.C.) by removal of five oral and parts of the adjacent lateral plates. The arm was then immobilized in a perspex clamp, oral side uppermost, and placed in fresh seawater maintained at 5°C. Recordings were made using conventional, polythene tipped suction electrodes with tip diameters of 100-300µm which were attached to the interganglionic region of the ectoneural portion of the R.N.C. Signals were amplified via a Neurolog A.C. preamp and then displayed on a Tektronix 565 oscilloscope. Permanent records were made by recording the signal onto a conventional stereo cassette and played back via a Sharp stereo cassette deck (RT.10). The signals were then filmed using a Nihon-Kohden oscilloscope camera.

2.2 Experimental procedure

The photic stimulus was produced by extinguishing a 1.5mm diameter light spot, transmitted via a fibre optic cable from a battery operated 1.2V lens tipped bulb on the distal portion of the arm. This was carried out in a number of different ways to produce a shadow of varying durations, intensities and lengths.

The vibration stimulus was in the form of a single pressure wave produced by the vertical movement in the water of a perspex rod driven off a waveform generator. The wave was produced 5cm from the tip of the unrestrained arm and could be varied in intensity and frequency during experimentation. Single spine studies were performed by moving a single arm spine after the attachment of the spine tip to a fine perspex rod driven by a mechanical transducer. The distance the spine was moved and the rate at which a spine was moved a given distance could be varied.

Chemoperception experiments were carried out with a range of concentrations of 10 amino acid solutions presented to the unrestrained distal portion of a brittlestar arm to ascertain the minimum threshold concentration of each amino acid detectable by the animals.

In all three cases behavioural experiments were carried out on laboratory maintained animals using similar stimuli. All experiments were carried out under carefully controlled conditions eliminating all sensory parameters except the stimulus being investigated.

3 RESULTS

Stimulation of the distal portion of a brittlestar's arm, by each of the three sensory modalities produced recordable activity within the R.N.C. The responses consisted of single unit potentials, ranging in amplitude between 10-50µv and varying in number, pattern and amplitude between the different stimuli.

3.1 Neurophysiology: Photic response

The photic response of brittlestars has been previously described in detail by Moore and Cobb (1984) after a preliminary study by Stubbs (1982). The greatest response was to a deep, fast moving shadow produced on the tip of the arm. The response was consistent and characteristic and is now standardly used as an indication as to the success of an electrode attachment and the suitability of a preparation for experimentation.

3.2 Vibration response

The stimulus presented was a pressure wave that produced a water movement relative to the position of the animal. The pressure wave could be varied to produce a range of intensities and frequencies but beyond these arbitary values, the stimulus was difficult to quantify because of the complexity of physics involved in measuring waveforms in a confined space.

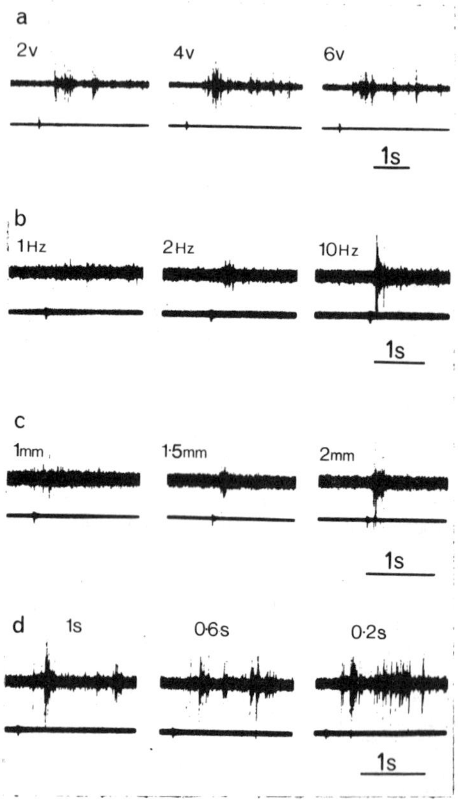

Figure 1. The responses recorded from the radial nerve cord to stimulation of the arm tip by pressure waves of different a) intensity and b) frequency and the movement of a single lateral spine c) different distances and d) at different rates through a given distance. The lower trace indicates the point of stimulation.

The response of the animal to movement in the water was a complex burst of single unit activity. Increasing both the intensity and frequency of the pressure wave elicited a corresponding increase in the size and number of recordable spikes (Fig. 1a and b). Studies using single spines where a lateral spine was mechanically moved produced similar recordable activity from the R.N.C. The responses increased, in terms of size and number of recordable spikes, as both the distance the spine was moved and the rate at which a spine was moved through a given distance increased (Fig. 1c and d). Removing the spines on a single arm preparation resulted in no recordable activity within the R.N.C. to a pressure wave of any frequency or intensity. These results indicated the mechanoreceptors were located at the spines. Removal of all surface structures and epithelium with pronase, confirmed by anatomical studies, had no effect on the animals ability to detect water movement. This evidence indicates that the receptors were not superficially located but were a type of interoreceptor, never before described in an echinoderm.

3.3 Chemical response

Ophiura ophiura was highly sensitive to low concentrations of the amino acid solutions (Table 1).

Table 1. The minimum threshold concentrations of 10 amino acids eliciting consistent recordable activity from the radial nerve cord.

Amino acid	Minimum Threshold concentration (M.)
L-Leucine	2×10^{-12}
L-Lysine	2×10^{-11}
L-Cysteine	2×10^{-11}
L-Lactic acid	2×10^{-11}
L-Valine	2×10^{-11}
L-Phenyl-alanine	2×10^{-11}
L-Malic acid	2×10^{-11}
N-Acetyl-L-Cysteine	2×10^{-11}
L-Serine	2×10^{-10}
L-Proline	2×10^{-10}

These minimum threshold concentrations produced consistent recordable activity from the R.N.C. The responses differed in the size and number of recordable spikes between the different amino acids

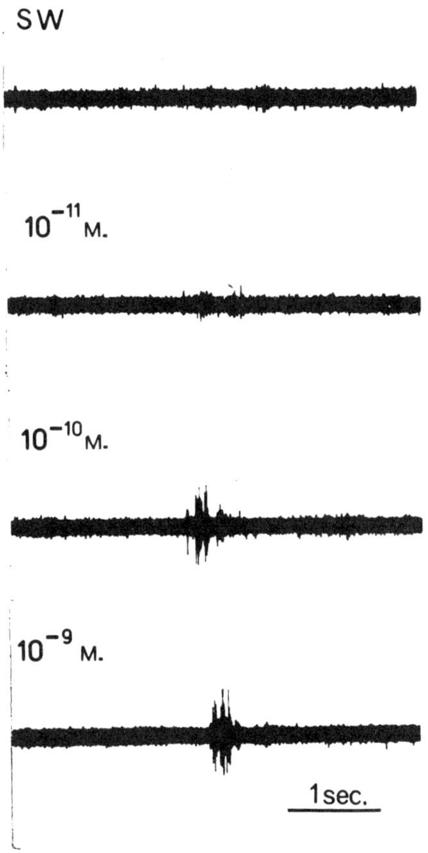

Figure 2. Extracellular recordings from the R.N.C. in response to stimulation of the arm tip by a range of concentrations of L-Valine. Control-seawater.

tested. In all experiments each response consisted of a complex burst of single unit activity (Fig. 2). The brittlestars were also sensitive to a range of foodstuffs commonly encountered in their natural environment.

3.4 BEHAVIOUR

The behaviour of brittlestars to photic stimuli has been fully described by Moore and Cobb (1984). This study showed that brittlestars performing any type of activity immediately ceased all movements at the onset of a shadow stimulus. This 'freezing' of the animal was both characteristic and consistent. Brittlestars starved for a period of 2 weeks responded to a chemical cue in their environment by an immediate movement towards the source of stimulus. In these circumstances a shadow stimulus had no effect on the behaviour of the animal. The response to the shadow was thus overridden by the response to the chemical stimuli.

The responses of Ophiura ophiura to a pressure wave stimulus in the aquatic medium were of two distinct types. A pressure wave of low intensity and frequency elicited a slow locomotory movement towards the source of stimulus. A sudden water movement created by an intense high frequency wave resulted in the immediate cessation of all activity in locomotory animals. This response was similar to the shadow response and resulted in the 'freezing' of the animal at the onset of the water movement.

The behavioural responses of brittlestars to chemical stimuli proved to be both inconclusive and inconsistent.

The minimum threshold concentrations of each amino acid solution that elicited recordable activity in the neurophysiological studies, produced no consistent activity in the experimental animals. The only consistency to emerge from the study was the attraction of responding brittlestars to all the amino acids tested.

4 DISCUSSION

The ophiuroids offer a unique opportunity within the echinoderms to record electrophysiologically the responses of the animals to a wide range of sensory modalities. Brehm (1977) demonstrated that unitary potentials could be detected in the R.N.C. of an ophiuroid in response to light, and Stubbs (1982) found that Ophiura ophiura was maximally sensitive to wavelengths and intensities of light normally encountered in marine habitats.

One aspect of the present research that requires clarification is exactly what type of neurone the responses are recorded from. Each unit potential is recorded from a type of interneurone within the ectoneural part of the radial nerve cord and not from the sensory receptor neurones themselves. The levels of stimulus recorded may therefore not be the absolute levels at which the animal is able to detect different sensory modalities in its environment. At present nothing is known of the mechanism by which activation of the primary receptors produces the interneurone spike burst.

The present study has shown that the responses to the various stimuli have important behavioural implications in the brittlestars. Ophiuroids form one of the main prey items of many gadoid and flatfish species. The detection of a rapidly moving shadow in its environment is of great importance and the intensity of shadow may well relate to the closeness of a possible predator. If the brittlestar ceases all movement at the onset of a deep, fast moving shadow, the animal may well escape detection by the predator. The shadow response is a simple antipredator device.

Vibration and water movement are also important sensory modalities providing information to the brittlestars about their aquatic environment. Ophiura ophiura occurs in dense populations on areas of the seabed where bottom currents are evident. The movement of the animal in response to water movements of low intensity may well relate to an animal's attempt to orientate itself towards a current flow. An intense water movement in the vicinity of an animal may be caused by the activity of a possible predator, resulting in the cessation of activity and the increased possibility of the animal avoiding detection. Anatomical studies of the arm spines have indicated the presence of structures similar to those described by Whitfield and Emson (1983) in Amphipholis squamata. The structures do not have projecting cilia and this fact coupled with the evidence from the removal of the epithelium without affecting the animals ability to detect water movement, suggests the presence of an interoreceptor located in the spine. It seems likely that the 'stabchen' described by Whitfield and Emson are chemoreceptors.

The levels of amino acids consistently detected by the brittlestars are extremely low. Reimer and Reimer (1975) have shown the ability of four tropical ophiuroids to detect a range of amino acids by monitoring the appetitive and feeding behaviour of the animals, but no electrophysiological recordings of an echinoderm's ability to detect chemicals within their environment have so far been reported.

One of the more significant findings of this study is the observation that the shadow response, although consistently recorded from the nervous system, is overridden behaviourly in the presence of food or other chemical stimuli. This result may suggest the presence of a behavioural hierarchy directly dependent upon the physiological and nutritional state of the particular animal. This has particular significance when work on the sensory discriminatory powers of echinoderms are based solely on behavioural changes in animals. It is important that the whole range of sensory modalities present are absolutely defined during experimentation to avoid many of the inconsistencies that occur under such conditions.

REFERENCES

Brehm, P. 1977. Electrophysiology and luminescence of an ophiuroid radial nerve. J. Exp. Biol. 77: 213-227.

Moore, A. and Cobb, J.L.S. 1984. Neurophysiological studies on photic responses in Ophiura ophiura (L.) (Echinodermata, Ophiuroidea). (In press).

Reimer, R.D. and Reimer, A.A. 1975. Chemical control of feeding in four species of tropical ophiuroids of the genus Ophioderma. Comp. Biochem. Physiol. 51A: 915-927.

Whitfield, P.J. and Emson, R.H. 1983. Presumptive ciliated receptors associated with the fibrillar glands of the spines of the echinoderm Amphipholis squamata. Cell Tissue Res. 232: 609-624.

Contrasting reproductive periodicities among north-eastern Pacific ophiuroids

STEVEN S.RUMRILL
University of Alberta, Edmonton, Canada

JOHN S.PEARSE
University of California, Santa Cruz, USA

ABSTRACT: Reproductive patterns were compared between 5 species of central California ophiuroids over a period of 2 years. Modes of development include the production of planktotrophic ophiopluteus larvae as well as viviparous and ovoviviparous brooding. Breeding was markedly periodic in the different broadcast-spawning species despite the absence of strong seasonal oscillations in sea temperature; Ophiopteris papillosa and Amphiodia occidentalis had restricted breeding cycles with respective peaks of spawning activity in the winter and spring. Spawning occurred continuously in Ophiothrix spiculata from late winter to early summer. Brooding ophiuroids also showed disparate reproductive patterns. Large-bodied ovoviviparous females of Ophioplocus esmarki brooded intermediate numbers of embryos only during the late spring and summer. Conversely, small-bodied viviparous hermaphroditic individuals of Axiognathus squamata continuously brooded fewer embryos throughout the year. Proximate cues other than sea temperature apparently synchronize the gametogenic cycles of these ophiuroids.

1 INTRODUCTION

The reproductive activities of echinoderms are regulated through endogenous pathways which are in turn coordinated by exogenous environmental stimuli. Annual reproductive cycles and their controlling factors have been investigated for several echinoids and asteroids along the temperate eastern Pacific coast of North America (Giese, 1958, 1959; Greenfield, et al., 1958; Gonor, 1973; Cochran & Engleman, 1975; Pearse, 1980; Pearse & Eernisse, 1982). By comparison, relatively little information has been gathered concerning the reproductive periodicities of ophiuroids from this same region (Patent, 1969; Cunningham, 1977). Widespread knowledge of echinoderm breeding patterns will aid in the identification of environmental variables that synchronize reproductive events.

The shallow-water ophiuroids of central California represent an abundant and ecologically diverse fauna (Austin & Hadfield, 1980). Members of this group display marked differences in morphology and habitat as well as a variety of contrasting reproductive patterns. Species having small eggs (<115 um) produce planktotrophic larvae and generally show seasonal reproductive cycles with spring spawning. Conversely, brooding species tend to have more broadly defined cycles or brood year-round. In this paper we document the modes of reproduction, fecundity, and seasonal reproductive periodicities of 5 species of ophiuroids from the central California coast.

2 MATERIALS AND METHODS

Ophiuroids were collected at monthly intervals from 2 sites on the Monterey Peninsula, California (36°38'N, 122°03'W). Specimens of Amphiodia occidentalis (Lyman, 1860) and Axiognathus (=Amphipholis) squamata (Delle Chiaje, 1829) were sampled intertidally from the Great Tide Pool from January 1981 to July 1982 (Fig. 1). Ophiothrix spiculata Le Conte, 1851, Ophiopteris papillosa (Lyman, 1875), and Ophioplocus esmarki Lyman, 1874 were gathered subtidally using SCUBA at Hopkins Marine Life Refuge from August 1980 to July 1982.

Each month between 10 and 20 specimens of each species were measured across the diameter of the disc and along the length of the longest arm. In broadcast-spawning species the arms were removed at the edge of the disc, and the disc was drained and weighed. An incision was made around the periphery of the disc with micropoint scissors, and the gonads were dissected from one inter-

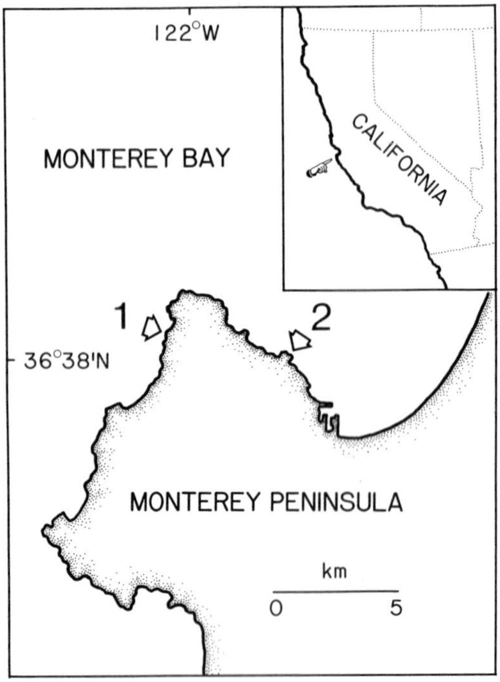

Figure 1. Map of Monterey Peninsula showing locations of collecting sites (arrows): 1, Great Tide Pool; 2, Hopkins Marine Life Refuge.

radius. The gonads were air dried and weighed to the nearest ±0.1 mg. Since gonadal tissues develop synchronously within the 5 interradii (Rumrill, 1984), a gonad index (G.I.) value was calculated for each specimen using the formula:

$$G.I. = \frac{5 \text{ x gonad wt.}}{\text{disc wt.}} (100)$$

Separate G.I. values were calculated for males and females of each species. For brooding species, the aboral portion of the disc and gut were removed from each specimen and the percentage of adults brooding was determined. Brooded embryos were dissected from the brooding cavity and counted under a stereomicroscope.

3 RESULTS

Body characteristics, types of development, and egg measurements are compared among the 5 species of ophiuroids in Table 1. O. spiculata, O. papillosa, and A. occidentalis exhibited broadcast-spawning and planktotrophic development. All 3 of these intermediate-sized ophiuroids produced large numbers (26,000-110,000) of small (<110 um) ova/female. Ovoviviparous females of O. esmarki brooded intermediate numbers (2,600) of large (>375 um) embryos, and viviparous hermaphroditic individuals of A. squamata brooded few (1-7) embryos which developed within the parent to large sizes (>900 um) from small eggs (<100 um).

Monthly changes in gonadal indices are plotted for the 3 broadcast-spawning ophiuroids along with annual sea temperatures in Figure 2. Sea temperatures in Monterey Bay fluctuated annually between 11 and 14°C but did not undergo large seasonal changes on a predictable schedule. Individuals of O. spiculata showed 2 broad periods of maximal gonad size, each lasting from 5 to 6 months in duration. The gonads of O. spiculata were smallest in late summer and consistently increased in size through the fall from August to October. Peak gonad index values were reached in November of 1980 and 1982 and the gonads remained large through the winter until March. Rapid decreases in gonad size occurred concomitantly

Table 1. Comparison of morphological and reproductive characteristics of California ophiuroids.

Species	Disc Dia. (mm) ($\bar{x}$ ±SD)	Arm/Disc Ratio	Mode of Reproduction	Max. Egg Dia. (um)	Max. no. Eggs/Female
Ophiothrix spiculata	6.0 ±1.3	6/1	planktotrophy	109	26,315
Ophiopteris papillosa	11.7 ±0.9	5/1	planktotrophy	102	109,080
Amphiodia occidentalis	4.9 ±0.9	13/1	planktotrophy	98	67,965
Axiognathus squamata	2.1 ±1.1	5/1	viviparous	105	7
Ophioplocus esmarki	18.4 ±2.2	3/1	ovoviviparous	328	2,360

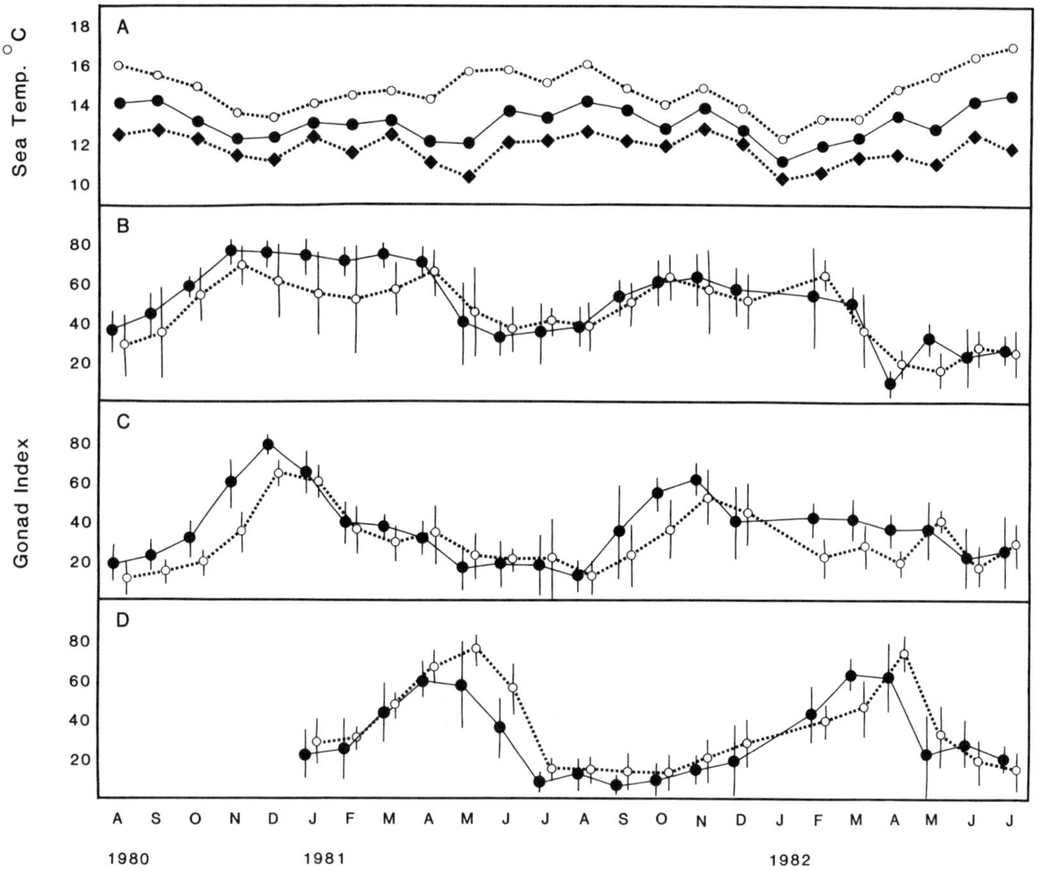

Figure 2. Annual sea temperature and gonadal indices of broadcast-spawning ophiuroids. A. Surface temperature (○ = monthly maximum, ● = mean, ◆ = mimimum). B-D. Monthly gonadal indices ($\overline{x}$ $\underline{+}$SD; ● = females, ○ = males). B. Ophiothrix spiculata. C. Ophiopteris papillosa. D. Amphiodia occidentalis.

with spawning in May 1981 and April 1982. O. papillosa and A. occidentalis each showed more distinctly seasonal reproductive cycles. In O. papillosa gonad index values generally increased through the fall and reached peaks in December 1980 and November 1981. The period of maximal gonad size lasted for 2 to 3 months in each season and was followed by an abrupt spawning in January. Oocyte proliferation and gonadal growth occurred in early spring in A. occidentalis and highest gonad index values were reached in May 1981 and April 1982. The gonads of A. occidentalis remained large between March and May and decreased rapidly in size following spawning in late May and June. In all 3 broadcast-spawning species the gonads of female and male specimens developed in relative synchrony throughout the year; however, male O. spiculata exhibited a depression in gonad size in the winter of 1981 which was not apparent in females, and the mean gonad index was consistently greater for males of A. occidentalis than females from January to December, 1981.

Brooding activity in A. squamata and O. esmarki are compared in Figure 3. A. squamata exhibited year-round brooding between February 1981 and July 1982 (Fig. 3A-B). An overall population mean of 71.3% were brooding throughout the year, and on 5 occasions the frequency of brooding equalled or exceeded 80%. Similarly, there was no seasonality in the numbers of embryos brooded. Mean brood size varied between 4 and 5 embryos/parent, although a few individuals brooded up to 7 embryos simultaneously within the bursae. Since variation in brood number was generally

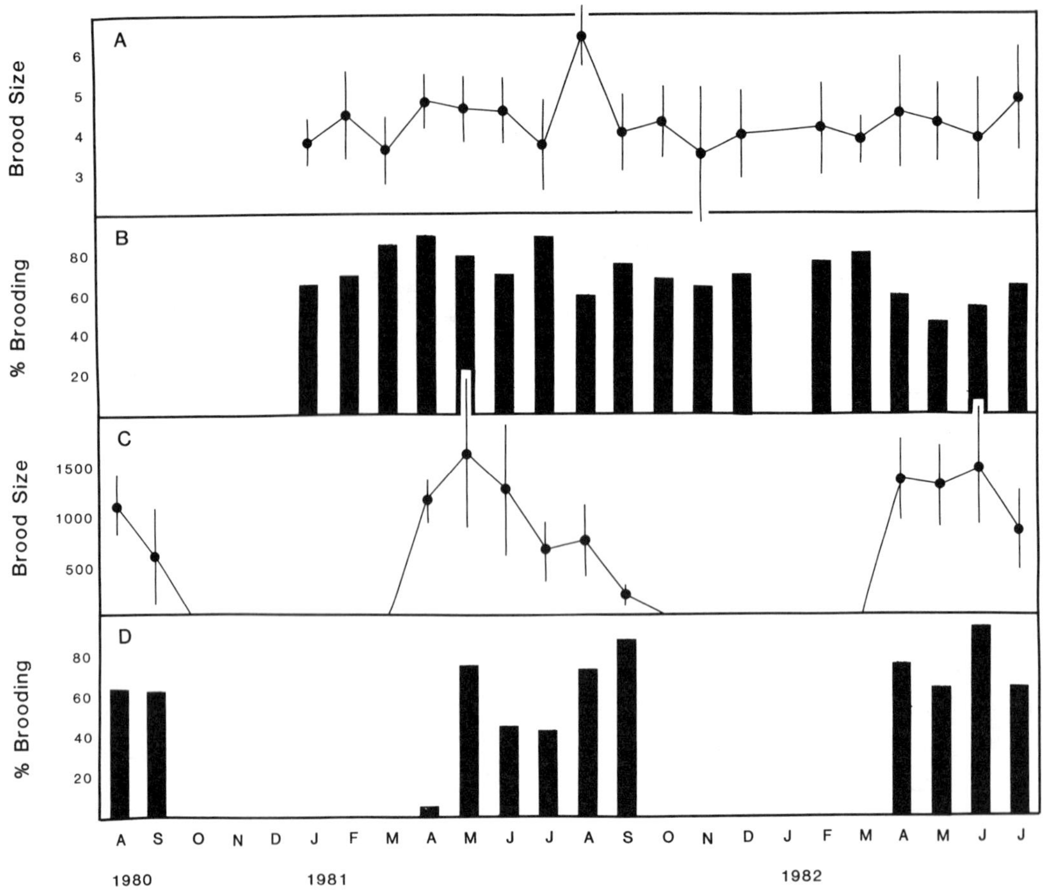

Figure 3. Annual brood size ($\bar{x}$ $\underline{+}$SD) and brooding frequencies of viviparous and ovoviviparous ophiuroids. A-B. Axiognathus squamata. C-D. Ophioplocus esmarki.

greater within months than the differences in brood size between months, the brooding of A. squamata does not appear to follow an annual cycle. Conversely, brooding activity followed a distinct seasonal cycle in O. esmarki (Fig. 3C-D). High frequencies of brooding females were observed in late summer 1980, from May to September 1981, and from April to July, 1982. No brooding specimens were observed in fall or winter months in 1980-81 or 1981-82. Brood size was generally highest in April and May at the onset of the brooding period and gradually declined as hatchlings were released over the summer.

4 DISCUSSION

In Monterey Bay, O. spiculata, O. papillosa, A. occidentalis, and O. esmarki were all found to breed on a rhythmic annual cycle. The marked reproductive seasonality in these species suggests that their breeding periods may be synchronized or entrained by predictable fluctuations in environmental parameters. However, differences in the timing of gametogenic events suggest that different species may be entrained by different environmental cues. For example, gametogenesis began 3 to 4 months prior to spawning in January and February in A. occidentalis while gonad growth began in August and September in O. spiculata and O. papillosa. Furthermore, simultaneous breeding of species with planktotrophic and ovoviviparous development, such as A. occidentalis and O. esmarki, suggests that factors which influence pelagic conditions for larvae are not necessarily of prime importance in the determination of reproductive patterns.

The extent to which ophiuroid reproductive cycles are coordinated by exogenous factors is not well understood. Changing ambient sea temperatures are commonly regarded as proximate cues that initiate gametogenic events. For example, Thorson (1934, 1946) suggested that summer breeding in Ophiocten sericeum in Kempe Fjord, Greenland was related to a period of warming associated with the break-up of ice, and Fenaux (1970) observed a correlation between spawning activity in Amphiura chiajei and increasing summer temperatures in Villefranche Bay, France. Similarly, Tyler (1977) and Bowmer (1982) noted that spawning took place in Ophiura albida and Amphiura filiformis during periods of high ambient sea temperatures in the north Atlantic. However, environmental factors other than temperature also may act as proximate cues that synchronize seasonal reproduction. Thorson (1934) could not explain asynchronous breeding between separated populations of O. sericeum from Kangerdlugsuak Fjord and Ella Island in the absence of a temperature gradient, and Hendler (1979) reported synchronous breeding between conspecific ophiuroid populations which experience different thermal conditions on the Caribbean and Pacific coasts of Panama. Furthermore, Schoener (1968) and Tyler and Gage (1979) documented a discrete seasonal reproductive cycle for Ophiura ljungmani in periodic samples from 2,900 m in the Rockall Trough, northeast Atlantic, where ambient sea temperatures are relatively constant throughout the year. Since nearshore sea temperatures do not undergo large seasonal changes in Monterey Bay, it is unlikely that the reproductive cycles of the ophiuroids reported here are directly influenced either by absolute or changing temperature conditions in the field. Other environmental factors which may be important as proximate cues in temperate ophiuroids include seasonally changing photoperiod, food and habitat availability, current regimes, and juvenile recruitment and survival. The role of proximate cues which influence benthic juvenile survival appear to be especially important for brooding species such as O. esmarki which has a well synchronized and distinctly seasonal brooding period. Proximate cues probably exert less influence on the reproduction of A. squamata in which the potential for opportunistic recruitment and rapid colonization may function as ultimate cues which select for aseasonal brooding.

Mladenov (1976) reported a tendency among tropical ophiuroids which have discrete annual reproductive cycles to spawn in late summer and fall. Conversely, ophiuroids from temperate waters show a propensity to spawn in spring; few species spawn in winter and summer (Patent, 1969; Fenaux, 1970; Hendler, 1973; Tyler, 1977; Tyler & Gage, 1979; Bowmer, 1982; this study). Our observations on the variable timing of reproductive seasons in California ophiuroids are in agreement with observations of breeding in other echinoderms from this region and support the trend identified by Thorson (1950) that breeding seasons become increasingly restricted to periods of longer daylengths and optimal phytoplankton growth with greater distance from the equator.

5 ACKNOWLEDGEMENTS

We thank W.T. Doyle, Director of the Center for Marine Studies, for providing facilities at the Joseph M. Long Marine Laboratory, and J.C. Cureton for preparing the figures and manuscript.

6 REFERENCES

Austin, W.C. & M.G. Hadfield. 1980. Ophiuroidea: the brittle stars. In R.H. Morris, D.P. Abbott & E.C. Haderlie (eds.), Intertidal invertebrates of California, p. 146-159. Stanford University Press, Stanford.

Bowmer, T. 1982. Reproduction in Amphiura filiformis (Echinodermata: Ophiuroidea): seasonality in gonad development. Mar. Biol. 69:281-290.

Cochran, R.C. & F. Engleman. 1975. Environmental regulation of the annual reproductive season of Strongylocentrotus purpuratus (Stimpson). Biol. Bull. 148: 393-401.

Cunningham, C.J.B. 1977. Aspects of the reproductive biology of intertidal amphiurids of Monterey Bay. M.A. Thesis, San Jose State University, San Jose. 65 p.

Fenaux, L. 1970. Maturation of the gonads and seasonal cycle of the planktonic larvae of the ophiuroid Amphiura chiajei Forbes. Biol. Bull. 138:262-271.

Giese, A.C. 1958. Comparative physiology: annual reproductive cycles of marine invertebrates. Ann. Rev. Physiol. 21: 547-576.

Giese, A.C. 1959. Reproductive cycles of some west coast invertebrates. In R.B. Withrow (ed.), Photoperiodism, p. 625-638. Amer. Assoc. Advanc. Sci., Wash. D.C.

Gonor, J.J. 1973. Reproductive cycles in Oregon populations of the echinoid Strongylocentrotus purpuratus (Stimpson). I. Annual gonad growth and ovarian gametogenic cycles. J. exp. mar. Biol. Ecol. 12:45-64.

Greenfield, L., A. Giese, A. Farmanfarmaian

& R.A. Boolootian. 1958. Cyclic biochemical changes in several echinoderms. J. exp. Zool. 139:507-524.
Hendler, G. 1973. Northwest Atlantic amphiurid brittle stars Amphioplus abditus (Verrill), Amphioplus macilentus (Verrill) and Amphioplus sepultus n. sp. (Ophiuroidea: Echinodermata): contributions on their systematics, zoogeography, annual periodicity, and larval adaptations. Ph.D. Thesis, University of Connecticut, Storrs. 255 p.
Hendler, G. 1979. Reproductive periodicity of ophiuroids (Echinodermata: Ophiuroidea) on the Atlantic and Pacific coasts of Panama. In S.E. Stancyk (ed.), Reproductive ecology of marine invertebrates, p. 145-156. University of South Carolina Press, Columbia.
Mladenov, P.V. 1976. Reproduction and larval development in ophiuroids (Echinodermata) of Barbados, West Indies. M.Sc. Thesis, McGill University, Montreal. 136 p.
Patent, D.H. 1969. The reproductive cycle of Gorgonocephalus cargi (Echinodermata: Ophiuroidea). Biol. Bull. 136:241-252.
Pearse, J.S. 1980. Synchronization of gametogenesis in the sea urchins Strongylocentrotus purpuratus and S. franciscanus. In W.H. Clark & T.S. Adams (eds.), Advances in invertebrate reproduction, p. 53-67.
Pearse, J.S. & D.J. Eernisse. 1982. Photoperiodic regulation of gametogenesis and gonadal growth in the sea star Pisaster ochraceus. Mar. Biol. 67:121-125.
Rumrill, S.S. 1984. Contrasting reproductive patterns among ophiuroids (Echinodermata) from southern Monterey Bay, USA. M.Sc. Thesis, University of California, Santa Cruz. 260 p.
Schoener, A. 1968. Evidence for reproductive periodicity in the deep sea. Ecology 49: 81-87.
Thorson, G. 1934. On the reproduction and larval stages of the brittle stars Ophiocten sericeum (Forbes) and Ophiura robusta Ayres in east Greenland. Meddr. Grønland. 100:1-21.
Thorson, G. 1946. Reproduction and larval development of Danish marine bottom invertebrates. Medd. Kommn Dan. Fisk. Havunders., Ser. Plankton 4:1-523.
Thorson, G. 1950. Reproductive and larval ecology of marine bottom invertebrates. Biol. Rev. 25:1-45.
Tyler, P.A. 1977. Seasonal variation and ecology of gametogenesis in the genus Ophiura (Ophiuroidea:Echinodermata) from the Bristol Channel. J. exp. mar. Biol. Ecol. 30:185-197.
Tyler, P.A. & J.D. Gage. 1979. Reproductive ecology of deep sea ophiuroids from the Rockall Trough. In E. Naylor & R.G. Hartnoll (eds.), Cyclic phenomenon in marine plants and animals, p. 215-222. Pergamon Press, Oxford.

Proceedings of the Fifth International Echinoderm Conference / Galway / 24-29 September 1984

Interference competition between brittle-stars?

E.M.SIDES
Department of Zoology and Discovery Bay Marine Laboratory, University of the West Indies, Jamaica

ABSTRACT: The brittle-stars Ophionereis reticulata, Ophiocoma pumila and Ophiolepis impressa, in addition to other species, were found under slabs of coral rubble lying on sand, in 1-2 m of water in Jamaica. Analysis of the species distribution under stones showed that O. reticulata, O. pumila and O. impressa were negatively associated with each other, i.e. did not generally occur under the same stone. Ophionereis reticulata and Ophiocoma pumila were shown to alter the crevice dimensions under a stone, by shifting sand, but Ophiolepis impressa did not. Experiments were run to determine if competition for space occurred between O. reticulata and O. impressa. Under experimental conditions competition for space was found to occur between the two species and O. reticulata was the competitively dominant species. The interaction took 2-8 days. Alteration of the crevice dimensions is suggested as the competitive mechanism.

1 INTRODUCTION

The ophiuroids Ophionereis reticulata (Say), Ophiocoma pumila Lütken, and Ophiolepis impressa Lütken are among the eight common species of brittle-stars found under rubble lying on sand in shallow water shoreward of the reef crest in Jamaica (Sides 1981). This overlap in habitat has been recorded previously (Grave 1898; Devaney 1974; Kissling and Taylor 1977; Lewis and Bray 1983). Grave (1898) and Fontaine (1953) recorded that some shallow-water species which occurred in the same habitat frequently occurred together under the same stones. However, detailed information of this kind was not available for O. reticulata, O. pumila, and O. impressa.

This paper quantifies the association of Ophionereis reticulata, Ophiocoma pumila, and Ophiolepis impressa under stones in the same area. Their association with other species will be reported elsewhere (Sides in prep). The ability of O. reticulata and O. pumila to alter their crevices by moving sand, and the apparent inability of Ophiolepis impressa to do this, is also described. In addition experiments were run to show that competition for space occurs between O. reticulata and O. impressa; and a competitive mechanism is suggested.

2 STUDY AREA

The study area was the lagoon on the west side of Discovery Bay, Jamaica, which has been previously described by Woodley and Robinson (1977) and Hammond (1980). The sites chosen were within the Transitional Zone of Hammond (1980), a distinct area shoreward of the Rear Zone (Kinzie 1973) in which dead and living coral becomes more widely spaced. Dead coral rubble is abundant in this zone and occurs as slabs lying on sand or in piles. The depth of the water is 1-2 m. On the shoreward side this zone merges in places with beds of turtle grass, Thalassia testudinum Koenig, in which sand patches devoid of rubble occur. Shrimps, Callianassa sp., and holothurians Holothuria arenicola Semper, are frequently present in these sand patches.

3 METHODS AND RESULTS

3.1 The species distribution under stones

The number of individuals of each species occurring under each piece of rubble, lying on sand, was recorded for 391 hapazardly chosen stones. The length and width of 161 of these stones were measured. Hulbert's

(1969) coefficient of interspecific association was calculated for the pairwise combinations of O. reticulata, O. pumila and O. impressa and showed that the three species were negatively associated with each other (Table 1). A two-tailed version of Fisher's exact test (Siegel 1956) was used to determine the significance level of the coefficients.

Table 1. Hulbert's coefficient of interspecific association. The coefficient ranges from +1 a positive association, through 0 no association, to -1 a negative association.

Species Pairs	Coefficient
O. reticulata X O. impressa	-0.39*
O. reticulata X O. pumila	-0.58**
O. pumila X O. impressa	-0.59**

* $p < 0.05$, ** $p < 0.001$

Table 2. The median size of stones under which each species was found in the field and the median number of individuals of these three species found.

Species	Median (sq cm)	50% Range	No. found	N
O.pumila	382	249- 600	1	64
O.reticulata	357	200- 604	1	43
O. impressa	950	275-1735	1+4*	27

* 1 O.impressa + 4 individuals of one or more species. O.reticulata and O.pumila typically occurred alone.

Analysis of the size of stones used by each species showed that the median stone size used by O. reticulata and O. pumila was similar and that only one individual was typically found under each stone (Table 2). The median stone size under which O. impressa was found was a little more than twice that used by the other two species. In spite of the large size of stone used by O. impressa the median number of individuals of this species found was again only one: however O. impressa rarely occurred alone under a stone, and a median number of 4 individuals of other species were found with it (Table 2).

3.2 Habitat alteration

Field observations indicated that O. reticulata and O. pumila often altered the shape of the openings, in particular the width of the openings, through which they extended their arms to feed, but that O. impressa did not. The following experiment was set up in the laboratory to examine the ability of each species to alter their crevice dimensions. The species were tested consecutively in a series of 10 trays (40 x 18 x 18 cm). Each tray was supplied with running sea-water, 2-3 cm of sand and a slab of rubble approximately 375 sq cm. One individual ophiuroid was placed in each tray and left for 5 days. The trays were photographed at the beginning and end of the experiment, the stones were then removed and notes were made on the amount of disturbance.

The results showed that both O. reticulata and O. pumila altered their crevice space as well as the openings, in particular O. reticulata, but that O. impressa did not (Table 3, Fig. 1).

3.3 Competition for space between Ophionereis reticulata and Ophiolepis impressa

The negative association between O. impressa, O. reticulata and O. pumila, and the ability of the latter two species to alter their crevices suggested that competition for space might occur between these species. This may be because O. reticulata and O. pumila make the crevices unsuitable for each other and for O. impressa. The experiments described here were run to determine whether or not competition for space occurred between O. reticulata and O. impressa.

The experiments were run in the lagoon in two adjacent sand patches, designated Area 1 and Area 2, which were devoid of rubble but contained the shrimp Callianassa sp. or the

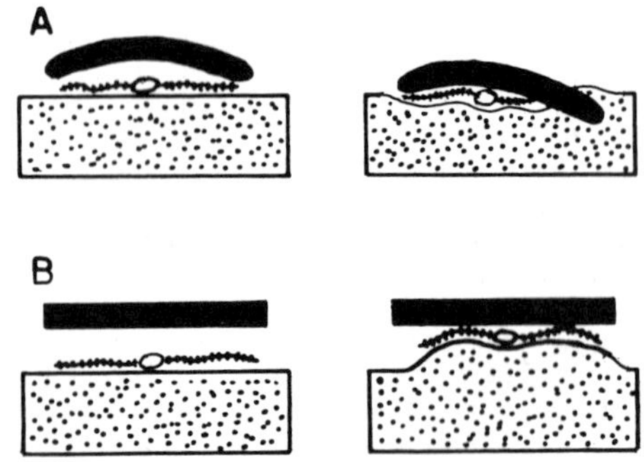

Figure 1. A diagrammatic sketch of how Ophionereis reticulata alters the crevice dimensions by (A) removing sand so that the stone sinks or, (B) building up the sand under the slab.

Table 3. A summary of the amount of crevice alteration by each species in the laboratory after 5 days (+++ major disturbance, ++ moderate, + slight, - none).

SPECIES	VISIBLE AMOUNT OF DISTURBANCE with stone in place	underneath stone	COMMENTS
O.reticulata	+++	+++	Parts of some stones lowered approximately 3 cm to the base of the tank. In one case the gap was narrowed by building up the sand.
O.pumila	++	+++	Parts of some stones lowered 1-2 cm into the sand, but rarely to the base of the tank.
O. impressa	-	+	Slight disturbance of the surface of the sand under the stone.

holothurian Holothuria arenicola, or possibly both. Experimental slabs of coral rubble were spaced 0.75 m apart, avoiding both ends of the shrimp or holothurian burrows. 15 pieces of rubble were placed in Area 1 and 20 in Area 2. The median stone size used was 423 sq cm (range 220-667). The median size of stone used was very close to that under which O. reticulata is found naturally (Table 2). The stones were well within the range used by O. impressa (Table 2), although the median size of the experimental stones was less than that under which O. impressa normally occurs. Thus the experimental stones were of a size expected to accomodate O. reticulata or O. impressa.

The experiment was divided into 3 steps.

1. To determine the probability of Ophiolepis remaining under a stone after it had been placed there.

This was done by placing 2 individuals of O. impressa under each stone in both areas, and examining the stones after 8 or 24 days. This species is positively associated with other brittle-stars, but rarely of the same species (Table 2 and section 3.1 above). However very few stones had 2 individuals under them after 8 days, suggesting that intraspecific competition between individuals of O. impressa may be occurring; this will be discussed elsewhere. The number of stones with 1 and 2 individuals were combined in the analysis. As the results were not significantly different in the two areas after 8 and 24 days (χ^2=2.357, df=2, p>0.3) the data were pooled to calculate the probability of O. impressa remaining under a stone after 8-24 days. Thus the probability of O. impressa remaining is 0.689 (i.e. 42/61) and O. impressa leaving is 0.311 (i.e. 19/61).

2. To determine the probability of Ophionereis reticulata remaining under a stone after it had been placed under it.

One experiment was run in Area 1 in which a single O. reticulata was placed under each of 12 stones and left for 8 days. All 12 O. reticulata were found after 8 days. However this sample size was considered too small on which to calculate the probability of O. reticulata remaining under a stone. To increase the sample size, the distribution of O. reticulata when it had been placed under a stone alone was compared with its distribution in 5 experiments when it had been placed under a stone with a single O. impressa. No significant difference was found between the experiments (χ^2=1.37, df=5, p>0.9) showing that the distribution of O. reticulata was not affected by the presence of O. impressa. Thus the probability of O. reticulata remaining under a stone was calculated from the pooled data, and found to be 0.851 (i.e. 63/74), and not remaining was 0.149 (i.e. 11/74).

If no competition occurs, then the distribution of each species should be independent of each other when they are both under the same stone. Combining the probabilities of O. reticulata and O. impressa being present or absent generates the probability of finding O. impressa alone, O. reticulata alone, O. impressa and O. reticulata together, and nothing under the stones.

3. To determine the distribution of Ophionereis reticulata and Ophiolepis impressa after one individual of each species had been simultaneously added to each stone.

If competition for space occurs between these two species it may be the result of alteration of the crevice dimensions under the stone by O. reticulata: however the degree of alteration necessary to force O. impressa to leave was not known, and the visible disturbance was not as obvious in the field as in the laboratory. Thus experiments were run for 30, 28, 18 and two experiments for 8 days. Only a few stones were found with O. impressa alone, both species, or nothing under them. Although unsatisfactory, it was necessary to lump these categories together for the analysis. As no significant difference was found between the experiments (χ^2=1.375, df=4, p> 0.8) the data for all experiments were pooled. A comparison of this observed distribution with the expected distribution showed that the two distributions were significantly different (Table 4). The results also showed that in these experimental conditions O. reticulata was the dominant species as approximately 2/3 of the stones had only O. reticulata under them. As the distribution of the two species did not differ between 30 and 8 days, the competitive interactions appear to have taken place by 8 days.

Table 4. A comparison of the expected and observed distribution in interspecific encounters between Ophionereis reticulata (O.R.) and Ophiolepis impressa (O.I.).

	O.I. alone	O.R. alone	O.R.+ O.I.	Zero	N Stones
Expected	6	16	37	3	62
Observed	4	43	8	7	62

χ^2 = 74.282, df = 3, p < 0.001

The comparison of an additional experiment terminated after two days with those terminated after 8 days showed the distribution of the species to be significantly different (Table 5), the majority of the stones still having both species after 2 days. Thus completion of the competitive interactions appears to require between 2 and 8 days. Also no aggressive interactions were observed.

4 DISCUSSION

Competition between brittle-stars "for suitable cryptic micro-habitats" has been suggested by Sloan et al (1979) as a mechanism which influences the distribution of species. Chartock (1972) and Sloan (1982) both suggested that the use of different sized crevices might also be a method of micro-habitat separation, particularly when

Table 5. A comparison of the distribution of Ophionereis reticulata (O.R.) and Ophiolepis impressa (O.I.) after 2 and 8 days.

	O.I.† alone	O.R.† alone	O.R.+ O.I.	Zero*	N
Day 2	2	1	11	1	15
Day 8	1	16	1	4	22

χ^2 = 17.92, df = 2, p < 0.001

† These two categories were combined in the analysis.
* Data omitted in the analysis.

there was a positive association between ophiuroid species occurring in the same coral head (Sloan 1982). Sloan (1982) found a negative association between two large and a small species of ophiocomids occurring in coral heads in Aldabra and suggested that the small species might be avoiding predation by the larger two. In contrast O. pumila, O. reticulata and, O. impressa are relatively similar in size but were found to be negatively associated with each other under slabs of rubble lying on sand. Although there is not the same size discrepancy between these three species compared to those described by Sloan (1982), the species differ in morphology. O. pumila and O. reticulata both have small discs and relatively long arms, whereas O. impressa has a slightly larger disc and shorter arms. The arms of O. reticulata are about twice the length of those of O. impressa but are very much finer.

O. pumila, O. reticulata, and O. impressa are all cryptic in their behaviour, even at night, which may be related to their susceptibility to predation (Sides 1981). Also they have been shown to have a broad overlap in the height and length of crevice opening through which they extend their arms to feed (Sides 1981; Sides in prep). Thus the negative association between these three species may be the result of competition for similar sized crevices. The ability of O. pumila and O. reticulata to alter their crevices means that they can maintain small crevice openings. The openings used by these two species in the field are frequently very small, particularly those of O. reticulata, such that other ophiuroids would have trouble entering them. The inability of O. impressa to alter its crevice may put it at a competitive disadvantage. But regardless of whether the competitive mechanism is the alteration of crevice space, or merely an avoidance reaction by the two species, the use of relatively small stones by O. reticulata means that they can dominate the available crevice space. The apparently poor competitiveness of O. impressa, at least with O. reticulata, and possibly also with O. pumila, may result in O. impressa selecting larger slabs with a much greater crevice space which neither O. reticulata or O. pumila can dominate or alter. This would explain the occurrence of O. impressa under larger slabs than O. reticulata and O. pumila.

The aggressive defense of crevice openings by a brittle-star has been described for Ophiocoma scolopendrina (Magnus 1967; Fishelson 1971). This type of behaviour has also been described for Ophiocoma echinata and O. wendtii and competition between them for crevice space has been experimentally shown to take place over a single night (Sides 1981; Sides in prep). Although no aggressive encounters were observed between O. reticulata and O. impressa avoidence interactions may have taken place within the crevices. The results of the experiments suggest that even when O. impressa occurs in a crevice that is also inhabited by O. reticulata, it is forced to leave by O. reticulata within 2-8 days (Table 5). The competitive mechanism suggested is the physical alteration of the crevice by O. reticulata. The latter makes the crevice too small for both species to co-habit by lowering the stone into the substrate, or increasing the volume of sand under the stone (Fig. 1), an activity that O. impressa cannot or does not try to counteract. Laboratory observations showed that O. reticulata was capable of considerably altering crevice dimensions within 5 days. However the amount of crevice alteration visible around the edges of the experimental stones in the field was not as obvious as it had been in the laboratory until after about two weeks. This may have been due to the water movement in the field removing some of the sand that was placed around the edge of the stone by O. reticulata. A certain amount of variability in the time taken to oust O. impressa was expected in the field, and so fewer individuals co-habiting were expected with an increase in the experimental time. This proved not to be the case.

The activities of sand gobies which frequently and rapidly, i.e. in a matter of minutes, colonised the stones, generally increased the gap under the stones. O. reticulata sometimes appeared unable to counteract this activity. This may have resulted in the sand gobies providing space in which O. impressa was able to remain. External factors such as this may affect the success with which O. reticulata forces O. impressa to leave. However some natural sinking of stones lying on sand takes place (Sides pers obs) which would enhance the activities of O. reticulata. These factors need to be separated, and the possibility of avoidance reactions investigated, before it can be conclusively demonstrated that it is the alteration of crevices by O. reticulata that is the mechanism by which it is competitively dominant over O. impressa.

5 CONCLUSIONS

In conclusion, the negative association between Ophionereis reticulata, Ophiocoma pumila and Ophiolepis impressa under slabs of coral rubble, lying on sand, in the same area is probably due to competition for crevice space. Competition for space, under experimental conditions, has been demonstrated between O. reticulata and O. impressa, and O. reticulata is the competitively dominant species. The competitive mechanism is not instantaneous but takes between 2 and 8 days. The mechanism suggested is the reduction of the crevice dimensions by the shifting of sand by O. reticulata.

6 ACKNOWLEDGEMENTS

I wish to thank Dr J. D. Woodley, B. Needleman, and T. Price for their assistance in the field. Drs J. D. Reynolds and A. B. West critised the manuscript. Prof. J. N. R. Grainger provided space in the Dept. of Zoology, Trinity College Dublin, where the paper was written. To all I am grateful. This is contribution number 341 from the Discovery Bay Marine Laboratory, University of the West Indies.

REFERENCES

Chartock, M.A. 1972. The role of detritus in a tropical marine ecosystem: niche separation in congeneric ophiuroids, food partitioning in cryptic invertebrates, and herbivore detritus production at Eniwetok, Marshall Islands. Ph.D. Thesis, University of Southern California.

Davaney, D.M. 1974. Shallow water echinoderms from British Honduras, with a description of a new genus of Ophiocoma (Ophiuoridea). Bull.Mar.Sci. 24:122-164.

Fishelson, L. 1971. Ecology and distribution of benthic fauna in the shallow waters of the Red Sea. Mar.Biol. 10:113-133.

Fontaine, A.R. 1953. The shallow water echinoderms of Jamaica, Part 2. The

brittle-stars (Class Ophiuroidea). Nat. Hist.Notes Jamaica, no 60:197-206.
Grave, C. 1898. Notes on the ophiuroids collected in Jamaica during June and July 1897. Johns Hopkins Univ.Circular. 18(no 137):7-8.
Hammond, L.S. 1980. The feeding ecology and substrate relations of sediment ingesting holothurians and echinoids (Echinodermata) from a shallow reef lagoon, Discovery Bay, Jamaica. Ph.D. Thesis. University of the West Indies.
Hulbert, S.H. 1969. A coefficient of inter-specific association. Ecology 50:1-9.
Kinzie, R.A. III. 1973. Zonation of West Indian gorgonians. Bull.Mar.Sci. 23:93-155.
Kissling, D.L. & G.T.Taylor. 1977. Habitat factors for reef-dwelling ophiuroids in the Florida Keys. Proc. 3rd Int.Symp.Coral Reefs 1:225-231.
Lewis, J.B. & R.D.Bray. 1983. Community structure of ophiuroids (Echinodermata) from three different habitats on a coral reef in Barbados, West Indies. Mar.Biol. 73:171-176.
Magnus, D.B.E. 1967. Ecological and ethological studies and experiments on the echinoderms of the Red Sea. Stud.Trop. Oceanogr. no. 5:635-664.
Sides, E.M. 1981. Aspects of space utilization in shallow-water brittle-stars (Echinodermata, Ophiuroidea) of Discovery Bay, Jamaica. Ph.D. Thesis. University of the West Indies.
Siegel, S. 1956. Nonparametric statistics for the behavioral sciences. N.Y. McGraw-Hill.
Sloan, N.A. 1982. Size and structure of echinoderm populations associated with different coexisting coral species at Aldabra Atoll, Seychelles. Mar.Biol. 66: 67-75.
Sloan, N.A. A.M.Clark, & J.D.Taylor. 1979. The echinoderms of Aldabra and their habitats. Bull.Br.Mus.Nat.Hist.(Zool.). 37:81-128.
Woodley, J.D. & E.Robinson. 1977. Field guidebook to the modern and ancient reefs of Jamaica. Miami:Atlantic Reef Committee.

Proceedings of the Fifth International Echinoderm Conference / Galway / 24-29 September 1984

The family Ophiuridae in the bathyal zone of the Indian Ocean: Origin and biogeography

C.VADON & A.GUILLE
Muséum national d'Histoire naturelle, Laboratoire de Biologie des Invertébrés Marins et Malacologie, Paris, France

ABSTRACT: An analysis of the biogeography of the family Ophiuridae from the deep Indian Ocean (200-5.000 m) is presented, based on data from all the expeditions since the "Investigator", including five recent French expeditions. At the present time 70 species are known, distributed in five faunistic stocks (endemic, indopacific, indo-atlantic, indo-antarctic and cosmopolitan). The high percentage (41 %) of endemic species indicates that the deep Indian Ocean is a distinct biogeographic area.

The purpose of this paper is to present an analysis of the biogeography and the bathymetric range of the eurybathic family Ophiuridae from the margins of the Indian Ocean.

The material used for this study has been sampled by five recent french expeditions (fig.1) :

- the trawlings of the "Vauban" (ORSTOM, 1972-1975), in the Mozambique channel, to a depth of 2000 meters,
- the "Benthedi" expedition of the "Suroît" (1977), in the north-eastern Mozambique channel, Glorieuses and Mayotte Islands, from the littoral zone to 3800 meters (Thomassin, 1978),
- the "Safari I and II" cruises of the "Marion-Dufresne" (1979 and 1981) in the Mozambique, Madagascar and Indian Central basins, from 1000 to 5350 meters (Monniot, in the press),
- the "MD 32" expedition of the "Marion-Dufresne" around the Reunion Island (1982), from the littoral zone to 4300 meters depth (Guille, 1982).

It is noteworthy to mention that the french expeditions of the last twelve years have equalled, in number of stations, all the benthic explorations that have been done in the Indian Ocean, since the "Investigator".

These recent collections have been supplemented by material collected during the following expeditions :

- the "Investigator", in the Bay of Bengal and in the Oman sea (R.Koehler, 1896),
- the "Deutschen Tiefsee-Expedition" by the "Valdivia", on the east African coast, Andaman islands and to the west of Sumatra (M.Hertz, 1927),
- the "John Murray expedition", by the "Mabahiss", in the north-western Indian Ocean (H.L.Clark, 1939),
- the expeditions of the "Vitiaz" and the "Academician Kurchatov" in different localities of the Indian Ocean (Belyaev and Litvinova, 1972; Litvinova, 1976).

This study covers a large number of continental and insular slopes of the Indian Ocean, up to 30°S, from 200 meters to more than 5000 meters depth, with the exception of the Red Sea, the Persian gulf and the west coast of Australia.

The heterogeneity of the distribution of stations and of the sampling methods does not allow quantitative comparison of populations ; it permits only the gathering of data on presence or absence of species in the different areas investigated.

Results (tab.1) :

34 species have been identified from the recent french expeditions. One new species has been described (Ophio-

Table 1. Geographic and bathymetric distribution of the 70 species of Ophiuridae sampled in the deep Indian Ocean.
x = material collected by the five recent French expeditions; + = material collected before these expeditions.

▬▬ bathymetric range in the Indian Ocean; ▬▬▬ bathymetric range in the Pacific Ocean;
▬·▬ bathymetric range in the Atlantic Ocean; ▬ ▬ ▬ bathymetric range in the Antarctic Ocean.

	Indian						Pacific			Atlantic	Antarctic
	Madagascar basin	Zanzibar	Réunion (Mascareignes basin)	Oman sea - Maldives	Indian Central basin	Sri Lanka to W.Sumatra	West	North	East (american coasts)		
ENDEMIC spp.											
Amphiophiura trifolium		+		+							
Amphiophiura concava		+									
Ophiomusium breve		+									
Ophiomusium ultima		+									
Ophiomusium nitidum		x		+							
Ophiomusium valdiviae						+					
Ophiozonella brachyactis				+							
Ophiomastus platydiscus		+	x								
Ophiomastus melaniae						+					
Anthophiura sp.		x			x						
Ophiopyrgus alcocki		+		+		+					
Ophiuraster perissus				+							
Ophiophysis sp.		x									
Ophiomisidium sp.		x									
Ophiophyllum borbonica			x								
Homalophiura euryplax				+							
Homalophiura glypta				+							
Anophiura simplex				+							
Anophiura planissima				+							
Ophiura podica				+							
Ophiura lanceolata				+							

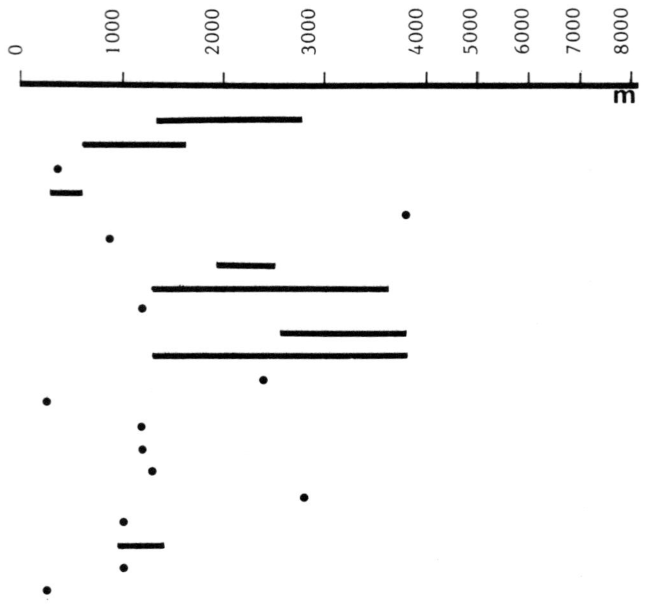

Gymnophiura chuni						+				
Gymnophiura micracantha		+				+				
Aspidophiura corone		+								
Ophiochorus granulatus		+		+						
Astrophiura permira		+								
Ophiuroglypha aequatoris					+					
Ophiuroglypha schmidt-otti		+				+				
Ophiurolepis inconveniens		+								
INDO-PACIFIC spp.										
Amphiophiura paupera		+	x	+		+	+			
Amphiophiura ornata						+	+	+		
Amphiophiura radiata		x		+			+	+		
Amphiophiura sordida						+	+			
Ophiomusium lunare	x	x					+			
Ophiomusium lutkeni	x	+					+			
Ophiomusium cancellatum		+					+			?
Ophiomusium familiare		+		+		+	+			
Ophiomusium elegans		x		+		+	+			
Ophiomusium properum		x					+			
Ophiomusium facetum		x					+			
Ophiomusium altum		+					+			
Ophiomusium relictum						+	+			
Ophiomusium diomedae		+				+			+	
Ophiomusium facundum		+					+			
Ophiomastus tumidus		+				+	+			
Ophiosphalma fimbriatum		+	x				+			
Ophiura micracantha		x					+			
Ophiura fluctuans		x					+			
Ophiura aequalis		+	x	+		+	+			
Ophiura imbecillis						+	+			
Ophiura palliata				+			+			
Ophiura undulata						+	+		+	
Ophiozonella longispina						+	+			
Ophiozonella molesta				+			+			
COSMOP. spp.										
Amphiophiura sculptilis	x	+	x	+	x	+	+		+	+
Amphiophiura convexa	x	+	x	+	x	+	+	+	+	+
Amphiophiura bullata	x				+			+		+
Ophiomusium validum		+		+			+			+
Ophiomusium lymani		+	x	+			+	+	+	+
Ophiomusium simplex		+	x	+		+	+		+	+
Ophiura irrorata		+	x	+	x	+	+	+	+	+
Ophiura loveni			x						+	+

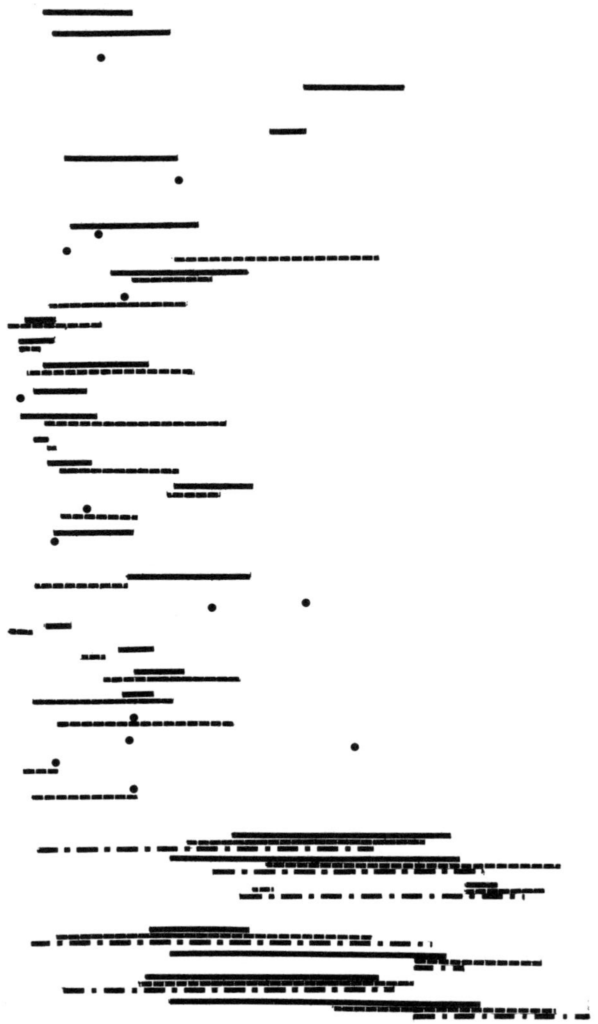

	Indian						Pacific				
	Madagascar basin	Zanzibar	Réunion (Mascareignes basin)	Oman sea - Maldives	Indian Central basin	Sri Lanka to W.Sumatra	West	North	East (american coasts)	Atlantic	Antarctic
Ophiotypa simplex	x	+	x	+	x	+	+		+	+	
Perlophiura profundissima	x	x	+				+	+	+	+	
INDO-ATLANTIC spp.											
Ophiosphalma planum	x	x	x	+	x	+				+	
Anthophiura ingolfi			x							+	
Uriopha ios		x	x		x					+	
Homalophiura abyssorum		x	?	+	x					+	
INDO-ANTARCTIC spp.											
Ophiurolepis mordax		x									+

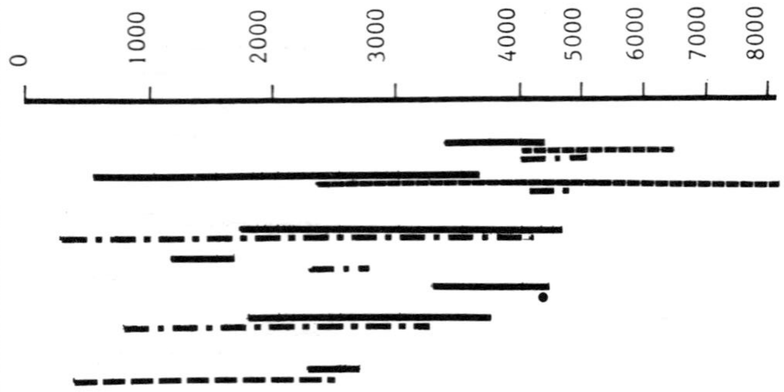

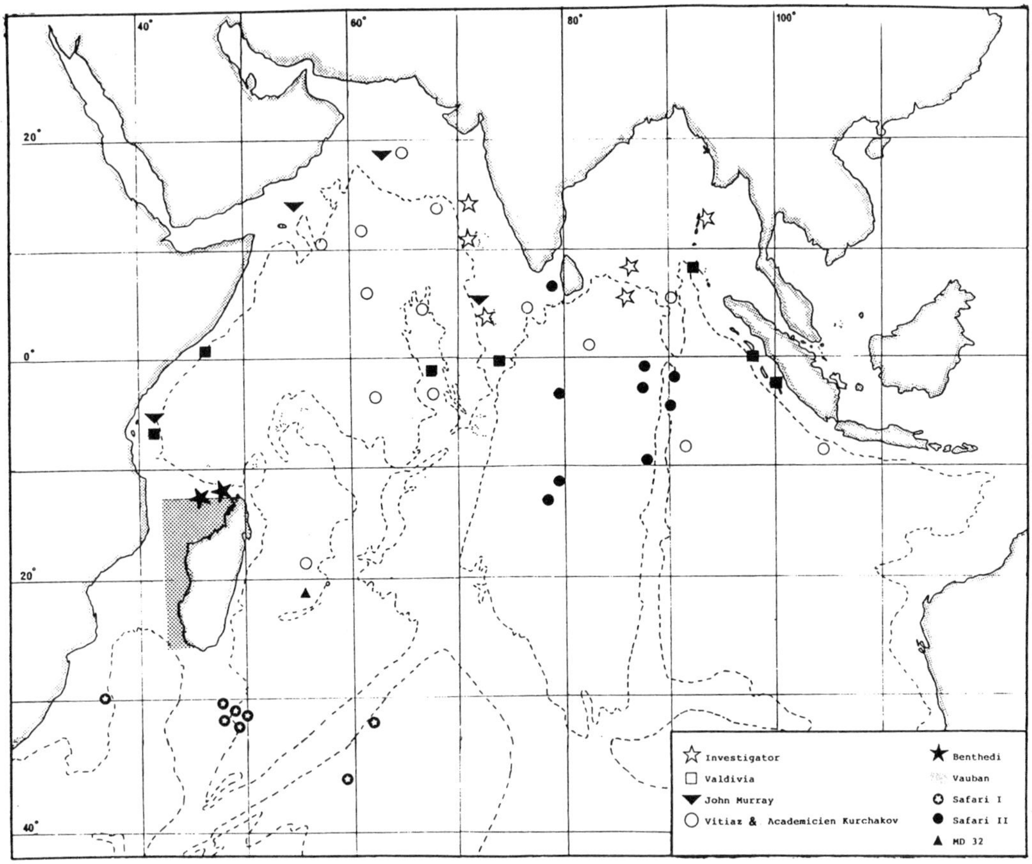

Figure 1. Areas investigated in the Indian ocean since the "Investigator" expedition.

phyllum borbonica Vadon and Guille, 1984) and three new species, belonging to the genera Ophiomisidium, Anthophiura and Ophiophysis are in the process of being described. Nine species were collected, for the first time, in the Indian Ocean : Ophiomusium lunare Lyman, O.properum Koehler, O.facetum Koehler, Ophiura micracantha H.L.Clark and O.fluctuans Koehler were known only from the west-pacific Ocean, Uriopha ios Paterson and Anthophiura ingolfi Fasmer from the north atlantic Ocean, Ophiura loveni from the Atlantic and Pacific (Kermadec islands) and south of the Indian Ocean ; Ophiurolepis mordax Koehler had been sampled only in the Antarctic Ocean.

At the present time, a total of 70 species (43 Ophiurinae and 27 Ophiolepidinae) belonging to 23 genera, are known from the deep Indian Ocean. These species have been arranged according to their geographic distribution (tables 1 and 2) in the Indian, Pacific, Atlantic and Antarctic Oceans. The bathymetric distribution of each species is given in each of four oceans. Five groups have been established :

1 - <u>endemic species</u>

29 species (41% of the total) are known only from the Indian Ocean. Ten belong to five genera represented worldwide by a large number of species : Ophiomusium, Amphiophiura, Ophiura, Homalophiura, Ophiozonella. 17 other species are distributed among 13 genera, also represented worldwide, but containing only a few species (tab.2). Among these 17 species are the four new species previously mentioned. The genera Ophiuraster and Ophiochorus are monotypic and represented by the species Ophiuraster perissus H.L.Clark and Ophiochorus granulatus H.L.Clark.

The 29 species are distribued from 200 meters to more than 4000 meters depth, but the majority were sampled between 1000 and 2000 meters.

Table 2. Relative proportion and bathymetric distribution of the five faunistic components of the Ophiuridae sampled in the deep Indian Ocean (dark grey : maximal frequency zone).

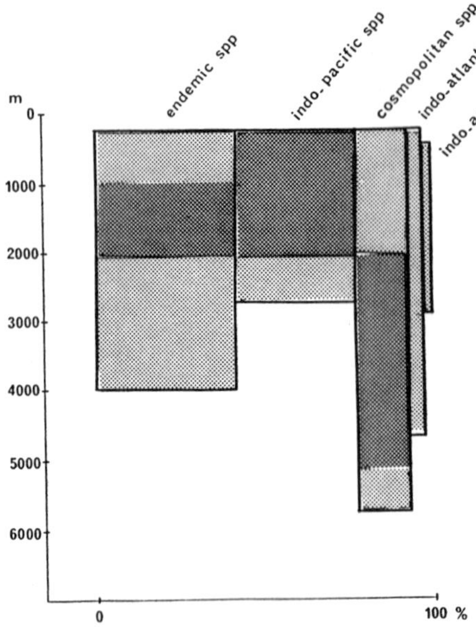

2 - indo-pacific species

25 species (36%) are known both from the Indian Ocean and the west-pacific Ocean. They belong to six genera Ophiomusium, Ophiosphalma, Ophiura, Amphiophiura, Ophiozonella and Ophiomastus.

With the exception of Ophiomisidium diomedae Lutken and Mortensen which occurs in the east-pacific, these species were described from the malayan archipelago, the north pacific (Amphiophiura ornata Lyman and A.radiata Lyman) and the american pacific margins (Ophiura undulata Lyman).

These indo-pacific species have been sampled, most frequently, from between 200 and 1500-2000 meters depth, though some of them reach 3000 meters (Ophiosphalma fimbriatum Koehler). 1ost show a wider bathymetric range n the Pacific Ocean.

3 - indo-atlantic species

Four species (5%) are known from the Indian and Atlantic Oceans. Their bathymetric distributions are bathyal and abyssal, and similar in the two oceans, with the exception of Ophiosphalma planum (Lyman), reported from the Atlantic at 540 meters (Lyman, 1878).

Some analogies between indian and atlantic abyssal faunas have already been shown, for Crustacea by Dahl (1954) and for Echinodermata by Madsen (1951). It is possible that these species will be collected during future deep pacific cruises and will be shown to be cosmopolitan species.

4 - indo-antarctic species

Ophiurolepis mordax Koehler, 1922, known from the Antarctic margin (Queen Mary, Princess Elisabeth and Wilkes Lands, 200-1600 meters) has been collected in the north of the Mozambique channel (Glorieuses islands), at 2500 meters depth.

This species constitutes the only exemple of the phenomenon of "equatorial submergence" (Ekman, 1953) or "polar emergence" (Menzies, George and Rowe, 1973), shown by cold stenothermal forms living in the same antarctic water layer under different latitudes.

5 - cosmopolitan species

Six genera and 10 species (14%) live in the Indian, Atlantic and Pacific Oceans.Their bathymetric distribution is very large, from 200 meters to the abyssal plain, and even in the very deep trenches (Perlophiura profundissima has been sampled at 8135 meters and Ophiura loveni at 7340 meters (Belyaev and Litvinova, 1972)). However, the major habitats of these species are between 1800-2000 meters and 5000 meters depth.

As for the indo-pacific species, the bathymetric ranges of the species observed in the Pacific and Atlantic Oceans are generally broader than in the Indian Ocean.

Below 4000 meters, in the Madagascar, Mascareignes and Indian Central basins, the specific diversity decreases. There, the Ophiuridae are represented by a few species (6 to 8), comprising very abundant populations (ex. : Madagascar basin, 3825 m, in one trawling sample, 74% of the fauna collected are Ophiuroids and among them, 150 specimens are Amphiophiura sculptilis).

DISCUSSION

Two faunistic stocks can be distinguished among the Ophiuridae living on the Indian Ocean margins : the indo-west-pacific species, belonging to some common and cosmopolitan genera, and distributed essentially between 200 and 2000 meters, and the endemic species, collected from 1000 to 2000 meters and belonging, generally, to genera comprising a small number of species.

Beginning at 2000-2500 meters, the abyssal and cosmopolitan species appear.

Similarly, Madsen (1961) notes that in the porcellanasterid Asteroids, which are essentially abyssal and cosmopolitan, the geographic distribution of the three bathyal genera (collected above 2000 m) is limited to the Indian Ocean and malayan archipelago.

The specific diversity of the indo-malayan area, thought to be the remains of the Tethys sea, is considered the highest of all the biogeographic areas (Ekman, 1953 ; Talbot, 1970), the number of species decreasing along two opposite gradients, towards east and west. This study shows (tab.3), in the case of the family Ophiuridae, that 50 and 22% respectively of indo-malayan species of the genera Ophiomusium and Amphiophiura are present in the Indian Ocean.

Table 3. List of the genera of the family Ophiuridae sampled in the deep Indian Ocean, the total number of species belonging to these genera, and specific distribution in the western Pacific.

	total	Indian	W.Pacif. + Indian	W.Pacif.
OPHIURA	75	2	8	11
OPHIOMUSIUM	70	3	14	29
AMPHIOPHIURA	66	2	7	32
HOMALOPHIURA	22	2	-	2
OPHIOZONELLA	20	1	2	5
OPHIUROLEPIS	19	1	-	-
OPHIOMASTUS	19	2	-	4
OPHIOSPHALMA	8	1	1	4
OPHIOMISIDIUM	7	1	-	1
OPHIOPYRGUS	6	1	-	4
ANTHOPHIURA	5	1	-	-
OPHIOPHYSIS	4	1	-	1
OPHIUROGLYPHA	4	2	-	-
ASPIDOPHIURA	4	1	-	-
OPHIOPHYLLUM	4	1	-	1
ANOPHIURA	3	2	-	-
GYMNOPHIURA	3	?	-	-
ASTROPHIURA	1	1	1	1
OPHIOTYPA	1	1	1	1
URIOPHA	1	1	-	-
PERLOPHIURA+	1	1	-	-
OPHIURASTER+	1	1	-	-
OPHIOCHORUS	1	1	-	-

(+ endemic genera of the Indian Ocean)

The high percentage (41%) of endemic species belonging to widespread indo-pacific genera (Ophiomusium breve H.L.Clark, O.ultima Hertz, O.valdiviae Hertz, Amphiophiura trifolium Hertz, A.concava Hertz,...) supports the suggestion that the Indian Ocean, at the bathyal level, is a distinct biogeographic area.

Cohen (1973) also notes high endemism (58%) in the Ophidioidei (benthic and benthopelagic fishes) of the deep Indian Ocean, in which he identifies two biogeographic sub-areas, western and eastern. No significant difference appears in the composition of Ophiuridae from all the insular and continental margins studied, demonstrating the homogeneity of this family in the Indian Ocean.

The apparent richness of the Zanzibar area must be viewed with great caution, and should be correlated with the abundance of samples.

In an analysis of the composition and the bathymetric range of the Ophiuroids fauna of the Philippines-Indonesia area, Guille (1980) defines, with the use of similarity coefficients, three faunistic discontinuities between 0 and 4000 meters depth of which one is at 1500 meters depth. This one is not correlated with environmental factors. The author emphasizes the role of heterogeneous sampling.

In our study, limited to one family of Ophiuroids (the Ophiuridae), a similar faunistic discontinuity appears at 2000 meters depth and corresponds to the gradual transition from the bathyal fauna to the abyssal fauna.

The data relating to the physical characteristics of the margin ecosystem are still very incomplete. However, the structure of Indian Ocean waters seem to be correlated with this faunistic discontinuity. As a matter of fact, its depth corresponds to a hydrological discontinuity : the lower limits, in the north part of the Indian Ocean, of the "north indian intermediate water", and in the south part, of the "intermediate antarctic water" (Wyrtki, 1973 ; Saetre and J.Silva, 1984). These two water layers have a great similarity in salinity (34.50/34.80‰) and temperature (4-8°C).

No correlation appears between substrata and the zonation of species. The transition from Pteropod ooze to Globigerina ooze appears at about 1200 meters (Guille, 1982) and the calcite compensation depth which delineates the red clays and calcareous oozes appears at about 5000 meters depth (Leclaire and Clocchiatti, 1976).

Finally, the high similarity between bathyal faunas of the Reunion island and the Glorieuses islands, of which the geological ages are very different (respectively 3.5 and almost 150 million years), emphasizes the major importance of hydrological factors in the distribution of marine species.

REFERENCES

Belyaev, G.M. & N.M.Litvinova 1972. New genera and species of deep-sea Ophiuroidea. Byull.mosk.Obshch.Ispyt.Prir. 77(3) : 5-20, 5 fig., 3 tab.

Clark, H.L. 1939. Ophiuroidea. Scient.Rep. John Murray Exped. 6 : 29-136, 62 fig.

Cohen, D.M. 1973. Zoogeography of the Fishes of the Indian Ocean. In B.Zeitzshel (ed.), Biology of the Indian Ocean, p. 451-463, 3 tab. Berlin, Springler-Verlag.

Dahl, E. 1954. The distribution of Deep Sea Crustacea. Union int.Sces Biol., sér. B, 16 (UNESCO).

Ekman, S. 1953. Zoogeography of the sea. London, Sidgwick and Jackson, 417 p.

Guille, A. 1980. Les Ophiures des marges continentales de la région Philippines-Indonésie. Distribution bathymétrique et étagement. In M.Jangoux (ed.), Echinoderms Present and Past, p. 97-105, 5 fig.,1 tab. Rotterdam, Balkema.

Guille, A. 1982. Les rapports des campagnes à la mer : MD 32/Réunion à bord du "Marion-Dufresne", 11 août-10 septembre 1982. Publications de la mission de recherche des Terres Australes et Antarctiques françaises, n° 82-03, 83 p.

Hertz, M. 1927. Die Ophiuriden der deutschen Tiefsee-Exped. I. Wiss.Ergebn.dt.Tiefsee-Exped."Valdivia" 22(3) : 59-122, 7 fig., pl. 6-9.

Koehler, R. 1897. Echinodermes recueillis par l'"Investigator" dans l'océan Indien. I. Les Ophiures de mer profonde. Ann.Sci. Nat. 8(4) : 277-372, 3 pl.

Koehler, R. 1922. Echinodermata Ophiuroidea. Sci.Rep.Aust.Antarct.Exped.C. 8(2) : 1-134, 15 pl.

Leclaire, L. & M.Clochiatti 1976. La dissolution des carbonates en milieu océanique Son rôle dans la genèse des dépôts pélagiques pendant le Cénozoïque. Bull.Soc. géol.France 7, t.XVIII(5) : 1315-1335.

Litvinova, N.M. 1971. The brittle-stars of the genus Amphiophiura of the Pacific and Indian oceans collected by soviet expeditions on the "Vitiaz" and "Akademik Kurchakov". Académie des Sciences d'URSS. Trav.Inst.Océanol. URSS 92 : 298-316, 8 fig., 2 tab.

Madsen, F.J. 1951. Ophiuroidea. Report Swed. Deep-Sea exp. 2(9) : 107-117, 2 fig.

Madsen , F.J. 1961. On the zoogeography and origin of the abyssal fauna, in view of the knowledge of the Porcellanasteridae. Galathea Rep. 4:177-215, 13 pl., 42 fig.

Menzies, R.J., R.Y.George & G.T.Rowe 1973. Abyssal environment and ecology of the world oceans. John Wiley and sons, 488p.

Monniot, C. sous presse. Composition des peuplements benthiques abyssaux : résultats des campagnes SAFARI dans l'océan Indien.C.N.F.R.A.

Saetre, R. & A.Jorge Da Silva 1984. The circulation of the Mozambique Channel. Deep-Sea Res. 31(5) :485-508, 17 fig., 4 tab.

Talbot, F.H. 1970. The South East Asian area as a centre of marine speciation : an ecological analysis of causes. Rep. Aust.Akad.Sci. 12:43-50.

Vadon, C. & A.Guille 1984. Les Ophiuridae (Ophiuroidea, Echinodermata) de la campagne MD 32 du "Marion-Dufresne" autour de l'île de la Réunion. Bull.Mus.natn.Hist. net., Paris 4ème sér., 6, sect.A, n°3: 583-615, 5 pl., 3 fig.

Wyrtki, K. 1973. Physical oceanography of the Indian Ocean. In B.Zeitzshel (ed.) Biology of the Indian Ocean, p. 18-36, 4 fig. Berlin, Springler-Verlag.

Acrocnida brachiata (Echinodermata: Ophiuroidea): Intertidal, subtidal populations in Brittany

ALLAIN BOURGOIN
Université de Bretagne Occidentale, Laboratoire d'Océanographie Biologique, Brest, France

The infaunal ophiuroid Acrocnida brachiata (Montagu)(Echinodermata: Ophiuroidea) buries 10 to 20 cm in fine sand substrata, and is known from several locations in European coastal waters. A comparison of an intertidal and a subtidal population has been in progress since January 1984 in the Bay of Douarnenez (Brittany, France). Both populations are sampled on a monthly basis: for the subtidal population, a diver-operated suction sampler with a 1mm mesh bag is used to collect 4 to 8 replicate samples of $0.064m^2$ at 18m depth; for the intertidal population, 15 to 20 samples of $0.25m^2$ are taken at 5m intervals and sieved on a 1mm screen. The population structures of Acrocnida brachiata are assessed by measuring the disc diameter, defined as the distance from the outer edge of the radial shields to the disc perimeter on the opposite side. Single readings are measured to the nearest 0.25mm with a drawing tube. In both the intertidal and subtidal populations, three well marked peaks constitute the size-frequency histograms, suggesting discrete year classes. For the same month, the mean disc sizes of the 2nd and 3rd peaks of each histogram are lower in the intertidal than those in the subtidal population. For example, in May, mean disc-sizes in the first population were 6.50mm and 9.50mm, the largest specimen reading 11.50mm. For the latter, the respective dimensions were 8.00, 11.00 and 14.50mm. In both cases, the first peak, i.e. juveniles, have the same mean disc-sizes. Thus, growth would become slower in the intertidal zone. New recruits were first observed in the subtidal population in May, with a mean disc-size of 2mm, but the mesh size used probably did not retain the earlier benthic stages. Young settlers were not sampled in the intertidal zone until July and had a mean disc size of 3mm. This sudden appearance of juveniles in the intertidal population is probably the result of a migration from the subtidal zone. Maturity stages were followed from the overall appearance of the gonads to the decrease in the mean disc size after gamete release, The shedding phase in the subtidal population occurred earlier, commencing in May, while in the intertidal population the breeding period began in June. A second maturity phase seems to occur in both populations.

The use of marine 'key' species in ecotoxicological testing

T.BOWMER, R.ROELENS* & B.F.KEEGAN
Zoology Department, School of Marine Sciences, University College, Galway, Ireland
** Institute of Industrial Research and Standards, Shannon, Ireland*

A major deficiency in marine toxicological testing procedures involves the choice of target organisms. There is a tendency for animals to be selected more for their amenability to laboratory experimentation than for any consideration of their ecological importance. This study investigates the suitability of so-called 'key' species (i.e. those which by their numbers, size, shape or activity control community organisation) in toxicological testing.

The burrowing ophiuroid Amphiura filiformis was chosen as one such species on the basis of its relatively stable, high density and longevous population in Galway Bay (west coast of Ireland), where it is suspected of 'regulating' both its sedimentary environment and associated macrofauna. In theory, at least, the disabling or removal of A. filiformis by toxic poisoning could totally change the character of the benthic community. The development of successful test protocols for such species may provide more environmentally relevant toxicity data for new chemicals and effluents to be discharged into the sea.

For this investigation, a test apparatus has been designed to accommodate a wide variety of sizes and types of animals. It consists of eight independent flow channels allowing seven concentrations of toxicant, and a control batch, to be run simultaneously in LC-50 (i.e. concentration lethal to 50% of the tested population) type experiments.

The performance of Amphiura filiformis has been examined in duplicate 96 hr. LC-50 tests using the following chemicals: unionised Ammonia, Cu^{++}, Dieldrin and Pentachlorophenol (PCP). These substances were chosen as being representative of a wide range of toxic chemical groupings. The ophiuroid's behaviour was monitored regularly to profile effects suitable for the compilation of EC-50 data. In this regard, the animal's tendency to come to the sediment surface under toxic stress is particularly useful.

Preliminary indications are that Amphiura filiformis is at least as sensitive as many comparable marine species (excluding zooplankton). It shows graded responses to toxicants, with time and concentration, and can be maintained for long periods in the laboratory.

This work is being carried out within the provisions of E.E.C. Contract No. ENV.695.EIR.(H). Contribution No. 208 - School of Marine Sciences, University College, Galway.

Proceedings of the Fifth International Echinoderm Conference / Galway / 24-29 September 1984

Occurrence of specialized papillae on the tube-feet of the ophiuroid *Amphipholis squamata*

LOUIS DE VOS
Laboratoire de Biologie animale et cellulaire, Université de Bruxelles, Belgium

The podial tips of the ophiuroid **Amphipholis squamata** have been observed in scanning and transmission electron microscopy. The podial knob is covered by a uniform microvillar fur interrupted by numerous cell assemblages slightly protruding above the knob surface. These cell clusters are made of two conical cells, one ciliated cell and one club-shaped cell.

Each conical cell bears a crown of very short microvilli. The two other cells are located between the conical cells. The ciliated cell has a short central cilium and a collar of 8 to 10 microvilli. The club-shaped cell has no special features. In section the conical cells appear to have a long cylindrical body. Their cytoplasm is stuffed with ovoid granules (about 0.6 um in diameter) having a very dense homogenous content. These granules are not membrane bound. The cytoplasm is, moreover, invaded by a bundle of microtubules located at least in the upper third of the cell and running parallel to its long axis. In most sections the ciliated cell is wedged between two conical cells. Its cell body is very slender and sinuous. The cytoplasm is very electron dense and contains numerous small spherical ,dense granules of 60 nm diameter. The cilium of this cell type posseses a long but narrow striated rootlet. Until now we failed identifying the characteristics of the club-shaped type cell.

Our observations show that there are striking homologies between the clusters (or papillae) of granular cells of the podia in **A. squamata** and the glandular elements described in some other echinoderms, viz. the holothuroids **Psolus chitinoides** and **Neopentadactylus mixta.** Presumably the big conical cells are responsible for the secretion of the tufts of slime-like material covering many papillae. On the other hand, the ciliated cell and perhaps the club-shaped cell could have a sensory function. We suggest that each cluster should be considered as a minute papilla forming a sensory and secretory unit. These papillae most likely play some key role in both food capture and locomotion.

Work supported by FRFC grant 2.4525.80.

Proceedings of the Fifth International Echinoderm Conference / Galway / 24-29 September 1984

Bioluminescence in deep and shallow water brittlestars

R.H.EMSON
Zoology Department, King's College, London, UK

P.J.HERRING
Institute of Oceanographic Sciences, Wormly, Godalming, Surrey, UK

Bioluminescence has been studied in four brittlestars, two from deep water and two from shallow. The two deepsea species were the ophiacanthid Ophiacantha abyssicola and an amphiurid Amphiura grandisquama, while the shallow water forms were the amphiurid, A. filiformis and an phiocomid Ophiopsila riisei. These represent four of the five brittlestar families in which luminescence is known, the fifth being the Ophioneridae. Field and laboratory observations showed two contrasting patterns of luminescence. In A. filiformis, the luminescence is confined to the arm spines, as it is in the closely related species Acrocnida brachiata (Bell, 1974). In the other species, the luminescence is a property of the arms. It is widespread in the arms but is concentrated in certain regions such as the area of the arm spine nerve ganglia. Another difference between A. filiformis and A. brachiata and other luminescent forms is that the sources of luminescence fluoresce, when suitably stimulated, in all species except A. filiformis and A. brachiata. Light and electron microscopy has revealed both the detailed distribution of potential light emitting cells and cell secretions, and the characteristics of the putative light emitting cells. In Amphiura filiformis, the light emission appears to be the function of basophilic gland cells in the spines which contain densely staining spherical granular bodies. These honeycomb-like structures appear identical to those presumed to be the source of light in Acrocnida by Bell (1974). In O. abyssicola, A. grandisquama and O. riisei, no such structures have been identified. The lateral areas of the arms where luminescence is concentrated contain only a few types of cell which could be potential sources of luminescence. Of these, one with secretions of finely granular material is both abundant, correctly positioned and closely associated with nerve axons containing presumed neurosecretory granules. These cells are tentatively suggested to be the source of luminescence. The abundance of these cells and of those containing photogenic granules in A. filiformis suggests either that luminescence has an important role in the life of these animals - or that these luminescent cells have several functions. This and other aspects of the phenomenon are being further investigated.

Proceedings of the Fifth International Echinoderm Conference / Galway / 24-29 September 1984

A study of ophiuroid size, abundance, and reproductive mode in relation to coral reef zones and substrata

G.HENDLER
Smithsonian Oceanographic Sorting Center, National Museum of Natural History, Smithsonian Institution, Washington, USA

This analysis of ophiuroid distribution is based on 3,980 specimens representing 33 species from the Belize Barrier Reef. Samples were collected from 4 reef zones: Back Reef (0.5m depth), Reef Crest (1m), Spur and Groove (5m), and Fore Reef Ridge (15m). In each zone, ophiuroids were isolated from quantitative samples of the calcareous alga (Halimeda opuntia), lettuce coral (Agaricia tenuifolia), finger coral (Porites porites and Madracis mirabilis), and from beneath pieces of coral rubble. The sizes (mean disc diameter) of 10 of 18 species varied significantly between reef zones. Also, the sizes of 9 of 15 species, and the density (mean number of individuals per volume of substratum) for 11 of 14 species, varied significantly between different substrata. In each case, the numbers of species tested differed because of the presence or absence of species (or the occurrence of invalid sample sizes) in certain test groups. It is important to note that the size and abundance of many species depended on the substratum sampled, and that equal volumes of each substratum were not sampled for each reef zone. Therefore, differences detected in brittlestar size and abundance between reef zones might be a sampling artifact (in this and other studies). One example of a probable artifact is that the larger specimens of some species tended to occur in shallow reef zones and smaller specimens in deep reef zones (e.g., Ophiocoma echinata, O. wendti, O. paucigranulata, Ophiothrix angulata, and O.oerstedii). In contrast, the larger specimens of Ophioderma rubicundum were sampled in the deeper zones. There was an apparent tendency for smaller ophiuroids to occur in Halimeda than in coral substrata, and in the cases where multiple range tests were possible this trend was borne out. Notably, for all 7 broadcast spawning species tested there usually was a significant relationship between body size and the type of substratum sampled. However, the mean sizes for 6 (of the 8 species collected) of both brooding, and fissiparous, brittlestars did not differ significantly among the substrata studied. In the discussion, the nature of the relationships between ophiuroid size, mode of reproduction, and substratum type are examined, and experiments are described that indicate the proximate cause of these trends.

Proceedings of the Fifth International Echinoderm Conference / Galway / 24-29 September 1984

Development and metamorphosis of the brittle star *Ophiocoma pumila:* Evolutionary and ecological implications

P.V.MLADENOV
Mount Allison University, Sackville, Canada

The development through to metamorphosis of the shallow-water tropical Atlantic ophiuroid Ophiocoma pumila is described for the first time. This species possesses an 8-armed planktotrophic ophiopluteus larva. This larva has a pelagic existence of about 3 months (26-27.5°C)(considerably longer than for other brittle star species so far studied) and is probably teleplanic and capable of trans-Atlantic dispersal along a number of routes. This likely accounts for the amphi-Atlantic distribution of O. pumila. During metamorphosis, all of the larval arms are resorbed into the developing ophiuroid rudiment although the right antero-lateral arm disappears last. Surprisingly, the metamorphosing larva transforms into a highly mobile vitellaria possessing transverse ciliary bands and tube feet. Ophiuroid vitellariae have hitherto been described only for species with abbreviated lecithotrophic development. The vitellaria of O. pumila can delay settlement for at least a week and probably functions as a pre-settling exploratory stage. The discovery of both an ophiopluteus and a vitellaria in the ontogeny of a brittle star is strong support for the proposal that ophiuroid vitellariae are related to ophioplutei and are not a divergent larval series. Abbreviated development in brittle stars with vitellariae may be the outcome of heterochrony caused by an acceleration of metamorphosis. Little is known about ophiuroid metamorphosis so it is possible that the ophioplutei of many other species pass through a vitellaria stage towards the end of their pelagic existence.

Proceedings of the Fifth International Echinoderm Conference / Galway / 24-29 September 1984

Annual recruitment in the brackish-water ophiuroid *Ophiophragmus filograneus*

R.L.TURNER
Harbor Branch Institution, Fort Pierce, and Florida Institute of Technology, Melbourne, USA

Ophiophragmus filograneus is endemic to brackish waters of Florida. Recruitment of juveniles was studied in a seagrass bed in the Banana River lagoon, an atidal site where seasonal changes in rainfall and evaporation produce wide, long-term changes in salinity. Ophiuroids were sampled monthly by two core/sieve methods--one for adults and one for juveniles. The population had small gonads, high percent autotomy, high density, and good recruitment. Spawn-out occurred in November, 1982 and corresponded with estimates of spawn-out for prior years and for other locations in Florida. Smallest specimens had oral-frame diameters of .43mm and only three free arm segments on arms of equal length. By November, juveniles formed a moderate-sized mode of the population, and some already had the 2+3 pattern of differential arm growth. In March, many specimens had a third arm that had begun to grow faster than the two remaining short arms. By late July, the juvenile mode had nearly merged with the adult mode of the population. It appears that members of a cohort grow to adult size within 1 year, spawn in late fall, and survive to spawn at least once more. The proximate spawning factor probably is temperature, which varies predictably in shallow waters of Florida. Salinity is unlikely to be the ultimate factor, for its variation is unpredictable within and between localities. A more likely factor regulating the timing of reproduction is predation or disturbance by stingrays, which retreat to deeper waters during the winter and return in spring.

Supported by a postdoctoral fellowship from HBI and sabbatical leave from FIT.

Proceedings of the Fifth International Echinoderm Conference / Galway / 24-29 September 1984

Use of dissolved organic matter during brooding in *Amphipholis squamata*: Implications for editing the developmental program of the ophiopluteus

CHARLES WALKER & FRANK SMITH
Department of Zoology, University of New Hampshire, USA

In order to investigate the possible use of dissolved organic matter (DOM) by the tissues of Amphipholis squamata, ultrastructural, autoradiographic and scintillation counting methods of analysis were applied to adult and embryonic tissues. During development, embryos are contained within 10 genital bursae where they grow from a fertilized ovum of 100 μm to a near-term juvenile of 800μm. Epithelial cells lining the cuticle-covered adult genital bursa are invariably associated with gram negative bacteria which lie in cytoplasmic pockets facing the lumen. Before metamorphosis, external epithelial cells of "vestigial plutei" do not form flagella or flagellated bands typical of the free-living ophiopluteus stage of most brittlestars. Instead, such cells prematurely secrete a cuticle and become associated with bacteria. Symbiosis of this kind is surprising because of the well-known bactericidal properties of echinoderm tissues. Abundant flagella in the genital slit of the adult bring continuous currents of sea water into each bursa. Grain counting of plastic sections indicates that bacteria absorb radiolabelled amino acids and glucose from such currents more readily than do cells of the adult or embryonic epithelia. Chloramphenicol (4mM) greatly reduces total incorporation of radiolabelled amino acids, suggesting that bacteria can quickly and effectively utilize our artificial DOM and that protein synthesis occurs principally in prokaryotic cytoplasm. Cycloheximide (4mM) shows little if any immediate effect on protein synthesis, demonstrating that direct utilization of DOM by eukaryotic cytoplasm is minimal. "Vestigial plutei" do not phenotypically express aspects of their developmental program leading to a swimming (e.g., spicule-supported larval arms) planktotrophic larva (mouth is absent prior to metamorphosis). In spite of this and their limited supply of vitellogenic nutrients, embryos regularly complete metamorphosis and grow to form large juveniles. We suggest that an additional source of nutrients to support such growth is DOM concentrated from sea water by symbiotic bacteria acting as efficient biological filters. Such resources might be used directly by the eukaryotic cells of the adult and embryos when bacteria are phagocytized or indirectly after prokaryotic processing and release. It seems highly probable that the generalized echinoderm phenomenon of epithelial absorption or so-called "skin digestion" of DOM may actually depend on the high transportation constants of bacteria associated with external epithelial cells of echinoderms.

Proceedings of the Fifth International Echinoderm Conference / Galway / 24-29 September 1984

Post-larval morphology of some N.W. European ophiuroids

CHRISTINE WEBB
Oceanography Department, University College, Swansea, UK

The early post-larval development of ophiuroids, in general, follows a similar pattern of disc and arm plate morphogenesis. Adult characteristics are developed at some stage in this period, but often well beyond metamorphosis. The transient post-larval morphology present in this intervening period, often differing markedly from subsequent stages, has generally poorly known taxonomic, ontogenetic and functional significance. A brief survey of the dorsal disc and arm morphology in the smallest free-living and settled post-larvae of a variety of N.W. European representatives of the order Ophiurae (Ophiura ophiura, O. albida, O. ljungmani, Ophiocten gracilis, Ophiomusium lymani, Acrocnida brachiata, Amphiura filiformis, A. chiajei, Amphipholis squamata, Ophiocomina nigra and Ophiacantha bidentata) serves to highlight and explore these aspects. The use of Scanning Electron Microscopy (SEM) proves invaluable in resolving such features. Primary plate stereom is an intriguing early post-larval feature which may prove to have significance at various taxonomic levels. For example, a distinctive stereom of a regular pattern of round, craterlike fenestrations extending to the plate edge, is observed in the three Ophiura species examined and appears to be confined to the genus. Similarly the rounded pentagonal central primary plate and rectangular terminal plate may be other generic features. Another plate stereom observed is the bordered type observed in Acrocnida brachiata, Amphiura filiformis and Ophiocten gracilis. In Acrocnida with 0.4-0.6mm discs, the thicker, central region has an open latticework while the region bordering the plate edge is thin, having its fenestrations partially infilled. Variations of this bordered plate stereom occur in the other two species but in all cases such borders are confined to the primary plates. Other features of plate development are specific such as the fenestration size in Ophiura or the particularly unique spined stereom of Ophiomusium lymani. The functional and adaptive value of ophiuroid primary plate morphology is generally unknown. The nature of the terminal plate of the recently metamorphosed post-larva exhibits greater variation between closely-related species, as do other aspects of arm development (i.e. arm segment number, spine form and equality of growth) whose adaptive and functional value may be more readily seen. That allied species with similar adult ecological habit remain distinguishable, may suggest that functional demands of the post-larval stage in a species's life history (possibly quite different to the adult period) are quite important. Differences in the relative rate of disc plate and arm development between some direct and indirect species and shallow and deep sea forms are tentatively suggested.

The tendons of *Ophiocomina nigra* and their role in arm autotomy: Evidence for variable tensility in a type IV collagenous structure

I.C.WILKIE
Department of Biological Sciences, Glasgow College of Technology, Scotland, UK

R.H.EMSON
Department of Zoology, King's College, University of London, UK

In the ophiuroid arm the vertebral ossicles of adjacent segments are interconnected by four muscle blocks. During arm autotomy in Ophiocomina nigra (Abildgaard) and most other species so far examined, the aboral (upper) intervertebral muscles in the autotomising segment always part cleanly from the ossicle at their distal insertion, and the oral (lower) muscles always part from the ossicle at their proximal insertion.

The muscles are linked to the vertebrae by connective tissue fibres which are strongly PAS-positive and show hyaluronidase-labile gamma metachromasia with toluidine blue, indicating a rich content of glycoproteins and glycosaminoglycans respectively. In the electron microscope they appear finely granular and contain aligned microfilaments 2 - 3 nm in diameter. The histochemistry and ultrastructure of these fibres, and their digestibility by purified collagenase, suggest that they consist of carbohydrate-rich, non-fibrillar, type IV collagen such as occurs in vertebrate basal laminae.

Examination of muscle-ossicle junctions in which the process of muscle separation had been initiated then arrested before complete detachment could occur, revealed that the tendons had an elongated, stretched out appearance, suggesting that they had undergone an increase in extensibility. Detachment of the muscles at autotomy thus seems to involve a change in the mechanical properties of the tendinous fibres resulting in their attenuation and rupture. This represents the first evidence for variable tensility in a non-fibrillar collagenous structure.

Granule-containing juxtaligamental cell processes are closely associated with the tendons of the 'autotomy insertions' but are absent from the 'non-autotomy insertions'. Further experimental evidence implicates these juxtaligamental cells in the mechanism of muscle detachment.